Atlas of historical eclipse maps

EAST ASIA 1500 BC – AD 1900

Translation of cover text
Part of a page from the Astronomical Treatise of
the *Chiu-t'ang-shu* ('Old History of the T'ang
Dynasty'). Among the astronomical records of
various kinds from around AD 760 is the
following detailed account of a total solar
eclipse:

'2nd year (of the Shang-yuan reign period), 7th
month, day *kuei-wei*, the first day of the month.
The Sun was eclipsed; the great stars were all
seen. The Astronomer Royal, Ch'u T'an,
reported [to the Emperor], "On *kuei-wei* day the
Sun was eclipsed. 6 *k'o* after the hour of *ch'en*
(8.12 a.m.) the eclipse began. It was total at 1 *k'o*
after the hour of *szŭ* (9.12 a.m.). The Sun was
fully seen 1*k'o* before the hour of *wu*
(10.48 a.m.). The eclipse was 4 degrees in
Chang (lunar mansion). This represents Chou
(state). The *Chin-tê* says that when the Sun is
eclipsed between (the hours) *szŭ* and *wu*, this
represents Chou. Now Chou is Ho-nan, which is
occupied by Shih Szu-ming and his rebels. The
I-szu-chan says that under a solar eclipse the
kingdom will be ruined".'

The recorded date of this eclipse corresponds
to AD 761 August 5 and it can be seen from the
computer-drawn map that on this day there was
a total eclipse visible at Ch'ang-an, the capital of
China at the time. The recorded times are only
approximately correct but they agree quite well
with calculation.

Atlas of historical eclipse maps

EAST ASIA
1500 BC – AD 1900

F. R. STEPHENSON
Department of Physics, University of Durham
M. A. HOULDEN
Department of Physics, University of Liverpool

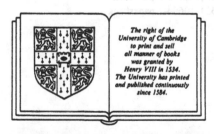

The right of the
University of Cambridge
to print and sell
all manner of books
was granted by
Henry VIII in 1534.
The University has printed
and published continuously
since 1584.

CAMBRIDGE UNIVERSITY PRESS
Cambridge
London   New York   New Rochelle
Melbourne   Sydney

CAMBRIDGE UNIVERSITY PRESS
Cambridge, New York, Melbourne, Madrid, Cape Town, Singapore, São Paulo, Delhi

Cambridge University Press
The Edinburgh Building, Cambridge CB2 8RU, UK

Published in the United States of America by Cambridge University Press, New York

www.cambridge.org
Information on this title: www.cambridge.org/9780521106948

First published 1986
This digitally printed version 2009

A catalogue record for this publication is available from the British Library

Library of Congress Catalogue Card Number: 85-47841

ISBN 978-0-521-26723-6 hardback
ISBN 978-0-521-10694-8 paperback

To Ellen and Penny

Contents

Introduction

Of the various kinds of astronomical phenomena recorded in history, eclipses have possibly the greatest interest and importance in present day research. Not only do they have considerable value in chronological studies, they have also contributed much to our understanding of changes in the Earth's rate of rotation, a major geophysical problem.

The astronomers of China have bequeathed to us the longest and most substantial series of eclipse observations from anywhere in the world. Before about 700 BC the extant records are very sporadic, but since that date roughly one thousand solar eclipses and several hundred lunar eclipses are preserved. The detailed histories of Korea and Japan, both cultural satellites of China, do not commence until much later – towards the end of the 1st millennium AD. However, from then on there are extensive lists of independent eclipse observations from both countries, in addition to those from China. For reasons given below, the present work is concerned only with *solar* eclipses and covers the period from 1500 BC (several centuries before the earliest known eclipse records in China) down to the end of last century.

We have restricted our attention to solar eclipses for two main reasons – historical and purely practical.

Although there are several extant references to lunar obscurations from the earliest historical period in China, the Shang Dynasty c. 1550 to 1045 BC (Keightley, 1978), at all subsequent periods down to the Sui Dynasty (AD 589–616) there are virtually no further records of these events. During this long interval, only solar eclipses were regarded as significant astrological omens and thus worthwhile reporting.

In studying historical eclipses, maps are very useful for representing solar eclipses since the local circumstances vary considerably from place to place. On account of both the lunar orbital motion and the terrestrial rotation, when the Moon passes in front of the Sun the lunar shadow sweeps across the Earth's surface in a few hours. The umbral shadow forms a narrow band, usually more than 10000 km in length but seldom more than 250 km wide. Within this band the eclipse may be described as central since the Moon covers the Sun more or less directly. For a considerable distance on either side of this zone a partial eclipse is visible. On the contrary, in the case of a lunar obscuration, mapping serves little purpose for the proportion of the Moon's disc in shadow is virtually independent of the observer's location; at any instant the appearance is much the same from the entire hemisphere of the Earth where the Moon is above the horizon. Brief tables such as those published by Oppolzer (1887) are thus quite satisfactory for investigating most records of these phenomena.

Central solar eclipses may be divided into three categories; total, annular, and 'annular-total'. A total obscuration, in which the Moon completely covers the Sun for a few brief minutes, producing a deep darkness with the appearance of stars, is certainly the most awe-inspiring. However, if the Moon is in the more distant part of its elliptical orbit when it passes in front of the Sun, an annular or ring eclipse is formed instead; here the diminution of light may be quite small. Very rarely, an eclipse may be annular near the sunrise and sunset position. Under such circumstances, the eclipse track is extremely narrow and over part of its extent the Sun is momentarily reduced to a broken ring of light. In this Atlas, all tracks of central eclipses crossing the Far East between 1500 BC and AD 1900 are mapped in detail, making use of recent extensive studies of the lunar motion and changes in the Earth's rotation in the past.

1. Calendar dates

All eclipse dates in this work are expressed in terms of either the Julian or Gregorian calendars, the transition date being AD 1582 Oct 5/15. On the BC/AD system there is, of course no year zero, 1 BC being immediately followed by AD 1. Hence in specifying years of eclipses before the Christian Era we have used negative numbers, equating the year 0 with 1 BC, −1 with 2 BC, etc. Thus the year of the earliest eclipse charted in this Atlas (−1496) corresponds to 1497 BC. This system is followed by Oppolzer (1887) and is in common use in astronomical practice. A number of other dates such as reference epochs are given in this introduction; for these we have adopted the BC/AD system since negative numbers could possibly be mistakenly interpreted as relating to quantities other than dates.

The calendar date which we assign to any particular eclipse is the local date and is independent of the Greenwich meridian. In the region covered by the maps, all places where a particular eclipse is visible share the same calendar date. This date will frequently be one day in advance of the Greenwich civil date (e.g. as listed by Oppolzer, 1887) since the selected range of longitude is far to the east of Greenwich. Eclipse dates as given here should be identical with the corresponding historical date as converted to the Western calendar, useful conversion tables being those of Hsüeh and Ou-yang (1956).

Following the calendar date of each eclipse we have given the Julian day number at the time of conjunction; this is rounded to the nearest integer. Julian *ephemeris* days, beginning at 12h ET (ephemeris time), are used here. The quoted day number thus may exceed by unity the appropriate value given by Oppolzer (1887), which is based on UT (universal time).

2. Fluctuations in the Earth's rotation and the accuracy of the eclipse maps

If the Earth's rate of rotation and the mean lunar motion were uniform, it would be possible to calculate with high accuracy the geographical co-ordinates of the zones of totality or annularity across the Earth's surface for eclipses at any time in the past (or future). In practice this is not the case, mainly on account of the tides produced by the Moon and Sun. During the course of a century, tidal friction in the oceans and − to a lesser extent − in the solid body of the Earth causes a simultaneous lengthening of the day by some 2.5 milliseconds (ms) and a recession of the Moon from the earth by some 3.5 metres. As the Moon's orbit expands in this way, its mean motion slows down (in accordance with Kepler's third law). The equivalent orbital acceleration is approximately −25 arcsec/century^2 ("/cy^2), producing a substantial drift in the lunar position over a period of a few millenia. A detailed discussion of the problem is given by Stephenson and Morrison (1984), to which the interested reader is referred. Unless otherwise indicated, the results given in this section are taken from this paper.

In all calculations of the Moon's position we have used the lunar ephemeris j=2 (IAU, 1968), but with a revised value for the lunar acceleration (ṅ). We have adopted a figure for ṅ of −26 "/cy^2, as deduced by Morrison and Ward (1975) from transits of Mercury observed over the last 250 years. This result is very similar to that determined from lunar laser ranging by Dickey and Williams (1982), i.e −25.1 ± 1.2. Over the period covered by this Atlas it is unlikely that tidal friction has changed significantly and we have thus assumed a constant ṅ throughout. The ephemeris j=2 is based on a value for ṅ of −22.44 "/cy^2. In order to fully incorporate our choice of ṅ we have applied the following additions to the mean lunar longitude (L) of the ephemeris:

$$\triangle L = -1".54 + 2".33\,T - 1".78\,T^2 \qquad (1)$$

Here T is measured in centuries from the epoch 1900.0.

Allowance for variations in the rate of rotation of the Earth is more complex, since both tidal and non-tidal changes are present. On the basis of conservation of angular momentum in the Earth-Moon system, the rate of lengthening of the day due to the combined effect of the lunar and solar tides may be calculated fairly accurately for a given value of the lunar acceleration. With $\dot{n} = -26$ "/cy^2, the result is 2.4 ms/cy. Although this may seem very small, the resultant accumulated clock drift ($\triangle T$) over the historical period is large; in T Julian centuries (each of 36525 days) it amounts to some 44 T^2 seconds. Hence by the beginning of the period covered by this Atlas (1500 BC) the tidal contribution to $\triangle T$ is roughly 0.5 day. Neglect of this would place eclipse tracks on the opposite side of the Earth from where they would otherwise be. The importance of carefully allowing for changes in the Earth's rotation when investigating past records of eclipses is thus obvious.

Both precise modern measurements and relatively crude pre-telescope observations (mainly of occultations and eclipses) reveal significant non-tidal fluctuations in the Earth's rotation which materially affect $\triangle T$. Irregular short-term changes, the so-called 'decade fluctuations', which can be accurately mapped over the last two centuries, are usually attributed to electromagnetic coupling between the core and mantle of the Earth. These produce deviations from the tidal $\triangle T$ parabola by up to 30 seconds of time. Pre-telescopic observations are much too inaccurate to reveal the decade fluctuations but they do provide clear evidence of significant non-tidal changes which are variable on the time-scale of millennia. Possible causes are (i) long-term electromagnetic coupling, (ii) redistribution of material in the interior of the Earth and (iii) global sea-level changes. Analysis of Babylonian observations (mean epoch around 400 BC) reveals that these non-tidal effects reduce the expected (tidal) value of $\triangle T$ by more than an hour in ancient times.

Above each map, beside the date, we have given the estimated value of $\triangle T$, based on a detailed investigation of (i) telescopic timings of occultations of stars by the Moon; (ii) medieval timings of lunar and solar eclipses by Arab astronomers; (iii) medieval and ancient sightings of total solar eclipses ob-

served mainly in Europe and China; and (iv) ancient timings of lunar eclipses by Babylonian astronomers. This analysis is based on the same value for $\dot{n}$ as adopted here (i.e. -26 "/cy^2). Our main source of reference is again Stephenson and Morrison (1984). It should be noted that the pre-telescopic values of $\triangle T$ are actually calculated on the basis of a preliminary study, the results of which differ only slightly from those of the main paper.

Between AD 1600 and 1900, we have deduced $\triangle T$ values from the data at yearly and 5-yearly intervals published by Stephenson and Morrison (1984). Before AD 1600, the following parabolic expressions were used instead, AD 948 being the mean epoch of the Arabian timings mentioned above.

(i) from AD 948 to 1600,

$$\triangle T = 22.5 \, t^2 \, \text{sec} \qquad (2)$$

(ii) at any time before AD 948,

$$\triangle T = 1830 - 405E + 46.5E^2 \, \text{sec} \qquad (3)$$

Here, t is measured in centuries from AD 1850 and E in centuries from AD 948.

The value of $\triangle T$ exceeds 1 minute during the whole of the period before AD 1640. Over this entire range of time we have rounded the individual values to the nearest 0.1 minute; later we have gradually increased the quoted precision. Some remarks are necessary on the real accuracy with which $\triangle T$ can be determined in the past since this has important bearing on the reliability of the maps and tables.

Reasonable estimates of the uncertainty in the adopted values of $\triangle T$ at century intervals back to AD 1600 are as follows: AD 1900, 0.1 sec; 1800, 1 sec; 1700, 5 sec; 1600, 30 sec. Because of the sporadic nature of the pre-telescopic data it would be impractical to quote values at similar intervals before AD 1600. However in the medieval period the uncertainty is not much more than a minute and even well into ancient times the value of $\triangle T$ is probably known to within a very few minutes. Thus the groups of Arabian and Babylonian timings give results for $\triangle T$ with standard errors of 1 min 20 seconds at AD 948 and approximately 7 minutes at 390 BC. Possibly a realistic estimate of the uncertainty in $\triangle T$ by 1500 BC is around 15 minutes. Obviously these latter estimates are much larger than the appropriate rounding errors (0.1 min). However, for purposes of reference we have felt it advisable to indicate the precise value of $\triangle T$ used in computing the data for each map.

With regard to the reliability of the eclipse maps and tabular data, the following remarks seem especially relevant. Uncertainties in $\triangle T$ are directly reflected in the computed local times of maximum eclipse. However, as the latter are rounded to the nearest minute, errors only become significant in the more ancient past. In practice, as eclipse times are seldom recorded in Chinese history before about AD 1500 – and then only to the nearest double hour – the calculated times may be regarded as of more than adequate precision. A detailed knowledge of the accuracy of the map tracks only becomes important when an observation alleges a very large eclipse. However, such records deserve special attention since they are of value in refining $\triangle T$. In calculating geographical co-ordinates, we have regarded longitude as the independent variable, deducing the latitudes of the edges of the belts of totality for discrete values of longitude. Uncertainties in $\triangle T$ produce errors in these latitude limits at each selected longitude. These errors depend also on the angle which a particular track makes with the equator at the appropriate longitude. Thus if a track of totality or annularity runs nearly parallel to the equator at some point, even quite a large uncertainty in $\triangle T$ will have negligible component in latitude. For other angles of inclination (I) to the equator, it is readily seen that the appropriate latitude error is approximately proportional to tan I. The Earth turns through 0.25 deg in 1 minute of time. Hence if we let M minutes be the estimated uncertainty in $\triangle T$ on a particular date, the corresponding error in the latitude limits is roughly (M tan I)/4. On a typical eclipse map, the mean value of I is some 25 deg (tangent close to 0.5). Thus around the epochs AD 1600, AD 1000 and 500 BC, likely latitude errors at any given longitude are respectively 0.06, 0.15 and 0.8 deg. Even this last amount represents only about 2 mm on the scale of the maps so that the mapping accuracy is expected to be high, even in ancient times.

On the question of investigating hitherto unused records of large solar eclipses from the Far East to deduce $\triangle T$ more accurately, possibly the best prospects would seem to relate to the more ancient observations; the material since about 200 BC has been extensively studied. At present, only three records of total solar eclipses are known before 200 BC, all from the chronicle of a single small state – Lu, the home of Confucius. The dates of these events as recorded in the *Ch'un-ch'iu* are as follows: 709, 601 and 549 BC. If any additional observations from the Shang or Chou Dynasties were to come to light they might revolutionise present knowledge of the history of $\triangle T$. It is hoped that the production of this Atlas will encourage research in this field.

3. Capital cities of China, Korea and Japan

(i) China

On each of the eclipse maps we have marked, for reference, the site of a major Chinese city – usually the capital of the time (the Wade-Giles system is used in the transliteration of names). Throughout much of the long history of China, because of the great importance attached to political astrology, the hub of astronomical activity was the metropolis. Most of the solar eclipses (as well as other celestial phenomena) recorded in the various dynastic histories can thus be assumed to have been observed at the appropriate capital. However, the earliest preserved series of solar eclipse observations (from any part of the world) probably does not come into this category. During the so-called 'Spring and Autumn Period' (722 to 481 BC), more than 30 solar eclipses are recorded in the *Ch'un-ch'iu*. It seems probable that most of the observations were made at the state capital, for the *Ch'un-ch'iu* notes several eclipse ceremonies taking place there.

In later times, we find occasional statements to the effect that a particular eclipse was not seen by the Astronomer Royal (e.g. on account of cloudy weather at the capital) but was reported from some provincial city instead. However, these appear to be only random sightings. Once the empire was centralised (221 BC) there is no evidence of provincial observers maintaining a systematic watch of the sky in the style of the astronomers at the imperial court.

The capital of China has been changed many times, any particular place seldom having held the honour for more than about two centuries. Often a move took place as the result of a new dynasty being established but not infrequently the change occurred on account of the loss of the original capital due to invasion or rebellion. At several periods the emperor and his court were in temporary exile, while the repeated movements of the T'ang court back and forth between Ch'ang-an and Lo-yang over several decades around AD 700 are well known to historians. We have only taken into account what we regard as the more important changes of capital. Nevertheless, there must be a degree of arbitrariness about our choice at certain critical times.

Several major Chinese cities have been known by different names at different periods. Thus, close to the site of the Former Han (202 BC – AD 9) metropolis of Ch'ang-an, the first Sui emperor founded his capital of Ta-hsing Ch'êng (AD 583). This was later (AD 618) renamed Ch'ang-an by the first T'ang ruler. At each period we have given the style by which the capital was usually known at the time, but we have avoided name changes during a dynasty.

Apart from minor periods, the capital of China has occupied a site near one of the following seven present-day cities: An-yang (pinyin spelling: Anyang), Hang-chou (Hangzhou), Hsi-an (Xian), K'ai-fêng (Kaifeng), Lo-yang (Luoyang), Nan-ching (Nanjing) and Pei-ching (Beijing). Table 1 lists in alphabetical order the various Chinese cities marked on the maps together with their geographical co-ordinates (in degrees and decimals). Note that the T'ang city of Ch'ang-an was located a little to the south-east of the Han city of the same name.

During the earliest truly historical dynasty – the Shang (c. 1550 BC to 1045 BC) – the capital appears to have changed several times. However, the details are obscure and the dates uncertain. Original Shang records – in the form of 'oracle bones' – have, however, been recovered only from a single site, the 'Wastes of Yin' near An-yang. The vast numbers of inscribed bones excavated here

Table 1. *Locations of Chinese cities marked on the various maps*

Name of City	Latitude	Longitude
Ch'ang-an (Han)	34.35°N	108.88°E
Ch'ang-an (T'ang)	34.27	108.90
Ch'êng-tu	30.62	104.10
Chien-k'ang	32.03	118.78
Hao	34.16	108.72
Hsien-yang	34.39	108.85
Hsü-chang	34.05	113.80
K'ai-fêng	34.78	114.33
Lin-an	30.25	120.17
Lo-i	34.75	112.50
Lo-yang	34.75	112.47
Pei-ching	39.92	116.42
Ta-hsing Ch'êng	34.27	108.90
Ta-tu	39.92	116.42
Yin	36.07	114.33
Ying-t'ien	32.03	118.78

(Keightley, 1978) indicate the importance of the place over a considerable period. It seems possible that the capital which was traditionally called Yin occupied this site and that this is recalled in the modern name for the locality. Lacking direct evidence to the contrary, we have assumed Yin to have been the capital during the entire period from the beginning of this Atlas (1500 BC) to the fall of the Shang Dynasty around 1045 BC (Pankenier, 1984). This is almost certainly an over-simplification, but any other Shang capitals were probably in the vicinity of this place.

During the subsequent Chou Dynasty (c. 1045 to 256 BC – if we include the Chan-kuo or Warring States Period) the capital was first at Hao, but for the whole of the period after 770 BC it was located at or near Lo-i. Around 516 BC the royal residence was transferred to Chêng-chou, some 10 km to the east of Lo-i, where it remained for two centuries. However, we have felt it unnecessary to incorporate such a trivial change in position. For reference, the geographical co-ordinates of the capital of the state of Lu (modern Chü-fu) where most of the extant observations between 722 and 481 BC were probably made are: lat 35.13 deg N; long 117.02 deg E.

Finally, we come to the question of the Chinese capital at other major periods of partition, notably the Northern and Southern Dynasties (AD 420 – 589) and the co-existence of the Kin and Sung Empires (AD 1127 – 1234). In each case we have marked the metropolis of the native Chinese Dynasty (respectively Chien-k'ang and Lin-an) on the appropriate maps. However, it should be emphasised that there was also much independent astronomical activity centred largely on the site of Ch'ang-an (AD 420 – 589) and K'ai-fêng (AD 1127 – 1234). In consequence, there are many parallel reports of eclipses and other astronomical phenomena at these times.

(ii) Korea and Japan

Eclipse records from these countries cover only about the last third of the time-span of this Atlas and in order to maintain a systematic format for the maps we have not marked the sites of the appropriate capitals. Instead, we have preferred to summarise the historical background. Because of the small extent of both Korea and Japan compared with China, the spatial movements of the various capitals are relatively minor on the scale of the maps.

A significant number of eclipse records are found in the Samguk Sagi, the earliest history of Korea, covering the period from the 1st century BC to AD 935. However, these records are generally acknowledged to be of dubious reliability, some possibly having originated in China. The detailed history of Korea does not begin until the establishment of the kingdom of Koryŏ in AD 918, although the first eclipse record in the Koryŏ-sa is as late as AD 1012. However, from this time onwards, both solar and lunar eclipses are fairly regularly noted in Korean history.

The capital of Korea was Songdo (lat 37.97 deg N; long 126.58 deg E) for almost the entire Koryŏ Dynasty (AD 918 – 1392) and apart from two periods of exile, we should expect the eclipse records to originate from there. It seems likely that eclipses charted in the periods AD 1236 to 1268 and 1290 to 1292 inclusive were observed on the island of Kangwha (37.73 deg N; 126.48 deg E), to which the court had fled during periods of Mongol invasion. Soon after the fall of the Koryŏ Dynasty, the newly established Yi Dynasty moved the capital to Hanyang

(modern Sŏul, 37.55 deg N; 126.97 deg E) in AD 1394. Here it has remained apart from brief periods during which no solar eclipses are charted.

The first solar eclipse record from Japan dates from AD 628 and from this time onwards eclipses are at first occasionally and later fairly frequently reported. However, the historical sources are by no means as systematic as those from China or Korea. Kanda (1935) compiled a valuable list from diaries of courtiers, temple records, etc as well as histories.

From very early times, the Japanese capital moved from one site to another in the Yamato Plain following the death of each emperor. Nara (34.68 deg N; 135.82 deg E) was established as the first fixed capital in AD 710, but the city held this position only until AD 784. For a further decade, Nagaoka (37.45 deg N; 138.83 deg E) became the imperial residence, after which the capital was established at Heian (modern Kyōto, 35.03 deg N; 135.75 deg E). Here it remained for more than a thousand years until it was replaced by Tōkyō (35.67 deg N; 139.75 deg E) in 1868.

In AD 1185, a Shogunate was founded at Kamakura (35.32 deg N; 139.55 deg E) and this became an important cultural centre. It is thus likely that a number of eclipse observations noted in Japanese history during this period were made here instead of Heian. Similar remarks apply to Edo (Tōkyō, 35.67 deg N; 139.75 deg E) between about AD 1590 and 1868, after which it became the capital.

4. Data represented on the maps

Above each map is given the following information: Julian ephemeris day number (rounded to the nearest integer), local calendar date (year, month and day) and value of $\triangle T$ used in making the various calculations. On the map itself is shown that portion of a track of totality or annularity which lies within the selected area: from 15 to 45 deg N and from 90 to 150 deg E. (The adopted sign convention for longitude is negative to the E of Greenwich). The edges of the zones of totality are shown by solid lines, those of annularity by broken lines. If an eclipse was annular for part of its trajectory and total for the remaining part the track edges are shown by solid lines. The actual phase at any particular longitude can be determined from the data listed below the map. Where appropriate, sunrise and sunset positions are denoted by short lines a little to the west or east of the main track and roughly perpendicular to it.

Below each map is given the degree of obscuration of the Sun at the E and W ends of the map track (over the range of longitudes listed to the right of the map). For a total eclipse, these figures will exceed 100 per cent, while for an annular eclipse the converse will be true. Where appropriate, the following additional information is supplied: (i) the greatest obscuration of the Sun along the eclipse track – if a maximum occurs in the region covered by the map; (ii) the approximate location at which the Sun rose or set centrally eclipsed, allowing for refraction and semi-diameter.

Accurate co-ordinates of an eclipse track are given in the space to the right of each map. Over a range of longitudes, usually from about 87 to 153 deg E, are given the following details: latitude of the N limit of the eclipse track, latitude of the S limit, approximate local time (in hours and minutes) of greatest phase and mean altitude of the Sun at this time. The latitude information will enable the user to draw the track on a more detailed map or investigate more carefully the local circumstances at a particular place, if required.

5. Comparison with historical records
(i) Total and near-total eclipses

Since all dates in this section directly concern eclipses, we shall use the convention of positive and negative numbers rather than the BC/AD system. Eclipses which were either total or nearly so are reported at almost all periods in Chinese history and there are occasional records from Korea and Japan. Such records are particularly prevalent in the astronomical treatises of the *Han-shu* and *Hou-han-shu* (which between them cover the period from about −200 to +220). It is instructive to compare these observations with the computed paths of totality and annularity as shown on the appropriate maps. The recorded dates, as converted to the Julian Calendar, and descriptions are given in Table 2.

Table 2. *Han records of large solar eclipses*

Julian Date	Description
−197 Aug 7	total
−187 July 17	almost complete
−180 March 4	total
−146 Nov 10	almost complete
−88 Sep 29	not complete, like a hook
−79 Sep 20	almost complete
−33 Aug 23	not complete, like a hook; it set
−27 Jun 19	not complete, like a hook
−1 Feb 5	not complete, like a hook
+ 2 Nov 23	total
+65 Dec 16	total
+120 Jan 18	almost complete, day became like evening

From Table 2 it will be seen that four total solar eclipses are recorded (−197, −180, +2 and +65) but without any details, although the *Shih-chi* states that on the occasion of the eclipse in −180 it became dark in the daytime. Reference to the maps shows that all but the eclipse of +2 are represented as central (only marginally so in +65) at the capital of the time (Ch'ang-an or Lo-yang) but in the case of −197 the phase was annular instead of total. It would seem that there was no separate term at this period to describe a ring eclipse; the annular obscuration of +516 is reported in the same way (in the *Nan-shih*). The observation in +2 is clearly discordant for a large change in △T would be necessary to render the eclipse central at Ch'ang-an. In fact, since the belt of totality was of virtually negligible width, the probability of the central phase actually being witnessed and reported to the capital is minute. Possibly an original report of an 'almost total' eclipse was later abbreviated.

Of the eclipses described as either 'not complete, like a hook' or 'almost complete' the relevant maps show that in −187, −146, −88, −79, −27 and −1 a large partial eclipse was indeed visible at the capital. The map for +120 makes this eclipse actually total at Lo-yang but only a small error in △T (some 3 minutes) would render the phase just partial – in keeping with the record, which alleges a very large obscuration. There was no eclipse visible in China in −33, but the fact that the text also indicates that the Sun set while the eclipse was still visible led Dubs (1941) to suggest an alternative date of −34 Nov 1. On this occasion the Sun would set eclipsed but the dating error is difficult to explain; it will be noted that the other dates of the eclipses discussed here agree exactly with the calculated dates.

(ii) Partial eclipses

Although the main emphasis in this work is on the accurate mapping of the paths of central eclipses, most large partial eclipses visible in the Far East should also be identifiable. A rough rule for estimating the magnitude of an eclipse outside the central zone is as follows. On the scale of the maps, a displacement of 1 mm perpendicular to the edge of a track of totality or annularity will decrease the degree of obscuration of the Sun by some 1 per cent. In the case of a total eclipse, the calculated magnitude on the edges of the map track is exactly 100 per cent, so that interpolation is straightforward. However, for an annular obscuration, allowance must be made for the fact that even the central obscuration may be much less than 100 per cent (see details given below map).

If an eclipse track passes a little to the N or S of the region covered by the maps, there will be no entry for this particular date despite the fact that a partial eclipse will be produced over much of our area of interest. Hence it is apparent that this Atlas is by no means complete for *partial* eclipses visible in the Far East.

As a practical illustration of the coverage of observed eclipses of *any* magnitude we may take the case of the *Ch'un-ch'iu* record already referred to above. Between −710 and −480, this chronicle notes 37 eclipses, mostly without any descriptive details. Four of these must be presumed to be either false sightings or possibly abortive predictions since on or near the stated dates no eclipse was visible on the Earth's surface. Of the remaining 33, as many as 28 eclipses are charted here, the exceptions being of fairly small computed magnitude in NE China. A similar degree of completeness seems likely at other periods in Far Eastern history.

References

Dickey, J. O. & Williams, J. G. (1982). Geophysical applications of lunar laser ranging. *Trans. Amer. Geophys. Union*, **163**, 301.

Hsueh Chung-san & Ou-yang I (1956). *A Sino-Western Calendar for Two Thousand Years: 1 – 2000 A.D.* Peking, San-lien Shu-tien.

IAU (1968). *Trans. Int. Astr. Union*, **13B**, 48.

Kanda Shigeru (1935). *Nihon Tenmon Shiryo*. Tokyo: Koseisha.

Keightley, D. K. (1978). *Sources of Shang History*. Berkeley: University of California Press.

Morrison, L. V. & Ward, C. G. (1975). An analysis of the transits of Mercury: 1677–1973. *Mon. Not. Roy. Astr. Soc.*, **173**, 183–206.

Oppolzer, T. R. von (1887). *Canon der Finsternisse*. Wien: Karl Gerold's Sohn. (Reprinted 1962 as *Canon of Eclipses*. New York: Dover Publications, Inc.)

Pankenier, D. W. (1982). Astronomical dates in Shang and Western Zhou. *Early China*, **7**, 1–37.

Stephenson, F. R. & Morrison, L. V. (1984). Long-term changes in the rotation of the Earth: 700 B.C. to A.D. 1980. *Phil. Trans. Roy. Soc. A*, **313**, 47–70.

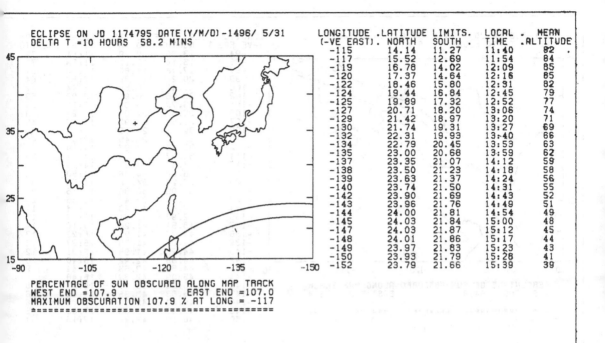

ECLIPSE ON JD 1174795 DATE(Y/M/D) -1496/ 5/31
DELTA T =10 HOURS 58.2 MINS

LONGITUDE (-VE EAST)	LATITUDE LIMITS. NORTH	SOUTH	LOCAL TIME	MEAN ALTITUDE
-115	14.14	11.27	11:40	82
-117	15.52	12.69	11:54	84
-119	16.78	14.02	12:09	85
-120	17.37	14.64	12:16	85
-122	18.46	15.80	12:31	82
-124	19.44	16.84	12:45	79
-125	19.89	17.32	12:52	77
-127	20.71	18.20	13:06	74
-129	21.42	18.97	13:20	71
-130	21.74	19.31	13:27	69
-132	22.31	19.93	13:40	66
-134	22.79	20.45	13:53	63
-135	23.00	20.68	13:59	62
-137	23.35	21.07	14:12	59
-138	23.50	21.23	14:18	58
-139	23.63	21.37	14:24	56
-140	23.74	21.50	14:31	55
-142	23.90	21.69	14:43	52
-143	23.96	21.76	14:49	51
-144	24.00	21.81	14:54	49
-145	24.03	21.84	15:00	48
-147	24.03	21.87	15:12	45
-148	24.01	21.86	15:17	44
-149	23.97	21.83	15:23	43
-150	23.93	21.79	15:28	41
-152	23.79	21.66	15:39	39

PERCENTAGE OF SUN OBSCURED ALONG MAP TRACK
WEST END =107.9   EAST END =107.0
MAXIMUM OBSCURATION 107.9 % AT LONG = -117
==

+ MARKS THE CITY OF YIN

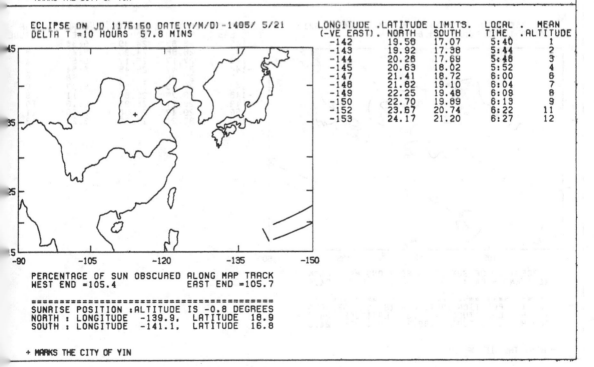

ECLIPSE ON JD 1176150 DATE(Y/M/D) -1405/ 5/21
DELTA T =10 HOURS 57.8 MINS

LONGITUDE (-VE EAST)	LATITUDE LIMITS. NORTH	SOUTH	LOCAL TIME	MEAN ALTITUDE
-142	19.58	17.07	5:40	1
-143	19.92	17.38	5:44	2
-144	20.28	17.69	5:48	3
-145	20.63	18.02	5:52	4
-147	21.41	18.72	6:00	6
-148	21.82	19.10	6:04	7
-149	22.25	19.48	6:09	8
-150	22.70	19.89	6:13	9
-152	23.67	20.74	6:22	11
-153	24.17	21.20	6:27	12

PERCENTAGE OF SUN OBSCURED ALONG MAP TRACK
WEST END =105.4   EAST END =105.7

==
SUNRISE POSITION :ALTITUDE IS -0.8 DEGREES
NORTH : LONGITUDE -139.9, LATITUDE 18.9
SOUTH : LONGITUDE -141.1, LATITUDE 16.8

+ MARKS THE CITY OF YIN

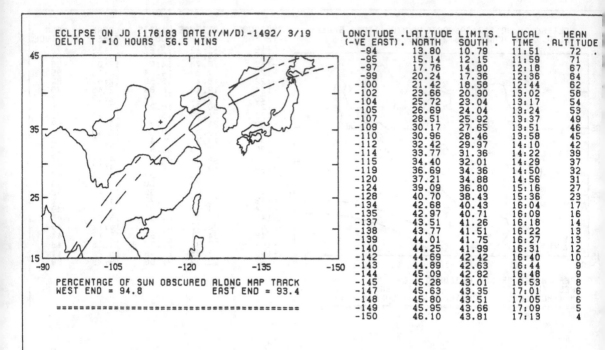

```
ECLIPSE ON JD 1176183 DATE(Y/M/D)-1492/ 3/19
DELTA T =10 HOURS  56.5 MINS
```

LONGITUDE (-VE EAST)	LATITUDE NORTH	LIMITS SOUTH	LOCAL TIME	MEAN ALTITUDE
-94	13.80	10.79	11:51	72
-95	15.14	12.15	11:59	71
-97	17.76	14.80	12:18	67
-99	20.24	17.36	12:36	64
-100	21.42	18.58	12:44	62
-102	23.66	20.90	13:02	58
-104	25.72	23.04	13:17	54
-105	26.69	24.04	13:24	53
-107	28.51	25.92	13:37	49
-109	30.17	27.65	13:51	46
-110	30.96	28.46	13:58	45
-112	32.42	29.97	14:10	42
-114	33.77	31.36	14:22	39
-115	34.40	32.01	14:29	37
-119	36.69	34.36	14:50	32
-120	37.21	34.88	14:56	31
-124	39.09	36.80	15:16	27
-128	40.70	38.43	15:36	23
-134	42.68	40.43	16:04	17
-135	42.97	40.71	16:09	16
-137	43.51	41.26	16:18	14
-138	43.77	41.51	16:22	13
-139	44.01	41.75	16:27	13
-140	44.25	41.99	16:31	12
-142	44.69	42.42	16:40	10
-143	44.89	42.63	16:44	9
-144	45.09	42.82	16:48	9
-145	45.28	43.01	16:53	8
-147	45.63	43.35	17:01	6
-148	45.80	43.51	17:05	6
-149	45.95	43.66	17:09	5
-150	46.10	43.81	17:13	4

```
PERCENTAGE OF SUN OBSCURED ALONG MAP TRACK
WEST END = 94.8            EAST END = 93.4
=============================================
```

+ MARKS THE CITY OF YIN

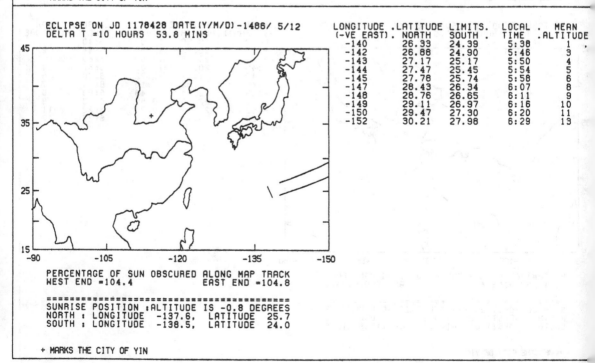

```
ECLIPSE ON JD 1178428 DATE(Y/M/D)-1486/ 5/12
DELTA T =10 HOURS  53.8 MINS
```

LONGITUDE (-VE EAST)	LATITUDE NORTH	LIMITS SOUTH	LOCAL TIME	MEAN ALTITUDE
-140	26.33	24.39	5:38	1
-142	26.88	24.90	5:46	3
-143	27.17	25.17	5:50	4
-144	27.47	25.45	5:54	5
-145	27.78	25.74	5:58	6
-147	28.43	26.34	6:07	8
-148	28.76	26.65	6:11	9
-149	29.11	26.97	6:16	10
-150	29.47	27.30	6:20	11
-152	30.21	27.98	6:29	13

```
PERCENTAGE OF SUN OBSCURED ALONG MAP TRACK
WEST END =104.4            EAST END =104.8
=============================================
SUNRISE POSITION :ALTITUDE IS -0.8 DEGREES
NORTH : LONGITUDE  -137.6,  LATITUDE  25.7
SOUTH : LONGITUDE  -138.5,  LATITUDE  24.0
```

+ MARKS THE CITY OF YIN

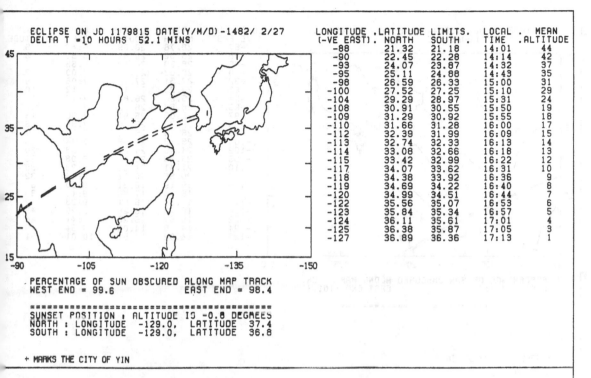

ECLIPSE ON JD 1179815 DATE(Y/M/D)-1482/ 2/27
DELTA T =10 HOURS 52.1 MINS

LONGITUDE (-VE EAST)	.LATITUDE LIMITS. NORTH	 SOUTH .	LOCAL . TIME	MEAN .ALTITUDE
-88	21.32	21.18	14:01	44
-90	22.45	22.28	14:14	42
-93	24.07	23.87	14:32	37
-95	25.11	24.88	14:43	35
-98	26.59	26.33	15:00	31
-100	27.52	27.25	15:10	29
-104	29.29	28.97	15:31	24
-108	30.91	30.55	15:50	19
-109	31.29	30.92	15:55	18
-110	31.66	31.28	16:00	17
-112	32.39	31.99	16:09	15
-113	32.74	32.33	16:13	14
-114	33.08	32.66	16:18	13
-115	33.42	32.99	16:22	12
-117	34.07	33.62	16:31	10
-118	34.38	33.92	16:36	9
-119	34.69	34.22	16:40	8
-120	34.99	34.51	16:44	7
-122	35.56	35.07	16:53	6
-123	35.84	35.34	16:57	5
-124	36.11	35.61	17:01	4
-125	36.38	35.87	17:05	3
-127	36.89	36.36	17:13	1

PERCENTAGE OF SUN OBSCURED ALONG MAP TRACK
WEST END = 99.6   EAST END = 98.4

===
SUNSET POSITION : ALTITUDE IS -0.8 DEGREES
NORTH : LONGITUDE -129.0, LATITUDE 37.4
SOUTH : LONGITUDE -129.0, LATITUDE 36.8

+ MARKS THE CITY OF YIN

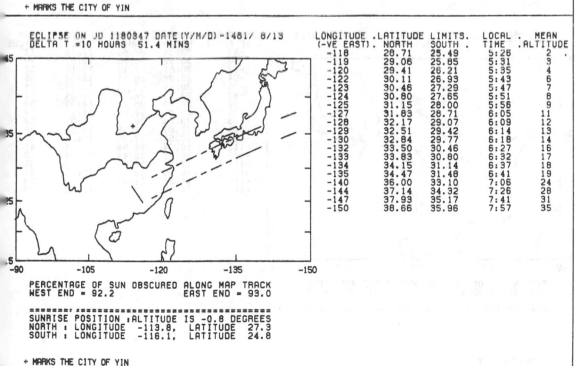

ECLIPSE ON JD 1180347 DATE(Y/M/D)-1481/ 8/13
DELTA T =10 HOURS 51.4 MINS

LONGITUDE (-VE EAST)	.LATITUDE LIMITS. NORTH	 SOUTH .	LOCAL . TIME	MEAN .ALTITUDE
-118	28.71	25.49	5:26	2
-119	29.06	25.85	5:31	3
-120	29.41	26.21	5:35	4
-122	30.11	26.93	5:43	6
-123	30.46	27.29	5:47	7
-124	30.80	27.65	5:51	8
-125	31.15	28.00	5:56	9
-127	31.83	28.71	6:05	11
-128	32.17	29.07	6:09	12
-129	32.51	29.42	6:14	13
-130	32.84	29.77	6:18	14
-132	33.50	30.46	6:27	16
-133	33.83	30.80	6:32	17
-134	34.15	31.14	6:37	18
-135	34.47	31.48	6:41	19
-140	36.00	33.10	7:06	24
-144	37.14	34.32	7:26	28
-147	37.93	35.17	7:41	31
-150	38.66	35.96	7:57	35

PERCENTAGE OF SUN OBSCURED ALONG MAP TRACK
WEST END = 92.2   EAST END = 93.0

===
SUNRISE POSITION :ALTITUDE IS -0.8 DEGREES
NORTH : LONGITUDE -113.8, LATITUDE 27.3
SOUTH : LONGITUDE -116.1, LATITUDE 24.8

+ MARKS THE CITY OF YIN

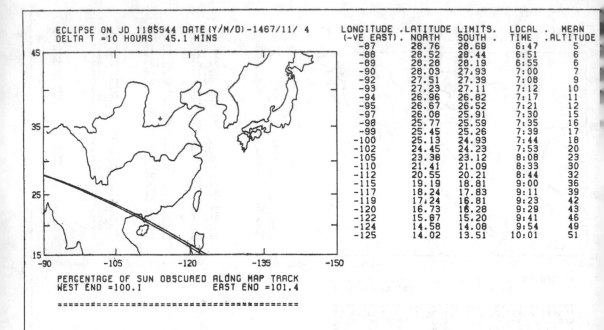

ECLIPSE ON JD 1185544 DATE(Y/M/D) -1467/11/ 4
DELTA T =10 HOURS   45.1 MINS

LONGITUDE (-VE EAST)	LATITUDE NORTH	LIMITS. SOUTH	LOCAL TIME	MEAN ALTITUDE
-87	28.76	28.69	6:47	5
-88	28.52	28.44	6:51	6
-89	28.28	28.19	6:55	6
-90	28.03	27.93	7:00	7
-92	27.51	27.39	7:08	9
-93	27.23	27.11	7:12	10
-94	26.96	26.82	7:17	11
-95	26.67	26.52	7:21	12
-97	26.08	25.91	7:30	15
-98	25.77	25.59	7:35	16
-99	25.45	25.26	7:39	17
-100	25.13	24.93	7:44	18
-102	24.45	24.23	7:53	20
-105	23.38	23.12	8:08	23
-110	21.41	21.09	8:33	30
-112	20.55	20.21	8:44	32
-115	19.19	18.81	9:00	36
-117	18.24	17.83	9:11	39
-119	17.24	16.81	9:23	42
-120	16.73	16.28	9:29	43
-122	15.87	15.20	9:41	46
-124	14.58	14.08	9:54	49
-125	14.02	13.51	10:01	51

PERCENTAGE OF SUN OBSCURED ALONG MAP TRACK
WEST END =100.1   EAST END =101.4

===

+ MARKS THE CITY OF YIN

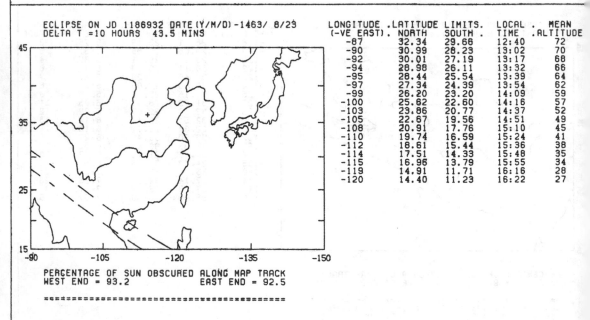

ECLIPSE ON JD 1186932 DATE(Y/M/D) -1463/ 8/23
DELTA T =10 HOURS   43.5 MINS

LONGITUDE (-VE EAST)	LATITUDE NORTH	LIMITS. SOUTH	LOCAL TIME	MEAN ALTITUDE
-87	32.34	29.66	12:40	72
-90	30.99	28.23	13:02	70
-92	30.01	27.19	13:17	68
-94	28.98	26.11	13:32	66
-95	28.44	25.54	13:39	64
-97	27.34	24.39	13:54	62
-99	26.20	23.20	14:09	59
-100	25.62	22.60	14:16	57
-103	23.86	20.77	14:37	52
-105	22.67	19.56	14:51	49
-108	20.91	17.76	15:10	45
-110	19.74	16.59	15:24	41
-112	18.61	15.44	15:36	38
-114	17.51	14.33	15:48	35
-115	16.96	13.79	15:55	34
-119	14.91	11.71	16:16	28
-120	14.40	11.23	16:22	27

PERCENTAGE OF SUN OBSCURED ALONG MAP TRACK
WEST END = 93.2   EAST END = 92.5

===

+ MARKS THE CITY OF YIN

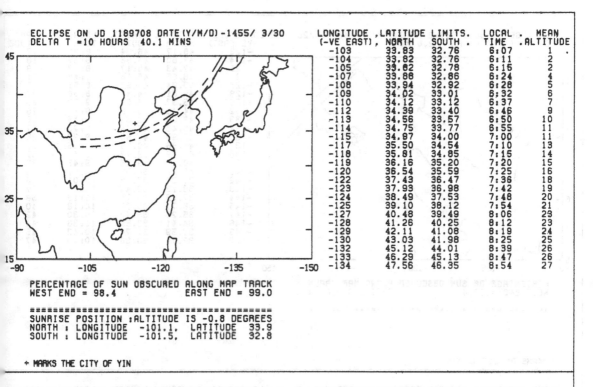

ECLIPSE ON JD 1189708 DATE(Y/M/D)-1455/ 3/30
DELTA T =10 HOURS 40.1 MINS

LONGITUDE (-VE EAST)	LATITUDE NORTH	LIMITS. SOUTH	LOCAL TIME	MEAN ALTITUDE
-103	33.83	32.76	6:07	1
-104	33.82	32.76	6:11	2
-105	33.82	32.78	6:15	2
-107	33.86	32.86	6:24	4
-108	33.94	32.92	6:28	5
-109	34.02	33.01	6:32	6
-110	34.12	33.12	6:37	7
-112	34.39	33.40	6:46	9
-113	34.56	33.57	6:50	10
-114	34.75	33.77	6:55	11
-115	34.97	34.00	7:00	11
-117	35.50	34.54	7:10	13
-118	35.81	34.85	7:15	14
-119	36.16	35.20	7:20	15
-120	36.54	35.59	7:25	16
-122	37.43	36.47	7:36	18
-123	37.93	36.98	7:42	19
-124	38.49	37.53	7:48	20
-125	39.10	38.12	7:54	21
-127	40.48	39.49	8:06	23
-128	41.26	40.25	8:12	23
-129	42.11	41.08	8:19	24
-130	43.03	41.98	8:25	25
-132	45.12	44.01	8:39	26
-133	46.29	45.13	8:47	26
-134	47.56	46.35	8:54	27

PERCENTAGE OF SUN OBSCURED ALONG MAP TRACK
WEST END = 98.4   EAST END = 99.0

===
SUNRISE POSITION :ALTITUDE IS -0.8 DEGREES
NORTH : LONGITUDE -101.1, LATITUDE 33.9
SOUTH : LONGITUDE -101.5, LATITUDE 32.8

+ MARKS THE CITY OF YIN

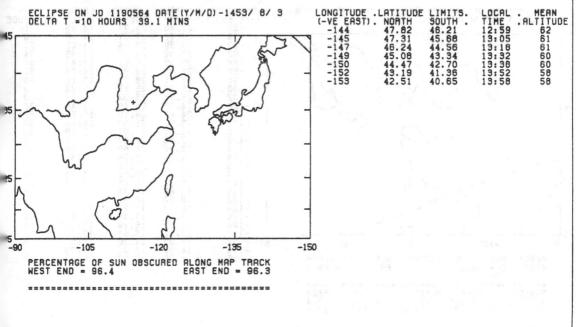

ECLIPSE ON JD 1190564 DATE(Y/M/D)-1453/ 8/ 3
DELTA T =10 HOURS 39.1 MINS

LONGITUDE (-VE EAST)	LATITUDE NORTH	LIMITS. SOUTH	LOCAL TIME	MEAN ALTITUDE
-144	47.82	46.21	12:59	62
-145	47.31	45.68	13:05	61
-147	46.24	44.56	13:18	61
-149	45.08	43.34	13:32	60
-150	44.47	42.70	13:38	60
-152	43.19	41.36	13:52	58
-153	42.51	40.65	13:58	58

PERCENTAGE OF SUN OBSCURED ALONG MAP TRACK
WEST END = 96.4   EAST END = 96.3

===

+ MARKS THE CITY OF YIN

5

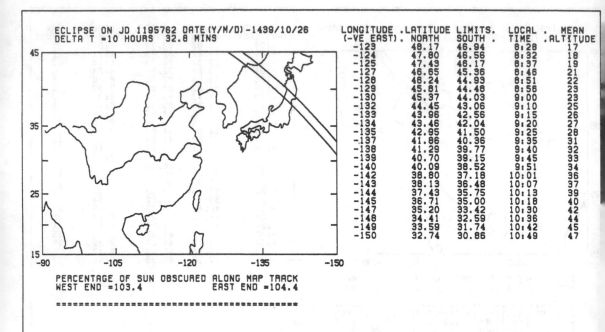

ECLIPSE ON JD 1195762 DATE(Y/M/D)-1439/10/26
DELTA T =10 HOURS 32.8 MINS

LONGITUDE (-VE EAST)	LATITUDE NORTH	LIMITS SOUTH	LOCAL TIME	MEAN ALTITUDE
-123	48.17	46.94	8:28	17
-124	47.80	46.56	8:32	18
-125	47.43	46.17	8:37	19
-127	46.65	45.36	8:46	21
-128	46.24	44.99	8:51	22
-129	45.81	44.48	8:56	23
-130	45.37	44.03	9:00	23
-132	44.45	43.06	9:10	25
-133	43.96	42.56	9:15	26
-134	43.46	42.04	9:20	27
-135	42.95	41.50	9:25	28
-137	41.86	40.36	9:35	31
-138	41.29	39.77	9:40	32
-139	40.70	39.15	9:45	33
-140	40.09	38.52	9:51	34
-142	38.80	37.18	10:01	36
-143	38.13	36.48	10:07	37
-144	37.43	35.75	10:13	39
-145	36.71	35.00	10:18	40
-147	35.20	33.42	10:30	42
-148	34.41	32.59	10:36	44
-149	33.59	31.74	10:42	45
-150	32.74	30.86	10:49	47

PERCENTAGE OF SUN OBSCURED ALONG MAP TRACK
WEST END =103.4   EAST END =104.4

==

+ MARKS THE CITY OF YIN

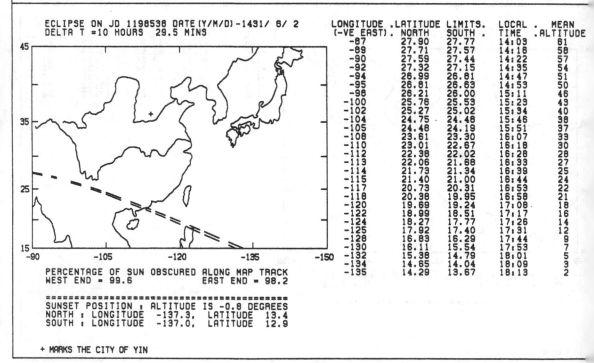

ECLIPSE ON JD 1198538 DATE(Y/M/D)-1431/ 6/ 2
DELTA T =10 HOURS 29.5 MINS

LONGITUDE (-VE EAST)	LATITUDE NORTH	LIMITS SOUTH	LOCAL TIME	MEAN ALTITUDE
-87	27.90	27.77	14:09	61
-89	27.71	27.57	14:16	58
-90	27.59	27.44	14:22	57
-92	27.32	27.15	14:35	54
-94	26.99	26.81	14:47	51
-95	26.81	26.63	14:53	50
-98	26.21	26.00	15:11	46
-100	25.76	25.53	15:23	43
-102	25.27	25.02	15:34	40
-104	24.75	24.48	15:46	38
-105	24.48	24.19	15:51	37
-108	23.61	23.30	16:07	33
-110	23.01	22.67	16:18	30
-112	22.38	22.02	16:28	28
-113	22.06	21.68	16:33	27
-114	21.73	21.34	16:39	25
-115	21.40	21.00	16:44	24
-117	20.73	20.31	16:53	22
-118	20.38	19.95	16:58	21
-120	19.69	19.24	17:08	18
-122	18.99	18.51	17:17	16
-124	18.27	17.77	17:26	14
-125	17.92	17.40	17:31	12
-128	16.83	16.29	17:44	9
-130	16.11	15.54	17:53	7
-132	15.38	14.79	18:01	5
-134	14.65	14.04	18:09	3
-135	14.29	13.67	18:13	2

PERCENTAGE OF SUN OBSCURED ALONG MAP TRACK
WEST END = 99.6   EAST END = 98.2

==
SUNSET POSITION : ALTITUDE IS -0.8 DEGREES
NORTH : LONGITUDE -137.3, LATITUDE 13.4
SOUTH : LONGITUDE -137.0, LATITUDE 12.9

+ MARKS THE CITY OF YIN

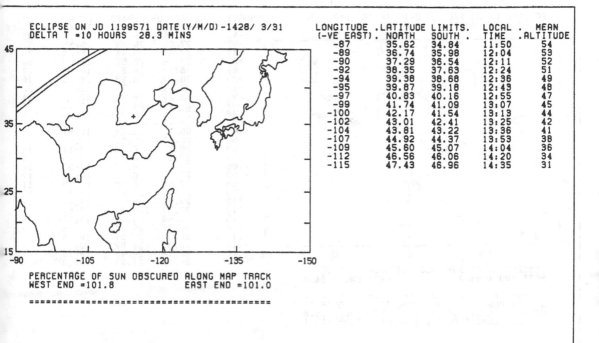

ECLIPSE ON JD 1199571 DATE(Y/M/D)-1428/ 3/31
DELTA T =10 HOURS 28.3 MINS

LONGITUDE (-VE EAST).	.LATITUDE LIMITS. NORTH	SOUTH .	LOCAL . TIME	MEAN .ALTITUDE
-87	35.62	34.84	11:50	54
-89	36.74	35.98	12:04	53
-90	37.29	36.54	12:11	52
-92	38.35	37.63	12:24	51
-94	39.38	38.68	12:36	49
-95	39.87	39.18	12:43	48
-97	40.83	40.16	12:55	47
-99	41.74	41.09	13:07	45
-100	42.17	41.54	13:13	44
-102	43.01	42.41	13:25	42
-104	43.81	43.22	13:36	41
-107	44.92	44.37	13:53	38
-109	45.60	45.07	14:04	36
-112	46.56	46.06	14:20	34
-115	47.43	46.96	14:35	31

PERCENTAGE OF SUN OBSCURED ALONG MAP TRACK
WEST END =101.8   EAST END =101.0

===

+ MARKS THE CITY OF YIN

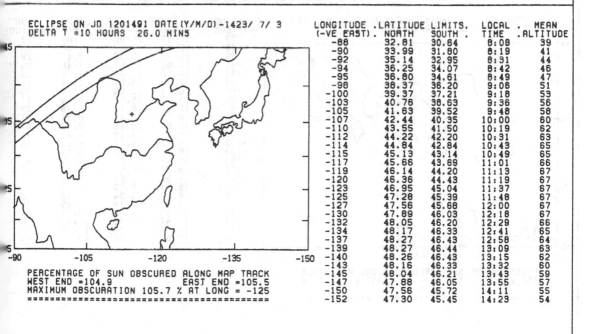

ECLIPSE ON JD 1201491 DATE(Y/M/D)-1423/ 7/ 3
DELTA T =10 HOURS 26.0 MINS

LONGITUDE (-VE EAST).	.LATITUDE LIMITS. NORTH	SOUTH .	LOCAL . TIME	MEAN .ALTITUDE
-88	32.81	30.64	8:08	39
-90	33.99	31.80	8:19	41
-92	35.14	32.95	8:31	44
-94	36.25	34.07	8:42	46
-95	36.80	34.61	8:49	47
-98	38.37	36.20	9:06	51
-100	39.37	37.21	9:18	53
-103	40.76	38.63	9:36	56
-105	41.63	39.52	9:48	58
-107	42.44	40.35	10:00	60
-110	43.55	41.50	10:19	62
-112	44.22	42.20	10:31	63
-114	44.84	42.84	10:43	65
-115	45.13	43.14	10:49	65
-117	45.66	43.69	11:01	66
-119	46.14	44.20	11:13	67
-120	46.36	44.43	11:19	67
-123	46.95	45.04	11:37	67
-125	47.28	45.39	11:48	67
-127	47.56	45.68	12:00	67
-130	47.89	46.03	12:18	67
-132	48.05	46.20	12:29	66
-134	48.17	46.33	12:41	65
-137	48.27	46.43	12:58	64
-139	48.27	46.44	13:09	63
-140	48.26	46.43	13:15	62
-143	48.16	46.33	13:32	60
-145	48.04	46.21	13:43	59
-147	47.88	46.05	13:55	57
-150	47.56	45.72	14:11	55
-152	47.30	45.45	14:23	54

PERCENTAGE OF SUN OBSCURED ALONG MAP TRACK
WEST END =104.9   EAST END =105.5
MAXIMUM OBSCURATION 105.7 % AT LONG = -125

===

+ MARKS THE CITY OF YIN

7

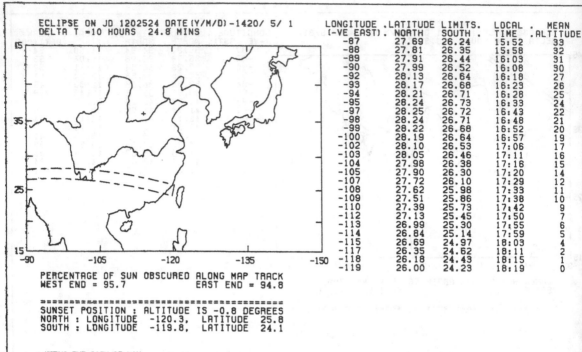

ECLIPSE ON JD 1202524 DATE(Y/M/D)−1420/ 5/ 1
DELTA T =10 HOURS 24.8 MINS

LONGITUDE (-VE EAST)	LATITUDE NORTH	LIMITS SOUTH	LOCAL TIME	MEAN ALTITUDE
-87	27.69	26.24	15:52	33
-88	27.81	26.35	15:58	32
-89	27.91	26.44	16:03	31
-90	27.99	26.52	16:08	30
-92	28.13	26.64	16:18	27
-93	28.17	26.68	16:23	26
-94	28.21	26.71	16:28	25
-95	28.24	26.73	16:33	24
-97	28.25	26.72	16:43	22
-98	28.24	26.71	16:48	21
-99	28.22	26.68	16:52	20
-100	28.19	26.64	16:57	19
-102	28.10	26.53	17:06	17
-103	28.05	26.46	17:11	16
-104	27.98	26.38	17:16	15
-105	27.90	26.30	17:20	14
-107	27.72	26.10	17:29	12
-108	27.62	25.98	17:33	11
-109	27.51	25.86	17:38	10
-110	27.39	25.73	17:42	9
-112	27.13	25.45	17:50	7
-113	26.99	25.30	17:55	6
-114	26.84	25.14	17:59	5
-115	26.69	24.97	18:03	4
-117	26.35	24.62	18:11	2
-118	26.18	24.43	18:15	1
-119	26.00	24.23	18:19	0

PERCENTAGE OF SUN OBSCURED ALONG MAP TRACK
WEST END = 95.7   EAST END = 94.8

===
SUNSET POSITION : ALTITUDE IS -0.8 DEGREES
NORTH : LONGITUDE -120.3, LATITUDE 25.8
SOUTH : LONGITUDE -119.8, LATITUDE 24.1

* MARKS THE CITY OF YIN

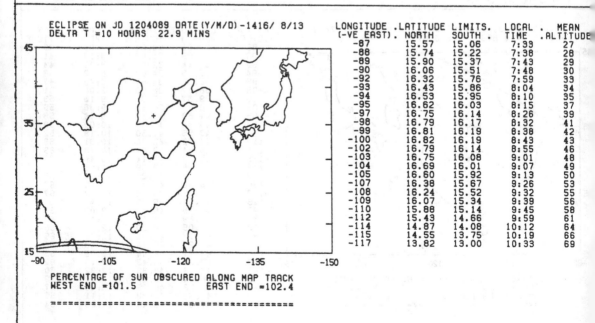

ECLIPSE ON JD 1204089 DATE(Y/M/D)−1416/ 8/13
DELTA T =10 HOURS 22.9 MINS

LONGITUDE (-VE EAST)	LATITUDE NORTH	LIMITS SOUTH	LOCAL TIME	MEAN ALTITUDE
-87	15.57	15.06	7:33	27
-88	15.74	15.22	7:38	28
-89	15.90	15.37	7:43	29
-90	16.06	15.51	7:48	30
-92	16.32	15.76	7:59	33
-93	16.43	15.86	8:04	34
-94	16.53	15.95	8:10	35
-95	16.62	16.03	8:15	37
-97	16.75	16.14	8:26	39
-98	16.79	16.17	8:32	41
-99	16.81	16.19	8:38	42
-100	16.82	16.19	8:43	43
-102	16.79	16.14	8:55	46
-103	16.75	16.08	9:01	48
-104	16.69	16.01	9:07	49
-105	16.60	15.92	9:13	50
-107	16.38	15.67	9:26	53
-108	16.24	15.52	9:32	55
-109	16.07	15.34	9:39	56
-110	15.88	15.14	9:45	58
-112	15.43	14.66	9:59	61
-114	14.87	14.08	10:12	64
-115	14.55	13.75	10:19	66
-117	13.82	13.00	10:33	69

PERCENTAGE OF SUN OBSCURED ALONG MAP TRACK
WEST END =101.5   EAST END =102.4

===

+ MARKS THE CITY OF YIN

ECLIPSE ON JD 1206688 DATE(Y/M/D)-1409/ 9/25
DELTA T =10 HOURS 19.8 MINS

LONGITUDE (-VE EAST)	LATITUDE LIMITS. NORTH	SOUTH	LOCAL TIME	MEAN ALTITUDE
-87	24.38	21.45	10:14	58
-89	23.54	20.57	10:28	62
-90	23.09	20.11	10:36	63
-92	22.13	19.11	10:51	67
-94	21.09	18.03	11:06	70
-95	20.54	17.45	11:14	72
-97	19.38	16.25	11:30	75
-99	18.15	14.97	11:46	77
-100	17.51	14.31	11:55	78
-102	16.18	12.94	12:12	79
-104	14.81	11.54	12:29	78
-105	14.12	10.83	12:37	77

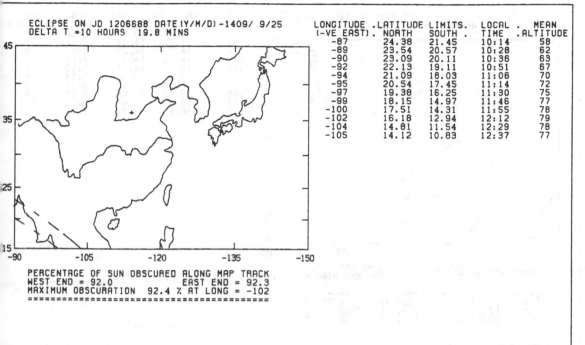

PERCENTAGE OF SUN OBSCURED ALONG MAP TRACK
WEST END = 92.0   EAST END = 92.3
MAXIMUM OBSCURATION 92.4 % AT LONG = -102
===

+ MARKS THE CITY OF YIN

ECLIPSE ON JD 1207545 DATE(Y/M/D)-1408/ 1/29
DELTA T =10 HOURS 18.8 MINS

LONGITUDE (-VE EAST)	LATITUDE LIMITS. NORTH	SOUTH	LOCAL TIME	MEAN ALTITUDE
-143	13.60	10.89	14:48	37
-145	14.71	11.98	15:00	34
-147	15.80	13.05	15:12	31
-150	17.41	14.64	15:28	27

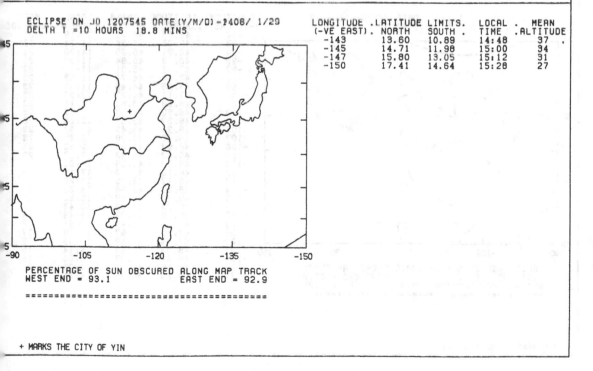

PERCENTAGE OF SUN OBSCURED ALONG MAP TRACK
WEST END = 93.1   EAST END = 92.9

===

+ MARKS THE CITY OF YIN

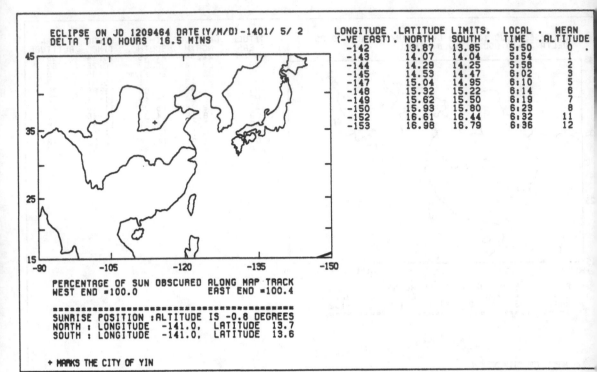

ECLIPSE ON JD 1209464 DATE(Y/M/D)-1401/ 5/ 2
DELTA T =10 HOURS 16.5 MINS

PERCENTAGE OF SUN OBSCURED ALONG MAP TRACK
WEST END =100.0   EAST END =100.4

==
SUNRISE POSITION :ALTITUDE IS -0.8 DEGREES
NORTH : LONGITUDE -141.0, LATITUDE 13.7
SOUTH : LONGITUDE -141.0, LATITUDE 13.6

+ MARKS THE CITY OF YIN

LONGITUDE (-VE EAST)	LATITUDE NORTH	LIMITS. SOUTH	LOCAL TIME	MEAN ALTITUDE
-142	13.87	13.85	5:50	0
-143	14.07	14.04	5:54	1
-144	14.29	14.25	5:58	2
-145	14.53	14.47	6:02	3
-147	15.04	14.95	6:10	5
-148	15.32	15.22	6:14	6
-149	15.62	15.50	6:19	7
-150	15.93	15.80	6:23	8
-152	16.61	16.44	6:32	11
-153	16.98	16.79	6:36	12

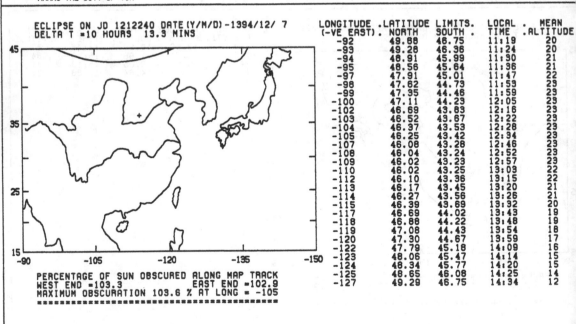

ECLIPSE ON JD 1212240 DATE(Y/M/D)-1394/12/ 7
DELTA T =10 HOURS 13.3 MINS

PERCENTAGE OF SUN OBSCURED ALONG MAP TRACK
WEST END =103.3   EAST END =102.9
MAXIMUM OBSCURATION 103.6 % AT LONG = -105
==

+ MARKS THE CITY OF YIN

LONGITUDE (-VE EAST)	LATITUDE NORTH	LIMITS. SOUTH	LOCAL TIME	MEAN ALTITUDE
-92	49.68	46.75	11:19	20
-93	49.28	46.36	11:24	20
-94	48.91	45.99	11:30	21
-95	48.56	45.64	11:36	21
-97	47.91	45.01	11:47	22
-98	47.62	44.73	11:59	23
-99	47.35	44.46	11:59	23
-100	47.11	44.29	12:05	23
-102	46.69	43.83	12:16	23
-103	46.52	43.67	12:22	23
-104	46.37	43.53	12:28	23
-105	46.25	43.42	12:34	23
-107	46.08	43.28	12:46	23
-108	46.04	43.24	12:52	23
-109	46.02	43.23	12:57	23
-110	46.02	43.25	13:03	22
-112	46.10	43.36	13:15	22
-113	46.17	43.45	13:20	21
-114	46.27	43.56	13:26	21
-115	46.39	43.69	13:32	20
-117	46.69	44.02	13:43	19
-118	46.88	44.22	13:48	19
-119	47.08	44.43	13:54	18
-120	47.30	44.67	13:59	17
-122	47.79	45.18	14:09	16
-123	48.06	45.47	14:14	15
-124	48.34	45.77	14:20	15
-125	48.65	46.08	14:25	14
-127	49.29	46.75	14:34	12

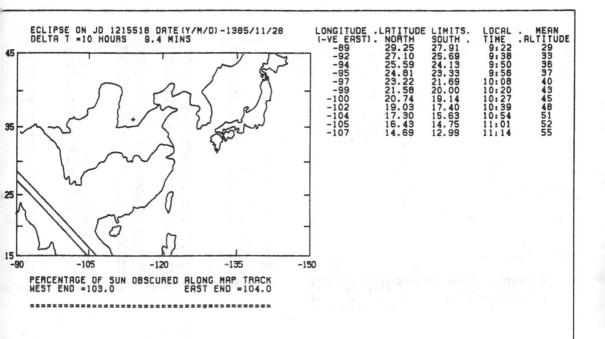

ECLIPSE ON JD 1215518 DATE (Y/M/D) -1385/11/28
DELTA T =10 HOURS   9.4 MINS

LONGITUDE (-VE EAST)	LATITUDE LIMITS NORTH	SOUTH	LOCAL TIME	MEAN ALTITUDE
-89	29.25	27.91	9:22	29
-92	27.10	25.69	9:38	33
-94	25.59	24.13	9:50	36
-95	24.81	23.33	9:56	37
-97	23.22	21.69	10:08	40
-99	21.58	20.00	10:20	43
-100	20.74	19.14	10:27	45
-102	19.03	17.40	10:39	48
-104	17.30	15.63	10:54	51
-105	16.43	14.75	11:01	52
-107	14.69	12.99	11:14	55

PERCENTAGE OF SUN OBSCURED ALONG MAP TRACK
WEST END =103.0   EAST END =104.0

=======================================

+ MARKS THE CITY OF YIN

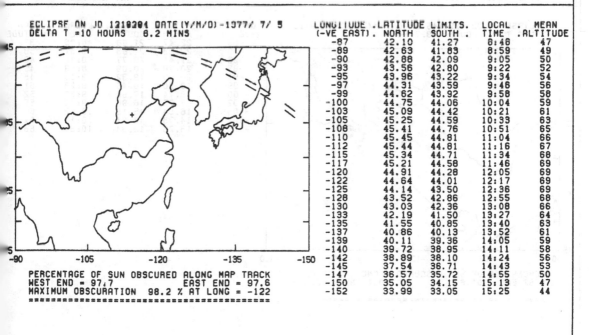

ECLIPSE ON JD 1219204 DATE (Y/M/D) -1377/ 7/ 5
DELTA T =10 HOURS   8.2 MINS

LONGITUDE (-VE EAST)	LATITUDE LIMITS NORTH	SOUTH	LOCAL TIME	MEAN ALTITUDE
-87	42.10	41.27	8:46	47
-89	42.63	41.83	8:59	49
-90	42.88	42.09	9:05	50
-93	43.56	42.80	9:22	52
-95	43.96	43.22	9:34	54
-97	44.31	43.59	9:46	56
-99	44.62	43.92	9:58	58
-100	44.75	44.06	10:04	59
-103	45.09	44.42	10:21	61
-105	45.25	44.59	10:33	63
-108	45.41	44.76	10:51	65
-110	45.45	44.81	11:04	66
-112	45.44	44.81	11:16	67
-115	45.34	44.71	11:34	68
-117	45.21	44.58	11:46	69
-120	44.91	44.28	12:05	69
-122	44.64	44.01	12:17	69
-125	44.14	43.50	12:36	69
-128	43.52	42.86	12:55	68
-130	43.03	42.36	13:08	66
-133	42.19	41.50	13:27	64
-135	41.55	40.85	13:40	63
-137	40.86	40.13	13:52	61
-139	40.11	39.36	14:05	59
-140	39.72	38.95	14:11	58
-142	38.89	38.10	14:24	56
-145	37.54	36.71	14:43	53
-147	36.57	35.72	14:55	50
-150	35.05	34.15	15:13	47
-152	33.99	33.05	15:25	44

PERCENTAGE OF SUN OBSCURED ALONG MAP TRACK
WEST END = 97.7   EAST END = 97.6
MAXIMUM OBSCURATION 98.2 % AT LONG = -122

=======================================

+ MARKS THE CITY OF YIN

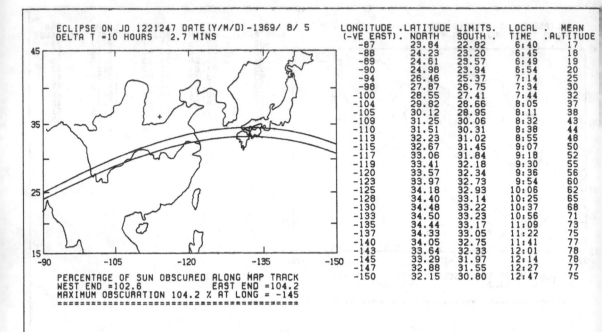

ECLIPSE ON JD 1221247 DATE(Y/M/D)-1369/ 8/ 5
DELTA T =10 HOURS   2.7 MINS

LONGITUDE (-VE EAST)	LATITUDE NORTH	LIMITS. SOUTH	LOCAL TIME	MEAN ALTITUDE
-87	23.84	22.82	6:40	17
-88	24.23	23.20	6:45	18
-89	24.61	23.57	6:49	19
-90	24.98	23.94	6:54	20
-94	26.46	25.37	7:14	25
-98	27.87	26.75	7:34	30
-100	28.55	27.41	7:44	32
-104	29.82	28.66	8:05	37
-105	30.12	28.95	8:11	38
-109	31.25	30.06	8:32	43
-110	31.51	30.31	8:38	44
-113	32.23	31.02	8:55	48
-115	32.67	31.45	9:07	50
-117	33.06	31.84	9:18	52
-119	33.41	32.18	9:30	55
-120	33.57	32.34	9:36	56
-123	33.97	32.73	9:54	60
-125	34.18	32.93	10:06	62
-128	34.40	33.14	10:25	65
-130	34.48	33.22	10:37	68
-133	34.50	33.23	10:56	71
-135	34.44	33.17	11:09	73
-137	34.33	33.05	11:22	75
-140	34.05	32.75	11:41	77
-143	33.64	32.33	12:01	78
-145	33.29	31.97	12:14	78
-147	32.88	31.55	12:27	77
-150	32.15	30.80	12:47	75

PERCENTAGE OF SUN OBSCURED ALONG MAP TRACK
WEST END =102.6   EAST END =104.2
MAXIMUM OBSCURATION 104.2 % AT LONG = -145
===

+ MARKS THE CITY OF YIN

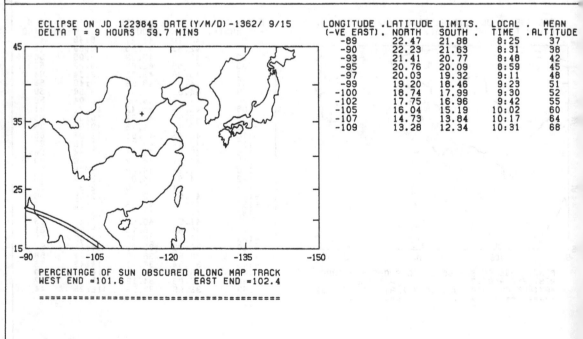

ECLIPSE ON JD 1223845 DATE(Y/M/D)-1362/ 9/15
DELTA T = 9 HOURS   59.7 MINS

LONGITUDE (-VE EAST)	LATITUDE NORTH	LIMITS. SOUTH	LOCAL TIME	MEAN ALTITUDE
-89	22.47	21.88	8:25	37
-90	22.23	21.63	8:31	38
-93	21.41	20.77	8:48	42
-95	20.76	20.09	8:59	45
-97	20.03	19.32	9:11	48
-99	19.20	18.46	9:23	51
-100	18.74	17.99	9:30	52
-102	17.75	16.96	9:42	55
-105	16.04	15.19	10:02	60
-107	14.73	13.84	10:17	64
-109	13.28	12.34	10:31	68

PERCENTAGE OF SUN OBSCURED ALONG MAP TRACK
WEST END =101.6   EAST END =102.4

===

+ MARKS THE CITY OF YIN

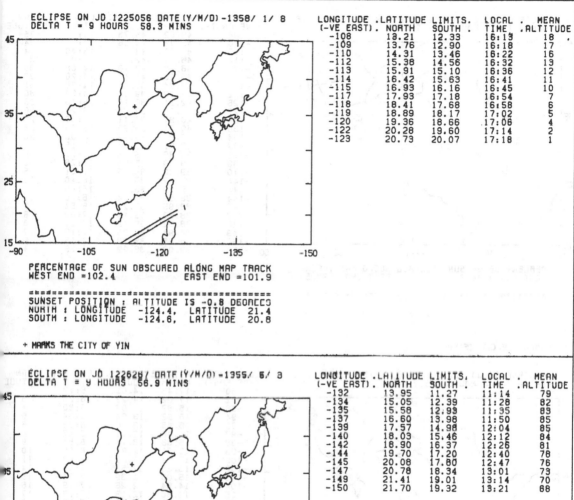

ECLIPSE ON JD 1225056 DATE (Y/M/D) -1358/ 1/ 8
DELTA T = 9 HOURS 58.3 MINS

LONGITUDE	.LATITUDE LIMITS.		LOCAL .	MEAN
(-VE EAST).	NORTH	SOUTH .	TIME	.ALTITUDE
-108	13.21	12.33	16:13	18
-109	13.76	12.90	16:18	17
-110	14.31	13.46	16:22	16
-112	15.38	14.56	16:32	13
-113	15.91	15.10	16:36	12
-114	16.42	15.63	16:41	11
-115	16.93	16.16	16:45	10
-117	17.93	17.18	16:54	7
-118	18.41	17.68	16:58	6
-119	18.89	18.17	17:02	5
-120	19.36	18.66	17:06	4
-122	20.28	19.60	17:14	2
-123	20.73	20.07	17:18	1

PERCENTAGE OF SUN OBSCURED ALONG MAP TRACK
WEST END =102.4   EAST END =101.9

=======================================
SUNSET POSITION : ALTITUDE IS -0.8 DEGREES
NORTH : LONGITUDE -124.4, LATITUDE 21.4
SOUTH : LONGITUDE -124.6, LATITUDE 20.8

+ MARKS THE CITY OF YIN

ECLIPSE ON JD 1226286/ DATE (Y/M/D) -1355/ 6/ 8
DELTA T = 9 HOURS 56.9 MINS

LONGITUDE	.LATITUDE LIMITS.		LOCAL .	MEAN
(-VE EAST).	NORTH	SOUTH .	TIME	.ALTITUDE
-132	13.95	11.27	11:14	79
-134	15.05	12.39	11:28	82
-135	15.58	12.93	11:35	83
-137	16.60	13.98	11:50	85
-139	17.57	14.98	12:04	85
-140	18.03	15.46	12:12	84
-142	18.90	16.37	12:26	81
-144	19.70	17.20	12:40	78
-145	20.08	17.80	12:47	76
-147	20.78	18.34	13:01	73
-149	21.41	19.01	13:14	70
-150	21.70	19.32	13:21	68

PERCENTAGE OF SUN OBSCURED ALONG MAP TRACK
WEST END =107.9   EAST END =107.8
MAXIMUM OBSCURATION 107.9 % AT LONG = -137
=======================================

+ MARKS THE CITY OF YIN

ECLIPSE ON JD 1226798 DATE(Y/M/D)-1354/10/16
DELTA T = 9 HOURS 56.2 MINS

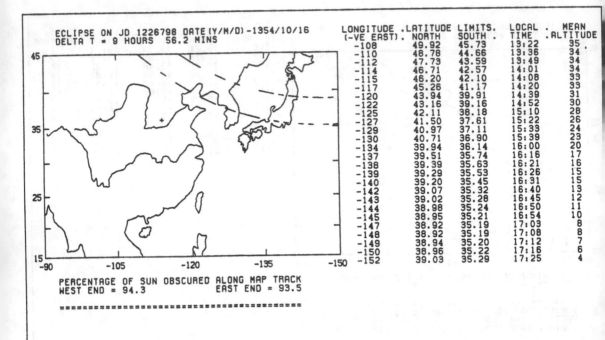

LONGITUDE (-VE EAST)	LATITUDE NORTH	LIMITS. SOUTH	LOCAL TIME	MEAN ALTITUDE
-108	49.92	45.73	13:22	35
-110	48.78	44.66	13:36	34
-112	47.73	43.59	13:49	34
-114	46.71	42.57	14:01	34
-115	46.20	42.10	14:08	33
-117	45.26	41.17	14:20	33
-120	43.94	39.91	14:39	31
-122	43.16	39.16	14:52	30
-125	42.11	38.18	15:10	28
-127	41.50	37.61	15:22	26
-129	40.97	37.11	15:33	24
-130	40.71	36.90	15:39	23
-134	39.94	36.14	16:00	20
-137	39.51	35.74	16:16	17
-138	39.39	35.63	16:21	16
-139	39.29	35.53	16:26	15
-140	39.20	35.45	16:31	15
-142	39.07	35.32	16:40	13
-143	39.02	35.28	16:45	12
-144	38.98	35.24	16:50	11
-145	38.95	35.21	16:54	10
-147	38.92	35.19	17:03	8
-148	38.92	35.19	17:08	8
-149	38.94	35.20	17:12	7
-150	38.96	35.22	17:16	6
-152	39.03	35.29	17:25	4

PERCENTAGE OF SUN OBSCURED ALONG MAP TRACK
WEST END = 94.3   EAST END = 93.5

==

+ MARKS THE CITY OF YIN

ECLIPSE ON JD 1227301 DATE(Y/M/D)-1352/ 3/ 2
DELTA T = 9 HOURS 55.7 MINS

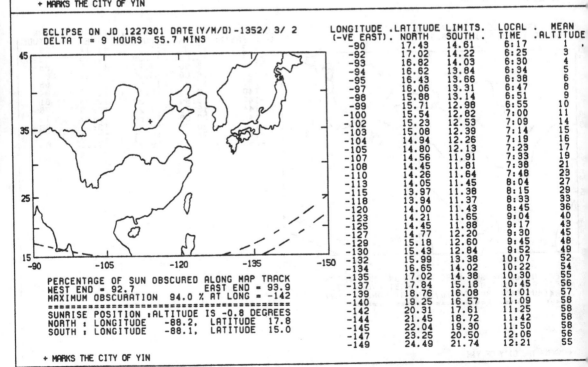

LONGITUDE (-VE EAST)	LATITUDE NORTH	LIMITS. SOUTH	LOCAL TIME	MEAN ALTITUDE
-90	17.43	14.61	6:17	1
-92	17.02	14.22	6:25	3
-93	16.82	14.03	6:30	4
-94	16.62	13.84	6:34	5
-95	16.43	13.66	6:38	6
-97	16.06	13.31	6:47	8
-98	15.88	13.14	6:51	9
-99	15.71	12.98	6:55	10
-100	15.54	12.82	7:00	11
-102	15.23	12.53	7:09	14
-103	15.08	12.39	7:14	15
-104	14.94	12.26	7:19	16
-105	14.80	12.13	7:23	17
-107	14.56	11.91	7:33	19
-108	14.45	11.81	7:38	21
-110	14.26	11.64	7:48	23
-113	14.05	11.45	8:04	27
-115	13.97	11.38	8:15	29
-118	13.94	11.37	8:33	33
-120	14.00	11.43	8:45	36
-123	14.21	11.65	9:04	40
-125	14.45	11.88	9:17	43
-127	14.77	12.20	9:30	45
-129	15.18	12.60	9:45	48
-130	15.43	12.84	9:52	49
-132	15.99	13.38	10:07	52
-134	16.65	14.02	10:22	54
-135	17.02	14.38	10:30	55
-137	17.84	15.18	10:45	56
-139	18.76	16.08	11:01	57
-140	19.25	16.57	11:09	58
-142	20.31	17.61	11:25	58
-144	21.45	18.72	11:42	58
-145	22.04	19.30	11:50	58
-147	23.25	20.50	12:06	56
-149	24.49	21.74	12:21	55

PERCENTAGE OF SUN OBSCURED ALONG MAP TRACK
WEST END = 92.7   EAST END = 93.9
MAXIMUM OBSCURATION 94.0 % AT LONG = -142
==
SUNRISE POSITION :ALTITUDE IS -0.8 DEGREES
NORTH : LONGITUDE -88.2, LATITUDE 17.8
SOUTH : LONGITUDE -88.1, LATITUDE 15.0

+ MARKS THE CITY OF YIN

14

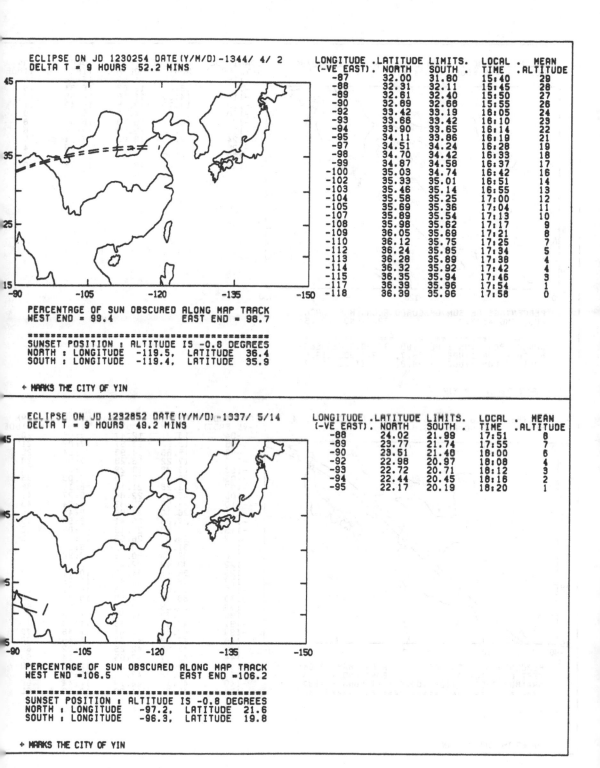

ECLIPSE ON JD 1230254 DATE(Y/M/D)-1344/ 4/ 2
DELTA T = 9 HOURS 52.2 MINS

LONGITUDE (-VE EAST)	LATITUDE NORTH	LIMITS. SOUTH	LOCAL TIME	MEAN ALTITUDE
-87	32.00	31.80	15:40	29
-88	32.31	32.11	15:45	28
-89	32.61	32.40	15:50	27
-90	32.89	32.68	15:55	26
-92	33.42	33.19	16:05	24
-93	33.66	33.42	16:10	23
-94	33.90	33.65	16:14	22
-95	34.11	33.86	16:19	21
-97	34.51	34.24	16:28	19
-98	34.70	34.42	16:33	18
-99	34.87	34.58	16:37	17
-100	35.03	34.74	16:42	16
-102	35.33	35.01	16:51	14
-103	35.46	35.14	16:55	13
-104	35.58	35.25	17:00	12
-105	35.69	35.36	17:04	11
-107	35.89	35.54	17:13	10
-108	35.98	35.62	17:17	9
-109	36.05	35.69	17:21	8
-110	36.12	35.75	17:25	7
-112	36.24	35.85	17:34	5
-113	36.28	35.89	17:38	4
-114	36.32	35.92	17:42	4
-115	36.35	35.94	17:46	3
-117	36.39	35.96	17:54	1
-118	36.39	35.96	17:58	0

PERCENTAGE OF SUN OBSCURED ALONG MAP TRACK
WEST END = 99.4   EAST END = 98.7

===
SUNSET POSITION : ALTITUDE IS -0.8 DEGREES
NORTH : LONGITUDE   -119.5, LATITUDE   36.4
SOUTH : LONGITUDE   -119.4, LATITUDE   35.9

+ MARKS THE CITY OF YIN

ECLIPSE ON JD 1292852 DATE(Y/M/D)-1337/ 5/14
DELTA T = 9 HOURS 48.2 MINS

LONGITUDE (-VE EAST)	LATITUDE NORTH	LIMITS. SOUTH	LOCAL TIME	MEAN ALTITUDE
-88	24.02	21.99	17:51	8
-89	23.77	21.74	17:55	7
-90	23.51	21.48	18:00	6
-92	22.98	20.97	18:08	4
-93	22.72	20.71	18:12	3
-94	22.44	20.45	18:16	2
-95	22.17	20.19	18:20	1

PERCENTAGE OF SUN OBSCURED ALONG MAP TRACK
WEST END =106.5   EAST END =106.2

===
SUNSET POSITION : ALTITUDE IS -0.8 DEGREES
NORTH : LONGITUDE   -97.2, LATITUDE   21.6
SOUTH : LONGITUDE   -96.3, LATITUDE   19.8

+ MARKS THE CITY OF YIN

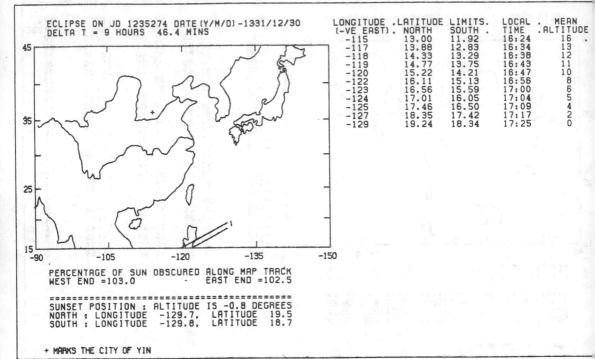

ECLIPSE ON JD 1235274 DATE(Y/M/D)-1331/12/30
DELTA T = 9 HOURS 46.4 MINS

LONGITUDE (-VE EAST)	.LATITUDE NORTH	LIMITS. SOUTH .	LOCAL TIME	MEAN .ALTITUDE
-115	13.00	11.92	16:24	16
-117	13.88	12.83	16:34	13
-118	14.33	13.29	16:38	12
-119	14.77	13.75	16:43	11
-120	15.22	14.21	16:47	10
-122	16.11	15.13	16:56	8
-123	16.56	15.59	17:00	6
-124	17.01	16.05	17:04	5
-125	17.46	16.50	17:09	4
-127	18.35	17.42	17:17	2
-129	19.24	18.34	17:25	0

PERCENTAGE OF SUN OBSCURED ALONG MAP TRACK
WEST END =103.0   -   EAST END =102.5

```
=================================================
SUNSET POSITION : ALTITUDE IS -0.8 DEGREES
NORTH : LONGITUDE  -129.7,  LATITUDE  19.5
SOUTH : LONGITUDE  -129.8,  LATITUDE  18.7
```

+ MARKS THE CITY OF YIN

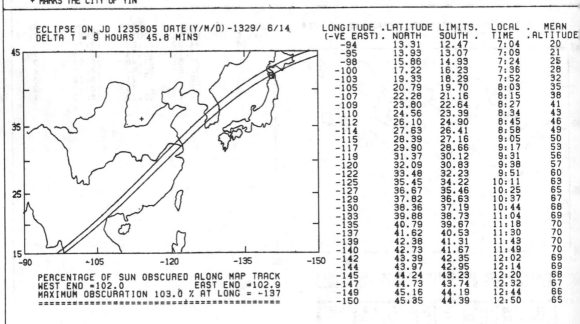

ECLIPSE ON JD 1235805 DATE(Y/M/D)-1329/ 6/14
DELTA T = 9 HOURS 45.8 MINS

LONGITUDE (-VE EAST)	.LATITUDE NORTH	LIMITS. SOUTH .	LOCAL TIME	MEAN .ALTITUDE
-94	13.31	12.47	7:04	20
-95	13.93	13.07	7:09	21
-98	15.86	14.93	7:24	25
-100	17.22	16.23	7:36	28
-103	19.33	18.29	7:52	32
-105	20.79	19.70	8:03	35
-107	22.28	21.16	8:15	38
-109	23.80	22.64	8:27	41
-110	24.56	23.39	8:34	43
-112	26.10	24.90	8:45	46
-114	27.63	26.41	8:58	49
-115	28.39	27.16	9:05	50
-117	29.90	28.66	9:17	53
-119	31.37	30.12	9:31	56
-120	32.09	30.83	9:38	57
-122	33.48	32.23	9:51	60
-125	35.45	34.22	10:11	63
-127	36.67	35.46	10:25	65
-129	37.82	36.63	10:37	67
-130	38.36	37.19	10:44	68
-133	39.88	38.73	11:04	69
-135	40.79	39.67	11:18	70
-137	41.62	40.53	11:30	70
-139	42.38	41.31	11:43	70
-140	42.73	41.67	11:49	70
-142	43.39	42.35	12:02	69
-144	43.97	42.95	12:14	69
-145	44.24	43.23	12:20	68
-147	44.73	43.74	12:32	67
-149	45.16	44.19	12:44	66
-150	45.35	44.39	12:50	65

PERCENTAGE OF SUN OBSCURED ALONG MAP TRACK
WEST END =102.0   EAST END =102.9
MAXIMUM OBSCURATION 103.0 % AT LONG = -137

```
==============================================
```

+ MARKS THE CITY OF YIN

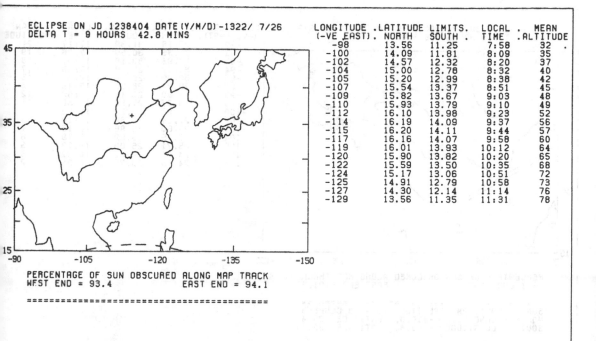

ECLIPSE ON JD 1238404 DATE(Y/M/D) -1322/ 7/26
DELTA T = 9 HOURS 42.8 MINS

LONGITUDE (-VE EAST).	LATITUDE NORTH	LIMITS. SOUTH .	LOCAL TIME	MEAN .ALTITUDE
-98	13.56	11.25	7:58	32
-100	14.09	11.81	8:09	35
-102	14.57	12.32	8:20	37
-104	15.00	12.78	8:32	40
-105	15.20	12.99	8:38	42
-107	15.54	13.37	8:51	45
-109	15.82	13.67	9:03	48
-110	15.93	13.79	9:10	49
-112	16.10	13.98	9:23	52
-114	16.19	14.09	9:37	56
-115	16.20	14.11	9:44	57
-117	16.16	14.07	9:58	60
-119	16.01	13.93	10:12	64
-120	15.90	13.82	10:20	65
-122	15.59	13.50	10:35	68
-124	15.17	13.06	10:51	72
-125	14.91	12.79	10:58	73
-127	14.30	12.14	11:14	76
-129	13.56	11.35	11:31	78

PERCENTAGE OF SUN OBSCURED ALONG MAP TRACK
WEST END = 93.4   EAST END = 94.1

===

+ MARKS THE CITY OF YIN

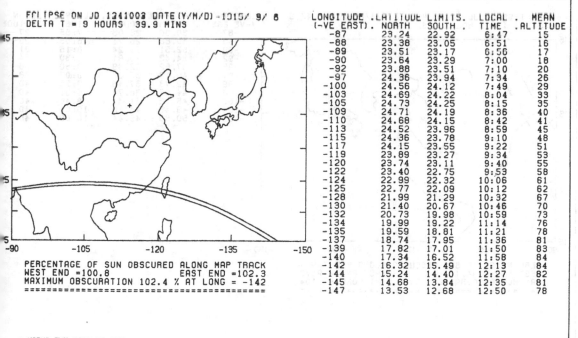

ECLIPSE ON JD 1241003 DATE(Y/M/D) -1315/ 9/ 8
DELTA T = 9 HOURS 39.9 MINS

LONGITUDE (-VE EAST).	LATITUDE NORTH	LIMITS. SOUTH .	LOCAL TIME	MEAN .ALTITUDE
-87	23.24	22.92	6:47	15
-88	23.38	23.05	6:51	16
-89	23.51	23.17	6:56	17
-90	23.64	23.29	7:00	18
-92	23.88	23.51	7:10	20
-97	24.36	23.94	7:34	26
-100	24.56	24.12	7:49	29
-103	24.69	24.22	8:04	33
-105	24.73	24.25	8:15	35
-109	24.71	24.19	8:36	40
-110	24.68	24.15	8:42	41
-113	24.52	23.96	8:59	45
-115	24.36	23.78	9:10	48
-117	24.15	23.55	9:22	51
-119	23.89	23.27	9:34	53
-120	23.74	23.11	9:40	55
-122	23.40	22.75	9:53	58
-124	22.99	22.32	10:06	61
-125	22.77	22.09	10:12	62
-128	21.99	21.29	10:32	67
-130	21.40	20.67	10:46	70
-132	20.73	19.98	10:59	73
-134	19.99	19.22	11:14	76
-135	19.59	18.81	11:21	78
-137	18.74	17.95	11:36	81
-139	17.82	17.01	11:50	83
-140	17.34	16.52	11:58	84
-142	16.32	15.49	12:13	84
-144	15.24	14.40	12:27	82
-145	14.68	13.84	12:35	81
-147	13.53	12.68	12:50	78

PERCENTAGE OF SUN OBSCURED ALONG MAP TRACK
WEST END =100.8   EAST END =102.3
MAXIMUM OBSCURATION 102.4 % AT LONG = -142
===

+ MARKS THE CITY OF YIN

17

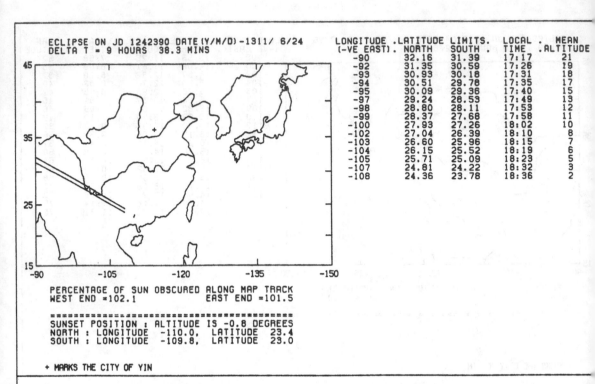

ECLIPSE ON JD 1242390 DATE(Y/M/D)-1311/ 6/24
DELTA T = 9 HOURS 38.3 MINS

LONGITUDE (-VE EAST)	LATITUDE NORTH	LIMITS. SOUTH	LOCAL TIME	MEAN ALTITUDE
-90	32.16	31.39	17:17	21
-92	31.35	30.59	17:26	19
-93	30.93	30.18	17:31	18
-94	30.51	29.78	17:35	17
-95	30.09	29.36	17:40	15
-97	29.24	28.53	17:49	13
-98	28.80	28.11	17:53	12
-99	28.37	27.68	17:58	11
-100	27.93	27.26	18:02	10
-102	27.04	26.39	18:10	8
-103	26.60	25.96	18:15	7
-104	26.15	25.52	18:19	6
-105	25.71	25.09	18:23	5
-107	24.81	24.22	18:32	3
-108	24.36	23.78	18:36	2

PERCENTAGE OF SUN OBSCURED ALONG MAP TRACK
WEST END =102.1   EAST END =101.5

===
SUNSET POSITION : ALTITUDE IS -0.8 DEGREES
NORTH : LONGITUDE -110.0, LATITUDE 23.4
SOUTH : LONGITUDE -109.8, LATITUDE 23.0

+ MARKS THE CITY OF YIN

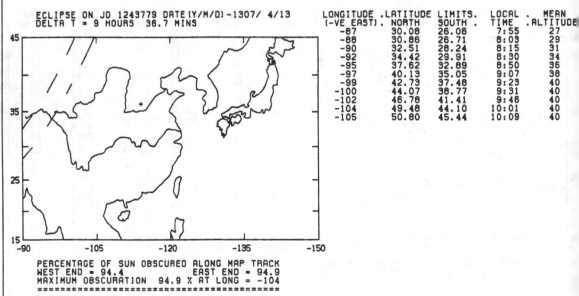

ECLIPSE ON JD 1243779 DATE(Y/M/D)-1307/ 4/13
DELTA T = 9 HOURS 36.7 MINS

LONGITUDE (-VE EAST)	LATITUDE NORTH	LIMITS. SOUTH	LOCAL TIME	MEAN ALTITUDE
-87	30.08	26.06	7:55	27
-88	30.86	26.71	8:03	29
-90	32.51	28.24	8:15	31
-92	34.42	29.91	8:30	34
-95	37.62	32.89	8:50	36
-97	40.13	35.05	9:07	38
-99	42.73	37.48	9:23	40
-100	44.07	38.77	9:31	40
-102	46.78	41.41	9:46	40
-104	49.48	44.10	10:01	40
-105	50.80	45.44	10:09	40

PERCENTAGE OF SUN OBSCURED ALONG MAP TRACK
WEST END = 94.4   EAST END = 94.9
MAXIMUM OBSCURATION 94.9 % AT LONG = -104
===

+ MARKS THE CITY OF YIN

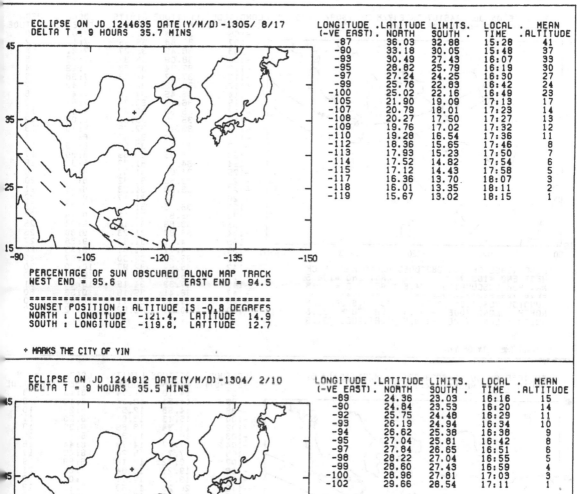

ECLIPSE ON JD 1244635 DATE(Y/M/D)-1305/ 8/17
DELTA T = 9 HOURS   35.7 MINS

LONGITUDE (-VE EAST)	.LATITUDE LIMITS. NORTH	SOUTH .	LOCAL TIME	. MEAN .ALTITUDE
-87	36.03	32.88	15:28	41
-90	33.18	30.05	15:48	37
-93	30.49	27.43	16:07	33
-95	28.82	25.79	16:19	30
-97	27.24	24.25	16:30	27
-99	25.76	22.83	16:42	24
-100	25.02	22.16	16:49	23
-105	21.90	19.09	17:13	17
-107	20.79	18.01	17:23	14
-108	20.27	17.50	17:27	13
-109	19.76	17.02	17:32	12
-110	19.28	16.54	17:36	11
-112	18.36	15.65	17:46	8
-113	17.93	15.23	17:50	7
-114	17.52	14.82	17:54	6
-115	17.12	14.43	17:58	5
-117	16.36	13.70	18:07	3
-118	16.01	13.35	18:11	2
-119	15.67	13.02	18:15	1

PERCENTAGE OF SUN OBSCURED ALONG MAP TRACK
WEST END = 95.6   EAST END = 94.5

===
SUNSET POSITION : ALTITUDE IS -0.8 DEGREES
NORTH : LONGITUDE -121.4, LATITUDE 14.9
SOUTH : LONGITUDE -119.8, LATITUDE 12.7

+ MARKS THE CITY OF YIN

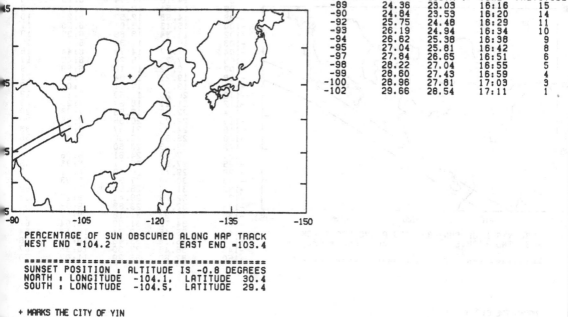

ECLIPSE ON JD 1244812 DATE(Y/M/D)-1304/ 2/10
DELTA T = 9 HOURS   35.5 MINS

LONGITUDE (-VE EAST)	.LATITUDE LIMITS. NORTH	SOUTH .	LOCAL TIME	. MEAN .ALTITUDE
-89	24.36	23.03	16:16	15
-90	24.84	23.53	16:20	14
-92	25.75	24.48	16:29	11
-93	26.19	24.94	16:34	10
-94	26.62	25.38	16:38	9
-95	27.04	25.81	16:42	8
-97	27.84	26.65	16:51	6
-98	28.22	27.04	16:55	5
-99	28.60	27.43	16:59	4
-100	28.96	27.81	17:03	3
-102	29.66	28.54	17:11	1

PERCENTAGE OF SUN OBSCURED ALONG MAP TRACK
WEST END =104.2   EAST END =103.4

===
SUNSET POSITION : ALTITUDE IS -0.8 DEGREES
NORTH : LONGITUDE -104.1, LATITUDE 30.4
SOUTH : LONGITUDE -104.5, LATITUDE 29.4

+ MARKS THE CITY OF YIN

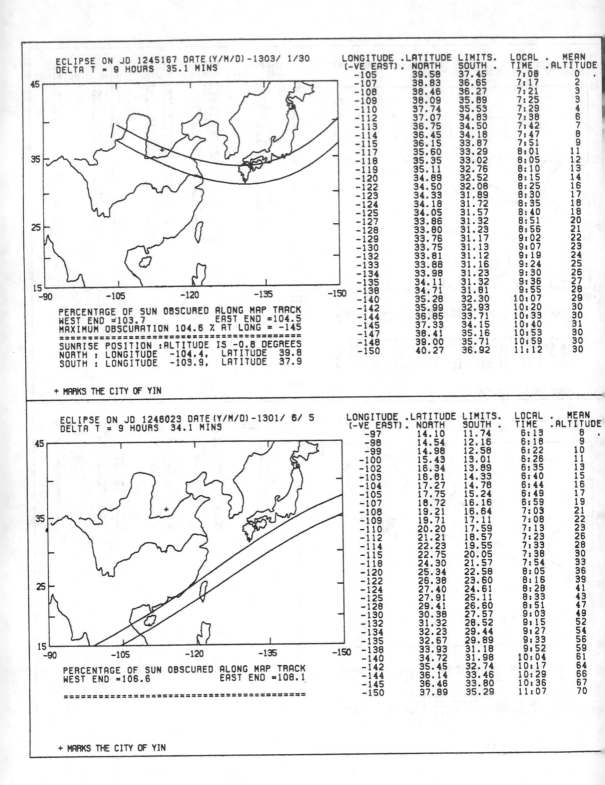

ECLIPSE ON JD 1245167 DATE(Y/M/D)-1303/ 1/30
DELTA T = 9 HOURS 35.1 MINS

LONGITUDE (-VE EAST)	LATITUDE NORTH	LIMITS. SOUTH	LOCAL TIME	MEAN ALTITUDE
-105	39.58	37.45	7:08	0
-107	38.83	36.65	7:17	2
-108	38.46	36.27	7:21	3
-109	38.09	35.89	7:25	3
-110	37.74	35.53	7:29	4
-112	37.07	34.83	7:38	6
-113	36.75	34.50	7:42	7
-114	36.45	34.18	7:47	8
-115	36.15	33.87	7:51	9
-117	35.60	33.29	8:01	11
-118	35.35	33.02	8:05	12
-119	35.11	32.76	8:10	13
-120	34.89	32.52	8:15	14
-122	34.50	32.08	8:25	16
-123	34.33	31.89	8:30	17
-124	34.18	31.72	8:35	18
-125	34.05	31.57	8:40	18
-127	33.86	31.32	8:51	20
-128	33.80	31.23	8:56	21
-129	33.76	31.17	9:02	22
-130	33.75	31.13	9:07	23
-132	33.81	31.12	9:19	24
-133	33.88	31.16	9:24	25
-134	33.98	31.23	9:30	26
-135	34.11	31.32	9:36	27
-138	34.71	31.81	9:55	28
-140	35.28	32.30	10:07	29
-142	35.99	32.93	10:20	30
-144	36.85	33.71	10:33	30
-145	37.33	34.15	10:40	31
-147	38.41	35.16	10:53	30
-148	39.00	35.71	10:59	30
-150	40.27	36.92	11:12	30

PERCENTAGE OF SUN OBSCURED ALONG MAP TRACK
WEST END =103.7   EAST END =104.5
MAXIMUM OBSCURATION 104.6 % AT LONG = -145
===
SUNRISE POSITION :ALTITUDE IS -0.8 DEGREES
NORTH : LONGITUDE -104.4, LATITUDE 39.8
SOUTH : LONGITUDE -103.9, LATITUDE 37.9

+ MARKS THE CITY OF YIN

ECLIPSE ON JD 1246023 DATE(Y/M/D)-1301/ 6/ 5
DELTA T = 9 HOURS 34.1 MINS

LONGITUDE (-VE EAST)	LATITUDE NORTH	LIMITS. SOUTH	LOCAL TIME	MEAN ALTITUDE
-97	14.10	11.74	6:13	8
-98	14.54	12.16	6:18	9
-99	14.98	12.58	6:22	10
-100	15.43	13.01	6:26	11
-102	16.34	13.89	6:35	13
-103	16.81	14.33	6:40	15
-104	17.27	14.78	6:44	16
-105	17.75	15.24	6:49	17
-107	18.72	16.16	6:59	19
-108	19.21	16.64	7:03	21
-109	19.71	17.11	7:08	22
-110	20.20	17.59	7:13	23
-112	21.21	18.57	7:23	26
-114	22.23	19.55	7:33	28
-115	22.75	20.05	7:38	30
-118	24.30	21.57	7:54	33
-120	25.34	22.58	8:05	36
-122	26.38	23.60	8:16	39
-124	27.40	24.61	8:28	41
-125	27.91	25.11	8:33	43
-128	29.41	26.60	8:51	47
-130	30.38	27.57	9:03	49
-132	31.32	28.52	9:15	52
-134	32.23	29.44	9:27	54
-135	32.67	29.89	9:33	56
-138	33.93	31.18	9:52	59
-140	34.72	31.98	10:04	61
-142	35.45	32.74	10:17	64
-144	36.14	33.46	10:29	66
-145	36.46	33.80	10:36	67
-150	37.89	35.29	11:07	70

PERCENTAGE OF SUN OBSCURED ALONG MAP TRACK
WEST END =106.6   EAST END =108.1

===

+ MARKS THE CITY OF YIN

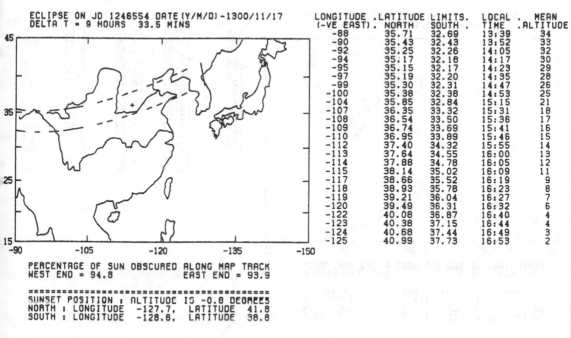

ECLIPSE ON JD 1246554 DATE(Y/M/D) -1300/11/17
DELTA T = 9 HOURS 33.5 MINS

LONGITUDE (-VE EAST)	LATITUDE LIMITS. NORTH	SOUTH	LOCAL TIME	MEAN ALTITUDE
-88	35.71	32.69	13:39	34
-90	35.43	32.43	13:52	33
-92	35.25	32.26	14:05	32
-94	35.17	32.18	14:17	30
-95	35.15	32.17	14:29	29
-97	35.19	32.20	14:35	28
-99	35.30	32.31	14:47	26
-100	35.38	32.38	14:53	25
-104	35.85	32.84	15:15	21
-107	36.35	33.32	15:31	18
-108	36.54	33.50	15:36	17
-109	36.74	33.69	15:41	16
-110	36.95	33.89	15:46	15
-112	37.40	34.32	15:55	14
-113	37.64	34.55	16:00	13
-114	37.88	34.78	16:05	12
-115	38.14	35.02	16:09	11
-117	38.66	35.52	16:19	9
-118	38.93	35.78	16:23	8
-119	39.21	36.04	16:27	7
-120	39.49	36.31	16:32	6
-122	40.08	36.87	16:40	4
-123	40.38	37.15	16:44	4
-124	40.68	37.44	16:49	3
-125	40.99	37.73	16:53	2

PERCENTAGE OF SUN OBSCURED ALONG MAP TRACK
WEST END = 94.8   EAST END = 93.9

===
SUNSET POSITION : ALTITUDE IS -0.8 DEGREES
NORTH : LONGITUDE -127.7, LATITUDE 41.8
SOUTH : LONGITUDE -128.6, LATITUDE 38.8

+ MARKS THE CITY OF YIN

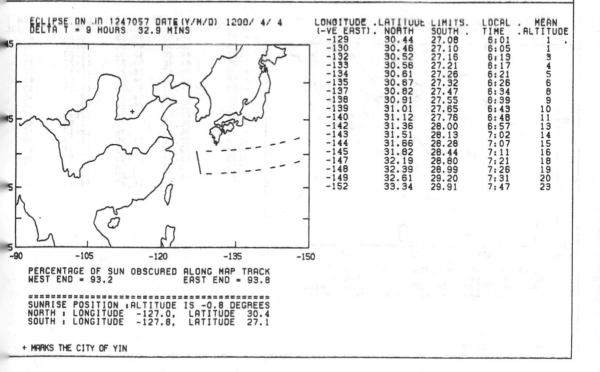

ECLIPSE ON JD 1247057 DATE(Y/M/D) 1200/ 4/ 4
DELTA T = 9 HOURS 32.9 MINS

LONGITUDE (-VE EAST)	LATITUDE LIMITS. NORTH	SOUTH	LOCAL TIME	MEAN ALTITUDE
-129	30.44	27.08	6:01	1
-130	30.46	27.10	6:05	1
-132	30.52	27.16	6:19	3
-133	30.56	27.21	6:17	4
-134	30.61	27.26	6:21	5
-135	30.67	27.32	6:26	6
-137	30.82	27.47	6:34	8
-138	30.91	27.55	6:39	9
-139	31.01	27.65	6:43	10
-140	31.12	27.76	6:48	11
-142	31.36	28.00	6:57	13
-143	31.51	28.13	7:02	14
-144	31.66	28.28	7:07	15
-145	31.82	28.44	7:11	16
-147	32.19	28.80	7:21	18
-148	32.39	28.99	7:26	19
-149	32.61	29.20	7:31	20
-152	33.34	29.91	7:47	23

PERCENTAGE OF SUN OBSCURED ALONG MAP TRACK
WEST END = 93.2   EAST END = 93.8

===
SUNRISE POSITION : ALTITUDE IS -0.8 DEGREES
NORTH : LONGITUDE -127.0, LATITUDE 30.4
SOUTH : LONGITUDE -127.8, LATITUDE 27.1

+ MARKS THE CITY OF YIN

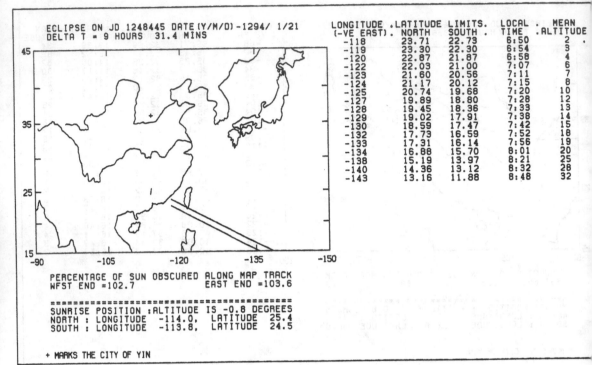

ECLIPSE ON JD 1248445 DATE(Y/M/D)-1294/ 1/21
DELTA T = 9 HOURS 31.4 MINS

LONGITUDE (-VE EAST)	.LATITUDE NORTH	LIMITS. SOUTH .	LOCAL TIME	MEAN .ALTITUDE
-118	23.71	22.73	6:50	2
-119	23.30	22.30	6:54	3
-120	22.87	21.87	6:58	4
-122	22.03	21.00	7:07	6
-123	21.60	20.56	7:11	7
-124	21.17	20.12	7:15	8
-125	20.74	19.68	7:20	10
-127	19.89	18.80	7:28	12
-128	19.45	18.36	7:33	13
-129	19.02	17.91	7:38	14
-130	18.59	17.47	7:42	15
-132	17.73	16.59	7:52	18
-133	17.31	16.14	7:56	19
-134	16.88	15.70	8:01	20
-138	15.19	13.97	8:21	25
-140	14.36	13.12	8:32	28
-143	13.16	11.88	8:48	32

PERCENTAGE OF SUN OBSCURED ALONG MAP TRACK
WEST END =102.7   EAST END =103.6

==
SUNRISE POSITION :ALTITUDE IS -0.8 DEGREES
NORTH : LONGITUDE -114.0, LATITUDE 25.4
SOUTH : LONGITUDE -113.8, LATITUDE 24.5

+ MARKS THE CITY OF YIN

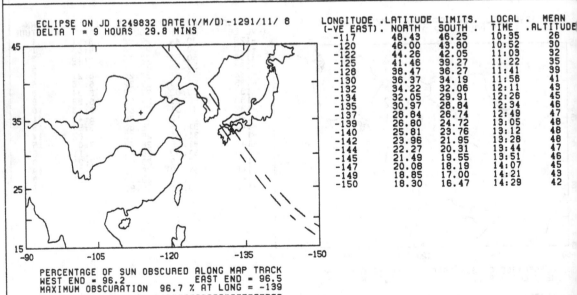

ECLIPSE ON JD 1249832 DATE(Y/M/D)-1291/11/ 8
DELTA T = 9 HOURS 29.8 MINS

LONGITUDE (-VE EAST)	.LATITUDE NORTH	LIMITS. SOUTH .	LOCAL TIME	MEAN .ALTITUDE
-117	48.43	46.25	10:35	26
-120	46.00	43.80	10:52	30
-122	44.26	42.05	11:09	32
-125	41.46	39.27	11:22	35
-128	38.47	36.27	11:41	39
-130	36.37	34.19	11:56	41
-132	34.22	32.06	12:11	43
-134	32.05	29.91	12:26	45
-135	30.97	28.84	12:34	46
-137	28.84	26.74	12:49	47
-139	26.80	24.72	13:05	48
-140	25.81	23.76	13:12	48
-142	23.96	21.95	13:28	48
-144	22.27	20.31	13:44	47
-145	21.49	19.55	13:51	46
-147	20.08	18.19	14:07	45
-149	18.85	17.00	14:21	43
-150	18.30	16.47	14:29	42

PERCENTAGE OF SUN OBSCURED ALONG MAP TRACK
WEST END = 96.2   EAST END = 96.5
MAXIMUM OBSCURATION 96.7 % AT LONG = -139
==

+ MARKS THE CITY OF YIN

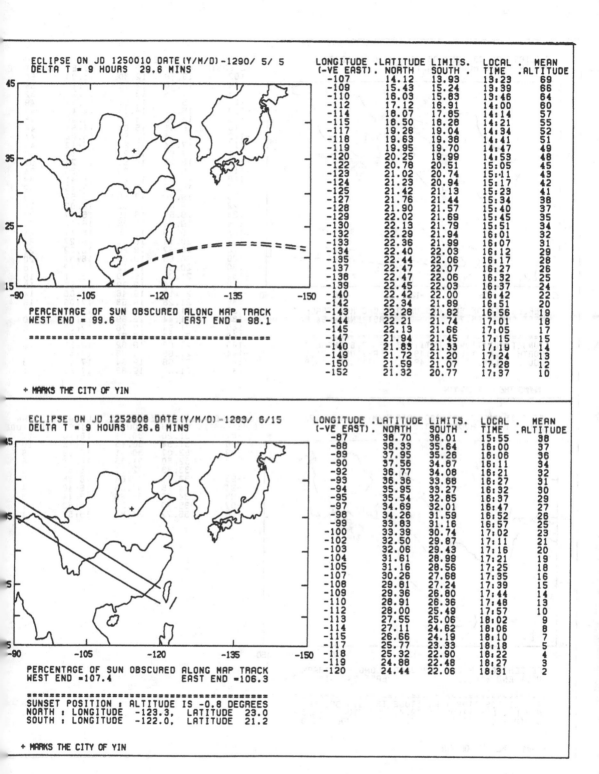

ECLIPSE ON JD 1250010 DATE(Y/M/D)-1290/ 5/ 5
DELTA T = 9 HOURS 29.6 MINS

LONGITUDE (-VE EAST).	.LATITUDE LIMITS. NORTH	SOUTH .	LOCAL . TIME	MEAN .ALTITUDE
-107	14.12	13.93	13:29	69
-109	15.43	15.24	13:39	66
-110	16.03	15.83	13:46	64
-112	17.12	16.91	14:00	60
-114	18.07	17.85	14:14	57
-115	18.50	18.28	14:21	55
-117	19.28	19.04	14:34	52
-118	19.63	19.38	14:41	51
-119	19.95	19.70	14:47	49
-120	20.25	19.99	14:53	48
-122	20.78	20.51	15:05	45
-123	21.02	20.74	15:11	43
-124	21.23	20.94	15:17	42
-125	21.42	21.13	15:23	41
-127	21.76	21.44	15:34	38
-128	21.90	21.57	15:40	37
-129	22.02	21.69	15:45	35
-130	22.13	21.79	15:51	34
-132	22.29	21.94	16:01	32
-133	22.36	21.99	16:07	31
-134	22.40	22.03	16:12	29
-135	22.44	22.06	16:17	28
-137	22.47	22.07	16:27	26
-138	22.47	22.06	16:32	25
-139	22.45	22.03	16:37	24
-140	22.42	22.00	16:42	22
-142	22.34	21.89	16:51	20
-143	22.28	21.82	16:56	19
-144	22.21	21.74	17:01	18
-145	22.13	21.66	17:05	17
-147	21.94	21.45	17:15	15
-148	21.85	21.33	17:19	14
-149	21.72	21.20	17:24	13
-150	21.59	21.07	17:28	12
-152	21.32	20.77	17:37	10

PERCENTAGE OF SUN OBSCURED ALONG MAP TRACK
WEST END = 99.6   EAST END = 98.1

==

+ MARKS THE CITY OF YIN

ECLIPSE ON JD 1252608 DATE(Y/M/D)-1283/ 6/15
DELTA T = 9 HOURS 26.6 MINS

LONGITUDE (-VE EAST).	.LATITUDE LIMITS. NORTH	SOUTH .	LOCAL . TIME	MEAN .ALTITUDE
-87	38.70	36.01	15:55	38
-88	38.33	35.64	16:00	37
-89	37.95	35.26	16:06	36
-90	37.56	34.87	16:11	34
-92	36.77	34.08	16:21	32
-93	36.36	33.68	16:27	31
-94	35.95	33.27	16:32	30
-95	35.54	32.85	16:37	29
-97	34.69	32.01	16:47	27
-98	34.26	31.59	16:52	26
-99	33.83	31.16	16:57	25
-100	33.39	30.74	17:02	23
-102	32.50	29.87	17:11	21
-103	32.06	29.43	17:16	20
-104	31.61	28.99	17:21	19
-105	31.16	28.56	17:25	18
-107	30.26	27.68	17:35	16
-108	29.81	27.24	17:39	15
-109	29.36	26.80	17:44	14
-110	28.91	26.36	17:48	13
-112	28.00	25.49	17:57	10
-113	27.55	25.06	18:02	9
-114	27.11	24.62	18:06	8
-115	26.66	24.19	18:10	7
-117	25.77	23.33	18:18	5
-118	25.32	22.90	18:22	4
-119	24.88	22.48	18:27	3
-120	24.44	22.06	18:31	2

PERCENTAGE OF SUN OBSCURED ALONG MAP TRACK
WEST END =107.4   EAST END =106.3

==
SUNSET POSITION : ALTITUDE IS -0.8 DEGREES
NORTH : LONGITUDE -123.3, LATITUDE 23.0
SOUTH : LONGITUDE -122.0, LATITUDE 21.2

+ MARKS THE CITY OF YIN

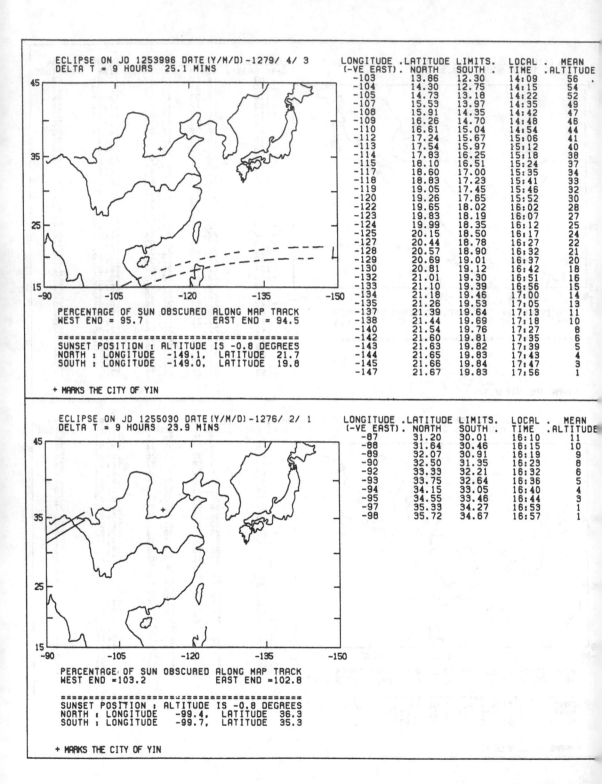

ECLIPSE ON JD 1253996 DATE(Y/M/D)-1279/ 4/ 3
DELTA T = 9 HOURS 25.1 MINS

LONGITUDE (-VE EAST)	LATITUDE NORTH	LIMITS. SOUTH	LOCAL TIME	MEAN ALTITUDE
-103	13.86	12.30	14:09	56
-104	14.30	12.75	14:15	54
-105	14.73	13.18	14:22	52
-107	15.53	13.97	14:35	49
-108	15.91	14.35	14:42	47
-109	16.26	14.70	14:48	46
-110	16.61	15.04	14:54	44
-112	17.24	15.67	15:06	41
-113	17.54	15.97	15:12	40
-114	17.83	16.25	15:18	38
-115	18.10	16.51	15:24	37
-117	18.60	17.00	15:35	34
-118	18.83	17.23	15:41	33
-119	19.05	17.45	15:46	32
-120	19.26	17.65	15:52	30
-122	19.65	18.02	16:02	28
-123	19.83	18.19	16:07	27
-124	19.99	18.35	16:12	25
-125	20.15	18.50	16:17	24
-127	20.44	18.78	16:27	22
-128	20.57	18.90	16:32	21
-129	20.69	19.01	16:37	20
-130	20.81	19.12	16:42	18
-132	21.01	19.30	16:51	16
-133	21.10	19.39	16:56	15
-134	21.18	19.46	17:00	14
-135	21.26	19.53	17:05	13
-137	21.39	19.64	17:13	11
-138	21.44	19.69	17:18	10
-140	21.54	19.76	17:27	8
-142	21.60	19.81	17:35	6
-143	21.63	19.82	17:39	5
-144	21.65	19.83	17:43	4
-145	21.66	19.84	17:47	3
-147	21.67	19.83	17:56	1

PERCENTAGE OF SUN OBSCURED ALONG MAP TRACK
WEST END = 95.7   EAST END = 94.5

===
SUNSET POSITION : ALTITUDE IS -0.8 DEGREES
NORTH : LONGITUDE   -149.1, LATITUDE   21.7
SOUTH : LONGITUDE   -149.0, LATITUDE   19.8

+ MARKS THE CITY OF YIN

ECLIPSE ON JD 1255030 DATE(Y/M/D)-1276/ 2/ 1
DELTA T = 9 HOURS 23.9 MINS

LONGITUDE (-VE EAST)	LATITUDE NORTH	LIMITS. SOUTH	LOCAL TIME	MEAN ALTITUDE
-87	31.20	30.01	16:10	11
-88	31.64	30.46	16:15	10
-89	32.07	30.91	16:19	9
-90	32.50	31.35	16:29	8
-92	33.33	32.21	16:32	6
-93	33.75	32.64	16:36	5
-94	34.15	33.05	16:40	4
-95	34.55	33.46	16:44	3
-97	35.33	34.27	16:53	1
-98	35.72	34.67	16:57	1

PERCENTAGE OF SUN OBSCURED ALONG MAP TRACK
WEST END =103.2   EAST END =102.8

===
SUNSET POSITION : ALTITUDE IS -0.8 DEGREES
NORTH : LONGITUDE   -99.4, LATITUDE   36.3
SOUTH : LONGITUDE   -99.7, LATITUDE   35.3

+ MARKS THE CITY OF YIN

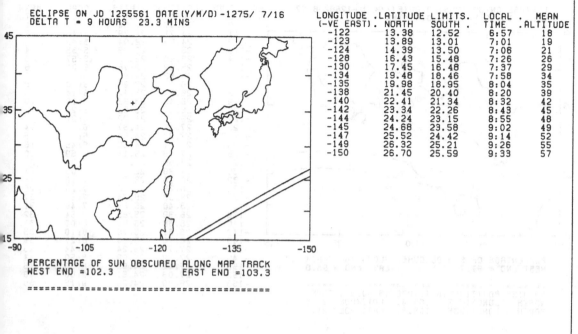

ECLIPSE ON JD 1255561 DATE (Y/M/D) -1275/ 7/16
DELTA T = 9 HOURS 23.3 MINS

LONGITUDE (-VE EAST)	LATITUDE NORTH	LIMITS SOUTH	LOCAL TIME	MEAN ALTITUDE
-122	13.38	12.52	6:57	18
-123	13.89	13.01	7:01	19
-124	14.39	13.50	7:06	21
-128	16.43	15.48	7:26	26
-130	17.45	16.48	7:37	29
-134	19.48	18.46	7:58	34
-135	19.98	18.95	8:04	35
-138	21.45	20.40	8:20	39
-140	22.41	21.34	8:32	42
-142	23.34	22.26	8:43	45
-144	24.24	23.15	8:55	48
-145	24.68	23.58	9:02	49
-147	25.52	24.42	9:14	52
-149	26.32	25.21	9:26	55
-150	26.70	25.59	9:33	57

PERCENTAGE OF SUN OBSCURED ALONG MAP TRACK
WEST END =102.3   EAST END =103.3

==

+ MARKS THE CITY OF YIN

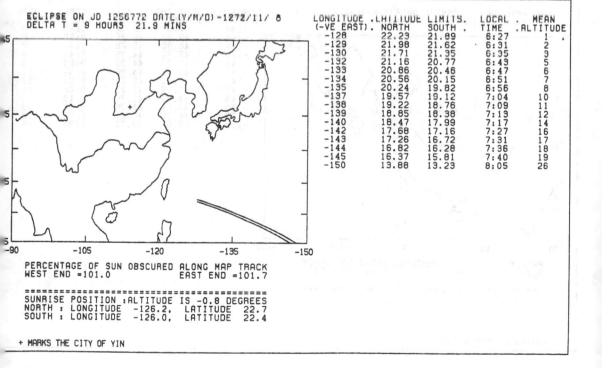

ECLIPSE ON JD 1256772 DATE (Y/M/D) -1272/11/ 8
DELTA T = 9 HOURS 21.9 MINS

LONGITUDE (-VE EAST)	LATITUDE NORTH	LIMITS SOUTH	LOCAL TIME	MEAN ALTITUDE
-128	22.23	21.89	6:27	1
-129	21.98	21.62	6:31	2
-130	21.71	21.35	6:35	3
-132	21.16	20.77	6:43	5
-133	20.86	20.46	6:47	6
-134	20.56	20.15	6:51	7
-135	20.24	19.82	6:56	8
-137	19.57	19.12	7:04	10
-138	19.22	18.76	7:09	11
-139	18.85	18.38	7:13	12
-140	18.47	17.99	7:17	14
-142	17.68	17.16	7:27	16
-143	17.26	16.72	7:31	17
-144	16.82	16.28	7:36	18
-145	16.37	15.81	7:40	19
-150	13.88	13.23	8:05	26

PERCENTAGE OF SUN OBSCURED ALONG MAP TRACK
WEST END =101.0   EAST END =101.7

==
SUNRISE POSITION :ALTITUDE IS -0.8 DEGREES
NORTH : LONGITUDE -126.2, LATITUDE 22.7
SOUTH : LONGITUDE -126.0, LATITUDE 22.4

+ MARKS THE CITY OF YIN

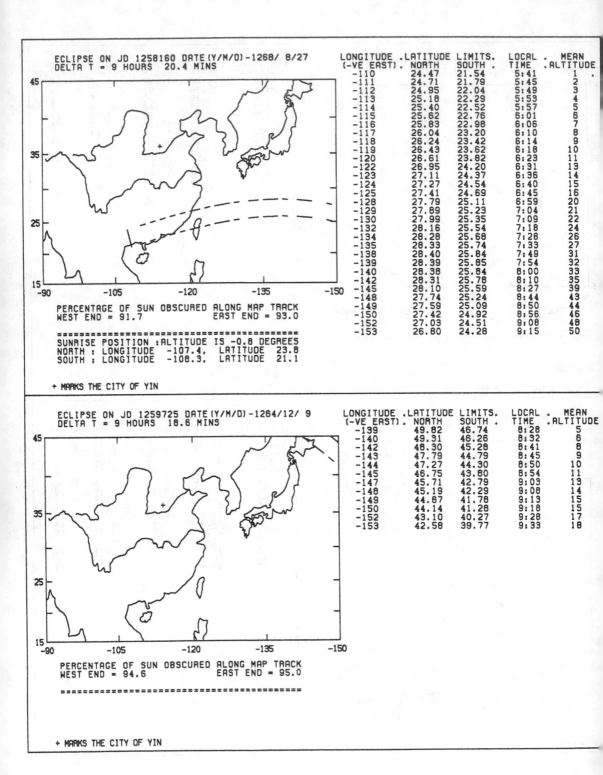

ECLIPSE ON JD 1258160 DATE(Y/M/D)-1268/ 8/27
DELTA T = 9 HOURS 20.4 MINS

LONGITUDE (-VE EAST)	LATITUDE NORTH	LIMITS. SOUTH	LOCAL TIME	MEAN ALTITUDE
-110	24.47	21.54	5:41	1
-111	24.71	21.79	5:45	2
-112	24.95	22.04	5:49	3
-113	25.18	22.29	5:53	4
-114	25.40	22.52	5:57	5
-115	25.62	22.76	6:01	6
-116	25.83	22.98	6:06	7
-117	26.04	23.20	6:10	8
-118	26.24	23.42	6:14	9
-119	26.43	23.62	6:18	10
-120	26.61	23.82	6:23	11
-122	26.95	24.20	6:31	13
-123	27.11	24.37	6:36	14
-124	27.27	24.54	6:40	15
-125	27.41	24.69	6:45	16
-128	27.79	25.11	6:59	20
-129	27.89	25.23	7:04	21
-130	27.99	25.35	7:09	22
-132	28.16	25.54	7:18	24
-134	28.28	25.68	7:28	26
-135	28.33	25.74	7:33	27
-138	28.40	25.84	7:49	31
-139	28.39	25.85	7:54	32
-140	28.38	25.84	8:00	33
-142	28.31	25.78	8:10	35
-145	28.10	25.59	8:27	39
-148	27.74	25.24	8:44	43
-149	27.59	25.09	8:50	44
-150	27.42	24.92	8:56	46
-152	27.03	24.51	9:08	48
-153	26.80	24.28	9:15	50

PERCENTAGE OF SUN OBSCURED ALONG MAP TRACK
WEST END = 91.7   EAST END = 93.0

==
SUNRISE POSITION :ALTITUDE IS -0.8 DEGREES
NORTH : LONGITUDE -107.4, LATITUDE 23.8
SOUTH : LONGITUDE -108.3, LATITUDE 21.1

+ MARKS THE CITY OF YIN

ECLIPSE ON JD 1259725 DATE(Y/M/D)-1264/12/ 9
DELTA T = 9 HOURS 18.6 MINS

LONGITUDE (-VE EAST)	LATITUDE NORTH	LIMITS. SOUTH	LOCAL TIME	MEAN ALTITUDE
-139	49.82	46.74	8:28	5
-140	49.31	46.26	8:32	6
-142	48.30	45.28	8:41	8
-143	47.79	44.79	8:45	9
-144	47.27	44.30	8:50	10
-145	46.75	43.80	8:54	11
-147	45.71	42.79	9:03	13
-148	45.19	42.29	9:08	14
-149	44.67	41.78	9:13	15
-150	44.14	41.28	9:18	15
-152	43.10	40.27	9:28	17
-153	42.58	39.77	9:33	18

PERCENTAGE OF SUN OBSCURED ALONG MAP TRACK
WEST END = 94.6   EAST END = 95.0

==

+ MARKS THE CITY OF YIN

ECLIPSE ON JD 1260759 DATE(Y/M/D)-1261/10/ 9
DELTA T = 9 HOURS 17.5 MINS

LONGITUDE (-VE EAST)	LATITUDE NORTH	LIMITS. SOUTH	LOCAL TIME	MEAN ALTITUDE
-87	16.30	16.24	8:22	33
-89	15.72	15.68	8:33	36
-90	15.41	15.39	8:39	38
-93	14.42	14.41	8:56	42
-95	13.71	13.68	9:08	45

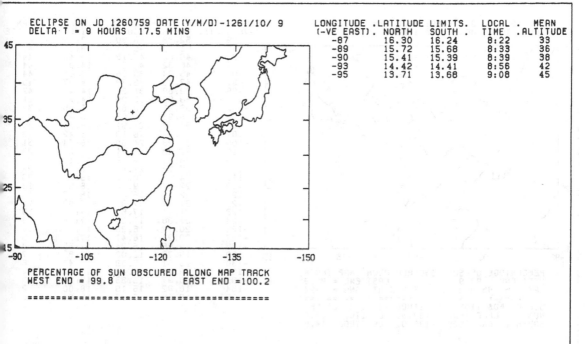

PERCENTAGE OF SUN OBSCURED ALONG MAP TRACK
WEST END = 99.8   EAST END =100.2

==

+ MARKS THE CITY OF YIN

ECLIPSE ON JD 1262148 DATE(Y/M/D)-1257/ 7/27
DELTA T = 9 HOURS 15.9 MINS

LONGITUDE (-VE EAST)	LATITUDE NORTH	LIMITS. SOUTH	LOCAL TIME	MEAN ALTITUDE
-88	16.98	15.93	15:16	44
-90	15.98	14.95	15:27	41
-92	14.96	13.96	15:39	38
-94	13.94	12.96	15:50	35

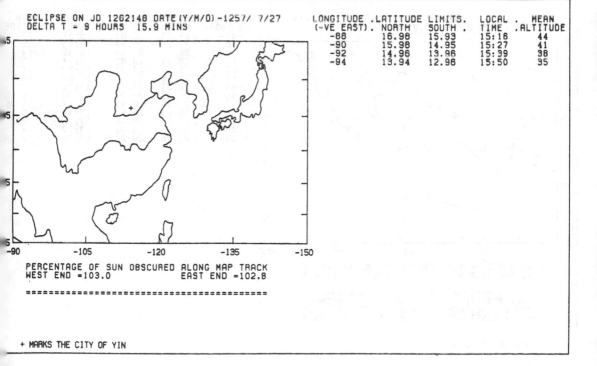

PERCENTAGE OF SUN OBSCURED ALONG MAP TRACK
WEST END =103.0   EAST END =102.8

==

+ MARKS THE CITY OF YIN

ECLIPSE ON JD 1264391 DATE(Y/M/D)-1251/ 9/18
DELTA T = 9 HOURS   13.4 MINS

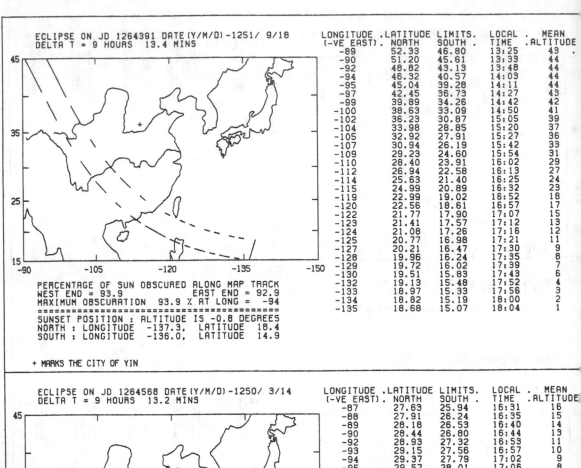

LONGITUDE (-VE EAST)	LATITUDE NORTH	LIMITS. SOUTH	LOCAL TIME	MEAN ALTITUDE
-89	52.33	46.80	13:25	43
-90	51.20	45.61	13:33	44
-92	48.82	43.13	13:48	44
-94	46.32	40.57	14:09	44
-95	45.04	39.28	14:11	44
-97	42.45	36.73	14:27	43
-99	39.89	34.26	14:42	42
-100	38.63	33.09	14:50	41
-102	36.23	30.87	15:05	39
-104	33.98	28.85	15:20	37
-105	32.92	27.91	15:27	36
-107	30.94	26.19	15:42	33
-109	29.23	24.60	15:54	31
-110	28.40	23.91	16:02	29
-112	26.94	22.58	16:13	27
-114	25.63	21.40	16:25	24
-115	24.99	20.89	16:32	23
-119	22.99	19.02	16:52	18
-120	22.56	18.61	16:57	17
-122	21.77	17.90	17:07	15
-123	21.41	17.57	17:12	13
-124	21.08	17.26	17:16	12
-125	20.77	16.98	17:21	11
-127	20.21	16.47	17:30	9
-128	19.96	16.24	17:35	8
-129	19.72	16.02	17:39	7
-130	19.51	15.83	17:43	6
-132	19.13	15.48	17:52	4
-133	18.97	15.33	17:56	3
-134	18.82	15.19	18:00	2
-135	18.68	15.07	18:04	1

PERCENTAGE OF SUN OBSCURED ALONG MAP TRACK
WEST END = 93.9   EAST END = 92.9
MAXIMUM OBSCURATION 93.9 % AT LONG = -94
==
SUNSET POSITION : ALTITUDE IS -0.8 DEGREES
NORTH : LONGITUDE -137.3, LATITUDE 18.4
SOUTH : LONGITUDE -136.0, LATITUDE 14.9

+ MARKS THE CITY OF YIN

ECLIPSE ON JD 1264568 DATE(Y/M/D)-1250/ 3/14
DELTA T = 9 HOURS   13.2 MINS

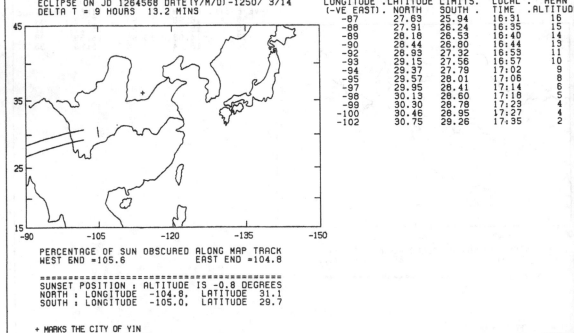

LONGITUDE (-VE EAST)	LATITUDE NORTH	LIMITS. SOUTH	LOCAL TIME	MEAN ALTITUDE
-87	27.63	25.94	16:31	16
-88	27.91	26.24	16:35	15
-89	28.18	26.53	16:40	14
-90	28.44	26.80	16:44	13
-92	28.93	27.32	16:53	11
-93	29.15	27.56	16:57	10
-94	29.37	27.79	17:02	9
-95	29.57	28.01	17:06	8
-97	29.95	28.41	17:14	6
-98	30.13	28.60	17:18	5
-99	30.30	28.78	17:23	4
-100	30.46	28.95	17:27	4
-102	30.75	29.26	17:35	2

PERCENTAGE OF SUN OBSCURED ALONG MAP TRACK
WEST END =105.6   EAST END =104.8

==
SUNSET POSITION : ALTITUDE IS -0.8 DEGREES
NORTH : LONGITUDE -104.8, LATITUDE 31.1
SOUTH : LONGITUDE -105.0, LATITUDE 29.7

+ MARKS THE CITY OF YIN

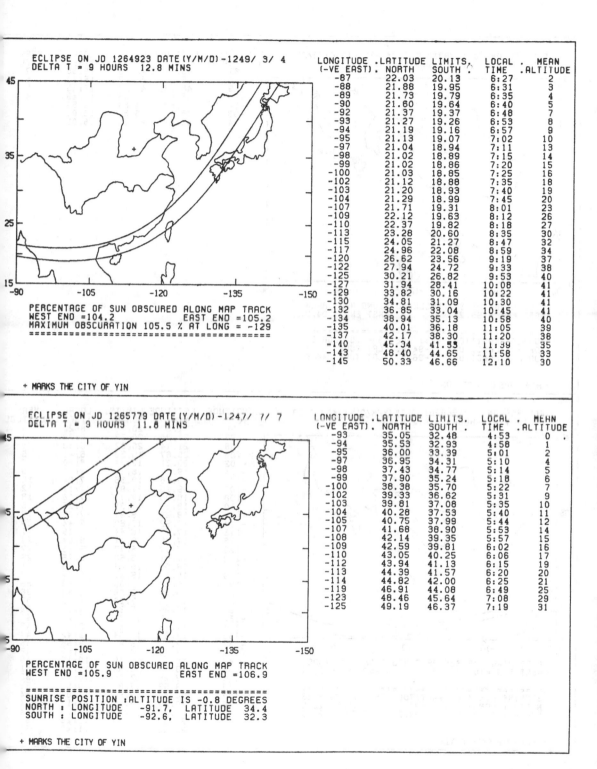

ECLIPSE ON JD 1264923 DATE(Y/M/D)-1249/ 3/ 4
DELTA T = 9 HOURS 12.8 MINS

LONGITUDE (-VE EAST)	.LATITUDE NORTH	LIMITS SOUTH	LOCAL TIME	MEAN ALTITUDE
-87	22.03	20.13	6:27	2
-88	21.88	19.95	6:31	3
-89	21.73	19.79	6:35	4
-90	21.60	19.64	6:40	5
-92	21.37	19.37	6:48	7
-93	21.27	19.26	6:53	8
-94	21.19	19.16	6:57	9
-95	21.13	19.07	7:02	10
-97	21.04	18.94	7:11	13
-98	21.02	18.89	7:15	14
-99	21.02	18.86	7:20	15
-100	21.03	18.85	7:25	16
-102	21.12	18.88	7:35	18
-103	21.20	18.93	7:40	19
-104	21.29	18.99	7:45	20
-107	21.71	19.31	8:01	23
-109	22.12	19.63	8:12	26
-110	22.37	19.82	8:18	27
-113	23.28	20.60	8:35	30
-115	24.05	21.27	8:47	32
-117	24.96	22.08	8:59	34
-120	26.62	23.56	9:19	37
-122	27.94	24.72	9:33	38
-125	30.21	26.82	9:53	40
-127	31.94	28.41	10:08	41
-129	33.82	30.16	10:22	41
-130	34.81	31.09	10:30	41
-132	36.85	33.04	10:45	41
-134	38.94	35.13	10:58	40
-135	40.01	36.18	11:05	39
-137	42.17	38.30	11:20	38
-140	45.34	41.53	11:39	35
-143	48.40	44.65	11:58	33
-145	50.33	46.66	12:10	30

PERCENTAGE OF SUN OBSCURED ALONG MAP TRACK
WEST END =104.2   EAST END =105.2
MAXIMUM OBSCURATION 105.5 % AT LONG = -129
===

+ MARKS THE CITY OF YIN

ECLIPSE ON JD 1265779 DATE(Y/M/D)-1247/ 7/ 7
DELTA T = 9 HOURS 11.8 MINS

LONGITUDE (-VE EAST)	.LATITUDE NORTH	LIMITS SOUTH	LOCAL TIME	MEAN ALTITUDE
-93	35.05	32.48	4:53	0
-94	35.53	32.93	4:58	1
-95	36.00	33.39	5:01	2
-97	36.95	34.31	5:10	4
-98	37.43	34.77	5:14	5
-99	37.90	35.24	5:18	6
-100	38.38	35.70	5:22	7
-102	39.33	36.62	5:31	9
-103	39.81	37.08	5:35	10
-104	40.28	37.53	5:40	11
-105	40.75	37.99	5:44	12
-107	41.68	38.90	5:53	14
-108	42.14	39.35	5:57	15
-109	42.59	39.81	6:02	16
-110	43.05	40.25	6:06	17
-112	43.94	41.13	6:15	19
-113	44.39	41.57	6:20	20
-114	44.82	42.00	6:25	21
-119	46.91	44.08	6:49	25
-123	48.46	45.64	7:08	29
-125	49.19	46.37	7:19	31

PERCENTAGE OF SUN OBSCURED ALONG MAP TRACK
WEST END =105.9   EAST END =106.9

===
SUNRISE POSITION :ALTITUDE IS -0.8 DEGREES
NORTH : LONGITUDE   -91.7, LATITUDE 34.4
SOUTH : LONGITUDE   -92.6, LATITUDE 32.3

+ MARKS THE CITY OF YIN

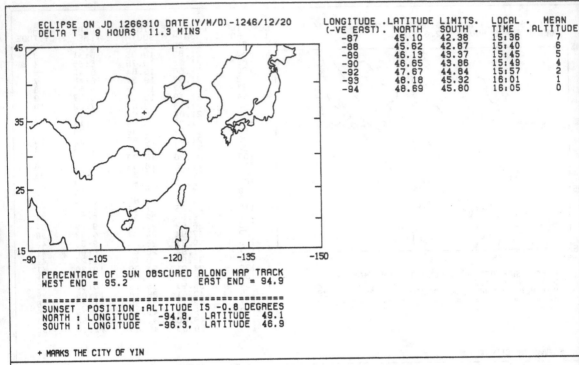

ECLIPSE ON JD 1266310 DATE(Y/M/D)-1246/12/20
DELTA T = 9 HOURS   11.3 MINS

LONGITUDE (-VE EAST)	LATITUDE NORTH	LIMITS. SOUTH	LOCAL TIME	MEAN ALTITUDE
-87	45.10	42.38	15:36	7
-88	45.62	42.87	15:40	6
-89	46.13	43.37	15:45	5
-90	46.65	43.86	15:49	4
-92	47.67	44.84	15:57	2
-93	48.18	45.32	16:01	1
-94	48.69	45.80	16:05	0

PERCENTAGE OF SUN OBSCURED ALONG MAP TRACK
WEST END = 95.2   EAST END = 94.9

==
SUNSET POSITION :ALTITUDE IS -0.8 DEGREES
NORTH : LONGITUDE   -94.8, LATITUDE   49.1
SOUTH : LONGITUDE   -96.3, LATITUDE   46.9

+ MARKS THE CITY OF YIN

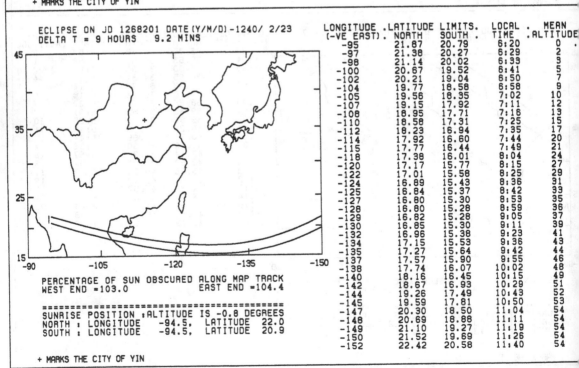

ECLIPSE ON JD 1268201 DATE(Y/M/D)-1240/ 2/23
DELTA T = 9 HOURS   9.2 MINS

LONGITUDE (-VE EAST)	LATITUDE NORTH	LIMITS. SOUTH	LOCAL TIME	MEAN ALTITUDE
-95	21.87	20.79	6:20	0
-97	21.38	20.27	6:29	2
-98	21.14	20.02	6:33	3
-100	20.67	19.52	6:41	5
-102	20.21	19.04	6:50	7
-104	19.77	18.58	6:58	9
-105	19.56	18.35	7:02	10
-107	19.15	17.92	7:11	12
-108	18.95	17.71	7:16	13
-110	18.58	17.31	7:25	15
-112	18.23	16.94	7:35	17
-114	17.92	16.60	7:44	20
-115	17.77	16.44	7:49	21
-118	17.38	16.01	8:04	24
-120	17.17	15.77	8:15	27
-122	17.01	15.58	8:25	29
-124	16.89	15.43	8:36	31
-125	16.84	15.37	8:42	33
-127	16.80	15.30	8:53	35
-128	16.80	15.28	8:59	36
-129	16.82	15.28	9:05	37
-130	16.85	15.30	9:11	39
-132	16.96	15.38	9:23	41
-134	17.15	15.53	9:36	43
-135	17.27	15.64	9:42	44
-137	17.57	15.90	9:55	46
-138	17.74	16.07	10:02	48
-140	18.16	16.45	10:15	49
-142	18.67	16.93	10:29	51
-144	19.26	17.49	10:43	52
-145	19.59	17.81	10:50	53
-147	20.30	18.50	11:04	54
-148	20.69	18.88	11:11	54
-149	21.10	19.27	11:19	54
-150	21.52	19.69	11:26	54
-152	22.42	20.58	11:40	54

PERCENTAGE OF SUN OBSCURED ALONG MAP TRACK
WEST END =103.0   EAST END =104.4

==
SUNRISE POSITION :ALTITUDE IS -0.8 DEGREES
NORTH : LONGITUDE   -94.5, LATITUDE   22.0
SOUTH : LONGITUDE   -94.5, LATITUDE   20.9

+ MARKS THE CITY OF YIN

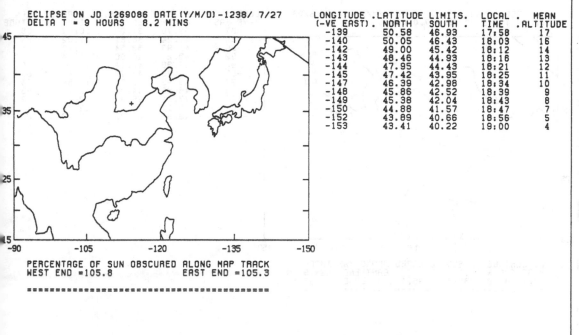

ECLIPSE ON JD 1269086 DATE(Y/M/D)-1238/ 7/27
DELTA T = 9 HOURS   8.2 MINS

LONGITUDE (-VE EAST)	LATITUDE NORTH	LIMITS SOUTH	LOCAL TIME	MEAN ALTITUDE
-139	50.58	46.93	17:58	17
-140	50.05	46.43	18:03	16
-142	49.00	45.42	18:12	14
-143	48.46	44.93	18:16	13
-144	47.95	44.43	18:21	12
-145	47.42	43.95	18:25	11
-147	46.39	42.98	18:34	10
-148	45.86	42.52	18:39	9
-149	45.38	42.04	18:43	8
-150	44.88	41.57	18:47	7
-152	43.89	40.66	18:56	5
-153	43.41	40.22	19:00	4

PERCENTAGE OF SUN OBSCURED ALONG MAP TRACK
WEST END =105.8   EAST END =105.3

==

+ MARKS THE CITY OF YIN

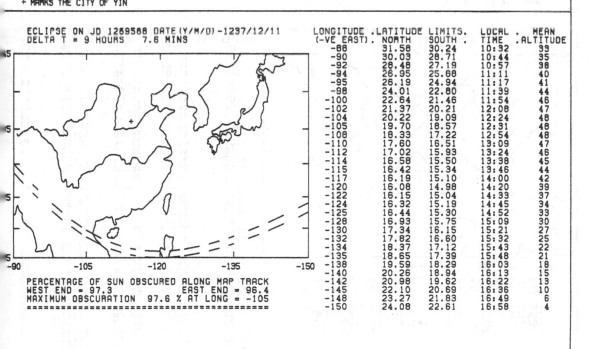

ECLIPSE ON JD 1269588 DATE(Y/M/D)-1237/12/11
DELTA T = 9 HOURS   7.6 MINS

LONGITUDE (-VE EAST)	LATITUDE NORTH	LIMITS SOUTH	LOCAL TIME	MEAN ALTITUDE
-88	31.58	30.24	10:32	39
-90	30.03	28.71	10:44	35
-92	28.48	27.19	10:57	38
-94	26.95	25.68	11:11	40
-95	26.19	24.94	11:17	41
-98	24.01	22.80	11:39	44
-100	22.64	21.46	11:54	46
-102	21.37	20.21	12:08	47
-104	20.22	19.09	12:24	48
-105	19.70	18.57	12:31	48
-108	18.33	17.22	12:54	48
-110	17.60	16.51	13:09	47
-112	17.02	15.93	13:24	46
-114	16.58	15.50	13:38	45
-115	16.42	15.34	13:46	44
-117	16.19	15.10	14:00	42
-120	16.08	14.98	14:20	39
-122	16.15	15.04	14:33	37
-124	16.32	15.19	14:45	34
-125	16.44	15.30	14:52	33
-128	16.93	15.75	15:09	30
-130	17.34	16.15	15:21	27
-132	17.82	16.60	15:32	25
-134	18.37	17.12	15:43	22
-135	18.65	17.39	15:48	21
-138	19.59	18.29	16:03	18
-140	20.26	18.94	16:13	15
-142	20.98	19.62	16:22	13
-145	22.10	20.69	16:36	10
-148	23.27	21.83	16:49	6
-150	24.08	22.61	16:58	4

PERCENTAGE OF SUN OBSCURED ALONG MAP TRACK
WEST END = 97.3   EAST END = 96.4
MAXIMUM OBSCURATION 97.6 % AT LONG = -105

==

+ MARKS THE CITY OF YIN

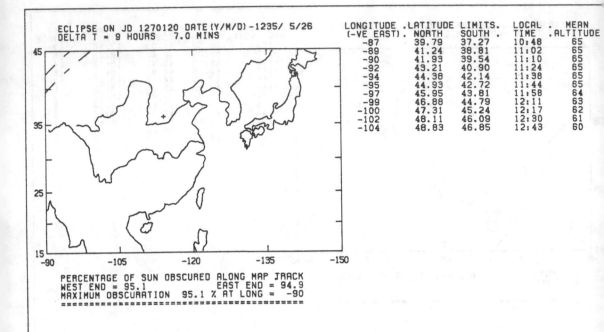

ECLIPSE ON JD 1270120 DATE(Y/M/D)-1235/ 5/26
DELTA T = 9 HOURS   7.0 MINS

LONGITUDE	.LATITUDE	LIMITS.	LOCAL	. MEAN
(-VE EAST).	NORTH	SOUTH .	TIME	.ALTITUDE
-87	39.79	37.27	10:48	65
-89	41.24	38.81	11:02	65
-90	41.93	39.54	11:10	65
-92	43.21	40.90	11:24	65
-94	44.38	42.14	11:38	65
-95	44.93	42.72	11:44	65
-97	45.95	43.81	11:58	64
-99	46.88	44.79	12:11	63
-100	47.31	45.24	12:17	62
-102	48.11	46.09	12:30	61
-104	48.83	46.85	12:43	60

PERCENTAGE OF SUN OBSCURED ALONG MAP TRACK
WEST END = 95.1   EAST END = 94.9
MAXIMUM OBSCURATION 95.1 % AT LONG = -90
===

+ MARKS THE CITY OF YIN

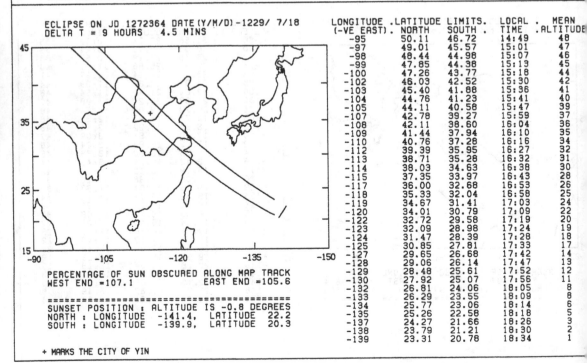

ECLIPSE ON JD 1272364 DATE(Y/M/D)-1229/ 7/18
DELTA T = 9 HOURS   4.5 MINS

LONGITUDE	.LATITUDE	LIMITS.	LOCAL	. MEAN
(-VE EAST).	NORTH	SOUTH .	TIME	.ALTITUDE
-95	50.11	46.72	14:49	48
-97	49.01	45.57	15:01	47
-98	48.44	44.98	15:07	46
-99	47.85	44.38	15:13	45
-100	47.26	43.77	15:18	44
-102	46.03	42.52	15:30	42
-103	45.40	41.88	15:36	41
-104	44.76	41.23	15:41	40
-105	44.11	40.58	15:47	39
-107	42.78	39.27	15:59	37
-108	42.11	38.60	16:04	36
-109	41.44	37.94	16:10	35
-110	40.76	37.28	16:16	34
-112	39.39	35.95	16:27	32
-113	38.71	35.28	16:32	31
-114	38.03	34.63	16:38	30
-115	37.35	33.97	16:43	28
-117	36.00	32.68	16:53	26
-118	35.33	32.04	16:58	25
-119	34.67	31.41	17:03	24
-120	34.01	30.79	17:09	22
-122	32.72	29.58	17:19	20
-123	32.09	28.98	17:24	19
-124	31.47	28.39	17:28	18
-125	30.85	27.81	17:33	17
-127	29.65	26.68	17:42	14
-128	29.06	26.14	17:47	13
-129	28.48	25.61	17:52	12
-130	27.92	25.07	17:56	11
-132	26.81	24.06	18:05	8
-133	26.29	23.55	18:09	8
-134	25.77	23.06	18:14	6
-135	25.26	22.58	18:18	5
-137	24.27	21.66	18:26	3
-138	23.79	21.21	18:30	2
-139	23.31	20.78	18:34	1

PERCENTAGE OF SUN OBSCURED ALONG MAP TRACK
WEST END =107.1   EAST END =105.6

===
SUNSET POSITION : ALTITUDE IS -0.8 DEGREES
NORTH : LONGITUDE -141.4, LATITUDE 22.2
SOUTH : LONGITUDE -139.9, LATITUDE 20.3

+ MARKS THE CITY OF YIN

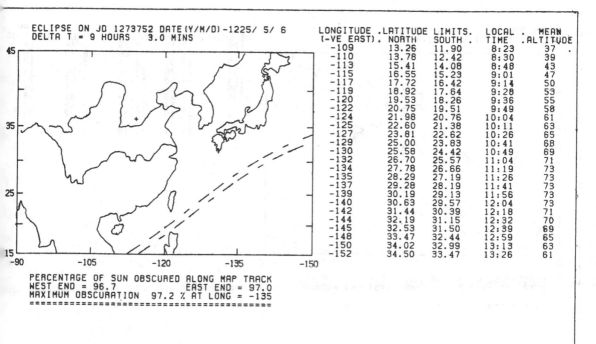

```
ECLIPSE ON JD 1273752 DATE(Y/M/D)-1225/ 5/ 6
DELTA T = 9 HOURS   3.0 MINS
```

LONGITUDE (-VE EAST)	.LATITUDE . NORTH	LIMITS. SOUTH .	LOCAL TIME	. MEAN .ALTITUDE
-109	13.26	11.90	8:23	37 .
-110	13.78	12.42	8:30	39
-113	15.41	14.08	8:48	43
-115	16.55	15.23	9:01	47
-117	17.72	16.42	9:14	50
-119	18.92	17.64	9:28	53
-120	19.53	18.26	9:36	55
-122	20.75	19.51	9:49	58
-124	21.98	20.76	10:04	61
-125	22.60	21.38	10:11	63
-127	23.81	22.62	10:26	65
-129	25.00	23.83	10:41	68
-130	25.58	24.42	10:49	69
-132	26.70	25.57	11:04	71
-134	27.78	26.66	11:19	73
-135	28.29	27.19	11:26	73
-137	29.28	28.19	11:41	73
-139	30.19	29.13	11:56	73
-140	30.63	29.57	12:04	73
-142	31.44	30.39	12:18	71
-144	32.19	31.15	12:32	70
-145	32.53	31.50	12:39	69
-148	33.47	32.44	12:59	65
-150	34.02	32.99	13:13	63
-152	34.50	33.47	13:26	61

```
PERCENTAGE OF SUN OBSCURED ALONG MAP TRACK
WEST END = 96.7              EAST END = 97.0
MAXIMUM OBSCURATION  97.2 % AT LONG = -135
================================================
```

+ MARKS THE CITY OF YIN

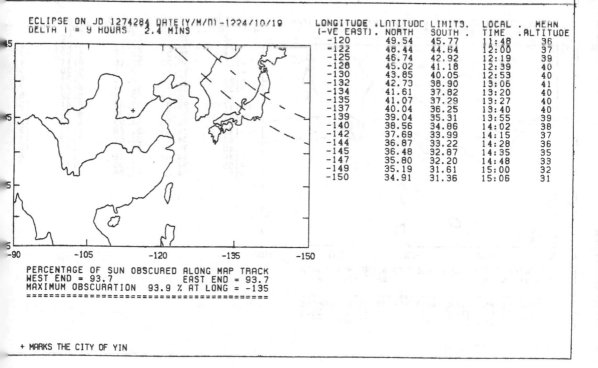

```
ECLIPSE ON JD 1274284 DATE(Y/M/D)-1224/10/19
DELTA T = 9 HOURS   2.4 MINS
```

LONGITUDE (-VE EAST)	.LATITUDE . NORTH	LIMITS. SOUTH .	LOCAL TIME	. MEAN .ALTITUDE
-120	49.54	45.77	11:48	36
-122	48.44	44.64	12:00	37
-125	46.74	42.92	12:19	39
-128	45.02	41.18	12:39	40
-130	43.85	40.05	12:53	40
-132	42.73	38.90	13:06	41
-134	41.61	37.82	13:20	40
-135	41.07	37.29	13:27	40
-137	40.04	36.25	13:40	40
-139	39.04	35.31	13:55	39
-140	38.56	34.86	14:02	38
-142	37.69	33.99	14:15	37
-144	36.87	33.22	14:28	36
-145	36.48	32.87	14:35	35
-147	35.80	32.20	14:48	33
-149	35.19	31.61	15:00	32
-150	34.91	31.36	15:06	31

```
PERCENTAGE OF SUN OBSCURED ALONG MAP TRACK
WEST END = 93.7              EAST END = 93.7
MAXIMUM OBSCURATION  93.9 % AT LONG = -135
================================================
```

+ MARKS THE CITY OF YIN

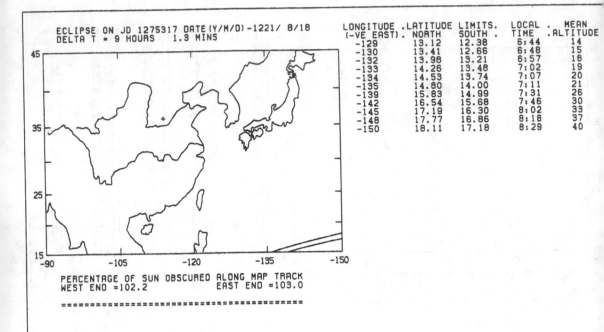

ECLIPSE ON JD 1275317 DATE(Y/M/D)-1221/ 8/18
DELTA T = 9 HOURS 1.3 MINS

LONGITUDE (-VE EAST)	LATITUDE NORTH	LIMITS SOUTH	LOCAL TIME	MEAN ALTITUDE
-129	13.12	12.38	6:44	14
-130	13.41	12.66	6:48	15
-132	13.98	13.21	6:57	18
-133	14.26	13.48	7:02	19
-134	14.53	13.74	7:07	20
-135	14.80	14.00	7:11	21
-139	15.83	14.99	7:31	26
-142	16.54	15.68	7:46	30
-145	17.19	16.30	8:02	33
-148	17.77	16.86	8:18	37
-150	18.11	17.18	8:29	40

PERCENTAGE OF SUN OBSCURED ALONG MAP TRACK
WEST END =102.2   EAST END =103.0

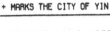

+ MARKS THE CITY OF YIN

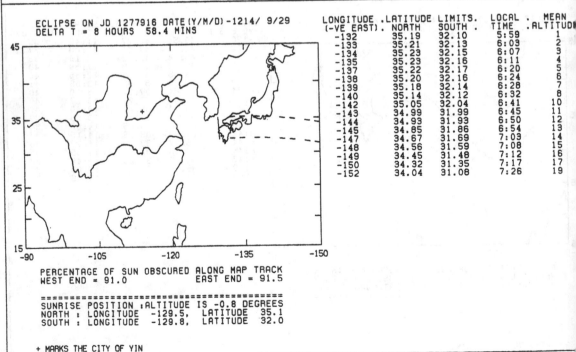

ECLIPSE ON JD 1277916 DATE(Y/M/D)-1214/ 9/29
DELTA T = 8 HOURS 58.4 MINS

LONGITUDE (-VE EAST)	LATITUDE NORTH	LIMITS SOUTH	LOCAL TIME	MEAN ALTITUDE
-132	35.19	32.10	5:59	1
-133	35.21	32.13	6:03	2
-134	35.23	32.15	6:07	3
-135	35.23	32.16	6:11	4
-137	35.22	32.17	6:20	5
-138	35.20	32.16	6:24	6
-139	35.18	32.14	6:28	7
-140	35.14	32.12	6:32	8
-142	35.05	32.04	6:41	10
-143	34.99	31.99	6:45	11
-144	34.93	31.93	6:50	12
-145	34.85	31.86	6:54	13
-147	34.67	31.69	7:03	14
-148	34.56	31.59	7:08	15
-149	34.45	31.48	7:12	16
-150	34.32	31.35	7:17	17
-152	34.04	31.08	7:26	19

PERCENTAGE OF SUN OBSCURED ALONG MAP TRACK
WEST END = 91.0   EAST END = 91.5

SUNRISE POSITION :ALTITUDE IS -0.8 DEGREES
NORTH : LONGITUDE -129.5, LATITUDE 35.1
SOUTH : LONGITUDE -129.8, LATITUDE 32.0

+ MARKS THE CITY OF YIN

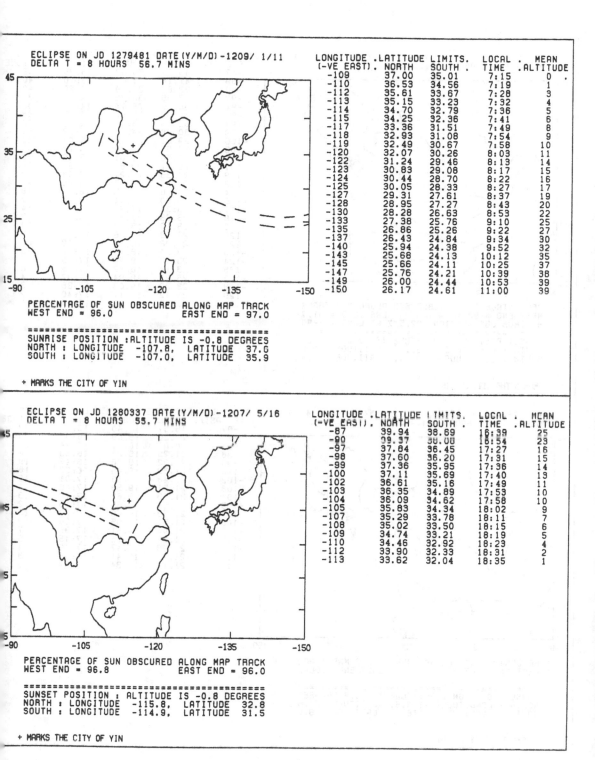

ECLIPSE ON JD 1279481 DATE(Y/M/D)-1209/ 1/11
DELTA T = 8 HOURS 56.7 MINS

LONGITUDE (-VE EAST).	LATITUDE NORTH	LIMITS. SOUTH .	LOCAL TIME	MEAN . ALTITUDE
-109	37.00	35.01	7:15	0 .
-110	36.53	34.56	7:19	1
-112	35.61	33.67	7:28	3
-113	35.15	33.23	7:32	4
-114	34.70	32.79	7:36	5
-115	34.25	32.36	7:41	6
-117	33.36	31.51	7:49	8
-118	32.93	31.08	7:54	9
-119	32.49	30.67	7:58	10
-120	32.07	30.26	8:03	11
-122	31.24	29.46	8:13	14
-123	30.83	29.08	8:17	15
-124	30.44	28.70	8:22	16
-125	30.05	28.33	8:27	17
-127	29.31	27.61	8:37	19
-128	28.95	27.27	8:43	20
-130	28.28	26.63	8:53	22
-133	27.38	25.76	9:10	25
-135	26.86	25.26	9:22	27
-137	26.43	24.84	9:34	30
-140	25.94	24.38	9:52	32
-143	25.68	24.13	10:12	35
-145	25.66	24.11	10:25	37
-147	25.76	24.21	10:39	38
-149	26.00	24.44	10:53	39
-150	26.17	24.61	11:00	39

PERCENTAGE OF SUN OBSCURED ALONG MAP TRACK
WEST END = 96.0   EAST END = 97.0

===
SUNRISE POSITION :ALTITUDE IS -0.8 DEGREES
NORTH : LONGITUDE -107.8, LATITUDE 37.0
SOUTH : LONGITUDE -107.0, LATITUDE 35.9

+ MARKS THE CITY OF YIN

ECLIPSE ON JD 1280337 DATE(Y/M/D)-1207/ 5/16
DELTA T = 8 HOURS 55.7 MINS

LONGITUDE (-VE EAST).	LATITUDE NORTH	LIMITS. SOUTH .	LOCAL TIME	MEAN . ALTITUDE
-87	39.94	38.69	16:39	25
-90	39.37	38.00	16:54	23
-97	37.84	36.45	17:27	16
-98	37.60	36.20	17:31	15
-99	37.36	35.95	17:36	14
-100	37.11	35.69	17:40	13
-102	36.61	35.16	17:49	11
-103	36.35	34.89	17:53	10
-104	36.09	34.62	17:58	10
-105	35.83	34.34	18:02	9
-107	35.29	33.78	18:11	7
-108	35.02	33.50	18:15	6
-109	34.74	33.21	18:19	5
-110	34.46	32.92	18:23	4
-112	33.90	32.33	18:31	2
-113	33.62	32.04	18:35	1

PERCENTAGE OF SUN OBSCURED ALONG MAP TRACK
WEST END = 96.8   EAST END = 96.0

===
SUNSET POSITION : ALTITUDE IS -0.8 DEGREES
NORTH : LONGITUDE -115.8, LATITUDE 32.8
SOUTH : LONGITUDE -114.9, LATITUDE 31.5

+ MARKS THE CITY OF YIN

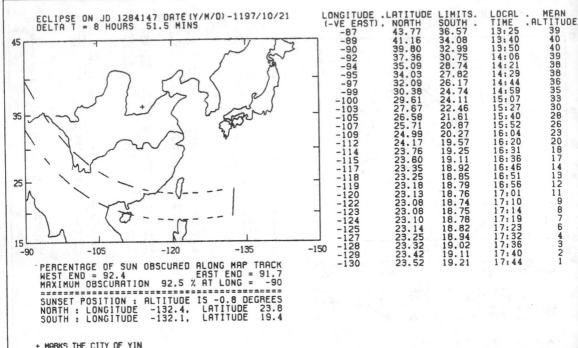

ECLIPSE ON JD 1284147 DATE(Y/M/D)-1197/10/21
DELTA T = 8 HOURS 51.5 MINS

LONGITUDE (-VE EAST)	LATITUDE LIMITS. NORTH	SOUTH .	LOCAL TIME	MEAN ALTITUDE
-87	43.77	36.57	13:25	39
-89	41.16	34.08	13:40	40
-90	39.80	32.99	13:50	40
-92	37.36	30.75	14:06	39
-94	35.09	28.74	14:21	38
-95	34.03	27.82	14:29	38
-97	32.09	26.17	14:44	36
-99	30.38	24.74	14:59	35
-100	29.61	24.11	15:07	33
-103	27.67	22.46	15:27	30
-105	26.58	21.61	15:40	28
-107	25.71	20.87	15:52	26
-109	24.99	20.27	16:04	23
-112	24.17	19.57	16:20	20
-114	23.76	19.25	16:31	18
-115	23.60	19.11	16:36	17
-117	23.35	18.92	16:46	14
-118	23.25	18.85	16:51	13
-119	23.18	18.79	16:56	12
-120	23.13	18.76	17:01	11
-122	23.08	18.74	17:10	9
-123	23.08	18.75	17:14	8
-124	23.10	18.78	17:19	7
-125	23.14	18.82	17:23	6
-127	23.25	18.94	17:32	4
-128	23.32	19.02	17:36	3
-129	23.42	19.11	17:40	2
-130	23.52	19.21	17:44	1

PERCENTAGE OF SUN OBSCURED ALONG MAP TRACK
WEST END = 92.4   EAST END = 91.7
MAXIMUM OBSCURATION 92.5 % AT LONG = -90
==
SUNSET POSITION : ALTITUDE IS -0.8 DEGREES
NORTH : LONGITUDE -132.4, LATITUDE 23.8
SOUTH : LONGITUDE -132.1, LATITUDE 19.4

+ MARKS THE CITY OF YIN

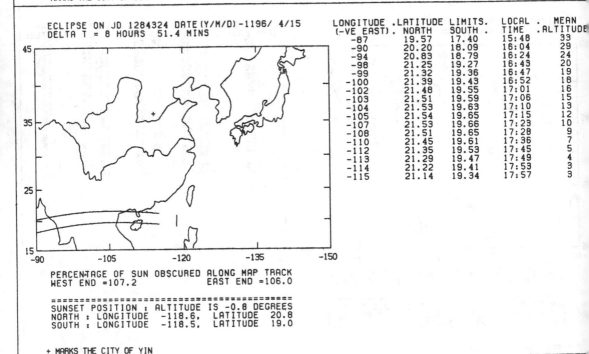

ECLIPSE ON JD 1284324 DATE(Y/M/D)-1196/ 4/15
DELTA T = 8 HOURS 51.4 MINS

LONGITUDE (-VE EAST)	LATITUDE NORTH	SOUTH .	LOCAL TIME	MEAN ALTITUDE
-87	19.57	17.40	15:48	33
-90	20.20	18.09	16:04	29
-94	20.83	18.79	16:24	24
-98	21.25	19.27	16:43	20
-99	21.32	19.36	16:47	19
-100	21.39	19.43	16:52	18
-102	21.48	19.55	17:01	16
-103	21.51	19.59	17:06	15
-104	21.53	19.63	17:10	13
-105	21.54	19.65	17:15	12
-107	21.53	19.66	17:23	10
-108	21.51	19.65	17:28	9
-110	21.45	19.61	17:36	7
-112	21.35	19.53	17:45	5
-113	21.29	19.47	17:49	4
-114	21.22	19.41	17:53	3
-115	21.14	19.34	17:57	3

PERCENTAGE OF SUN OBSCURED ALONG MAP TRACK
WEST END =107.2   EAST END =106.0

==
SUNSET POSITION : ALTITUDE IS -0.8 DEGREES
NORTH : LONGITUDE -118.6, LATITUDE 20.8
SOUTH : LONGITUDE -118.5, LATITUDE 19.0

+ MARKS THE CITY OF YIN

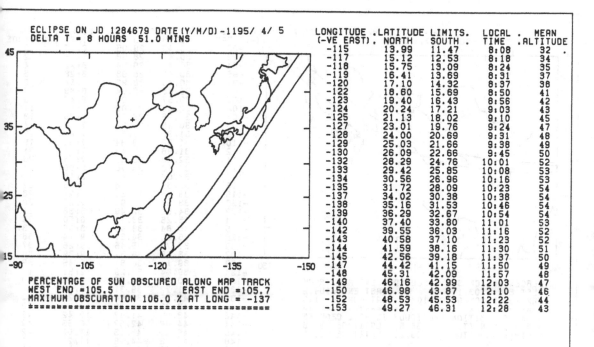

ECLIPSE ON JD 1284679 DATE(Y/M/D)-1195/ 4/ 5
DELTA T = 8 HOURS 51.0 MINS

LONGITUDE (-VE EAST)	LATITUDE NORTH	LIMITS. SOUTH	LOCAL TIME	MEAN ALTITUDE
-115	13.99	11.47	8:08	32
-117	15.12	12.53	8:18	34
-118	15.75	13.09	8:24	35
-119	16.41	13.69	8:31	37
-120	17.10	14.32	8:37	38
-122	18.60	15.69	8:50	41
-123	19.40	16.43	8:56	42
-124	20.24	17.21	9:03	43
-125	21.13	18.02	9:10	45
-127	23.01	19.76	9:24	47
-128	24.00	20.69	9:31	48
-129	25.03	21.66	9:38	49
-130	26.09	22.66	9:45	50
-132	28.29	24.76	10:01	52
-133	29.42	25.85	10:08	53
-134	30.56	26.96	10:16	53
-135	31.72	28.09	10:23	54
-137	34.02	30.38	10:38	54
-138	35.16	31.53	10:46	54
-139	36.29	32.67	10:54	54
-140	37.40	33.80	11:01	53
-142	39.55	36.03	11:16	52
-143	40.58	37.10	11:23	52
-144	41.59	38.16	11:30	51
-145	42.56	39.18	11:37	50
-147	44.42	41.15	11:50	49
-148	45.31	42.09	11:57	48
-149	46.16	42.99	12:03	47
-150	46.98	43.87	12:10	46
-152	48.53	45.53	12:22	44
-153	49.27	46.31	12:28	43

PERCENTAGE OF SUN OBSCURED ALONG MAP TRACK
WEST END =105.5   EAST END =105.7
MAXIMUM OBSCURATION 106.0 % AT LONG = -137
===

+ MARKS THE CITY OF YIN

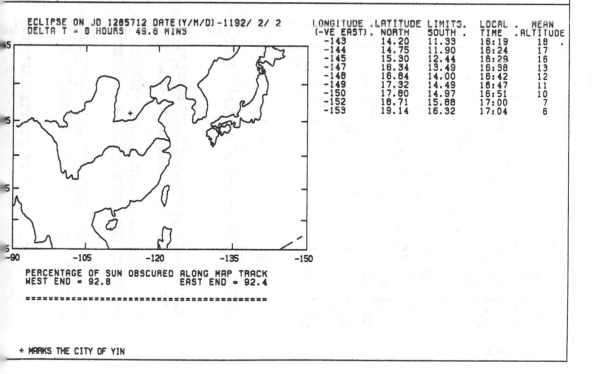

ECLIPSE ON JD 1285712 DATE(Y/M/D)-1192/ 2/ 2
DELTA T = 8 HOURS 49.8 MINS

LONGITUDE (-VE EAST)	LATITUDE NORTH	LIMITS. SOUTH	LOCAL TIME	MEAN ALTITUDE
-143	14.20	11.33	16:19	18
-144	14.75	11.90	16:24	17
-145	15.30	12.44	16:29	16
-147	16.34	13.49	16:38	13
-148	16.84	14.00	16:42	12
-149	17.32	14.49	16:47	11
-150	17.80	14.97	16:51	10
-152	18.71	15.88	17:00	7
-153	19.14	16.32	17:04	6

PERCENTAGE OF SUN OBSCURED ALONG MAP TRACK
WEST END = 92.8   EAST END = 92.4

===

+ MARKS THE CITY OF YIN

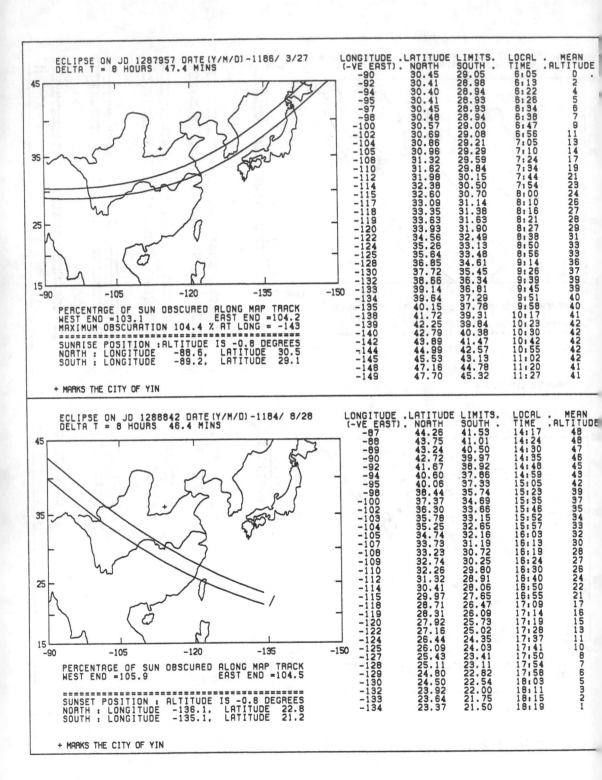

ECLIPSE ON JD 1287957 DATE(Y/M/D)-1186/ 3/27
DELTA T = 8 HOURS 47.4 MINS

LONGITUDE (-VE EAST)	LATITUDE LIMITS NORTH	SOUTH	LOCAL TIME	MEAN ALTITUDE
-90	30.45	29.05	6:05	0
-92	30.41	28.98	6:13	2
-94	30.40	28.94	6:22	4
-95	30.41	28.93	6:26	5
-97	30.45	28.93	6:34	6
-98	30.48	28.94	6:38	7
-100	30.57	29.00	6:47	9
-102	30.69	29.08	6:56	11
-104	30.86	29.21	7:05	13
-105	30.96	29.29	7:10	14
-108	31.32	29.59	7:24	17
-110	31.62	29.84	7:34	19
-112	31.98	30.15	7:44	21
-114	32.38	30.50	7:54	23
-115	32.60	30.70	8:00	24
-117	33.09	31.14	8:10	26
-118	33.35	31.38	8:16	27
-119	33.63	31.63	8:21	28
-120	33.93	31.90	8:27	29
-122	34.56	32.49	8:38	31
-124	35.26	33.13	8:50	33
-125	35.64	33.48	8:56	33
-128	36.85	34.61	9:14	36
-130	37.72	35.45	9:26	37
-132	38.66	36.34	9:39	39
-133	39.14	36.81	9:45	39
-134	39.64	37.29	9:51	40
-135	40.15	37.78	9:58	40
-138	41.72	39.31	10:17	41
-139	42.25	39.84	10:23	42
-140	42.79	40.38	10:30	42
-142	43.89	41.47	10:42	42
-144	44.99	42.57	10:55	42
-145	45.53	43.13	11:02	42
-148	47.16	44.78	11:20	41
-149	47.70	45.32	11:27	41

PERCENTAGE OF SUN OBSCURED ALONG MAP TRACK
WEST END =103.1   EAST END =104.2
MAXIMUM OBSCURATION 104.4 % AT LONG = -143
===
SUNRISE POSITION :ALTITUDE IS -0.8 DEGREES
NORTH : LONGITUDE -88.6, LATITUDE 30.5
SOUTH : LONGITUDE -89.2, LATITUDE 29.1

+ MARKS THE CITY OF YIN

ECLIPSE ON JD 1288842 DATE(Y/M/D)-1184/ 8/28
DELTA T = 8 HOURS 46.4 MINS

LONGITUDE (-VE EAST)	LATITUDE LIMITS NORTH	SOUTH	LOCAL TIME	MEAN ALTITUDE
-87	44.26	41.53	14:17	48
-88	43.75	41.01	14:24	48
-89	43.24	40.50	14:30	47
-90	42.72	39.97	14:35	46
-92	41.67	38.92	14:48	45
-94	40.60	37.86	14:59	43
-95	40.06	37.33	15:05	42
-98	38.44	35.74	15:23	39
-100	37.37	34.69	15:35	37
-102	36.30	33.66	15:46	35
-103	35.78	33.15	15:52	34
-104	35.25	32.65	15:57	33
-105	34.74	32.16	16:03	32
-107	33.73	31.19	16:13	30
-108	33.23	30.72	16:19	28
-109	32.74	30.25	16:24	27
-110	32.26	29.80	16:30	26
-112	31.32	28.91	16:40	24
-114	30.41	28.06	16:50	22
-115	29.97	27.65	16:55	21
-118	28.71	26.47	17:09	17
-119	28.31	26.09	17:14	16
-120	27.92	25.73	17:19	15
-122	27.16	25.02	17:28	13
-124	26.44	24.35	17:37	11
-125	26.09	24.03	17:41	10
-127	25.43	23.41	17:50	8
-128	25.11	23.11	17:54	7
-129	24.80	22.82	17:58	6
-130	24.50	22.54	18:03	5
-132	23.92	22.00	18:11	3
-133	23.64	21.75	18:15	2
-134	23.37	21.50	18:19	1

PERCENTAGE OF SUN OBSCURED ALONG MAP TRACK
WEST END =105.9   EAST END =104.5

===
SUNSET POSITION : ALTITUDE IS -0.8 DEGREES
NORTH : LONGITUDE -136.1, LATITUDE 22.8
SOUTH : LONGITUDE -135.1, LATITUDE 21.2

+ MARKS THE CITY OF YIN

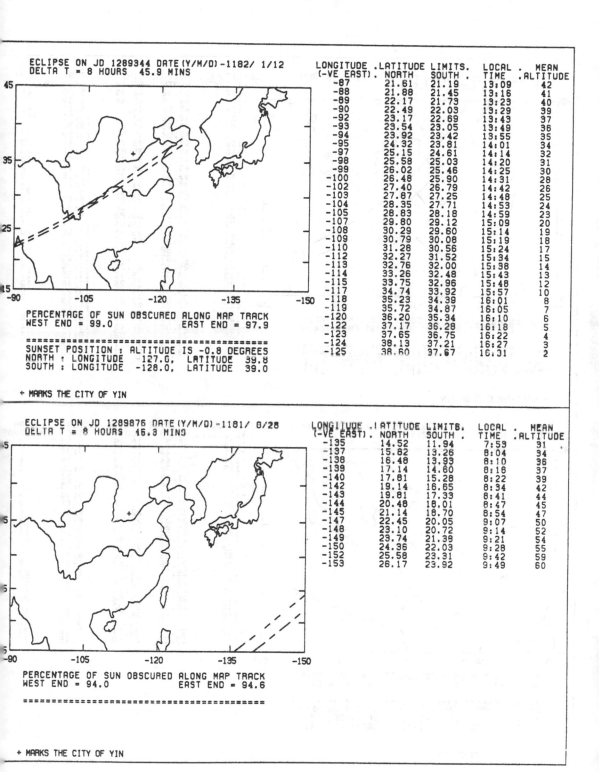

ECLIPSE ON JD 1289344 DATE (Y/M/D) -1182/ 1/12
DELTA T = 8 HOURS   45.9 MINS

LONGITUDE (-VE EAST)	LATITUDE LIMITS. NORTH	SOUTH	LOCAL TIME	MEAN ALTITUDE
-87	21.61	21.19	13:09	42
-88	21.88	21.45	13:16	41
-89	22.17	21.73	13:23	40
-90	22.49	22.03	13:29	39
-92	23.17	22.69	13:43	37
-93	23.54	23.05	13:49	36
-94	23.92	23.42	13:55	35
-95	24.32	23.81	14:01	34
-97	25.15	24.61	14:14	32
-98	25.58	25.03	14:20	31
-99	26.02	25.46	14:25	30
-100	26.48	25.90	14:31	28
-102	27.40	26.79	14:42	26
-103	27.87	27.25	14:48	25
-104	28.35	27.71	14:53	24
-105	28.83	28.18	14:59	23
-107	29.80	29.12	15:09	20
-108	30.29	29.60	15:14	19
-109	30.79	30.08	15:19	18
-110	31.28	30.56	15:24	17
-112	32.27	31.52	15:34	15
-113	32.76	32.00	15:38	14
-114	33.26	32.48	15:43	13
-115	33.75	32.96	15:48	12
-117	34.74	33.92	15:57	10
-118	35.23	34.39	16:01	8
-119	35.72	34.87	16:05	7
-120	36.20	35.34	16:10	6
-122	37.17	36.28	16:18	5
-123	37.65	36.75	16:22	4
-124	38.13	37.21	16:27	3
-125	38.60	37.67	16:31	2

PERCENTAGE OF SUN OBSCURED ALONG MAP TRACK
WEST END = 99.0   EAST END = 97.9

===
SUNSET POSITION : ALTITUDE IS -0.8 DEGREES
NORTH : LONGITUDE 127.0, LATITUDE 39.8
SOUTH : LONGITUDE -128.0, LATITUDE 39.0

+ MARKS THE CITY OF YIN

ECLIPSE ON JD 1289876 DATE (Y/M/D) -1181/ 8/28
DELTA T = 8 HOURS   16.3 MINS

LONGITUDE (-VE EAST)	LATITUDE LIMITS. NORTH	SOUTH	LOCAL TIME	MEAN ALTITUDE
-135	14.52	11.94	7:53	31
-137	15.82	13.26	8:04	34
-138	16.48	13.93	8:10	36
-139	17.14	14.60	8:16	37
-140	17.81	15.28	8:22	39
-142	19.14	16.65	8:34	42
-143	19.81	17.33	8:41	44
-144	20.48	18.01	8:47	45
-145	21.14	18.70	8:54	47
-147	22.45	20.05	9:07	50
-148	23.10	20.72	9:14	52
-149	23.74	21.38	9:21	54
-150	24.36	22.03	9:28	55
-152	25.58	23.31	9:42	59
-153	26.17	23.92	9:49	60

PERCENTAGE OF SUN OBSCURED ALONG MAP TRACK
WEST END = 94.0   EAST END = 94.6

===

+ MARKS THE CITY OF YIN

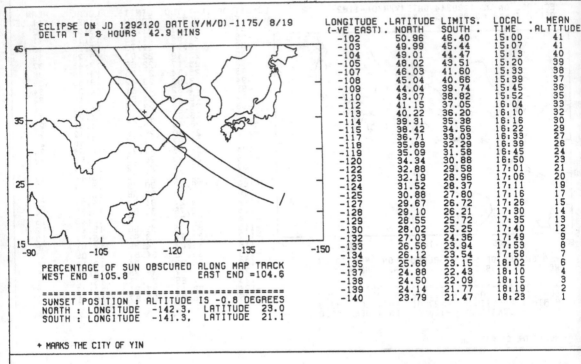

```
ECLIPSE ON JD 1292120 DATE(Y/M/D)-1175/ 8/19
DELTA T = 8 HOURS  42.9 MINS
```

LONGITUDE (-VE EAST)	LATITUDE NORTH	LIMITS. SOUTH	LOCAL TIME	MEAN ALTITUDE
-102	50.96	46.40	15:00	41
-103	49.99	45.44	15:07	41
-104	49.01	44.47	15:13	40
-105	48.02	43.51	15:20	39
-107	46.03	41.60	15:33	38
-108	45.04	40.66	15:39	37
-109	44.04	39.74	15:45	36
-110	43.07	38.82	15:52	35
-112	41.15	37.05	16:04	33
-113	40.22	36.20	16:10	32
-114	39.31	35.38	16:16	30
-115	38.42	34.56	16:22	29
-117	36.71	33.03	16:33	27
-118	35.89	32.29	16:39	26
-119	35.09	31.58	16:45	24
-120	34.34	30.88	16:50	23
-122	32.88	29.58	17:01	21
-123	32.19	28.96	17:06	20
-124	31.52	28.37	17:11	19
-125	30.88	27.80	17:16	17
-127	29.67	26.72	17:26	15
-128	29.10	26.21	17:30	14
-129	28.55	25.72	17:35	13
-130	28.02	25.25	17:40	12
-132	27.03	24.36	17:49	9
-133	26.56	23.94	17:53	8
-134	26.12	23.54	17:58	7
-135	25.68	23.15	18:02	6
-137	24.88	22.43	18:10	4
-138	24.50	22.09	18:15	3
-139	24.14	21.77	18:19	2
-140	23.79	21.47	18:23	1

```
PERCENTAGE OF SUN OBSCURED ALONG MAP TRACK
WEST END =105.8              EAST END =104.6

==============================================
SUNSET POSITION : ALTITUDE IS -0.8 DEGREES
NORTH : LONGITUDE -142.3,  LATITUDE  23.0
SOUTH : LONGITUDE -141.3,  LATITUDE  21.1

 + MARKS THE CITY OF YIN
```

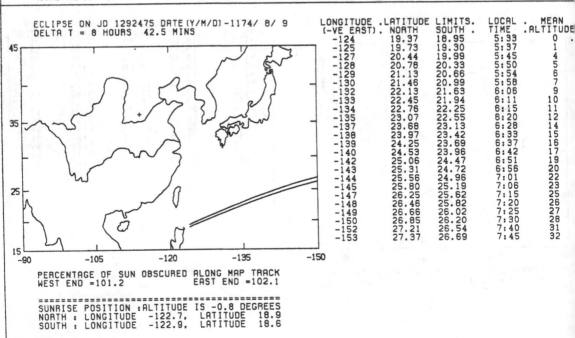

```
ECLIPSE ON JD 1292475 DATE(Y/M/D)-1174/ 8/ 9
DELTA T = 8 HOURS  42.5 MINS
```

LONGITUDE (-VE EAST)	LATITUDE NORTH	LIMITS. SOUTH	LOCAL TIME	MEAN ALTITUDE
-124	19.37	18.95	5:33	0
-125	19.73	19.30	5:37	1
-127	20.44	19.99	5:45	4
-128	20.78	20.33	5:50	5
-129	21.13	20.66	5:54	6
-130	21.46	20.99	5:58	7
-132	22.13	21.63	6:06	9
-133	22.45	21.94	6:11	10
-134	22.76	22.25	6:15	11
-135	23.07	22.55	6:20	12
-137	23.68	23.13	6:28	14
-138	23.97	23.42	6:33	15
-139	24.25	23.69	6:37	16
-140	24.53	23.96	6:42	17
-142	25.06	24.47	6:51	19
-143	25.31	24.72	6:56	20
-144	25.56	24.96	7:01	22
-145	25.80	25.19	7:06	23
-147	26.25	25.62	7:15	25
-148	26.46	25.82	7:20	26
-149	26.66	26.02	7:25	27
-150	26.85	26.20	7:30	28
-152	27.21	26.54	7:40	31
-153	27.37	26.69	7:45	32

```
PERCENTAGE OF SUN OBSCURED ALONG MAP TRACK
WEST END =101.2              EAST END =102.1

==============================================
SUNRISE POSITION :ALTITUDE IS -0.8 DEGREES
NORTH : LONGITUDE -122.7,  LATITUDE  18.9
SOUTH : LONGITUDE -122.9,  LATITUDE  18.6

 + MARKS THE CITY OF YIN
```

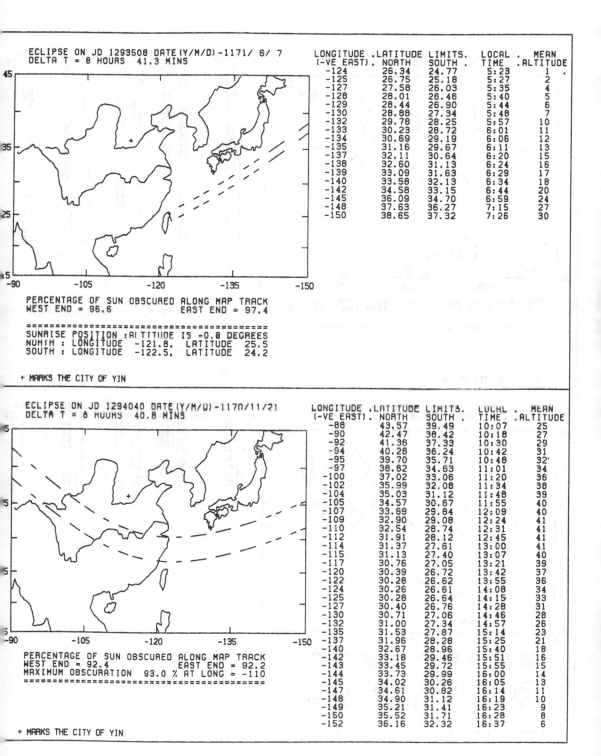

ECLIPSE ON JD 1293508 DATE(Y/M/D) -1171/ 6/ 7
DELTA T = 8 HOURS 41.3 MINS

LONGITUDE	.LATITUDE	LIMITS.	LOCAL	. MEAN
(-VE EAST).	NORTH	SOUTH .	TIME	.ALTITUDE
-124	26.34	24.77	5:23	1
-125	26.75	25.18	5:27	2
-127	27.58	26.03	5:35	4
-128	28.01	26.46	5:40	5
-129	28.44	26.90	5:44	6
-130	28.88	27.34	5:48	7
-132	29.78	28.25	5:57	10
-133	30.23	28.72	6:01	11
-134	30.69	29.19	6:06	12
-135	31.16	29.67	6:11	13
-137	32.11	30.64	6:20	15
-138	32.60	31.13	6:24	16
-139	33.09	31.63	6:29	17
-140	33.58	32.13	6:34	18
-142	34.58	33.15	6:44	20
-145	36.09	34.70	6:59	24
-148	37.63	36.27	7:15	27
-150	38.65	37.32	7:26	30

PERCENTAGE OF SUN OBSCURED ALONG MAP TRACK
WEST END = 96.6   EAST END = 97.4

==
SUNRISE POSITION :ALTITUDE IS -0.8 DEGREES
NORTH : LONGITUDE -121.8, LATITUDE 25.5
SOUTH : LONGITUDE -122.5, LATITUDE 24.2

+ MARKS THE CITY OF YIN

ECLIPSE ON JD 1294040 DATE(Y/M/D) -1170/11/21
DELTA T = 8 HOURS 40.8 MINS

LONGITUDE	.LATITUDE	LIMITS.	LOCAL	. MEAN
(-VE EAST).	NORTH	SOUTH .	TIME	.ALTITUDE
-88	43.57	39.49	10:07	25
-90	42.47	38.42	10:18	27
-92	41.36	37.33	10:30	29
-94	40.28	36.24	10:42	31
-95	39.70	35.71	10:48	32
-97	38.62	34.63	11:01	34
-100	37.02	33.06	11:20	36
-102	35.99	32.08	11:34	38
-104	35.03	31.12	11:48	39
-105	34.57	30.67	11:55	40
-107	33.69	29.84	12:09	40
-109	32.90	29.08	12:24	41
-110	32.54	28.74	12:31	41
-112	31.91	28.12	12:45	41
-114	31.37	27.61	13:00	41
-115	31.13	27.40	13:07	40
-117	30.76	27.05	13:21	39
-120	30.39	26.72	13:42	37
-122	30.28	26.62	13:55	36
-124	30.26	26.61	14:08	34
-125	30.28	26.64	14:15	33
-127	30.40	26.76	14:28	31
-130	30.71	27.06	14:46	28
-132	31.00	27.34	14:57	26
-135	31.53	27.87	15:14	23
-137	31.96	28.28	15:25	21
-140	32.67	28.96	15:40	18
-142	33.18	29.46	15:51	16
-143	33.45	29.72	15:55	15
-144	33.73	29.99	16:00	14
-145	34.02	30.26	16:05	13
-147	34.61	30.82	16:14	11
-148	34.90	31.12	16:19	10
-149	35.21	31.41	16:23	9
-150	35.52	31.71	16:28	8
-152	36.16	32.32	16:37	6

PERCENTAGE OF SUN OBSCURED ALONG MAP TRACK
WEST END = 92.4   EAST END = 92.2
MAXIMUM OBSCURATION 93.0 % AT LONG = -110
==

+ MARKS THE CITY OF YIN

41

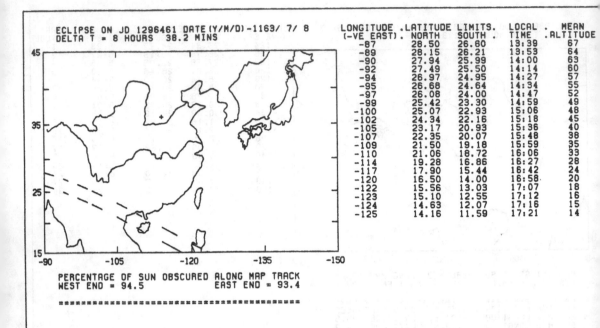

ECLIPSE ON JD 1296461 DATE(Y/M/D) -1163/ 7/ 8
DELTA T = 8 HOURS 38.2 MINS

LONGITUDE	.LATITUDE	LIMITS.	LOCAL	. MEAN
(-VE EAST).	NORTH	SOUTH .	TIME	.ALTITUDE
-87	28.50	26.80	13:39	67
-89	28.15	26.21	13:53	64
-90	27.94	25.99	14:00	63
-92	27.49	25.50	14:14	60
-94	26.97	24.95	14:27	57
-95	26.68	24.64	14:34	55
-97	26.08	24.00	14:47	52
-99	25.42	23.30	14:59	49
-100	25.07	22.93	15:06	48
-102	24.34	22.16	15:18	45
-105	23.17	20.93	15:36	40
-107	22.35	20.07	15:48	38
-109	21.50	19.18	15:59	35
-110	21.06	18.72	16:06	33
-114	19.28	16.86	16:27	28
-117	17.90	15.44	16:42	24
-120	16.50	14.00	16:58	20
-122	15.56	13.03	17:07	18
-123	15.10	12.55	17:12	16
-124	14.63	12.07	17:16	15
-125	14.16	11.59	17:21	14

PERCENTAGE OF SUN OBSCURED ALONG MAP TRACK
WEST END = 94.5   EAST END = 93.4

==

+ MARKS THE CITY OF YIN

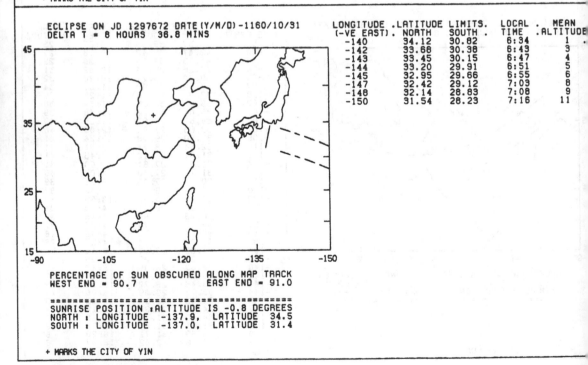

ECLIPSE ON JD 1297672 DATE(Y/M/D) -1160/10/31
DELTA T = 8 HOURS 38.8 MINS

LONGITUDE	.LATITUDE	LIMITS.	LOCAL	. MEAN
(-VE EAST).	NORTH	SOUTH .	TIME	.ALTITUDE
-140	34.12	30.82	6:34	1
-142	33.68	30.38	6:43	3
-143	33.45	30.15	6:47	4
-144	33.20	29.91	6:51	5
-145	32.95	29.66	6:55	6
-147	32.42	29.12	7:03	8
-148	32.14	28.83	7:08	9
-150	31.54	28.23	7:16	11

PERCENTAGE OF SUN OBSCURED ALONG MAP TRACK
WEST END = 90.7   EAST END = 91.0

==
SUNRISE POSITION :ALTITUDE IS -0.8 DEGREES
NORTH : LONGITUDE -137.9, LATITUDE 34.5
SOUTH : LONGITUDE -137.0, LATITUDE 31.4

+ MARKS THE CITY OF YIN

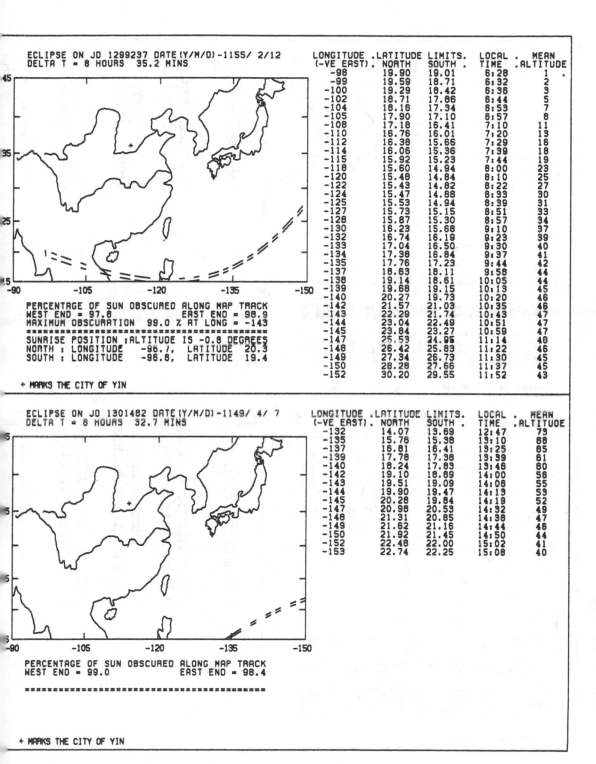

ECLIPSE ON JD 1299237 DATE(Y/M/D)-1155/ 2/12
DELTA T = 8 HOURS 35.2 MINS

45
35
25
15
-90   -105   -120   -135   -150

PERCENTAGE OF SUN OBSCURED ALONG MAP TRACK
WEST END = 97.8   EAST END = 98.9
MAXIMUM OBSCURATION 99.0 % AT LONG = -143
===
SUNRISE POSITION :ALTITUDE IS -0.8 DEGREES
NORTH : LONGITUDE -96.7, LATITUDE 20.3
SOUTH : LONGITUDE -96.8, LATITUDE 19.4

+ MARKS THE CITY OF YIN

LONGITUDE	.LATITUDE	LIMITS.	LOCAL	MEAN
(-VE EAST).	NORTH	SOUTH .	TIME	.ALTITUDE
-98	19.90	19.01	6:28	1
-99	19.59	18.71	6:32	2
-100	19.29	18.42	6:36	3
-102	18.71	17.86	6:44	5
-104	18.16	17.34	6:53	7
-105	17.90	17.10	6:57	8
-108	17.18	16.41	7:10	11
-110	16.76	16.01	7:20	13
-112	16.38	15.66	7:29	16
-114	16.06	15.36	7:39	18
-115	15.92	15.23	7:44	19
-118	15.60	14.94	8:00	23
-120	15.48	14.84	8:10	25
-122	15.43	14.82	8:22	27
-124	15.47	14.88	8:33	30
-125	15.53	14.94	8:39	31
-127	15.73	15.15	8:51	33
-128	15.87	15.30	8:57	34
-130	16.23	15.68	9:10	37
-132	16.74	16.19	9:23	39
-133	17.04	16.50	9:30	40
-134	17.38	16.84	9:37	41
-135	17.76	17.23	9:44	42
-137	18.63	18.11	9:58	44
-138	19.14	18.61	10:05	44
-139	19.68	19.15	10:13	45
-140	20.27	19.73	10:20	46
-142	21.57	21.03	10:35	46
-143	22.29	21.74	10:43	47
-144	23.04	22.49	10:51	47
-145	23.84	23.27	10:59	47
-147	25.53	24.95	11:14	48
-148	26.42	25.83	11:22	46
-149	27.34	26.73	11:30	45
-150	28.28	27.66	11:37	45
-152	30.20	29.55	11:52	43

ECLIPSE ON JD 1301482 DATE(Y/M/D)-1149/ 4/ 7
DELTA T = 8 HOURS 32.7 MINS

PERCENTAGE OF SUN OBSCURED ALONG MAP TRACK
WEST END = 99.0   EAST END = 98.4

===

+ MARKS THE CITY OF YIN

LONGITUDE	.LATITUDE	LIMITS.	LOCAL	MEAN
(-VE EAST).	NORTH	SOUTH .	TIME	.ALTITUDE
-132	14.07	13.69	12:47	79
-135	15.76	15.38	13:10	68
-137	16.81	16.41	13:25	65
-139	17.78	17.38	13:39	61
-140	18.24	17.83	13:46	60
-142	19.10	18.69	14:00	56
-143	19.51	19.09	14:06	55
-144	19.90	19.47	14:13	53
-145	20.28	19.84	14:19	52
-147	20.98	20.53	14:32	49
-148	21.31	20.85	14:38	47
-149	21.62	21.16	14:44	46
-150	21.92	21.45	14:50	44
-152	22.48	22.00	15:02	41
-153	22.74	22.25	15:08	40

ECLIPSE ON JD 1302515 DATE(Y/M/D)-1146/ 2/ 3
DELTA T = 8 HOURS   31.6 MINS

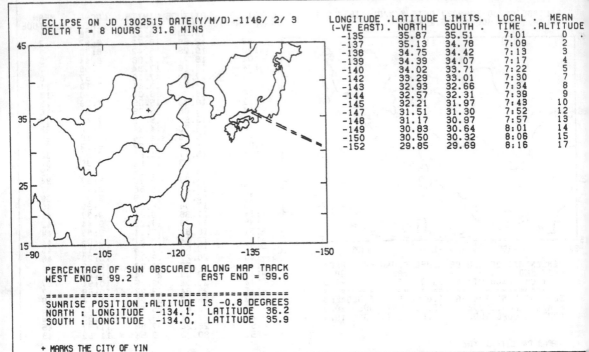

LONGITUDE (-VE EAST)	.LATITUDE LIMITS. NORTH	SOUTH .	LOCAL TIME	MEAN .ALTITUDE
-135	35.87	35.51	7:01	0
-137	35.13	34.78	7:09	2
-138	34.75	34.42	7:13	3
-139	34.39	34.07	7:17	4
-140	34.02	33.71	7:22	5
-142	33.29	33.01	7:30	7
-143	32.93	32.66	7:34	8
-144	32.57	32.31	7:39	9
-145	32.21	31.97	7:43	10
-147	31.51	31.30	7:52	12
-148	31.17	30.97	7:57	13
-149	30.83	30.64	8:01	14
-150	30.50	30.32	8:06	15
-152	29.85	29.69	8:16	17

PERCENTAGE OF SUN OBSCURED ALONG MAP TRACK
WEST END = 99.2   EAST END = 99.6

===
SUNRISE POSITION :ALTITUDE IS -0.8 DEGREES
NORTH : LONGITUDE -134.1, LATITUDE 36.2
SOUTH : LONGITUDE -134.0, LATITUDE 35.9

+ MARKS THE CITY OF YIN

ECLIPSE ON JD 1303903 DATE(Y/M/D)-1143/11/22
DELTA T = 8 HOURS   30.1 MINS

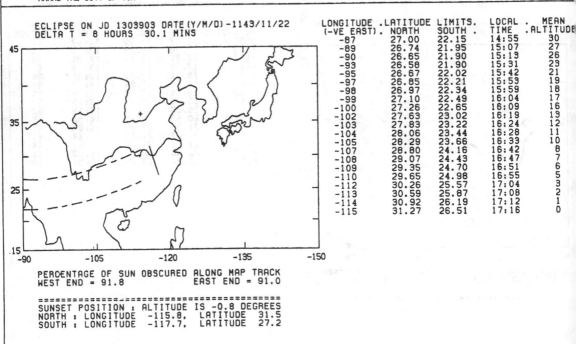

LONGITUDE (-VE EAST)	.LATITUDE NORTH	LIMITS. SOUTH .	LOCAL TIME	MEAN .ALTITUDE
-87	27.00	22.15	14:55	30
-89	26.74	21.95	15:07	27
-90	26.65	21.90	15:19	26
-93	26.58	21.90	15:31	23
-95	26.67	22.02	15:42	21
-97	26.85	22.21	15:53	19
-98	26.97	22.34	15:59	18
-99	27.10	22.49	16:04	17
-100	27.26	22.65	16:09	16
-102	27.63	23.02	16:19	13
-103	27.83	23.22	16:24	12
-104	28.06	23.44	16:28	11
-105	28.29	23.66	16:33	10
-107	28.80	24.16	16:42	8
-108	29.07	24.43	16:47	7
-109	29.35	24.70	16:51	6
-110	29.65	24.98	16:55	5
-112	30.26	25.57	17:04	3
-113	30.59	25.87	17:08	2
-114	30.92	26.19	17:12	1
-115	31.27	26.51	17:16	0

PERCENTAGE OF SUN OBSCURED ALONG MAP TRACK
WEST END = 91.8   EAST END = 91.0

===
SUNSET POSITION : ALTITUDE IS -0.8 DEGREES
NORTH : LONGITUDE -115.8, LATITUDE 31.5
SOUTH : LONGITUDE -117.7, LATITUDE 27.2

+ MARKS THE CITY OF YIN

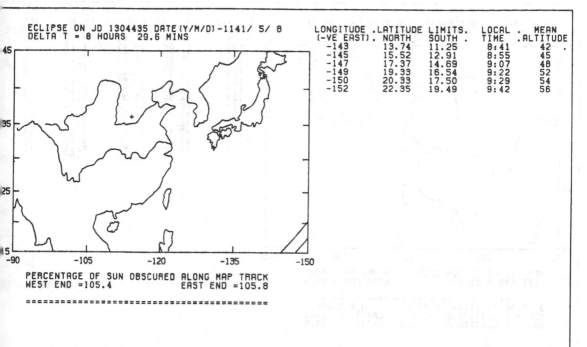

ECLIPSE ON JD 1304435 DATE(Y/M/D) -1141/ 5/ 8
DELTA T = 8 HOURS 29.6 MINS

LONGITUDE (-VE EAST)	LATITUDE NORTH	LIMITS. SOUTH	LOCAL TIME	MEAN ALTITUDE
-143	13.74	11.25	8:41	42
-145	15.52	12.91	8:55	45
-147	17.37	14.69	9:07	48
-149	19.33	16.54	9:22	52
-150	20.33	17.50	9:29	54
-152	22.35	19.49	9:42	56

PERCENTAGE OF SUN OBSCURED ALONG MAP TRACK
WEST END =105.4   EAST END =105.8

===

+ MARKS THE CITY OF YIN

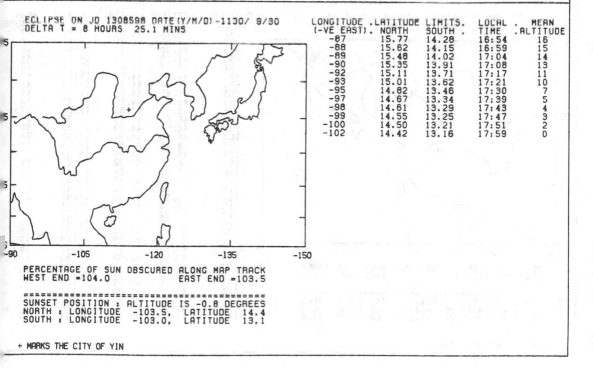

ECLIPSE ON JD 1308598 DATE(Y/M/D) -1130/ 9/30
DELTA T = 8 HOURS 25.1 MINS

LONGITUDE (-VE EAST)	LATITUDE NORTH	LIMITS. SOUTH	LOCAL TIME	MEAN ALTITUDE
-87	15.77	14.28	16:54	16
-88	15.62	14.15	16:59	15
-89	15.48	14.02	17:04	14
-90	15.35	13.91	17:08	13
-92	15.11	13.71	17:17	11
-93	15.01	13.62	17:21	10
-95	14.82	13.46	17:30	7
-97	14.67	13.34	17:39	5
-98	14.61	13.29	17:43	4
-99	14.55	13.25	17:47	3
-100	14.50	13.21	17:51	2
-102	14.42	13.16	17:59	0

PERCENTAGE OF SUN OBSCURED ALONG MAP TRACK
WEST END =104.0   EAST END =103.5

===
SUNSET POSITION : ALTITUDE IS -0.8 DEGREES
NORTH : LONGITUDE -103.5, LATITUDE 14.4
SOUTH : LONGITUDE -103.0, LATITUDE 13.1

+ MARKS THE CITY OF YIN

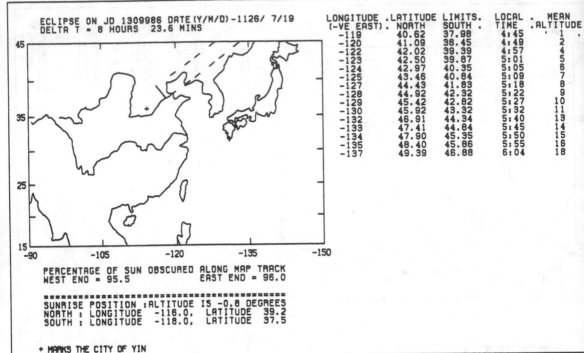

ECLIPSE ON JD 1309986 DATE(Y/M/D)-1126/ 7/19
DELTA T = 8 HOURS 23.6 MINS

LONGITUDE (-VE EAST)	LATITUDE NORTH	LIMITS. SOUTH	LOCAL TIME	MEAN ALTITUDE
-119	40.62	37.98	4:45	1
-120	41.09	38.45	4:49	2
-122	42.02	39.39	4:57	4
-123	42.50	39.87	5:01	5
-124	42.97	40.35	5:05	6
-125	43.46	40.84	5:09	7
-127	44.43	41.83	5:18	8
-128	44.92	42.32	5:22	9
-129	45.42	42.82	5:27	10
-130	45.92	43.32	5:32	11
-132	46.91	44.34	5:40	13
-133	47.41	44.84	5:45	14
-134	47.90	45.35	5:50	15
-135	48.40	45.86	5:55	16
-137	49.39	46.88	6:04	18

PERCENTAGE OF SUN OBSCURED ALONG MAP TRACK
WEST END = 95.5   EAST END = 96.0

===
SUNRISE POSITION : ALTITUDE IS -0.8 DEGREES
NORTH : LONGITUDE -116.0, LATITUDE 39.2
SOUTH : LONGITUDE -118.0, LATITUDE 37.5

+ MARKS THE CITY OF YIN

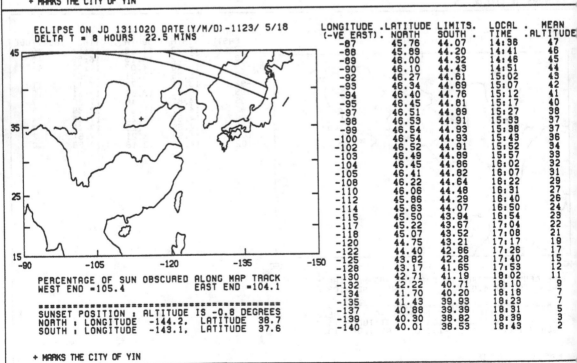

ECLIPSE ON JD 1311020 DATE(Y/M/D)-1123/ 5/18
DELTA T = 8 HOURS 22.5 MINS

LONGITUDE (-VE EAST)	LATITUDE NORTH	LIMITS. SOUTH	LOCAL TIME	MEAN ALTITUDE
-87	45.76	44.07	14:36	47
-88	45.89	44.20	14:41	46
-89	46.00	44.32	14:46	45
-90	46.10	44.43	14:51	44
-92	46.27	44.61	15:02	49
-93	46.34	44.69	15:07	42
-94	46.40	44.76	15:12	41
-95	46.45	44.81	15:17	40
-97	46.51	44.89	15:27	38
-98	46.53	44.91	15:33	37
-99	46.54	44.93	15:38	37
-100	46.54	44.93	15:43	36
-102	46.52	44.91	15:52	34
-103	46.49	44.89	15:57	33
-104	46.45	44.86	16:02	32
-105	46.41	44.82	16:07	31
-108	46.22	44.64	16:22	29
-110	46.06	44.48	16:31	27
-112	45.86	44.29	16:40	26
-114	45.63	44.07	16:50	24
-115	45.50	43.94	16:54	23
-117	45.22	43.67	17:04	22
-118	45.07	43.52	17:08	21
-120	44.75	43.21	17:17	19
-122	44.40	42.86	17:26	17
-125	43.82	42.28	17:40	15
-128	43.17	41.65	17:53	12
-130	42.71	41.19	18:02	11
-132	42.22	40.71	18:10	9
-134	41.70	40.20	18:18	7
-135	41.43	39.93	18:23	7
-137	40.88	39.39	18:31	5
-139	40.30	38.82	18:39	3
-140	40.01	38.53	18:43	2

PERCENTAGE OF SUN OBSCURED ALONG MAP TRACK
WEST END =105.4   EAST END =104.1

===
SUNSET POSITION : ALTITUDE IS -0.8 DEGREES
NORTH : LONGITUDE -144.2, LATITUDE 38.7
SOUTH : LONGITUDE -143.1, LATITUDE 37.6

+ MARKS THE CITY OF YIN

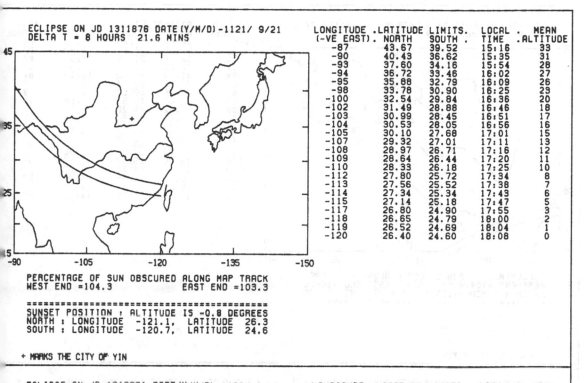

ECLIPSE ON JD 1311876 DATE(Y/M/D)-1121/ 9/21
DELTA T = 8 HOURS   21.6 MINS

LONGITUDE (-VE EAST)	LATITUDE LIMITS. NORTH	SOUTH	LOCAL TIME	MEAN ALTITUDE
-87	43.67	39.52	15:16	33
-90	40.43	36.62	15:35	31
-93	37.60	34.16	15:54	28
-94	36.72	33.46	16:02	27
-95	35.88	32.79	16:09	26
-98	33.78	30.90	16:25	23
-100	32.54	29.84	16:36	20
-102	31.49	28.88	16:46	18
-103	30.99	28.45	16:51	17
-104	30.53	28.05	16:56	16
-105	30.10	27.68	17:01	15
-107	29.32	27.01	17:11	13
-108	28.97	26.71	17:16	12
-109	28.64	26.44	17:20	11
-110	28.33	26.18	17:25	10
-112	27.80	25.72	17:34	8
-113	27.56	25.52	17:38	7
-114	27.34	25.34	17:43	6
-115	27.14	25.18	17:47	5
-117	26.80	24.90	17:55	3
-118	26.65	24.79	18:00	2
-119	26.52	24.69	18:04	1
-120	26.40	24.60	18:08	0

PERCENTAGE OF SUN OBSCURED ALONG MAP TRACK
WEST END =104.3   EAST END =103.3

```
=================================================
SUNSET POSITION : ALTITUDE IS -0.8 DEGREES
NORTH : LONGITUDE  -121.1,  LATITUDE  26.3
SOUTH : LONGITUDE  -120.7,  LATITUDE  24.6
```

+ MARKS THE CITY OF YIN

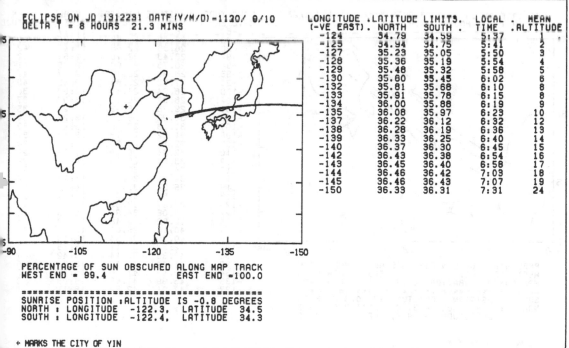

ECLIPSE ON JD 1312231 DATE(Y/M/D)-1120/ 8/10
DELTA T = 8 HOURS   21.3 MINS

LONGITUDE (-VE EAST)	LATITUDE LIMITS. NORTH	SOUTH	LOCAL TIME	MEAN ALTITUDE
-124	34.79	34.59	5:37	1
-125	34.94	34.75	5:41	2
-127	35.23	35.05	5:50	3
-128	35.36	35.19	5:54	4
-129	35.48	35.32	5:58	5
-130	35.60	35.45	6:02	6
-132	35.81	35.68	6:10	8
-133	35.91	35.78	6:15	8
-134	36.00	35.88	6:19	9
-135	36.08	35.97	6:23	10
-137	36.22	36.12	6:32	12
-138	36.28	36.19	6:36	13
-139	36.33	36.25	6:40	14
-140	36.37	36.30	6:45	15
-142	36.43	36.38	6:54	16
-143	36.45	36.40	6:58	17
-144	36.46	36.42	7:03	18
-145	36.46	36.43	7:07	19
-150	36.33	36.31	7:31	24

PERCENTAGE OF SUN OBSCURED ALONG MAP TRACK
WEST END = 99.4   EAST END =100.0

```
=================================================
SUNRISE POSITION : ALTITUDE IS -0.8 DEGREES
NORTH : LONGITUDE  -122.3,  LATITUDE  34.5
SOUTH : LONGITUDE  -122.4,  LATITUDE  34.3
```

+ MARKS THE CITY OF YIN

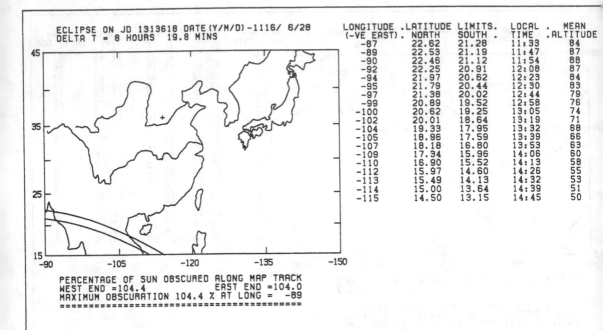

ECLIPSE ON JD 1313618 DATE (Y/M/D) -1116/ 6/28
DELTA T = 8 HOURS   19.8 MINS

LONGITUDE (-VE EAST)	LATITUDE LIMITS. NORTH	SOUTH	LOCAL TIME	MEAN ALTITUDE
-87	22.62	21.28	11:33	84
-89	22.53	21.19	11:47	87
-90	22.46	21.12	11:54	88
-92	22.25	20.91	12:08	87
-94	21.97	20.62	12:23	84
-95	21.79	20.44	12:30	83
-97	21.38	20.02	12:44	79
-99	20.89	19.52	12:58	76
-100	20.62	19.25	13:05	74
-102	20.01	18.64	13:19	71
-104	19.33	17.95	13:32	68
-105	18.96	17.59	13:39	66
-107	18.18	16.80	13:53	63
-109	17.34	15.96	14:06	60
-110	16.90	15.52	14:13	58
-112	15.97	14.60	14:26	55
-113	15.49	14.13	14:32	53
-114	15.00	13.64	14:39	51
-115	14.50	13.15	14:45	50

PERCENTAGE OF SUN OBSCURED ALONG MAP TRACK
WEST END =104.4   EAST END =104.0
MAXIMUM OBSCURATION 104.4 % AT LONG = -89
===

+ MARKS THE CITY OF YIN

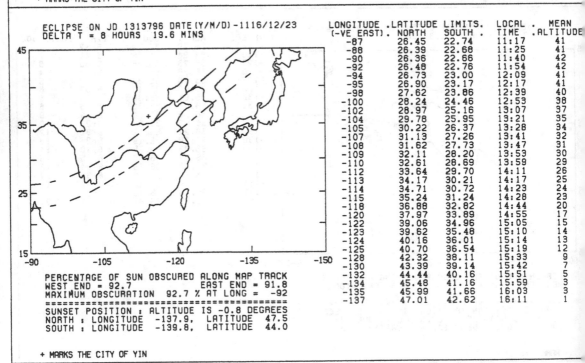

ECLIPSE ON JD 1313796 DATE (Y/M/D) -1116/12/23
DELTA T = 8 HOURS   19.6 MINS

LONGITUDE (-VE EAST)	LATITUDE LIMITS. NORTH	SOUTH	LOCAL TIME	MEAN ALTITUDE
-87	26.45	22.74	11:17	41
-88	26.39	22.68	11:25	41
-90	26.36	22.66	11:40	42
-92	26.48	22.76	11:54	42
-94	26.73	23.00	12:09	41
-95	26.90	23.17	12:17	41
-98	27.62	23.86	12:39	40
-100	28.24	24.46	12:53	38
-102	28.97	25.16	13:07	37
-104	29.78	25.95	13:21	35
-105	30.22	26.37	13:28	34
-107	31.13	27.26	13:41	32
-108	31.62	27.73	13:47	31
-109	32.11	28.20	13:53	30
-110	32.61	28.69	13:59	29
-112	33.64	29.70	14:11	26
-113	34.17	30.21	14:17	25
-114	34.71	30.72	14:23	24
-115	35.24	31.24	14:28	23
-118	36.88	32.82	14:44	20
-120	37.97	33.89	14:55	17
-122	39.06	34.96	15:05	15
-123	39.62	35.48	15:10	14
-124	40.16	36.01	15:14	13
-125	40.70	36.54	15:19	12
-128	42.32	38.11	15:33	9
-130	43.39	39.14	15:42	7
-132	44.44	40.16	15:51	5
-134	45.48	41.16	15:59	3
-135	45.99	41.66	16:03	3
-137	47.01	42.62	16:11	1

PERCENTAGE OF SUN OBSCURED ALONG MAP TRACK
WEST END = 92.7   EAST END = 91.8
MAXIMUM OBSCURATION 92.7 % AT LONG = -92
===
SUNSET POSITION : ALTITUDE IS -0.8 DEGREES
NORTH : LONGITUDE -137.9, LATITUDE 47.5
SOUTH : LONGITUDE -139.8, LATITUDE 44.0

+ MARKS THE CITY OF YIN

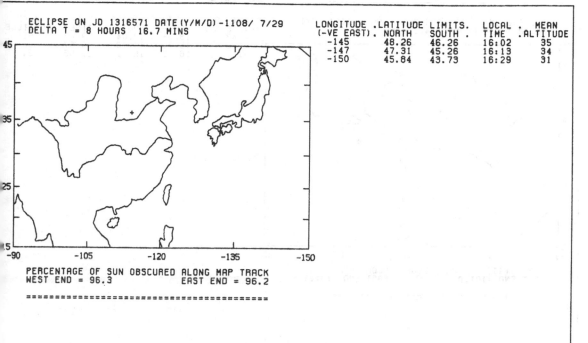

ECLIPSE ON JD 1316571 DATE(Y/M/D)-1108/ 7/29
DELTA T = 8 HOURS 16.7 MINS

LONGITUDE	.LATITUDE LIMITS.		LOCAL	. MEAN
(-VE EAST).	NORTH	SOUTH .	TIME	.ALTITUDE
-145	48.26	46.26	16:02	35
-147	47.31	45.26	16:13	34
-150	45.84	43.73	16:29	31

PERCENTAGE OF SUN OBSCURED ALONG MAP TRACK
WEST END = 96.3   EAST END = 96.2

===

+ MARKS THE CITY OF YIN

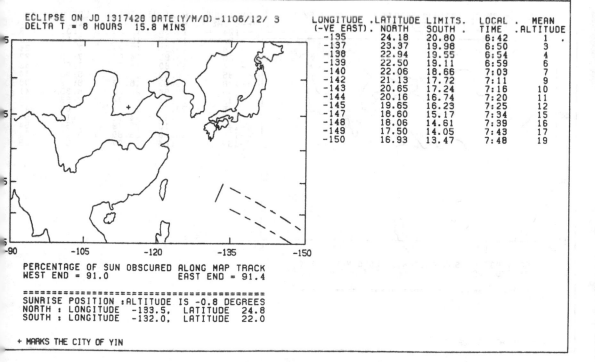

ECLIPSE ON JD 1317420 DATE(Y/M/D)-1106/12/ 3
DELTA T = 8 HOURS 15.8 MINS

LONGITUDE	.LATITUDE LIMITS.		LOCAL	. MEAN
(-VE EAST).	NORTH	SOUTH .	TIME	.ALTITUDE
-135	24.18	20.80	6:42	1 .
-137	23.37	19.98	6:50	3
-138	22.94	19.55	6:54	4
-139	22.50	19.11	6:59	6
-140	22.06	18.66	7:03	7
-142	21.13	17.72	7:11	9
-143	20.65	17.24	7:16	10
-144	20.16	16.74	7:20	11
-145	19.65	16.23	7:25	12
-147	18.60	15.17	7:34	15
-148	18.06	14.61	7:39	16
-149	17.50	14.05	7:43	17
-150	16.93	13.47	7:48	19

PERCENTAGE OF SUN OBSCURED ALONG MAP TRACK
WEST END = 91.0   EAST END = 91.4

===
SUNRISE POSITION :ALTITUDE IS -0.8 DEGREES
NORTH : LONGITUDE -133.5, LATITUDE 24.8
SOUTH : LONGITUDE -132.0, LATITUDE 22.0

+ MARKS THE CITY OF YIN

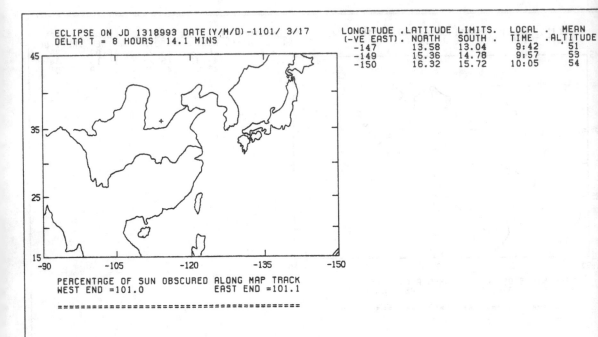

ECLIPSE ON JD 1318993 DATE(Y/M/D)-1101/ 3/17
DELTA T = 8 HOURS   14.1 MINS

LONGITUDE (-VE EAST)	LATITUDE NORTH	LIMITS SOUTH	LOCAL TIME	MEAN ALTITUDE
-147	13.58	13.04	9:42	51
-149	15.36	14.78	9:57	53
-150	16.32	15.72	10:05	54

PERCENTAGE OF SUN OBSCURED ALONG MAP TRACK
WEST END =101.0   EAST END =101.1

==

+ MARKS THE CITY OF YIN

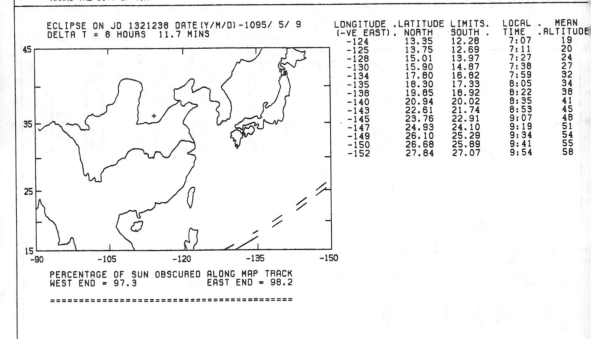

ECLIPSE ON JD 1321238 DATE(Y/M/D)-1095/ 5/ 9
DELTA T = 8 HOURS   11.7 MINS

LONGITUDE (-VE EAST)	LATITUDE NORTH	LIMITS SOUTH	LOCAL TIME	MEAN ALTITUDE
-124	13.35	12.28	7:07	19
-125	13.75	12.69	7:11	20
-128	15.01	13.97	7:27	24
-130	15.90	14.87	7:38	27
-134	17.80	16.82	7:59	32
-135	18.30	17.33	8:05	34
-138	19.85	18.92	8:22	38
-140	20.94	20.02	8:35	41
-143	22.61	21.74	8:53	45
-145	23.76	22.91	9:07	48
-147	24.93	24.10	9:19	51
-149	26.10	25.29	9:34	54
-150	26.68	25.89	9:41	55
-152	27.84	27.07	9:54	58

PERCENTAGE OF SUN OBSCURED ALONG MAP TRACK
WEST END = 97.3   EAST END = 98.2

==

+ MARKS THE CITY OF YIN

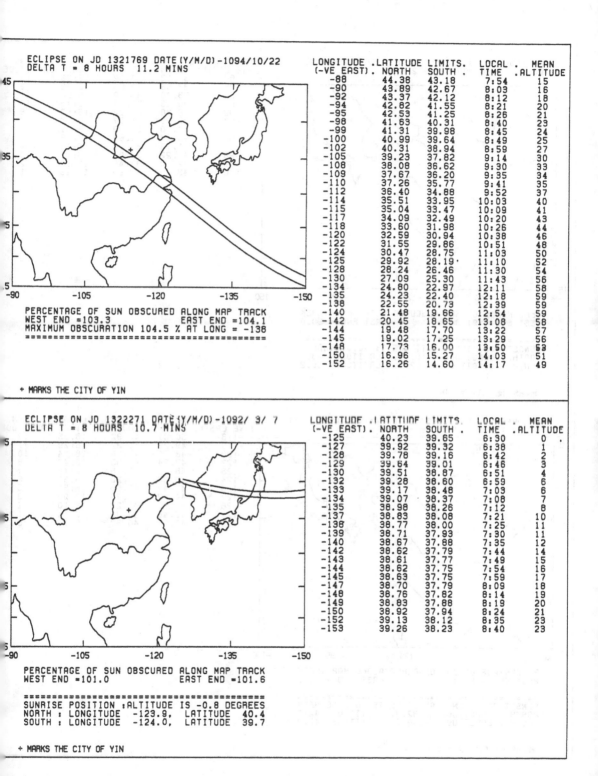

ECLIPSE ON JD 1321769 DATE(Y/M/D)-1094/10/22
DELTA T = 8 HOURS 11.2 MINS

LONGITUDE (-VE EAST)	LATITUDE NORTH	LIMITS SOUTH	LOCAL TIME	MEAN ALTITUDE
-88	44.38	43.18	7:54	15
-90	43.89	42.67	8:03	16
-92	43.37	42.12	8:12	18
-94	42.82	41.55	8:21	20
-95	42.53	41.25	8:26	21
-98	41.63	40.31	8:40	23
-99	41.31	39.98	8:45	24
-100	40.99	39.64	8:49	25
-102	40.31	38.94	8:59	27
-105	39.23	37.82	9:14	30
-108	38.08	36.62	9:30	33
-109	37.67	36.20	9:35	34
-110	37.26	35.77	9:41	35
-112	36.40	34.88	9:52	37
-114	35.51	33.95	10:03	40
-115	35.04	33.47	10:09	41
-117	34.09	32.49	10:20	43
-118	33.60	31.98	10:26	44
-120	32.59	30.94	10:38	46
-122	31.55	29.86	10:51	48
-124	30.47	28.75	11:03	50
-125	29.92	28.19·	11:10	52
-128	28.24	26.46	11:30	54
-130	27.09	25.30	11:43	56
-134	24.80	22.97	12:11	58
-135	24.23	22.40	12:18	59
-138	22.55	20.73	12:39	59
-140	21.48	19.66	12:54	59
-142	20.45	18.65	13:08	58
-144	19.48	17.70	13:22	57
-145	19.02	17.25	13:29	56
-148	17.73	16.00	13:50	53
-150	16.96	15.27	14:03	51
-152	16.26	14.60	14:17	49

PERCENTAGE OF SUN OBSCURED ALONG MAP TRACK
WEST END =103.3   EAST END =104.1
MAXIMUM OBSCURATION 104.5 % AT LONG = -138
==

+ MARKS THE CITY OF YIN

ECLIPSE ON JD 1322271 DATE(Y/M/D)-1092/ 3/ 7
DELTA T = 8 HOURS 10.7 MINS

LONGITUDE (-VE EAST)	LATITUDE NORTH	LIMITS SOUTH	LOCAL TIME	MEAN ALTITUDE
-125	40.23	39.65	6:30	0
-127	39.92	39.32	6:38	1
-128	39.78	39.16	6:42	2
-129	39.64	39.01	6:46	3
-130	39.51	38.87	6:51	4
-132	39.28	38.60	6:59	6
-133	39.17	38.48	7:03	6
-134	39.07	38.37	7:08	7
-135	38.98	38.26	7:12	8
-137	38.83	38.08	7:21	10
-138	38.77	38.00	7:25	11
-139	38.71	37.93	7:30	11
-140	38.67	37.88	7:35	12
-142	38.62	37.79	7:44	14
-143	38.61	37.77	7:49	15
-144	38.62	37.75	7:54	16
-145	38.63	37.75	7:59	17
-147	38.70	37.79	8:09	18
-148	38.76	37.82	8:14	19
-149	38.83	37.88	8:19	20
-150	38.92	37.94	8:24	21
-152	39.13	38.12	8:35	23
-153	39.26	38.23	8:40	23

PERCENTAGE OF SUN OBSCURED ALONG MAP TRACK
WEST END =101.0   EAST END =101.6

==
SUNRISE POSITION :ALTITUDE IS -0.8 DEGREES
NORTH : LONGITUDE -123.9, LATITUDE 40.4
SOUTH : LONGITUDE -124.0, LATITUDE 39.7

+ MARKS THE CITY OF YIN

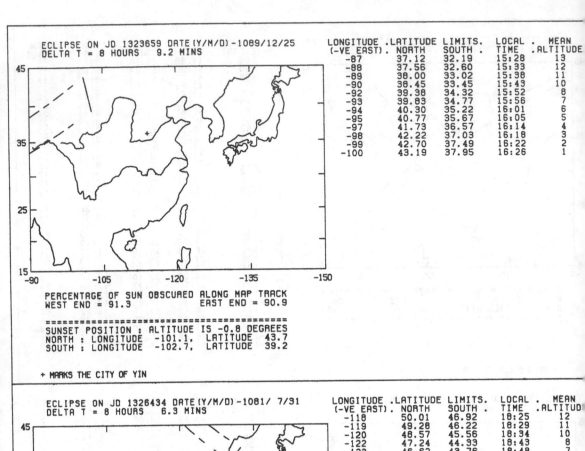

ECLIPSE ON JD 1323659 DATE(Y/M/D)-1089/12/25
DELTA T = 8 HOURS   9.2 MINS

LONGITUDE (-VE EAST)	LATITUDE LIMITS. NORTH	SOUTH	LOCAL TIME	MEAN ALTITUDE
-87	37.12	32.19	15:28	13
-88	37.56	32.60	15:33	12
-89	38.00	33.02	15:38	11
-90	38.45	33.45	15:43	10
-92	39.38	34.32	15:52	8
-93	39.83	34.77	15:56	7
-94	40.30	35.22	16:01	6
-95	40.77	35.67	16:05	5
-97	41.73	36.57	16:14	4
-98	42.22	37.03	16:18	3
-99	42.70	37.49	16:22	2
-100	43.19	37.95	16:26	1

PERCENTAGE OF SUN OBSCURED ALONG MAP TRACK
WEST END = 91.3   EAST END = 90.9

==
SUNSET POSITION : ALTITUDE IS -0.8 DEGREES
NORTH : LONGITUDE -101.1, LATITUDE 43.7
SOUTH : LONGITUDE -102.7, LATITUDE 39.2

+ MARKS THE CITY OF YIN

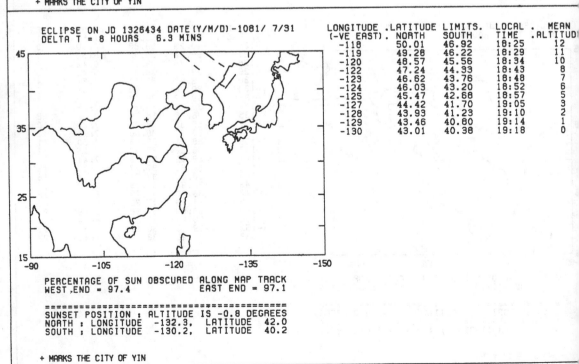

ECLIPSE ON JD 1326434 DATE(Y/M/D)-1081/ 7/31
DELTA T = 8 HOURS   6.3 MINS

LONGITUDE (-VE EAST)	LATITUDE LIMITS. NORTH	SOUTH	LOCAL TIME	MEAN ALTITUD
-118	50.01	46.92	18:25	12
-119	49.28	46.22	18:29	11
-120	48.57	45.56	18:34	10
-122	47.24	44.33	18:43	8
-123	46.62	43.76	18:48	7
-124	46.03	43.20	18:52	6
-125	45.47	42.68	18:57	5
-127	44.42	41.70	19:05	3
-128	43.93	41.23	19:10	2
-129	43.46	40.80	19:14	1
-130	43.01	40.38	19:18	0

PERCENTAGE OF SUN OBSCURED ALONG MAP TRACK
WEST ,END = 97.4   EAST END = 97.1

==
SUNSET POSITION : ALTITUDE IS -0.8 DEGREES
NORTH : LONGITUDE -132.3, LATITUDE 42.0
SOUTH : LONGITUDE -130.2, LATITUDE 40.2

+ MARKS THE CITY OF YIN

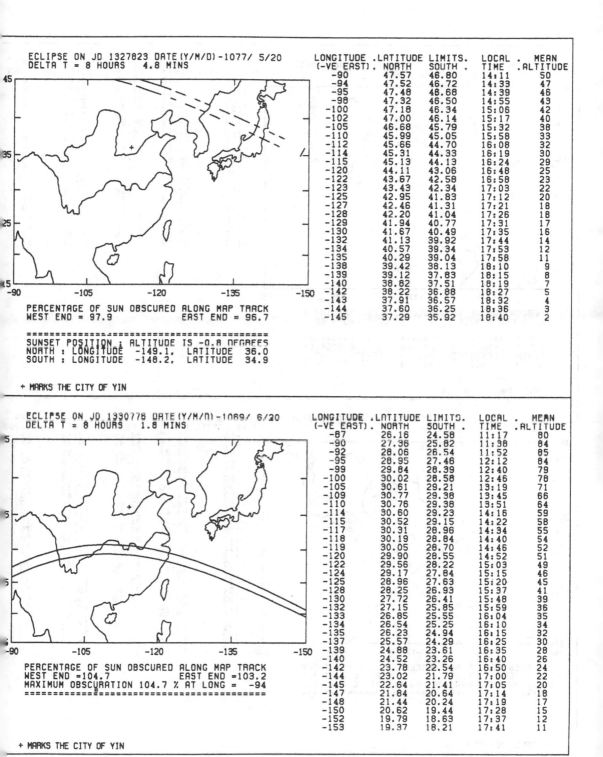

ECLIPSE ON JD 1327823 DATE(Y/M/D)-1077/ 5/20
DELTA T = 8 HOURS   4.8 MINS

LONGITUDE (-VE EAST)	LATITUDE NORTH	LIMITS SOUTH	LOCAL TIME	MEAN ALTITUDE
-90	47.57	46.80	14:11	50
-94	47.52	46.72	14:33	47
-95	47.48	46.68	14:39	46
-98	47.32	46.50	14:55	43
-100	47.18	46.34	15:06	42
-102	47.00	46.14	15:17	40
-105	46.68	45.79	15:32	38
-110	45.99	45.05	15:58	33
-112	45.66	44.70	16:08	32
-114	45.31	44.33	16:19	30
-115	45.13	44.13	16:24	29
-120	44.11	43.06	16:48	25
-122	43.67	42.58	16:58	23
-123	43.43	42.34	17:03	22
-125	42.95	41.83	17:12	20
-127	42.46	41.31	17:21	18
-128	42.20	41.04	17:26	18
-129	41.94	40.77	17:31	17
-130	41.67	40.49	17:35	16
-132	41.13	39.92	17:44	14
-134	40.57	39.34	17:53	12
-135	40.29	39.04	17:58	11
-138	39.42	38.13	18:10	9
-139	39.12	37.83	18:15	8
-140	38.82	37.51	18:19	7
-142	38.22	36.88	18:27	5
-143	37.91	36.57	18:32	4
-144	37.60	36.25	18:36	3
-145	37.29	35.92	18:40	2

PERCENTAGE OF SUN OBSCURED ALONG MAP TRACK
WEST END = 97.9   EAST END = 96.7

==

SUNSET POSITION : ALTITUDE IS -0.8 DEGREES
NORTH : LONGITUDE -149.1,   LATITUDE   36.0
SOUTH : LONGITUDE -148.2,   LATITUDE   34.9

+ MARKS THE CITY OF YIN

ECLIPSE ON JD 1330776 DATE(Y/M/D)-1069/ 6/20
DELTA T = 8 HOURS   1.8 MINS

LONGITUDE (-VE EAST)	LATITUDE NORTH	LIMITS SOUTH	LOCAL TIME	MEAN ALTITUDE
-87	26.16	24.58	11:17	80
-90	27.36	25.82	11:38	84
-92	28.06	26.54	11:52	85
-95	28.95	27.46	12:12	84
-99	29.84	28.39	12:40	79
-100	30.02	28.58	12:46	78
-105	30.61	29.21	13:19	71
-109	30.77	29.38	13:45	66
-110	30.76	29.38	13:51	64
-114	30.60	29.23	14:16	59
-115	30.52	29.15	14:22	58
-117	30.31	28.96	14:34	55
-118	30.19	28.84	14:40	54
-119	30.05	28.70	14:46	52
-120	29.90	28.55	14:52	51
-122	29.56	28.22	15:03	49
-124	29.17	27.84	15:15	46
-125	28.96	27.63	15:20	45
-128	28.25	26.93	15:37	41
-130	27.72	26.41	15:48	39
-132	27.15	25.85	15:59	36
-133	26.85	25.55	16:04	35
-134	26.54	25.25	16:10	34
-135	26.23	24.94	16:15	32
-137	25.57	24.29	16:25	30
-139	24.88	23.61	16:35	28
-140	24.52	23.26	16:40	26
-142	23.78	22.54	16:50	24
-144	23.02	21.79	17:00	22
-145	22.64	21.41	17:05	20
-147	21.84	20.64	17:14	18
-148	21.44	20.24	17:19	17
-150	20.62	19.44	17:28	15
-152	19.79	18.63	17:37	12
-153	19.37	18.21	17:41	11

PERCENTAGE OF SUN OBSCURED ALONG MAP TRACK
WEST END =104.7   EAST END =103.2
MAXIMUM OBSCURATION 104.7 % AT LONG = -94
==

+ MARKS THE CITY OF YIN

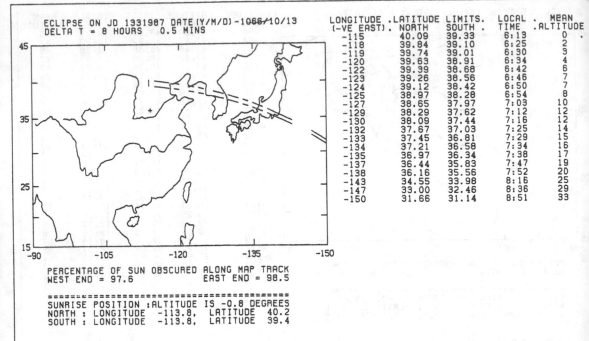

ECLIPSE ON JD 1331987 DATE(Y/M/D)-1066/10/13
DELTA T = 8 HOURS   0.5 MINS

LONGITUDE (-VE EAST)	LATITUDE NORTH	LIMITS. SOUTH	LOCAL TIME	MEAN ALTITUDE
-115	40.09	39.33	6:13	0
-118	39.84	39.10	6:25	2
-119	39.74	39.01	6:30	3
-120	39.63	38.91	6:34	4
-122	39.39	38.68	6:42	6
-123	39.26	38.56	6:46	7
-124	39.12	38.42	6:50	7
-125	38.97	38.28	6:54	8
-127	38.65	37.97	7:03	10
-129	38.29	37.62	7:12	12
-130	38.09	37.44	7:16	12
-132	37.67	37.03	7:25	14
-133	37.45	36.81	7:29	15
-134	37.21	36.58	7:34	16
-135	36.97	36.34	7:38	17
-137	36.44	35.83	7:47	19
-138	36.16	35.56	7:52	20
-143	34.55	33.98	8:16	25
-147	33.00	32.46	8:36	29
-150	31.66	31.14	8:51	33

PERCENTAGE OF SUN OBSCURED ALONG MAP TRACK
WEST END = 97.6   EAST END = 98.5

===
SUNRISE POSITION :ALTITUDE IS -0.8 DEGREES
NORTH : LONGITUDE -113.8, LATITUDE 40.2
SOUTH : LONGITUDE -113.8, LATITUDE 39.4

+ MARKS THE CITY OF YIN

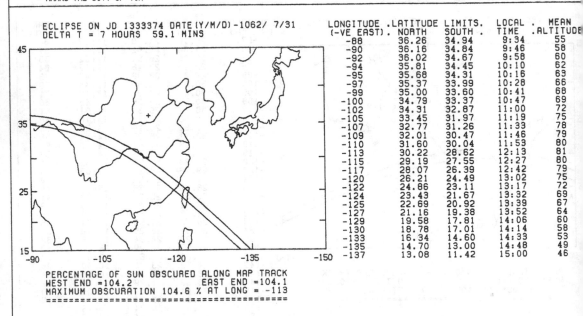

ECLIPSE ON JD 1333374 DATE(Y/M/D)-1062/ 7/31
DELTA T = 7 HOURS 59.1 MINS

LONGITUDE (-VE EAST)	LATITUDE NORTH	LIMITS. SOUTH	LOCAL TIME	MEAN ALTITUDE
-88	36.26	34.94	9:34	55
-90	36.16	34.84	9:46	58
-92	36.02	34.67	9:58	60
-94	35.81	34.45	10:10	62
-95	35.68	34.31	10:16	63
-97	35.37	33.99	10:28	66
-99	35.00	33.60	10:41	68
-100	34.79	33.37	10:47	69
-102	34.31	32.87	11:00	72
-105	33.45	31.97	11:19	75
-107	32.77	31.26	11:33	78
-109	32.01	30.47	11:46	79
-110	31.60	30.04	11:53	80
-113	30.22	28.62	12:13	81
-115	29.19	27.55	12:27	80
-117	28.07	26.39	12:42	79
-120	26.21	24.49	13:02	75
-122	24.86	23.11	13:17	72
-124	23.43	21.67	13:32	69
-125	22.69	20.92	13:39	67
-127	21.16	19.38	13:52	64
-129	19.58	17.81	14:06	60
-130	18.78	17.01	14:14	58
-133	16.34	14.60	14:33	53
-135	14.70	13.00	14:48	49
-137	13.08	11.42	15:00	46

PERCENTAGE OF SUN OBSCURED ALONG MAP TRACK
WEST END =104.2   EAST END =104.1
MAXIMUM OBSCURATION 104.6 % AT LONG = -113
===

+ MARKS THE CITY OF YIN

ECLIPSE ON JD 1333552 DATE(Y/M/D)-1061/ 1/25
DELTA T = 7 HOURS 58.9 MINS

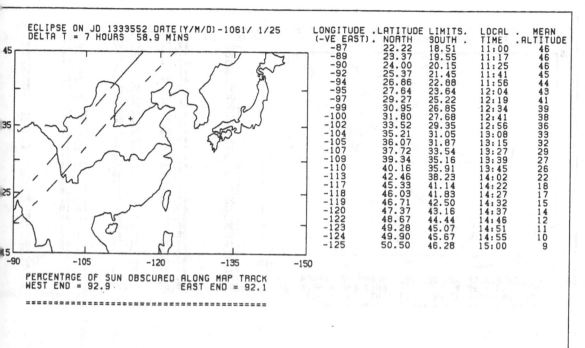

LONGITUDE	.LATITUDE	LIMITS.	LOCAL	MEAN
(-VE EAST).	NORTH	SOUTH .	TIME	.ALTITUDE
-87	22.22	18.51	11:00	46
-89	23.37	19.55	11:17	46
-90	24.00	20.15	11:25	46
-92	25.37	21.45	11:41	45
-94	26.86	22.88	11:56	44
-95	27.64	23.64	12:04	43
-97	29.27	25.22	12:19	41
-99	30.95	26.85	12:34	39
-100	31.80	27.68	12:41	38
-102	33.52	29.35	12:56	36
-104	35.21	31.05	13:08	33
-105	36.07	31.87	13:15	32
-107	37.72	33.54	13:27	29
-109	39.34	35.16	13:39	27
-110	40.16	35.91	13:45	26
-113	42.46	38.23	14:02	22
-117	45.33	41.14	14:22	18
-118	46.03	41.83	14:27	17
-119	46.71	42.50	14:32	15
-120	47.37	43.16	14:37	14
-122	48.67	44.44	14:46	12
-123	49.28	45.07	14:51	11
-124	49.90	45.67	14:55	10
-125	50.50	46.28	15:00	9

PERCENTAGE OF SUN OBSCURED ALONG MAP TRACK
WEST END = 92.9   EAST END = 92.1

==

+ MARKS THE CITY OF YIN

ECLIPSE ON JD 1335786 DATE(Y/M/D)-1055/ 3/18
DELTA T = 7 HOURS 58.8 MINS

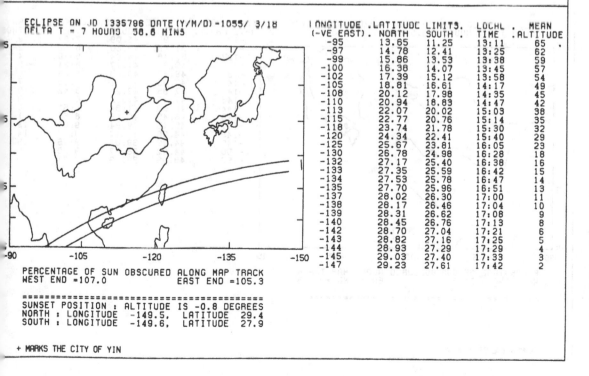

LONGITUDE	.LATITUDE	LIMITS.	LOCAL	MEAN
(-VE EAST).	NORTH	SOUTH .	TIME	.ALTITUDE
-95	13.65	11.25	13:11	65
-97	14.78	12.41	13:25	62
-99	15.86	13.53	13:38	59
-100	16.38	14.07	13:45	57
-102	17.39	15.12	13:58	54
-105	18.81	16.61	14:17	49
-108	20.12	17.98	14:35	45
-110	20.94	18.83	14:47	42
-113	22.07	20.02	15:03	38
-115	22.77	20.76	15:14	35
-118	23.74	21.78	15:30	32
-120	24.34	22.41	15:40	29
-125	25.67	23.81	16:05	23
-130	26.78	24.98	16:28	18
-132	27.17	25.40	16:38	16
-133	27.35	25.59	16:42	15
-134	27.53	25.78	16:47	14
-135	27.70	25.96	16:51	13
-137	28.02	26.30	17:00	11
-138	28.17	26.46	17:04	10
-139	28.31	26.62	17:08	9
-140	28.45	26.76	17:13	8
-142	28.70	27.04	17:21	6
-143	28.82	27.16	17:25	5
-144	28.93	27.29	17:29	4
-145	29.03	27.40	17:33	3
-147	29.23	27.61	17:42	2

PERCENTAGE OF SUN OBSCURED ALONG MAP TRACK
WEST END =107.0   EAST END =105.3

==
SUNSET POSITION : ALTITUDE IS -0.8 DEGREES
NORTH : LONGITUDE -149.5, LATITUDE 29.4
SOUTH : LONGITUDE -149.6, LATITUDE 27.9

+ MARKS THE CITY OF YIN

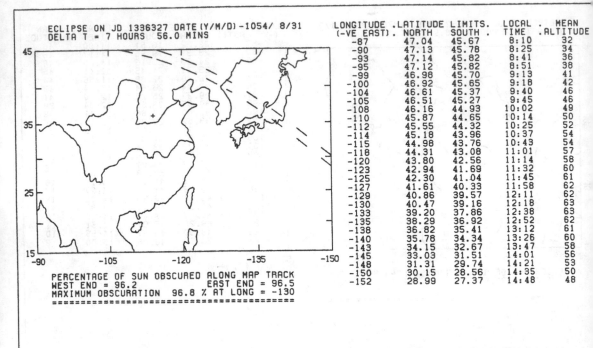

ECLIPSE ON JD 1396327 DATE(Y/M/D)-1054/ 8/31
DELTA T = 7 HOURS 56.0 MINS

LONGITUDE (-VE EAST)	LATITUDE NORTH	LIMITS SOUTH	LOCAL TIME	MEAN ALTITUDE
-87	47.04	45.67	8:10	32
-90	47.13	45.78	8:25	34
-93	47.14	45.82	8:41	36
-95	47.12	45.82	8:51	38
-99	46.98	45.70	9:13	41
-100	46.92	45.65	9:18	42
-104	46.61	45.37	9:40	46
-105	46.51	45.27	9:45	46
-108	46.16	44.93	10:02	49
-110	45.87	44.65	10:14	50
-112	45.55	44.32	10:25	52
-114	45.18	43.96	10:37	54
-115	44.98	43.76	10:43	54
-118	44.31	43.08	11:01	57
-120	43.80	42.56	11:14	58
-123	42.94	41.69	11:32	60
-125	42.30	41.04	11:45	61
-127	41.61	40.33	11:58	62
-129	40.86	39.57	12:11	62
-130	40.47	39.16	12:18	63
-133	39.20	37.86	12:38	63
-135	38.29	36.92	12:52	62
-138	36.82	35.41	13:12	61
-140	35.78	34.34	13:26	60
-143	34.15	32.67	13:47	58
-145	33.03	31.51	14:01	56
-148	31.31	29.74	14:21	53
-150	30.15	28.56	14:35	50
-152	28.99	27.37	14:48	48

PERCENTAGE OF SUN OBSCURED ALONG MAP TRACK
WEST END = 96.2   EAST END = 96.5
MAXIMUM OBSCURATION 96.8 % AT LONG = -130
===

+ MARKS THE CITY OF YIN

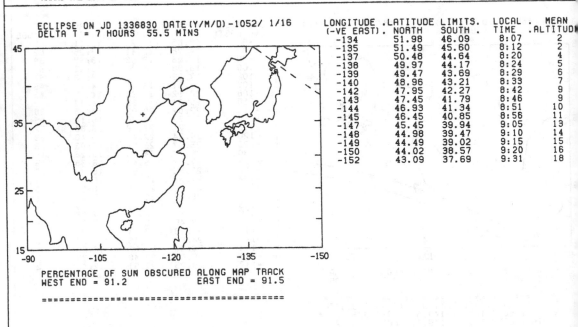

ECLIPSE ON JD 1336830 DATE(Y/M/D)-1052/ 1/16
DELTA T = 7 HOURS 55.5 MINS

LONGITUDE (-VE EAST)	LATITUDE NORTH	LIMITS SOUTH	LOCAL TIME	MEAN ALTITUDE
-134	51.98	46.09	8:07	2
-135	51.49	45.60	8:12	2
-137	50.48	44.64	8:20	4
-138	49.97	44.17	8:24	5
-139	49.47	43.69	8:29	6
-140	48.96	43.21	8:33	7
-142	47.95	42.27	8:42	9
-143	47.45	41.79	8:46	9
-144	46.93	41.34	8:51	10
-145	46.45	40.85	8:56	11
-147	45.45	39.94	9:05	13
-148	44.98	39.47	9:10	14
-149	44.49	39.02	9:15	15
-150	44.02	38.57	9:20	16
-152	43.09	37.69	9:31	18

PERCENTAGE OF SUN OBSCURED ALONG MAP TRACK
WEST END = 91.2   EAST END = 91.5

===

+ MARKS THE CITY OF YIN

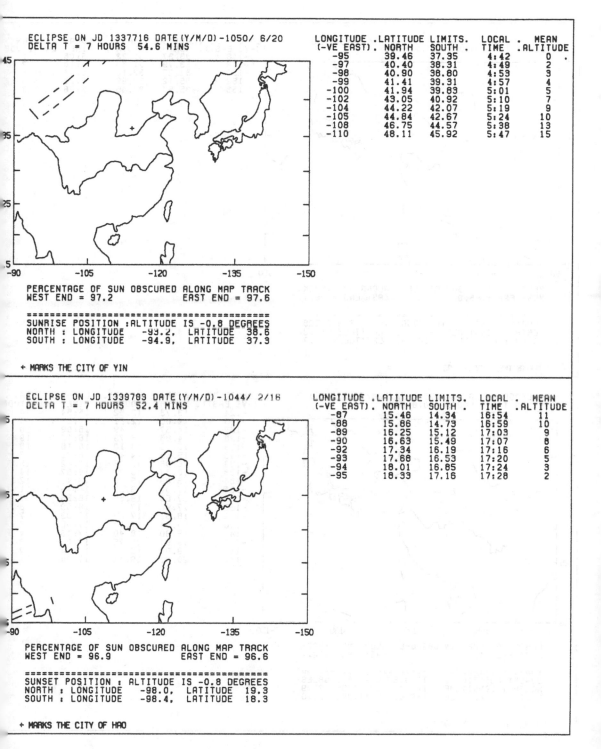

ECLIPSE ON JD 1337716 DATE(Y/M/D)-1050/ 6/20
DELTA T = 7 HOURS 54.6 MINS

LONGITUDE	.LATITUDE	LIMITS.	LOCAL	. MEAN
(-VE EAST).	NORTH	SOUTH .	TIME	.ALTITUDE
-95	39.46	37.35	4:42	0 .
-97	40.40	38.31	4:49	2
-98	40.90	38.80	4:53	3
-99	41.41	39.31	4:57	4
-100	41.94	39.83	5:01	5
-102	43.05	40.92	5:10	7
-104	44.22	42.07	5:19	9
-105	44.84	42.67	5:24	10
-108	46.75	44.57	5:38	13
-110	48.11	45.92	5:47	15

PERCENTAGE OF SUN OBSCURED ALONG MAP TRACK
WEST END = 97.2   EAST END = 97.6

==
SUNRISE POSITION :ALTITUDE IS -0.8 DEGREES
NORTH : LONGITUDE   -93.2, LATITUDE 38.6
SOUTH : LONGITUDE   -94.9, LATITUDE 37.3

+ MARKS THE CITY OF YIN

ECLIPSE ON JD 1339783 DATE(Y/M/D)-1044/ 2/16
DELTA T = 7 HOURS 52.4 MINS

LONGITUDE	.LATITUDE	LIMITS.	LOCAL	. MEAN
(-VE EAST).	NORTH	SOUTH .	TIME	.ALTITUDE
-87	15.46	14.34	16:54	11
-88	15.86	14.73	16:59	10
-89	16.25	15.12	17:03	9
-90	16.63	15.49	17:07	8
-92	17.34	16.19	17:16	6
-93	17.68	16.53	17:20	5
-94	18.01	16.85	17:24	3
-95	18.33	17.16	17:28	2

PERCENTAGE OF SUN OBSCURED ALONG MAP TRACK
WEST END = 96.9   EAST END = 96.6

==
SUNSET POSITION : ALTITUDE IS -0.8 DEGREES
NORTH : LONGITUDE   -98.0, LATITUDE 19.3
SOUTH : LONGITUDE   -98.4, LATITUDE 18.3

+ MARKS THE CITY OF HAO

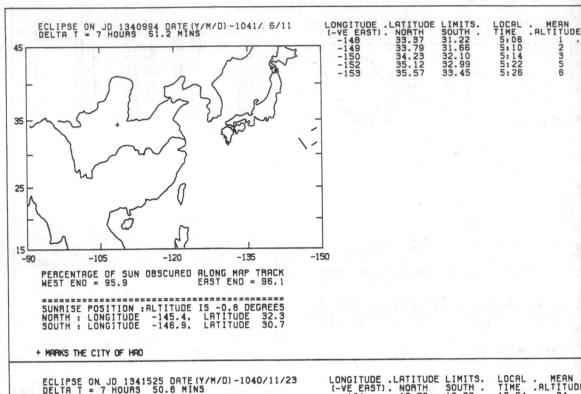

ECLIPSE ON JD 1340994 DATE(Y/M/D)-1041/. 6/11
DELTA T = 7 HOURS 51.2 MINS

LONGITUDE (-VE EAST).	LATITUDE NORTH	LIMITS. SOUTH .	LOCAL TIME	MEAN ALTITUDE
-148	33.37	31.22	5:06	1
-149	33.79	31.66	5:10	2
-150	34.23	32.10	5:14	3
-152	35.12	32.99	5:22	5
-153	35.57	33.45	5:26	6

PERCENTAGE OF SUN OBSCURED ALONG MAP TRACK
WEST END = 95.9   EAST END = 96.1

==
SUNRISE POSITION :ALTITUDE IS -0.8 DEGREES
NORTH : LONGITUDE -145.4, LATITUDE 32.3
SOUTH : LONGITUDE -146.9, LATITUDE 30.7

+ MARKS THE CITY OF HAO

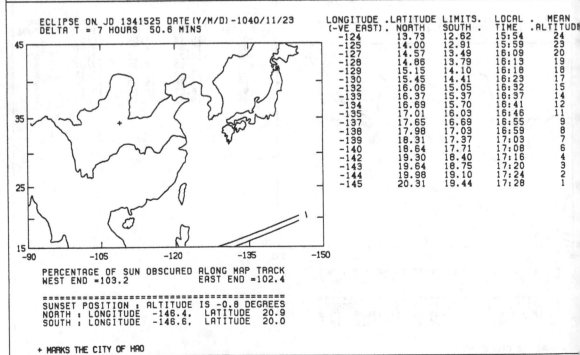

ECLIPSE ON JD 1341525 DATE(Y/M/D)-1040/11/23
DELTA T = 7 HOURS 50.6 MINS

LONGITUDE (-VE EAST).	LATITUDE NORTH	LIMITS. SOUTH .	LOCAL TIME	MEAN ALTITUDE
-124	13.73	12.62	15:54	24
-125	14.00	12.91	15:59	23
-127	14.57	13.49	16:09	20
-128	14.86	13.79	16:13	19
-129	15.15	14.10	16:18	18
-130	15.45	14.41	16:23	17
-132	16.06	15.05	16:32	15
-133	16.37	15.37	16:37	14
-134	16.69	15.70	16:41	12
-135	17.01	16.03	16:46	11
-137	17.65	16.69	16:55	9
-138	17.98	17.03	16:59	8
-139	18.31	17.37	17:03	7
-140	18.64	17.71	17:08	6
-142	19.30	18.40	17:16	4
-143	19.64	18.75	17:20	3
-144	19.98	19.10	17:24	2
-145	20.31	19.44	17:28	1

PERCENTAGE OF SUN OBSCURED ALONG MAP TRACK
WEST END =103.2   EAST END =102.4

==
SUNSET POSITION : ALTITUDE IS -0.8 DEGREES
NORTH : LONGITUDE -146.4, LATITUDE 20.9
SOUTH : LONGITUDE -146.6, LATITUDE 20.0

+ MARKS THE CITY OF HAO

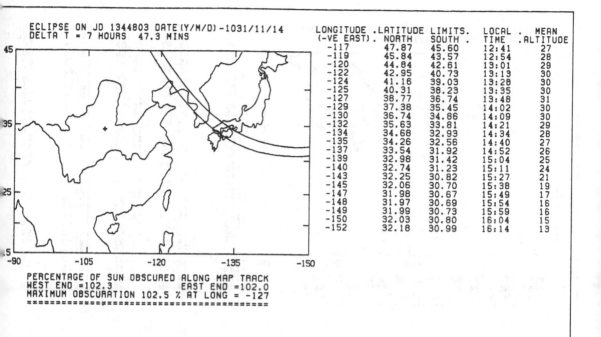

ECLIPSE ON JD 1344803 DATE(Y/M/D) -1031/11/14
DELTA T = 7 HOURS 47.3 MINS

LONGITUDE (-VE EAST)	LATITUDE NORTH	LIMITS SOUTH	LOCAL TIME	MEAN ALTITUDE
-117	47.87	45.60	12:41	27
-119	45.84	43.57	12:54	28
-120	44.84	42.61	13:01	29
-122	42.95	40.73	13:13	30
-124	41.16	39.03	13:28	30
-125	40.31	38.23	13:35	30
-127	38.77	36.74	13:48	31
-129	37.38	35.45	14:02	30
-130	36.74	34.86	14:09	30
-132	35.63	33.81	14:21	29
-134	34.68	32.93	14:34	28
-135	34.26	32.56	14:40	27
-137	33.54	31.92	14:52	26
-139	32.98	31.42	15:04	25
-140	32.74	31.23	15:11	24
-143	32.25	30.82	15:27	21
-145	32.06	30.70	15:38	19
-147	31.98	30.67	15:49	17
-148	31.97	30.69	15:54	16
-149	31.99	30.73	15:59	16
-150	32.03	30.80	16:04	15
-152	32.18	30.99	16:14	13

PERCENTAGE OF SUN OBSCURED ALONG MAP TRACK
WEST END =102.3   EAST END =102.0
MAXIMUM OBSCURATION 102.5 % AT LONG = -127
==

+ MARKS THE CITY OF HAO

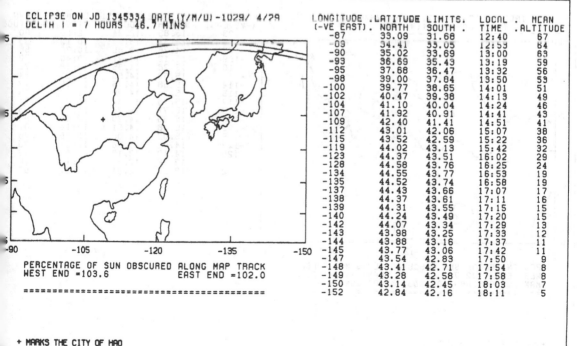

ECLIPSE ON JD 1345334 DATE(Y/M/D) -1029/ 4/29
DELTA T = 7 HOURS 46.7 MINS

LONGITUDE (-VE EAST)	LATITUDE NORTH	LIMITS SOUTH	LOCAL TIME	MEAN ALTITUDE
-87	33.09	31.68	12:40	67
-89	34.41	33.05	12:53	64
-90	35.02	33.69	13:00	63
-93	36.69	35.43	13:19	59
-95	37.68	36.47	13:32	56
-98	39.00	37.84	13:50	59
-100	39.77	38.65	14:01	51
-102	40.47	39.38	14:13	49
-104	41.10	40.04	14:24	46
-107	41.92	40.91	14:41	43
-109	42.40	41.41	14:51	41
-112	43.01	42.06	15:07	38
-115	43.52	42.59	15:22	36
-119	44.02	43.13	15:42	32
-123	44.37	43.51	16:02	29
-128	44.58	43.76	16:25	24
-134	44.55	43.77	16:53	19
-135	44.52	43.74	16:58	19
-137	44.43	43.66	17:07	17
-138	44.37	43.61	17:11	16
-139	44.31	43.55	17:15	15
-140	44.24	43.49	17:20	15
-142	44.07	43.34	17:29	13
-143	43.98	43.25	17:33	12
-144	43.88	43.16	17:37	11
-145	43.77	43.06	17:42	11
-147	43.54	42.83	17:50	9
-148	43.41	42.71	17:54	8
-149	43.28	42.58	17:58	8
-150	43.14	42.45	18:03	7
-152	42.84	42.16	18:11	5

PERCENTAGE OF SUN OBSCURED ALONG MAP TRACK
WEST END =103.6   EAST END =102.0

==

+ MARKS THE CITY OF HAO

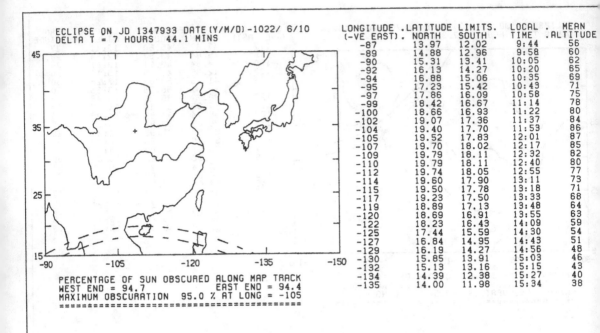

ECLIPSE ON JD 1347933 DATE(Y/M/D)-1022/ 6/10
DELTA T = 7 HOURS 44.1 MINS

LONGITUDE (-VE EAST)	LATITUDE NORTH	LIMITS. SOUTH	LOCAL TIME	MEAN ALTITUDE
-87	13.97	12.02	9:44	56
-89	14.88	12.96	9:58	60
-90	15.31	13.41	10:05	62
-92	16.13	14.27	10:20	65
-94	16.88	15.06	10:35	69
-95	17.23	15.42	10:43	71
-97	17.86	16.09	10:58	75
-99	18.42	16.67	11:14	78
-100	18.66	16.93	11:22	80
-102	19.07	17.36	11:37	84
-104	19.40	17.70	11:53	86
-105	19.52	17.83	12:01	87
-107	19.70	18.02	12:17	85
-109	19.79	18.11	12:32	82
-110	19.79	18.11	12:40	80
-112	19.74	18.05	12:55	77
-114	19.60	17.90	13:11	73
-115	19.50	17.78	13:18	71
-117	19.23	17.50	13:33	68
-119	18.89	17.13	13:48	64
-120	18.69	16.91	13:55	63
-122	18.23	16.43	14:09	59
-125	17.44	15.59	14:30	54
-127	16.84	14.95	14:43	51
-129	16.19	14.27	14:56	48
-130	15.85	13.91	15:03	46
-132	15.13	13.16	15:15	43
-134	14.39	12.38	15:27	40
-135	14.00	11.98	15:34	38

PERCENTAGE OF SUN OBSCURED ALONG MAP TRACK
WEST END = 94.7   EAST END = 94.4
MAXIMUM OBSCURATION 95.0 % AT LONG = -105
===

+ MARKS THE CITY OF HAO

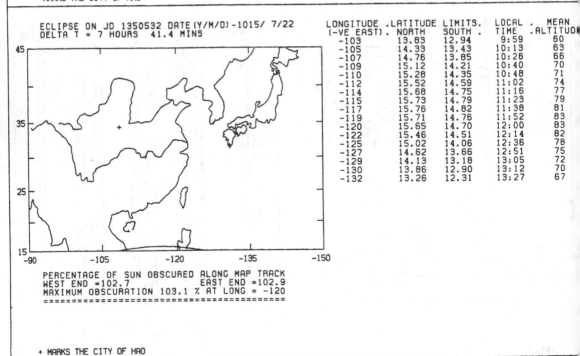

ECLIPSE ON JD 1350532 DATE(Y/M/D)-1015/ 7/22
DELTA T = 7 HOURS 41.4 MINS

LONGITUDE (-VE EAST)	LATITUDE NORTH	LIMITS. SOUTH	LOCAL TIME	MEAN ALTITUDE
-103	13.83	12.94	9:59	60
-105	14.33	13.43	10:13	63
-107	14.76	13.85	10:26	66
-109	15.12	14.21	10:40	70
-110	15.28	14.35	10:48	71
-112	15.52	14.59	11:02	74
-114	15.68	14.75	11:16	77
-115	15.73	14.79	11:23	79
-117	15.76	14.82	11:38	81
-119	15.71	14.76	11:52	83
-120	15.65	14.70	12:00	83
-122	15.46	14.51	12:14	82
-125	15.02	14.06	12:36	78
-127	14.62	13.66	12:51	75
-129	14.13	13.18	13:05	72
-130	13.86	12.90	13:12	70
-132	13.26	12.31	13:27	67

PERCENTAGE OF SUN OBSCURED ALONG MAP TRACK
WEST END =102.7   EAST END =102.9
MAXIMUM OBSCURATION 103.1 % AT LONG = -120
===

+ MARKS THE CITY OF HAO

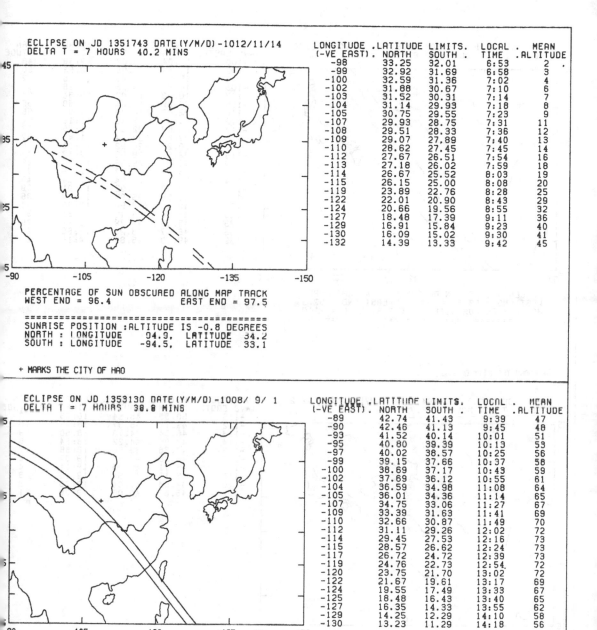

ECLIPSE ON JD 1351743 DATE(Y/M/D)-1012/11/14
DELTA T = 7 HOURS 40.2 MINS

LONGITUDE (-VE EAST)	LATITUDE LIMITS. NORTH	SOUTH	LOCAL TIME	MEAN ALTITUDE
-98	33.25	32.01	6:53	2
-99	32.92	31.69	6:58	3
-100	32.59	31.36	7:02	4
-102	31.88	30.67	7:10	6
-103	31.52	30.31	7:14	7
-104	31.14	29.93	7:18	8
-105	30.75	29.55	7:23	9
-107	29.93	28.75	7:31	11
-108	29.51	28.33	7:36	12
-109	29.07	27.89	7:40	13
-110	28.62	27.45	7:45	14
-112	27.67	26.51	7:54	16
-113	27.18	26.02	7:59	18
-114	26.67	25.52	8:03	19
-115	26.15	25.00	8:08	20
-119	23.89	22.76	8:28	25
-122	22.01	20.90	8:43	29
-124	20.66	19.56	8:55	32
-127	18.48	17.39	9:11	36
-129	16.91	15.84	9:23	40
-130	16.09	15.02	9:30	41
-132	14.39	13.33	9:42	45

PERCENTAGE OF SUN OBSCURED ALONG MAP TRACK
WEST END = 96.4   EAST END = 97.5

===
SUNRISE POSITION :ALTITUDE IS -0.8 DEGREES
NORTH : LONGITUDE   94.9, LATITUDE 34.2
SOUTH : LONGITUDE   -94.5, LATITUDE 33.1

+ MARKS THE CITY OF HAO

ECLIPSE ON JD 1353130 DATE(Y/M/D)-1008/ 9/ 1
DELTA T = 7 HOURS 38.8 MINS

LONGITUDE (-VE EAST)	LATITUDE LIMITS. NORTH	SOUTH	LOCAL TIME	MEAN ALTITUDE
-89	42.74	41.43	9:39	47
-90	42.46	41.13	9:45	48
-93	41.52	40.14	10:01	51
-95	40.80	39.39	10:13	53
-97	40.02	38.57	10:25	56
-99	39.15	37.66	10:37	58
-100	38.69	37.17	10:43	59
-102	37.69	36.12	10:55	61
-104	36.59	34.98	11:08	64
-105	36.01	34.36	11:14	65
-107	34.75	33.06	11:27	67
-109	33.39	31.63	11:41	69
-110	32.66	30.87	11:49	70
-112	31.11	29.26	12:02	72
-114	29.45	27.53	12:16	73
-115	28.57	26.62	12:24	73
-117	26.72	24.72	12:39	73
-119	24.76	22.73	12:54	72
-120	23.75	21.70	13:02	72
-122	21.67	19.61	13:17	69
-124	19.55	17.49	13:33	67
-125	18.48	16.43	13:40	65
-127	16.35	14.33	13:55	62
-129	14.25	12.29	14:10	58
-130	13.23	11.29	14:18	56

PERCENTAGE OF SUN OBSCURED ALONG MAP TRACK
WEST END =103.8   EAST END =104.1
MAXIMUM OBSCURATION 104.3 % AT LONG = -115
===

+ MARKS THE CITY OF HAO

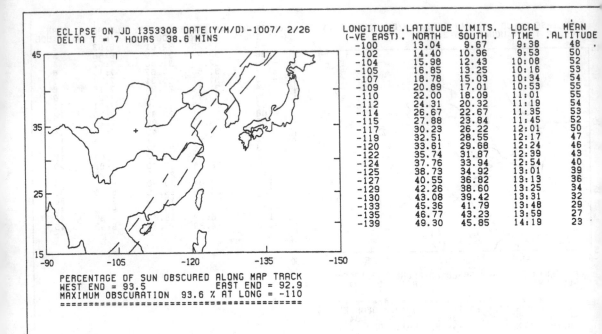

ECLIPSE ON JD 1353308 DATE(Y/M/D)-1007/ 2/26
DELTA T = 7 HOURS 38.6 MINS

LONGITUDE	.LATITUDE	LIMITS.	LOCAL .	MEAN
(-VE EAST).	NORTH	SOUTH .	TIME	.ALTITUDE
-100	13.04	9.67	9:38	48
-102	14.40	10.96	9:53	50
-104	15.98	12.43	10:08	52
-105	16.85	13.25	10:16	53
-107	18.78	15.03	10:34	54
-109	20.89	17.01	10:53	55
-110	22.00	18.09	11:01	55
-112	24.31	20.32	11:19	54
-114	26.67	22.67	11:35	53
-115	27.88	23.84	11:45	52
-117	30.23	26.22	12:01	50
-119	32.51	28.55	12:17	47
-120	33.61	29.68	12:24	46
-122	35.74	31.87	12:39	43
-124	37.76	33.94	12:54	40
-125	38.73	34.92	13:01	39
-127	40.55	36.82	13:13	36
-129	42.26	38.60	13:25	34
-130	43.08	39.42	13:31	32
-133	45.36	41.79	13:48	29
-135	46.77	43.23	13:59	27
-139	49.30	45.85	14:19	23

PERCENTAGE OF SUN OBSCURED ALONG MAP TRACK
WEST END = 93.5   EAST END = 92.9
MAXIMUM OBSCURATION 93.6 % AT LONG = -110
===

+ MARKS THE CITY OF HAO

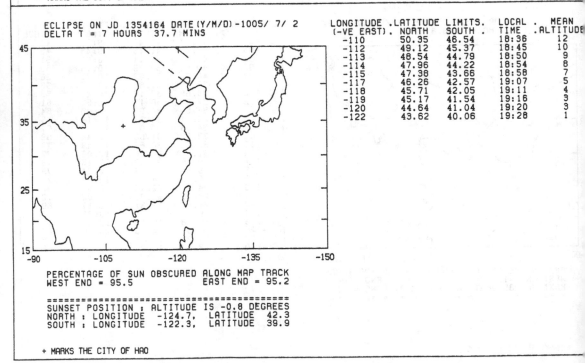

ECLIPSE ON JD 1354164 DATE(Y/M/D)-1005/ 7/ 2
DELTA T = 7 HOURS 37.7 MINS

LONGITUDE	.LATITUDE	LIMITS.	LOCAL .	MEAN
(-VE EAST).	NORTH	SOUTH .	TIME	.ALTITUDE
-110	50.35	46.54	18:36	12
-112	49.12	45.37	18:45	10
-113	48.54	44.79	18:50	9
-114	47.96	44.22	18:54	8
-115	47.38	43.66	18:58	7
-117	46.26	42.57	19:07	5
-118	45.71	42.05	19:11	4
-119	45.17	41.54	19:16	3
-120	44.64	41.04	19:20	3
-122	43.62	40.06	19:28	1

PERCENTAGE OF SUN OBSCURED ALONG MAP TRACK
WEST END = 95.5   EAST END = 95.2

===
SUNSET POSITION : ALTITUDE IS -0.8 DEGREES
NORTH : LONGITUDE -124.7, LATITUDE 42.3
SOUTH : LONGITUDE -122.3, LATITUDE 39.9

+ MARKS THE CITY OF HAO

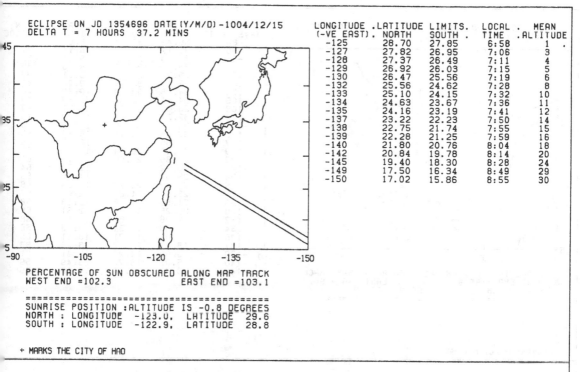

ECLIPSE ON JD 1354696 DATE(Y/M/D)-1004/12/15
DELTA T = 7 HOURS 37.2 MINS

LONGITUDE (-VE EAST)	LATITUDE LIMITS. NORTH	SOUTH	LOCAL TIME	MEAN ALTITUDE
-125	28.70	27.85	6:58	1
-127	27.82	26.95	7:06	3
-128	27.37	26.49	7:11	4
-129	26.92	26.03	7:15	5
-130	26.47	25.56	7:19	6
-132	25.56	24.62	7:28	8
-133	25.10	24.15	7:32	10
-134	24.63	23.67	7:36	11
-135	24.16	23.19	7:41	12
-137	23.22	22.23	7:50	14
-138	22.75	21.74	7:55	15
-139	22.28	21.25	7:59	16
-140	21.80	20.76	8:04	18
-142	20.84	19.78	8:14	20
-145	19.40	18.30	8:28	24
-149	17.50	16.34	8:49	29
-150	17.02	15.86	8:55	30

PERCENTAGE OF SUN OBSCURED ALONG MAP TRACK
WEST END =102.3   EAST END =103.1

===
SUNRISE POSITION :ALTITUDE IS -0.8 DEGREES
NORTH : LONGITUDE -123.0, LATITUDE 29.6
SOUTH : LONGITUDE -122.9, LATITUDE 28.8

+ MARKS THE CITY OF HAO

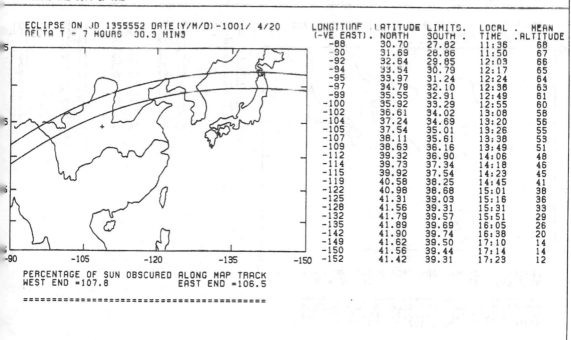

ECLIPSE ON JD 1355552 DATE(Y/M/D)-1001/ 4/20
DELTA T = 7 HOURS 30.3 MINS

LONGITUDE (-VE EAST)	LATITUDE LIMITS. NORTH	SOUTH	LOCAL TIME	MEAN ALTITUDE
-88	30.70	27.82	11:36	68
-90	31.69	28.86	11:50	67
-92	32.64	29.85	12:03	66
-94	33.54	30.79	12:17	65
-95	33.97	31.24	12:24	64
-97	34.79	32.10	12:36	63
-99	35.55	32.91	12:49	61
-100	35.92	33.29	12:55	60
-102	36.61	34.02	13:08	58
-104	37.24	34.69	13:20	56
-105	37.54	35.01	13:26	55
-107	38.11	35.61	13:38	53
-109	38.63	36.16	13:49	51
-112	39.32	36.90	14:06	48
-114	39.73	37.34	14:18	46
-115	39.92	37.54	14:23	45
-119	40.58	38.25	14:45	41
-122	40.98	38.68	15:01	38
-125	41.31	39.03	15:16	36
-128	41.56	39.31	15:31	33
-132	41.79	39.57	15:51	29
-135	41.89	39.69	16:05	26
-142	41.90	39.74	16:38	20
-149	41.62	39.50	17:10	14
-150	41.56	39.44	17:14	14
-152	41.42	39.31	17:23	12

PERCENTAGE OF SUN OBSCURED ALONG MAP TRACK
WEST END =107.8   EAST END =106.5

===

+ MARKS THE CITY OF HAO

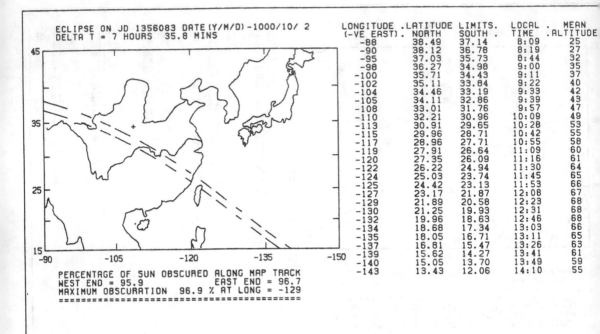

ECLIPSE ON JD 1356083 DATE(Y/M/D) -1000/10/ 2
DELTA T = 7 HOURS   35.8 MINS

LONGITUDE (-VE EAST)	LATITUDE NORTH	LIMITS. SOUTH	LOCAL TIME	MEAN ALTITUDE
-88	38.49	37.14	8:09	25
-90	38.12	36.78	8:19	27
-95	37.03	35.73	8:44	32
-98	36.27	34.98	9:00	35
-100	35.71	34.43	9:11	37
-102	35.11	33.84	9:22	40
-104	34.46	33.19	9:33	42
-105	34.11	32.86	9:39	43
-108	33.01	31.76	9:57	47
-110	32.21	30.96	10:09	49
-113	30.91	29.65	10:28	53
-115	29.96	28.71	10:42	55
-117	28.96	27.71	10:55	58
-119	27.91	26.64	11:09	60
-120	27.35	26.09	11:16	61
-122	26.22	24.94	11:30	64
-124	25.03	23.74	11:45	65
-125	24.42	23.13	11:53	66
-127	23.17	21.87	12:08	67
-129	21.89	20.58	12:23	68
-130	21.25	19.93	12:31	68
-132	19.96	18.63	12:46	68
-134	18.68	17.34	13:03	66
-135	18.05	16.71	13:11	65
-137	16.81	15.47	13:26	63
-139	15.62	14.27	13:41	61
-140	15.05	13.70	13:49	59
-143	13.43	12.06	14:10	55

PERCENTAGE OF SUN OBSCURED ALONG MAP TRACK
WEST END = 95.9   EAST END = 96.7
MAXIMUM OBSCURATION   96.9 % AT LONG = -129
==

+ MARKS THE CITY OF HAO

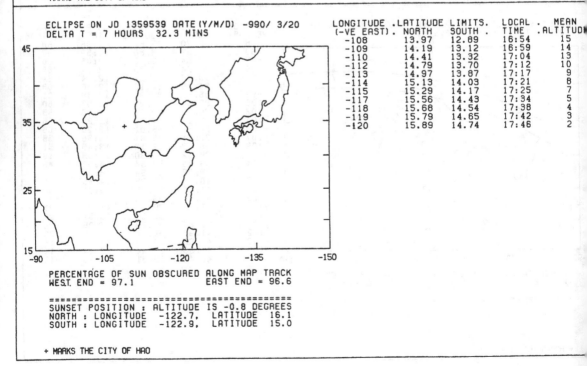

ECLIPSE ON JD 1359539 DATE(Y/M/D) -990/ 3/20
DELTA T = 7 HOURS   32.3 MINS

LONGITUDE (-VE EAST)	LATITUDE NORTH	LIMITS. SOUTH	LOCAL TIME	MEAN ALTITUDE
-108	13.97	12.89	16:54	15
-109	14.19	13.12	16:59	14
-110	14.41	13.32	17:04	13
-112	14.79	13.70	17:12	10
-113	14.97	13.87	17:17	9
-114	15.13	14.03	17:21	8
-115	15.29	14.17	17:25	7
-117	15.56	14.43	17:34	5
-118	15.68	14.54	17:38	4
-119	15.79	14.65	17:42	3
-120	15.89	14.74	17:46	2

PERCENTAGE OF SUN OBSCURED ALONG MAP TRACK
WEST END = 97.1   EAST END = 96.6

==
SUNSET POSITION : ALTITUDE IS -0.8 DEGREES
NORTH : LONGITUDE -122.7,   LATITUDE   16.1
SOUTH : LONGITUDE -122.9,   LATITUDE   15.0

+ MARKS THE CITY OF HAO

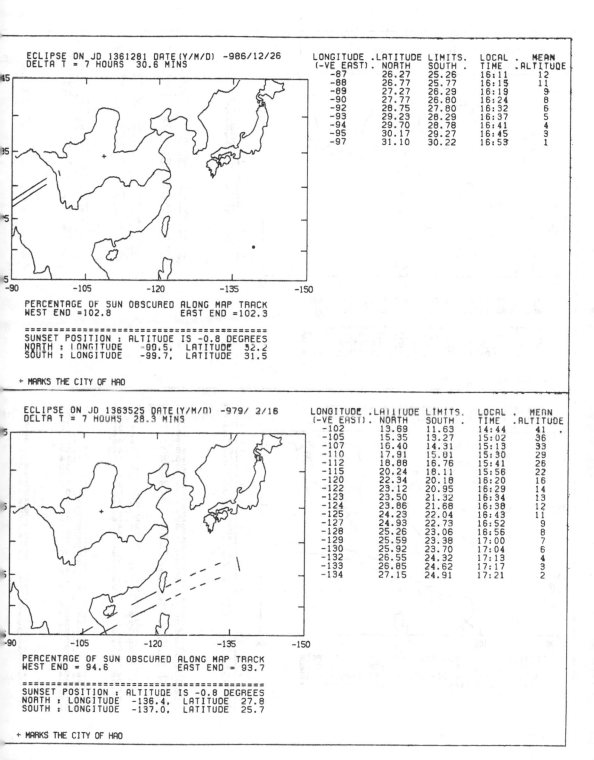

ECLIPSE ON JD 1361281 DATE(Y/M/D) -986/12/26
DELTA T = 7 HOURS 30.6 MINS

LONGITUDE (-VE EAST)	LATITUDE LIMITS		LOCAL	MEAN
	NORTH	SOUTH	TIME	ALTITUDE
-87	26.27	25.26	16:11	12
-88	26.77	25.77	16:15	11
-89	27.27	26.29	16:19	9
-90	27.77	26.80	16:24	8
-92	28.75	27.80	16:32	6
-93	29.23	28.29	16:37	5
-94	29.70	28.78	16:41	4
-95	30.17	29.27	16:45	3
-97	31.10	30.22	16:53	1

PERCENTAGE OF SUN OBSCURED ALONG MAP TRACK
WEST END =102.8   EAST END =102.3

===
SUNSET POSITION : ALTITUDE IS -0.8 DEGREES
NORTH : LONGITUDE   -90.5, LATITUDE 52.2
SOUTH : LONGITUDE -99.7, LATITUDE 31.5

+ MARKS THE CITY OF HAO

ECLIPSE ON JD 1363525 DATE(Y/M/D) -979/ 2/16
DELTA T = 7 HOURS 28.3 MINS

LONGITUDE (-VE EAST)	LATITUDE LIMITS		LOCAL	MEAN
	NORTH	SOUTH	TIME	ALTITUDE
-102	13.69	11.63	14:44	41
-105	15.35	13.27	15:02	36
-107	16.40	14.31	15:13	33
-110	17.91	15.81	15:30	29
-112	18.88	16.76	15:41	26
-115	20.24	18.11	15:56	22
-120	22.34	20.18	16:20	16
-122	23.12	20.95	16:29	14
-123	23.50	21.32	16:34	13
-124	23.86	21.68	16:38	12
-125	24.23	22.04	16:43	11
-127	24.93	22.73	16:52	9
-128	25.26	23.06	16:56	8
-129	25.59	23.38	17:00	7
-130	25.92	23.70	17:04	6
-132	26.55	24.32	17:13	4
-133	26.85	24.62	17:17	3
-134	27.15	24.91	17:21	2

PERCENTAGE OF SUN OBSCURED ALONG MAP TRACK
WEST END = 94.6   EAST END = 93.7

===
SUNSET POSITION : ALTITUDE IS -0.8 DEGREES
NORTH : LONGITUDE -136.4, LATITUDE 27.8
SOUTH : LONGITUDE -137.0, LATITUDE 25.7

+ MARKS THE CITY OF HAO

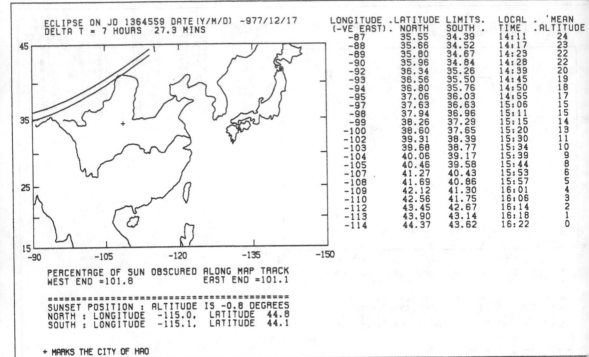

ECLIPSE ON JD 1364559 DATE(Y/M/D) -977/12/17
DELTA T = 7 HOURS 27.3 MINS

LONGITUDE (-VE EAST)	LATITUDE LIMITS. NORTH	SOUTH	LOCAL TIME	MEAN ALTITUDE
-87	35.55	34.39	14:11	24
-88	35.66	34.52	14:17	23
-89	35.80	34.67	14:23	22
-90	35.96	34.84	14:28	22
-92	36.34	35.26	14:39	20
-93	36.56	35.50	14:45	19
-94	36.80	35.76	14:50	18
-95	37.06	36.03	14:55	17
-97	37.63	36.63	15:06	15
-98	37.94	36.96	15:11	15
-99	38.26	37.29	15:15	14
-100	38.60	37.65	15:20	13
-102	39.31	38.39	15:30	11
-103	39.68	38.77	15:34	10
-104	40.06	39.17	15:39	9
-105	40.46	39.58	15:44	8
-107	41.27	40.43	15:53	6
-108	41.69	40.86	15:57	5
-109	42.12	41.30	16:01	4
-110	42.56	41.75	16:06	3
-112	43.45	42.67	16:14	2
-113	43.90	43.14	16:18	1
-114	44.37	43.62	16:22	0

PERCENTAGE OF SUN OBSCURED ALONG MAP TRACK
WEST END =101.8   EAST END =101.1

==

SUNSET POSITION : ALTITUDE IS -0.8 DEGREES
NORTH : LONGITUDE -115.0, LATITUDE 44.8
SOUTH : LONGITUDE -115.1, LATITUDE 44.1

+ MARKS THE CITY OF HAO

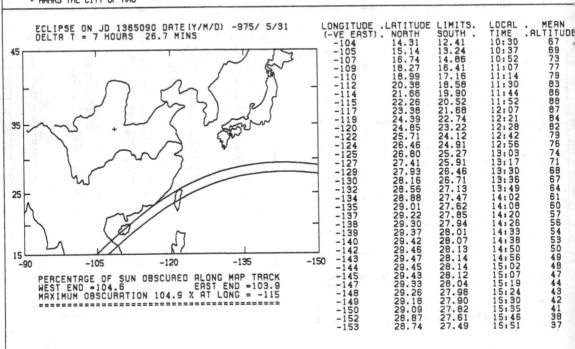

ECLIPSE ON JD 1365090 DATE(Y/M/D) -975/ 5/31
DELTA T = 7 HOURS 26.7 MINS

LONGITUDE (-VE EAST)	LATITUDE LIMITS. NORTH	SOUTH	LOCAL TIME	MEAN ALTITUDE
-104	14.31	12.41	10:30	67
-105	15.14	13.24	10:37	69
-107	16.74	14.86	10:52	73
-109	18.27	16.41	11:07	77
-110	18.99	17.16	11:14	79
-112	20.38	18.58	11:30	83
-114	21.66	19.90	11:44	86
-115	22.26	20.52	11:52	88
-117	23.38	21.68	12:07	87
-119	24.39	22.74	12:21	84
-120	24.85	23.22	12:28	82
-122	25.71	24.12	12:42	79
-124	26.46	24.91	12:56	76
-125	26.80	25.27	13:03	74
-127	27.41	25.91	13:17	71
-129	27.93	26.46	13:30	68
-130	28.16	26.71	13:36	67
-132	28.56	27.13	13:49	64
-134	28.88	27.47	14:02	61
-135	29.01	27.62	14:08	60
-137	29.22	27.85	14:20	57
-138	29.30	27.94	14:26	56
-139	29.37	28.01	14:33	54
-140	29.42	28.07	14:38	53
-142	29.46	28.13	14:50	50
-143	29.47	28.14	14:56	49
-144	29.45	28.14	15:02	48
-145	29.43	28.12	15:07	47
-147	29.33	28.04	15:19	44
-148	29.26	27.98	15:24	43
-149	29.18	27.90	15:30	42
-150	29.09	27.82	15:35	41
-152	28.87	27.61	15:46	38
-153	28.74	27.49	15:51	37

PERCENTAGE OF SUN OBSCURED ALONG MAP TRACK
WEST END =104.6   EAST END =103.9
MAXIMUM OBSCURATION 104.9 % AT LONG = -115
==

+ MARKS THE CITY OF HAO

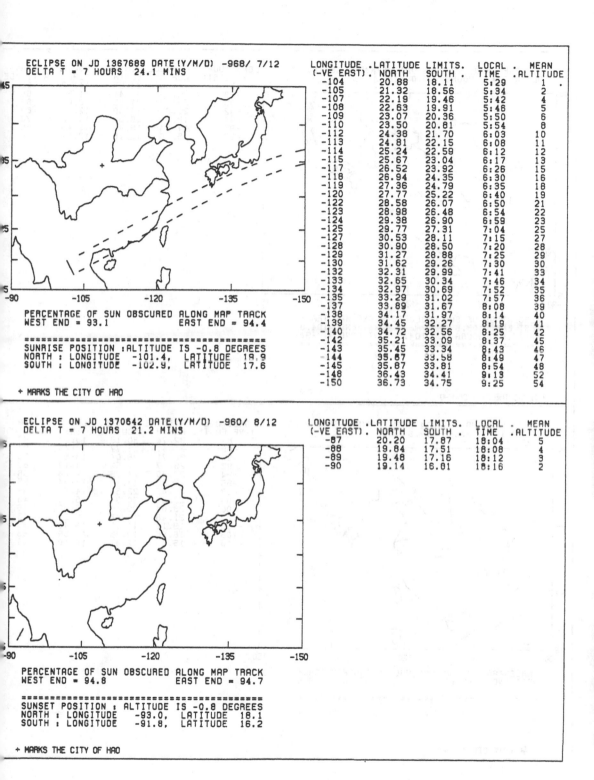

ECLIPSE ON JD 1367689 DATE(Y/M/D) -968/ 7/12
DELTA T = 7 HOURS 24.1 MINS

LONGITUDE (-VE EAST)	LATITUDE NORTH	LIMITS. SOUTH	LOCAL TIME	MEAN ALTITUDE
-104	20.88	18.11	5:29	1
-105	21.32	18.56	5:34	2
-107	22.19	19.46	5:42	4
-108	22.63	19.91	5:46	5
-109	23.07	20.36	5:50	6
-110	23.50	20.81	5:54	8
-112	24.38	21.70	6:03	10
-113	24.81	22.15	6:08	11
-114	25.24	22.59	6:12	12
-115	25.67	23.04	6:17	13
-117	26.52	23.92	6:26	15
-118	26.94	24.35	6:30	16
-119	27.36	24.79	6:35	18
-120	27.77	25.22	6:40	19
-122	28.58	26.07	6:50	21
-123	28.98	26.48	6:54	22
-124	29.38	26.90	6:59	23
-125	29.77	27.31	7:04	25
-127	30.53	28.11	7:15	27
-128	30.90	28.50	7:20	28
-129	31.27	28.88	7:25	29
-130	31.62	29.26	7:30	30
-132	32.31	29.99	7:41	33
-133	32.65	30.34	7:46	34
-134	32.97	30.69	7:52	35
-135	33.29	31.02	7:57	36
-137	33.89	31.67	8:08	39
-138	34.17	31.97	8:14	40
-139	34.45	32.27	8:19	41
-140	34.72	32.56	8:25	42
-142	35.21	33.09	8:37	45
-143	35.45	33.34	8:43	46
-144	35.67	33.58	8:49	47
-145	35.87	33.81	8:54	48
-148	36.43	34.41	9:19	52
-150	36.73	34.75	9:25	54

PERCENTAGE OF SUN OBSCURED ALONG MAP TRACK
WEST END = 93.1   EAST END = 94.4

=======================================
SUNRISE POSITION :ALTITUDE IS -0.8 DEGREES
NORTH : LONGITUDE -101.4, LATITUDE 19.9
SOUTH : LONGITUDE -102.9, LATITUDE 17.6

+ MARKS THE CITY OF HAO

ECLIPSE ON JD 1370642 DATE(Y/M/D) -960/ 8/12
DELTA T = 7 HOURS 21.2 MINS

LONGITUDE (-VE EAST)	LATITUDE NORTH	LIMITS. SOUTH	LOCAL TIME	MEAN ALTITUDE
-87	20.20	17.87	18:04	5
-88	19.84	17.51	18:08	4
-89	19.48	17.16	18:12	3
-90	19.14	16.81	18:16	2

PERCENTAGE OF SUN OBSCURED ALONG MAP TRACK
WEST END = 94.8   EAST END = 94.7

=======================================
SUNSET POSITION : ALTITUDE IS -0.8 DEGREES
NORTH : LONGITUDE -93.0, LATITUDE 18.1
SOUTH : LONGITUDE -91.8, LATITUDE 16.2

+ MARKS THE CITY OF HAO

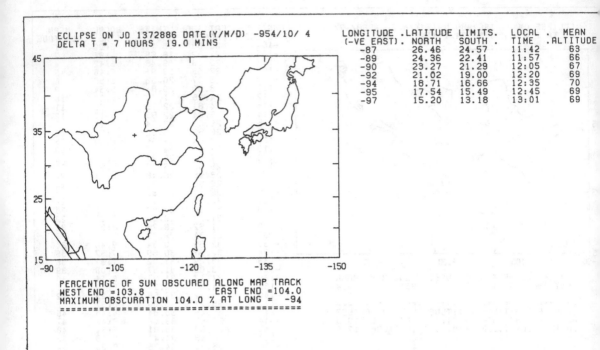

ECLIPSE ON JD 1372886 DATE(Y/M/D) -954/10/ 4
DELTA T = 7 HOURS 19.0 MINS

LONGITUDE	.LATITUDE	LIMITS.	LOCAL	. MEAN
(-VE EAST).	NORTH	SOUTH .	TIME	.ALTITUDE
-87	26.46	24.57	11:42	63
-89	24.36	22.41	11:57	66
-90	23.27	21.29	12:05	67
-92	21.02	19.00	12:20	69
-94	18.71	16.66	12:35	70
-95	17.54	15.49	12:45	69
-97	15.20	13.18	13:01	69

PERCENTAGE OF SUN OBSCURED ALONG MAP TRACK
WEST END =103.8   EAST END =104.0
MAXIMUM OBSCURATION 104.0 % AT LONG = -94
===

+ MARKS THE CITY OF HAO

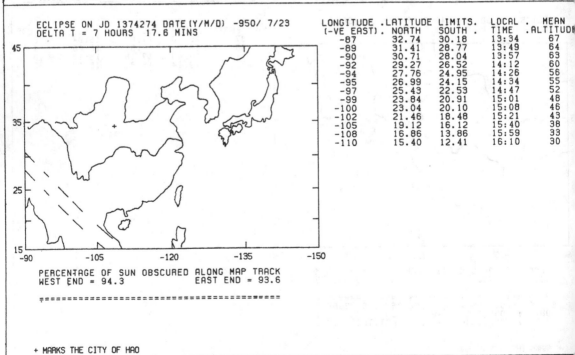

ECLIPSE ON JD 1374274 DATE(Y/M/D) -950/ 7/23
DELTA T = 7 HOURS 17.6 MINS

LONGITUDE	.LATITUDE	LIMITS.	LOCAL	. MEAN
(-VE EAST).	NORTH	SOUTH .	TIME	.ALTITUDE
-87	32.74	30.18	13:34	67
-89	31.41	28.77	13:49	64
-90	30.71	28.04	13:57	63
-92	29.27	26.52	14:12	60
-94	27.76	24.95	14:26	56
-95	26.99	24.15	14:34	55
-97	25.43	22.53	14:47	52
-99	23.84	20.91	15:01	48
-100	23.04	20.10	15:08	46
-102	21.46	18.48	15:21	43
-105	19.12	16.12	15:40	38
-108	16.86	13.86	15:59	33
-110	15.40	12.41	16:10	30

PERCENTAGE OF SUN OBSCURED ALONG MAP TRACK
WEST END = 94.3   EAST END = 93.6

===

+ MARKS THE CITY OF HAO

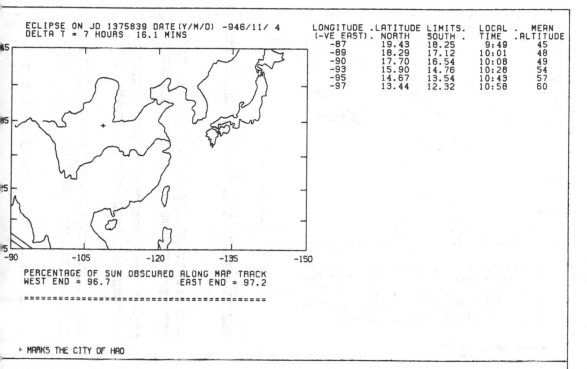

ECLIPSE ON JD 1375839 DATE(Y/M/D) -946/11/ 4
DELTA T = 7 HOURS 16.1 MINS

LONGITUDE	.LATITUDE	LIMITS.	LOCAL	. MEAN
(-VE EAST).	NORTH	SOUTH .	TIME	.ALTITUDE
-87	19.43	18.25	9:49	45
-89	18.29	17.12	10:01	48
-90	17.70	16.54	10:08	49
-93	15.90	14.76	10:28	54
-95	14.67	13.54	10:43	57
-97	13.44	12.32	10:58	60

PERCENTAGE OF SUN OBSCURED ALONG MAP TRACK
WEST END = 96.7   EAST END = 97.2

==

+ MARKS THE CITY OF HAO

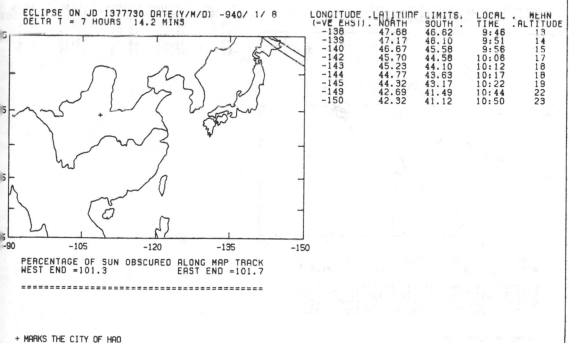

ECLIPSE ON JD 1377730 DATE(Y/M/D) -940/ 1/ 8
DELTA T = 7 HOURS 14.2 MINS

LONGITUDE	.LATITUDE	LIMITS.	LOCAL	. MEAN
(-VE EAST).	NORTH	SOUTH .	TIME	.ALTITUDE
-138	47.68	46.62	9:46	13
-139	47.17	46.10	9:51	14
-140	46.67	45.58	9:56	15
-142	45.70	44.58	10:06	17
-143	45.23	44.10	10:12	18
-144	44.77	43.63	10:17	18
-145	44.32	43.17	10:22	19
-149	42.69	41.49	10:44	22
-150	42.32	41.12	10:50	23

PERCENTAGE OF SUN OBSCURED ALONG MAP TRACK
WEST END =101.3   EAST END =101.7

==

+ MARKS THE CITY OF HAO

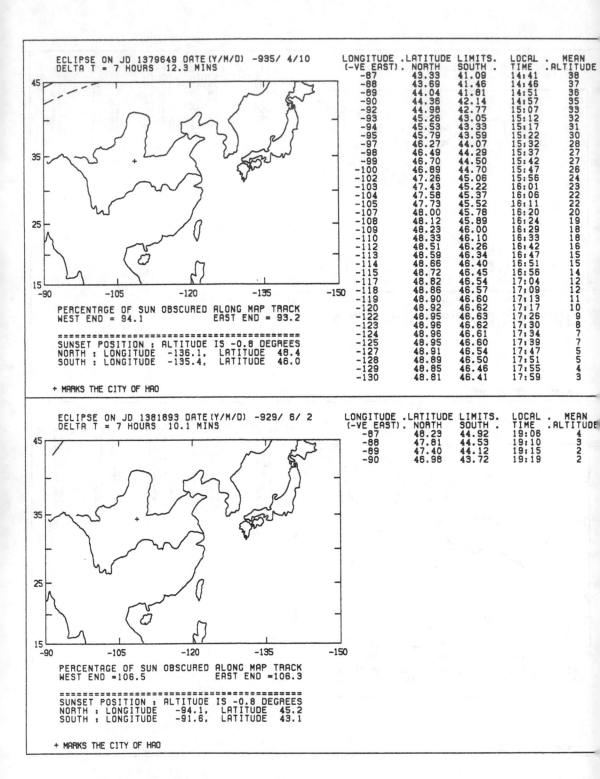

ECLIPSE ON JD 1379649 DATE(Y/M/D) -935/ 4/10
DELTA T = 7 HOURS 12.3 MINS

LONGITUDE (-VE EAST)	LATITUDE NORTH	LIMITS. SOUTH	LOCAL TIME	MEAN ALTITUDE
-87	43.33	41.09	14:41	38
-88	43.69	41.46	14:46	37
-89	44.04	41.81	14:51	36
-90	44.36	42.14	14:57	35
-92	44.98	42.77	15:07	33
-93	45.26	43.05	15:12	32
-94	45.53	43.33	15:17	31
-95	45.79	43.59	15:22	30
-97	46.27	44.07	15:32	28
-98	46.49	44.29	15:37	27
-99	46.70	44.50	15:42	27
-100	46.89	44.70	15:47	26
-102	47.26	45.06	15:56	24
-103	47.43	45.22	16:01	23
-104	47.58	45.37	16:06	22
-105	47.73	45.52	16:11	22
-107	48.00	45.78	16:20	20
-108	48.12	45.89	16:24	19
-109	48.23	46.00	16:29	18
-110	48.33	46.10	16:33	18
-112	48.51	46.26	16:42	16
-113	48.59	46.34	16:47	15
-114	48.66	46.40	16:51	15
-115	48.72	46.45	16:56	14
-117	48.82	46.54	17:04	12
-118	48.86	46.57	17:09	12
-119	48.90	46.60	17:13	11
-120	48.92	46.62	17:17	10
-122	48.95	46.63	17:26	9
-123	48.96	46.62	17:30	8
-124	48.96	46.61	17:34	7
-125	48.95	46.60	17:39	7
-127	48.91	46.54	17:47	5
-128	48.89	46.50	17:51	5
-129	48.85	46.46	17:55	4
-130	48.81	46.41	17:59	3

PERCENTAGE OF SUN OBSCURED ALONG MAP TRACK
WEST END = 94.1   EAST END = 93.2

==
SUNSET POSITION : ALTITUDE IS -0.8 DEGREES
NORTH : LONGITUDE -136.1, LATITUDE 48.4
SOUTH : LONGITUDE -135.4, LATITUDE 46.0

+ MARKS THE CITY OF HAO

ECLIPSE ON JD 1381893 DATE(Y/M/D) -929/ 6/ 2
DELTA T = 7 HOURS 10.1 MINS

LONGITUDE (-VE EAST)	LATITUDE NORTH	LIMITS. SOUTH	LOCAL TIME	MEAN ALTITUDE
-87	48.23	44.92	19:06	4
-88	47.81	44.53	19:10	3
-89	47.40	44.12	19:15	2
-90	46.98	43.72	19:19	2

PERCENTAGE OF SUN OBSCURED ALONG MAP TRACK
WEST END =106.5   EAST END =106.3

==
SUNSET POSITION : ALTITUDE IS -0.8 DEGREES
NORTH : LONGITUDE   -94.1, LATITUDE 45.2
SOUTH : LONGITUDE   -91.6, LATITUDE 43.1

+ MARKS THE CITY OF HAO

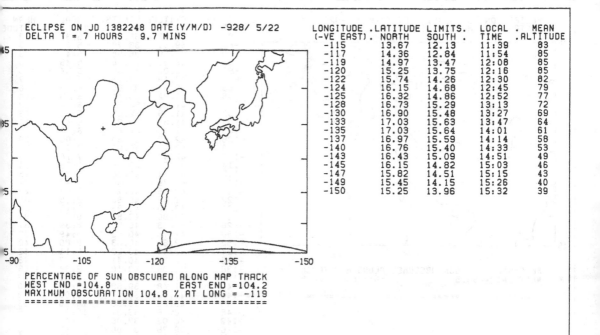

ECLIPSE ON JD 1382248 DATE(Y/M/D) -928/ 5/22
DELTA T = 7 HOURS   9.7 MINS

LONGITUDE (-VE EAST)	LATITUDE LIMITS. NORTH	SOUTH	LOCAL TIME	MEAN ALTITUDE
-115	13.67	12.13	11:39	83
-117	14.36	12.84	11:54	85
-119	14.97	13.47	12:08	85
-120	15.25	13.75	12:16	85
-122	15.74	14.26	12:30	82
-124	16.15	14.68	12:45	79
-125	16.32	14.86	12:52	77
-128	16.73	15.29	13:13	72
-130	16.90	15.48	13:27	69
-133	17.03	15.63	13:47	64
-135	17.03	15.64	14:01	61
-137	16.97	15.59	14:14	58
-140	16.76	15.40	14:33	53
-143	16.43	15.09	14:51	49
-145	16.15	14.82	15:03	46
-147	15.82	14.51	15:15	43
-149	15.45	14.15	15:26	40
-150	15.25	13.96	15:32	39

PERCENTAGE OF SUN OBSCURED ALONG MAP TRACK
WEST END =104.8   EAST END =104.2
MAXIMUM OBSCURATION 104.8 % AT LONG = -119
==

+ MARKS THE CITY OF HAO

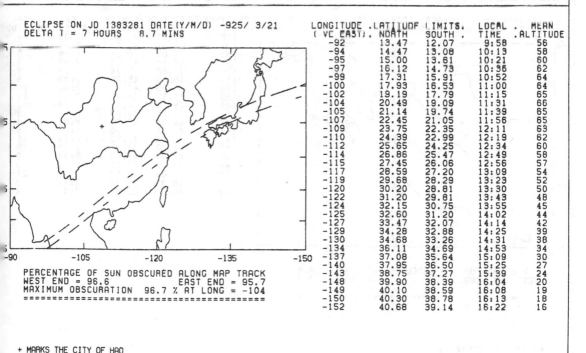

ECLIPSE ON JD 1383281 DATE(Y/M/D) -925/ 3/21
DELTA T = 7 HOURS   8.7 MINS

LONGITUDE (-VE EAST)	LATITUDE LIMITS. NORTH	SOUTH	LOCAL TIME	MEAN ALTITUDE
-92	13.47	12.07	9:58	56
-94	14.47	13.08	10:13	58
-95	15.00	13.61	10:21	60
-97	16.12	14.73	10:36	62
-99	17.31	15.91	10:52	64
-100	17.93	16.53	11:00	64
-102	19.19	17.79	11:15	65
-104	20.49	19.09	11:31	66
-105	21.14	19.74	11:39	65
-107	22.45	21.05	11:56	65
-109	23.75	22.35	12:11	63
-110	24.39	22.99	12:19	62
-112	25.65	24.25	12:34	60
-114	26.86	25.47	12:49	58
-115	27.45	26.06	12:56	57
-117	28.59	27.20	13:09	54
-119	29.68	28.29	13:23	52
-120	30.20	28.81	13:30	50
-122	31.20	29.81	13:43	48
-124	32.15	30.75	13:55	45
-125	32.60	31.20	14:02	44
-127	33.47	32.07	14:14	42
-129	34.28	32.88	14:25	39
-130	34.68	33.26	14:31	38
-134	36.11	34.69	14:53	34
-137	37.08	35.64	15:09	30
-140	37.95	36.50	15:25	27
-143	38.75	37.27	15:39	24
-148	39.90	38.39	16:04	20
-149	40.10	38.59	16:08	19
-150	40.30	38.78	16:13	18
-152	40.68	39.14	16:22	16

PERCENTAGE OF SUN OBSCURED ALONG MAP TRACK
WEST END = 96.6   EAST END = 95.7
MAXIMUM OBSCURATION 96.7 % AT LONG = -104
==

+ MARKS THE CITY OF HAO

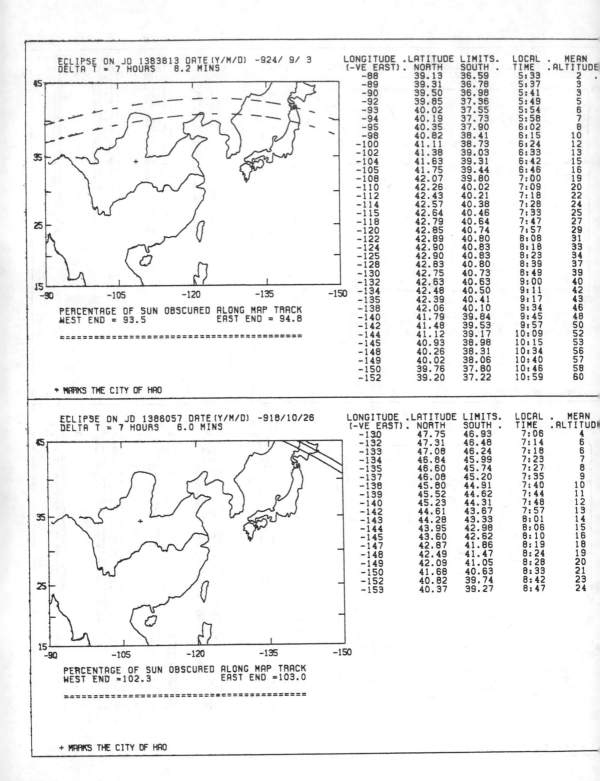

ECLIPSE ON JD 1383813 DATE(Y/M/D) -924/ 9/ 3
DELTA T = 7 HOURS   8.2 MINS

LONGITUDE (-VE EAST)	.LATITUDE NORTH	LIMITS. SOUTH .	LOCAL TIME	. MEAN .ALTITUDE
-88	39.13	36.59	5:33	2
-89	39.31	36.78	5:37	3
-90	39.50	36.98	5:41	3
-92	39.85	37.36	5:49	5
-93	40.02	37.55	5:54	6
-94	40.19	37.73	5:58	7
-95	40.35	37.90	6:02	8
-98	40.82	38.41	6:15	10
-100	41.11	38.73	6:24	12
-102	41.38	39.03	6:33	13
-104	41.63	39.31	6:42	15
-105	41.75	39.44	6:46	16
-108	42.07	39.80	7:00	19
-110	42.26	40.02	7:09	20
-112	42.43	40.21	7:18	22
-114	42.57	40.38	7:28	24
-115	42.64	40.46	7:33	25
-118	42.79	40.64	7:47	27
-120	42.85	40.74	7:57	29
-122	42.89	40.80	8:08	31
-124	42.90	40.83	8:18	33
-125	42.90	40.83	8:23	34
-128	42.83	40.80	8:39	37
-130	42.75	40.73	8:49	39
-132	42.63	40.63	9:00	40
-134	42.48	40.50	9:11	42
-135	42.39	40.41	9:17	43
-138	42.06	40.10	9:34	46
-140	41.79	39.84	9:45	48
-142	41.48	39.53	9:57	50
-144	41.12	39.17	10:09	52
-145	40.93	38.98	10:15	53
-148	40.26	38.31	10:34	56
-149	40.02	38.06	10:40	57
-150	39.76	37.80	10:46	58
-152	39.20	37.22	10:59	60

PERCENTAGE OF SUN OBSCURED ALONG MAP TRACK
WEST END = 93.5   EAST END = 94.8

+ MARKS THE CITY OF HAO

ECLIPSE ON JD 1386057 DATE(Y/M/D) -918/10/26
DELTA T = 7 HOURS   6.0 MINS

LONGITUDE (-VE EAST)	.LATITUDE NORTH	LIMITS. SOUTH .	LOCAL TIME	. MEAN .ALTITUDE
-130	47.75	46.93	7:06	4
-132	47.31	46.48	7:14	6
-133	47.08	46.24	7:18	6
-134	46.84	45.99	7:23	7
-135	46.60	45.74	7:27	8
-137	46.08	45.20	7:35	9
-138	45.80	44.91	7:40	10
-139	45.52	44.62	7:44	11
-140	45.23	44.31	7:48	12
-142	44.61	43.67	7:57	13
-143	44.28	43.33	8:01	14
-144	43.95	42.98	8:06	15
-145	43.60	42.62	8:10	16
-147	42.87	41.86	8:19	18
-148	42.49	41.47	8:24	19
-149	42.09	41.05	8:28	20
-150	41.68	40.63	8:33	21
-152	40.82	39.74	8:42	23
-153	40.37	39.27	8:47	24

PERCENTAGE OF SUN OBSCURED ALONG MAP TRACK
WEST END =102.3   EAST END =103.0

+ MARKS THE CITY OF HAO

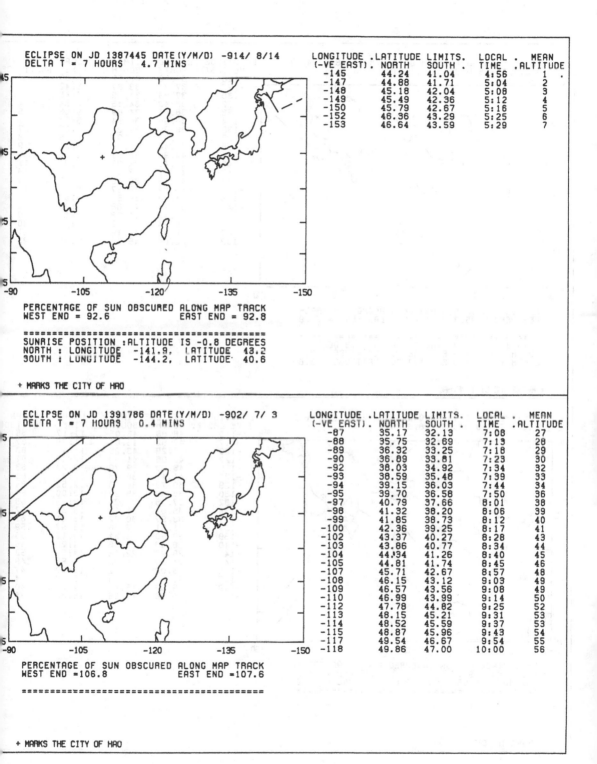

ECLIPSE ON JD 1387445 DATE(Y/M/D) -914/ 8/14
DELTA T = 7 HOURS   4.7 MINS

LONGITUDE (-VE EAST)	.LATITUDE . NORTH	LIMITS. SOUTH .	LOCAL . TIME	MEAN .ALTITUDE
-145	44.24	41.04	4:56	1 .
-147	44.88	41.71	5:04	2
-148	45.18	42.04	5:08	3
-149	45.49	42.36	5:12	4
-150	45.79	42.67	5:16	5
-152	46.36	43.29	5:25	6
-153	46.64	43.59	5:29	7

-90   -105   -120   -135   -150

PERCENTAGE OF SUN OBSCURED ALONG MAP TRACK
WEST END = 92.6   EAST END = 92.8

===
SUNRISE POSITION :ALTITUDE IS -0.8 DEGREES
NORTH : LONGITUDE -141.9,   LATITUDE 43.2
SOUTH : LONGITUDE -144.2,   LATITUDE 40.6

+ MARKS THE CITY OF HAO

ECLIPSE ON JD 1391786 DATE(Y/M/D) -902/ 7/ 3
DELTA T = 7 HOURS   0.4 MINS

LONGITUDE (-VE EAST)	.LATITUDE . NORTH	LIMITS. SOUTH .	LOCAL TIME	MEAN .ALTITUDE
-87	35.17	32.13	7:08	27
-88	35.75	32.69	7:13	28
-89	36.32	33.25	7:18	29
-90	36.89	33.81	7:23	30
-92	38.03	34.92	7:34	32
-93	38.59	35.48	7:39	33
-94	39.15	36.03	7:44	34
-95	39.70	36.58	7:50	36
-97	40.79	37.66	8:01	38
-98	41.32	38.20	8:06	39
-99	41.85	38.73	8:12	40
-100	42.36	39.25	8:17	41
-102	43.37	40.27	8:28	43
-103	43.86	40.77	8:34	44
-104	44.34	41.26	8:40	45
-105	44.81	41.74	8:45	46
-107	45.71	42.67	8:57	48
-108	46.15	43.12	9:03	49
-109	46.57	43.56	9:08	49
-110	46.99	43.99	9:14	50
-112	47.78	44.82	9:25	52
-113	48.15	45.21	9:31	53
-114	48.52	45.59	9:37	53
-115	48.87	45.96	9:43	54
-117	49.54	46.67	9:54	55
-118	49.86	47.00	10:00	56

-90   -105   -120   -135   -150

PERCENTAGE OF SUN OBSCURED ALONG MAP TRACK
WEST END =106.8   EAST END =107.6

==

+ MARKS THE CITY OF HAO

ECLIPSE ON JD 1393174 DATE (Y/M/D) -898/ 4/21
DELTA T = 6 HOURS 59.1 MINS

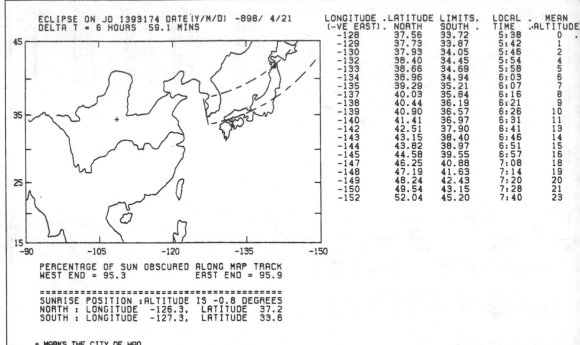

LONGITUDE (-VE EAST).	LATITUDE NORTH	LIMITS. SOUTH .	LOCAL . TIME	MEAN .ALTITUDE
-128	37.56	33.72	5:38	0
-129	37.73	33.87	5:42	1
-130	37.93	34.05	5:46	2
-132	38.40	34.45	5:54	4
-133	38.66	34.69	5:58	5
-134	38.96	34.94	6:03	6
-135	39.29	35.21	6:07	7
-137	40.03	35.84	6:16	8
-138	40.44	36.19	6:21	9
-139	40.90	36.57	6:26	10
-140	41.41	36.97	6:31	11
-142	42.51	37.90	6:41	13
-143	43.15	38.40	6:46	14
-144	43.82	38.97	6:51	15
-145	44.58	39.55	6:57	16
-147	46.25	40.88	7:08	18
-148	47.19	41.63	7:14	19
-149	48.24	42.43	7:20	20
-150	49.54	43.15	7:28	21
-152	52.04	45.20	7:40	23

PERCENTAGE OF SUN OBSCURED ALONG MAP TRACK
WEST END = 95.3   EAST END = 95.9

==
SUNRISE POSITION :ALTITUDE IS -0.8 DEGREES
NORTH : LONGITUDE -126.3, LATITUDE 37.2
SOUTH : LONGITUDE -127.3, LATITUDE 33.6

+ MARKS THE CITY OF HAO

ECLIPSE ON JD 1394030 DATE (Y/M/D) -896/ 8/24
DELTA T = 6 HOURS 58.3 MINS

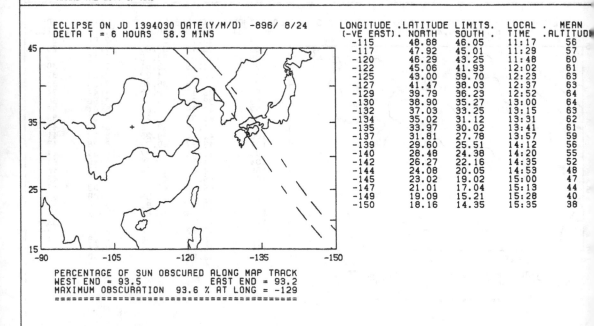

LONGITUDE (-VE EAST).	LATITUDE NORTH	LIMITS. SOUTH .	LOCAL . TIME	MEAN .ALTITUDE
-115	48.88	46.05	11:17	56
-117	47.92	45.01	11:29	57
-120	46.29	43.25	11:48	60
-122	45.06	41.93	12:02	61
-125	43.00	39.70	12:23	63
-127	41.47	38.03	12:37	63
-129	39.79	36.23	12:52	64
-130	38.90	35.27	13:00	64
-132	37.03	33.25	13:15	63
-134	35.02	31.12	13:31	62
-135	33.97	30.02	13:41	61
-137	31.81	27.78	13:57	59
-139	29.60	25.51	14:12	56
-140	28.48	24.38	14:20	55
-142	26.27	22.16	14:35	52
-144	24.08	20.05	14:53	48
-145	23.02	19.02	15:00	47
-147	21.01	17.04	15:13	44
-149	19.09	15.21	15:28	40
-150	18.16	14.35	15:35	38

PERCENTAGE OF SUN OBSCURED ALONG MAP TRACK
WEST END = 93.5   EAST END = 93.2
MAXIMUM OBSCURATION 93.6 % AT LONG = -129
==

+ MARKS THE CITY OF HAO

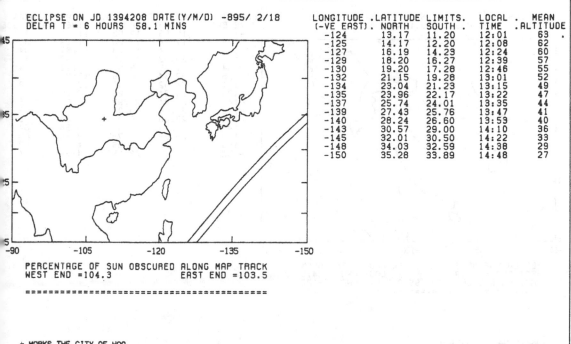

ECLIPSE ON JD 1394208 DATE(Y/M/D) -895/ 2/18
DELTA T = 6 HOURS 58.1 MINS

LONGITUDE (-VE EAST)	LATITUDE LIMITS. NORTH	SOUTH	LOCAL TIME	MEAN ALTITUDE
-124	13.17	11.20	12:01	63
-125	14.17	12.20	12:08	62
-127	16.19	14.23	12:24	60
-129	18.20	16.27	12:39	57
-130	19.20	17.28	12:46	55
-132	21.15	19.28	13:01	52
-134	23.04	21.23	13:15	49
-135	23.96	22.17	13:22	47
-137	25.74	24.01	13:35	44
-139	27.43	25.76	13:47	41
-140	28.24	26.60	13:59	40
-143	30.57	29.00	14:10	36
-145	32.01	30.50	14:22	33
-148	34.03	32.59	14:38	29
-150	35.28	33.89	14:48	27

PERCENTAGE OF SUN OBSCURED ALONG MAP TRACK
WEST END =104.3   EAST END =103.5

===

+ MARKS THE CITY OF HAO

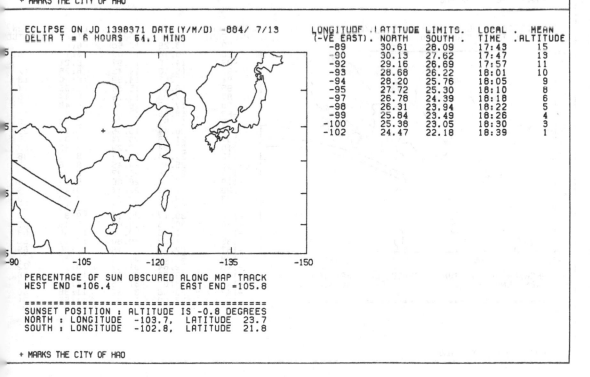

ECLIPSE ON JD 1398371 DATE(Y/M/D) -884/ 7/13
DELTA T = 6 HOURS 54.1 MINS

LONGITUDE (-VE EAST)	LATITUDE LIMITS. NORTH	SOUTH	LOCAL TIME	MEAN ALTITUDE
-89	30.61	28.09	17:43	15
-90	30.13	27.62	17:47	13
-92	29.16	26.69	17:57	11
-93	28.68	26.22	18:01	10
-94	28.20	25.76	18:05	9
-95	27.72	25.30	18:10	8
-97	26.78	24.39	18:18	6
-98	26.31	23.94	18:22	5
-99	25.84	23.49	18:26	4
-100	25.38	23.05	18:30	3
-102	24.47	22.18	18:39	1

PERCENTAGE OF SUN OBSCURED ALONG MAP TRACK
WEST END =106.4   EAST END =105.8

===
SUNSET POSITION : ALTITUDE IS -0.8 DEGREES
NORTH : LONGITUDE -103.7, LATITUDE 23.7
SOUTH : LONGITUDE -102.8, LATITUDE 21.8

+ MARKS THE CITY OF HAO

ECLIPSE ON JD 1399405 DATE(Y/M/D) -881/ 5/13
DELTA T = 6 HOURS 53.1 MINS

LONGITUDE (-VE EAST)	.LATITUDE . NORTH	LIMITS. SOUTH .	LOCAL . TIME	MEAN .ALTITUDE
-127	15.05	12.52	11:11	78
-129	17.14	14.69	11:29	83
-130	18.14	15.73	11:37	84
-132	20.02	17.70	11:56	87
-134	21.75	19.51	12:12	84
-135	22.54	20.36	12:21	82
-137	24.02	21.91	12:37	78
-139	25.34	23.29	12:53	75
-140	25.94	23.92	13:01	73
-142	27.04	25.08	13:16	69
-144	28.00	26.09	13:30	66
-145	28.44	26.54	13:38	64
-148	29.59	27.73	13:58	60
-150	30.23	28.39	14:12	57
-152	30.78	28.95	14:24	54

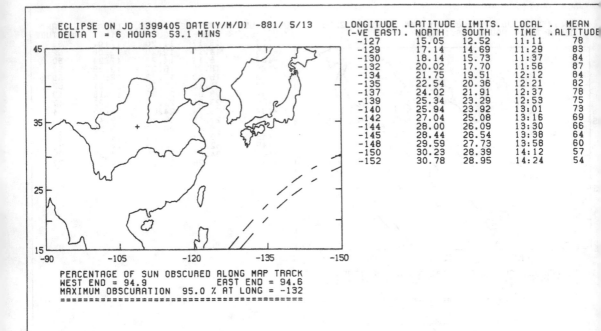

PERCENTAGE OF SUN OBSCURED ALONG MAP TRACK
WEST END = 94.9   EAST END = 94.6
MAXIMUM OBSCURATION 95.0 % AT LONG = -132
===

+ MARKS THE CITY OF HAO

ECLIPSE ON JD 1402004 DATE(Y/M/D) -874/ 6/24
DELTA T = 6 HOURS 50.6 MINS

LONGITUDE (-VE EAST)	.LATITUDE . NORTH	LIMITS. SOUTH .	LOCAL . TIME	MEAN .ALTITUD
-89	13.29	12.60	6:08	7
-90	13.76	13.05	6:12	8
-92	14.70	13.97	6:21	11
-93	15.18	14.43	6:26	12
-94	15.65	14.89	6:30	13
-95	16.13	15.36	6:35	14
-97	17.09	16.29	6:44	16
-98	17.57	16.76	6:48	18
-99	18.05	17.22	6:53	19
-100	18.53	17.69	6:58	20
-104	20.44	19.56	7:18	25
-107	21.86	20.94	7:33	29
-110	23.24	22.29	7:49	33
-112	24.14	23.17	8:00	36
-115	25.45	24.45	8:16	39
-117	26.28	25.27	8:28	42
-119	27.08	26.06	8:39	45
-120	27.47	26.44	8:45	46
-123	28.57	27.53	9:03	50
-125	29.25	28.20	9:15	53
-128	30.17	29.12	9:33	57
-130	30.73	29.67	9:46	60
-133	31.45	30.39	10:05	63
-135	31.87	30.80	10:17	66
-137	32.22	31.16	10:30	69
-139	32.52	31.45	10:43	71
-140	32.64	31.57	10:50	72
-143	32.92	31.85	11:09	76
-145	33.02	31.95	11:22	78
-147	33.06	31.99	11:36	80
-150	33.00	31.92	11:55	81

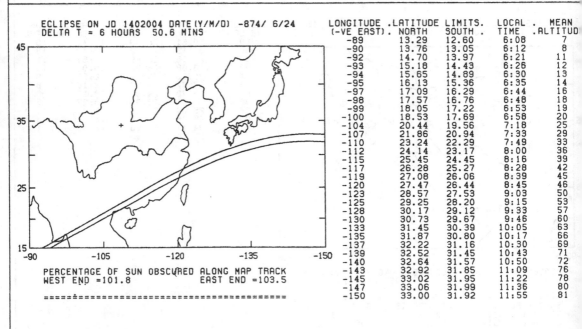

PERCENTAGE OF SUN OBSCURED ALONG MAP TRACK
WEST END =101.8   EAST END =103.5

===

+ MARKS THE CITY OF HAO

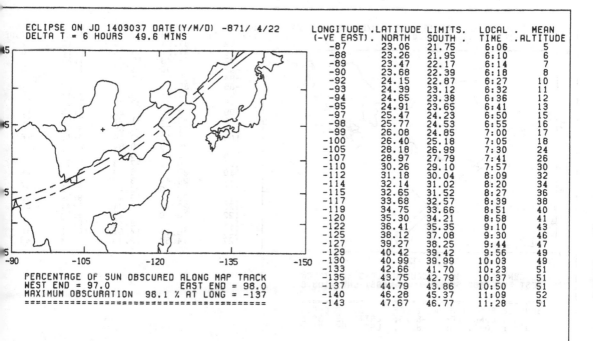

ECLIPSE ON JD 1403037 DATE(Y/M/D) -871/ 4/22
DELTA T = 6 HOURS 49.6 MINS

PERCENTAGE OF SUN OBSCURED ALONG MAP TRACK
WEST END = 97.0   EAST END = 98.0
MAXIMUM OBSCURATION 98.1 % AT LONG = -137
===

+ MARKS THE CITY OF HAO

LONGITUDE (-VE EAST)	LATITUDE NORTH	LIMITS SOUTH	LOCAL TIME	MEAN ALTITUDE
-87	23.06	21.75	6:06	5
-88	23.26	21.95	6:10	6
-89	23.47	22.17	6:14	7
-90	23.68	22.39	6:18	8
-92	24.15	22.87	6:27	10
-93	24.39	23.12	6:32	11
-94	24.65	23.38	6:36	12
-95	24.91	23.65	6:41	13
-97	25.47	24.23	6:50	15
-98	25.77	24.53	6:55	16
-99	26.08	24.85	7:00	17
-100	26.40	25.18	7:05	18
-105	28.18	26.99	7:30	24
-107	28.97	27.79	7:41	26
-110	30.26	29.10	7:57	30
-112	31.18	30.04	8:09	32
-114	32.14	31.02	8:20	34
-115	32.65	31.52	8:27	36
-117	33.68	32.57	8:39	38
-119	34.75	33.66	8:51	40
-120	35.30	34.21	8:58	41
-122	36.41	35.35	9:10	43
-125	38.12	37.08	9:30	46
-127	39.27	38.25	9:44	47
-129	40.42	39.42	9:56	49
-130	40.99	39.99	10:03	49
-133	42.66	41.70	10:23	51
-135	43.75	42.79	10:37	51
-137	44.79	43.86	10:50	51
-140	46.28	45.37	11:09	52
-143	47.67	46.77	11:28	51

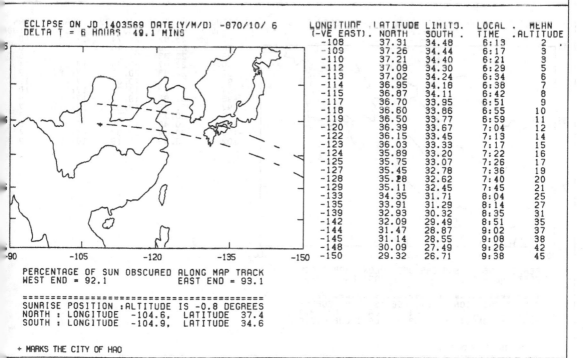

ECLIPSE ON JD 1403569 DATE(Y/M/D) -870/10/ 6
DELTA T = 6 HOURS 49.1 MINS

PERCENTAGE OF SUN OBSCURED ALONG MAP TRACK
WEST END = 92.1   EAST END = 93.1

===
SUNRISE POSITION :ALTITUDE IS -0.8 DEGREES
NORTH : LONGITUDE -104.6, LATITUDE 37.4
SOUTH : LONGITUDE -104.9, LATITUDE 34.6

+ MARKS THE CITY OF HAO

LONGITUDE (-VE EAST)	LATITUDE NORTH	LIMITS SOUTH	LOCAL TIME	MEAN ALTITUDE
-108	37.31	34.48	6:13	2
-109	37.26	34.44	6:17	3
-110	37.21	34.40	6:21	3
-112	37.09	34.30	6:29	5
-113	37.02	34.24	6:34	6
-114	36.95	34.18	6:38	7
-115	36.87	34.11	6:42	8
-117	36.70	33.95	6:51	9
-118	36.60	33.86	6:55	10
-119	36.50	33.77	6:59	11
-120	36.39	33.67	7:04	12
-122	36.15	33.45	7:13	14
-123	36.03	33.33	7:17	15
-124	35.89	33.20	7:22	16
-125	35.75	33.07	7:26	17
-127	35.45	32.78	7:36	19
-128	35.28	32.62	7:40	20
-129	35.11	32.45	7:45	21
-133	34.35	31.71	8:04	25
-135	33.91	31.29	8:14	27
-139	32.93	30.32	8:35	31
-142	32.09	29.49	8:51	35
-144	31.47	28.87	9:02	37
-145	31.14	28.55	9:08	38
-148	30.09	27.49	9:26	42
-150	29.32	26.71	9:38	45

ECLIPSE ON JD 1405813 DATE(Y/M/D) -864/11/27
DELTA T = 6 HOURS 47.0 MINS

LONGITUDE (-VE EAST)	.LATITUDE LIMITS.		LOCAL TIME	.MEAN ALTITUDE
	NORTH	SOUTH		
-89	37.64	36.76	7:12	2
-90	37.24	36.35	7:16	3
-92	36.43	35.51	7:25	4
-93	36.00	35.07	7:29	5
-94	35.57	34.62	7:33	6
-95	35.13	34.17	7:37	7
-97	34.21	33.22	7:46	9
-98	33.73	32.73	7:50	10
-99	33.25	32.23	7:55	11
-100	32.75	31.71	7:59	12
-102	31.72	30.65	8:08	15
-103	31.19	30.10	8:13	16
-104	30.65	29.54	8:17	17
-105	30.09	28.97	8:22	18
-107	28.94	27.78	8:31	20
-110	27.11	25.90	8:46	24
-114	24.49	23.19	9:07	29
-115	23.80	22.48	9:13	31
-119	20.89	19.49	9:34	36
-120	20.14	18.71	9:41	38
-122	18.58	17.11	9:53	41
-124	16.97	15.46	10:05	44
-125	16.16	14.62	10:12	46
-127	14.50	12.93	10:24	49

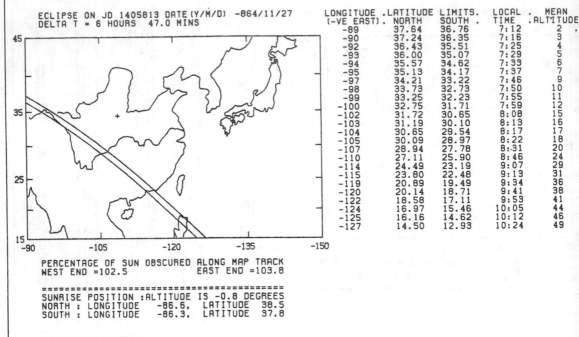

PERCENTAGE OF SUN OBSCURED ALONG MAP TRACK
WEST END =102.5   EAST END =103.8

==
SUNRISE POSITION :ALTITUDE IS -0.8 DEGREES
NORTH : LONGITUDE -86.6, LATITUDE 38.5
SOUTH : LONGITUDE -86.3, LATITUDE 37.8

+ MARKS THE CITY OF HAO

ECLIPSE ON JD 1405990 DATE(Y/M/D) -863/ 5/23
DELTA T = 6 HOURS 46.8 MINS

LONGITUDE (-VE EAST)	.LATITUDE LIMITS.		LOCAL TIME	.MEAN ALTITUDE
	NORTH	SOUTH		
-87	23.82	21.74	17:13	18
-88	23.58	21.48	17:17	16
-89	23.33	21.21	17:22	15
-90	23.07	20.94	17:27	14
-92	22.53	20.37	17:36	12
-93	22.25	20.08	17:40	11
-94	21.97	19.78	17:44	10
-95	21.68	19.48	17:49	9
-97	21.08	18.85	17:57	7
-98	20.77	18.53	18:02	6
-99	20.46	18.20	18:06	5
-100	20.14	17.87	18:10	4
-102	19.49	17.19	18:18	2

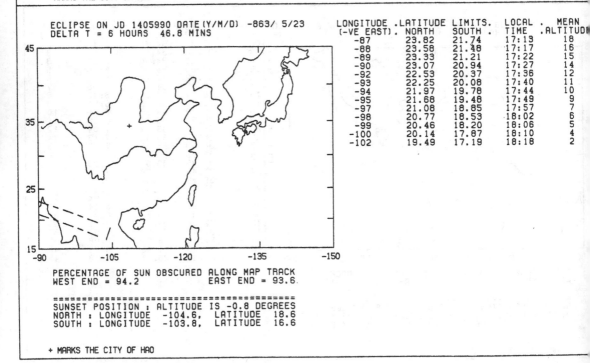

PERCENTAGE OF SUN OBSCURED ALONG MAP TRACK
WEST END = 94.2   EAST END = 93.6

==
SUNSET POSITION : ALTITUDE IS -0.8 DEGREES
NORTH : LONGITUDE -104.6, LATITUDE 18.6
SOUTH : LONGITUDE -103.8, LATITUDE 16.6

+ MARKS THE CITY OF HAO

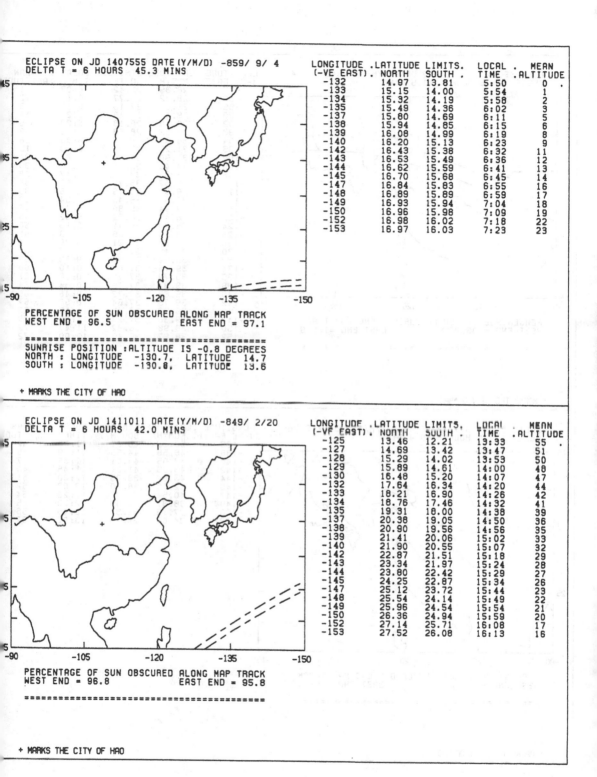

ECLIPSE ON JD 1407555 DATE (Y/M/D) -859/ 9/ 4
DELTA T = 6 HOURS 45.3 MINS

LONGITUDE (-VE EAST)	LATITUDE LIMITS. NORTH	SOUTH	LOCAL TIME	MEAN ALTITUDE
-132	14.97	13.81	5:50	0
-133	15.15	14.00	5:54	1
-134	15.32	14.19	5:58	2
-135	15.49	14.36	6:02	3
-137	15.80	14.69	6:11	5
-138	15.94	14.85	6:15	6
-139	16.08	14.99	6:19	8
-140	16.20	15.13	6:23	9
-142	16.43	15.38	6:32	11
-143	16.53	15.49	6:36	12
-144	16.62	15.59	6:41	13
-145	16.70	15.68	6:45	14
-147	16.84	15.83	6:55	16
-148	16.89	15.89	6:59	17
-149	16.93	15.94	7:04	18
-150	16.96	15.98	7:09	19
-152	16.98	16.02	7:18	22
-153	16.97	16.03	7:23	23

PERCENTAGE OF SUN OBSCURED ALONG MAP TRACK
WEST END = 96.5   EAST END = 97.1

===
SUNRISE POSITION :ALTITUDE IS -0.8 DEGREES
NORTH : LONGITUDE -130.7, LATITUDE 14.7
SOUTH : LONGITUDE -130.8, LATITUDE 13.6

+ MARKS THE CITY OF HAO

ECLIPSE ON JD 1411011 DATE (Y/M/D) -849/ 2/20
DELTA T = 6 HOURS 42.0 MINS

LONGITUDE (-VE EAST)	LATITUDE LIMITS. NORTH	SOUTH	LOCAL TIME	MEAN ALTITUDE
-125	13.46	12.21	13:33	55
-127	14.69	13.42	13:47	51
-128	15.29	14.02	13:53	50
-129	15.89	14.61	14:00	48
-130	16.48	15.20	14:07	47
-132	17.64	16.34	14:20	44
-133	18.21	16.90	14:26	42
-134	18.76	17.46	14:32	41
-135	19.31	18.00	14:38	39
-137	20.38	19.05	14:50	36
-138	20.90	19.56	14:56	35
-139	21.41	20.06	15:02	33
-140	21.90	20.55	15:07	32
-142	22.87	21.51	15:18	29
-143	23.34	21.97	15:24	28
-144	23.80	22.42	15:29	27
-145	24.25	22.87	15:34	26
-147	25.12	23.72	15:44	23
-148	25.54	24.14	15:49	22
-149	25.96	24.54	15:54	21
-150	26.36	24.94	15:59	20
-152	27.14	25.71	16:08	17
-153	27.52	26.08	16:13	16

PERCENTAGE OF SUN OBSCURED ALONG MAP TRACK
WEST END = 96.8   EAST END = 95.8

===

+ MARKS THE CITY OF HAO

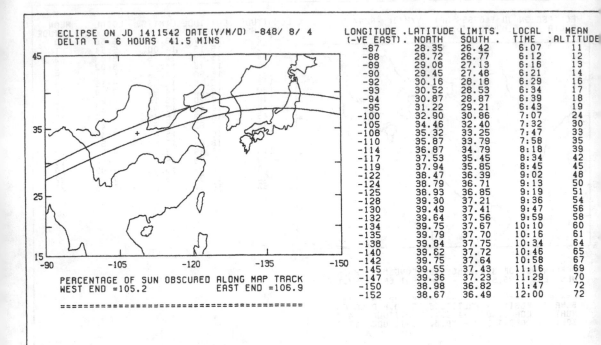

ECLIPSE ON JD 1411542 DATE(Y/M/D) -848/ 8/ 4
DELTA T = 6 HOURS 41.5 MINS

LONGITUDE (-VE EAST)	LATITUDE NORTH	LIMITS. SOUTH	LOCAL TIME	MEAN ALTITUDE
-87	28.35	26.42	6:07	11
-88	28.72	26.77	6:12	12
-89	29.08	27.13	6:16	13
-90	29.45	27.48	6:21	14
-92	30.16	28.18	6:29	16
-93	30.52	28.53	6:34	17
-94	30.87	28.87	6:39	18
-95	31.22	29.21	6:43	19
-100	32.90	30.86	7:07	24
-105	34.46	32.40	7:32	30
-108	35.32	33.25	7:47	33
-110	35.87	33.79	7:58	35
-114	36.87	34.79	8:18	39
-117	37.53	35.45	8:34	42
-119	37.94	35.85	8:45	45
-122	38.47	36.39	9:02	48
-124	38.79	36.71	9:13	50
-125	38.93	36.85	9:19	51
-128	39.30	37.21	9:36	54
-130	39.49	37.41	9:47	56
-132	39.64	37.56	9:59	58
-134	39.75	37.67	10:10	60
-135	39.79	37.70	10:16	61
-138	39.84	37.75	10:34	64
-140	39.82	37.72	10:46	65
-142	39.75	37.64	10:58	67
-145	39.55	37.43	11:16	69
-147	39.36	37.23	11:29	70
-150	38.98	36.82	11:47	72
-152	38.67	36.49	12:00	72

PERCENTAGE OF SUN OBSCURED ALONG MAP TRACK
WEST END =105.2   EAST END =106.9

+ MARKS THE CITY OF HAO

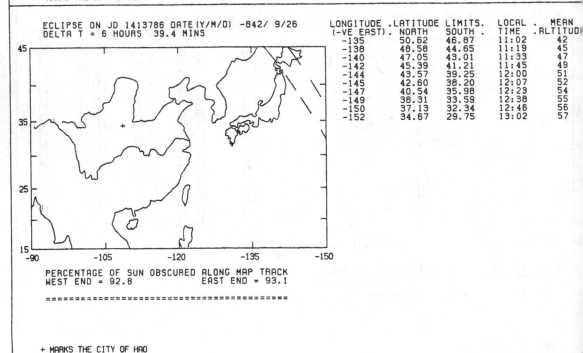

ECLIPSE ON JD 1413786 DATE(Y/M/D) -842/ 9/26
DELTA T = 6 HOURS 39.4 MINS

LONGITUDE (-VE EAST)	LATITUDE NORTH	LIMITS. SOUTH	LOCAL TIME	MEAN ALTITUDE
-135	50.62	46.87	11:02	42
-138	48.58	44.65	11:19	45
-140	47.05	43.01	11:33	47
-142	45.39	41.21	11:45	49
-144	43.57	39.25	12:00	51
-145	42.60	38.20	12:07	52
-147	40.54	35.98	12:23	54
-149	38.31	33.59	12:38	55
-150	37.13	32.34	12:46	56
-152	34.67	29.75	13:02	57

PERCENTAGE OF SUN OBSCURED ALONG MAP TRACK
WEST END = 92.8   EAST END = 93.1

+ MARKS THE CITY OF HAO

ECLIPSE ON JD 1413964 DATE(Y/M/D) -841/ 3/23
DELTA T = 6 HOURS 39.2 MINS

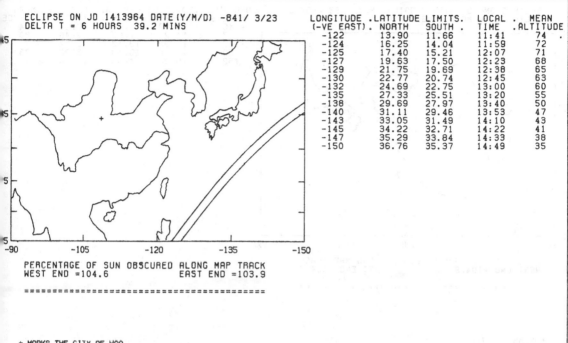

PERCENTAGE OF SUN OBSCURED ALONG MAP TRACK
WEST END =104.6   EAST END =103.9

===

LONGITUDE (-VE EAST)	LATITUDE LIMITS. NORTH	SOUTH .	LOCAL TIME	MEAN ALTITUDE
-122	13.90	11.66	11:41	74
-124	16.25	14.04	11:59	72
-125	17.40	15.21	12:07	71
-127	19.63	17.50	12:23	68
-129	21.75	19.69	12:38	65
-130	22.77	20.74	12:45	63
-132	24.69	22.75	13:00	60
-135	27.33	25.51	13:20	55
-138	29.69	27.97	13:40	50
-140	31.11	29.46	13:53	47
-143	33.05	31.49	14:10	43
-145	34.22	32.71	14:22	41
-147	35.29	33.84	14:33	38
-150	36.76	35.37	14:49	35

+ MARKS THE CITY OF HAO

ECLIPSE ON JD 1415351 DATE(Y/M/D) -837/ 1/ 8
DELTA T = 6 HOURS 37.9 MINS

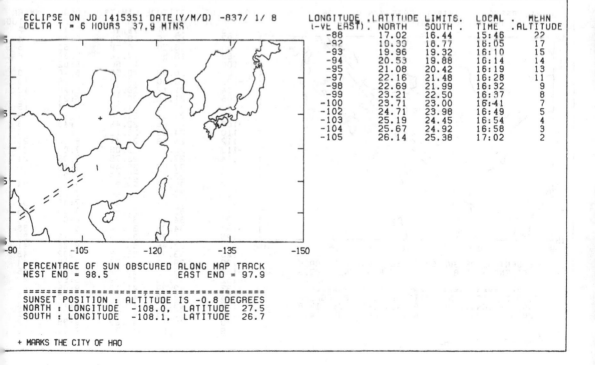

PERCENTAGE OF SUN OBSCURED ALONG MAP TRACK
WEST END = 98.5   EAST END = 97.9

===
SUNSET POSITION : ALTITUDE IS -0.8 DEGREES
NORTH : LONGITUDE -108.0, LATITUDE 27.5
SOUTH : LONGITUDE -108.1, LATITUDE 26.7

+ MARKS THE CITY OF HAO

LONGITUDE (-VE EAST)	LATITUDE LIMITS. NORTH	SOUTH .	LOCAL TIME	MEAN ALTITUDE
-88	17.02	16.44	15:46	22
-92	18.30	18.77	16:05	17
-93	19.96	19.32	16:10	15
-94	20.53	19.88	16:14	14
-95	21.08	20.42	16:19	13
-97	22.16	21.48	16:28	11
-98	22.69	21.99	16:32	9
-99	23.21	22.50	16:37	8
-100	23.71	23.00	16:41	7
-102	24.71	23.98	16:49	5
-103	25.19	24.45	16:54	4
-104	25.67	24.92	16:58	3
-105	26.14	25.38	17:02	2

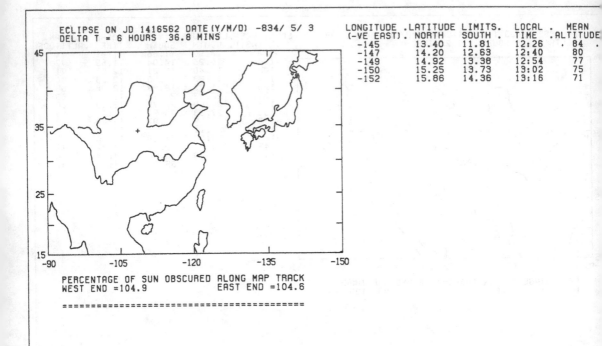

ECLIPSE ON JD 1416562 DATE(Y/M/D) -834/ 5/ 3
DELTA T = 6 HOURS .36.8 MINS

LONGITUDE	.LATITUDE	LIMITS.	LOCAL	.	MEAN
(-VE EAST).	NORTH	SOUTH .	TIME	.	ALTITUDE
-145	13.40	11.81	12:26	.	84 .
-147	14.20	12.63	12:40		80
-149	14.92	13.38	12:54		77
-150	15.25	13.73	13:02		75
-152	15.86	14.36	13:16		71

PERCENTAGE OF SUN OBSCURED ALONG MAP TRACK
WEST END =104.9   EAST END =104.6

==

+ MARKS THE CITY OF HAO

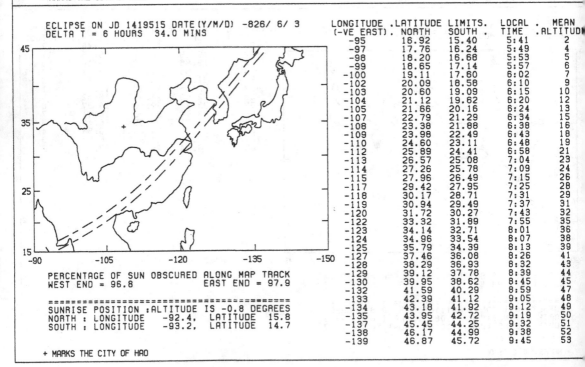

ECLIPSE ON JD 1419515 DATE(Y/M/D) -826/ 6/ 3
DELTA T = 6 HOURS 34.0 MINS

LONGITUDE	.LATITUDE	LIMITS.	LOCAL	.	MEAN
(-VE EAST).	NORTH	SOUTH .	TIME	.	ALTITUDE
-95	16.92	15.40	5:41		2
-97	17.76	16.24	5:49		4
-98	18.20	16.68	5:53		5
-99	18.65	17.14	5:57		6
-100	19.11	17.60	6:02		7
-102	20.09	18.58	6:10		9
-103	20.60	19.09	6:15		10
-104	21.12	19.62	6:20		12
-105	21.66	20.16	6:24		13
-107	22.79	21.29	6:34		15
-108	23.38	21.88	6:38		16
-109	23.98	22.49	6:43		18
-110	24.60	23.11	6:48		19
-112	25.89	24.41	6:58		21
-113	26.57	25.08	7:04		23
-114	27.26	25.78	7:09		24
-115	27.96	26.49	7:15		26
-117	29.42	27.95	7:25		28
-118	30.17	28.71	7:31		29
-119	30.94	29.49	7:37		31
-120	31.72	30.27	7:43		32
-122	33.32	31.89	7:55		35
-123	34.14	32.71	8:01		36
-124	34.96	33.54	8:07		38
-125	35.79	34.39	8:13		39
-127	37.46	36.08	8:26		41
-128	38.29	36.93	8:32		43
-129	39.12	37.78	8:39		44
-130	39.95	38.62	8:45		45
-132	41.59	40.29	8:59		47
-133	42.39	41.12	9:05		48
-134	43.18	41.92	9:12		49
-135	43.95	42.72	9:19		50
-137	45.45	44.25	9:32		51
-138	46.17	44.99	9:38		52
-139	46.87	45.72	9:45		53

PERCENTAGE OF SUN OBSCURED ALONG MAP TRACK
WEST END = 96.8   EAST END = 97.9

==
SUNRISE POSITION :ALTITUDE IS -0.8 DEGREES
NORTH : LONGITUDE -92.4, LATITUDE 15.8
SOUTH : LONGITUDE -93.2, LATITUDE 14.7

+ MARKS THE CITY OF HAO

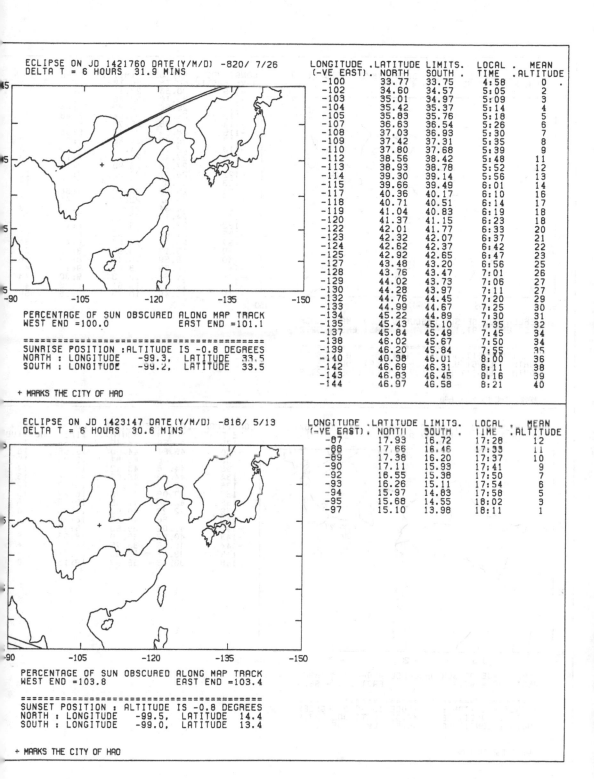

ECLIPSE ON JD 1421760 DATE(Y/M/D) -820/ 7/26
DELTA T = 6 HOURS 31.9 MINS

LONGITUDE (-VE EAST)	LATITUDE LIMITS. NORTH	SOUTH	LOCAL TIME	MEAN ALTITUDE
-100	33.77	33.75	4:58	0
-102	34.60	34.57	5:05	2
-103	35.01	34.97	5:09	3
-104	35.42	35.37	5:14	4
-105	35.83	35.76	5:18	5
-107	36.63	36.54	5:26	6
-108	37.03	36.93	5:30	7
-109	37.42	37.31	5:35	8
-110	37.80	37.68	5:39	9
-112	38.56	38.42	5:48	11
-113	38.93	38.78	5:52	12
-114	39.30	39.14	5:56	13
-115	39.66	39.49	6:01	14
-117	40.36	40.17	6:10	16
-118	40.71	40.51	6:14	17
-119	41.04	40.83	6:19	18
-120	41.37	41.15	6:23	18
-122	42.01	41.77	6:33	20
-123	42.32	42.07	6:37	21
-124	42.62	42.37	6:42	22
-125	42.92	42.65	6:47	23
-127	43.48	43.20	6:56	25
-128	43.76	43.47	7:01	26
-129	44.02	43.73	7:06	27
-130	44.28	43.97	7:11	27
-132	44.76	44.45	7:20	29
-133	44.99	44.67	7:25	30
-134	45.22	44.89	7:30	31
-135	45.43	45.10	7:35	32
-137	45.84	45.49	7:45	34
-138	46.02	45.67	7:50	34
-139	46.20	45.84	7:55	35
-140	46.38	46.01	8:00	36
-142	46.69	46.31	8:11	38
-143	46.83	46.45	8:16	39
-144	46.97	46.58	8:21	40

PERCENTAGE OF SUN OBSCURED ALONG MAP TRACK
WEST END =100.0   EAST END =101.1

==
SUNRISE POSITION :ALTITUDE IS -0.8 DEGREES
NORTH : LONGITUDE -99.3, LATITUDE 33.5
SOUTH : LONGITUDE -99.2, LATITUDE 33.5

+ MARKS THE CITY OF HAO

ECLIPSE ON JD 1423147 DATE(Y/M/D) -816/ 5/13
DELTA T = 6 HOURS 30.6 MINS

LONGITUDE (-VE EAST)	LATITUDE LIMITS. NORTH	SOUTH	LOCAL TIME	MEAN ALTITUDE
-87	17.93	16.72	17:28	12
-88	17.66	16.46	17:33	11
-89	17.38	16.20	17:37	10
-90	17.11	15.93	17:41	9
-92	16.55	15.38	17:50	7
-93	16.26	15.11	17:54	6
-94	15.97	14.83	17:58	5
-95	15.68	14.55	18:02	3
-97	15.10	13.98	18:11	1

PERCENTAGE OF SUN OBSCURED ALONG MAP TRACK
WEST END =103.8   EAST END =103.4

==
SUNSET POSITION : ALTITUDE IS -0.8 DEGREES
NORTH : LONGITUDE -99.5, LATITUDE 14.4
SOUTH : LONGITUDE -99.0, LATITUDE 13.4

+ MARKS THE CITY OF HAO

ECLIPSE ON JD 1423325 DATE(Y/M/D) -816/11/ 7
DELTA T = 6 HOURS 30.4 MINS

LONGITUDE (-VE EAST).	LATITUDE NORTH	LIMITS. SOUTH .	LOCAL TIME	MEAN .ALTITUDE
-109	30.36	27.12	6:32	0
-110	30.10	26.86	6:36	1
-112	29.56	26.32	6:44	3
-113	29.27	26.04	6:48	4
-114	28.98	25.76	6:52	5
-115	28.69	25.47	6:56	6
-117	28.07	24.86	7:05	8
-118	27.76	24.55	7:09	9
-119	27.43	24.23	7:14	10
-120	27.10	23.90	7:18	11
-122	26.41	23.22	7:27	13
-123	26.05	22.86	7:32	14
-124	25.69	22.50	7:36	15
-125	25.32	22.13	7:41	16
-127	24.54	21.37	7:50	19
-128	24.15	20.97	7:55	20
-132	22.46	19.29	8:15	25
-135	21.09	17.92	8:31	29
-137	20.12	16.96	8:42	31
-139	19.12	15.95	8:53	34
-140	18.59	15.44	8:59	36
-143	16.98	13.81	9:17	40
-145	15.84	12.68	9:30	43
-147	14.67	11.51	9:43	47
-149	13.46	10.29	9:56	50

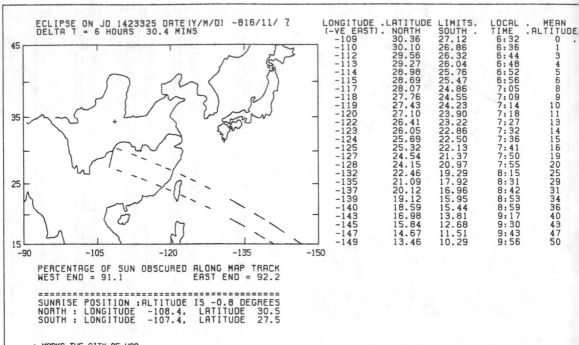

PERCENTAGE OF SUN OBSCURED ALONG MAP TRACK
WEST END = 91.1   EAST END = 92.2

==
SUNRISE POSITION :ALTITUDE IS -0.8 DEGREES
NORTH : LONGITUDE -108.4, LATITUDE 30.5
SOUTH : LONGITUDE -107.4, LATITUDE 27.5

+ MARKS THE CITY OF HAO

ECLIPSE ON JD 1426100 DATE(Y/M/D) -808/ 6/13
DELTA T = 6 HOURS 27.9 MINS

LONGITUDE (-VE EAST).	LATITUDE NORTH	LIMITS. SOUTH .	LOCAL TIME	MEAN .ALTITUDE
-125	47.59	46.45	17:33	21
-130	45.81	44.60	17:56	16
-132	45.06	43.82	18:05	15
-133	44.68	43.42	18:09	14
-134	44.29	43.02	18:14	13
-135	43.90	42.61	18:18	12
-137	43.09	41.78	18:27	10
-138	42.68	41.35	18:32	9
-139	42.27	40.93	18:36	8
-140	41.86	40.50	18:40	8
-142	41.02	39.63	18:48	6
-143	40.59	39.19	18:53	5
-144	40.16	38.75	18:57	4
-145	39.73	38.30	19:01	3
-148	38.43	36.97	19:13	0

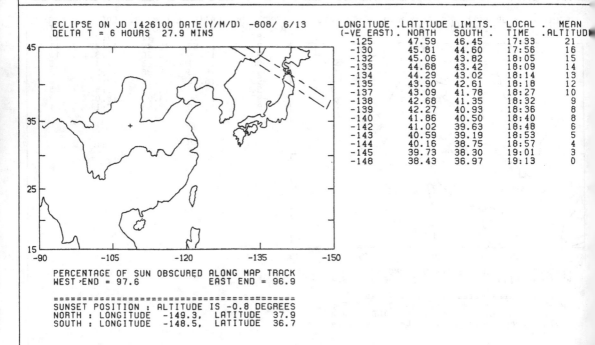

PERCENTAGE OF SUN OBSCURED ALONG MAP TRACK
WEST END = 97.6   EAST END = 96.9

==
SUNSET POSITION : ALTITUDE IS -0.8 DEGREES
NORTH : LONGITUDE -149.3, LATITUDE 37.9
SOUTH : LONGITUDE -148.5, LATITUDE 36.7

+ MARKS THE CITY OF HAO

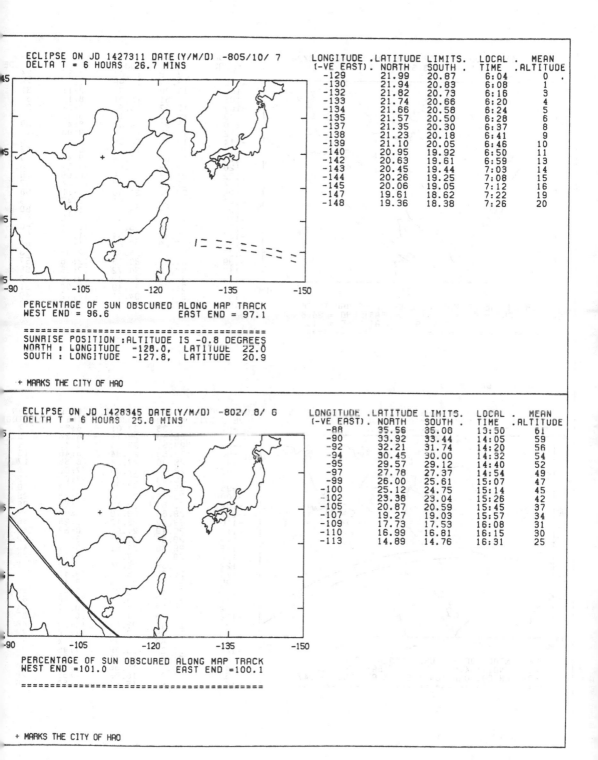

ECLIPSE ON JD 1427311 DATE(Y/M/D) -805/10/ 7
DELTA T = 6 HOURS 26.7 MINS

LONGITUDE	.LATITUDE	LIMITS.	LOCAL	. MEAN
(-VE EAST).	NORTH	SOUTH .	TIME	.ALTITUDE
-129	21.99	20.87	6:04	0 .
-130	21.94	20.83	6:08	1
-132	21.82	20.73	6:16	3
-133	21.74	20.66	6:20	4
-134	21.66	20.58	6:24	5
-135	21.57	20.50	6:28	6
-137	21.35	20.30	6:37	8
-138	21.23	20.18	6:41	9
-139	21.10	20.05	6:46	10
-140	20.95	19.92	6:50	11
-142	20.63	19.61	6:59	13
-143	20.45	19.44	7:03	14
-144	20.26	19.25	7:08	15
-145	20.06	19.05	7:12	16
-147	19.61	18.62	7:22	19
-148	19.36	18.38	7:26	20

PERCENTAGE OF SUN OBSCURED ALONG MAP TRACK
WEST END = 96.6   EAST END = 97.1

===
SUNRISE POSITION :ALTITUDE IS -0.8 DEGREES
NORTH : LONGITUDE -128.0, LATITUDE 22.0
SOUTH : LONGITUDE -127.8, LATITUDE 20.9

+ MARKS THE CITY OF HAO

ECLIPSE ON JD 1428345 DATE(Y/M/D) -802/ 8/ 6
DELTA T = 6 HOURS 25.8 MINS

LONGITUDE	.LATITUDE	LIMITS.	LOCAL	. MEAN
(-VE EAST).	NORTH	SOUTH .	TIME	.ALTITUDE
-88	35.56	35.00	13:50	61
-90	33.92	33.44	14:05	59
-92	32.21	31.74	14:20	56
-94	30.45	30.00	14:32	54
-95	29.57	29.12	14:40	52
-97	27.78	27.37	14:54	49
-99	26.00	25.61	15:07	47
-100	25.12	24.75	15:14	45
-102	23.38	23.04	15:26	42
-105	20.87	20.59	15:45	37
-107	19.27	19.03	15:57	34
-109	17.73	17.53	16:08	31
-110	16.99	16.81	16:15	30
-113	14.89	14.76	16:31	25

PERCENTAGE OF SUN OBSCURED ALONG MAP TRACK
WEST END =101.0   EAST END =100.1

===

+ MARKS THE CITY OF HAO

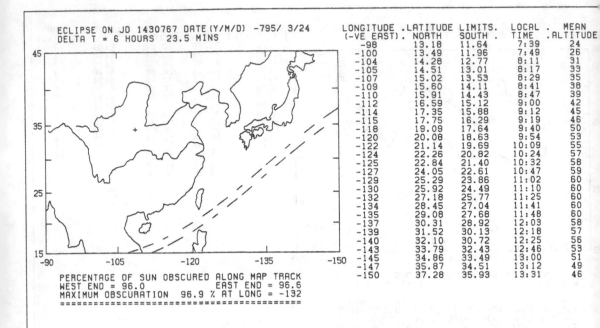

ECLIPSE ON JD 1430767 DATE(Y/M/D) -795/ 3/24
DELTA T = 6 HOURS 23.5 MINS

LONGITUDE (-VE EAST).	LATITUDE NORTH	LIMITS. SOUTH .	LOCAL TIME	MEAN .ALTITUDE
-98	13.18	11.64	7:39	24
-100	13.49	11.96	7:49	26
-104	14.28	12.77	8:11	31
-105	14.51	13.01	8:17	33
-107	15.02	13.53	8:29	35
-109	15.60	14.11	8:41	38
-110	15.91	14.43	8:47	39
-112	16.59	15.12	9:00	42
-114	17.35	15.88	9:12	45
-115	17.75	16.29	9:19	46
-118	19.09	17.64	9:40	50
-120	20.08	18.63	9:54	53
-122	21.14	19.69	10:09	55
-124	22.26	20.82	10:24	57
-125	22.84	21.40	10:32	58
-127	24.05	22.61	10:47	59
-129	25.29	23.86	11:02	60
-130	25.92	24.49	11:10	60
-132	27.18	25.77	11:25	60
-134	28.45	27.04	11:41	60
-135	29.08	27.68	11:48	60
-137	30.31	28.92	12:03	58
-139	31.52	30.13	12:18	57
-140	32.10	30.72	12:25	56
-143	33.79	32.43	12:46	53
-145	34.86	33.49	13:00	51
-147	35.87	34.51	13:12	49
-150	37.28	35.93	13:31	46

PERCENTAGE OF SUN OBSCURED ALONG MAP TRACK
WEST END = 96.0   EAST END = 96.6
MAXIMUM OBSCURATION 96.9 % AT LONG = -132
==

+ MARKS THE CITY OF HAO

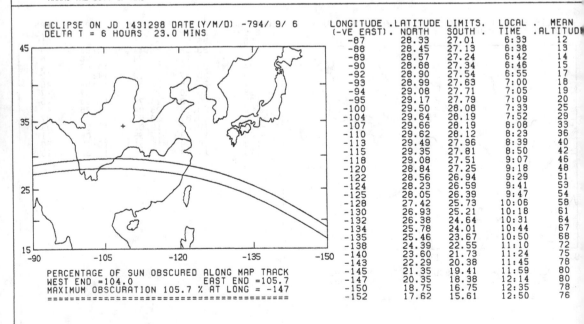

ECLIPSE ON JD 1431298 DATE(Y/M/D) -794/ 9/ 6
DELTA T = 6 HOURS 23.0 MINS

LONGITUDE (-VE EAST).	LATITUDE NORTH	LIMITS. SOUTH .	LOCAL TIME	MEAN .ALTITUDE
-87	28.33	27.01	6:33	12
-88	28.45	27.13	6:38	13
-89	28.57	27.24	6:42	14
-90	28.68	27.34	6:46	15
-92	28.90	27.54	6:55	17
-93	28.99	27.63	7:00	18
-94	29.08	27.71	7:05	19
-95	29.17	27.79	7:09	20
-100	29.50	28.08	7:33	25
-104	29.64	28.19	7:52	29
-107	29.66	28.19	8:08	33
-110	29.62	28.12	8:23	36
-113	29.49	27.96	8:39	40
-115	29.35	27.81	8:50	42
-118	29.08	27.51	9:07	46
-120	28.84	27.25	9:18	48
-122	28.56	26.94	9:29	51
-124	28.23	26.59	9:41	53
-125	28.05	26.39	9:47	54
-128	27.42	25.73	10:06	58
-130	26.93	25.21	10:18	61
-132	26.38	24.64	10:31	64
-134	25.78	24.01	10:44	67
-135	25.46	23.67	10:50	68
-138	24.39	22.55	11:10	72
-140	23.60	21.73	11:24	75
-143	22.29	20.38	11:45	78
-145	21.35	19.41	11:59	80
-147	20.35	18.38	12:14	80
-150	18.75	16.75	12:35	78
-152	17.62	15.61	12:50	76

PERCENTAGE OF SUN OBSCURED ALONG MAP TRACK
WEST END =104.0   EAST END =105.7
MAXIMUM OBSCURATION 105.7 % AT LONG = -147
==

+ MARKS THE CITY OF HAO

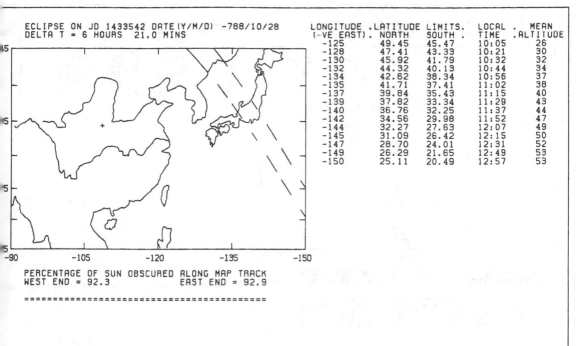

ECLIPSE ON JD 1433542 DATE(Y/M/D) -788/10/28
DELTA T = 6 HOURS 21.0 MINS

LONGITUDE (-VE EAST).	LATITUDE LIMITS. NORTH	SOUTH .	LOCAL TIME	MEAN .ALTITUDE
-125	49.45	45.47	10:05	26
-128	47.41	43.33	10:21	30
-130	45.92	41.79	10:32	32
-132	44.32	40.13	10:44	34
-134	42.62	38.34	10:56	37
-135	41.71	37.41	11:02	38
-137	39.84	35.43	11:15	40
-139	37.82	33.34	11:29	43
-140	36.76	32.25	11:37	44
-142	34.56	29.98	11:52	47
-144	32.27	27.63	12:07	49
-145	31.09	26.42	12:15	50
-147	28.70	24.01	12:31	52
-149	26.29	21.65	12:49	53
-150	25.11	20.49	12:57	53

PERCENTAGE OF SUN OBSCURED ALONG MAP TRACK
WEST END = 92.3   EAST END = 92.9

+ MARKS THE CITY OF HAO

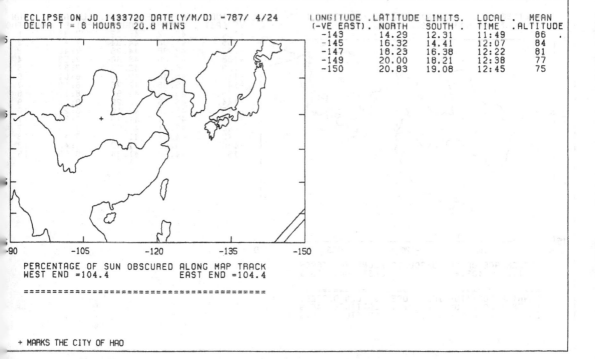

ECLIPSE ON JD 1433720 DATE(Y/M/D) -787/ 4/24
DELTA T = 8 HOURS 20.8 MINS

LONGITUDE (-VE EAST).	LATITUDE LIMITS. NORTH	SOUTH .	LOCAL TIME	MEAN .ALTITUDE
-143	14.29	12.31	11:49	86 .
-145	16.32	14.41	12:07	84
-147	18.23	16.38	12:22	81
-149	20.00	18.21	12:38	77
-150	20.83	19.08	12:45	75

PERCENTAGE OF SUN OBSCURED ALONG MAP TRACK
WEST END =104.4   EAST END =104.4

+ MARKS THE CITY OF HAO

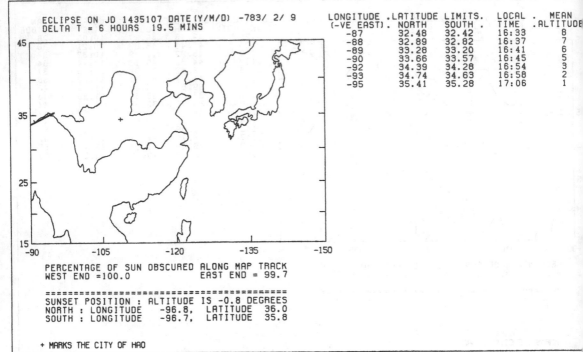

ECLIPSE ON JD 1435107 DATE(Y/M/D) -783/ 2/ 9
DELTA T = 6 HOURS 19.5 MINS

LONGITUDE (-VE EAST)	.LATITUDE. NORTH	LIMITS. SOUTH .	LOCAL TIME	. MEAN .ALTITUDE
-87	32.48	32.42	16:33	8
-88	32.89	32.82	16:37	7
-89	33.28	33.20	16:41	6
-90	33.66	33.57	16:45	5
-92	34.39	34.28	16:54	3
-93	34.74	34.63	16:58	2
-95	35.41	35.28	17:06	1

PERCENTAGE OF SUN OBSCURED ALONG MAP TRACK
WEST END =100.0   EAST END = 99.7

===
SUNSET POSITION : ALTITUDE IS -0.8 DEGREES
NORTH : LONGITUDE   -96.8, LATITUDE 36.0
SOUTH : LONGITUDE   -96.7, LATITUDE 35.8

+ MARKS THE CITY OF HAO

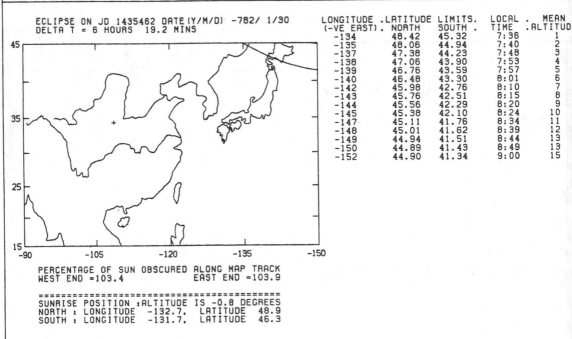

ECLIPSE ON JD 1435462 DATE(Y/M/D) -782/ 1/30
DELTA T = 6 HOURS 19.2 MINS

LONGITUDE (-VE EAST)	.LATITUDE. NORTH	LIMITS. SOUTH .	LOCAL TIME	. MEAN .ALTITUD
-134	48.42	45.32	7:36	1
-135	48.06	44.94	7:40	2
-137	47.38	44.23	7:48	3
-138	47.06	43.90	7:53	4
-139	46.76	43.59	7:57	5
-140	46.48	43.30	8:01	6
-142	45.98	42.76	8:10	7
-143	45.76	42.51	8:15	8
-144	45.56	42.29	8:20	9
-145	45.38	42.10	8:24	10
-147	45.11	41.76	8:34	11
-148	45.01	41.62	8:39	12
-149	44.94	41.51	8:44	13
-150	44.89	41.43	8:49	13
-152	44.90	41.34	9:00	15

PERCENTAGE OF SUN OBSCURED ALONG MAP TRACK
WEST END =103.4   EAST END =103.9

===
SUNRISE POSITION :ALTITUDE IS -0.8 DEGREES
NORTH : LONGITUDE -132.7, LATITUDE 48.9
SOUTH : LONGITUDE -131.7, LATITUDE 46.3

+ MARKS THE CITY OF HAO

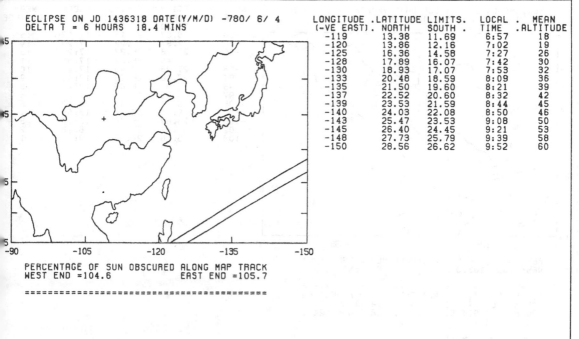

ECLIPSE ON JD 1436318 DATE(Y/M/D) -780/ 6/ 4
DELTA T = 6 HOURS 18.4 MINS

LONGITUDE (-VE EAST)	LATITUDE LIMITS. NORTH	SOUTH	LOCAL TIME	MEAN ALTITUDE
-119	13.38	11.69	6:57	18
-120	13.86	12.16	7:02	19
-125	16.36	14.58	7:27	26
-128	17.89	16.07	7:42	30
-130	18.93	17.07	7:53	32
-133	20.48	18.59	8:09	36
-135	21.50	19.60	8:21	39
-137	22.52	20.60	8:32	42
-139	23.53	21.59	8:44	45
-140	24.03	22.08	8:50	46
-143	25.47	23.53	9:08	50
-145	26.40	24.45	9:21	53
-148	27.73	25.79	9:39	58
-150	28.56	26.62	9:52	60

PERCENTAGE OF SUN OBSCURED ALONG MAP TRACK
WEST END =104.6   EAST END =105.7

===

+ MARKS THE CITY OF HAO

ECLIPSE ON JD 1438740 DATE(Y/M/D) -773/ 1/21
DELTA T = 6 HOURS 16.2 MINS

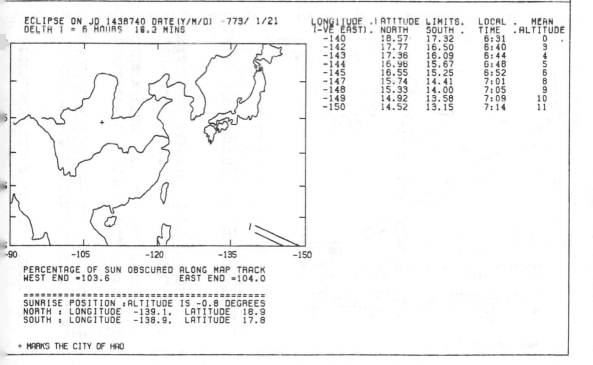

LONGITUDE (-VE EAST)	LATITUDE LIMITS. NORTH	SOUTH	LOCAL TIME	MEAN ALTITUDE
-140	18.57	17.32	6:31	0
-142	17.77	16.50	6:40	3
-143	17.36	16.09	6:44	4
-144	16.96	15.67	6:48	5
-145	16.55	15.25	6:52	6
-147	15.74	14.41	7:01	8
-148	15.33	14.00	7:05	9
-149	14.92	13.58	7:09	10
-150	14.52	13.15	7:14	11

PERCENTAGE OF SUN OBSCURED ALONG MAP TRACK
WEST END =103.6   EAST END =104.0

===
SUNRISE POSITION :ALTITUDE IS -0.8 DEGREES
NORTH : LONGITUDE -139.1,   LATITUDE 18.9
SOUTH : LONGITUDE -138.9,   LATITUDE 17.8

+ MARKS THE CITY OF HAO

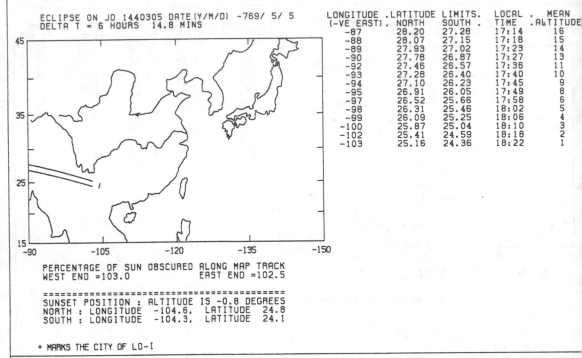

ECLIPSE ON JD 1440305 DATE(Y/M/D) -769/ 5/ 5
DELTA T = 6 HOURS 14.8 MINS

LONGITUDE (-VE EAST)	LATITUDE LIMITS. NORTH	SOUTH	LOCAL TIME	MEAN ALTITUDE
-87	28.20	27.28	17:14	16
-88	28.07	27.15	17:18	15
-89	27.93	27.02	17:23	14
-90	27.78	26.87	17:27	13
-92	27.46	26.57	17:36	11
-93	27.28	26.40	17:40	10
-94	27.10	26.23	17:45	9
-95	26.91	26.05	17:49	8
-97	26.52	25.66	17:58	6
-98	26.31	25.46	18:02	5
-99	26.09	25.25	18:06	4
-100	25.87	25.04	18:10	3
-102	25.41	24.59	18:18	2
-103	25.16	24.36	18:22	1

PERCENTAGE OF SUN OBSCURED ALONG MAP TRACK
WEST END =103.0   EAST END =102.5

=======================================
SUNSET POSITION : ALTITUDE IS -0.8 DEGREES
NORTH : LONGITUDE -104.6, LATITUDE 24.8
SOUTH : LONGITUDE -104.3, LATITUDE 24.1

+ MARKS THE CITY OF LO-I

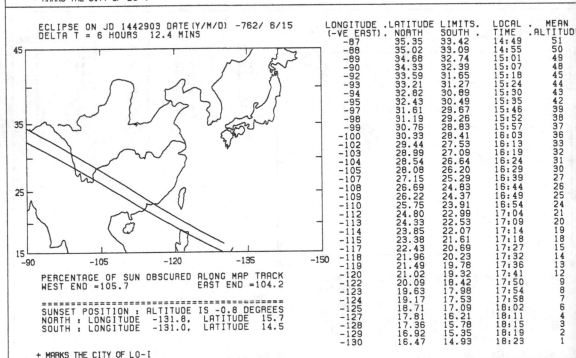

ECLIPSE ON JD 1442903 DATE(Y/M/D) -762/ 6/15
DELTA T = 6 HOURS 12.4 MINS

LONGITUDE (-VE EAST)	LATITUDE LIMITS. NORTH	SOUTH	LOCAL TIME	MEAN ALTITUDE
-87	35.35	33.42	14:49	51
-88	35.02	33.09	14:55	50
-89	34.68	32.74	15:01	49
-90	34.33	32.39	15:07	48
-92	33.59	31.65	15:18	45
-93	33.21	31.27	15:24	44
-94	32.82	30.89	15:30	43
-95	32.43	30.49	15:35	42
-97	31.61	29.67	15:46	39
-98	31.19	29.26	15:52	38
-99	30.76	28.83	15:57	37
-100	30.33	28.41	16:03	36
-102	29.44	27.53	16:13	33
-103	28.99	27.09	16:19	32
-104	28.54	26.64	16:24	31
-105	28.08	26.20	16:29	30
-107	27.15	25.29	16:39	27
-108	26.69	24.83	16:44	26
-109	26.22	24.37	16:49	25
-110	25.75	23.91	16:54	24
-112	24.80	22.99	17:04	21
-113	24.33	22.53	17:09	20
-114	23.85	22.07	17:14	19
-115	23.38	21.61	17:18	18
-117	22.43	20.69	17:27	15
-118	21.96	20.23	17:32	14
-119	21.49	19.78	17:36	13
-120	21.02	19.32	17:41	12
-122	20.09	18.42	17:50	9
-123	19.63	17.98	17:54	8
-124	19.17	17.53	17:58	7
-125	18.71	17.09	18:02	6
-127	17.81	16.21	18:11	4
-128	17.36	15.78	18:15	3
-129	16.92	15.35	18:19	2
-130	16.47	14.93	18:23	1

PERCENTAGE OF SUN OBSCURED ALONG MAP TRACK
WEST END =105.7   EAST END =104.2

=======================================
SUNSET POSITION : ALTITUDE IS -0.8 DEGREES
NORTH : LONGITUDE -131.8, LATITUDE 15.7
SOUTH : LONGITUDE -131.0, LATITUDE 14.5

+ MARKS THE CITY OF LO-I

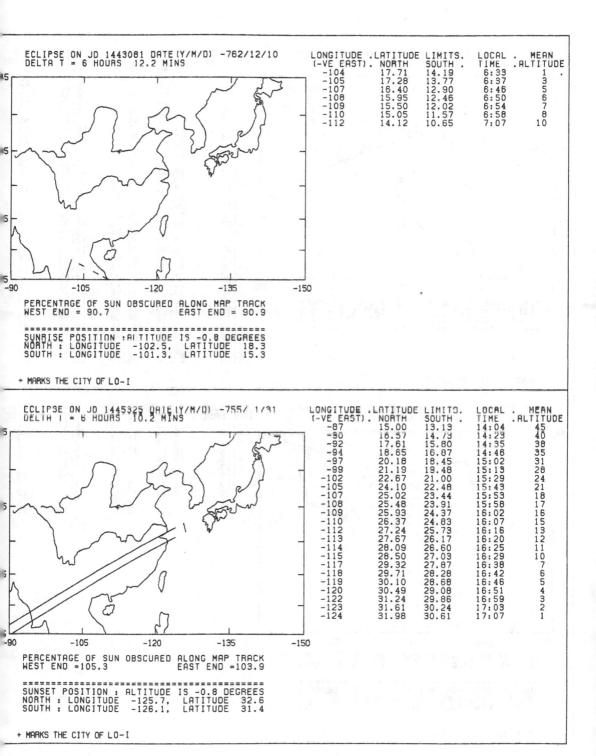

ECLIPSE ON JD 1443081 DATE(Y/M/D) -762/12/10
DELTA T = 6 HOURS 12.2 MINS

LONGITUDE (-VE EAST)	LATITUDE NORTH	LIMITS. SOUTH	LOCAL TIME	MEAN ALTITUDE
-104	17.71	14.19	6:33	1
-105	17.28	13.77	6:37	3
-107	16.40	12.90	6:46	5
-108	15.95	12.46	6:50	6
-109	15.50	12.02	6:54	7
-110	15.05	11.57	6:58	8
-112	14.12	10.65	7:07	10

PERCENTAGE OF SUN OBSCURED ALONG MAP TRACK
WEST END = 90.7   EAST END = 90.9

=======================================
SUNRISE POSITION : ALTITUDE IS -0.8 DEGREES
NORTH : LONGITUDE -102.5, LATITUDE 18.3
SOUTH : LONGITUDE -101.3, LATITUDE 15.3

+ MARKS THE CITY OF LO-I

ECLIPSE ON JD 1445325 DATE(Y/M/D) -755/ 1/31
DELTA T = 6 HOURS 10.2 MINS

LONGITUDE (-VE EAST)	LATITUDE NORTH	LIMITS. SOUTH	LOCAL TIME	MEAN ALTITUDE
-87	15.00	13.13	14:04	45
-90	16.57	14.73	14:23	40
-92	17.61	15.80	14:35	38
-94	18.65	16.87	14:48	35
-97	20.18	18.45	15:02	31
-99	21.19	19.48	15:19	28
-102	22.67	21.00	15:29	24
-105	24.10	22.48	15:43	21
-107	25.02	23.44	15:53	18
-108	25.48	23.91	15:58	17
-109	25.93	24.37	16:02	16
-110	26.37	24.83	16:07	15
-112	27.24	25.73	16:16	13
-113	27.67	26.17	16:20	12
-114	28.09	26.60	16:25	11
-115	28.50	27.03	16:29	10
-117	29.32	27.87	16:38	7
-118	29.71	28.28	16:42	6
-119	30.10	28.68	16:46	5
-120	30.49	29.08	16:51	4
-122	31.24	29.86	16:59	3
-123	31.61	30.24	17:03	2
-124	31.98	30.61	17:07	1

PERCENTAGE OF SUN OBSCURED ALONG MAP TRACK
WEST END =105.3   EAST END =103.9

=======================================
SUNSET POSITION : ALTITUDE IS -0.8 DEGREES
NORTH : LONGITUDE -125.7, LATITUDE 32.6
SOUTH : LONGITUDE -126.1, LATITUDE 31.4

+ MARKS THE CITY OF LO-I

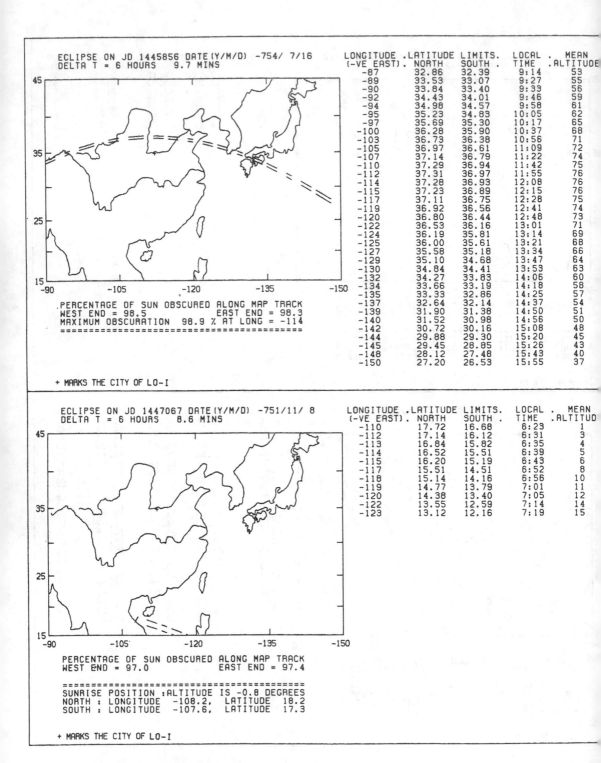

ECLIPSE ON JD 1445856 DATE(Y/M/D) -754/ 7/16
DELTA T = 6 HOURS   9.7 MINS

LONGITUDE	.LATITUDE	LIMITS.	LOCAL	. MEAN
(-VE EAST).	NORTH	SOUTH .	TIME	.ALTITUDE
-87	32.86	32.39	9:14	53
-89	33.53	33.07	9:27	55
-90	33.84	33.40	9:33	56
-92	34.43	34.01	9:46	59
-94	34.98	34.57	9:58	61
-95	35.23	34.83	10:05	62
-97	35.69	35.30	10:17	65
-100	36.28	35.90	10:37	68
-103	36.73	36.38	10:56	71
-105	36.97	36.61	11:09	72
-107	37.14	36.79	11:22	74
-110	37.29	36.94	11:42	75
-112	37.31	36.97	11:55	76
-114	37.28	36.93	12:08	76
-115	37.23	36.89	12:15	76
-117	37.11	36.75	12:28	75
-119	36.92	36.56	12:41	74
-120	36.80	36.44	12:48	73
-122	36.53	36.16	13:01	71
-124	36.19	35.81	13:14	69
-125	36.00	35.61	13:21	68
-127	35.58	35.18	13:34	66
-129	35.10	34.68	13:47	64
-130	34.84	34.41	13:53	63
-132	34.27	33.83	14:06	60
-134	33.66	33.19	14:18	58
-135	33.33	32.86	14:25	57
-137	32.64	32.14	14:37	54
-139	31.90	31.38	14:50	51
-140	31.52	30.98	14:56	50
-142	30.72	30.16	15:08	48
-144	29.88	29.30	15:20	45
-145	29.45	28.85	15:26	43
-148	28.12	27.48	15:43	40
-150	27.20	26.53	15:55	37

.PERCENTAGE OF SUN OBSCURED ALONG MAP TRACK
WEST END = 98.5   EAST END = 98.3
MAXIMUM OBSCURATION 98.9 % AT LONG = -114
======================================

+ MARKS THE CITY OF LO-I

ECLIPSE ON JD 1447067 DATE(Y/M/D) -751/11/ 8
DELTA T = 6 HOURS   8.6 MINS

LONGITUDE	.LATITUDE	LIMITS.	LOCAL	. MEAN
(-VE EAST).	NORTH	SOUTH .	TIME	.ALTITUD
-110	17.72	16.68	6:23	1
-112	17.14	16.12	6:31	3
-113	16.84	15.82	6:35	4
-114	16.52	15.51	6:39	5
-115	16.20	15.19	6:43	6
-117	15.51	14.51	6:52	8
-118	15.14	14.16	6:56	10
-119	14.77	13.79	7:01	11
-120	14.38	13.40	7:05	12
-122	13.55	12.59	7:14	14
-123	13.12	12.16	7:19	15

PERCENTAGE OF SUN OBSCURED ALONG MAP TRACK
WEST END = 97.0   EAST END = 97.4

======================================
SUNRISE POSITION :ALTITUDE IS -0.8 DEGREES
NORTH : LONGITUDE -108.2, LATITUDE 18.2
SOUTH : LONGITUDE -107.6, LATITUDE 17.3

+ MARKS THE CITY OF LO-I

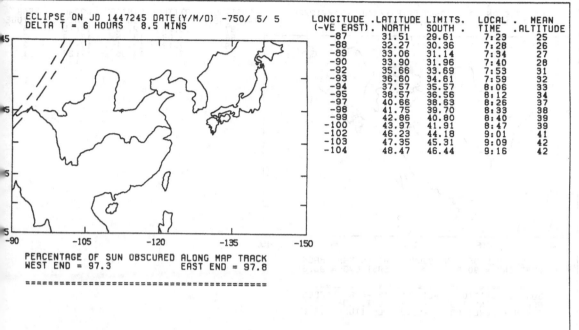

ECLIPSE ON JD 1447245 DATE(Y/M/D) -750/ 5/ 5
DELTA T = 6 HOURS   8.5 MINS

LONGITUDE (-VE EAST)	LATITUDE LIMITS. NORTH	SOUTH	LOCAL TIME	MEAN ALTITUDE
-87	31.51	29.61	7:23	25
-88	32.27	30.36	7:28	26
-89	33.06	31.14	7:34	27
-90	33.90	31.96	7:40	28
-92	35.66	33.69	7:53	31
-93	36.60	34.61	7:59	32
-94	37.57	35.57	8:06	33
-95	38.57	36.56	8:12	34
-97	40.66	38.63	8:26	37
-98	41.75	39.70	8:33	38
-99	42.86	40.80	8:40	39
-100	43.97	41.91	8:47	39
-102	46.23	44.18	9:01	41
-103	47.35	45.31	9:09	42
-104	48.47	46.44	9:16	42

PERCENTAGE OF SUN OBSCURED ALONG MAP TRACK
WEST END = 97.3   EAST END = 97.8

==

+ MARKS THE CITY OF LO-I

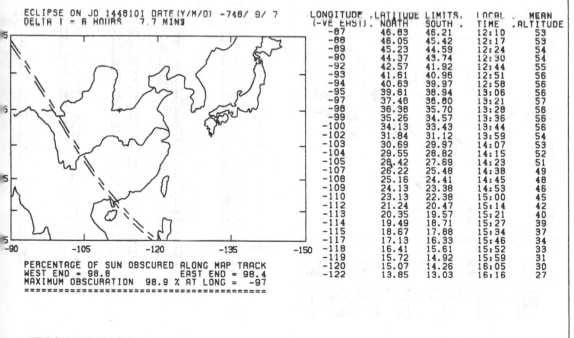

ECLIPSE ON JD 1448101 DATE(Y/M/D) -748/ 9/ 7
DELTA T = 6 HOURS   7.7 MINS

LONGITUDE (-VE EAST)	LATITUDE LIMITS. NORTH	SOUTH	LOCAL TIME	MEAN ALTITUDE
-87	46.83	46.21	12:10	53
-88	46.05	45.42	12:17	53
-89	45.23	44.59	12:24	54
-90	44.37	43.74	12:30	54
-92	42.57	41.92	12:44	55
-93	41.61	40.96	12:51	56
-94	40.63	39.97	12:58	56
-95	39.61	38.94	13:06	56
-97	37.48	36.80	13:21	57
-98	36.38	35.70	13:28	56
-99	35.26	34.57	13:36	56
-100	34.13	33.43	13:44	56
-102	31.84	31.12	13:59	54
-103	30.69	29.97	14:07	53
-104	29.55	28.82	14:15	52
-105	28.42	27.69	14:23	51
-107	26.22	25.48	14:38	49
-108	25.16	24.41	14:45	48
-109	24.13	23.38	14:53	46
-110	23.13	22.38	15:00	45
-112	21.24	20.47	15:14	42
-113	20.35	19.57	15:21	40
-114	19.49	18.71	15:27	39
-115	18.67	17.88	15:34	37
-117	17.13	16.33	15:46	34
-118	16.41	15.61	15:52	33
-119	15.72	14.92	15:59	31
-120	15.07	14.26	16:05	30
-122	13.85	13.03	16:16	27

PERCENTAGE OF SUN OBSCURED ALONG MAP TRACK
WEST END = 98.8   EAST END = 98.4
MAXIMUM OBSCURATION 98.9 % AT LONG = -97

==

+ MARKS THE CITY OF LO-I

93

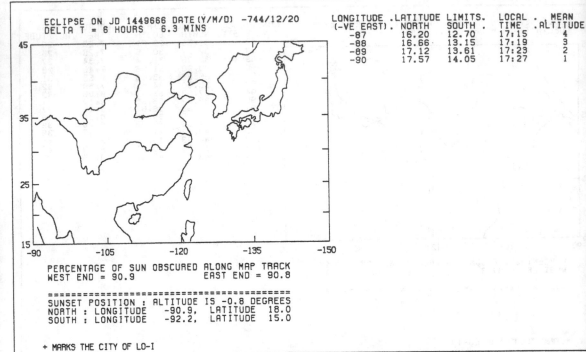

ECLIPSE ON JD 1449666 DATE(Y/M/D) -744/12/20
DELTA T = 6 HOURS   6.3 MINS

LONGITUDE (-VE EAST)	LATITUDE NORTH	LIMITS. SOUTH	LOCAL TIME	MEAN ALTITUDE
-87	16.20	12.70	17:15	4
-88	16.66	13.15	17:19	3
-89	17.12	13.61	17:23	2
-90	17.57	14.05	17:27	1

PERCENTAGE OF SUN OBSCURED ALONG MAP TRACK
WEST END = 90.9   EAST END = 90.8

==
SUNSET POSITION : ALTITUDE IS -0.8 DEGREES
NORTH : LONGITUDE   -90.9, LATITUDE 18.0
SOUTH : LONGITUDE   -92.2, LATITUDE 15.0

+ MARKS THE CITY OF LO-I

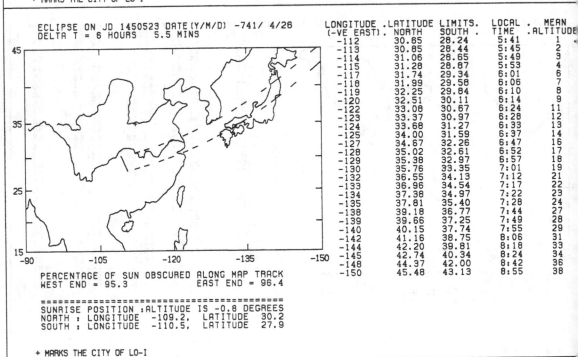

ECLIPSE ON JD 1450523 DATE(Y/M/D) -741/ 4/26
DELTA T = 6 HOURS   5.5 MINS

LONGITUDE (-VE EAST)	LATITUDE NORTH	LIMITS. SOUTH	LOCAL TIME	MEAN ALTITUDE
-112	30.65	28.24	5:41	1
-113	30.85	28.44	5:45	2
-114	31.06	28.65	5:49	3
-115	31.28	28.87	5:53	4
-117	31.74	29.34	6:01	6
-118	31.99	29.58	6:06	7
-119	32.25	29.84	6:10	8
-120	32.51	30.11	6:14	9
-122	33.08	30.67	6:24	11
-123	33.37	30.97	6:28	12
-124	33.68	31.27	6:33	13
-125	34.00	31.59	6:37	14
-127	34.67	32.26	6:47	16
-128	35.02	32.61	6:52	17
-129	35.38	32.97	6:57	18
-130	35.76	33.35	7:01	19
-132	36.55	34.13	7:12	21
-133	36.96	34.54	7:17	22
-134	37.38	34.97	7:22	23
-135	37.81	35.40	7:28	24
-138	39.18	36.77	7:44	27
-139	39.66	37.25	7:49	28
-140	40.15	37.74	7:55	29
-142	41.16	38.75	8:06	31
-144	42.20	39.81	8:18	33
-145	42.74	40.34	8:24	34
-148	44.37	42.00	8:42	36
-150	45.48	43.13	8:55	38

PERCENTAGE OF SUN OBSCURED ALONG MAP TRACK
WEST END = 95.3   EAST END = 96.4

==
SUNRISE POSITION : ALTITUDE IS -0.8 DEGREES
NORTH : LONGITUDE -109.2, LATITUDE 30.2
SOUTH : LONGITUDE -110.5, LATITUDE 27.9

+ MARKS THE CITY OF LO-I

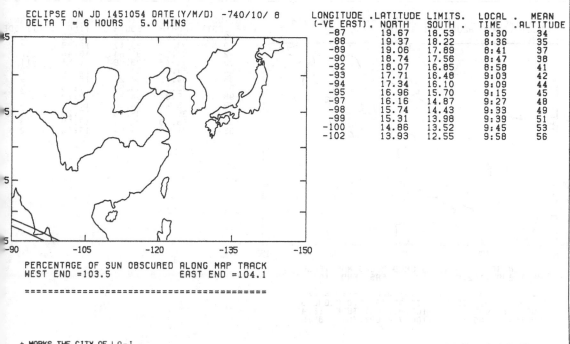

ECLIPSE ON JD 1451054 DATE(Y/M/D) -740/10/ 8
DELTA T = 6 HOURS   5.0 MINS

LONGITUDE	.LATITUDE	LIMITS.	LOCAL	. MEAN
(-VE EAST).	NORTH	SOUTH .	TIME	.ALTITUDE
-87	19.67	18.53	8:30	34
-88	19.37	18.22	8:36	35
-89	19.06	17.89	8:41	37
-90	18.74	17.56	8:47	38
-92	18.07	16.85	8:58	41
-93	17.71	16.48	9:03	42
-94	17.34	16.10	9:09	44
-95	16.96	15.70	9:15	45
-97	16.16	14.87	9:27	48
-98	15.74	14.43	9:33	49
-99	15.31	13.98	9:39	51
-100	14.86	13.52	9:45	53
-102	13.93	12.55	9:58	56

PERCENTAGE OF SUN OBSCURED ALONG MAP TRACK
WEST END =103.5   EAST END =104.1

==

+ MARKS THE CITY OF LO-I

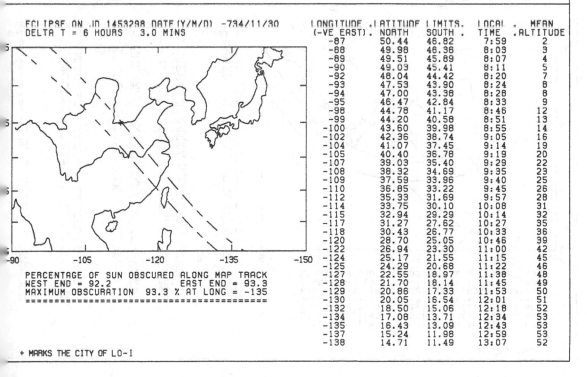

ECLIPSE ON JD 1453298 DATE(Y/M/D) -734/11/30
DELTA T = 6 HOURS   3.0 MINS

LONGITUDE	.LATITUDE	LIMITS.	LOCAL	. MEAN
(-VE EAST).	NORTH	SOUTH .	TIME	.ALTITUDE
-87	50.44	46.82	7:59	2
-88	49.98	46.36	8:03	3
-89	49.51	45.89	8:07	4
-90	49.03	45.41	8:11	5
-92	48.04	44.42	8:20	7
-93	47.53	43.90	8:24	8
-94	47.00	43.38	8:28	8
-95	46.47	42.84	8:33	9
-98	44.78	41.17	8:46	12
-99	44.20	40.58	8:51	13
-100	43.60	39.98	8:55	14
-102	42.36	38.74	9:05	16
-104	41.07	37.45	9:14	19
-105	40.40	36.78	9:19	20
-107	39.03	35.40	9:29	22
-108	38.32	34.69	9:35	23
-109	37.59	33.96	9:40	25
-110	36.85	33.22	9:45	26
-112	35.33	31.69	9:57	28
-114	33.75	30.10	10:08	31
-115	32.94	29.29	10:14	32
-117	31.27	27.62	10:27	35
-118	30.43	26.77	10:33	36
-120	28.70	25.05	10:46	39
-122	26.94	23.30	11:00	42
-124	25.17	21.55	11:15	45
-125	24.29	20.68	11:22	46
-127	22.55	18.97	11:38	48
-128	21.70	18.14	11:45	49
-129	20.86	17.33	11:53	50
-130	20.05	16.54	12:01	51
-132	18.50	15.06	12:18	52
-134	17.08	13.71	12:34	53
-135	16.43	13.09	12:43	53
-137	15.24	11.98	12:59	53
-138	14.71	11.49	13:07	52

PERCENTAGE OF SUN OBSCURED ALONG MAP TRACK
WEST END = 92.2   EAST END = 93.3
MAXIMUM OBSCURATION 93.3 % AT LONG = -135

==

+ MARKS THE CITY OF LO-I

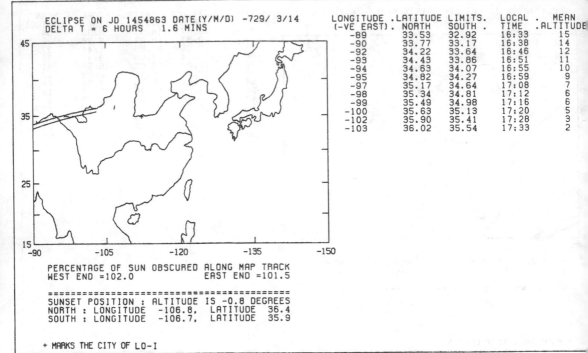

ECLIPSE ON JD 1454863 DATE(Y/M/D) -729/ 3/14
DELTA T = 6 HOURS   1.6 MINS

LONGITUDE (-VE EAST)	LATITUDE NORTH	LIMITS. SOUTH	LOCAL TIME	MEAN ALTITUDE
-89	33.53	32.92	16:33	15
-90	33.77	33.17	16:38	14
-92	34.22	33.64	16:46	12
-93	34.43	33.86	16:51	11
-94	34.63	34.07	16:55	10
-95	34.82	34.27	16:59	9
-97	35.17	34.64	17:08	7
-98	35.34	34.81	17:12	6
-99	35.49	34.98	17:16	6
-100	35.63	35.13	17:20	5
-102	35.90	35.41	17:28	3
-103	36.02	35.54	17:33	2

PERCENTAGE OF SUN OBSCURED ALONG MAP TRACK
WEST END =102.0   EAST END =101.5

===
SUNSET POSITION : ALTITUDE IS -0.8 DEGREES
NORTH : LONGITUDE -106.8, LATITUDE 36.4
SOUTH : LONGITUDE -106.7, LATITUDE 35.9

+ MARKS THE CITY OF LO-I

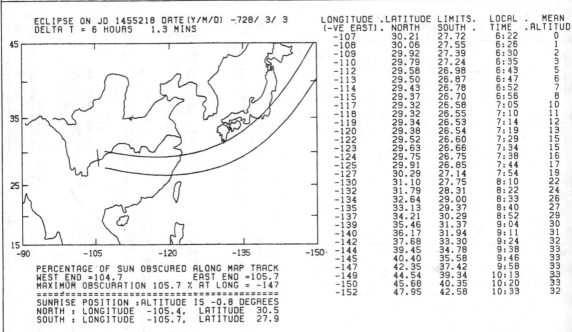

ECLIPSE ON JD 1455218 DATE(Y/M/D) -728/ 3/ 3
DELTA T = 6 HOURS   1.3 MINS

LONGITUDE (-VE EAST)	LATITUDE NORTH	LIMITS. SOUTH	LOCAL TIME	MEAN ALTITUD
-107	30.21	27.72	6:22	0
-108	30.06	27.55	6:26	1
-109	29.92	27.39	6:30	2
-110	29.79	27.24	6:35	3
-112	29.58	26.98	6:43	5
-113	29.50	26.87	6:47	6
-114	29.43	26.78	6:52	7
-115	29.37	26.70	6:56	8
-117	29.32	26.58	7:05	10
-118	29.32	26.55	7:10	11
-119	29.34	26.53	7:14	12
-120	29.38	26.54	7:19	13
-122	29.52	26.60	7:29	15
-123	29.63	26.66	7:34	15
-124	29.75	26.75	7:38	16
-125	29.91	26.85	7:44	17
-127	30.29	27.14	7:54	19
-130	31.10	27.75	8:10	22
-132	31.79	28.31	8:22	24
-134	32.64	29.00	8:33	26
-135	33.13	29.37	8:40	27
-137	34.21	30.29	8:52	29
-139	35.46	31.37	9:04	30
-140	36.17	31.94	9:11	31
-142	37.68	33.30	9:24	32
-144	39.45	34.78	9:38	33
-145	40.40	35.58	9:46	33
-147	42.35	37.42	9:58	33
-149	44.54	39.34	10:13	33
-150	45.68	40.35	10:20	33
-152	47.95	42.58	10:33	32

PERCENTAGE OF SUN OBSCURED ALONG MAP TRACK
WEST END =104.7   EAST END =105.7
MAXIMUM OBSCURATION 105.7 % AT LONG = -147
===
SUNRISE POSITION :ALTITUDE IS -0.8 DEGREES
NORTH : LONGITUDE -105.4, LATITUDE 30.5
SOUTH : LONGITUDE -105.7, LATITUDE 27.9

+ MARKS THE CITY OF LO-I

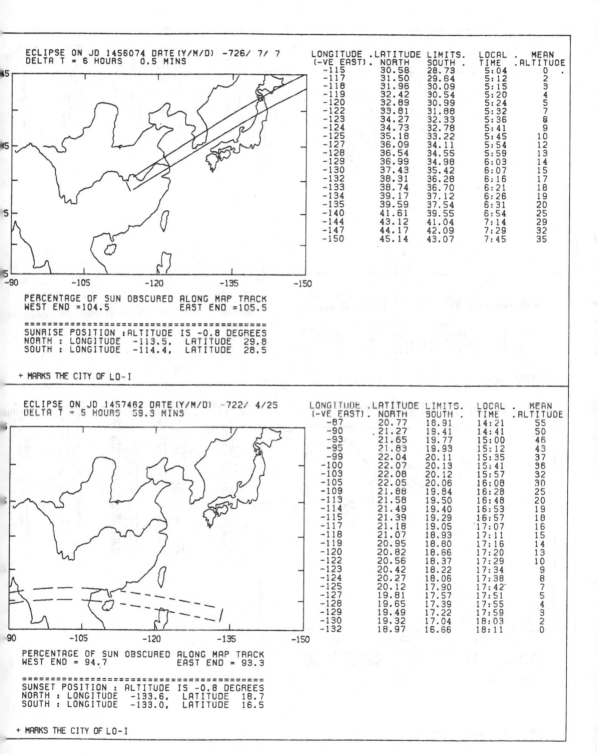

ECLIPSE ON JD 1456074 DATE (Y/M/D) -726/ 7/ 7
DELTA T = 6 HOURS   0.5 MINS

LONGITUDE (-VE EAST)	LATITUDE NORTH	LIMITS SOUTH	LOCAL TIME	MEAN ALTITUDE
-115	30.58	28.73	5:04	0
-117	31.50	29.64	5:12	2
-118	31.96	30.09	5:15	3
-119	32.42	30.54	5:20	4
-120	32.89	30.99	5:24	5
-122	33.81	31.88	5:32	7
-123	34.27	32.33	5:36	8
-124	34.73	32.78	5:41	9
-125	35.18	33.22	5:45	10
-127	36.09	34.11	5:54	12
-128	36.54	34.55	5:59	13
-129	36.99	34.98	6:03	14
-130	37.43	35.42	6:07	15
-132	38.31	36.28	6:16	17
-133	38.74	36.70	6:21	18
-134	39.17	37.12	6:26	19
-135	39.59	37.54	6:31	20
-140	41.61	39.55	6:54	25
-144	43.12	41.04	7:14	29
-147	44.17	42.09	7:29	32
-150	45.14	43.07	7:45	35

PERCENTAGE OF SUN OBSCURED ALONG MAP TRACK
WEST END =104.5   EAST END =105.5

==
SUNRISE POSITION :ALTITUDE IS -0.8 DEGREES
NORTH : LONGITUDE -113.5,   LATITUDE   29.8
SOUTH : LONGITUDE -114.4,   LATITUDE   28.5

+ MARKS THE CITY OF LO-I

ECLIPSE ON JD 1457462 DATE(Y/M/D) -722/ 4/25
DELTA T = 5 HOURS 59.3 MINS

LONGITUDE (-VE EAST)	LATITUDE NORTH	LIMITS SOUTH	LOCAL TIME	MEAN ALTITUDE
-87	20.77	18.91	14:21	55
-90	21.27	19.41	14:41	50
-93	21.65	19.77	15:00	46
-95	21.83	19.93	15:12	43
-99	22.04	20.11	15:35	37
-100	22.07	20.13	15:41	36
-103	22.08	20.12	15:57	32
-105	22.05	20.06	16:08	30
-109	21.88	19.84	16:28	25
-113	21.58	19.50	16:48	20
-114	21.49	19.40	16:53	19
-115	21.39	19.29	16:57	18
-117	21.18	19.05	17:07	16
-118	21.07	18.93	17:11	15
-119	20.95	18.80	17:16	14
-120	20.82	18.66	17:20	13
-122	20.56	18.37	17:29	10
-123	20.42	18.22	17:34	9
-124	20.27	18.06	17:38	8
-125	20.12	17.90	17:42	7
-127	19.81	17.57	17:51	5
-128	19.65	17.39	17:55	4
-129	19.49	17.22	17:59	3
-130	19.32	17.04	18:03	2
-132	18.97	16.66	18:11	0

PERCENTAGE OF SUN OBSCURED ALONG MAP TRACK
WEST END = 94.7   EAST END = 93.3

==
SUNSET POSITION : ALTITUDE IS -0.8 DEGREES
NORTH : LONGITUDE -133.6,   LATITUDE   18.7
SOUTH : LONGITUDE -133.0,   LATITUDE   16.5

+ MARKS THE CITY OF LO-I

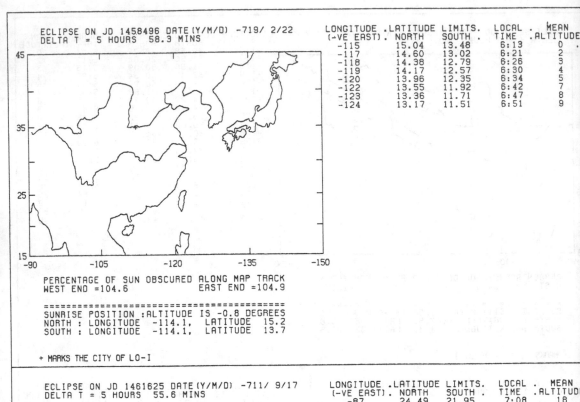

ECLIPSE ON JD 1458496 DATE(Y/M/D) -719/ 2/22
DELTA T = 5 HOURS 58.3 MINS

LONGITUDE	.LATITUDE	LIMITS.	LOCAL .	MEAN
(-VE EAST).	NORTH	SOUTH .	TIME	.ALTITUDE
-115	15.04	13.48	6:13	0 .
-117	14.60	13.02	6:21	2
-118	14.38	12.79	6:26	3
-119	14.17	12.57	6:30	4
-120	13.96	12.35	6:34	5
-122	13.55	11.92	6:42	7
-123	13.36	11.71	6:47	8
-124	13.17	11.51	6:51	9

PERCENTAGE OF SUN OBSCURED ALONG MAP TRACK
WEST END =104.6   EAST END =104.9

===
SUNRISE POSITION :ALTITUDE IS -0.8 DEGREES
NORTH : LONGITUDE -114.1,   LATITUDE   15.2
SOUTH : LONGITUDE -114.1,   LATITUDE   13.7

+ MARKS THE CITY OF LO-I

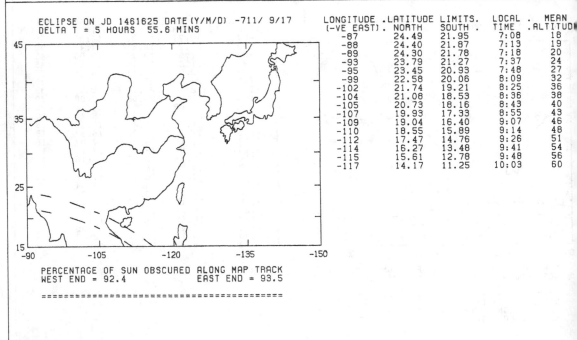

ECLIPSE ON JD 1461625 DATE(Y/M/D) -711/ 9/17
DELTA T = 5 HOURS 55.6 MINS

LONGITUDE	.LATITUDE	LIMITS.	LOCAL .	MEAN
(-VE EAST).	NORTH	SOUTH .	TIME	.ALTITUDE
-87	24.49	21.95	7:08	18
-88	24.40	21.87	7:13	19
-89	24.30	21.78	7:18	20
-93	23.79	21.27	7:37	24
-95	23.45	20.93	7:48	27
-99	22.58	20.06	8:09	32
-102	21.74	19.21	8:25	36
-104	21.08	18.53	8:36	38
-105	20.73	18.16	8:43	40
-107	19.93	17.33	8:55	43
-109	19.04	16.40	9:07	46
-110	18.55	15.89	9:14	48
-112	17.47	14.76	9:26	51
-114	16.27	13.48	9:41	54
-115	15.61	12.78	9:48	56
-117	14.17	11.25	10:03	60

PERCENTAGE OF SUN OBSCURED ALONG MAP TRACK
WEST END = 92.4   EAST END = 93.5

===

+ MARKS THE CITY OF LO-I

ECLIPSE ON JD 1462659 DATE(Y/M/D) -708/ 7/17
DELTA T = 5 HOURS 54.6 MINS

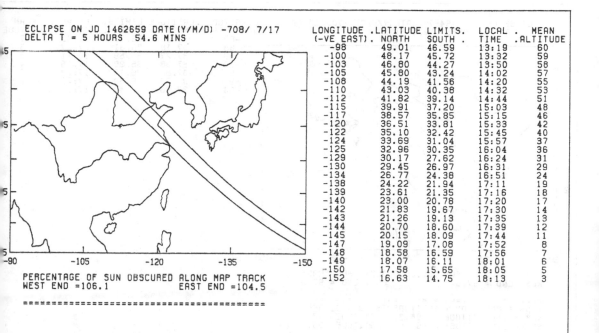

PERCENTAGE OF SUN OBSCURED ALONG MAP TRACK
WEST END =106.1   EAST END =104.5

===

+ MARKS THE CITY OF LO-I

LONGITUDE (-VE EAST)	LATITUDE NORTH	LIMITS SOUTH	LOCAL TIME	MEAN ALTITUDE
-98	49.01	46.59	13:19	60
-100	48.17	45.72	13:32	59
-103	46.80	44.27	13:50	58
-105	45.80	43.24	14:02	57
-108	44.19	41.56	14:20	55
-110	43.03	40.38	14:32	53
-112	41.82	39.14	14:44	51
-115	39.91	37.20	15:03	48
-117	38.57	35.85	15:15	46
-120	36.51	33.81	15:33	42
-122	35.10	32.42	15:45	40
-124	33.69	31.04	15:57	37
-125	32.96	30.35	16:04	36
-129	30.17	27.62	16:24	31
-130	29.45	26.97	16:31	29
-134	26.77	24.38	16:51	24
-138	24.22	21.94	17:11	19
-139	23.61	21.35	17:16	18
-140	23.00	20.78	17:20	17
-142	21.83	19.67	17:30	14
-143	21.26	19.13	17:35	13
-144	20.70	18.60	17:39	12
-145	20.15	18.09	17:44	11
-147	19.09	17.08	17:52	8
-148	18.58	16.59	17:56	7
-149	18.07	16.11	18:01	6
-150	17.58	15.65	18:05	5
-152	16.63	14.75	18:13	3

ECLIPSE ON JD 1463191 DATE(Y/M/D) -707/12/31
DELTA T = 5 HOURS 54.2 MINS

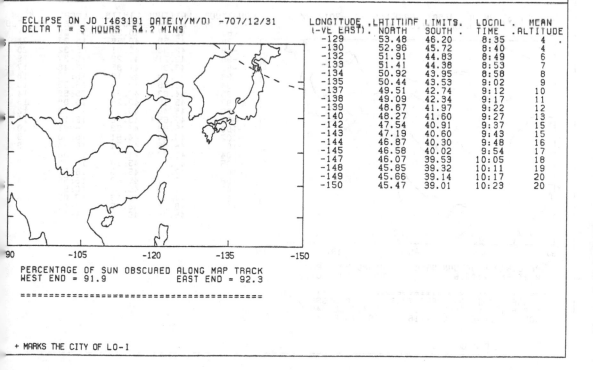

PERCENTAGE OF SUN OBSCURED ALONG MAP TRACK
WEST END = 91.9   EAST END = 92.3

===

+ MARKS THE CITY OF LO-I

LONGITUDE (-VE EAST)	LATITUDE NORTH	LIMITS SOUTH	LOCAL TIME	MEAN ALTITUDE
-129	53.48	46.20	8:35	4
-130	52.96	45.72	8:40	4
-132	51.91	44.83	8:49	6
-133	51.41	44.38	8:53	7
-134	50.92	43.95	8:58	8
-135	50.44	43.53	9:02	9
-137	49.51	42.74	9:12	10
-138	49.09	42.34	9:17	11
-139	48.67	41.97	9:22	12
-140	48.27	41.60	9:27	13
-142	47.54	40.91	9:37	15
-143	47.19	40.60	9:43	15
-144	46.87	40.30	9:48	16
-145	46.58	40.02	9:54	17
-147	46.07	39.53	10:05	18
-148	45.85	39.32	10:11	19
-149	45.66	39.14	10:17	20
-150	45.47	39.01	10:23	20

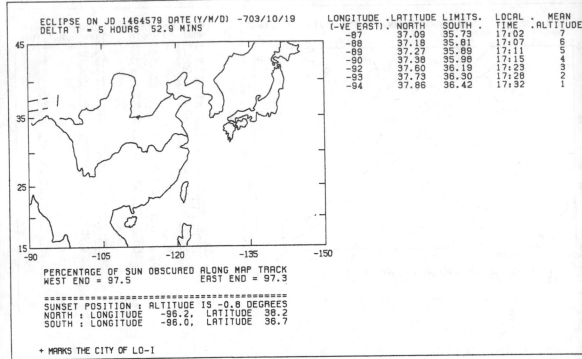

ECLIPSE ON JD 1464579 DATE(Y/M/D) -703/10/19
DELTA T = 5 HOURS 52.9 MINS

LONGITUDE	.LATITUDE	LIMITS.	LOCAL .	MEAN
(-VE EAST).	NORTH	SOUTH .	TIME	.ALTITUDE
-87	37.09	35.73	17:02	7
-88	37.18	35.81	17:07	6
-89	37.27	35.89	17:11	5
-90	37.38	35.98	17:15	4
-92	37.60	36.19	17:23	3
-93	37.73	36.30	17:28	2
-94	37.86	36.42	17:32	1

PERCENTAGE OF SUN OBSCURED ALONG MAP TRACK
WEST END = 97.5   EAST END = 97.3

===
SUNSET POSITION : ALTITUDE IS -0.8 DEGREES
NORTH : LONGITUDE -96.2, LATITUDE 38.2
SOUTH : LONGITUDE -96.0, LATITUDE 36.7

+ MARKS THE CITY OF LO-I

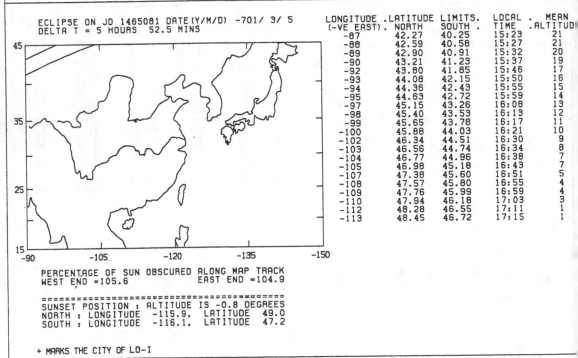

ECLIPSE ON JD 1465081 DATE(Y/M/D) -701/ 3/ 5
DELTA T = 5 HOURS 52.5 MINS

LONGITUDE	.LATITUDE	LIMITS.	LOCAL .	MEAN
(-VE EAST).	NORTH	SOUTH .	TIME	.ALTITUD
-87	42.27	40.25	15:23	21
-88	42.59	40.58	15:27	21
-89	42.90	40.91	15:32	20
-90	43.21	41.23	15:37	19
-92	43.80	41.85	15:46	17
-93	44.08	42.15	15:50	16
-94	44.36	42.43	15:55	15
-95	44.63	42.72	15:59	14
-97	45.15	43.26	16:08	13
-98	45.40	43.53	16:13	12
-99	45.65	43.78	16:17	11
-100	45.88	44.03	16:21	10
-102	46.34	44.51	16:30	9
-103	46.56	44.74	16:34	8
-104	46.77	44.96	16:38	7
-105	46.98	45.18	16:43	7
-107	47.38	45.60	16:51	5
-108	47.57	45.80	16:55	4
-109	47.76	45.99	16:59	4
-110	47.94	46.18	17:03	3
-112	48.28	46.55	17:11	1
-113	48.45	46.72	17:15	1

PERCENTAGE OF SUN OBSCURED ALONG MAP TRACK
WEST END =105.6   EAST END =104.9

===
SUNSET POSITION : ALTITUDE IS -0.8 DEGREES
NORTH : LONGITUDE -115.9, LATITUDE 49.0
SOUTH : LONGITUDE -116.1, LATITUDE 47.2

+ MARKS THE CITY OF LO-I

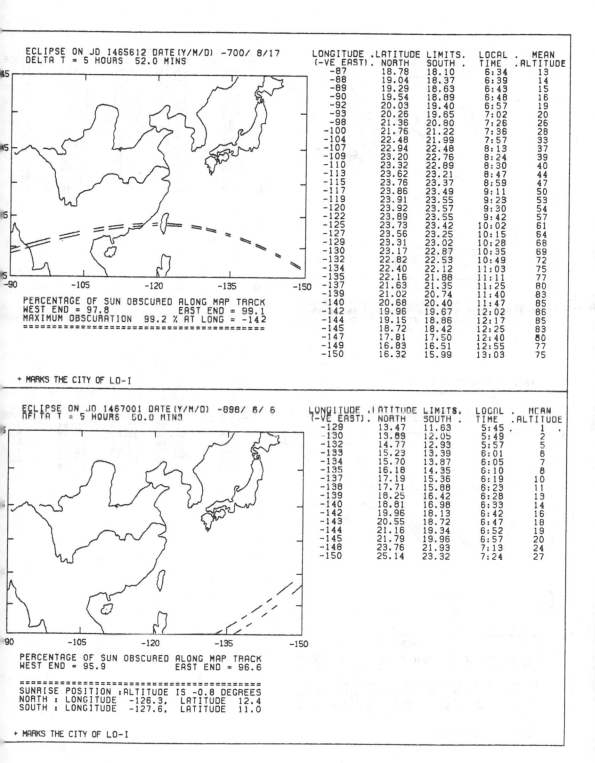

ECLIPSE ON JD 1465612 DATE(Y/M/D) -700/ 8/17
DELTA T = 5 HOURS 52.0 MINS

LONGITUDE (-VE EAST)	LATITUDE NORTH	LIMITS SOUTH	LOCAL TIME	MEAN ALTITUDE
-87	18.78	18.10	6:34	13
-88	19.04	18.37	6:39	14
-89	19.29	18.63	6:43	15
-90	19.54	18.89	6:48	16
-92	20.03	19.40	6:57	19
-93	20.26	19.65	7:02	20
-98	21.36	20.80	7:26	26
-100	21.76	21.22	7:36	28
-104	22.48	21.99	7:57	33
-107	22.94	22.48	8:13	37
-109	23.20	22.76	8:24	39
-110	23.32	22.89	8:30	40
-113	23.62	23.21	8:47	44
-115	23.76	23.37	8:59	47
-117	23.86	23.49	9:11	50
-119	23.91	23.55	9:23	53
-120	23.92	23.57	9:30	54
-122	23.89	23.55	9:42	57
-125	23.73	23.42	10:02	61
-127	23.56	23.25	10:15	64
-129	23.31	23.02	10:28	68
-130	23.17	22.87	10:35	69
-132	22.82	22.53	10:49	72
-134	22.40	22.12	11:03	75
-135	22.16	21.88	11:11	77
-137	21.63	21.35	11:25	80
-139	21.02	20.74	11:40	83
-140	20.68	20.40	11:47	85
-142	19.96	19.67	12:02	86
-144	19.15	18.86	12:17	85
-145	18.72	18.42	12:25	83
-147	17.81	17.50	12:40	80
-149	16.83	16.51	12:55	77
-150	16.32	15.99	13:03	75

PERCENTAGE OF SUN OBSCURED ALONG MAP TRACK
WEST END = 97.8   EAST END = 99.1
MAXIMUM OBSCURATION 99.2 % AT LONG = -142
==

+ MARKS THE CITY OF LO-I

ECLIPSE ON JD 1467001 DATE(Y/M/D) -898/ 6/ 6
DELTA T = 5 HOURS 50.0 MINS

LONGITUDE (-VE EAST)	LATITUDE NORTH	LIMITS SOUTH	LOCAL TIME	MEAN ALTITUDE
-129	13.47	11.63	5:45	1
-130	13.89	12.05	5:49	2
-132	14.77	12.93	5:57	5
-133	15.23	13.39	6:01	6
-134	15.70	13.87	6:05	7
-135	16.18	14.35	6:10	8
-137	17.19	15.36	6:19	10
-138	17.71	15.88	6:23	11
-139	18.25	16.42	6:28	13
-140	18.81	16.98	6:33	14
-142	19.96	18.13	6:42	16
-143	20.55	18.72	6:47	18
-144	21.16	19.34	6:52	19
-145	21.79	19.96	6:57	20
-148	23.76	21.93	7:13	24
-150	25.14	23.32	7:24	27

PERCENTAGE OF SUN OBSCURED ALONG MAP TRACK
WEST END = 95.9   EAST END = 96.6

==
SUNRISE POSITION :ALTITUDE IS -0.8 DEGREES
NORTH : LONGITUDE -126.3, LATITUDE 12.4
SOUTH : LONGITUDE -127.6, LATITUDE 11.0

+ MARKS THE CITY OF LO-I

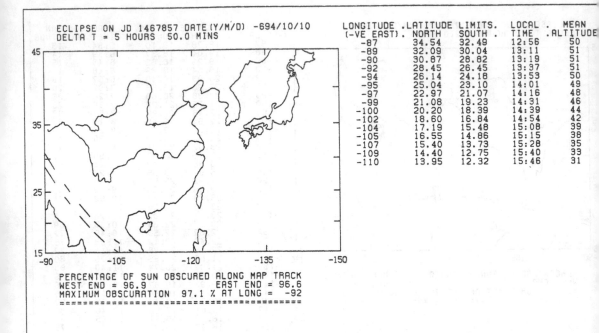

ECLIPSE ON JD 1467857 DATE (Y/M/D) -694/10/10
DELTA T = 5 HOURS 50.0 MINS

LONGITUDE (-VE EAST)	.LATITUDE NORTH	LIMITS. SOUTH .	LOCAL TIME	. MEAN .ALTITUDE
-87	34.54	32.49	12:56	50
-89	32.09	30.04	13:11	51
-90	30.87	28.82	13:19	51
-92	28.45	26.45	13:37	51
-94	26.14	24.18	13:53	50
-95	25.04	23.10	14:01	49
-97	22.97	21.07	14:16	48
-99	21.08	19.23	14:31	46
-100	20.20	18.39	14:39	44
-102	18.60	16.84	14:54	42
-104	17.19	15.48	15:08	39
-105	16.55	14.86	15:15	38
-107	15.40	13.73	15:28	35
-109	14.40	12.75	15:40	33
-110	13.95	12.32	15:46	31

PERCENTAGE OF SUN OBSCURED ALONG MAP TRACK
WEST END = 96.9   EAST END = 96.6
MAXIMUM OBSCURATION 97.1 % AT LONG = -92
==

+ MARKS THE CITY OF LO-I

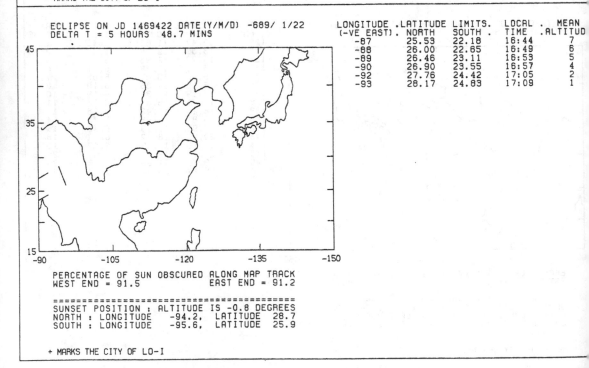

ECLIPSE ON JD 1469422 DATE (Y/M/D) -689/ 1/22
DELTA T = 5 HOURS 48.7 MINS

LONGITUDE (-VE EAST)	.LATITUDE NORTH	LIMITS. SOUTH .	LOCAL TIME	. MEAN .ALTITUD
-87	25.53	22.18	16:44	7
-88	26.00	22.65	16:49	6
-89	26.46	23.11	16:53	5
-90	26.90	23.55	16:57	4
-92	27.76	24.42	17:05	2
-93	28.17	24.83	17:09	1

PERCENTAGE OF SUN OBSCURED ALONG MAP TRACK
WEST END = 91.5   EAST END = 91.2

==
SUNSET POSITION : ALTITUDE IS -0.8 DEGREES
NORTH : LONGITUDE -94.2, LATITUDE 28.7
SOUTH : LONGITUDE -95.6, LATITUDE 25.9

+ MARKS THE CITY OF LO-I

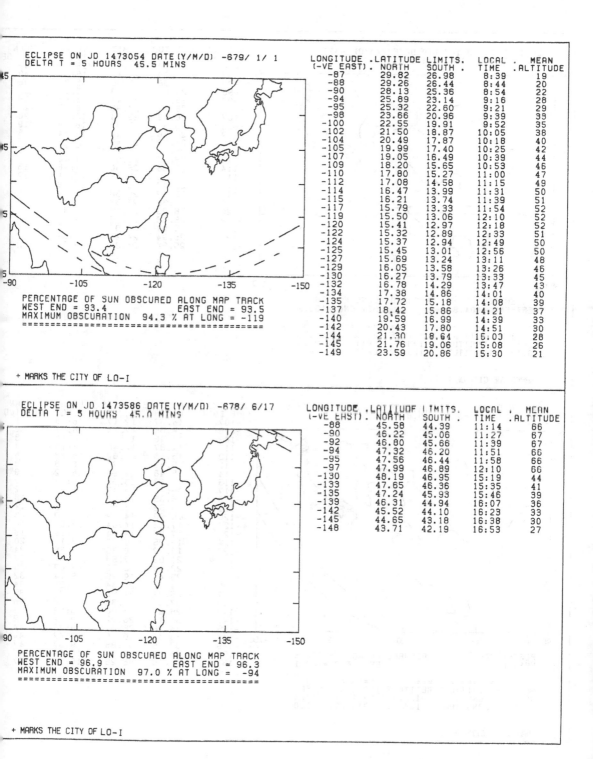

ECLIPSE ON JD 1473054 DATE(Y/M/D) -679/ 1/ 1
DELTA T = 5 HOURS 45.5 MINS

LONGITUDE (-VE EAST)	LATITUDE LIMITS.		LOCAL TIME	MEAN ALTITUDE
	NORTH	SOUTH		
-87	29.82	26.98	8:39	19
-88	29.26	26.44	8:44	20
-90	28.13	25.36	8:54	22
-94	25.89	23.14	9:16	28
-95	25.32	22.60	9:21	29
-98	23.66	20.96	9:39	33
-100	22.55	19.91	9:52	35
-102	21.50	18.87	10:05	38
-104	20.49	17.87	10:18	40
-105	19.99	17.40	10:25	42
-107	19.05	16.49	10:39	44
-109	18.20	15.65	10:53	46
-110	17.80	15.27	11:00	47
-112	17.08	14.58	11:15	49
-114	16.47	13.99	11:31	50
-115	16.21	13.74	11:39	51
-117	15.79	13.33	11:54	52
-119	15.50	13.06	12:10	52
-120	15.41	12.97	12:18	52
-122	15.32	12.89	12:33	51
-124	15.37	12.94	12:49	50
-125	15.45	13.01	12:56	50
-127	15.69	13.24	13:11	48
-129	16.05	13.58	13:26	46
-130	16.27	13.79	13:33	45
-132	16.78	14.29	13:47	43
-134	17.38	14.86	14:01	40
-135	17.72	15.18	14:08	39
-137	18.42	15.86	14:21	37
-140	19.59	16.99	14:39	33
-142	20.43	17.80	14:51	30
-144	21.30	18.64	15:03	28
-145	21.76	19.06	15:08	26
-149	23.59	20.86	15:30	21

PERCENTAGE OF SUN OBSCURED ALONG MAP TRACK
WEST END = 93.4   EAST END = 93.5
MAXIMUM OBSCURATION 94.3 % AT LONG = -119
==

+ MARKS THE CITY OF LO-I

ECLIPSE ON JD 1473586 DATE(Y/M/D) -678/ 6/17
DELTA T = 5 HOURS 45.0 MINS

LONGITUDE (-VE EAST)	LATITUDE LIMITS.		LOCAL TIME	MEAN ALTITUDE
	NORTH	SOUTH		
-88	45.58	44.39	11:14	66
-90	46.22	45.06	11:27	67
-92	46.80	45.66	11:39	67
-94	47.32	46.20	11:51	66
-95	47.56	46.44	11:58	66
-97	47.99	46.89	12:10	66
-130	48.19	46.95	15:19	44
-133	47.65	46.36	15:35	41
-135	47.24	45.93	15:46	39
-139	46.31	44.94	16:07	36
-142	45.52	44.10	16:23	33
-145	44.65	43.18	16:38	30
-148	43.71	42.19	16:53	27

PERCENTAGE OF SUN OBSCURED ALONG MAP TRACK
WEST END = 96.9   EAST END = 96.3
MAXIMUM OBSCURATION 97.0 % AT LONG = -94
==

+ MARKS THE CITY OF LO-I

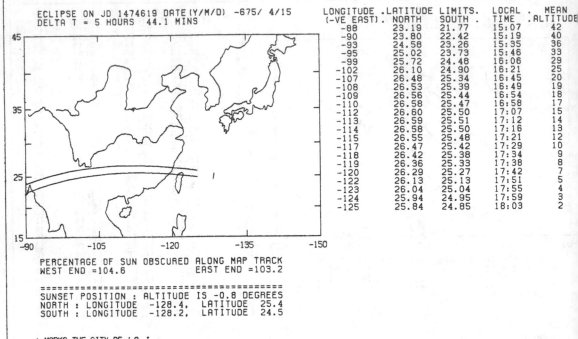

ECLIPSE ON JD 1474619 DATE(Y/M/D) -675/ 4/15
DELTA T = 5 HOURS 44.1 MINS

LONGITUDE (-VE EAST)	.LATITUDE LIMITS. NORTH	SOUTH .	LOCAL TIME	MEAN .ALTITUDE
-88	23.19	21.77	15:07	42
-90	23.80	22.42	15:19	40
-93	24.58	23.26	15:35	36
-95	25.02	23.73	15:46	33
-99	25.72	24.48	16:06	29
-102	26.10	24.90	16:21	25
-107	26.48	25.34	16:45	20
-108	26.53	25.39	16:49	19
-109	26.56	25.44	16:54	18
-110	26.58	25.47	16:58	17
-112	26.60	25.50	17:07	15
-113	26.59	25.51	17:12	14
-114	26.58	25.50	17:16	13
-115	26.55	25.48	17:21	12
-117	26.47	25.42	17:29	10
-118	26.42	25.38	17:34	9
-119	26.36	25.33	17:38	8
-120	26.29	25.27	17:42	7
-122	26.13	25.13	17:51	5
-123	26.04	25.04	17:55	4
-124	25.94	24.95	17:59	3
-125	25.84	24.85	18:03	2

PERCENTAGE OF SUN OBSCURED ALONG MAP TRACK
WEST END =104.6   EAST END =103.2

===
SUNSET POSITION : ALTITUDE IS -0.8 DEGREES
NORTH : LONGITUDE -128.4, LATITUDE 25.4
SOUTH : LONGITUDE -128.2, LATITUDE 24.5

+ MARKS THE CITY OF LO-I

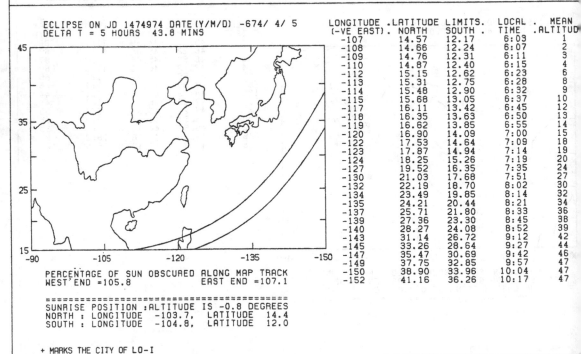

ECLIPSE ON JD 1474974 DATE(Y/M/D) -674/ 4/ 5
DELTA T = 5 HOURS 43.8 MINS

LONGITUDE (-VE EAST)	.LATITUDE LIMITS. NORTH	SOUTH .	LOCAL TIME	MEAN .ALTITUDE
-107	14.57	12.17	6:09	1
-108	14.66	12.24	6:07	2
-109	14.76	12.31	6:11	3
-110	14.87	12.40	6:15	4
-112	15.15	12.62	6:23	6
-113	15.31	12.75	6:28	8
-114	15.48	12.90	6:32	9
-115	15.68	13.05	6:37	10
-117	16.11	13.42	6:45	12
-118	16.35	13.63	6:50	13
-119	16.62	13.85	6:55	14
-120	16.90	14.09	7:00	15
-122	17.53	14.64	7:09	18
-123	17.87	14.94	7:14	19
-124	18.25	15.26	7:19	20
-127	19.52	16.35	7:35	24
-130	21.03	17.68	7:51	27
-132	22.19	18.70	8:02	30
-134	23.49	19.85	8:14	32
-135	24.21	20.44	8:21	34
-137	25.71	21.80	8:33	36
-139	27.36	23.30	8:45	38
-140	28.27	24.08	8:52	39
-143	31.14	26.72	9:12	42
-145	33.26	28.64	9:27	44
-147	35.47	30.69	9:42	46
-149	37.75	32.85	9:57	47
-150	38.90	33.96	10:04	47
-152	41.16	36.26	10:17	47

PERCENTAGE OF SUN OBSCURED ALONG MAP TRACK
WEST END =105.8   EAST END =107.1

===
SUNRISE POSITION :ALTITUDE IS -0.8 DEGREES
NORTH : LONGITUDE -103.7, LATITUDE 14.4
SOUTH : LONGITUDE -104.8, LATITUDE 12.0

+ MARKS THE CITY OF LO-I

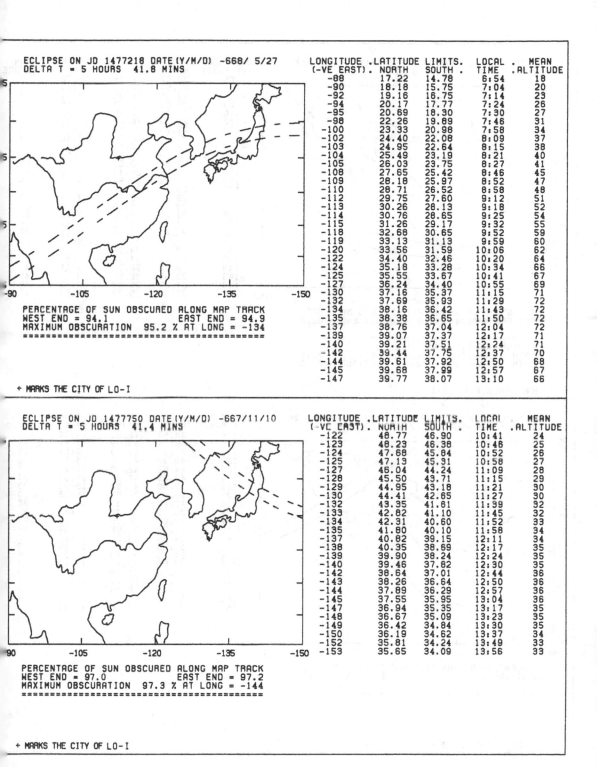

ECLIPSE ON JD 1477218 DATE(Y/M/D) -668/ 5/27
DELTA T = 5 HOURS 41.8 MINS

LONGITUDE (-VE EAST)	LATITUDE LIMITS. NORTH	SOUTH	LOCAL TIME	MEAN ALTITUDE
-88	17.22	14.78	6:54	18
-90	18.18	15.75	7:04	20
-92	19.16	16.75	7:14	23
-94	20.17	17.77	7:24	26
-95	20.69	18.30	7:30	27
-98	22.26	19.89	7:46	31
-100	23.33	20.98	7:58	34
-102	24.40	22.08	8:09	37
-103	24.95	22.64	8:15	38
-104	25.49	23.19	8:21	40
-105	26.03	23.75	8:27	41
-108	27.65	25.42	8:46	45
-109	28.18	25.97	8:52	47
-110	28.71	26.52	8:58	48
-112	29.75	27.60	9:12	51
-113	30.26	28.13	9:18	52
-114	30.76	28.65	9:25	54
-115	31.26	29.17	9:32	55
-118	32.68	30.65	9:52	59
-119	33.13	31.13	9:59	60
-120	33.56	31.59	10:06	62
-122	34.40	32.46	10:20	64
-124	35.18	33.28	10:34	66
-125	35.55	33.67	10:41	67
-127	36.24	34.40	10:55	69
-130	37.16	35.37	11:15	71
-132	37.69	35.93	11:29	72
-134	38.16	36.42	11:43	72
-135	38.38	36.65	11:50	72
-137	38.76	37.04	12:04	72
-139	39.07	37.37	12:17	71
-140	39.21	37.51	12:24	71
-142	39.44	37.75	12:37	70
-144	39.61	37.92	12:50	68
-145	39.68	37.99	12:57	67
-147	39.77	38.07	13:10	66

PERCENTAGE OF SUN OBSCURED ALONG MAP TRACK
WEST END = 94.1   EAST END = 94.9
MAXIMUM OBSCURATION 95.2 % AT LONG = -134
==

+ MARKS THE CITY OF LO-I

ECLIPSE ON JD 1477750 DATE(Y/M/D) -667/11/10
DELTA T = 5 HOURS 41.4 MINS

LONGITUDE (-VE EAST)	LATITUDE LIMITS. NORTH	SOUTH	LOCAL TIME	MEAN ALTITUDE
-122	48.77	46.90	10:41	24
-123	48.23	46.38	10:46	25
-124	47.68	45.84	10:52	26
-125	47.13	45.31	10:58	27
-127	46.04	44.24	11:09	28
-128	45.50	43.71	11:15	29
-129	44.95	43.18	11:21	30
-130	44.41	42.65	11:27	30
-132	43.35	41.61	11:39	32
-133	42.82	41.10	11:45	32
-134	42.31	40.60	11:52	33
-135	41.80	40.10	11:58	34
-137	40.82	39.15	12:11	34
-138	40.35	38.69	12:17	35
-139	39.90	38.24	12:24	35
-140	39.46	37.82	12:30	35
-142	38.64	37.01	12:44	36
-143	38.26	36.64	12:50	36
-144	37.89	36.29	12:57	36
-145	37.55	35.95	13:04	36
-147	36.94	35.35	13:17	35
-148	36.67	35.09	13:23	35
-149	36.42	34.84	13:30	35
-150	36.19	34.62	13:37	34
-152	35.81	34.24	13:49	33
-153	35.65	34.09	13:56	33

PERCENTAGE OF SUN OBSCURED ALONG MAP TRACK
WEST END = 97.0   EAST END = 97.2
MAXIMUM OBSCURATION 97.3 % AT LONG = -144
==

+ MARKS THE CITY OF LO-I

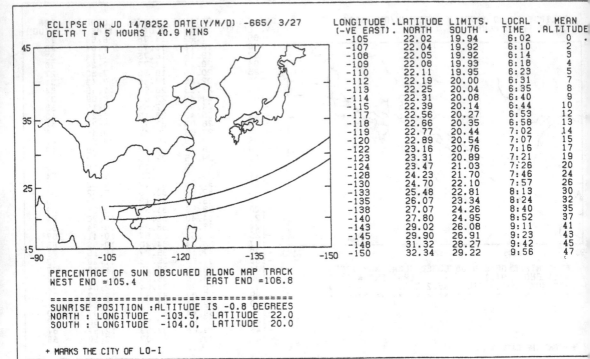

ECLIPSE ON JD 1478252 DATE(Y/M/D) -665/ 3/27
DELTA T = 5 HOURS 40.9 MINS

LONGITUDE (-VE EAST)	.LATITUDE NORTH	LIMITS. SOUTH .	LOCAL TIME	MEAN .ALTITUDE
-105	22.02	19.94	6:02	0
-107	22.04	19.92	6:10	2
-108	22.05	19.92	6:14	3
-109	22.08	19.93	6:18	4
-110	22.11	19.95	6:23	5
-112	22.19	20.00	6:31	7
-113	22.25	20.04	6:35	8
-114	22.31	20.08	6:40	9
-115	22.39	20.14	6:44	10
-117	22.56	20.27	6:53	12
-118	22.66	20.35	6:58	13
-119	22.77	20.44	7:02	14
-120	22.89	20.54	7:07	15
-122	23.16	20.76	7:16	17
-123	23.31	20.89	7:21	19
-124	23.47	21.03	7:26	20
-128	24.23	21.70	7:46	24
-130	24.70	22.10	7:57	26
-133	25.48	22.81	8:13	30
-135	26.07	23.34	8:24	32
-138	27.07	24.26	8:40	35
-140	27.80	24.95	8:52	37
-143	29.02	26.08	9:11	41
-145	29.90	26.91	9:23	43
-148	31.32	28.27	9:42	45
-150	32.34	29.22	9:56	47

PERCENTAGE OF SUN OBSCURED ALONG MAP TRACK
WEST END =105.4   EAST END =106.8

==
SUNRISE POSITION :ALTITUDE IS -0.8 DEGREES
NORTH : LONGITUDE -103.5, LATITUDE 22.0
SOUTH : LONGITUDE -104.0, LATITUDE 20.0

+ MARKS THE CITY OF LO-I

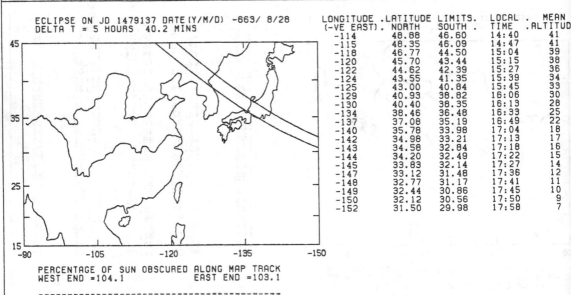

ECLIPSE ON JD 1479137 DATE(Y/M/D) -663/ 8/28
DELTA T = 5 HOURS 40.2 MINS

LONGITUDE (-VE EAST)	.LATITUDE NORTH	LIMITS. SOUTH .	LOCAL TIME	MEAN .ALTITUDE
-114	48.88	46.60	14:40	41
-115	48.35	46.09	14:47	41
-118	46.77	44.50	15:04	39
-120	45.70	43.44	15:15	38
-122	44.62	42.39	15:27	36
-124	43.55	41.35	15:39	34
-125	43.00	40.84	15:45	33
-129	40.93	38.82	16:06	30
-130	40.40	38.35	16:13	28
-134	38.46	36.48	16:33	25
-137	37.08	35.19	16:49	22
-140	35.78	33.98	17:04	18
-142	34.98	33.21	17:13	17
-143	34.58	32.84	17:18	16
-144	34.20	32.49	17:22	15
-145	33.83	32.14	17:27	14
-147	33.12	31.48	17:36	12
-148	32.77	31.17	17:41	11
-149	32.44	30.86	17:45	10
-150	32.12	30.56	17:50	9
-152	31.50	29.98	17:58	7

PERCENTAGE OF SUN OBSCURED ALONG MAP TRACK
WEST END =104.1   EAST END =103.1

==

+ MARKS THE CITY OF LO-I

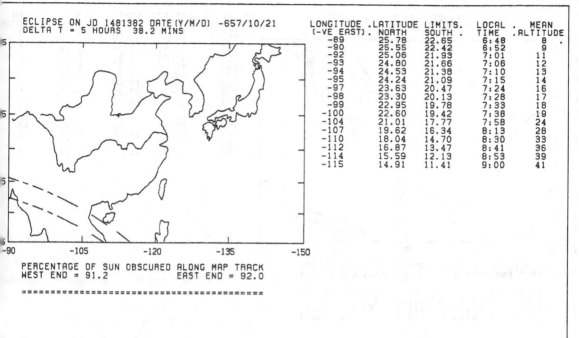

ECLIPSE ON JD 1481382 DATE(Y/M/D) -657/10/21
DELTA T = 5 HOURS 38.2 MINS

LONGITUDE (-VE EAST)	LATITUDE LIMITS. NORTH	SOUTH	LOCAL TIME	MEAN ALTITUDE
-89	25.78	22.65	6:48	8
-90	25.55	22.42	6:52	9
-92	25.06	21.93	7:01	11
-93	24.80	21.66	7:06	12
-94	24.53	21.38	7:10	13
-95	24.24	21.09	7:15	14
-97	23.63	20.47	7:24	16
-98	23.30	20.13	7:28	17
-99	22.95	19.78	7:33	18
-100	22.60	19.42	7:38	19
-104	21.01	17.77	7:58	24
-107	19.62	16.34	8:13	28
-110	18.04	14.70	8:30	33
-112	16.87	13.47	8:41	36
-114	15.59	12.13	8:53	39
-115	14.91	11.41	9:00	41

PERCENTAGE OF SUN OBSCURED ALONG MAP TRACK
WEST END = 91.2   EAST END = 92.0

===

+ MARKS THE CITY OF LO-I

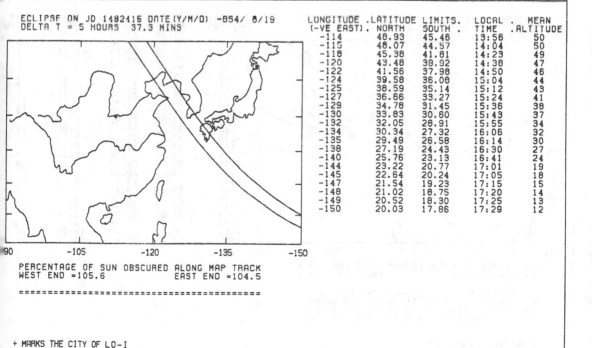

ECLIPSE ON JD 1482415 DATE(Y/M/D) -654/ 8/19
DELTA T = 5 HOURS 37.3 MINS

LONGITUDE (-VE EAST)	LATITUDE LIMITS. NORTH	SOUTH	LOCAL TIME	MEAN ALTITUDE
-114	48.93	45.46	13:56	50
-115	48.07	44.57	14:04	50
-118	45.38	41.81	14:23	49
-120	43.48	39.92	14:38	47
-122	41.56	37.98	14:50	46
-124	39.58	36.08	15:04	44
-125	38.59	35.14	15:12	43
-127	36.66	33.27	15:24	41
-129	34.78	31.45	15:36	38
-130	33.83	30.60	15:43	37
-132	32.05	28.91	15:55	34
-134	30.34	27.32	16:06	32
-135	29.49	26.58	16:14	30
-138	27.19	24.43	16:30	27
-140	25.76	23.13	16:41	24
-144	23.22	20.77	17:01	19
-145	22.64	20.24	17:05	18
-147	21.54	19.23	17:15	15
-148	21.02	18.75	17:20	14
-149	20.52	18.30	17:25	13
-150	20.03	17.86	17:29	12

PERCENTAGE OF SUN OBSCURED ALONG MAP TRACK
WEST END =105.6   EAST END =104.5

===

+ MARKS THE CITY OF LO-I

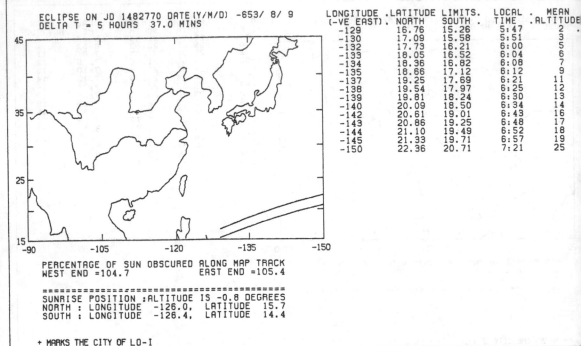

ECLIPSE ON JD 1482770 DATE(Y/M/D) -653/ 8/ 9
DELTA T = 5 HOURS 37.0 MINS

LONGITUDE (-VE EAST)	LATITUDE NORTH	LIMITS SOUTH	LOCAL TIME	MEAN ALTITUDE
-129	16.76	15.26	5:47	2
-130	17.09	15.58	5:51	3
-132	17.73	16.21	6:00	5
-133	18.05	16.52	6:04	6
-134	18.36	16.82	6:08	7
-135	18.66	17.12	6:12	9
-137	19.25	17.69	6:21	11
-138	19.54	17.97	6:25	12
-139	19.81	18.24	6:30	13
-140	20.09	18.50	6:34	14
-142	20.61	19.01	6:43	16
-143	20.86	19.25	6:48	17
-144	21.10	19.49	6:52	18
-145	21.33	19.71	6:57	19
-150	22.36	20.71	7:21	25

PERCENTAGE OF SUN OBSCURED ALONG MAP TRACK
WEST END =104.7   EAST END =105.4

==
SUNRISE POSITION :ALTITUDE IS -0.8 DEGREES
NORTH : LONGITUDE -126.0, LATITUDE 15.7
SOUTH : LONGITUDE -126.4, LATITUDE 14.4

+ MARKS THE CITY OF LO-I

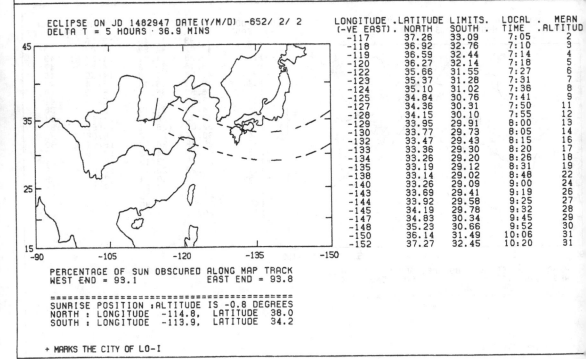

ECLIPSE ON JD 1482947 DATE(Y/M/D) -652/ 2/ 2
DELTA T = 5 HOURS 36.9 MINS

LONGITUDE (-VE EAST)	LATITUDE NORTH	LIMITS SOUTH	LOCAL TIME	MEAN ALTITUD
-117	37.26	33.09	7:05	2
-118	36.92	32.76	7:10	3
-119	36.59	32.44	7:14	4
-120	36.27	32.14	7:18	5
-122	35.66	31.55	7:27	6
-123	35.37	31.28	7:31	7
-124	35.10	31.02	7:36	8
-125	34.84	30.76	7:41	9
-127	34.36	30.31	7:50	11
-128	34.15	30.10	7:55	12
-129	33.95	29.91	8:00	13
-130	33.77	29.73	8:05	14
-132	33.47	29.43	8:15	16
-133	33.36	29.30	8:20	17
-134	33.26	29.20	8:26	18
-135	33.19	29.12	8:31	19
-138	33.14	29.02	8:48	22
-140	33.26	29.09	9:00	24
-143	33.69	29.41	9:19	26
-144	33.92	29.58	9:25	27
-145	34.19	29.78	9:32	28
-147	34.83	30.34	9:45	29
-148	35.23	30.66	9:52	30
-150	36.14	31.49	10:06	31
-152	37.27	32.45	10:20	31

PERCENTAGE OF SUN OBSCURED ALONG MAP TRACK
WEST END = 93.1   EAST END = 93.8

==
SUNRISE POSITION :ALTITUDE IS -0.8 DEGREES
NORTH : LONGITUDE -114.8, LATITUDE 38.0
SOUTH : LONGITUDE -113.9, LATITUDE 34.2

+ MARKS THE CITY OF LO-I

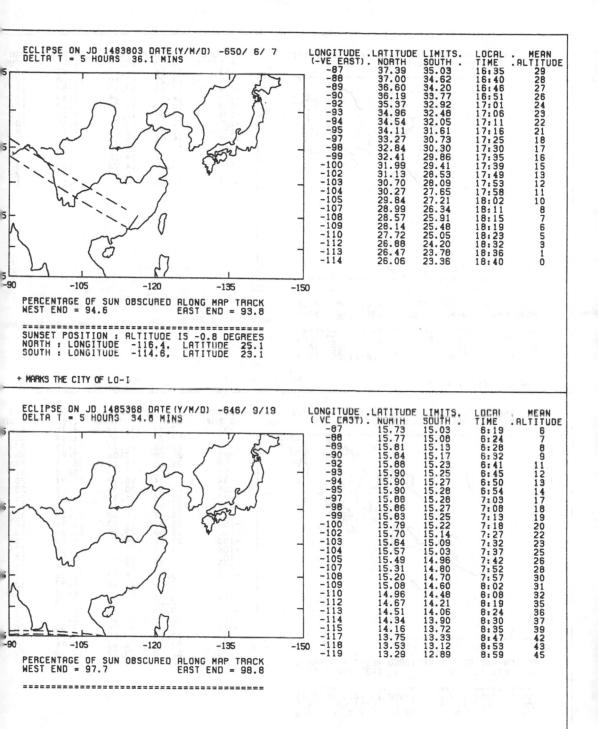

ECLIPSE ON JD 1483803 DATE (Y/M/D) -650/ 6/ 7
DELTA T = 5 HOURS 36.1 MINS

LONGITUDE (-VE EAST)	LATITUDE LIMITS. NORTH	SOUTH	LOCAL TIME	MEAN ALTITUDE
-87	37.39	35.03	16:35	29
-88	37.00	34.62	16:40	28
-89	36.60	34.20	16:46	27
-90	36.19	33.77	16:51	26
-92	35.37	32.92	17:01	24
-93	34.96	32.48	17:06	23
-94	34.54	32.05	17:11	22
-95	34.11	31.61	17:16	21
-97	33.27	30.73	17:25	18
-98	32.84	30.30	17:30	17
-99	32.41	29.86	17:35	16
-100	31.99	29.41	17:39	15
-102	31.13	28.53	17:49	13
-103	30.70	28.09	17:53	12
-104	30.27	27.65	17:58	11
-105	29.84	27.21	18:02	10
-107	28.99	26.34	18:11	8
-108	28.57	25.91	18:15	7
-109	28.14	25.48	18:19	6
-110	27.72	25.05	18:23	5
-112	26.88	24.20	18:32	3
-113	26.47	23.78	18:36	1
-114	26.06	23.36	18:40	0

PERCENTAGE OF SUN OBSCURED ALONG MAP TRACK
WEST END = 94.6   EAST END = 93.8

===
SUNSET POSITION : ALTITUDE IS -0.8 DEGREES
NORTH : LONGITUDE -116.4, LATITUDE 25.1
SOUTH : LONGITUDE -114.6, LATITUDE 23.1

+ MARKS THE CITY OF LO-I

ECLIPSE ON JD 1485368 DATE (Y/M/D) -646/ 9/19
DELTA T = 5 HOURS 34.8 MINS

LONGITUDE (-VE EAST)	LATITUDE LIMITS. NORTH	SOUTH	LOCAL TIME	MEAN ALTITUDE
-87	15.73	15.03	6:19	6
-88	15.77	15.08	6:24	7
-89	15.81	15.13	6:28	8
-90	15.84	15.17	6:32	9
-92	15.88	15.23	6:41	11
-93	15.90	15.25	6:45	12
-94	15.90	15.27	6:50	13
-95	15.90	15.28	6:54	14
-97	15.88	15.28	7:03	17
-98	15.86	15.27	7:08	18
-99	15.83	15.25	7:13	19
-100	15.79	15.22	7:18	20
-102	15.70	15.14	7:27	22
-103	15.64	15.09	7:32	23
-104	15.57	15.03	7:37	25
-105	15.49	14.96	7:42	26
-107	15.31	14.80	7:52	28
-108	15.20	14.70	7:57	30
-109	15.08	14.60	8:02	31
-110	14.96	14.48	8:08	32
-112	14.67	14.21	8:19	35
-113	14.51	14.06	8:24	36
-114	14.34	13.90	8:30	37
-115	14.16	13.72	8:35	39
-117	13.75	13.33	8:47	42
-118	13.53	13.12	8:53	43
-119	13.29	12.89	8:59	45

PERCENTAGE OF SUN OBSCURED ALONG MAP TRACK
WEST END = 97.7   EAST END = 98.8

===

+ MARKS THE CITY OF LO-I

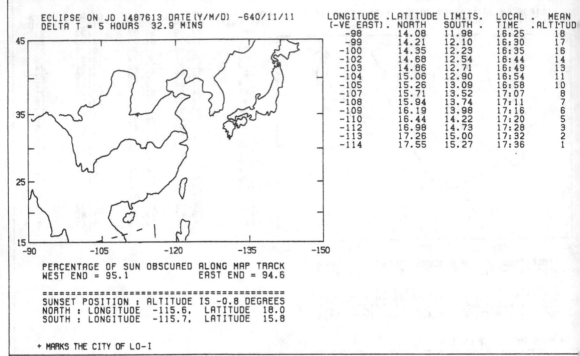

ECLIPSE ON JD 1487613 DATE(Y/M/D) -640/11/11
DELTA T = 5 HOURS 32.9 MINS

LONGITUDE (-VE EAST)	LATITUDE LIMITS. NORTH	SOUTH	LOCAL TIME	MEAN ALTITUD
-98	14.08	11.98	16:25	18
-99	14.21	12.10	16:30	17
-100	14.35	12.23	16:35	16
-102	14.68	12.54	16:44	14
-103	14.86	12.71	16:49	13
-104	15.06	12.90	16:54	11
-105	15.26	13.09	16:58	10
-107	15.71	13.52	17:07	8
-108	15.94	13.74	17:11	7
-109	16.19	13.98	17:16	6
-110	16.44	14.22	17:20	5
-112	16.98	14.73	17:28	3
-113	17.26	15.00	17:32	2
-114	17.55	15.27	17:36	1

PERCENTAGE OF SUN OBSCURED ALONG MAP TRACK
WEST END = 95.1   EAST END = 94.6

==
SUNSET POSITION : ALTITUDE IS -0.8 DEGREES
NORTH : LONGITUDE -115.6, LATITUDE 18.0
SOUTH : LONGITUDE -115.7, LATITUDE 15.8

+ MARKS THE CITY OF LO-I

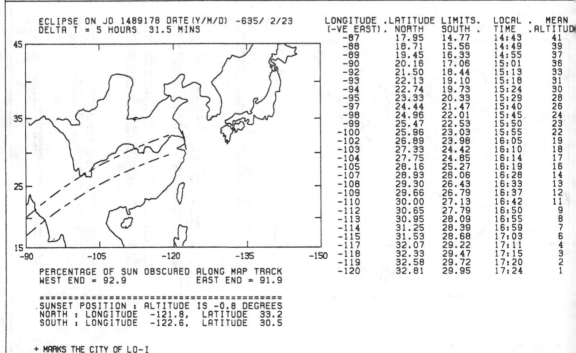

ECLIPSE ON JD 1489178 DATE(Y/M/D) -635/ 2/23
DELTA T = 5 HOURS 31.5 MINS

LONGITUDE (-VE EAST)	LATITUDE LIMITS. NORTH	SOUTH	LOCAL TIME	MEAN ALTITUD
-87	17.95	14.77	14:43	41
-88	18.71	15.56	14:49	39
-89	19.45	16.33	14:55	37
-90	20.16	17.06	15:01	36
-92	21.50	18.44	15:13	33
-93	22.13	19.10	15:18	31
-94	22.74	19.73	15:24	30
-95	23.33	20.33	15:29	28
-97	24.44	21.47	15:40	26
-98	24.96	22.01	15:45	24
-99	25.47	22.53	15:50	23
-100	25.96	23.03	15:55	22
-102	26.89	23.98	16:05	19
-103	27.33	24.42	16:10	18
-104	27.75	24.85	16:14	17
-105	28.16	25.27	16:19	16
-107	28.93	26.06	16:28	14
-108	29.30	26.43	16:33	13
-109	29.66	26.79	16:37	12
-110	30.00	27.13	16:42	11
-112	30.65	27.79	16:50	9
-113	30.95	28.09	16:55	8
-114	31.25	28.39	16:59	7
-115	31.53	28.68	17:03	6
-117	32.07	29.22	17:11	4
-118	32.33	29.47	17:15	3
-119	32.58	29.72	17:20	2
-120	32.81	29.95	17:24	1

PERCENTAGE OF SUN OBSCURED ALONG MAP TRACK
WEST END = 92.9   EAST END = 91.9

==
SUNSET POSITION : ALTITUDE IS -0.8 DEGREES
NORTH : LONGITUDE -121.8, LATITUDE 33.2
SOUTH : LONGITUDE -122.6, LATITUDE 30.5

+ MARKS THE CITY OF LO-I

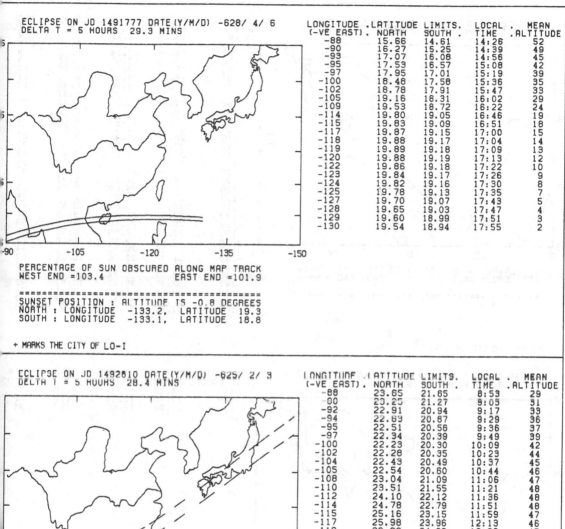

ECLIPSE ON JD 1491777 DATE(Y/M/D) -628/ 4/ 6
DELTA T = 5 HOURS 29.3 MINS

LONGITUDE (-VE EAST)	LATITUDE NORTH	LIMITS SOUTH	LOCAL TIME	MEAN ALTITUDE
-88	15.66	14.61	14:26	52
-90	16.27	15.25	14:39	49
-93	17.07	16.08	14:56	45
-95	17.53	16.57	15:08	42
-97	17.95	17.01	15:19	39
-100	18.48	17.58	15:36	35
-102	18.78	17.91	15:47	33
-105	19.16	18.31	16:02	29
-109	19.53	18.72	16:22	24
-114	19.80	19.05	16:46	19
-115	19.83	19.09	16:51	18
-117	19.87	19.15	17:00	15
-118	19.88	19.17	17:04	14
-119	19.89	19.18	17:09	13
-120	19.88	19.19	17:13	12
-122	19.86	19.18	17:22	10
-123	19.84	19.17	17:26	9
-124	19.82	19.16	17:30	8
-125	19.78	19.13	17:35	7
-127	19.70	19.07	17:43	5
-128	19.65	19.03	17:47	4
-129	19.60	18.99	17:51	3
-130	19.54	18.94	17:55	2

PERCENTAGE OF SUN OBSCURED ALONG MAP TRACK
WEST END =103.4   EAST END =101.9

===
SUNSET POSITION : ALTITUDE IS -0.8 DEGREES
NORTH : LONGITUDE -133.2, LATITUDE 19.3
SOUTH : LONGITUDE -133.1, LATITUDE 18.8

+ MARKS THE CITY OF LO-I

ECLIPSE ON JD 1492810 DATE(Y/M/D) -625/ 2/ 3
DELTA T = 5 HOURS 28.4 MINS

LONGITUDE (-VE EAST)	LATITUDE NORTH	LIMITS SOUTH	LOCAL TIME	MEAN ALTITUDE
-88	23.65	21.65	8:53	29
-90	23.25	21.27	9:05	51
-92	22.91	20.94	9:17	33
-94	22.69	20.67	9:29	36
-95	22.51	20.56	9:36	37
-97	22.34	20.39	9:49	39
-100	22.23	20.30	10:09	42
-102	22.28	20.35	10:23	44
-104	22.43	20.49	10:37	45
-105	22.54	20.60	10:44	46
-108	23.04	21.09	11:06	47
-110	23.51	21.55	11:21	48
-112	24.10	22.12	11:36	48
-114	24.78	22.79	11:51	48
-115	25.16	23.15	11:59	47
-117	25.98	23.96	12:13	46
-119	26.88	24.83	12:28	45
-120	27.36	25.30	12:35	44
-123	28.86	26.77	12:56	42
-125	29.92	27.80	13:10	40
-127	30.99	28.86	13:23	38
-130	32.63	30.47	13:42	34
-133	34.29	32.08	14:00	31
-135	35.38	33.15	14:11	29
-138	36.98	34.74	14:28	25
-140	38.04	35.76	14:38	23
-144	40.07	37.77	14:59	19
-145	40.57	38.26	15:04	18
-147	41.54	39.21	15:13	16
-148	42.02	39.68	15:18	15
-149	42.49	40.14	15:23	14
-150	42.95	40.59	15:27	13
-152	43.86	41.48	15:36	11

PERCENTAGE OF SUN OBSCURED ALONG MAP TRACK
WEST END = 95.1   EAST END = 94.8
MAXIMUM OBSCURATION 95.7 % AT LONG = -110
===

+ MARKS THE CITY OF LO-I

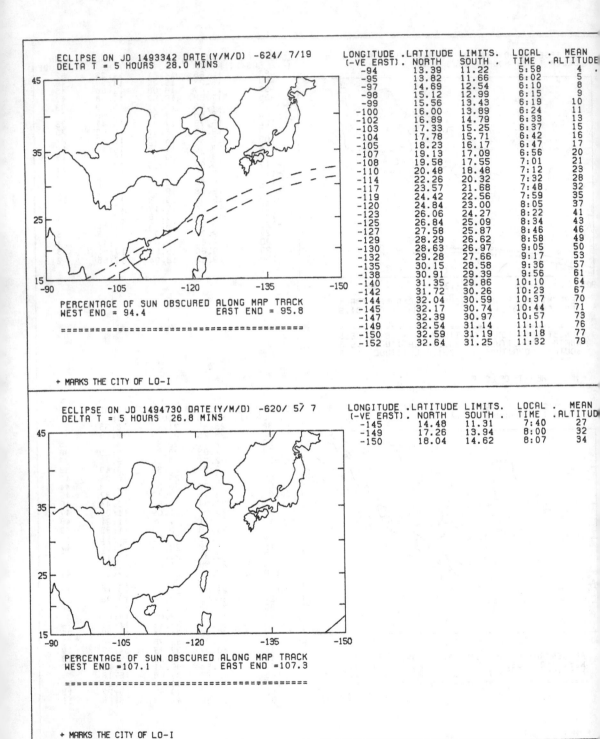

ECLIPSE ON JD 1493342 DATE(Y/M/D) -624/ 7/19
DELTA T = 5 HOURS 28.0 MINS

LONGITUDE (-VE EAST)	LATITUDE NORTH	LIMITS SOUTH	LOCAL TIME	MEAN ALTITUDE
-94	13.39	11.22	5:58	4
-95	13.82	11.66	6:02	5
-97	14.69	12.54	6:10	8
-98	15.12	12.99	6:15	9
-99	15.56	13.43	6:19	10
-100	16.00	13.89	6:24	11
-102	16.89	14.79	6:33	13
-103	17.33	15.25	6:37	15
-104	17.78	15.71	6:42	16
-105	18.23	16.17	6:47	17
-107	19.13	17.09	6:56	20
-108	19.58	17.55	7:01	21
-110	20.48	18.48	7:12	23
-114	22.26	20.32	7:32	28
-117	23.57	21.68	7:48	32
-119	24.42	22.56	7:59	35
-120	24.84	23.00	8:05	37
-123	26.06	24.27	8:22	41
-125	26.84	25.09	8:34	43
-127	27.58	25.87	8:46	46
-129	28.29	26.62	8:58	49
-130	28.63	26.97	9:05	50
-132	29.28	27.66	9:17	53
-135	30.15	28.58	9:36	57
-138	30.91	29.39	9:56	61
-140	31.35	29.86	10:10	64
-142	31.72	30.26	10:23	67
-144	32.04	30.59	10:37	70
-145	32.17	30.74	10:44	71
-147	32.39	30.97	10:57	73
-149	32.54	31.14	11:11	76
-150	32.59	31.19	11:18	77
-152	32.64	31.25	11:32	79

PERCENTAGE OF SUN OBSCURED ALONG MAP TRACK
WEST END = 94.4   EAST END = 95.8

==

+ MARKS THE CITY OF LO-I

ECLIPSE ON JD 1494730 DATE(Y/M/D) -620/ 5/ 7
DELTA T = 5 HOURS 26.8 MINS

LONGITUDE (-VE EAST)	LATITUDE NORTH	LIMITS SOUTH	LOCAL TIME	MEAN ALTITUDE
-145	14.48	11.31	7:40	27
-149	17.26	13.94	8:00	32
-150	18.04	14.62	8:07	34

PERCENTAGE OF SUN OBSCURED ALONG MAP TRACK
WEST END =107.1   EAST END =107.3

==

+ MARKS THE CITY OF LO-I

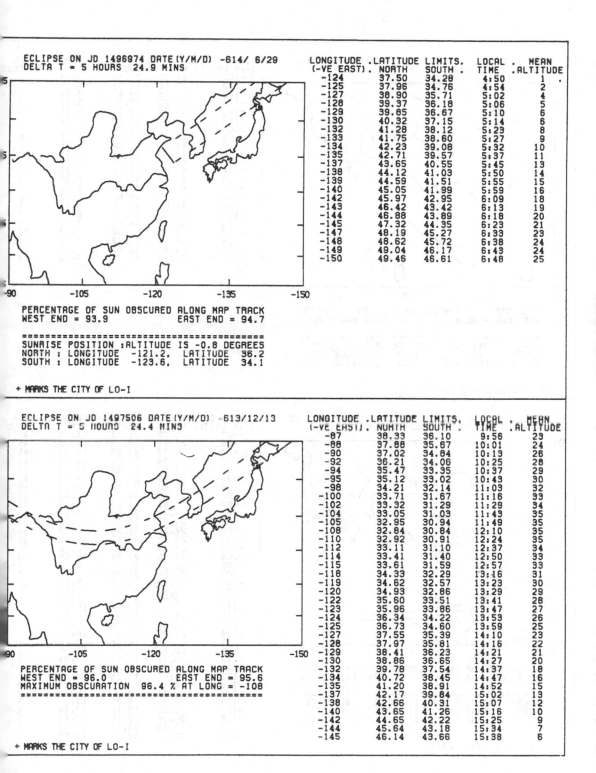

ECLIPSE ON JD 1496974 DATE(Y/M/D) -614/ 6/29
DELTA T = 5 HOURS 24.9 MINS

LONGITUDE (-VE EAST)	LATITUDE LIMITS. NORTH	SOUTH	LOCAL TIME	MEAN ALTITUDE
-124	37.50	34.28	4:50	1
-125	37.96	34.76	4:54	2
-127	38.90	35.71	5:02	4
-128	39.37	36.18	5:06	5
-129	39.85	36.67	5:10	6
-130	40.32	37.15	5:14	6
-132	41.28	38.12	5:23	8
-133	41.75	38.60	5:27	9
-134	42.23	39.08	5:32	10
-135	42.71	39.57	5:37	11
-137	43.65	40.55	5:45	13
-138	44.12	41.03	5:50	14
-139	44.59	41.51	5:55	15
-140	45.05	41.99	5:59	16
-142	45.97	42.95	6:09	18
-143	46.42	43.42	6:13	19
-144	46.88	43.89	6:18	20
-145	47.32	44.35	6:23	21
-147	48.19	45.27	6:93	23
-148	48.62	45.72	6:38	24
-149	49.04	46.17	6:43	24
-150	49.46	46.61	6:48	25

PERCENTAGE OF SUN OBSCURED ALONG MAP TRACK
WEST END = 93.9   EAST END = 94.7

=======================================
SUNRISE POSITION :ALTITUDE IS -0.8 DEGREES
NORTH : LONGITUDE -121.2, LATITUDE 36.2
SOUTH : LONGITUDE -123.6, LATITUDE 34.1

+ MARKS THE CITY OF LO-I

ECLIPSE ON JD 1497506 DATE(Y/M/D) -613/12/13
DELTA T = 5 HOURS 24.4 MINS

LONGITUDE (-VE EAST)	LATITUDE LIMITS. NORTH	SOUTH	LOCAL TIME	MEAN ALTITUDE
-87	38.33	36.10	9:56	23
-88	37.88	35.67	10:01	24
-90	37.02	34.84	10:13	26
-92	36.21	34.06	10:25	28
-94	35.47	33.35	10:37	29
-95	35.12	33.02	10:43	30
-98	34.21	32.14	11:03	32
-100	33.71	31.67	11:16	33
-102	33.32	31.29	11:29	34
-104	33.05	31.03	11:43	35
-105	32.95	30.94	11:49	35
-108	32.84	30.84	12:10	35
-110	32.92	30.91	12:24	35
-112	33.11	31.10	12:37	34
-114	33.41	31.40	12:50	33
-115	33.61	31.59	12:57	33
-118	34.33	32.29	13:16	31
-119	34.62	32.57	13:23	30
-120	34.93	32.86	13:29	29
-122	35.60	33.51	13:41	28
-123	35.96	33.86	13:47	27
-124	36.34	34.22	13:53	26
-125	36.73	34.60	13:59	25
-127	37.55	35.39	14:10	23
-128	37.97	35.81	14:16	22
-129	38.41	36.23	14:21	21
-130	38.86	36.65	14:27	20
-132	39.78	37.54	14:37	18
-134	40.72	38.45	14:47	16
-135	41.20	38.91	14:52	15
-137	42.17	39.84	15:02	13
-138	42.66	40.31	15:07	12
-140	43.65	41.26	15:16	10
-142	44.65	42.22	15:25	9
-144	45.64	43.18	15:34	7
-145	46.14	43.66	15:38	6

PERCENTAGE OF SUN OBSCURED ALONG MAP TRACK
WEST END = 96.0   EAST END = 95.6
MAXIMUM OBSCURATION 96.4 % AT LONG = -108
=======================================

+ MARKS THE CITY OF LO-I

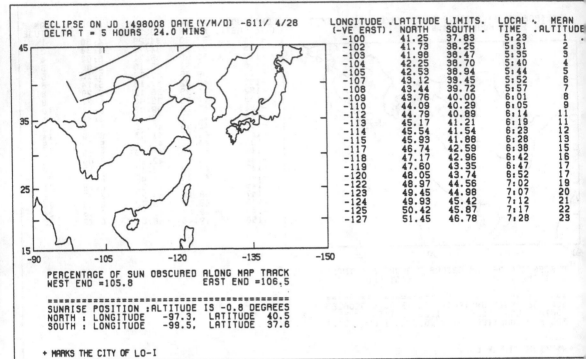

ECLIPSE ON JD 1498008 DATE(Y/M/D) -611/ 4/28
DELTA T = 5 HOURS 24.0 MINS

LONGITUDE (-VE EAST)	LATITUDE LIMITS NORTH	SOUTH	LOCAL TIME	MEAN ALTITUDE
-100	41.25	37.83	5:23	1
-102	41.73	38.25	5:31	2
-103	41.98	38.47	5:35	3
-104	42.25	38.70	5:40	4
-105	42.53	38.94	5:44	5
-107	43.12	39.45	5:52	6
-108	43.44	39.72	5:57	7
-109	43.76	40.00	6:01	8
-110	44.09	40.29	6:05	9
-112	44.79	40.89	6:14	11
-113	45.17	41.21	6:19	11
-114	45.54	41.54	6:23	12
-115	45.93	41.88	6:28	13
-117	46.74	42.59	6:38	15
-118	47.17	42.96	6:42	16
-119	47.60	43.35	6:47	17
-120	48.05	43.74	6:52	17
-122	48.97	44.56	7:02	19
-123	49.45	44.98	7:07	20
-124	49.93	45.42	7:12	21
-125	50.42	45.87	7:17	22
-127	51.45	46.78	7:28	23

PERCENTAGE OF SUN OBSCURED ALONG MAP TRACK
WEST END =105.8   EAST END =106.5

===
SUNRISE POSITION : ALTITUDE IS -0.8 DEGREES
NORTH : LONGITUDE -97.3, LATITUDE 40.5
SOUTH : LONGITUDE -99.5, LATITUDE 37.6

+ MARKS THE CITY OF LO-I

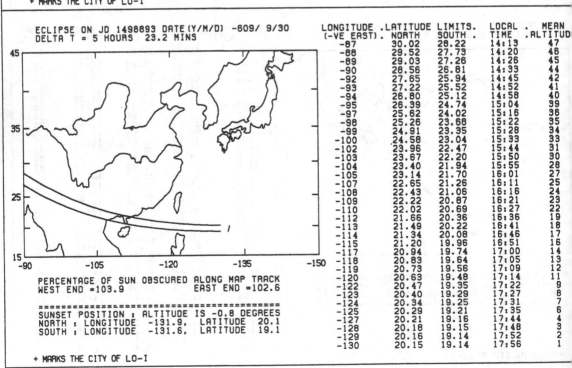

ECLIPSE ON JD 1498893 DATE(Y/M/D) -609/ 9/30
DELTA T = 5 HOURS 23.2 MINS

LONGITUDE (-VE EAST)	LATITUDE LIMITS NORTH	SOUTH	LOCAL TIME	MEAN ALTITUDE
-87	30.02	28.22	14:13	47
-88	29.52	27.73	14:20	46
-89	29.03	27.26	14:26	45
-90	28.56	26.81	14:33	44
-92	27.65	25.94	14:45	42
-93	27.22	25.52	14:52	41
-94	26.80	25.12	14:58	40
-95	26.39	24.74	15:04	39
-97	25.62	24.02	15:16	36
-98	25.26	23.68	15:22	35
-99	24.91	23.35	15:28	34
-100	24.58	23.04	15:33	33
-102	23.96	22.47	15:44	31
-103	23.67	22.20	15:50	30
-104	23.40	21.94	15:55	28
-105	23.14	21.70	16:01	27
-107	22.65	21.26	16:11	25
-108	22.43	21.06	16:16	24
-109	22.22	20.87	16:21	23
-110	22.02	20.69	16:27	22
-112	21.66	20.36	16:36	19
-113	21.49	20.22	16:41	18
-114	21.34	20.08	16:46	17
-115	21.20	19.96	16:51	16
-117	20.94	19.74	17:00	14
-118	20.83	19.64	17:05	13
-119	20.73	19.56	17:09	12
-120	20.63	19.48	17:14	11
-122	20.47	19.35	17:22	9
-123	20.40	19.29	17:27	8
-124	20.34	19.25	17:31	7
-125	20.29	19.21	17:35	6
-127	20.21	19.16	17:44	4
-128	20.18	19.15	17:48	3
-129	20.16	19.14	17:52	2
-130	20.15	19.14	17:56	1

PERCENTAGE OF SUN OBSCURED ALONG MAP TRACK
WEST END =103.9   EAST END =102.6

===
SUNSET POSITION : ALTITUDE IS -0.8 DEGREES
NORTH : LONGITUDE -131.9, LATITUDE 20.1
SOUTH : LONGITUDE -131.6, LATITUDE 19.1

+ MARKS THE CITY OF LO-I

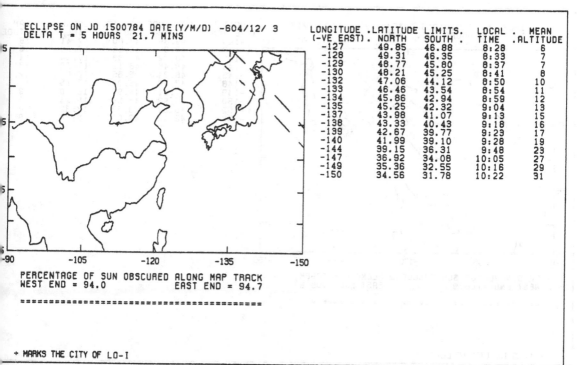

ECLIPSE ON JD 1500784 DATE(Y/M/D) -604/12/ 3
DELTA T = 5 HOURS 21.7 MINS

LONGITUDE (-VE EAST)	LATITUDE LIMITS. NORTH	SOUTH .	LOCAL TIME	MEAN ALTITUDE
-127	49.85	46.88	8:28	6
-128	49.31	46.35	8:33	7
-129	48.77	45.80	8:37	7
-130	48.21	45.25	8:41	8
-132	47.06	44.12	8:50	10
-133	46.46	43.54	8:54	11
-134	45.86	42.94	8:59	12
-135	45.25	42.32	9:04	13
-137	43.98	41.07	9:13	15
-138	43.33	40.43	9:18	16
-139	42.67	39.77	9:23	17
-140	41.99	39.10	9:28	19
-144	99.15	36.31	9:48	23
-147	36.92	34.08	10:05	27
-149	35.36	32.55	10:16	29
-150	34.56	31.78	10:22	31

PERCENTAGE OF SUN OBSCURED ALONG MAP TRACK
WEST END = 94.0   EAST END = 94.7

===

+ MARKS THE CITY OF LO-I

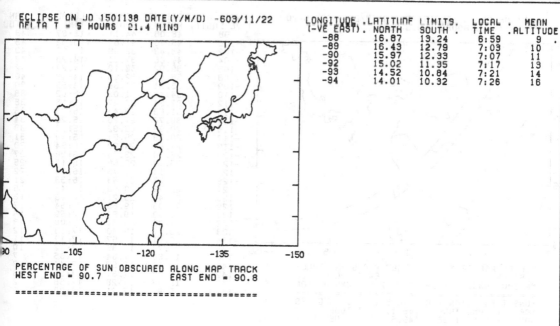

ECLIPSE ON JD 1501138 DATE(Y/M/D) -603/11/22
DELTA T = 5 HOURS 21.4 MINS

LONGITUDE (-VE EAST)	LATITUDE LIMITS. NORTH	SOUTH .	LOCAL TIME	MEAN ALTITUDE
-88	16.87	13.24	6:59	9
-89	16.43	12.79	7:03	10
-90	15.97	12.33	7:07	11
-92	15.02	11.35	7:17	13
-93	14.52	10.84	7:21	14
-94	14.01	10.32	7:26	16

PERCENTAGE OF SUN OBSCURED ALONG MAP TRACK
WEST END = 90.7   EAST END = 90.8

===

+ MARKS THE CITY OF LO-I

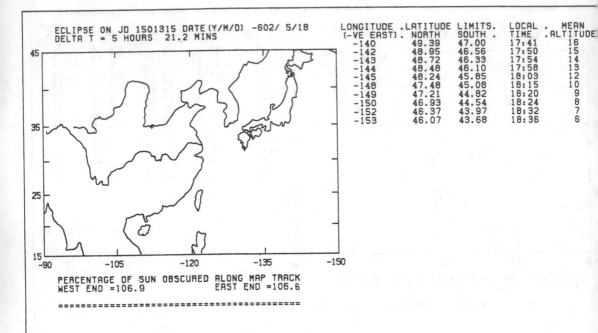

ECLIPSE ON JD 1501315 DATE(Y/M/D) -602/ 5/18
DELTA T = 5 HOURS 21.2 MINS

LONGITUDE	.LATITUDE	LIMITS.	LOCAL .	MEAN
(-VE EAST).	NORTH	SOUTH .	TIME	.ALTITUDE
-140	49.39	47.00	17:41	16
-142	48.95	46.56	17:50	15
-143	48.72	46.33	17:54	14
-144	48.48	46.10	17:58	13
-145	48.24	45.85	18:03	12
-148	47.48	45.08	18:15	10
-149	47.21	44.82	18:20	9
-150	46.93	44.54	18:24	8
-152	46.37	43.97	18:32	7
-153	46.07	43.68	18:36	6

PERCENTAGE OF SUN OBSCURED ALONG MAP TRACK
WEST END =106.9   EAST END =106.6

==

+ MARKS THE CITY OF LO-I

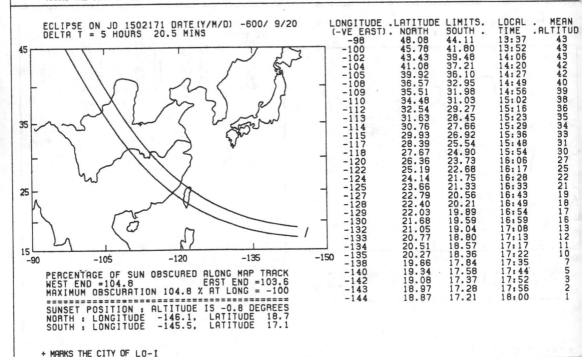

ECLIPSE ON JD 1502171 DATE(Y/M/D) -600/ 9/20
DELTA T = 5 HOURS 20.5 MINS

LONGITUDE	.LATITUDE	LIMITS.	LOCAL .	MEAN
(-VE EAST).	NORTH	SOUTH .	TIME	.ALTITUD
-98	48.08	44.11	13:37	43
-100	45.78	41.80	13:52	43
-102	43.43	39.48	14:06	43
-104	41.08	37.21	14:20	42
-105	39.92	36.10	14:27	42
-108	36.57	32.95	14:49	40
-109	35.51	31.98	14:56	39
-110	34.48	31.03	15:02	38
-112	32.54	29.27	15:16	36
-113	31.63	28.45	15:23	35
-114	30.76	27.66	15:29	34
-115	29.93	26.92	15:36	33
-117	28.39	25.54	15:48	31
-118	27.67	24.90	15:54	30
-120	26.36	23.73	16:06	27
-122	25.19	22.68	16:17	25
-124	24.14	21.75	16:28	22
-125	23.66	21.33	16:33	21
-127	22.79	20.56	16:43	19
-128	22.40	20.21	16:49	18
-129	22.03	19.89	16:54	17
-130	21.68	19.59	16:59	16
-132	21.05	19.04	17:08	13
-133	20.77	18.80	17:13	12
-134	20.51	18.57	17:17	11
-135	20.27	18.36	17:22	10
-138	19.66	17.84	17:35	7
-140	19.34	17.58	17:44	5
-142	19.08	17.37	17:52	3
-143	18.97	17.28	17:56	2
-144	18.87	17.21	18:00	1

PERCENTAGE OF SUN OBSCURED ALONG MAP TRACK
WEST END =104.8   EAST END =103.6
MAXIMUM OBSCURATION 104.8 % AT LONG = -100
==
SUNSET POSITION : ALTITUDE IS -0.8 DEGREES
NORTH : LONGITUDE -146.1, LATITUDE 18.7
SOUTH : LONGITUDE -145.5, LATITUDE 17.1

+ MARKS THE CITY OF LO-I

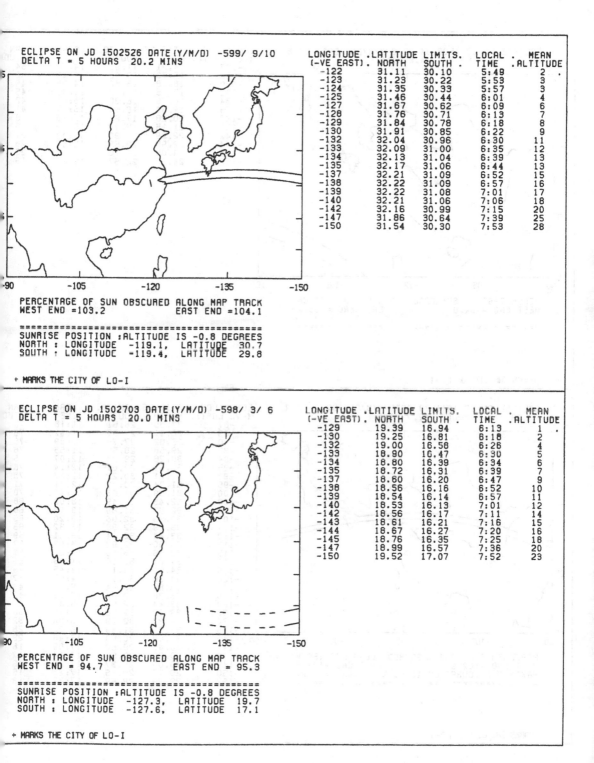

ECLIPSE ON JD 1502526 DATE(Y/M/D) -599/ 9/10
DELTA T = 5 HOURS 20.2 MINS

LONGITUDE (-VE EAST)	LATITUDE LIMITS. NORTH	SOUTH	LOCAL TIME	MEAN ALTITUDE
-122	31.11	30.10	5:49	2
-123	31.23	30.22	5:53	3
-124	31.35	30.33	5:57	3
-125	31.46	30.44	6:01	4
-127	31.67	30.62	6:09	6
-128	31.76	30.71	6:13	7
-129	31.84	30.78	6:18	8
-130	31.91	30.85	6:22	9
-132	32.04	30.96	6:30	11
-133	32.09	31.00	6:35	12
-134	32.13	31.04	6:39	13
-135	32.17	31.06	6:44	13
-137	32.21	31.09	6:52	15
-138	32.22	31.09	6:57	16
-139	32.22	31.08	7:01	17
-140	32.21	31.06	7:06	18
-142	32.16	30.99	7:15	20
-147	31.86	30.64	7:39	25
-150	31.54	30.30	7:53	28

PERCENTAGE OF SUN OBSCURED ALONG MAP TRACK
WEST END =103.2   EAST END =104.1

===
SUNRISE POSITION :ALTITUDE IS -0.8 DEGREES
NORTH : LONGITUDE -119.1, LATITUDE 30.7
SOUTH : LONGITUDE -119.4, LATITUDE 29.8

+ MARKS THE CITY OF LO-I

ECLIPSE ON JD 1502703 DATE(Y/M/D) -598/ 3/ 6
DELTA T = 5 HOURS 20.0 MINS

LONGITUDE (-VE EAST)	LATITUDE LIMITS. NORTH	SOUTH	LOCAL TIME	MEAN ALTITUDE
-129	19.39	16.94	6:13	1
-130	19.25	16.81	6:18	2
-132	19.00	16.58	6:26	4
-133	18.90	16.47	6:30	5
-134	18.80	16.39	6:34	6
-135	18.72	16.31	6:39	7
-137	18.60	16.20	6:47	9
-138	18.56	16.16	6:52	10
-139	18.54	16.14	6:57	11
-140	18.53	16.13	7:01	12
-142	18.56	16.17	7:11	14
-143	18.61	16.21	7:16	15
-144	18.67	16.27	7:20	16
-145	18.76	16.35	7:25	18
-147	18.99	16.57	7:36	20
-150	19.52	17.07	7:52	23

PERCENTAGE OF SUN OBSCURED ALONG MAP TRACK
WEST END = 94.7   EAST END = 95.3

===
SUNRISE POSITION :ALTITUDE IS -0.8 DEGREES
NORTH : LONGITUDE -127.3, LATITUDE 19.7
SOUTH : LONGITUDE -127.6, LATITUDE 17.1

+ MARKS THE CITY OF LO-I

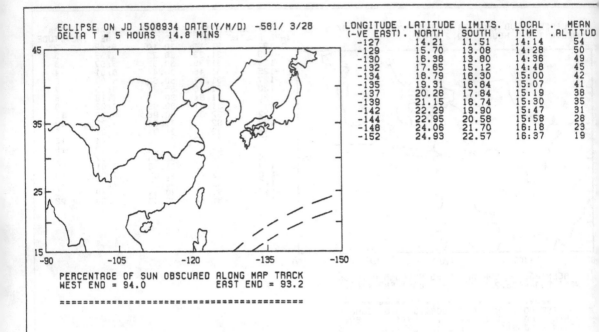

ECLIPSE ON JD 1508934 DATE(Y/M/D) -581/ 3/28
DELTA T = 5 HOURS 14.8 MINS

LONGITUDE (-VE EAST)	.LATITUDE NORTH	LIMITS. SOUTH	LOCAL TIME	. MEAN .ALTITUD
-127	14.21	11.51	14:14	54
-129	15.70	13.08	14:28	50
-130	16.38	13.80	14:36	49
-132	17.65	15.12	14:48	45
-134	18.79	16.30	15:00	42
-135	19.31	16.84	15:07	41
-137	20.28	17.84	15:19	38
-139	21.15	18.74	15:30	35
-142	22.29	19.90	15:47	31
-144	22.95	20.58	15:58	28
-148	24.06	21.70	16:18	23
-152	24.93	22.57	16:37	19

PERCENTAGE OF SUN OBSCURED ALONG MAP TRACK
WEST END = 94.0   EAST END = 93.2

===

+ MARKS THE CITY OF LO-I

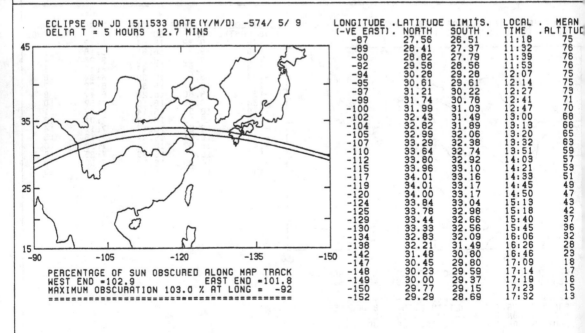

ECLIPSE ON JD 1511533 DATE(Y/M/D) -574/ 5/ 9
DELTA T = 5 HOURS 12.7 MINS

LONGITUDE (-VE EAST)	.LATITUDE NORTH	LIMITS. SOUTH	LOCAL TIME	. MEAN .ALTITUD
-87	27.56	26.51	11:18	75
-89	28.41	27.37	11:32	76
-90	28.82	27.79	11:39	76
-92	29.58	28.56	11:53	76
-94	30.28	29.28	12:07	75
-95	30.61	29.61	12:14	75
-97	31.21	30.22	12:27	73
-99	31.74	30.78	12:41	71
-100	31.99	31.03	12:47	70
-102	32.43	31.49	13:00	68
-104	32.82	31.89	13:13	66
-105	32.99	32.06	13:20	65
-107	33.29	32.38	13:32	63
-110	33.64	32.74	13:51	59
-112	33.80	32.92	14:03	57
-115	33.96	33.10	14:21	53
-117	34.01	33.16	14:33	51
-119	34.01	33.17	14:45	49
-120	34.00	33.17	14:50	47
-124	33.84	33.04	15:13	43
-125	33.78	32.98	15:18	42
-129	33.44	32.66	15:40	37
-130	33.33	32.56	15:45	36
-134	32.83	32.09	16:06	32
-138	32.21	31.49	16:26	28
-142	31.48	30.80	16:46	23
-147	30.45	29.80	17:09	18
-148	30.23	29.59	17:14	17
-149	30.00	29.37	17:19	16
-150	29.77	29.15	17:23	15
-152	29.29	28.69	17:32	13

PERCENTAGE OF SUN OBSCURED ALONG MAP TRACK
WEST END =102.9   EAST END =101.8
MAXIMUM OBSCURATION 103.0 % AT LONG = -92
===

+ MARKS THE CITY OF LO-I

ECLIPSE ON JD 1512064 DATE(Y/M/D) -573/10/22
DELTA T = 5 HOURS 12.2 MINS

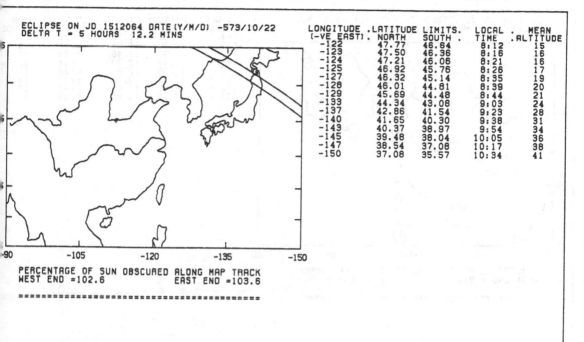

LONGITUDE (-VE EAST)	LATITUDE NORTH	LIMITS SOUTH	LOCAL TIME	MEAN ALTITUDE
-122	47.77	46.64	8:12	15
-123	47.50	46.36	8:16	16
-124	47.21	46.06	8:21	16
-125	46.92	45.76	8:26	17
-127	46.32	45.14	8:35	19
-128	46.01	44.81	8:39	20
-129	45.69	44.48	8:44	21
-133	44.34	43.08	9:03	24
-137	42.86	41.54	9:23	28
-140	41.65	40.30	9:38	31
-143	40.37	38.97	9:54	34
-145	39.48	38.04	10:05	36
-147	38.54	37.08	10:17	38
-150	37.08	35.57	10:34	41

PERCENTAGE OF SUN OBSCURED ALONG MAP TRACK
WEST END =102.6   EAST END =103.6

==

+ MARKS THE CITY OF LO-I

ECLIPSE ON JD 1512566 DATE(Y/M/D) -571/ 3/ 7
DELTA T = 5 HOURS 11.8 MINS

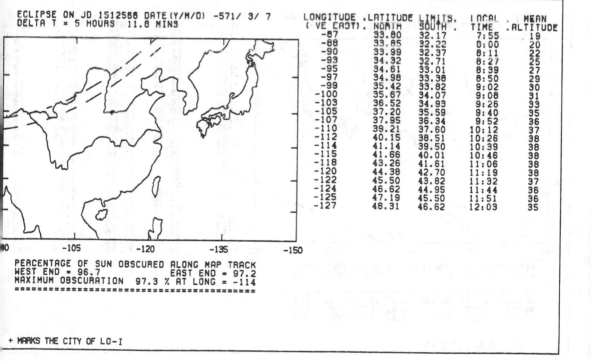

LONGITUDE (-VE EAST)	LATITUDE NORTH	LIMITS SOUTH	LOCAL TIME	MEAN ALTITUDE
-87	33.80	32.17	7:55	19
-88	33.85	32.22	8:00	20
-90	33.99	32.37	8:11	22
-93	34.32	32.71	8:27	25
-95	34.61	33.01	8:39	27
-97	34.98	33.38	8:50	29
-99	35.42	33.82	9:02	30
-100	35.67	34.07	9:08	31
-103	36.52	34.93	9:26	33
-105	37.20	35.59	9:40	35
-107	37.95	36.34	9:52	36
-110	39.21	37.60	10:12	37
-112	40.15	38.51	10:26	38
-114	41.14	39.50	10:39	38
-115	41.66	40.01	10:46	38
-118	43.26	41.61	11:06	38
-120	44.38	42.70	11:19	38
-122	45.50	43.82	11:32	37
-124	46.62	44.95	11:44	36
-125	47.19	45.50	11:51	36
-127	48.31	46.62	12:03	35

PERCENTAGE OF SUN OBSCURED ALONG MAP TRACK
WEST END = 96.7   EAST END = 97.2
MAXIMUM OBSCURATION 97.3 % AT LONG = -114

==

+ MARKS THE CITY OF LO-I

ECLIPSE ON JD 1513098 DATE(Y/M/D) -570/ 8/21
DELTA T = 5 HOURS 11.4 MINS

LONGITUDE	.LATITUDE	LIMITS.	LOCAL .	MEAN
(-VE EAST) .	NORTH	SOUTH .	TIME	.ALTITUDE
-144	14.11	11.70	6:17	7
-145	14.36	11.95	6:21	8
-147	14.83	12.46	6:30	11
-148	15.07	12.70	6:34	12
-149	15.30	12.94	6:39	13
-150	15.52	13.18	6:43	14
-152	15.96	13.65	6:52	16
-153	16.18	13.88	6:57	17

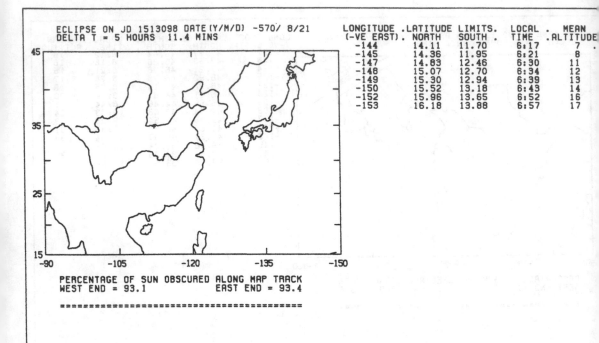

PERCENTAGE OF SUN OBSCURED ALONG MAP TRACK
WEST END = 93.1   EAST END = 93.4

===

+ MARKS THE CITY OF LO-I

ECLIPSE ON JD 1517084 DATE(Y/M/D) -559/ 7/20
DELTA T = 5 HOURS   8.1 MINS

LONGITUDE	.LATITUDE	LIMITS.	LOCAL .	MEAN
(-VE EAST) .	NORTH	SOUTH .	TIME	.ALTITUD
-120	13.56	13.06	5:41	1
-122	14.37	13.90	5:50	3
-123	14.78	14.31	5:54	4
-125	15.57	15.13	6:02	6
-127	16.36	15.95	6:11	9
-128	16.75	16.35	6:15	10
-129	17.14	16.75	6:19	11
-130	17.52	17.14	6:24	12
-132	18.28	17.92	6:33	14
-133	18.65	18.30	6:37	15
-134	19.01	18.68	6:42	16
-135	19.37	19.05	6:47	18
-137	20.08	19.77	6:56	20
-138	20.42	20.13	7:01	21
-139	20.75	20.48	7:06	22
-140	21.08	20.82	7:11	24
-142	21.72	21.48	7:21	26
-143	22.03	21.80	7:26	27
-144	22.33	22.11	7:31	28
-145	22.62	22.41	7:36	30
-147	23.18	22.99	7:46	32
-148	23.44	23.26	7:52	33
-149	23.70	23.53	7:57	35
-150	23.94	23.78	8:02	36
-152	24.40	24.26	8:13	38
-153	24.61	24.48	8:19	40

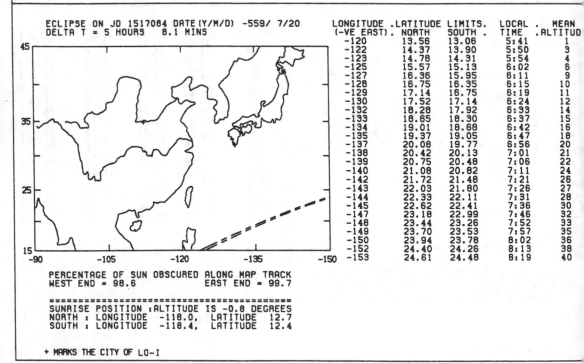

PERCENTAGE OF SUN OBSCURED ALONG MAP TRACK
WEST END = 98.6   EAST END = 99.7

===
SUNRISE POSITION :ALTITUDE IS -0.8 DEGREES
NORTH : LONGITUDE -118.0, LATITUDE 12.7
SOUTH : LONGITUDE -118.4, LATITUDE 12.4

+ MARKS THE CITY OF LO-I

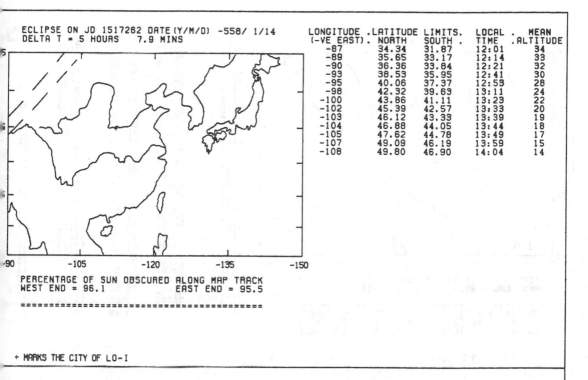

ECLIPSE ON JD 1517262 DATE (Y/M/D) -558/ 1/14
DELTA T = 5 HOURS 7.9 MINS

LONGITUDE (-VE EAST)	LATITUDE NORTH	LIMITS. SOUTH	LOCAL TIME	MEAN ALTITUDE
-87	34.34	31.87	12:01	34
-89	35.65	33.17	12:14	33
-90	36.36	33.84	12:21	32
-93	38.53	35.95	12:41	30
-95	40.06	37.37	12:53	28
-98	42.32	39.63	13:11	24
-100	43.86	41.11	13:23	22
-102	45.39	42.57	13:33	20
-103	46.12	43.33	13:39	19
-104	46.88	44.05	13:44	18
-105	47.62	44.78	13:49	17
-107	49.09	46.19	13:59	15
-108	49.80	46.90	14:04	14

PERCENTAGE OF SUN OBSCURED ALONG MAP TRACK
WEST END = 96.1   EAST END = 95.5

===

+ MARKS THE CITY OF LO-I

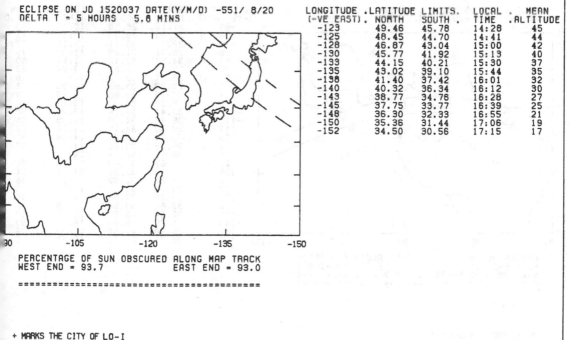

ECLIPSE ON JD 1520037 DATE (Y/M/D) -551/ 8/20
DELTA T = 5 HOURS 5.8 MINS

LONGITUDE (-VE EAST)	LATITUDE NORTH	LIMITS. SOUTH	LOCAL TIME	MEAN ALTITUDE
-123	49.46	45.78	14:28	45
-125	48.45	44.70	14:41	44
-128	46.87	43.04	15:00	42
-130	45.77	41.92	15:13	40
-133	44.15	40.21	15:30	37
-135	43.02	39.10	15:44	35
-138	41.40	37.42	16:01	32
-140	40.32	36.34	16:12	30
-143	38.77	34.78	16:28	27
-145	37.75	33.77	16:39	25
-148	36.30	32.33	16:55	21
-150	35.36	31.44	17:06	19
-152	34.50	30.56	17:15	17

PERCENTAGE OF SUN OBSCURED ALONG MAP TRACK
WEST END = 93.7   EAST END = 93.0

===

+ MARKS THE CITY OF LO-I

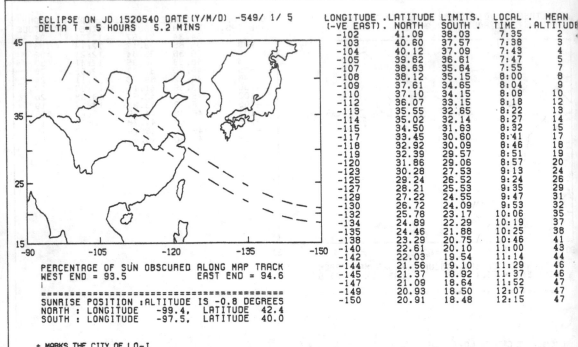

ECLIPSE ON JD 1520540 DATE(Y/M/D) -549/ 1/ 5
DELTA T = 5 HOURS   5.2 MINS

LONGITUDE (-VE EAST)	.LATITUDE LIMITS. NORTH	SOUTH .	LOCAL . TIME	MEAN .ALTITUDE
-102	41.09	38.03	7:35	2
-103	40.60	37.57	7:38	3
-104	40.12	37.09	7:43	4
-105	39.62	36.61	7:47	5
-107	38.63	35.64	7:55	7
-108	38.12	35.15	8:00	8
-109	37.61	34.65	8:04	9
-110	37.10	34.15	8:09	10
-112	36.07	33.15	8:18	12
-113	35.55	32.65	8:22	13
-114	35.02	32.14	8:27	14
-115	34.50	31.63	8:32	15
-117	33.45	30.60	8:41	17
-118	32.92	30.09	8:46	18
-119	32.39	29.57	8:51	19
-120	31.86	29.06	8:57	20
-123	30.28	27.53	9:13	24
-125	29.24	26.52	9:24	26
-127	28.21	25.53	9:35	29
-129	27.22	24.55	9:47	31
-130	26.72	24.09	9:53	32
-132	25.78	23.17	10:06	35
-134	24.89	22.29	10:19	37
-135	24.46	21.88	10:25	38
-138	23.29	20.75	10:46	41
-140	22.61	20.10	11:00	43
-142	22.03	19.54	11:14	44
-144	21.56	19.10	11:29	46
-145	21.37	18.92	11:37	46
-147	21.09	18.64	11:52	47
-149	20.93	18.50	12:07	47
-150	20.91	18.48	12:15	47

PERCENTAGE OF SUN OBSCURED ALONG MAP TRACK
WEST END = 93.5   EAST END = 94.6

===
SUNRISE POSITION :ALTITUDE IS -0.8 DEGREES
NORTH : LONGITUDE -99.4, LATITUDE 42.4
SOUTH : LONGITUDE -97.5, LATITUDE 40.0

+ MARKS THE CITY OF LO-I

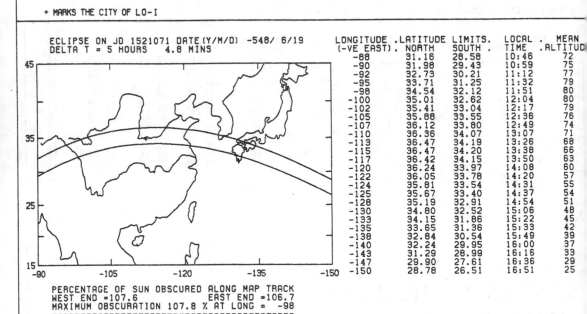

ECLIPSE ON JD 1521071 DATE(Y/M/D) -548/ 6/19
DELTA T = 5 HOURS   4.8 MINS

LONGITUDE (-VE EAST)	.LATITUDE LIMITS. NORTH	SOUTH .	LOCAL . TIME	MEAN .ALTITUDE
-88	31.16	28.58	10:46	72
-90	31.98	29.43	10:59	75
-92	32.73	30.21	11:12	77
-95	33.71	31.25	11:32	79
-98	34.54	32.12	11:51	80
-100	35.01	32.62	12:04	80
-102	35.41	33.04	12:17	79
-105	35.88	33.55	12:36	76
-107	36.12	33.80	12:49	74
-110	36.36	34.07	13:07	71
-113	36.47	34.19	13:26	68
-115	36.47	34.20	13:38	66
-117	36.42	34.15	13:50	63
-120	36.24	33.97	14:08	60
-122	36.05	33.78	14:20	57
-124	35.81	33.54	14:31	55
-125	35.67	33.40	14:37	54
-128	35.19	32.91	14:54	51
-130	34.80	32.52	15:06	48
-133	34.15	31.86	15:22	45
-135	33.65	31.36	15:33	42
-138	32.84	30.54	15:49	39
-140	32.24	29.95	16:00	37
-143	31.29	28.99	16:16	33
-147	29.90	27.61	16:36	29
-150	28.78	26.51	16:51	25

PERCENTAGE OF SUN OBSCURED ALONG MAP TRACK
WEST END =107.6   EAST END =106.7
MAXIMUM OBSCURATION 107.8 % AT LONG = -98
===

+ MARKS THE CITY OF LO-I

ECLIPSE ON JD 1521927 DATE(Y/M/D) -546/10/23
DELTA T = 5 HOURS   4.1 MINS

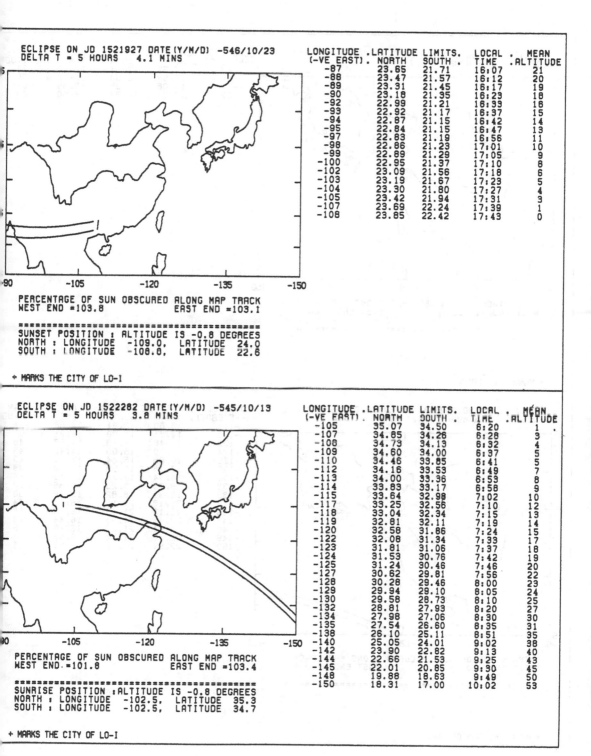

LONGITUDE (-VE EAST)	LATITUDE LIMITS. NORTH	SOUTH	LOCAL TIME	MEAN ALTITUDE
-87	23.65	21.71	16:07	21
-88	23.47	21.57	16:12	20
-89	23.31	21.45	16:17	19
-90	23.18	21.35	16:23	18
-92	22.99	21.21	16:33	16
-93	22.92	21.17	16:37	15
-94	22.87	21.15	16:42	14
-95	22.84	21.15	16:47	13
-97	22.83	21.19	16:56	11
-98	22.86	21.23	17:01	10
-99	22.89	21.29	17:05	9
-100	22.95	21.37	17:10	8
-102	23.09	21.56	17:18	6
-103	23.19	21.67	17:23	5
-104	23.30	21.80	17:27	4
-105	23.42	21.94	17:31	3
-107	23.69	22.24	17:39	1
-108	23.85	22.42	17:43	0

PERCENTAGE OF SUN OBSCURED ALONG MAP TRACK
WEST END =103.8   EAST END =103.1

===
SUNSET POSITION : ALTITUDE IS -0.8 DEGREES
NORTH : LONGITUDE -109.0, LATITUDE 24.0
SOUTH : LONGITUDE -108.8, LATITUDE 22.6

+ MARKS THE CITY OF LO-I

ECLIPSE ON JD 1522282 DATE(Y/M/D) -545/10/13
DELTA T = 5 HOURS   3.8 MINS

LONGITUDE (-VE EAST)	LATITUDE LIMITS. NORTH	SOUTH	LOCAL TIME	MEAN ALTITUDE
-105	35.07	34.50	6:20	1
-107	34.85	34.26	6:28	3
-108	34.73	34.13	6:32	4
-109	34.60	34.00	6:37	5
-110	34.46	33.85	6:41	5
-112	34.16	33.53	6:49	7
-113	34.00	33.36	6:53	8
-114	33.83	33.17	6:58	9
-115	33.64	32.98	7:02	10
-117	33.25	32.56	7:10	12
-118	33.04	32.34	7:15	13
-119	32.81	32.11	7:19	14
-120	32.58	31.86	7:24	15
-122	32.08	31.34	7:33	17
-123	31.81	31.06	7:37	18
-124	31.53	30.76	7:42	19
-125	31.24	30.46	7:46	20
-127	30.62	29.81	7:56	22
-128	30.28	29.46	8:00	23
-129	29.94	29.10	8:05	24
-130	29.58	28.73	8:10	25
-132	28.81	27.93	8:20	27
-134	27.98	27.06	8:30	30
-135	27.54	26.60	8:35	31
-138	26.10	25.11	8:51	35
-140	25.05	24.01	9:02	38
-142	23.90	22.82	9:13	40
-144	22.66	21.53	9:25	43
-145	22.01	20.85	9:30	45
-148	19.88	18.63	9:49	50
-150	18.31	17.00	10:02	53

PERCENTAGE OF SUN OBSCURED ALONG MAP TRACK
WEST END =101.8   EAST END =103.4

===
SUNRISE POSITION : ALTITUDE IS -0.8 DEGREES
NORTH : LONGITUDE -102.5, LATITUDE 35.3
SOUTH : LONGITUDE -102.5, LATITUDE 34.7

+ MARKS THE CITY OF LO-I

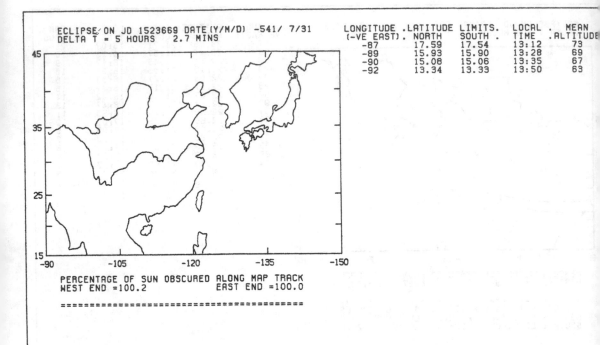

ECLIPSE ON JD 1523669 DATE(Y/M/D) -541/ 7/31
DELTA T = 5 HOURS   2.7 MINS

LONGITUDE	.LATITUDE	LIMITS.	LOCAL	. MEAN
(-VE EAST).	NORTH	SOUTH .	TIME	.ALTITUDE
-87	17.59	17.54	13:12	73
-89	15.93	15.90	13:28	69
-90	15.08	15.06	13:35	67
-92	13.34	13.33	13:50	63

PERCENTAGE OF SUN OBSCURED ALONG MAP TRACK
WEST END =100.2   EAST END =100.0

+ MARKS THE CITY OF LO-I

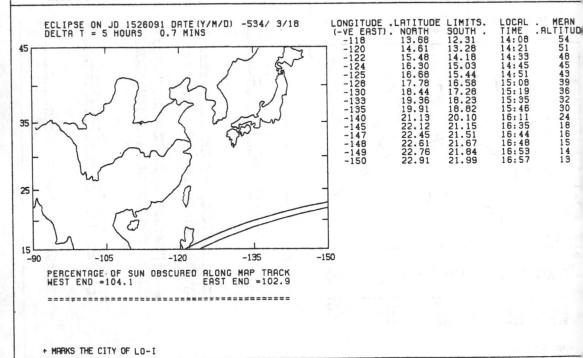

ECLIPSE ON JD 1526091 DATE(Y/M/D) -534/ 3/18
DELTA T = 5 HOURS   0.7 MINS

LONGITUDE	.LATITUDE	LIMITS.	LOCAL	. MEAN
(-VE EAST).	NORTH	SOUTH .	TIME	.ALTITUDE
-118	13.68	12.31	14:08	54
-120	14.61	13.28	14:21	51
-122	15.48	14.18	14:33	48
-124	16.30	15.03	14:45	45
-125	16.68	15.44	14:51	43
-128	17.78	16.58	15:08	39
-130	18.44	17.28	15:19	36
-133	19.36	18.23	15:35	32
-135	19.91	18.82	15:46	30
-140	21.13	20.10	16:11	24
-145	22.12	21.15	16:35	18
-147	22.45	21.51	16:44	16
-148	22.61	21.67	16:48	15
-149	22.76	21.84	16:53	14
-150	22.91	21.99	16:57	13

PERCENTAGE OF SUN OBSCURED ALONG MAP TRACK
WEST END =104.1   EAST END =102.9

+ MARKS THE CITY OF LO-I

ECLIPSE ON JD 1529044 DATE(Y/M/D) -526/ 4/18
DELTA T = 4 HOURS 58.3 MINS

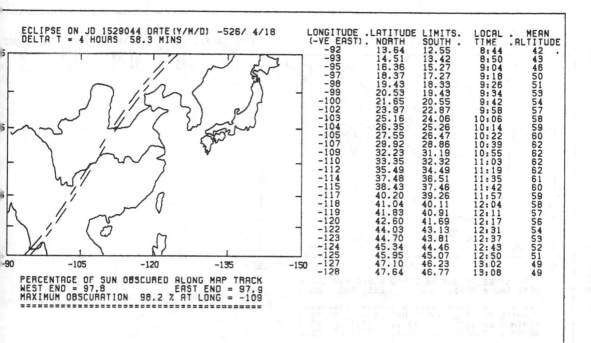

LONGITUDE (-VE EAST)	LATITUDE NORTH	LIMITS SOUTH	LOCAL TIME	MEAN ALTITUDE
-92	13.64	12.55	8:44	42
-93	14.51	13.42	8:50	43
-95	16.36	15.27	9:04	46
-97	18.37	17.27	9:18	50
-98	19.43	18.33	9:26	51
-99	20.53	19.43	9:34	53
-100	21.65	20.55	9:42	54
-102	23.97	22.87	9:58	57
-103	25.16	24.06	10:06	58
-104	26.35	25.26	10:14	59
-105	27.55	26.47	10:22	60
-107	29.92	28.86	10:39	62
-109	32.23	31.19	10:55	62
-110	33.35	32.32	11:03	62
-112	35.49	34.49	11:19	62
-114	37.48	36.51	11:35	61
-115	38.43	37.46	11:42	60
-117	40.20	39.26	11:57	59
-118	41.04	40.11	12:04	58
-119	41.83	40.91	12:11	57
-120	42.60	41.69	12:17	56
-122	44.03	43.13	12:31	54
-123	44.70	43.81	12:37	53
-124	45.34	44.46	12:43	52
-125	45.95	45.07	12:50	51
-127	47.10	46.23	13:02	49
-128	47.64	46.77	13:08	49

PERCENTAGE OF SUN OBSCURED ALONG MAP TRACK
WEST END = 97.8   EAST END = 97.9
MAXIMUM OBSCURATION 98.2 % AT LONG = -109
===

+ MARKS THE CITY OF LO-I

ECLIPSE ON JD 1529900 DATE(Y/M/D) -524/ 8/21
DELTA T = 4 HOURS 57.6 MINS

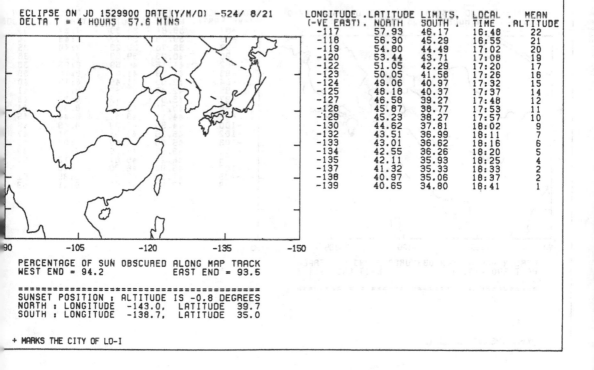

LONGITUDE (-VE EAST)	LATITUDE NORTH	LIMITS SOUTH	LOCAL TIME	MEAN ALTITUDE
-117	57.93	46.17	16:48	22
-118	56.30	45.29	16:55	21
-119	54.80	44.49	17:02	20
-120	53.44	43.71	17:08	19
-122	51.05	42.29	17:20	17
-123	50.05	41.58	17:26	16
-124	49.06	40.97	17:32	15
-125	48.18	40.37	17:37	14
-127	46.58	39.27	17:48	12
-128	45.87	38.77	17:53	11
-129	45.23	38.27	17:57	10
-130	44.62	37.81	18:02	9
-132	43.51	36.99	18:11	7
-133	43.01	36.62	18:16	6
-134	42.55	36.26	18:20	5
-135	42.11	35.93	18:25	4
-137	41.32	35.33	18:33	2
-138	40.97	35.06	18:37	2
-139	40.65	34.80	18:41	1

PERCENTAGE OF SUN OBSCURED ALONG MAP TRACK
WEST END = 94.2   EAST END = 93.5

===
SUNSET POSITION : ALTITUDE IS -0.8 DEGREES
NORTH : LONGITUDE -143.0, LATITUDE 39.7
SOUTH : LONGITUDE -138.7, LATITUDE 35.0

+ MARKS THE CITY OF LO-I

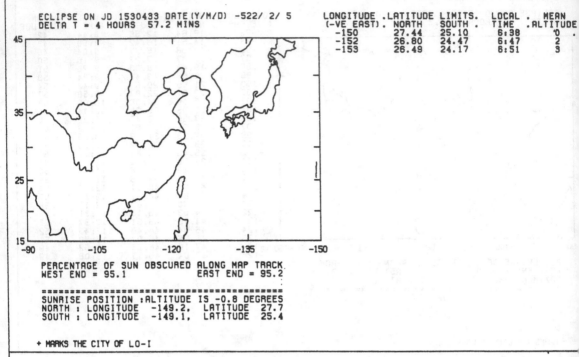

ECLIPSE ON JD 1530433 DATE(Y/M/D) -522/ 2/ 5
DELTA T = 4 HOURS 57.2 MINS

LONGITUDE (-VE EAST)	.LATITUDE NORTH	LIMITS. SOUTH	LOCAL TIME	MEAN .ALTITUDE
-150	27.44	25.10	6:38	0
-152	26.80	24.47	6:47	2
-153	26.49	24.17	6:51	3

PERCENTAGE OF SUN OBSCURED ALONG MAP TRACK.
WEST END = 95.1   EAST END = 95.2

==
SUNRISE POSITION :ALTITUDE IS -0.8 DEGREES
NORTH : LONGITUDE -149.2, LATITUDE 27.7
SOUTH : LONGITUDE -149.1, LATITUDE 25.4

+ MARKS THE CITY OF LO-I

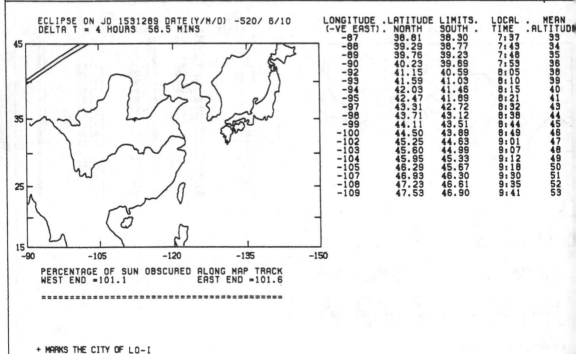

ECLIPSE ON JD 1531289 DATE(Y/M/D) -520/ 6/10
DELTA T = 4 HOURS 56.5 MINS

LONGITUDE (-VE EAST)	.LATITUDE NORTH	LIMITS. SOUTH	LOCAL TIME	MEAN .ALTITUDE
-87	38.81	38.90	7:37	33
-88	39.29	38.77	7:43	34
-89	39.76	39.23	7:48	35
-90	40.23	39.69	7:53	36
-92	41.15	40.59	8:05	38
-93	41.59	41.03	8:10	39
-94	42.03	41.46	8:15	40
-95	42.47	41.89	8:21	41
-97	43.31	42.72	8:32	43
-98	43.71	43.12	8:38	44
-99	44.11	43.51	8:44	45
-100	44.50	43.89	8:49	46
-102	45.25	44.63	9:01	47
-103	45.60	44.99	9:07	48
-104	45.95	45.33	9:12	49
-105	46.29	45.67	9:18	50
-107	46.93	46.30	9:30	51
-108	47.23	46.61	9:35	52
-109	47.53	46.90	9:41	53

PERCENTAGE OF SUN OBSCURED ALONG MAP TRACK
WEST END =101.1   EAST END =101.6

==

+ MARKS THE CITY OF LO-I

ECLIPSE ON JD 1531820 DATE(Y/M/D) -519/11/23
DELTA T = 4 HOURS 56.0 MINS

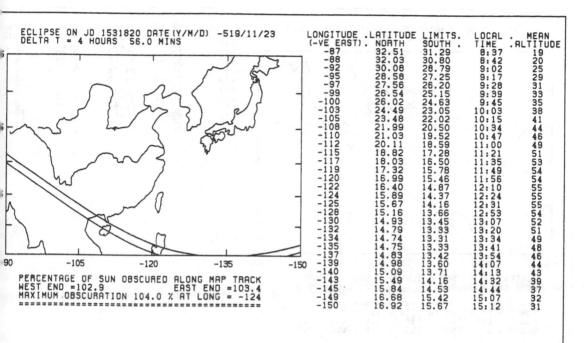

LONGITUDE (-VE EAST)	LATITUDE NORTH	LIMITS SOUTH	LOCAL TIME	MEAN ALTITUDE
-87	32.51	31.29	8:37	19
-88	32.03	30.80	8:42	20
-92	30.08	28.79	9:02	25
-95	28.58	27.25	9:17	29
-97	27.56	26.20	9:28	31
-99	26.54	25.15	9:39	33
-100	26.02	24.63	9:45	35
-103	24.49	23.05	10:03	38
-105	23.48	22.02	10:15	41
-108	21.99	20.50	10:34	44
-110	21.03	19.52	10:47	46
-112	20.11	18.59	11:00	49
-115	18.82	17.28	11:21	51
-117	18.03	16.50	11:35	53
-119	17.32	15.78	11:49	54
-120	16.99	15.46	11:56	54
-122	16.40	14.87	12:10	55
-124	15.89	14.37	12:24	55
-125	15.67	14.16	12:31	55
-128	15.16	13.66	12:53	54
-130	14.93	13.45	13:07	52
-132	14.79	13.33	13:20	51
-134	14.74	13.31	13:34	49
-135	14.75	13.33	13:41	48
-137	14.83	13.42	13:54	46
-139	14.98	13.60	14:07	44
-140	15.09	13.71	14:13	43
-143	15.49	14.16	14:32	39
-145	15.84	14.53	14:44	37
-149	16.68	15.42	15:07	32
-150	16.92	15.67	15:12	31

PERCENTAGE OF SUN OBSCURED ALONG MAP TRACK
WEST END =102.9   EAST END =103.4
MAXIMUM OBSCURATION 104.0 % AT LONG = -124
===

+ MARKS THE CITY OF LO-I

ECLIPSE ON JD 1535098 DATE(Y/M/D) -510/11/14
DELTA T = 4 HOURS 53.4 MINS

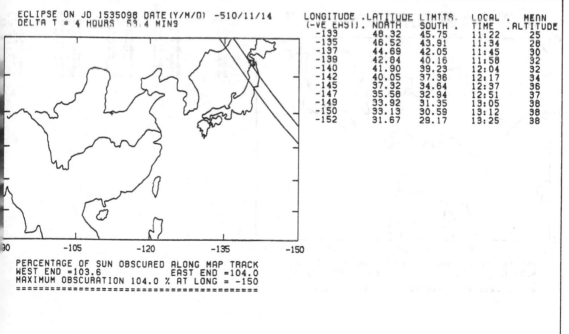

LONGITUDE (-VE EAST)	LATITUDE NORTH	LIMITS SOUTH	LOCAL TIME	MEAN ALTITUDE
-133	48.32	45.75	11:22	25
-135	46.52	43.91	11:34	28
-137	44.69	42.05	11:45	30
-138	42.84	40.16	11:58	32
-140	41.90	39.23	12:04	32
-142	40.05	37.36	12:17	34
-145	37.32	34.64	12:37	36
-147	35.58	32.94	12:51	37
-149	33.92	31.35	13:05	38
-150	33.13	30.59	13:12	38
-152	31.67	29.17	13:25	38

PERCENTAGE OF SUN OBSCURED ALONG MAP TRACK
WEST END =103.6   EAST END =104.0
MAXIMUM OBSCURATION 104.0 % AT LONG = -150
===

+ MARKS THE CITY OF LO-I

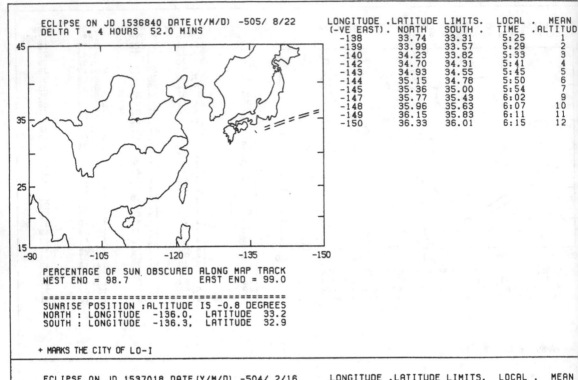

ECLIPSE ON JD 1536840 DATE(Y/M/D) -505/ 8/22
DELTA T = 4 HOURS 52.0 MINS

LONGITUDE (-VE EAST)	LATITUDE NORTH	LIMITS SOUTH	LOCAL TIME	MEAN ALTITUD
-138	33.74	33.31	5:25	1
-139	33.99	33.57	5:29	2
-140	34.23	33.82	5:33	3
-142	34.70	34.31	5:41	4
-143	34.93	34.55	5:45	5
-144	35.15	34.78	5:50	6
-145	35.36	35.00	5:54	7
-147	35.77	35.43	6:02	9
-148	35.96	35.63	6:07	10
-149	36.15	35.83	6:11	11
-150	36.33	36.01	6:15	12

PERCENTAGE OF SUN OBSCURED ALONG MAP TRACK
WEST END = 98.7   EAST END = 99.0

===
SUNRISE POSITION :ALTITUDE IS -0.8 DEGREES
NORTH : LONGITUDE -136.0, LATITUDE 33.2
SOUTH : LONGITUDE -136.3, LATITUDE 32.9

+ MARKS THE CITY OF LO-I

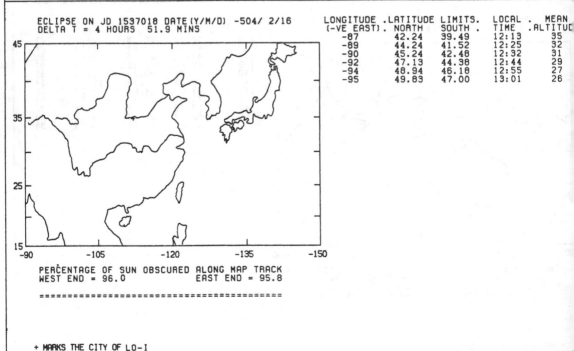

ECLIPSE ON JD 1537018 DATE(Y/M/D) -504/ 2/16
DELTA T = 4 HOURS 51.9 MINS

LONGITUDE (-VE EAST)	LATITUDE NORTH	LIMITS SOUTH	LOCAL TIME	MEAN ALTITUD
-87	42.24	39.49	12:13	35
-89	44.24	41.52	12:25	32
-90	45.24	42.48	12:32	31
-92	47.13	44.38	12:44	29
-94	48.94	46.18	12:55	27
-95	49.83	47.00	13:01	26

PERCENTAGE OF SUN OBSCURED ALONG MAP TRACK
WEST END = 96.0   EAST END = 95.8

===

+ MARKS THE CITY OF LO-I

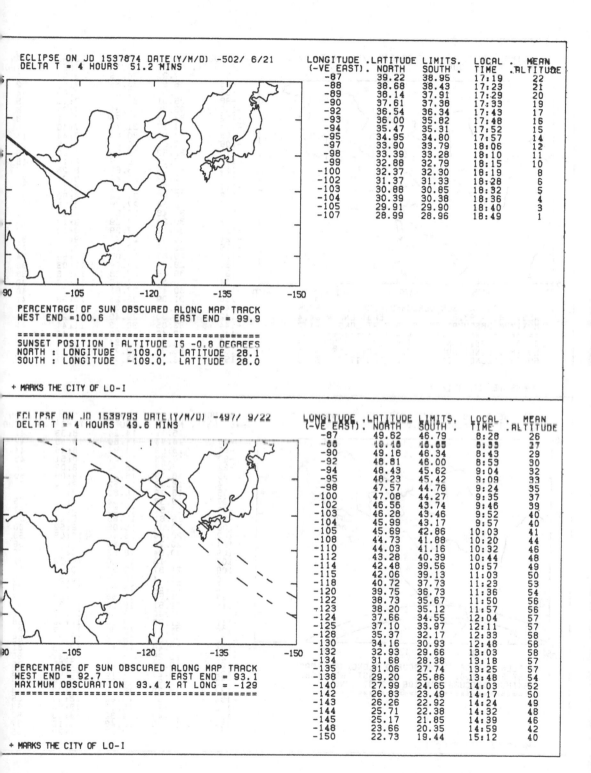

ECLIPSE ON JD 1537874 DATE(Y/M/D) -502/ 6/21
DELTA T = 4 HOURS 51.2 MINS

LONGITUDE (-VE EAST)	LATITUDE LIMITS. NORTH	SOUTH	LOCAL TIME	MEAN ALTITUDE
-87	39.22	38.95	17:19	22
-88	38.68	38.43	17:23	21
-89	38.14	37.91	17:29	20
-90	37.61	37.38	17:33	19
-92	36.54	36.34	17:43	17
-93	36.00	35.82	17:48	16
-94	35.47	35.31	17:52	15
-95	34.95	34.80	17:57	14
-97	33.90	33.79	18:06	12
-98	33.39	33.28	18:10	11
-99	32.88	32.79	18:15	10
-100	32.37	32.30	18:19	8
-102	31.37	31.33	18:28	6
-103	30.88	30.85	18:32	5
-104	30.39	30.38	18:36	4
-105	29.91	29.90	18:40	3
-107	28.99	28.96	18:49	1

PERCENTAGE OF SUN OBSCURED ALONG MAP TRACK
WEST END =100.6   EAST END = 99.9

==
SUNSET POSITION : ALTITUDE IS -0.8 DEGREES
NORTH : LONGITUDE -109.0, LATITUDE 28.1
SOUTH : LONGITUDE -109.0, LATITUDE 28.0

+ MARKS THE CITY OF LO-I

ECLIPSE ON JD 1538793 DATE(Y/M/D) -497/ 9/22
DELTA T = 4 HOURS 49.6 MINS

LONGITUDE (-VE EAST)	LATITUDE LIMITS. NORTH	SOUTH	LOCAL TIME	MEAN ALTITUDE
-87	49.62	46.79	8:28	26
-88	49.16	46.65	8:33	27
-90	49.16	46.34	8:43	29
-92	48.81	46.00	8:59	30
-94	48.43	45.62	9:04	32
-95	48.23	45.42	9:09	33
-98	47.57	44.76	9:24	35
-100	47.08	44.27	9:35	37
-102	46.56	43.74	9:46	39
-103	46.28	43.46	9:52	40
-104	45.99	43.17	9:57	40
-105	45.69	42.86	10:03	41
-108	44.73	41.88	10:20	44
-110	44.03	41.16	10:32	46
-112	43.28	40.39	10:44	48
-114	42.48	39.56	10:57	49
-115	42.06	39.13	11:03	50
-118	40.72	37.73	11:23	53
-120	39.75	36.73	11:36	54
-122	38.73	35.67	11:50	56
-123	38.20	35.12	11:57	56
-124	37.66	34.55	12:04	57
-125	37.10	33.97	12:11	57
-128	35.37	32.17	12:33	58
-130	34.16	30.93	12:48	58
-132	32.93	29.66	13:03	58
-134	31.68	28.38	13:18	57
-135	31.06	27.74	13:25	57
-138	29.20	25.86	13:48	54
-140	27.99	24.65	14:03	52
-142	26.83	23.49	14:17	50
-143	26.26	22.92	14:24	49
-144	25.71	22.38	14:32	48
-145	25.17	21.85	14:39	46
-148	23.66	20.35	14:59	42
-150	22.73	19.44	15:12	40

PERCENTAGE OF SUN OBSCURED ALONG MAP TRACK
WEST END = 92.7   EAST END = 93.1
MAXIMUM OBSCURATION 93.4 % AT LONG = -129
==

+ MARKS THE CITY OF LO-I

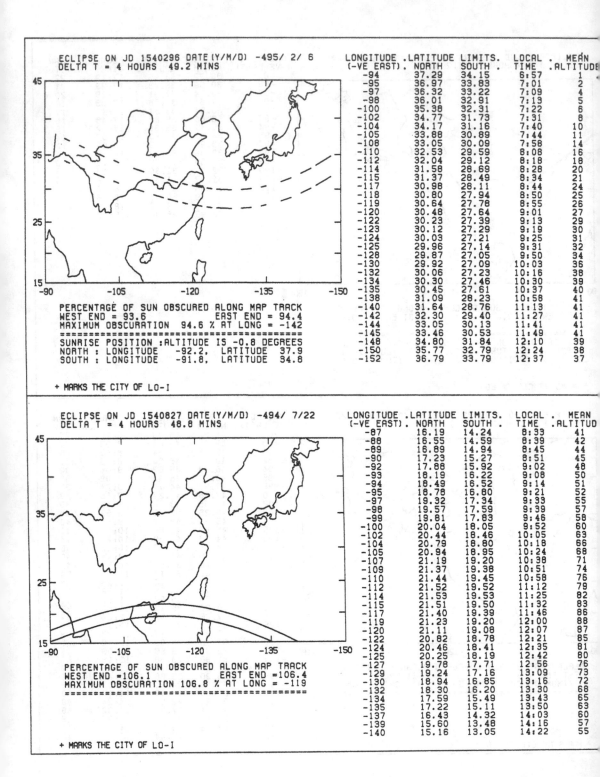

ECLIPSE ON JD 1540296 DATE (Y/M/D) -495/ 2/ 6
DELTA T = 4 HOURS 49.2 MINS

PERCENTAGE OF SUN OBSCURED ALONG MAP TRACK
WEST END = 93.6   EAST END = 94.4
MAXIMUM OBSCURATION 94.6 % AT LONG = -142
===
SUNRISE POSITION :ALTITUDE IS -0.8 DEGREES
NORTH : LONGITUDE -92.2, LATITUDE 37.9
SOUTH : LONGITUDE -91.8, LATITUDE 34.8

+ MARKS THE CITY OF LO-I

LONGITUDE (-VE EAST)	LATITUDE LIMITS. NORTH	SOUTH	LOCAL TIME	MEAN ALTITUDE
-94	37.29	34.15	6:57	1
-95	36.97	33.83	7:01	2
-97	36.32	33.22	7:09	4
-98	36.01	32.91	7:13	5
-100	35.38	32.31	7:22	6
-102	34.77	31.73	7:31	8
-104	34.17	31.16	7:40	10
-105	33.88	30.89	7:44	11
-108	33.05	30.09	7:58	14
-110	32.53	29.59	8:08	16
-112	32.04	29.12	8:18	18
-114	31.58	28.69	8:28	20
-115	31.37	28.49	8:34	21
-117	30.98	28.11	8:44	24
-118	30.80	27.94	8:50	25
-119	30.64	27.78	8:55	26
-120	30.48	27.64	9:01	27
-122	30.23	27.39	9:13	29
-123	30.12	27.29	9:19	30
-124	30.03	27.21	9:25	31
-125	29.96	27.14	9:31	32
-128	29.87	27.05	9:50	34
-130	29.92	27.09	10:03	36
-132	30.06	27.23	10:16	38
-134	30.30	27.46	10:30	39
-135	30.45	27.61	10:37	40
-138	31.09	28.23	10:58	41
-140	31.64	28.76	11:13	41
-142	32.30	29.40	11:27	41
-144	33.05	30.13	11:41	41
-145	33.46	30.53	11:49	41
-148	34.80	31.84	12:10	39
-150	35.77	32.79	12:24	38
-152	36.79	33.79	12:37	37

ECLIPSE ON JD 1540827 DATE (Y/M/D) -494/ 7/22
DELTA T = 4 HOURS 48.8 MINS

PERCENTAGE OF SUN OBSCURED ALONG MAP TRACK
WEST END =106.1   EAST END =106.4
MAXIMUM OBSCURATION 106.8 % AT LONG = -119
===

+ MARKS THE CITY OF LO-I

LONGITUDE (-VE EAST)	LATITUDE LIMITS. NORTH	SOUTH	LOCAL TIME	MEAN ALTITUD
-87	16.19	14.24	8:33	41
-88	16.55	14.59	8:39	42
-89	16.89	14.94	8:45	44
-90	17.23	15.27	8:51	45
-92	17.88	15.92	9:02	48
-93	18.19	16.22	9:08	50
-94	18.49	16.52	9:14	51
-95	18.78	16.80	9:21	52
-97	19.32	17.34	9:33	55
-98	19.57	17.59	9:39	57
-99	19.81	17.83	9:46	58
-100	20.04	18.05	9:52	60
-102	20.44	18.46	10:05	63
-104	20.79	18.80	10:18	66
-105	20.94	18.95	10:24	68
-107	21.19	19.20	10:38	71
-109	21.37	19.38	10:51	74
-110	21.44	19.45	10:58	76
-112	21.52	19.52	11:12	79
-114	21.59	19.53	11:25	82
-115	21.51	19.50	11:32	83
-117	21.40	19.39	11:46	86
-119	21.23	19.20	12:00	88
-120	21.11	19.08	12:07	87
-122	20.82	18.78	12:21	85
-124	20.46	18.41	12:35	81
-125	20.25	18.19	12:42	80
-127	19.78	17.71	12:56	76
-129	19.24	17.16	13:09	73
-130	18.94	16.85	13:16	72
-132	18.30	16.20	13:30	68
-134	17.59	15.49	13:43	65
-135	17.22	15.11	13:50	63
-137	16.43	14.32	14:03	60
-139	15.60	13.48	14:16	57
-140	15.16	13.05	14:22	55

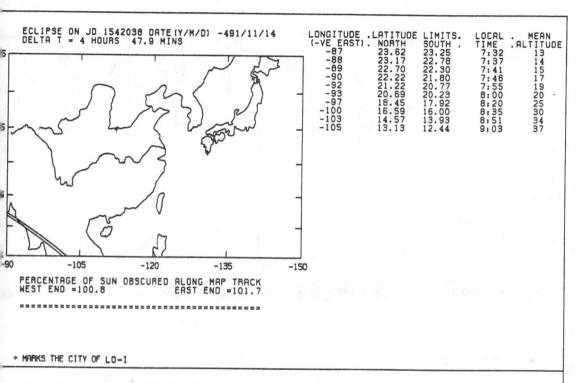

ECLIPSE ON JD 1542038 DATE(Y/M/D) -491/11/14
DELTA T = 4 HOURS 47.9 MINS

LONGITUDE (-VE EAST)	LATITUDE NORTH	LIMITS. SOUTH	LOCAL TIME	MEAN ALTITUDE
-87	23.62	23.25	7:32	13
-88	23.17	22.78	7:37	14
-89	22.70	22.30	7:41	15
-90	22.22	21.80	7:46	17
-92	21.22	20.77	7:55	19
-93	20.69	20.23	8:00	20
-97	18.45	17.92	8:20	25
-100	16.59	16.00	8:35	30
-103	14.57	13.93	8:51	34
-105	13.13	12.44	9:03	37

PERCENTAGE OF SUN OBSCURED ALONG MAP TRACK
WEST END =100.8   EAST END =101.7

===

+ MARKS THE CITY OF LO-I

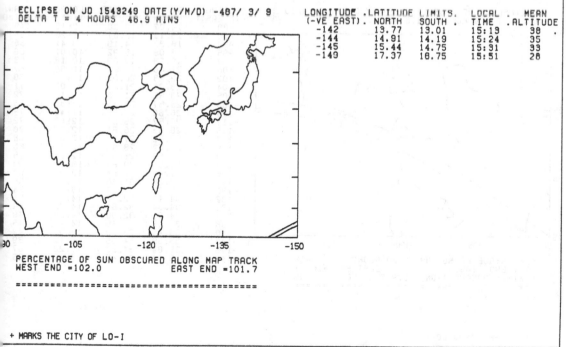

ECLIPSE ON JD 1543249 DATE(Y/M/D) -487/ 3/ 9
DELTA T = 4 HOURS 46.9 MINS

LONGITUDE (-VE EAST)	LATITUDE NORTH	LIMITS. SOUTH	LOCAL TIME	MEAN ALTITUDE
-142	13.77	13.01	15:19	38
-144	14.91	14.19	15:24	35
-145	15.44	14.75	15:31	33
-149	17.37	16.75	15:51	28

PERCENTAGE OF SUN OBSCURED ALONG MAP TRACK
WEST END =102.0   EAST END =101.7

===

+ MARKS THE CITY OF LO-I

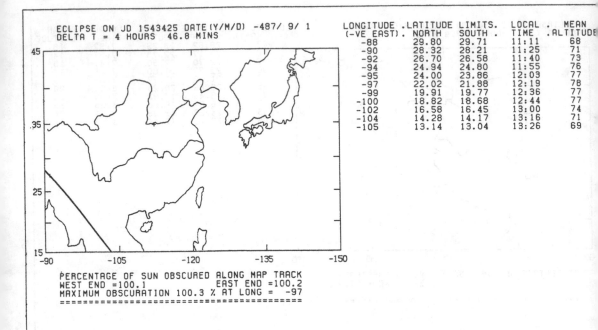

ECLIPSE ON JD 1543425 DATE(Y/M/D) -487/ 9/ 1
DELTA T = 4 HOURS   46.8 MINS

LONGITUDE (-VE EAST)	LATITUDE NORTH	LIMITS. SOUTH	LOCAL TIME	MEAN ALTITUDE
-88	29.80	29.71	11:11	68
-90	28.32	28.21	11:25	71
-92	26.70	26.58	11:40	73
-94	24.94	24.80	11:55	76
-95	24.00	23.86	12:03	77
-97	22.02	21.88	12:19	78
-99	19.91	19.77	12:36	77
-100	18.82	18.68	12:44	77
-102	16.58	16.45	13:00	74
-104	14.28	14.17	13:16	71
-105	13.14	13.04	13:26	69

PERCENTAGE OF SUN OBSCURED ALONG MAP TRACK
WEST END =100.1   EAST END =100.2
MAXIMUM OBSCURATION 100.3 % AT LONG = -97
==

+ MARKS THE CITY OF LO-I

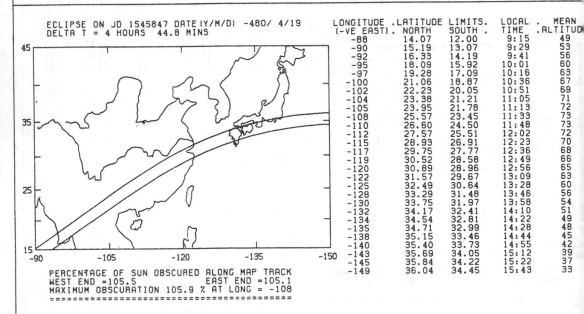

ECLIPSE ON JD 1545847 DATE(Y/M/D) -480/ 4/19
DELTA T = 4 HOURS   44.8 MINS

LONGITUDE (-VE EAST)	LATITUDE NORTH	LIMITS. SOUTH	LOCAL TIME	MEAN ALTITUDE
-88	14.07	12.00	9:15	49
-90	15..19	13.07	9:29	53
-92	16.33	14.19	9:41	56
-95	18.09	15.92	10:01	60
-97	19.28	17.09	10:16	63
-100	21.06	18.87	10:36	67
-102	22.23	20.05	10:51	69
-104	23.38	21.21	11:05	71
-105	23.95	21.78	11:13	72
-108	25.57	23.45	11:33	73
-110	26.60	24.50	11:48	73
-112	27.57	25.51	12:02	72
-115	28.93	26.91	12:23	70
-117	29.75	27.77	12:36	68
-119	30.52	28.58	12:49	66
-120	30.89	28.96	12:56	65
-122	31.57	29.67	13:09	63
-125	32.49	30.64	13:28	60
-128	33.29	31.48	13:46	56
-130	33.75	31.97	13:58	54
-132	34.17	32.41	14:10	51
-134	34.54	32.81	14:22	49
-135	34.71	32.99	14:28	48
-138	35.15	33.46	14:44	45
-140	35.40	33.73	14:55	42
-143	35.69	34.05	15:12	39
-145	35.84	34.22	15:22	37
-149	36.04	34.45	15:43	33

PERCENTAGE OF SUN OBSCURED ALONG MAP TRACK
WEST END =105.5   EAST END =105.1
MAXIMUM OBSCURATION 105.9 % AT LONG = -108
==

+ MARKS THE CITY OF LO-I

ECLIPSE ON JD 1548800 DATE(Y/M/D) -472/ 5/20
DELTA T = 4 HOURS 42.5 MINS

LONGITUDE	.LATITUDE	LIMITS.	LOCAL	. MEAN
(-VE EAST).	NORTH	SOUTH .	TIME	.ALTITUDE .
-147	13.67	13.42	9:12	49
-149	15.51	15.28	9:26	53
-150	16.44	16.23	9:34	55

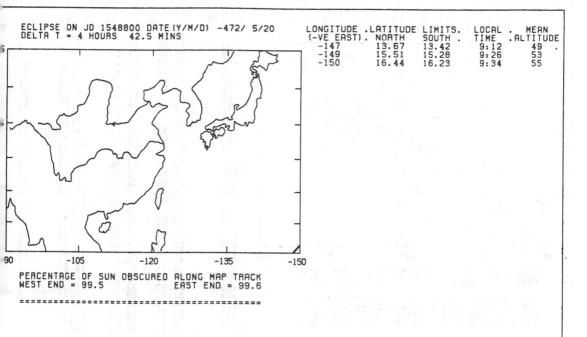

PERCENTAGE OF SUN OBSCURED ALONG MAP TRACK
WEST END = 99.5   EAST END = 99.6

==

+ MARKS THE CITY OF LO-I

ECLIPSE ON JD 1551399 DATE(Y/M/D) -465/ 7/ 2
DELTA T = 4 HOURS 40.5 MINS

LONGITUDE	.LATITUDE	LIMITS.	LOCAL	. MEAN
(-VE EAST).	NORTH	SOUTH .	TIME	.ALTITUDE
-87	21.86	20.49	11:29	82
-89	21.58	20.19	11:44	85
-90	21.38	20.00	11:52	86
-92	20.94	19.55	12:07	86
-94	20.39	18.98	12:23	83
-95	20.08	18.66	12:31	82
-97	19.39	17.93	12:46	78
-99	18.60	17.11	13:01	74
-100	18.17	16.66	13:09	73
-102	17.24	15.70	13:24	69
-104	16.23	14.65	13:40	65
-105	15.70	14.10	13:47	63
-107	14.60	12.96	14:02	59
-109	13.43	11.75	14:16	56

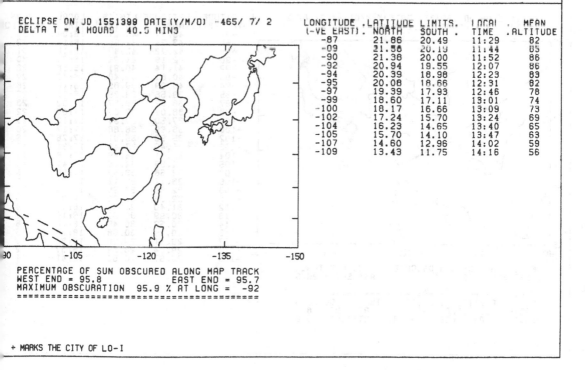

PERCENTAGE OF SUN OBSCURED ALONG MAP TRACK
WEST END = 95.8   EAST END = 95.7
MAXIMUM OBSCURATION 95.9 % AT LONG = -92

==

+ MARKS THE CITY OF LO-I

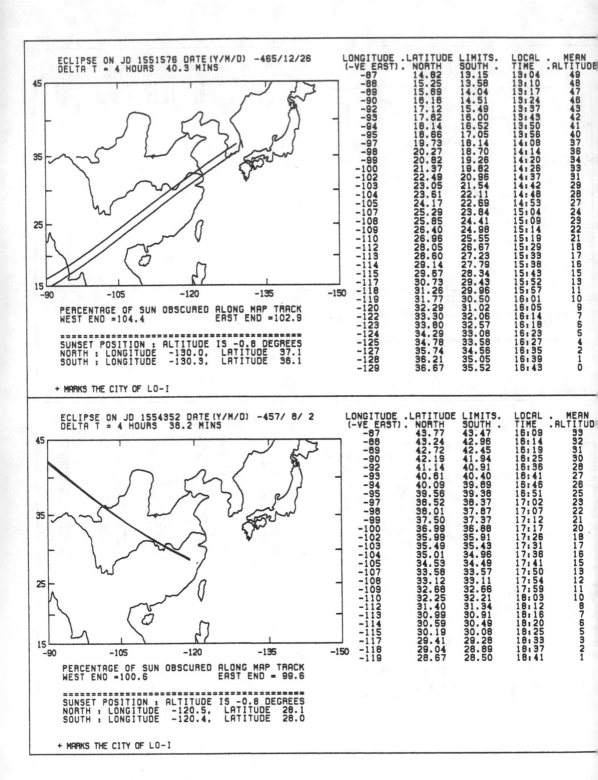

ECLIPSE ON JD 1551576 DATE(Y/M/D) -465/12/26
DELTA T = 4 HOURS 40.3 MINS

PERCENTAGE OF SUN OBSCURED ALONG MAP TRACK
WEST END =104.4   EAST END =102.9

SUNSET POSITION : ALTITUDE IS -0.8 DEGREES
NORTH : LONGITUDE -130.0, LATITUDE 37.1
SOUTH : LONGITUDE -130.3, LATITUDE 36.1

+ MARKS THE CITY OF LO-I

LONGITUDE (-VE EAST)	LATITUDE LIMITS. NORTH	SOUTH	LOCAL TIME	MEAN ALTITUDE
-87	14.82	13.15	13:04	49
-88	15.25	13.58	13:10	48
-89	15.69	14.04	13:17	47
-90	16.16	14.51	13:24	46
-92	17.12	15.49	13:37	43
-93	17.62	16.00	13:43	42
-94	18.14	16.52	13:50	41
-95	18.66	17.05	13:56	40
-97	19.73	18.14	14:08	37
-98	20.27	18.70	14:14	36
-99	20.82	19.26	14:20	34
-100	21.37	19.82	14:26	33
-102	22.49	20.96	14:37	31
-103	23.05	21.54	14:42	29
-104	23.61	22.11	14:48	28
-105	24.17	22.69	14:53	27
-107	25.29	23.84	15:04	24
-108	25.85	24.41	15:09	23
-109	26.40	24.98	15:14	22
-110	26.96	25.55	15:19	21
-112	28.05	26.67	15:29	18
-113	28.60	27.23	15:33	17
-114	29.14	27.79	15:38	16
-115	29.67	28.34	15:43	15
-117	30.73	29.43	15:52	13
-118	31.26	29.96	15:57	11
-119	31.77	30.50	16:01	10
-120	32.29	31.02	16:05	9
-122	33.30	32.06	16:14	7
-123	33.80	32.57	16:18	6
-124	34.29	33.08	16:23	5
-125	34.78	33.58	16:27	4
-127	35.74	34.56	16:35	2
-128	36.21	35.05	16:39	1
-129	36.67	35.52	16:43	0

ECLIPSE ON JD 1554352 DATE(Y/M/D) -457/ 8/ 2
DELTA T = 4 HOURS 38.2 MINS

PERCENTAGE OF SUN OBSCURED ALONG MAP TRACK
WEST END =100.6   EAST END = 99.6

SUNSET POSITION : ALTITUDE IS -0.8 DEGREES
NORTH : LONGITUDE -120.5, LATITUDE 28.1
SOUTH : LONGITUDE -120.4, LATITUDE 28.0

+ MARKS THE CITY OF LO-I

LONGITUDE (-VE EAST)	LATITUDE LIMITS. NORTH	SOUTH	LOCAL TIME	MEAN ALTITUD
-87	43.77	43.47	16:09	33
-88	43.24	42.96	16:14	32
-89	42.72	42.45	16:19	31
-90	42.19	41.94	16:25	30
-92	41.14	40.91	16:36	28
-93	40.61	40.40	16:41	27
-94	40.09	39.89	16:46	26
-95	39.56	39.38	16:51	25
-97	38.52	38.37	17:02	23
-98	38.01	37.87	17:07	22
-99	37.50	37.37	17:12	21
-100	36.99	36.88	17:17	20
-102	35.99	35.91	17:26	18
-103	35.49	35.43	17:31	17
-104	35.01	34.96	17:36	16
-105	34.53	34.49	17:41	15
-107	33.58	33.57	17:50	13
-108	33.12	33.11	17:54	12
-109	32.68	32.66	17:59	11
-110	32.25	32.21	18:03	10
-112	31.40	31.34	18:12	8
-113	30.99	30.91	18:16	7
-114	30.59	30.49	18:20	6
-115	30.19	30.08	18:25	5
-117	29.41	29.28	18:33	3
-118	29.04	28.89	18:37	2
-119	28.67	28.50	18:41	1

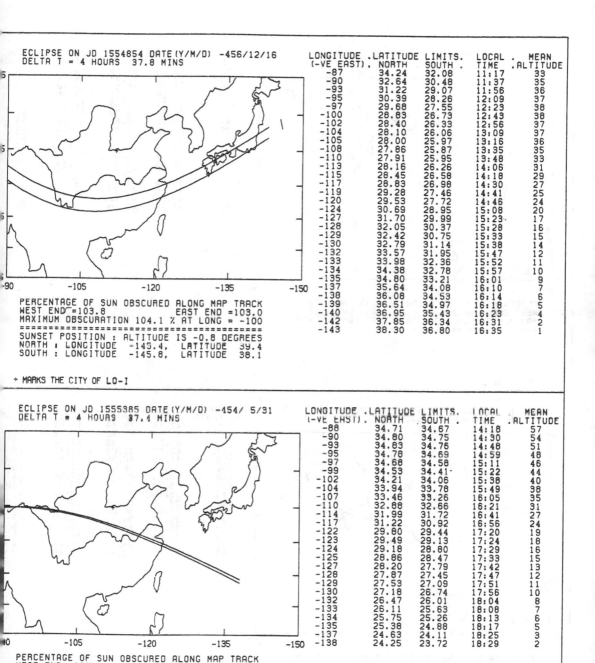

ECLIPSE ON JD 1554854 DATE(Y/M/D) -456/12/16
DELTA T = 4 HOURS 37.8 MINS

LONGITUDE (-VE EAST)	LATITUDE NORTH	LIMITS. SOUTH	LOCAL TIME	MEAN ALTITUDE
-87	34.24	32.08	11:17	33
-90	32.64	30.48	11:37	35
-93	31.22	29.07	11:56	36
-95	30.39	28.26	12:09	37
-97	29.68	27.55	12:23	38
-100	28.83	26.73	12:43	38
-102	28.40	26.33	12:56	37
-104	28.10	26.06	13:09	37
-105	28.00	25.97	13:16	36
-108	27.86	25.87	13:35	35
-110	27.91	25.95	13:48	33
-113	28.16	26.26	14:06	31
-115	28.45	26.58	14:18	29
-117	28.83	26.98	14:30	27
-119	29.28	27.46	14:41	25
-120	29.53	27.72	14:46	24
-124	30.69	28.95	15:08	20
-127	31.70	29.99	15:23	17
-128	32.05	30.37	15:28	16
-129	32.42	30.75	15:33	15
-130	32.79	31.14	15:38	14
-132	33.57	31.95	15:47	12
-133	33.98	32.36	15:52	11
-134	34.38	32.78	15:57	10
-135	34.80	33.21	16:01	9
-137	35.64	34.08	16:10	7
-138	36.08	34.53	16:14	6
-139	36.51	34.97	16:18	5
-140	36.95	35.43	16:23	4
-142	37.85	36.34	16:31	2
-143	38.30	36.80	16:35	1

PERCENTAGE OF SUN OBSCURED ALONG MAP TRACK
WEST END =103.8   EAST END =103.0
MAXIMUM OBSCURATION 104.1 % AT LONG = -100
===
SUNSET POSITION : ALTITUDE IS -0.8 DEGREES
NORTH : LONGITUDE -145.4,   LATITUDE   39.4
SOUTH : LONGITUDE -145.8,   LATITUDE   38.1

+ MARKS THE CITY OF LO-I

ECLIPSE ON JD 1555385 DATE(Y/M/D) -454/ 5/31
DELTA T = 4 HOURS 37.1 MINS

LONGITUDE (-VE EAST)	LATITUDE NORTH	LIMITS. SOUTH	LOCAL TIME	MEAN ALTITUDE
-88	34.71	34.67	14:18	57
-90	34.80	34.75	14:30	54
-93	34.83	34.76	14:48	51
-95	34.78	34.69	14:59	48
-97	34.68	34.58	15:11	46
-99	34.53	34.41	15:22	44
-102	34.21	34.06	15:38	40
-104	33.94	33.78	15:49	38
-107	33.46	33.26	16:05	35
-110	32.88	32.66	16:21	31
-114	31.99	31.72	16:41	27
-117	31.22	30.92	16:56	24
-122	29.80	29.44	17:20	19
-123	29.49	29.13	17:24	18
-124	29.18	28.80	17:29	16
-125	28.86	28.47	17:33	15
-127	28.20	27.79	17:42	13
-128	27.87	27.45	17:47	12
-129	27.53	27.09	17:51	11
-130	27.18	26.74	17:56	10
-132	26.47	26.01	18:04	8
-133	26.11	25.63	18:08	7
-134	25.75	25.26	18:13	6
-135	25.38	24.88	18:17	5
-137	24.63	24.11	18:25	3
-138	24.25	23.72	18:29	2

PERCENTAGE OF SUN OBSCURED ALONG MAP TRACK
WEST END =100.0   EAST END = 98.5

===
SUNSET POSITION : ALTITUDE IS -0.8 DEGREES
NORTH : LONGITUDE -140.6,   LATITUDE   23.3
SOUTH : LONGITUDE -140.3,   LATITUDE   22.8

+ MARKS THE CITY OF LO-I

ECLIPSE ON JD 1556774 DATE(Y/M/D) -450/ 3/20
DELTA T = 4 HOURS 36.3 MINS

LONGITUDE (-VE EAST)	.LATITUDE LIMITS. NORTH	SOUTH .	LOCAL TIME .	MEAN ALTITUDE
-87	18.23	16.08	9:17	45
-89	20.09	17.87	9:32	48
-90	21.09	18.84	9:40	49
-92	23.25	20.91	9:58	51
-94	25.57	23.19	10:14	53
-95	26.77	24.39	10:22	53
-97	29.28	26.85	10:40	54
-99	31.80	29.39	10:56	54
-100	33.05	30.65	11:04	53
-102	35.52	33.13	11:21	52
-104	37.87	35.53	11:37	50
-105	39.00	36.69	11:44	49
-107	41.15	38.88	11:59	47
-110	44.09	41.91	12:19	44
-112	45.88	43.73	12:33	42
-114	47.52	45.42	12:45	40
-115	48.30	46.20	12:52	39

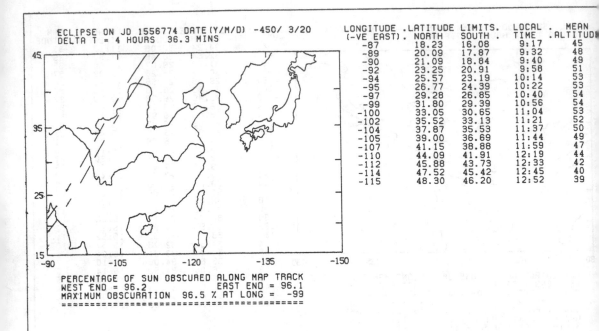

PERCENTAGE OF SUN OBSCURED ALONG MAP TRACK
WEST END = 96.2   EAST END = 96.1
MAXIMUM OBSCURATION 96.5 % AT LONG = -99
==

+ MARKS THE CITY OF LO-I

ECLIPSE ON JD 1557630 DATE(Y/M/D) -448/ 7/23
DELTA T = 4 HOURS 35.6 MINS

LONGITUDE (-VE EAST)	.LATITUDE LIMITS. NORTH	SOUTH .	LOCAL TIME .	MEAN ALTITUDE
-130	47.43	46.49	17:03	25
-132	45.81	44.86	17:15	23
-135	43.52	42.53	17:31	20
-137	42.07	41.07	17:41	18
-138	41.38	40.36	17:46	16
-139	40.71	39.68	17:50	15
-140	40.05	39.02	17:56	14
-142	38.79	37.75	18:05	12
-143	38.20	37.15	18:10	11
-144	37.62	36.56	18:14	10
-145	37.06	36.00	18:19	9
-147	35.99	34.91	18:28	7
-148	35.48	34.40	18:32	6
-149	34.99	33.90	18:36	5
-150	34.51	33.42	18:41	4
-152	33.61	32.51	18:49	2

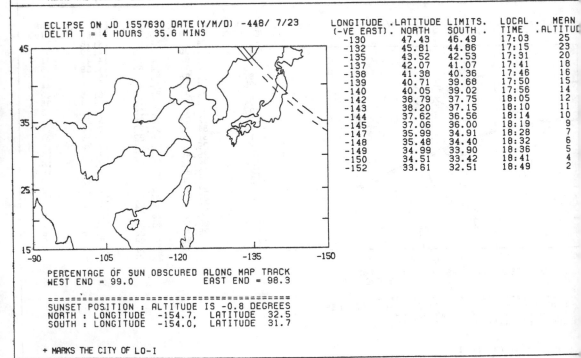

PERCENTAGE OF SUN OBSCURED ALONG MAP TRACK
WEST END = 99.0   EAST END = 98.3

==
SUNSET POSITION : ALTITUDE IS -0.8 DEGREES
NORTH : LONGITUDE -154.7, LATITUDE 32.5
SOUTH : LONGITUDE -154.0, LATITUDE 31.7

+ MARKS THE CITY OF LO-I

ECLIPSE ON JD 1559549 DATE(Y/M/D) -443/10/24
DELTA T = 4 HOURS 34.1 MINS

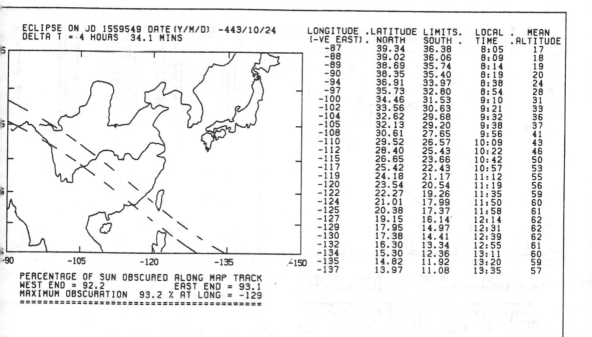

LONGITUDE (-VE EAST)	LATITUDE NORTH	LIMITS. SOUTH	LOCAL TIME	MEAN ALTITUDE
-87	39.34	36.38	8:05	17
-88	39.02	36.06	8:09	18
-89	38.69	35.74	8:14	19
-90	38.35	35.40	8:19	20
-94	36.91	33.97	8:38	24
-97	35.73	32.80	8:54	28
-100	34.46	31.53	9:10	31
-102	33.56	30.63	9:21	33
-104	32.62	29.68	9:32	36
-105	32.13	29.20	9:38	37
-108	30.61	27.65	9:56	41
-110	29.52	26.57	10:09	43
-112	28.40	25.43	10:22	46
-115	26.65	23.66	10:42	50
-117	25.42	22.43	10:57	53
-119	24.18	21.17	11:12	55
-120	23.54	20.54	11:19	56
-122	22.27	19.26	11:35	59
-124	21.01	17.99	11:50	60
-125	20.38	17.37	11:58	61
-127	19.15	16.14	12:14	62
-129	17.95	14.97	12:31	62
-130	17.38	14.41	12:39	62
-132	16.30	13.34	12:55	61
-134	15.30	12.36	13:11	60
-135	14.82	11.92	13:20	59
-137	13.97	11.08	13:35	57

PERCENTAGE OF SUN OBSCURED ALONG MAP TRACK
WEST END = 92.2   EAST END = 93.1
MAXIMUM OBSCURATION 93.2 % AT LONG = -129
===

+ MARKS THE CITY OF LO-I

ECLIPSE ON JD 1560052 DATE(Y/M/D) -441/ 3/11
DELTA T = 4 HOURS 33.7 MINS

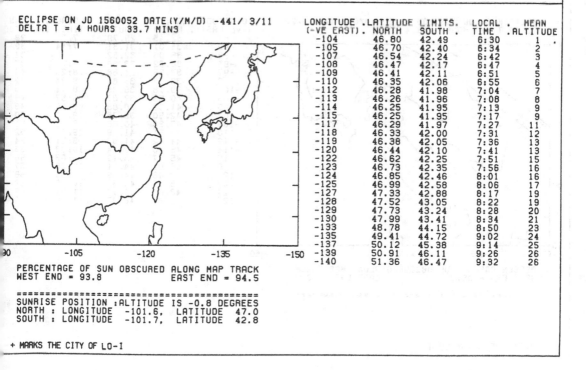

LONGITUDE (-VE EAST)	LATITUDE NORTH	LIMITS. SOUTH	LOCAL TIME	MEAN ALTITUDE
-104	46.80	42.49	6:30	1
-105	46.70	42.40	6:34	2
-107	46.54	42.24	6:42	3
-108	46.47	42.17	6:47	4
-109	46.41	42.11	6:51	5
-110	46.35	42.06	6:55	6
-112	46.28	41.98	7:04	7
-113	46.26	41.96	7:08	8
-114	46.25	41.95	7:13	9
-115	46.25	41.95	7:17	9
-117	46.29	41.97	7:27	11
-118	46.33	42.00	7:31	12
-119	46.38	42.05	7:36	13
-120	46.44	42.10	7:41	13
-122	46.62	42.25	7:51	15
-123	46.73	42.35	7:56	16
-124	46.85	42.46	8:01	16
-125	46.99	42.58	8:06	17
-127	47.33	42.88	8:17	19
-128	47.52	43.05	8:22	19
-129	47.73	43.24	8:28	20
-130	47.99	43.41	8:34	21
-133	48.78	44.15	8:50	23
-135	49.41	44.72	9:02	24
-137	50.12	45.38	9:14	25
-139	50.91	46.11	9:26	26
-140	51.36	46.47	9:32	26

PERCENTAGE OF SUN OBSCURED ALONG MAP TRACK
WEST END = 93.8   EAST END = 94.5

==
SUNRISE POSITION :ALTITUDE IS -0.8 DEGREES
NORTH : LONGITUDE -101.6, LATITUDE 47.0
SOUTH : LONGITUDE -101.7, LATITUDE 42.8

+ MARKS THE CITY OF LO-I

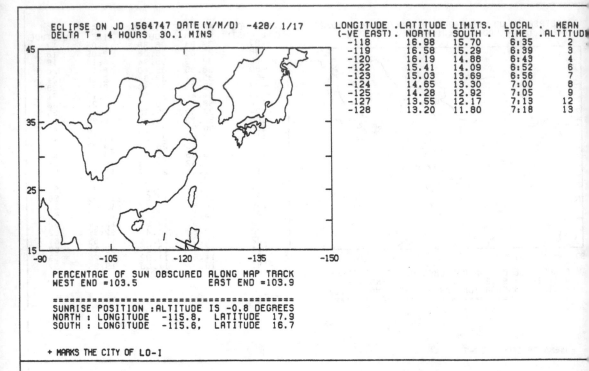

ECLIPSE ON JD 1564747 DATE(Y/M/D) -428/ 1/17
DELTA T = 4 HOURS 30.1 MINS

LONGITUDE (-VE EAST)	LATITUDE NORTH	LIMITS. SOUTH	LOCAL TIME	MEAN ALTITUDE
-118	16.98	15.70	6:35	2
-119	16.58	15.29	6:39	3
-120	16.19	14.88	6:43	4
-122	15.41	14.09	6:52	6
-123	15.03	13.69	6:56	7
-124	14.65	13.30	7:00	8
-125	14.28	12.92	7:05	9
-127	13.55	12.17	7:13	12
-128	13.20	11.80	7:18	13

PERCENTAGE OF SUN OBSCURED ALONG MAP TRACK
WEST END =103.5   EAST END =103.9

===
SUNRISE POSITION :ALTITUDE IS -0.8 DEGREES
NORTH : LONGITUDE -115.8, LATITUDE 17.9
SOUTH : LONGITUDE -115.6, LATITUDE 16.7

+ MARKS THE CITY OF LO-I

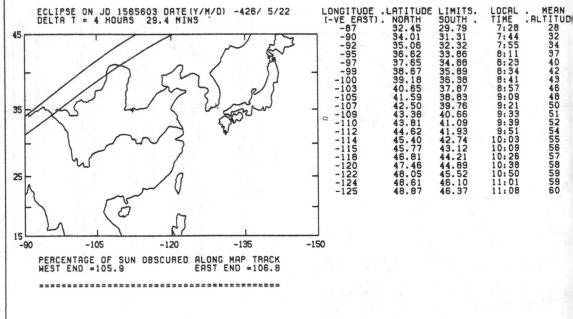

ECLIPSE ON JD 1565603 DATE(Y/M/D) -426/ 5/22
DELTA T = 4 HOURS 29.4 MINS

LONGITUDE (-VE EAST)	LATITUDE NORTH	LIMITS. SOUTH	LOCAL TIME	MEAN ALTITUDE
-87	32.45	29.79	7:28	28
-90	34.01	31.31	7:44	32
-92	35.06	32.32	7:55	34
-95	36.62	33.86	8:11	37
-97	37.65	34.88	8:23	40
-99	38.67	35.89	8:34	42
-100	39.18	36.38	8:41	43
-103	40.65	37.87	8:57	46
-105	41.59	38.83	9:09	48
-107	42.50	39.76	9:21	50
-109	43.38	40.66	9:33	51
-110	43.81	41.09	9:39	52
-112	44.62	41.93	9:51	54
-114	45.40	42.74	10:03	55
-115	45.77	43.12	10:09	56
-118	46.81	44.21	10:26	57
-120	47.46	44.89	10:38	58
-122	48.05	45.52	10:50	59
-124	48.61	46.10	11:01	59
-125	48.87	46.37	11:08	60

PERCENTAGE OF SUN OBSCURED ALONG MAP TRACK
WEST END =105.9   EAST END =106.8

===

+ MARKS THE CITY OF LO-I

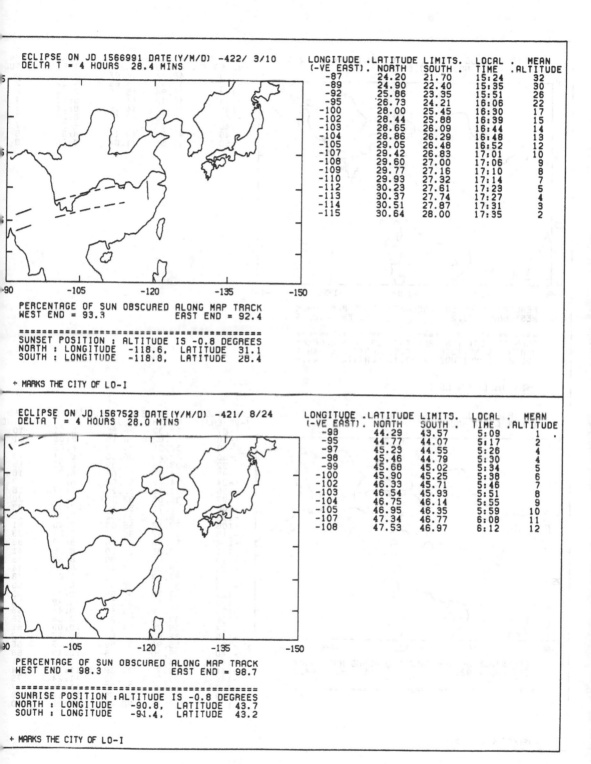

ECLIPSE ON JD 1566991 DATE(Y/M/D) -422/ 3/10
DELTA T = 4 HOURS 28.4 MINS

LONGITUDE (-VE EAST)	LATITUDE NORTH	LIMITS. SOUTH	LOCAL TIME	MEAN ALTITUDE
-87	24.20	21.70	15:24	32
-89	24.90	22.40	15:35	30
-92	25.86	23.35	15:51	26
-95	26.73	24.21	16:06	22
-100	28.00	25.45	16:30	17
-102	28.44	25.88	16:39	15
-103	28.65	26.09	16:44	14
-104	28.86	26.29	16:48	13
-105	29.05	26.48	16:52	12
-107	29.42	26.83	17:01	10
-108	29.60	27.00	17:06	9
-109	29.77	27.16	17:10	8
-110	29.93	27.32	17:14	7
-112	30.23	27.61	17:23	5
-113	30.37	27.74	17:27	4
-114	30.51	27.87	17:31	3
-115	30.64	28.00	17:35	2

PERCENTAGE OF SUN OBSCURED ALONG MAP TRACK
WEST END = 93.3   EAST END = 92.4

==
SUNSET POSITION : ALTITUDE IS -0.8 DEGREES
NORTH : LONGITUDE -118.6, LATITUDE 31.1
SOUTH : LONGITUDE -118.8, LATITUDE 28.4

+ MARKS THE CITY OF LO-I

ECLIPSE ON JD 1567523 DATE(Y/M/D) -421/ 8/24
DELTA T = 4 HOURS 28.0 MINS

LONGITUDE (-VE EAST)	LATITUDE NORTH	LIMITS. SOUTH	LOCAL TIME	MEAN ALTITUDE
-93	44.29	43.57	5:09	1
-95	44.77	44.07	5:17	2
-97	45.23	44.55	5:26	4
-98	45.46	44.79	5:30	4
-99	45.68	45.02	5:34	5
-100	45.90	45.25	5:38	6
-102	46.33	45.71	5:46	7
-103	46.54	45.93	5:51	8
-104	46.75	46.14	5:55	9
-105	46.95	46.35	5:59	10
-107	47.34	46.77	6:08	11
-108	47.53	46.97	6:12	12

PERCENTAGE OF SUN OBSCURED ALONG MAP TRACK
WEST END = 98.3   EAST END = 98.7

==
SUNRISE POSITION : ALTITUDE IS -0.8 DEGREES
NORTH : LONGITUDE -90.8, LATITUDE 43.7
SOUTH : LONGITUDE -91.4, LATITUDE 43.2

+ MARKS THE CITY OF LO-I

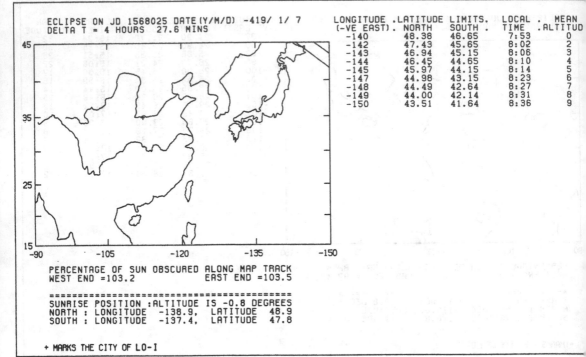

ECLIPSE ON JD 1568025 DATE(Y/M/D) -419/ 1/ 7
DELTA T = 4 HOURS 27.6 MINS

LONGITUDE	.LATITUDE	LIMITS.	LOCAL	. MEAN
(-VE EAST).	NORTH	SOUTH .	TIME	.ALTITUD
-140	48.38	46.65	7:53	0
-142	47.43	45.65	8:02	2
-143	46.94	45.15	8:06	3
-144	46.45	44.65	8:10	4
-145	45.97	44.15	8:14	5
-147	44.98	43.15	8:23	6
-148	44.49	42.64	8:27	7
-149	44.00	42.14	8:31	8
-150	43.51	41.64	8:36	9

PERCENTAGE OF SUN OBSCURED ALONG MAP TRACK
WEST END =103.2   EAST END =103.5

==
SUNRISE POSITION :ALTITUDE IS -0.8 DEGREES
NORTH : LONGITUDE -138.9,   LATITUDE 48.9
SOUTH : LONGITUDE -137.4,   LATITUDE 47.8

+ MARKS THE CITY OF LO-I

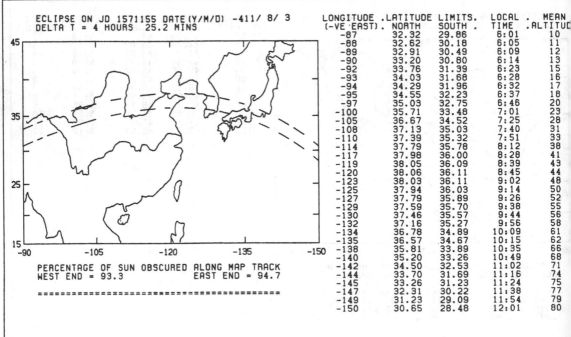

ECLIPSE ON JD 1571155 DATE(Y/M/D) -411/ 8/ 3
DELTA T = 4 HOURS 25.2 MINS

LONGITUDE	.LATITUDE	LIMITS.	LOCAL	. MEAN
(-VE EAST).	NORTH	SOUTH .	TIME	.ALTITUD
-87	32.32	29.86	6:01	10
-88	32.62	30.18	6:05	11
-89	32.91	30.49	6:09	12
-90	33.20	30.80	6:14	13
-92	33.76	31.39	6:23	15
-93	34.03	31.68	6:28	16
-94	34.29	31.96	6:32	17
-95	34.55	32.23	6:37	18
-97	35.03	32.75	6:46	20
-100	35.71	33.48	7:01	23
-105	36.67	34.52	7:25	28
-108	37.13	35.03	7:40	31
-110	37.39	35.32	7:51	33
-114	37.79	35.78	8:12	38
-117	37.98	36.00	8:28	41
-119	38.05	36.09	8:39	43
-120	38.06	36.11	8:45	44
-123	38.03	36.11	9:02	48
-125	37.94	36.03	9:14	50
-127	37.79	35.89	9:26	52
-129	37.59	35.70	9:38	55
-130	37.46	35.57	9:44	56
-132	37.16	35.27	9:56	58
-134	36.78	34.89	10:09	61
-135	36.57	34.67	10:15	62
-138	35.81	33.89	10:35	66
-140	35.20	33.26	10:49	68
-142	34.50	32.53	11:02	71
-144	33.70	31.69	11:16	74
-145	33.26	31.23	11:24	75
-147	32.31	30.22	11:38	77
-149	31.23	29.09	11:54	79
-150	30.65	28.48	12:01	80

PERCENTAGE OF SUN OBSCURED ALONG MAP TRACK
WEST END = 93.3   EAST END = 94.7

==

+ MARKS THE CITY OF LO-I

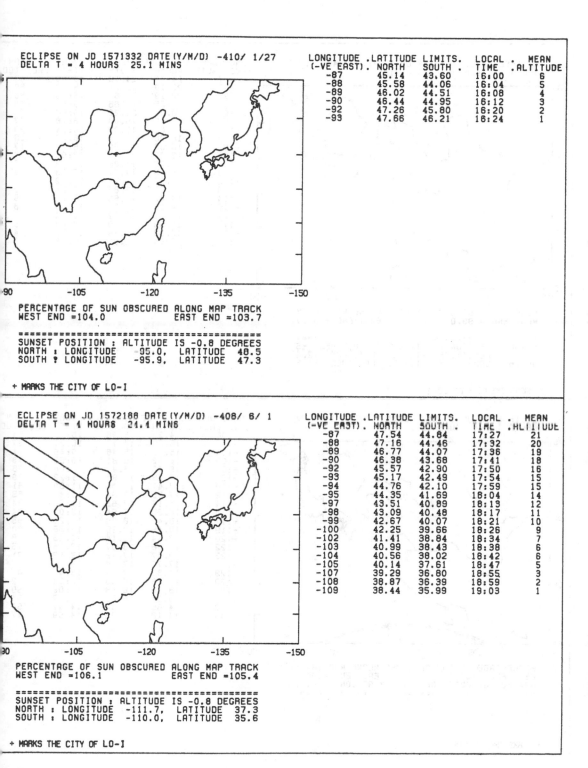

ECLIPSE ON JD 1571332 DATE(Y/M/D) -410/ 1/27
DELTA T = 4 HOURS 25.1 MINS

LONGITUDE	.LATITUDE	LIMITS.	LOCAL .	MEAN
(-VE EAST).	NORTH	SOUTH .	TIME	.ALTITUDE
-87	45.14	43.60	16:00	6
-88	45.58	44.06	16:04	5
-89	46.02	44.51	16:08	4
-90	46.44	44.95	16:12	3
-92	47.26	45.80	16:20	2
-93	47.66	46.21	16:24	1

PERCENTAGE OF SUN OBSCURED ALONG MAP TRACK
WEST END =104.0   EAST END =103.7

==
SUNSET POSITION : ALTITUDE IS -0.8 DEGREES
NORTH : LONGITUDE   -95.0, LATITUDE 48.5
SOUTH : LONGITUDE   -95.9, LATITUDE 47.3

+ MARKS THE CITY OF LO-I

ECLIPSE ON JD 1572188 DATE(Y/M/D) -408/ 6/ 1
DELTA T = 4 HOURS 24.1 MINS

LONGITUDE	.LATITUDE	LIMITS.	LOCAL .	MEAN
(-VE EAST).	NORTH	SOUTH .	TIME	.ALTITUDE
-87	47.54	44.84	17:27	21
-88	47.16	44.46	17:32	20
-89	46.77	44.07	17:36	19
-90	46.38	43.68	17:41	18
-92	45.57	42.90	17:50	16
-93	45.17	42.49	17:54	15
-94	44.76	42.10	17:59	15
-95	44.35	41.69	18:04	14
-97	43.51	40.89	18:13	12
-98	43.09	40.48	18:17	11
-99	42.67	40.07	18:21	10
-100	42.25	39.66	18:26	9
-102	41.41	38.84	18:34	7
-103	40.99	38.43	18:38	6
-104	40.56	38.02	18:42	6
-105	40.14	37.61	18:47	5
-107	39.29	36.80	18:55	3
-108	38.87	36.39	18:59	2
-109	38.44	35.99	19:03	1

PERCENTAGE OF SUN OBSCURED ALONG MAP TRACK
WEST END =106.1   EAST END =105.4

==
SUNSET POSITION : ALTITUDE IS -0.8 DEGREES
NORTH : LONGITUDE -111.7, LATITUDE 37.3
SOUTH : LONGITUDE -110.0, LATITUDE 35.6

+ MARKS THE CITY OF LO-I

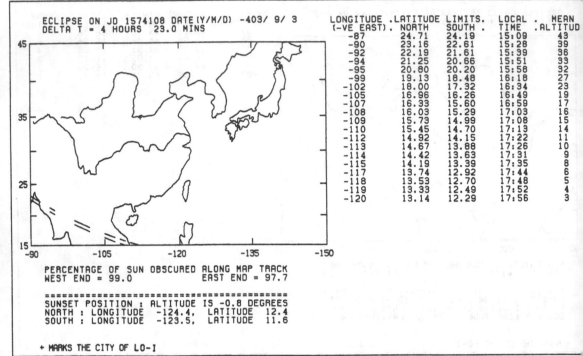

ECLIPSE ON JD 1574108 DATE(Y/M/D) -403/ 9/ 3
DELTA T = 4 HOURS 23.0 MINS

LONGITUDE	.LATITUDE	LIMITS.	LOCAL	. MEAN
(-VE EAST).	NORTH	SOUTH .	TIME	.ALTITUD
-87	24.71	24.19	15:09	43
-90	23.16	22.61	15:28	39
-92	22.19	21.61	15:39	36
-94	21.25	20.66	15:51	33
-95	20.80	20.20	15:58	32
-99	19.13	18.48	16:18	27
-102	18.00	17.32	16:34	23
-105	16.96	16.26	16:49	19
-107	16.33	15.60	16:59	17
-108	16.03	15.29	17:03	16
-109	15.73	14.99	17:08	15
-110	15.45	14.70	17:13	14
-112	14.92	14.15	17:22	11
-113	14.67	13.88	17:26	10
-114	14.42	13.63	17:31	9
-115	14.19	13.39	17:35	8
-117	13.74	12.92	17:44	6
-118	13.53	12.70	17:48	5
-119	13.33	12.49	17:52	4
-120	13.14	12.29	17:56	3

PERCENTAGE OF SUN OBSCURED ALONG MAP TRACK
WEST END = 99.0   EAST END = 97.7

==
SUNSET POSITION : ALTITUDE IS -0.8 DEGREES
NORTH : LONGITUDE -124.4, LATITUDE 12.4
SOUTH : LONGITUDE -123.5, LATITUDE 11.6

+ MARKS THE CITY OF LO-I

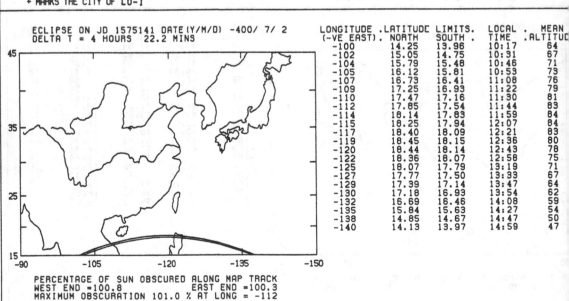

ECLIPSE ON JD 1575141 DATE(Y/M/D) -400/ 7/ 2
DELTA T = 4 HOURS 22.2 MINS

LONGITUDE	.LATITUDE	LIMITS.	LOCAL	. MEAN
(-VE EAST).	NORTH	SOUTH .	TIME	.ALTITUD
-100	14.25	13.96	10:17	64
-102	15.05	14.75	10:31	67
-104	15.79	15.48	10:46	71
-105	16.12	15.81	10:53	73
-107	16.73	16.41	11:08	76
-109	17.25	16.93	11:22	79
-110	17.47	17.16	11:30	81
-112	17.85	17.54	11:44	83
-114	18.14	17.83	11:59	84
-115	18.25	17.94	12:07	84
-117	18.40	18.09	12:21	83
-119	18.45	18.15	12:36	80
-120	18.44	18.14	12:43	78
-122	18.36	18.07	12:58	75
-125	18.07	17.79	13:19	71
-127	17.77	17.50	13:33	67
-129	17.39	17.14	13:47	64
-130	17.18	16.93	13:54	62
-132	16.69	16.46	14:08	59
-135	15.84	15.63	14:27	54
-138	14.85	14.67	14:47	50
-140	14.13	13.97	14:59	47

PERCENTAGE OF SUN OBSCURED ALONG MAP TRACK
WEST END =100.8   EAST END =100.3
MAXIMUM OBSCURATION 101.0 % AT LONG = -112
==

+ MARKS THE CITY OF LO-I

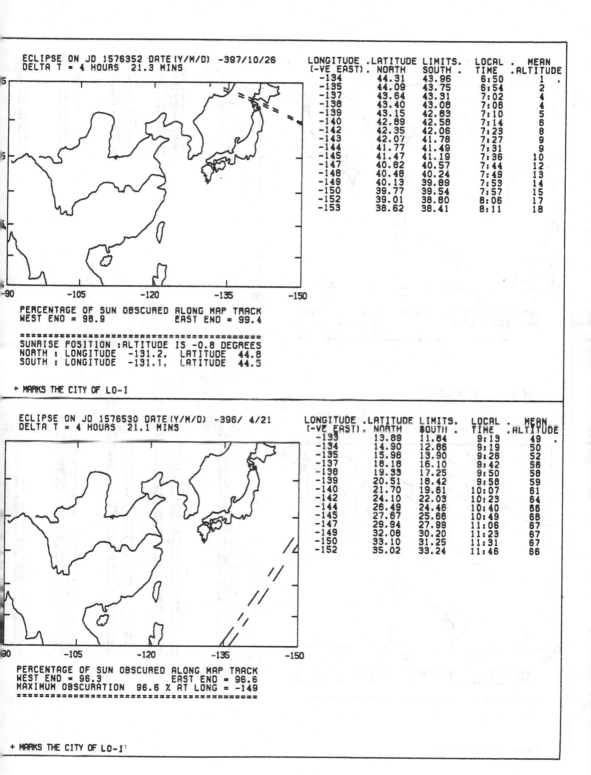

ECLIPSE ON JD 1576352 DATE(Y/M/D) -397/10/26
DELTA T = 4 HOURS 21.3 MINS

LONGITUDE (-VE EAST)	LATITUDE LIMITS. NORTH	SOUTH	LOCAL TIME	MEAN ALTITUDE
-134	44.31	43.96	6:50	1
-135	44.09	43.75	6:54	2
-137	43.64	43.31	7:02	4
-138	43.40	43.08	7:06	4
-139	43.15	42.83	7:10	5
-140	42.89	42.58	7:14	6
-142	42.35	42.06	7:23	8
-143	42.07	41.78	7:27	9
-144	41.77	41.49	7:31	9
-145	41.47	41.19	7:36	10
-147	40.82	40.57	7:44	12
-148	40.48	40.24	7:49	13
-149	40.13	39.89	7:53	14
-150	39.77	39.54	7:57	15
-152	39.01	38.80	8:06	17
-153	38.62	38.41	8:11	18

PERCENTAGE OF SUN OBSCURED ALONG MAP TRACK
WEST END = 98.9   EAST END = 99.4

SUNRISE POSITION :ALTITUDE IS -0.8 DEGREES
NORTH : LONGITUDE -131.2, LATITUDE 44.8
SOUTH : LONGITUDE -131.1, LATITUDE 44.5

+ MARKS THE CITY OF LO-I

ECLIPSE ON JD 1576530 DATE(Y/M/D) -396/ 4/21
DELTA T = 4 HOURS 21.1 MINS

LONGITUDE (-VE EAST)	LATITUDE LIMITS. NORTH	SOUTH	LOCAL TIME	MEAN ALTITUDE
-133	13.89	11.84	9:13	49
-134	14.90	12.86	9:19	50
-135	15.96	13.90	9:26	52
-137	18.18	16.10	8:42	56
-138	19.33	17.25	9:50	58
-139	20.51	18.42	9:58	59
-140	21.70	19.61	10:07	61
-142	24.10	22.03	10:23	64
-144	26.49	24.46	10:40	66
-145	27.67	25.66	10:49	66
-147	29.94	27.99	11:06	67
-149	32.08	30.20	11:23	67
-150	33.10	31.25	11:31	67
-152	35.02	33.24	11:46	66

PERCENTAGE OF SUN OBSCURED ALONG MAP TRACK
WEST END = 96.3   EAST END = 96.6
MAXIMUM OBSCURATION 96.6 % AT LONG = -149

+ MARKS THE CITY OF LO-I

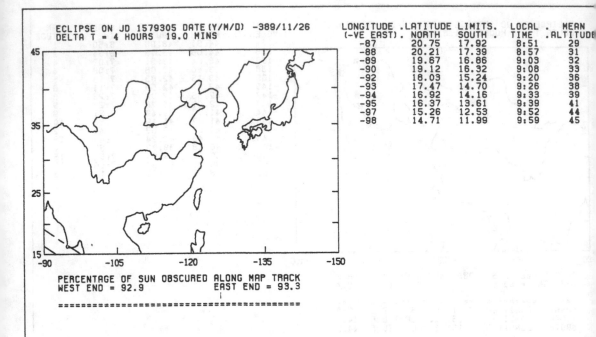

ECLIPSE ON JD 1579305 DATE(Y/M/D) -389/11/26
DELTA T = 4 HOURS 19.0 MINS

LONGITUDE (-VE EAST)	LATITUDE NORTH	LIMITS. SOUTH	LOCAL TIME	MEAN ALTITUDE
-87	20.75	17.92	8:51	29
-88	20.21	17.39	8:57	31
-89	19.67	16.86	9:03	32
-90	19.12	16.32	9:08	33
-92	18.03	15.24	9:20	36
-93	17.47	14.70	9:26	38
-94	16.92	14.16	9:33	39
-95	16.37	13.61	9:39	41
-97	15.26	12.53	9:52	44
-98	14.71	11.99	9:59	45

PERCENTAGE OF SUN OBSCURED ALONG MAP TRACK
WEST END = 92.9   EAST END = 93.3

===

+ MARKS THE CITY OF LO-I

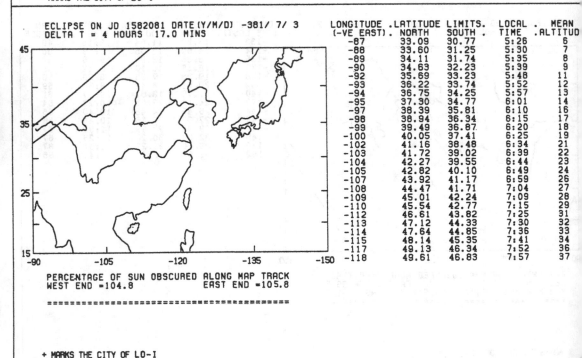

ECLIPSE ON JD 1582081 DATE(Y/M/D) -381/ 7/ 3
DELTA T = 4 HOURS 17.0 MINS

LONGITUDE (-VE EAST)	LATITUDE NORTH	LIMITS. SOUTH	LOCAL TIME	MEAN ALTITUD
-87	33.09	30.77	5:26	6
-88	33.60	31.25	5:30	7
-89	34.11	31.74	5:35	8
-90	34.63	32.23	5:39	9
-92	35.69	33.23	5:48	11
-93	36.22	33.74	5:52	12
-94	36.75	34.25	5:57	13
-95	37.30	34.77	6:01	14
-97	38.39	35.81	6:10	16
-98	38.94	36.34	6:15	17
-99	39.49	36.87	6:20	18
-100	40.05	37.41	6:25	19
-102	41.16	38.48	6:34	21
-103	41.72	39.02	6:39	22
-104	42.27	39.55	6:44	23
-105	42.82	40.10	6:49	24
-107	43.92	41.17	6:59	26
-108	44.47	41.71	7:04	27
-109	45.01	42.24	7:09	28
-110	45.54	42.77	7:15	29
-112	46.61	43.82	7:25	31
-113	47.12	44.33	7:30	32
-114	47.64	44.85	7:36	33
-115	48.14	45.35	7:41	34
-117	49.13	46.34	7:52	36
-118	49.61	46.83	7:57	37

PERCENTAGE OF SUN OBSCURED ALONG MAP TRACK
WEST END =104.8   EAST END =105.8

===

+ MARKS THE CITY OF LO-I

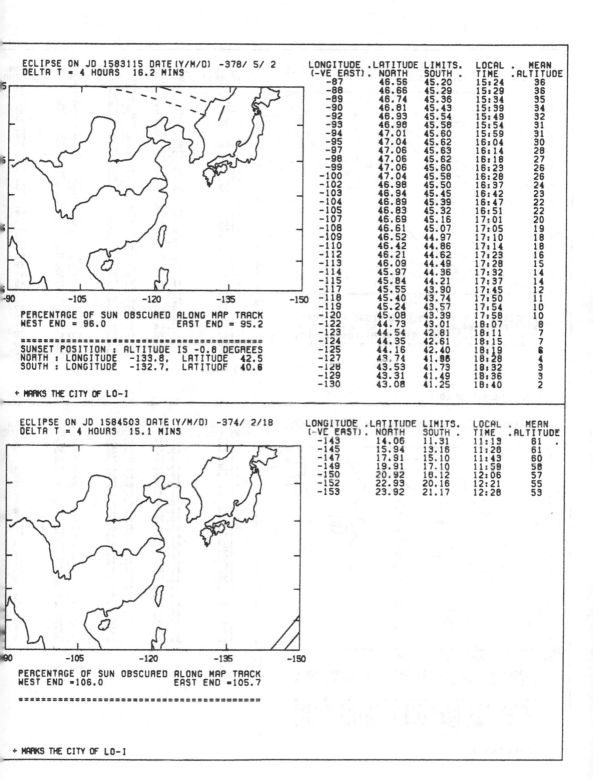

ECLIPSE ON JD 1583115 DATE(Y/M/D) -378/ 5/ 2
DELTA T = 4 HOURS 16.2 MINS

LONGITUDE (-VE EAST)	LATITUDE LIMITS. NORTH	SOUTH	LOCAL TIME	MEAN ALTITUDE
-87	46.56	45.20	15:24	36
-88	46.66	45.29	15:29	36
-89	46.74	45.36	15:34	35
-90	46.81	45.43	15:39	34
-92	46.93	45.54	15:49	32
-93	46.98	45.58	15:54	31
-94	47.01	45.60	15:59	31
-95	47.04	45.62	16:04	30
-97	47.06	45.63	16:14	28
-98	47.06	45.62	16:18	27
-99	47.06	45.60	16:23	26
-100	47.04	45.58	16:28	26
-102	46.98	45.50	16:37	24
-103	46.94	45.45	16:42	23
-104	46.89	45.39	16:47	22
-105	46.83	45.32	16:51	22
-107	46.69	45.16	17:01	20
-108	46.61	45.07	17:05	19
-109	46.52	44.97	17:10	18
-110	46.42	44.86	17:14	18
-112	46.21	44.62	17:23	16
-113	46.09	44.49	17:28	15
-114	45.97	44.36	17:32	14
-115	45.84	44.21	17:37	14
-117	45.55	43.90	17:45	12
-118	45.40	43.74	17:50	11
-119	45.24	43.57	17:54	10
-120	45.08	43.39	17:58	10
-122	44.73	43.01	18:07	8
-123	44.54	42.81	18:11	7
-124	44.35	42.61	18:15	7
-125	44.16	42.40	18:19	6
-127	43.74	41.98	18:28	4
-128	43.53	41.73	18:32	3
-129	43.31	41.49	18:36	3
-130	43.08	41.25	18:40	2

PERCENTAGE OF SUN OBSCURED ALONG MAP TRACK
WEST END = 96.0   EAST END = 95.2

===
SUNSET POSITION : ALTITUDE IS -0.8 DEGREES
NORTH : LONGITUDE -133.8, LATITUDE 42.5
SOUTH : LONGITUDE -132.7, LATITUDE 40.6

+ MARKS THE CITY OF LO-I

ECLIPSE ON JD 1584503 DATE(Y/M/D) -374/ 2/18
DELTA T = 4 HOURS 15.1 MINS

LONGITUDE (-VE EAST)	LATITUDE NORTH	SOUTH	LOCAL TIME	MEAN ALTITUDE
-143	14.06	11.31	11:19	61
-145	15.94	13.16	11:28	61
-147	17.91	15.10	11:43	60
-149	19.91	17.10	11:58	58
-150	20.92	18.12	12:06	57
-152	22.93	20.16	12:21	55
-153	23.92	21.17	12:28	53

PERCENTAGE OF SUN OBSCURED ALONG MAP TRACK
WEST END =106.0   EAST END =105.7

===

+ MARKS THE CITY OF LO-I

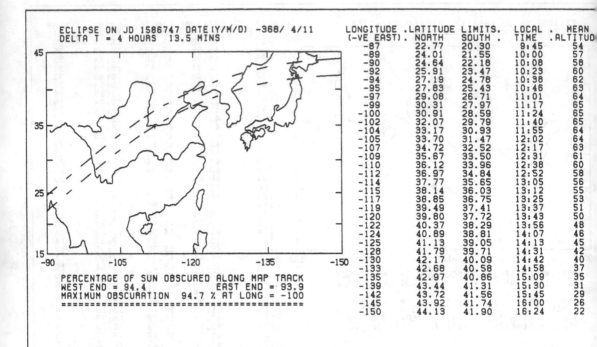

ECLIPSE ON JD 1586747 DATE(Y/M/D) -368/ 4/11
DELTA T = 4 HOURS 13.5 MINS

LONGITUDE (-VE EAST)	LATITUDE LIMITS. NORTH	SOUTH	LOCAL TIME	MEAN ALTITUDE
-87	22.77	20.30	9:45	54
-89	24.01	21.55	10:00	57
-90	24.64	22.18	10:08	58
-92	25.91	23.47	10:23	60
-94	27.19	24.78	10:38	62
-95	27.83	25.43	10:46	63
-97	29.08	26.71	11:01	64
-99	30.31	27.97	11:17	65
-100	30.91	28.59	11:24	65
-102	32.07	29.79	11:40	65
-104	33.17	30.93	11:55	64
-105	33.70	31.47	12:02	64
-107	34.72	32.52	12:17	63
-109	35.67	33.50	12:31	61
-110	36.12	33.96	12:38	60
-112	36.97	34.84	12:52	58
-114	37.77	35.65	13:05	56
-115	38.14	36.03	13:12	55
-117	38.85	36.75	13:25	53
-119	39.49	37.41	13:37	51
-120	39.80	37.72	13:43	50
-122	40.37	38.29	13:56	48
-124	40.89	38.81	14:07	46
-125	41.13	39.05	14:13	45
-128	41.79	39.71	14:31	42
-130	42.17	40.09	14:42	40
-133	42.68	40.58	14:58	37
-135	42.97	40.86	15:09	35
-139	43.44	41.31	15:30	31
-142	43.72	41.56	15:45	29
-145	43.92	41.74	16:00	26
-150	44.13	41.90	16:24	22

PERCENTAGE OF SUN OBSCURED ALONG MAP TRACK
WEST END = 94.4   EAST END = 93.9
MAXIMUM OBSCURATION 94.7 % AT LONG = -100
===

+ MARKS THE CITY OF LO-I

ECLIPSE ON JD 1587279 DATE(Y/M/D) -367/ 9/25
DELTA T = 4 HOURS 13.1 MINS

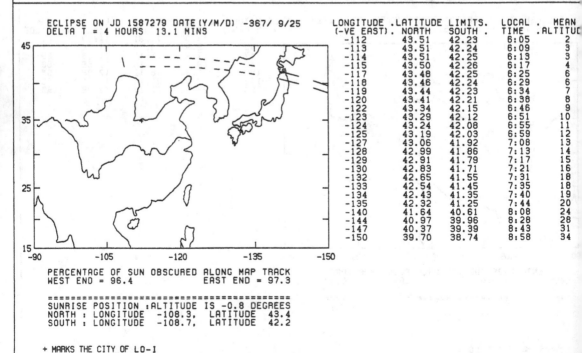

LONGITUDE (-VE EAST)	LATITUDE LIMITS. NORTH	SOUTH	LOCAL TIME	MEAN ALTITUDE
-112	43.51	42.23	6:05	2
-113	43.51	42.24	6:09	3
-114	43.51	42.25	6:13	3
-115	43.50	42.26	6:17	4
-117	43.48	42.25	6:25	6
-118	43.46	42.24	6:29	6
-119	43.44	42.23	6:34	7
-120	43.41	42.21	6:38	8
-122	43.34	42.15	6:46	9
-123	43.29	42.12	6:51	10
-124	43.24	42.08	6:55	11
-125	43.19	42.03	6:59	12
-127	43.06	41.92	7:08	13
-128	42.99	41.86	7:13	14
-129	42.91	41.79	7:17	15
-130	42.83	41.71	7:21	16
-132	42.65	41.55	7:31	18
-133	42.54	41.45	7:35	18
-134	42.43	41.35	7:40	19
-135	42.32	41.25	7:44	20
-140	41.64	40.61	8:08	24
-144	40.97	39.96	8:28	28
-147	40.37	39.39	8:43	31
-150	39.70	38.74	8:58	34

PERCENTAGE OF SUN OBSCURED ALONG MAP TRACK
WEST END = 96.4   EAST END = 97.3

===
SUNRISE POSITION :ALTITUDE IS -0.8 DEGREES
NORTH : LONGITUDE -108.3, LATITUDE 43.4
SOUTH : LONGITUDE -108.7, LATITUDE 42.2

+ MARKS THE CITY OF LO-I

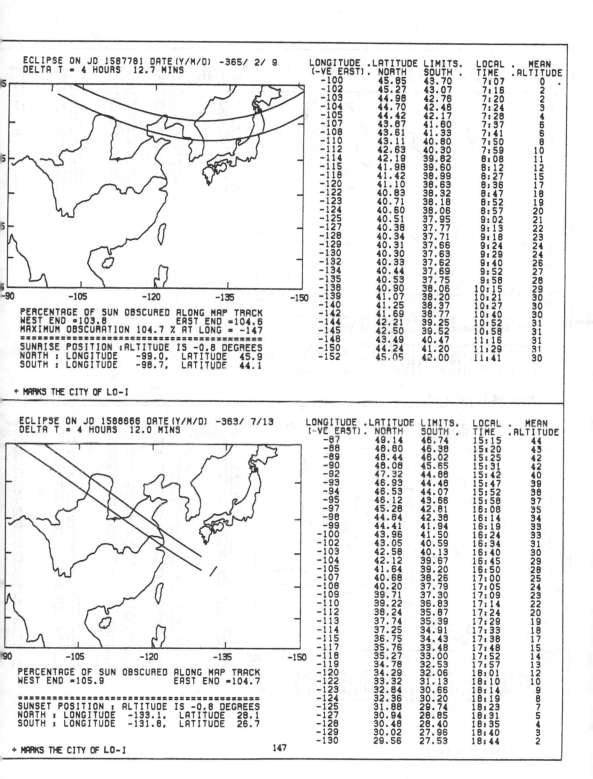

ECLIPSE ON JD 1587781 DATE(Y/M/D) -365/ 2/ 9
DELTA T = 4 HOURS 12.7 MINS

LONGITUDE (-VE EAST)	LATITUDE NORTH	LIMITS. SOUTH	LOCAL TIME	MEAN ALTITUDE
-100	45.85	43.70	7:07	0
-102	45.27	43.07	7:16	2
-103	44.98	42.76	7:20	2
-104	44.70	42.46	7:24	3
-105	44.42	42.17	7:28	4
-107	43.87	41.60	7:97	6
-108	43.61	41.33	7:41	6
-110	43.11	40.80	7:50	8
-112	42.63	40.30	7:59	10
-114	42.19	39.82	8:08	11
-115	41.98	39.60	8:12	12
-118	41.42	38.99	8:27	15
-120	41.10	38.63	8:36	17
-122	40.83	38.32	8:47	18
-123	40.71	38.18	8:52	19
-124	40.60	38.06	8:57	20
-125	40.51	37.95	9:02	21
-127	40.38	37.77	9:13	22
-128	40.34	37.71	9:18	23
-129	40.31	37.66	9:24	24
-130	40.30	37.63	9:29	24
-132	40.33	37.62	9:40	26
-134	40.44	37.69	9:52	27
-135	40.53	37.75	9:58	28
-138	40.90	38.06	10:15	29
-139	41.07	38.20	10:21	30
-140	41.25	38.37	10:27	30
-142	41.69	38.77	10:40	30
-144	42.21	39.25	10:52	31
-145	42.50	39.52	10:58	31
-148	43.49	40.47	11:16	31
-150	44.24	41.20	11:29	31
-152	45.05	42.00	11:41	30

PERCENTAGE OF SUN OBSCURED ALONG MAP TRACK
WEST END =103.8   EAST END =104.6
MAXIMUM OBSCURATION 104.7 % AT LONG = -147
===
SUNRISE POSITION :ALTITUDE IS -0.8 DEGREES
NORTH : LONGITUDE -99.0, LATITUDE 45.9
SOUTH : LONGITUDE -98.7, LATITUDE 44.1

+ MARKS THE CITY OF LO-I

ECLIPSE ON JD 1588666 DATE(Y/M/D) -363/ 7/13
DELTA T = 4 HOURS 12.0 MINS

LONGITUDE (-VE EAST)	LATITUDE NORTH	LIMITS. SOUTH	LOCAL TIME	MEAN ALTITUDE
-87	49.14	46.74	15:15	44
-88	48.80	46.38	15:20	43
-89	48.44	46.02	15:25	42
-90	48.08	45.65	15:31	42
-92	47.32	44.88	15:42	40
-93	46.93	44.48	15:47	39
-94	46.53	44.07	15:52	38
-95	46.12	43.66	15:58	37
-97	45.28	42.81	16:08	35
-98	44.84	42.38	16:14	34
-99	44.41	41.94	16:19	33
-100	43.96	41.50	16:24	33
-102	43.05	40.59	16:34	31
-103	42.58	40.13	16:40	30
-104	42.12	39.67	16:45	29
-105	41.64	39.20	16:50	28
-107	40.68	38.26	17:00	25
-108	40.20	37.79	17:05	24
-109	39.71	37.30	17:09	23
-110	39.22	36.83	17:14	22
-112	38.24	35.87	17:24	20
-113	37.74	35.39	17:29	19
-114	37.25	34.91	17:33	18
-115	36.75	34.43	17:38	17
-117	35.76	33.48	17:48	15
-118	35.27	33.00	17:52	14
-119	34.78	32.53	17:57	13
-120	34.29	32.06	18:01	12
-122	33.32	31.13	18:10	10
-123	32.84	30.66	18:14	9
-124	32.36	30.20	18:19	8
-125	31.88	29.74	18:23	7
-127	30.94	28.85	18:31	5
-128	30.48	28.40	18:35	4
-129	30.02	27.96	18:40	3
-130	29.56	27.53	18:44	2

PERCENTAGE OF SUN OBSCURED ALONG MAP TRACK
WEST END =105.9   EAST END =104.7

===
SUNSET POSITION : ALTITUDE IS -0.8 DEGREES
NORTH : LONGITUDE -133.1, LATITUDE 28.1
SOUTH : LONGITUDE -131.8, LATITUDE 26.7

+ MARKS THE CITY OF LO-I   147

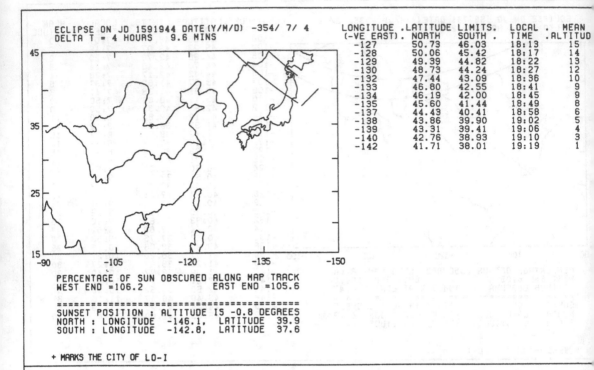

ECLIPSE ON JD 1591944 DATE(Y/M/D) -354/ 7/ 4
DELTA T = 4 HOURS   9.6 MINS

| LONGITUDE .LATITUDE LIMITS. | | LOCAL . | MEAN |
(-VE EAST). NORTH	SOUTH .	TIME	.ALTITUD
-127 50.73	46.03	18:13	15
-128 50.06	45.42	18:17	14
-129 49.39	44.82	18:22	13
-130 48.73	44.24	18:27	12
-132 47.44	43.09	18:36	10
-133 46.80	42.55	18:41	9
-134 46.19	42.00	18:45	9
-135 45.60	41.44	18:49	8
-137 44.43	40.41	18:58	6
-138 43.86	39.90	19:02	5
-139 43.31	39.41	19:06	4
-140 42.76	38.93	19:10	3
-142 41.71	38.01	19:19	1

PERCENTAGE OF SUN OBSCURED ALONG MAP TRACK
WEST END =106.2   EAST END =105.6

==
SUNSET POSITION : ALTITUDE IS -0.8 DEGREES
NORTH : LONGITUDE -146.1, LATITUDE   39.9
SOUTH : LONGITUDE -142.8, LATITUDE   37.6

+ MARKS THE CITY OF LO-I

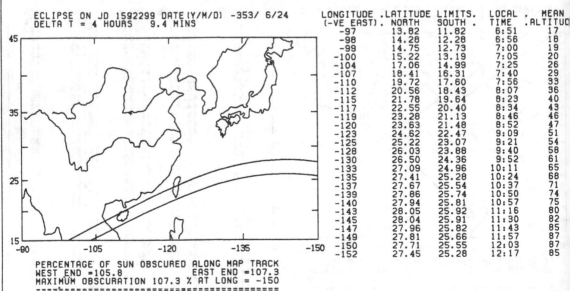

ECLIPSE ON JD 1592299 DATE(Y/M/D) -353/ 6/24
DELTA T = 4 HOURS   9.4 MINS

| LONGITUDE .LATITUDE LIMITS. | | LOCAL | MEAN |
(-VE EAST). NORTH	SOUTH .	TIME	.ALTITUD
-97 13.82	11.82	6:51	17
-98 14.28	12.28	6:56	18
-99 14.75	12.73	7:00	19
-100 15.22	13.19	7:05	20
-104 17.06	14.99	7:25	26
-107 18.41	16.31	7:40	29
-110 19.72	17.60	7:56	33
-112 20.56	18.43	8:07	36
-115 21.78	19.64	8:23	40
-117 22.55	20.40	8:34	43
-119 23.28	21.13	8:46	46
-120 23.63	21.48	8:52	47
-123 24.62	22.47	9:09	51
-125 25.22	23.07	9:21	54
-128 26.03	23.88	9:40	58
-130 26.50	24.36	9:52	61
-133 27.09	24.96	10:11	65
-135 27.41	25.28	10:24	68
-137 27.67	25.54	10:37	71
-139 27.86	25.74	10:50	74
-140 27.94	25.81	10:57	75
-143 28.05	25.92	11:16	80
-145 28.04	25.91	11:30	82
-147 27.96	25.82	11:43	85
-149 27.81	25.66	11:57	87
-150 27.71	25.55	12:03	87
-152 27.45	25.28	12:17	85

PERCENTAGE OF SUN OBSCURED ALONG MAP TRACK
WEST END =105.8   EAST END =107.3
MAXIMUM OBSCURATION 107.3 % AT LONG = -150
==

+ MARKS THE CITY OF LO-I

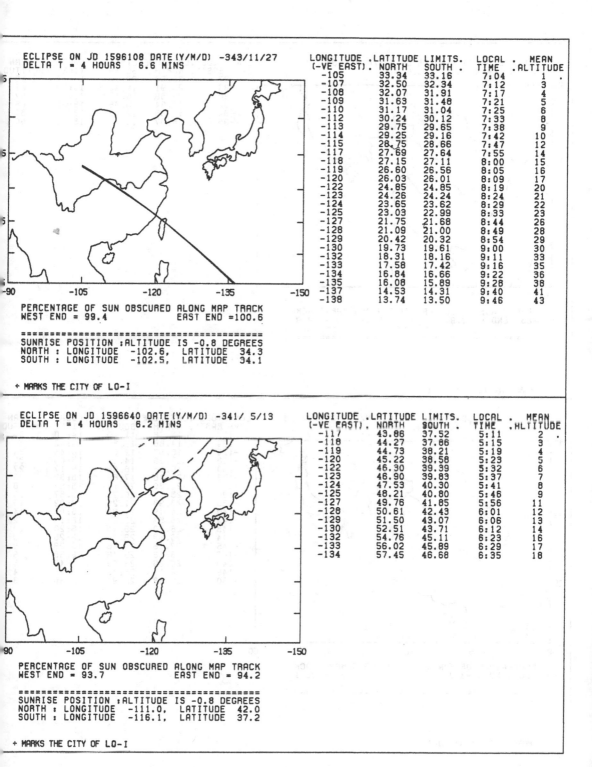

ECLIPSE ON JD 1596108 DATE(Y/M/D) -343/11/27
DELTA T = 4 HOURS   6.6 MINS

LONGITUDE (-VE EAST)	LATITUDE LIMITS. NORTH	SOUTH	LOCAL TIME	MEAN ALTITUDE
-105	33.34	33.16	7:04	1
-107	32.50	32.34	7:12	3
-108	32.07	31.91	7:17	4
-109	31.63	31.48	7:21	5
-110	31.17	31.04	7:25	6
-112	30.24	30.12	7:33	8
-113	29.75	29.65	7:38	9
-114	29.25	29.16	7:42	10
-115	28.75	28.66	7:47	12
-117	27.69	27.64	7:55	14
-118	27.15	27.11	8:00	15
-119	26.60	26.56	8:05	16
-120	26.03	26.01	8:09	17
-122	24.85	24.85	8:19	20
-123	24.26	24.24	8:24	21
-124	23.65	23.62	8:29	22
-125	23.03	22.99	8:33	23
-127	21.75	21.68	8:44	26
-128	21.09	21.00	8:49	28
-129	20.42	20.32	8:54	29
-130	19.73	19.61	9:00	30
-132	18.31	18.16	9:11	33
-133	17.58	17.42	9:16	35
-134	16.84	16.66	9:22	36
-135	16.08	15.89	9:28	38
-137	14.53	14.31	9:40	41
-138	13.74	13.50	9:46	43

PERCENTAGE OF SUN OBSCURED ALONG MAP TRACK
WEST END = 99.4   EAST END =100.6

=======================================
SUNRISE POSITION :ALTITUDE IS -0.8 DEGREES
NORTH : LONGITUDE -102.6, LATITUDE 34.3
SOUTH : LONGITUDE -102.5, LATITUDE 34.1

+ MARKS THE CITY OF LO-I

ECLIPSE ON JD 1596640 DATE(Y/M/D) -341/ 5/13
DELTA T = 4 HOURS   6.2 MINS

LONGITUDE (-VE EAST)	LATITUDE LIMITS. NORTH	SOUTH	LOCAL TIME	MEAN ALTITUDE
-117	43.86	37.52	5:11	2
-118	44.27	37.86	5:15	3
-119	44.73	38.21	5:19	4
-120	45.22	38.58	5:23	5
-122	46.30	39.39	5:32	6
-123	46.90	39.83	5:37	7
-124	47.53	40.30	5:41	8
-125	48.21	40.80	5:46	9
-127	49.76	41.85	5:56	11
-128	50.61	42.43	6:01	12
-129	51.50	43.07	6:06	13
-130	52.51	43.71	6:12	14
-132	54.76	45.11	6:23	16
-133	56.02	45.89	6:29	17
-134	57.45	46.68	6:35	18

PERCENTAGE OF SUN OBSCURED ALONG MAP TRACK
WEST END = 93.7   EAST END = 94.2

===
SUNRISE POSITION :ALTITUDE IS -0.8 DEGREES
NORTH : LONGITUDE -111.0, LATITUDE 42.0
SOUTH : LONGITUDE -116.1, LATITUDE 37.2

+ MARKS THE CITY OF LO-I

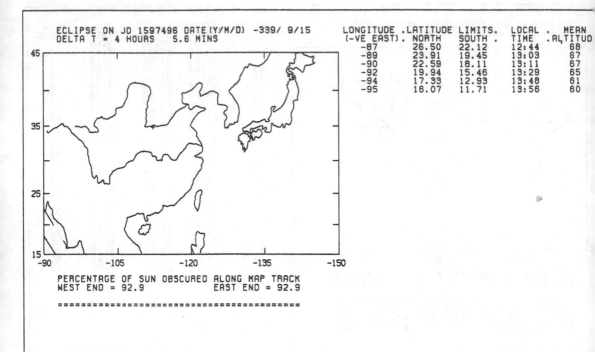

ECLIPSE ON JD 1597496 DATE (Y/M/D) -339/ 9/15
DELTA T = 4 HOURS   5.6 MINS

LONGITUDE	.LATITUDE	LIMITS.	LOCAL	. MEAN
(-VE EAST).	NORTH	SOUTH .	TIME	.ALTITUD
-87	26.50	22.12	12:44	68
-89	23.91	19.45	13:03	67
-90	22.59	18.11	13:11	67
-92	19.94	15.46	13:29	65
-94	17.93	12.93	13:48	61
-95	16.07	11.71	13:56	60

PERCENTAGE OF SUN OBSCURED ALONG MAP TRACK
WEST END = 92.9   EAST END = 92.9

==

+ MARKS THE CITY OF LO-I

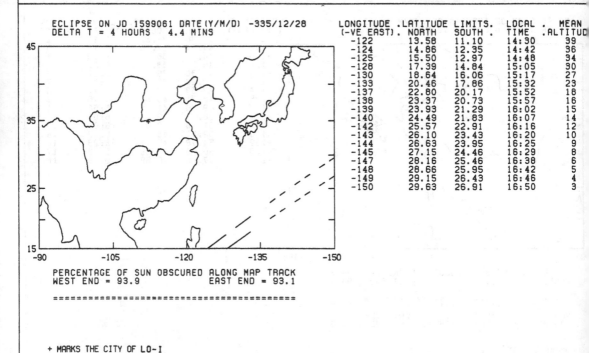

ECLIPSE ON JD 1599061 DATE (Y/M/D) -335/12/28
DELTA T = 4 HOURS   4.4 MINS

LONGITUDE	.LATITUDE	LIMITS.	LOCAL	. MEAN
(-VE EAST).	NORTH	SOUTH .	TIME	.ALTITUD
-122	13.58	11.10	14:30	39
-124	14.86	12.35	14:42	36
-125	15.50	12.97	14:48	34
-128	17.39	14.84	15:05	30
-130	18.64	16.06	15:17	27
-133	20.46	17.86	15:32	23
-137	22.80	20.17	15:52	18
-138	23.37	20.73	15:57	16
-139	23.93	21.29	16:02	15
-140	24.49	21.83	16:07	14
-142	25.57	22.91	16:16	12
-143	26.10	23.43	16:20	10
-144	26.63	23.95	16:25	9
-145	27.15	24.46	16:29	8
-147	28.16	25.46	16:38	6
-148	28.66	25.95	16:42	5
-149	29.15	26.43	16:46	4
-150	29.63	26.91	16:50	3

PERCENTAGE OF SUN OBSCURED ALONG MAP TRACK
WEST END = 93.9   EAST END = 93.1

==

+ MARKS THE CITY OF LO-I

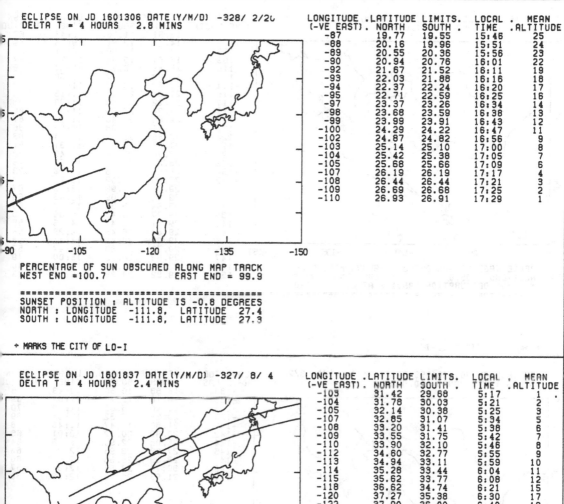

ECLIPSE ON JD 1601306 DATE(Y/M/D) -328/ 2/20
DELTA T = 4 HOURS   2.8 MINS

LONGITUDE	.LATITUDE	LIMITS.	LOCAL	. MEAN
(-VE EAST).	NORTH	SOUTH .	TIME	.ALTITUDE
-87	19.77	19.55	15:46	25
-88	20.16	19.96	15:51	24
-89	20.55	20.36	15:56	23
-90	20.94	20.76	16:01	22
-92	21.67	21.52	16:11	19
-93	22.03	21.88	16:16	18
-94	22.37	22.24	16:20	17
-95	22.71	22.59	16:25	16
-97	23.37	23.26	16:34	14
-98	23.68	23.59	16:38	13
-99	23.99	23.91	16:43	12
-100	24.29	24.22	16:47	11
-102	24.87	24.82	16:56	9
-103	25.14	25.10	17:00	8
-104	25.42	25.38	17:05	7
-105	25.68	25.66	17:09	6
-107	26.19	26.19	17:17	4
-108	26.44	26.44	17:21	3
-109	26.69	26.68	17:25	2
-110	26.93	26.91	17:29	1

PERCENTAGE OF SUN OBSCURED ALONG MAP TRACK
WEST END =100.7   EAST END = 99.9

==
SUNSET POSITION : ALTITUDE IS -0.8 DEGREES
NORTH : LONGITUDE -111.8, LATITUDE 27.4
SOUTH : LONGITUDE -111.8, LATITUDE 27.3

+ MARKS THE CITY OF LO-I

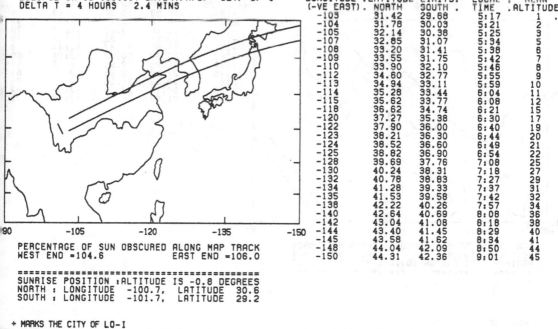

ECLIPSE ON JD 1601837 DATE(Y/M/D) -327/ 8/ 4
DELTA T = 4 HOURS   2.4 MINS

LONGITUDE	.LATITUDE	LIMITS.	LOCAL	. MEAN
(-VE EAST).	NORTH	SOUTH .	TIME	.ALTITUDE
-103	31.42	29.68	5:17	1 .
-104	31.78	30.03	5:21	2
-105	32.14	30.38	5:25	3
-107	32.85	31.07	5:34	5
-108	33.20	31.41	5:38	6
-109	33.55	31.75	5:42	7
-110	33.90	32.10	5:46	8
-112	34.60	32.77	5:55	9
-113	34.94	33.11	5:59	10
-114	35.28	33.44	6:04	11
-115	35.62	33.77	6:08	12
-118	36.62	34.74	6:21	15
-120	37.27	35.38	6:30	17
-122	37.90	36.00	6:40	19
-123	38.21	36.30	6:44	20
-124	38.52	36.60	6:49	21
-125	38.82	36.90	6:54	22
-128	39.69	37.76	7:08	25
-130	40.24	38.31	7:18	27
-132	40.78	38.83	7:27	29
-134	41.28	39.33	7:37	31
-135	41.53	39.58	7:42	32
-138	42.22	40.26	7:57	34
-140	42.64	40.69	8:08	36
-142	43.04	41.08	8:18	38
-144	43.40	41.45	8:29	40
-145	43.58	41.62	8:34	41
-148	44.04	42.09	8:50	44
-150	44.31	42.36	9:01	45

PERCENTAGE OF SUN OBSCURED ALONG MAP TRACK
WEST END =104.6   EAST END =106.0

==
SUNRISE POSITION :ALTITUDE IS -0.8 DEGREES
NORTH : LONGITUDE -100.7, LATITUDE 30.6
SOUTH : LONGITUDE -101.7, LATITUDE 29.2

+ MARKS THE CITY OF LO-I

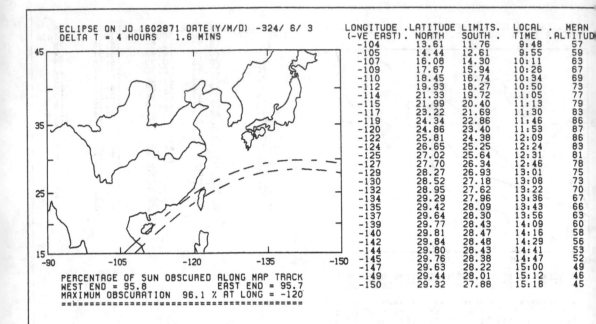

ECLIPSE ON JD 1602871 DATE(Y/M/D) -324/ 6/ 3
DELTA T = 4 HOURS   1.6 MINS

LONGITUDE (-VE EAST)	.LATITUDE NORTH	LIMITS. SOUTH .	LOCAL TIME	. MEAN .ALTITUDE
-104	13.61	11.76	9:48	57
-105	14.44	12.61	9:55	59
-107	16.08	14.30	10:11	63
-109	17.67	15.94	10:26	67
-110	18.45	16.74	10:34	69
-112	19.93	18.27	10:50	73
-114	21.33	19.72	11:05	77
-115	21.99	20.40	11:13	79
-117	23.22	21.69	11:30	83
-119	24.34	22.86	11:46	86
-120	24.86	23.40	11:53	87
-122	25.81	24.38	12:09	86
-124	26.65	25.25	12:24	83
-125	27.02	25.64	12:31	81
-127	27.70	26.34	12:46	78
-129	28.27	26.93	13:01	75
-130	28.52	27.18	13:08	73
-132	28.95	27.62	13:22	70
-134	29.29	27.96	13:36	67
-135	29.42	28.09	13:43	66
-137	29.64	28.30	13:56	63
-139	29.77	28.43	14:09	60
-140	29.81	28.47	14:16	58
-142	29.84	28.48	14:29	56
-144	29.80	28.43	14:41	53
-145	29.76	28.38	14:47	52
-147	29.63	28.22	15:00	49
-149	29.44	28.01	15:12	46
-150	29.32	27.88	15:18	45

PERCENTAGE OF SUN OBSCURED ALONG MAP TRACK
WEST END = 95.8   EAST END = 95.7
MAXIMUM OBSCURATION 96.1 % AT LONG = -120
==

+ MARKS THE CITY OF LO-I

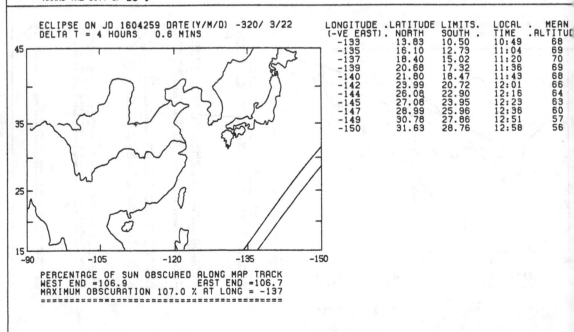

ECLIPSE ON JD 1604259 DATE(Y/M/D) -320/ 3/22
DELTA T = 4 HOURS   0.6 MINS

LONGITUDE (-VE EAST)	.LATITUDE NORTH	LIMITS. SOUTH .	LOCAL TIME	. MEAN .ALTITUDE
-133	13.83	10.50	10:49	68
-135	16.10	12.73	11:04	69
-137	18.40	15.02	11:20	70
-139	20.68	17.32	11:36	69
-140	21.80	18.47	11:43	68
-142	23.99	20.72	12:01	66
-144	26.08	22.90	12:16	64
-145	27.08	23.95	12:23	63
-147	28.99	25.96	12:36	60
-149	30.78	27.86	12:51	57
-150	31.63	28.76	12:58	56

PERCENTAGE OF SUN OBSCURED ALONG MAP TRACK
WEST END =106.9   EAST END =106.7
MAXIMUM OBSCURATION 107.0 % AT LONG = -137
==

+ MARKS THE CITY OF LO-I

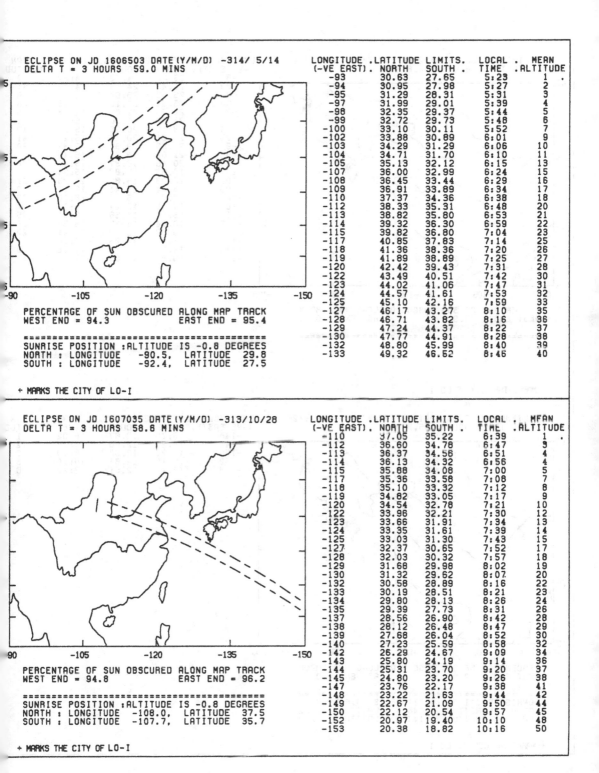

ECLIPSE ON JD 1606503 DATE(Y/M/D) -314/ 5/14
DELTA T = 3 HOURS 59.0 MINS

LONGITUDE (-VE EAST)	LATITUDE NORTH	LIMITS. SOUTH	LOCAL TIME	MEAN ALTITUDE
-93	30.63	27.65	5:23	1
-94	30.95	27.98	5:27	2
-95	31.29	28.31	5:31	3
-97	31.99	29.01	5:39	4
-98	32.35	29.37	5:44	5
-99	32.72	29.73	5:48	6
-100	33.10	30.11	5:52	7
-102	33.88	30.89	6:01	9
-103	34.29	31.29	6:06	10
-104	34.71	31.70	6:10	11
-105	35.13	32.12	6:15	13
-107	36.00	32.99	6:24	15
-108	36.45	33.44	6:29	16
-109	36.91	33.89	6:34	17
-110	37.37	34.36	6:38	18
-112	38.33	35.31	6:48	20
-113	38.82	35.80	6:53	21
-114	39.32	36.30	6:59	22
-115	39.82	36.80	7:04	23
-117	40.85	37.83	7:14	25
-118	41.36	38.36	7:20	26
-119	41.89	38.89	7:25	27
-120	42.42	39.43	7:31	28
-122	43.49	40.51	7:42	30
-123	44.02	41.06	7:47	31
-124	44.57	41.61	7:53	32
-125	45.10	42.16	7:59	33
-127	46.17	43.27	8:10	35
-128	46.71	43.82	8:16	36
-129	47.24	44.37	8:22	37
-130	47.77	44.91	8:28	38
-132	48.80	45.99	8:40	39
-133	49.32	46.62	8:46	40

PERCENTAGE OF SUN OBSCURED ALONG MAP TRACK
WEST END = 94.3   EAST END = 95.4

===
SUNRISE POSITION :ALTITUDE IS -0.8 DEGREES
NORTH : LONGITUDE -90.5, LATITUDE 29.8
SOUTH : LONGITUDE -92.4, LATITUDE 27.5

+ MARKS THE CITY OF LO-I

ECLIPSE ON JD 1607035 DATE(Y/M/D) -313/10/28
DELTA T = 3 HOURS 58.6 MINS

LONGITUDE (-VE EAST)	LATITUDE NORTH	LIMITS. SOUTH	LOCAL TIME	MEAN ALTITUDE
-110	37.05	35.22	6:39	1
-112	36.60	34.78	6:47	3
-113	36.37	34.56	6:51	4
-114	36.13	34.32	6:56	4
-115	35.88	34.08	7:00	5
-117	35.36	33.58	7:08	7
-118	35.10	33.32	7:12	8
-119	34.82	33.05	7:17	9
-120	34.54	32.78	7:21	10
-122	33.96	32.21	7:30	12
-123	33.66	31.91	7:34	13
-124	33.35	31.61	7:39	14
-125	33.03	31.30	7:43	15
-127	32.37	30.65	7:52	17
-128	32.03	30.32	7:57	18
-129	31.68	29.98	8:02	19
-130	31.32	29.62	8:07	20
-132	30.58	28.89	8:16	22
-133	30.19	28.51	8:21	23
-134	29.80	28.13	8:26	24
-135	29.39	27.73	8:31	26
-137	28.56	26.90	8:42	28
-138	28.12	26.48	8:47	29
-139	27.68	26.04	8:52	30
-140	27.23	25.59	8:58	32
-142	26.29	24.67	9:09	34
-143	25.80	24.19	9:14	36
-144	25.31	23.70	9:20	37
-145	24.80	23.20	9:26	38
-147	23.76	22.17	9:38	41
-148	23.22	21.63	9:44	42
-149	22.67	21.09	9:50	44
-150	22.12	20.54	9:57	45
-152	20.97	19.40	10:10	48
-153	20.38	18.82	10:16	50

PERCENTAGE OF SUN OBSCURED ALONG MAP TRACK
WEST END = 94.8   EAST END = 96.2

===
SUNRISE POSITION :ALTITUDE IS -0.8 DEGREES
NORTH : LONGITUDE -108.0, LATITUDE 37.5
SOUTH : LONGITUDE -107.7, LATITUDE 35.7

+ MARKS THE CITY OF LO-I

ECLIPSE ON JD 1608422 DATE(Y/M/D) -309/ 8/15
DELTA T = 3 HOURS 57.6 MINS

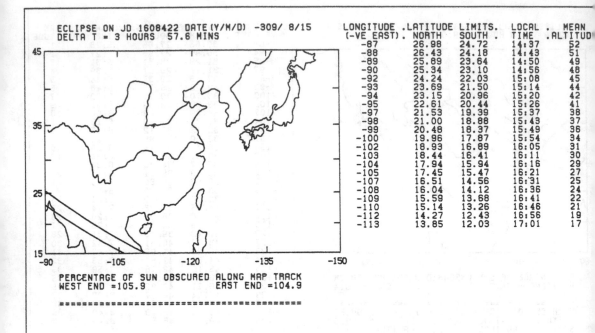

LONGITUDE (-VE EAST)	LATITUDE NORTH	LIMITS. SOUTH	LOCAL TIME	MEAN ALTITUD
-87	26.98	24.72	14:37	52
-88	26.43	24.18	14:43	51
-89	25.89	23.64	14:50	49
-90	25.34	23.10	14:56	48
-92	24.24	22.03	15:08	45
-93	23.69	21.50	15:14	44
-94	23.15	20.96	15:20	42
-95	22.61	20.44	15:26	41
-97	21.53	19.39	15:37	38
-98	21.00	18.88	15:43	37
-99	20.48	18.37	15:49	36
-100	19.96	17.87	15:54	34
-102	18.93	16.89	16:05	31
-103	18.44	16.41	16:11	30
-104	17.94	15.94	16:16	29
-105	17.45	15.47	16:21	27
-107	16.51	14.56	16:31	25
-108	16.04	14.12	16:36	24
-109	15.59	13.68	16:41	22
-110	15.14	13.26	16:46	21
-112	14.27	12.43	16:56	19
-113	13.85	12.03	17:01	17

PERCENTAGE OF SUN OBSCURED ALONG MAP TRACK
WEST END =105.9   EAST END =104.9

==

+ MARKS THE CITY OF LO-I

ECLIPSE ON JD 1611021 DATE(Y/M/D) -302/ 9/26
DELTA T = 3 HOURS 55.8 MINS

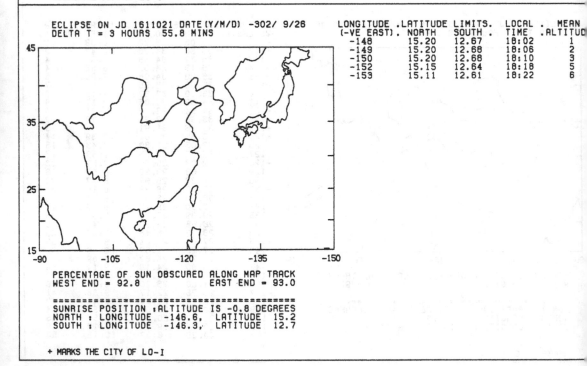

LONGITUDE (-VE EAST)	LATITUDE NORTH	LIMITS. SOUTH	LOCAL TIME	MEAN ALTITUD
-148	15.20	12.67	18:02	1
-149	15.20	12.68	18:06	2
-150	15.20	12.68	18:10	3
-152	15.15	12.64	18:18	5
-153	15.11	12.61	18:22	6

PERCENTAGE OF SUN OBSCURED ALONG MAP TRACK
WEST END = 92.8   EAST END = 93.0

==
SUNRISE POSITION :ALTITUDE IS -0.8 DEGREES
NORTH : LONGITUDE -146.6, LATITUDE 15.2
SOUTH : LONGITUDE -146.3, LATITUDE 12.7

+ MARKS THE CITY OF LO-I

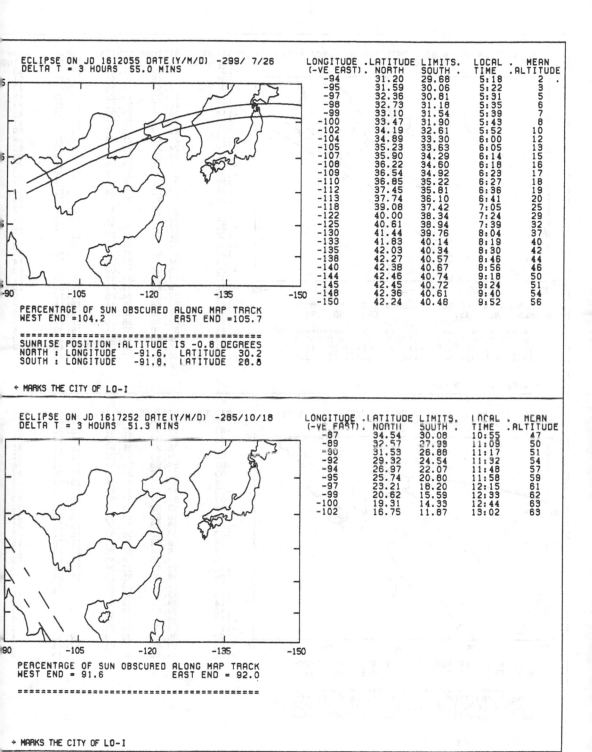

ECLIPSE ON JD 1612055 DATE (Y/M/D) -299/ 7/26
DELTA T = 3 HOURS 55.0 MINS

PERCENTAGE OF SUN OBSCURED ALONG MAP TRACK
WEST END =104.2   EAST END =105.7

==
SUNRISE POSITION :ALTITUDE IS -0.8 DEGREES
NORTH : LONGITUDE   -91.6. LATITUDE 30.2
SOUTH : LONGITUDE   -91.8. LATITUDE 28.8

+ MARKS THE CITY OF LO-I

LONGITUDE (-VE EAST)	LATITUDE NORTH	LIMITS. SOUTH	LOCAL TIME	MEAN ALTITUDE
-94	31.20	29.68	5:18	2
-95	31.59	30.06	5:22	3
-97	32.36	30.81	5:31	5
-98	32.73	31.18	5:35	6
-99	33.10	31.54	5:39	7
-100	33.47	31.90	5:43	8
-102	34.19	32.61	5:52	10
-104	34.89	33.30	6:00	12
-105	35.23	33.63	6:05	13
-107	35.90	34.29	6:14	15
-108	36.22	34.60	6:18	16
-109	36.54	34.92	6:23	17
-110	36.85	35.22	6:27	18
-112	37.45	35.81	6:36	19
-113	37.74	36.10	6:41	20
-118	39.08	37.42	7:05	25
-122	40.00	38.34	7:24	29
-125	40.61	38.94	7:39	32
-130	41.44	39.76	8:04	37
-133	41.83	40.14	8:19	40
-135	42.03	40.34	8:30	42
-138	42.27	40.57	8:46	44
-140	42.38	40.67	8:56	46
-144	42.46	40.74	9:18	50
-145	42.45	40.72	9:24	51
-148	42.36	40.61	9:40	54
-150	42.24	40.48	9:52	56

ECLIPSE ON JD 1617252 DATE (Y/M/D) -285/10/18
DELTA T = 3 HOURS 51.3 MINS

PERCENTAGE OF SUN OBSCURED ALONG MAP TRACK
WEST END = 91.6   EAST END = 92.0

==

+ MARKS THE CITY OF LO-I

LONGITUDE (-VE EAST)	LATITUDE NORTH	LIMITS. SOUTH	LOCAL TIME	MEAN ALTITUDE
-87	34.54	30.08	10:55	47
-89	32.57	27.99	11:09	50
-90	31.53	26.88	11:17	51
-92	29.32	24.54	11:32	54
-94	26.97	22.07	11:48	57
-95	25.74	20.80	11:58	59
-97	23.21	18.20	12:15	61
-99	20.62	15.59	12:33	62
-100	19.31	14.33	12:44	63
-102	16.75	11.87	13:02	63

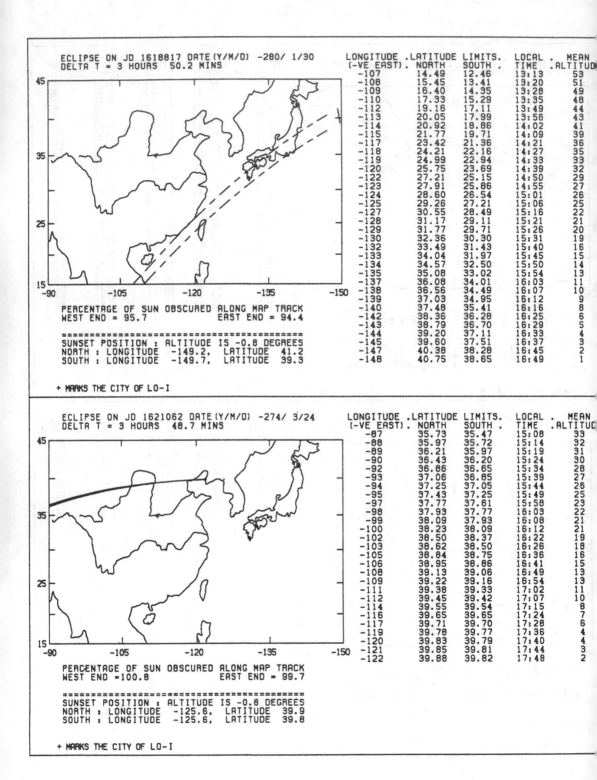

ECLIPSE ON JD 1618817 DATE(Y/M/D) -280/ 1/30
DELTA T = 3 HOURS 50.2 MINS

LONGITUDE (-VE EAST)	LATITUDE NORTH	LIMITS SOUTH	LOCAL TIME	MEAN ALTITUD
-107	14.49	12.46	13:13	53
-108	15.45	13.41	13:20	51
-109	16.40	14.35	13:28	49
-110	17.33	15.29	13:35	48
-112	19.16	17.11	13:49	44
-113	20.05	17.99	13:56	43
-114	20.92	18.86	14:02	41
-115	21.77	19.71	14:09	39
-117	23.42	21.36	14:21	36
-118	24.21	22.16	14:27	35
-119	24.99	22.94	14:33	33
-120	25.75	23.69	14:39	32
-122	27.21	25.15	14:50	29
-123	27.91	25.86	14:55	27
-124	28.60	26.54	15:01	26
-125	29.26	27.21	15:06	25
-127	30.55	28.49	15:16	22
-128	31.17	29.11	15:21	21
-129	31.77	29.71	15:26	20
-130	32.36	30.30	15:31	19
-132	33.49	31.43	15:40	16
-133	34.04	31.97	15:45	15
-134	34.57	32.50	15:50	14
-135	35.08	33.02	15:54	13
-137	36.08	34.01	16:03	11
-138	36.56	34.49	16:07	10
-139	37.03	34.95	16:12	9
-140	37.48	35.41	16:16	8
-142	38.36	36.28	16:25	6
-143	38.79	36.70	16:29	5
-144	39.20	37.11	16:33	4
-145	39.60	37.51	16:37	3
-147	40.38	38.28	16:45	2
-148	40.75	38.65	16:49	1

PERCENTAGE OF SUN OBSCURED ALONG MAP TRACK
WEST END = 95.7   EAST END = 94.4

===
SUNSET POSITION : ALTITUDE IS -0.8 DEGREES
NORTH : LONGITUDE -149.2, LATITUDE 41.2
SOUTH : LONGITUDE -149.7, LATITUDE 39.3

+ MARKS THE CITY OF LO-I

ECLIPSE ON JD 1621062 DATE(Y/M/D) -274/ 3/24
DELTA T = 3 HOURS 48.7 MINS

LONGITUDE (-VE EAST)	LATITUDE NORTH	LIMITS SOUTH	LOCAL TIME	MEAN ALTITUD
-87	35.73	35.47	15:08	33
-88	35.97	35.72	15:14	32
-89	36.21	35.97	15:19	31
-90	36.43	36.20	15:24	30
-92	36.86	36.65	15:34	28
-93	37.06	36.85	15:39	27
-94	37.25	37.05	15:44	26
-95	37.43	37.25	15:49	25
-97	37.77	37.61	15:58	23
-98	37.93	37.77	16:03	22
-99	38.09	37.93	16:08	21
-100	38.23	38.09	16:12	21
-102	38.50	38.37	16:22	19
-103	38.62	38.50	16:26	18
-105	38.84	38.75	16:36	16
-106	38.95	38.86	16:41	15
-108	39.13	39.06	16:49	13
-109	39.22	39.16	16:54	13
-111	39.38	39.33	17:02	11
-112	39.45	39.42	17:07	10
-114	39.55	39.54	17:15	8
-116	39.65	39.65	17:24	7
-117	39.71	39.70	17:28	6
-119	39.78	39.77	17:36	4
-120	39.83	39.79	17:40	4
-121	39.85	39.81	17:44	3
-122	39.88	39.82	17:48	2

PERCENTAGE OF SUN OBSCURED ALONG MAP TRACK
WEST END =100.8   EAST END = 99.7

===
SUNSET POSITION : ALTITUDE IS -0.8 DEGREES
NORTH : LONGITUDE -125.6, LATITUDE 39.9
SOUTH : LONGITUDE -125.6, LATITUDE 39.8

+ MARKS THE CITY OF LO-I

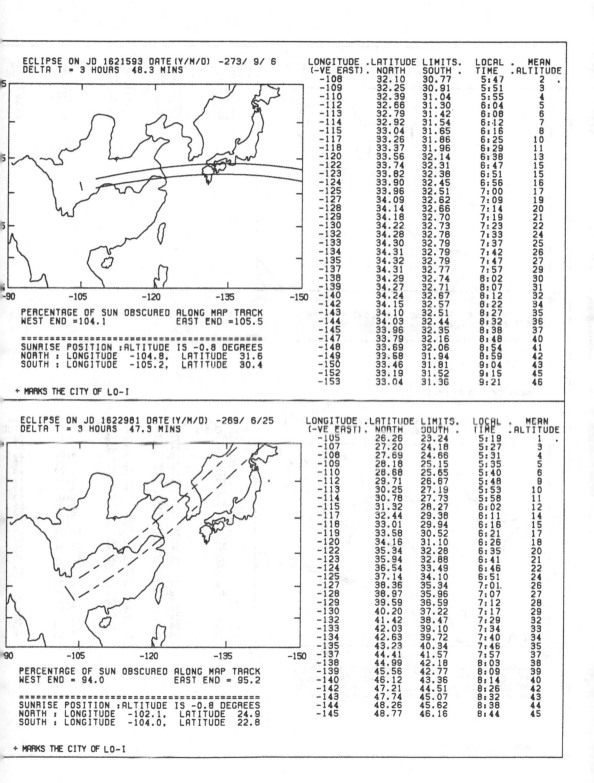

ECLIPSE ON JD 1621593 DATE(Y/M/D) -273/ 9/ 6
DELTA T = 3 HOURS 48.3 MINS

LONGITUDE (-VE EAST)	LATITUDE NORTH	LIMITS SOUTH	LOCAL TIME	MEAN ALTITUDE
-108	32.10	30.77	5:47	2
-109	32.25	30.91	5:51	3
-110	32.39	31.04	5:55	4
-112	32.66	31.30	6:04	5
-113	32.79	31.42	6:08	6
-114	32.92	31.54	6:12	7
-115	33.04	31.65	6:16	8
-117	33.26	31.86	6:25	10
-118	33.37	31.96	6:29	11
-120	33.56	32.14	6:38	13
-122	33.74	32.31	6:47	15
-123	33.82	32.38	6:51	15
-124	33.90	32.45	6:56	16
-125	33.96	32.51	7:00	17
-127	34.09	32.62	7:09	19
-128	34.14	32.66	7:14	20
-129	34.18	32.70	7:19	21
-130	34.22	32.73	7:23	22
-132	34.28	32.78	7:33	24
-133	34.30	32.79	7:37	25
-134	34.31	32.79	7:42	26
-135	34.32	32.79	7:47	27
-137	34.31	32.77	7:57	29
-138	34.29	32.74	8:02	30
-139	34.27	32.71	8:07	31
-140	34.24	32.67	8:12	32
-142	34.15	32.57	8:22	34
-143	34.10	32.51	8:27	35
-144	34.03	32.44	8:32	36
-145	33.96	32.35	8:38	37
-147	33.79	32.16	8:48	40
-148	33.69	32.06	8:54	41
-149	33.58	31.94	8:59	42
-150	33.46	31.81	9:04	43
-152	33.19	31.52	9:15	45
-153	33.04	31.36	9:21	46

PERCENTAGE OF SUN OBSCURED ALONG MAP TRACK
WEST END =104.1   EAST END =105.5

===
SUNRISE POSITION :ALTITUDE IS -0.8 DEGREES
NORTH : LONGITUDE -104.8, LATITUDE 31.6
SOUTH : LONGITUDE -105.2, LATITUDE 30.4

+ MARKS THE CITY OF LO-I

ECLIPSE ON JD 1622981 DATE(Y/M/D) -269/ 6/25
DELTA T = 3 HOURS 47.3 MINS

LONGITUDE (-VE EAST)	LATITUDE NORTH	LIMITS SOUTH	LOCAL TIME	MEAN ALTITUDE
-105	26.26	23.24	5:19	1
-107	27.20	24.18	5:27	3
-108	27.69	24.66	5:31	4
-109	28.18	25.15	5:35	5
-110	28.68	25.65	5:40	6
-112	29.71	26.67	5:48	8
-113	30.25	27.19	5:53	10
-114	30.78	27.73	5:58	11
-115	31.32	28.27	6:02	12
-117	32.44	29.38	6:11	14
-118	33.01	29.94	6:16	15
-119	33.58	30.52	6:21	17
-120	34.16	31.10	6:26	18
-122	35.34	32.28	6:35	20
-123	35.94	32.88	6:41	21
-124	36.54	33.49	6:46	22
-125	37.14	34.10	6:51	24
-127	38.36	35.34	7:01	26
-128	38.97	35.96	7:07	27
-129	39.59	36.59	7:12	28
-130	40.20	37.22	7:17	29
-132	41.42	38.47	7:29	32
-133	42.03	39.10	7:34	33
-134	42.63	39.72	7:40	34
-135	43.23	40.34	7:46	35
-137	44.41	41.57	7:57	37
-138	44.99	42.18	8:03	38
-139	45.56	42.77	8:09	39
-140	46.12	43.36	8:14	40
-142	47.21	44.51	8:26	42
-143	47.74	45.07	8:32	43
-144	48.26	45.62	8:38	44
-145	48.77	46.16	8:44	45

PERCENTAGE OF SUN OBSCURED ALONG MAP TRACK
WEST END = 94.0   EAST END = 95.2

===
SUNRISE POSITION :ALTITUDE IS -0.8 DEGREES
NORTH : LONGITUDE -102.1, LATITUDE 24.9
SOUTH : LONGITUDE -104.0, LATITUDE 22.8

+ MARKS THE CITY OF LO-I

157

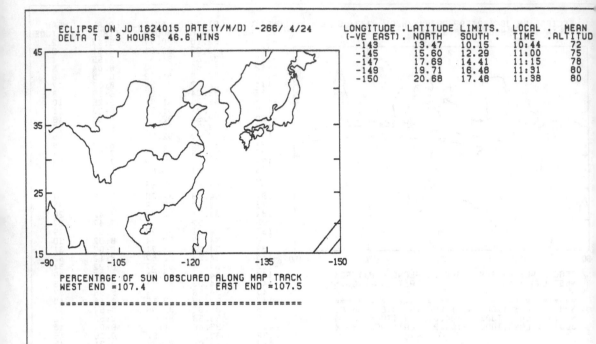

ECLIPSE ON JD 1624015 DATE(Y/M/D) -266/ 4/24
DELTA T = 3 HOURS 46.6 MINS

LONGITUDE	.LATITUDE	LIMITS.	LOCAL	. MEAN
(-VE EAST).	NORTH	SOUTH .	TIME	.ALTITUD
-143	13.47	10.15	10:44	72
-145	15.60	12.29	11:00	75
-147	17.69	14.41	11:15	78
-149	19.71	16.48	11:31	80
-150	20.68	17.48	11:38	80

PERCENTAGE OF SUN OBSCURED ALONG MAP TRACK
WEST END =107.4   EAST END =107.5

+ MARKS THE CITY OF LO-I

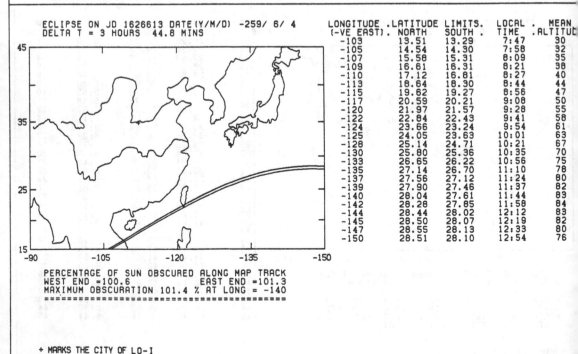

ECLIPSE ON JD 1626613 DATE(Y/M/D) -259/ 6/ 4
DELTA T = 3 HOURS 44.8 MINS

LONGITUDE	.LATITUDE	LIMITS.	LOCAL	. MEAN
(-VE EAST).	NORTH	SOUTH .	TIME	.ALTITUD
-103	13.51	13.29	7:47	30
-105	14.54	14.30	7:58	32
-107	15.58	15.31	8:09	35
-109	16.61	16.31	8:21	38
-110	17.12	16.81	8:27	40
-113	18.64	18.30	8:44	44
-115	19.62	19.27	8:56	47
-117	20.59	20.21	9:08	50
-120	21.97	21.57	9:28	55
-122	22.84	22.43	9:41	58
-124	23.66	23.24	9:54	61
-125	24.05	23.63	10:01	63
-128	25.14	24.71	10:21	67
-130	25.80	25.36	10:35	70
-133	26.65	26.22	10:56	75
-135	27.14	26.70	11:10	78
-137	27.56	27.12	11:24	80
-139	27.90	27.46	11:37	82
-140	28.04	27.61	11:44	83
-142	28.28	27.85	11:58	84
-144	28.44	28.02	12:12	83
-145	28.50	28.07	12:19	82
-147	28.55	28.13	12:33	80
-150	28.51	28.10	12:54	76

PERCENTAGE OF SUN OBSCURED ALONG MAP TRACK
WEST END =100.6   EAST END =101.3
MAXIMUM OBSCURATION 101.4 % AT LONG = -140

+ MARKS THE CITY OF LO-I

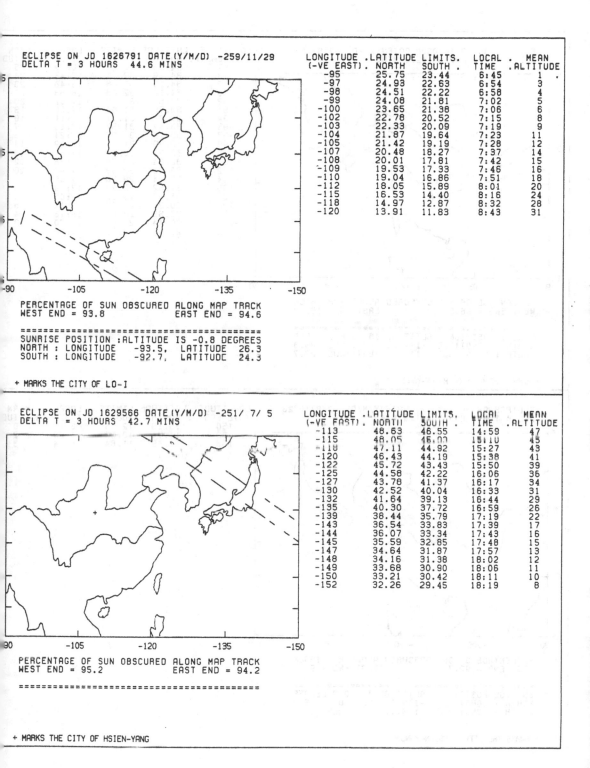

ECLIPSE ON JD 1626791 DATE(Y/M/D) -259/11/29
DELTA T = 3 HOURS 44.6 MINS

LONGITUDE (-VE EAST)	LATITUDE NORTH	LIMITS. SOUTH	LOCAL TIME	MEAN ALTITUDE
-95	25.75	23.44	6:45	1
-97	24.93	22.63	6:54	3
-98	24.51	22.22	6:58	4
-99	24.08	21.81	7:02	5
-100	23.65	21.38	7:06	6
-102	22.78	20.52	7:15	8
-103	22.33	20.09	7:19	9
-104	21.87	19.64	7:23	11
-105	21.42	19.19	7:28	12
-107	20.48	18.27	7:37	14
-108	20.01	17.81	7:42	15
-109	19.53	17.33	7:46	16
-110	19.04	16.86	7:51	18
-112	18.05	15.89	8:01	20
-115	16.53	14.40	8:16	24
-118	14.97	12.87	8:32	28
-120	13.91	11.83	8:43	31

PERCENTAGE OF SUN OBSCURED ALONG MAP TRACK
WEST END = 93.8   EAST END = 94.6

==
SUNRISE POSITION :ALTITUDE IS -0.8 DEGREES
NORTH : LONGITUDE   -93.5, LATITUDE 26.3
SOUTH : LONGITUDE   -92.7, LATITUDE 24.3

+ MARKS THE CITY OF LO-I

ECLIPSE ON JD 1629566 DATE(Y/M/D) -251/ 7/ 5
DELTA T = 3 HOURS 42.7 MINS

LONGITUDE (-VE EAST)	LATITUDE NORTH	LIMITS. SOUTH	LOCAL TIME	MEAN ALTITUDE
-113	48.63	46.55	14:59	47
-115	48.05	46.00	15:10	45
-118	47.11	44.92	15:27	43
-120	46.43	44.19	15:38	41
-122	45.72	43.43	15:50	39
-125	44.58	42.22	16:06	36
-127	43.78	41.37	16:17	34
-130	42.52	40.04	16:33	31
-132	41.64	39.13	16:44	29
-135	40.30	37.72	16:59	26
-139	38.44	35.79	17:19	22
-143	36.54	33.83	17:39	17
-144	36.07	33.34	17:43	16
-145	35.59	32.85	17:48	15
-147	34.64	31.87	17:57	13
-148	34.16	31.38	18:02	12
-149	33.68	30.90	18:06	11
-150	33.21	30.42	18:11	10
-152	32.26	29.45	18:19	8

PERCENTAGE OF SUN OBSCURED ALONG MAP TRACK
WEST END = 95.2   EAST END = 94.2

==

+ MARKS THE CITY OF HSIEN-YANG

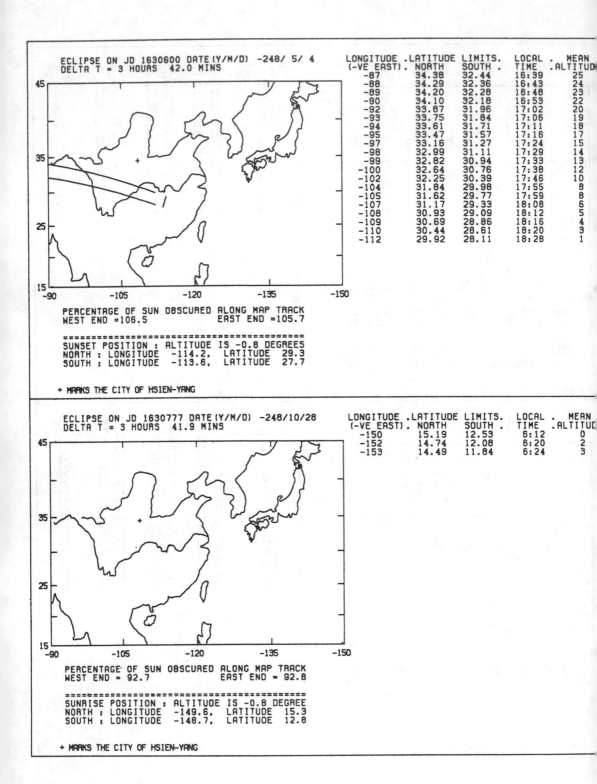

ECLIPSE ON JD 1630600 DATE(Y/M/D) -248/ 5/ 4
DELTA T = 3 HOURS 42.0 MINS

LONGITUDE (-VE EAST)	LATITUDE NORTH	LIMITS. SOUTH	LOCAL TIME	MEAN ALTITUDE
-87	34.38	32.44	16:39	25
-88	34.29	32.36	16:43	24
-89	34.20	32.28	16:48	23
-90	34.10	32.18	16:53	22
-92	33.87	31.96	17:02	20
-93	33.75	31.84	17:06	19
-94	33.61	31.71	17:11	18
-95	33.47	31.57	17:16	17
-97	33.16	31.27	17:24	15
-98	32.99	31.11	17:29	14
-99	32.82	30.94	17:33	13
-100	32.64	30.76	17:38	12
-102	32.25	30.39	17:46	10
-104	31.84	29.98	17:55	8
-105	31.62	29.77	17:59	8
-107	31.17	29.33	18:08	6
-108	30.93	29.09	18:12	5
-109	30.69	28.86	18:16	4
-110	30.44	28.61	18:20	3
-112	29.92	28.11	18:28	1

PERCENTAGE OF SUN OBSCURED ALONG MAP TRACK
WEST END =106.5   EAST END =105.7

===
SUNSET POSITION : ALTITUDE IS -0.8 DEGREES
NORTH : LONGITUDE -114.2, LATITUDE 29.3
SOUTH : LONGITUDE -113.6, LATITUDE 27.7

+ MARKS THE CITY OF HSIEN-YANG

ECLIPSE ON JD 1630777 DATE(Y/M/D) -248/10/28
DELTA T = 3 HOURS 41.9 MINS

LONGITUDE (-VE EAST)	LATITUDE NORTH	LIMITS. SOUTH	LOCAL TIME	MEAN ALTITUDE
-150	15.19	12.53	6:12	0
-152	14.74	12.08	6:20	2
-153	14.49	11.84	6:24	3

PERCENTAGE OF SUN OBSCURED ALONG MAP TRACK
WEST END = 92.7   EAST END = 92.8

===
SUNRISE POSITION : ALTITUDE IS -0.8 DEGREE
NORTH : LONGITUDE -149.6, LATITUDE 15.3
SOUTH : LONGITUDE -148.7, LATITUDE 12.8

+ MARKS THE CITY OF HSIEN-YANG

ECLIPSE ON JD 1630955 DATE(Y/M/D) -247/ 4/24
DELTA T = 3 HOURS   41.8 MINS

LONGITUDE	.LATITUDE	LIMITS.	LOCAL	. MEAN
(-VE EAST)	. NORTH	SOUTH .	TIME	.ALTITUDE
-92	33.62	32.71	5:31	0 .
-93	33.86	32.91	5:35	1
-94	34.10	33.13	5:39	2
-95	34.37	33.37	5:43	3
-97	34.98	33.91	5:52	5
-98	35.31	34.21	5:56	6
-99	35.67	34.53	6:01	7
-100	36.05	34.87	6:05	8
-102	36.91	35.62	6:14	10
-103	37.36	36.05	6:19	11
-104	37.86	36.49	6:23	12
-105	38.39	36.96	6:29	13
-107	39.55	38.00	6:38	15
-108	40.19	38.57	6:43	16
-109	40.86	39.17	6:49	17
-110	41.58	39.81	6:54	18
-112	43.16	41.22	7:05	20
-114	44.99	42.77	7:18	22
-115	46.00	43.58	7:25	23
-117	48.14	45.49	7:37	25

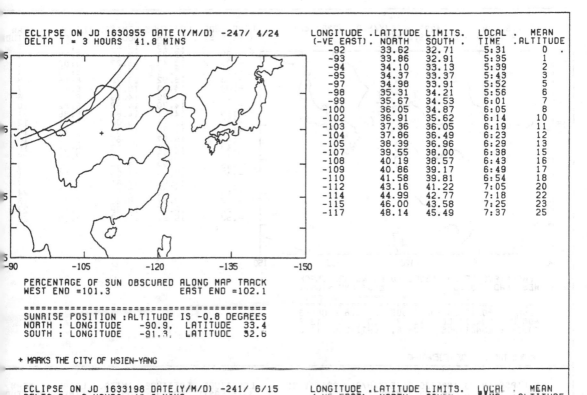

PERCENTAGE OF SUN OBSCURED ALONG MAP TRACK
WEST END =101.3   EAST END =102.1

===
SUNRISE POSITION :ALTITUDE IS -0.8 DEGREES
NORTH : LONGITUDE -90.9, LATITUDE 33.4
SOUTH : LONGITUDE -91.3, LATITUDE 32.6

+ MARKS THE CITY OF HSIEN-YANG

ECLIPSE ON JD 1633198 DATE(Y/M/D) -241/ 6/15
DELTA T = 3 HOURS   40.2 MINS

LONGITUDE	.LATITUDE	LIMITS.	LOCAL	. MEAN
(-VE EAST)	. NORTH	SOUTH .	TIME	.ALTITUDE
-90	15.62	15.47	17:15	16
-92	14.64	14.52	17:24	14
-93	14.15	14.04	17:29	12
-94	13.67	13.57	17:33	11
-95	13.19	13.10	17:37	10

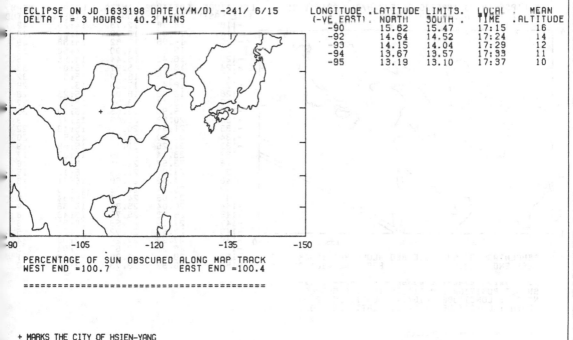

PERCENTAGE OF SUN OBSCURED ALONG MAP TRACK
WEST END =100.7   EAST END =100.4

===

+ MARKS THE CITY OF HSIEN-YANG

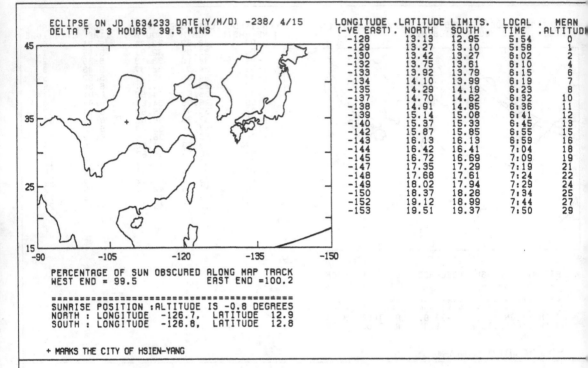

ECLIPSE ON JD 1634233 DATE(Y/M/D) -238/ 4/15
DELTA T = 3 HOURS 39.5 MINS

LONGITUDE (-VE EAST)	LATITUDE LIMITS. NORTH	SOUTH	LOCAL TIME	MEAN ALTITUDE
-128	13.13	12.95	5:54	0
-129	13.27	13.10	5:58	1
-130	13.42	13.27	6:02	2
-132	13.75	13.61	6:10	4
-133	13.92	13.79	6:15	6
-134	14.10	13.99	6:19	7
-135	14.29	14.19	6:23	8
-137	14.70	14.62	6:32	10
-138	14.91	14.85	6:36	11
-139	15.14	15.08	6:41	12
-140	15.37	15.33	6:45	13
-142	15.87	15.85	6:55	15
-143	16.13	16.13	6:59	16
-144	16.42	16.41	7:04	18
-145	16.72	16.69	7:09	19
-147	17.35	17.29	7:19	21
-148	17.68	17.61	7:24	22
-149	18.02	17.94	7:29	24
-150	18.37	18.28	7:34	25
-152	19.12	18.99	7:44	27
-153	19.51	19.37	7:50	29

PERCENTAGE OF SUN OBSCURED ALONG MAP TRACK
WEST END = 99.5   EAST END =100.2

===
SUNRISE POSITION :ALTITUDE IS -0.8 DEGREES
NORTH : LONGITUDE -126.7, LATITUDE 12.9
SOUTH : LONGITUDE -126.8, LATITUDE 12.8

+ MARKS THE CITY OF HSIEN-YANG

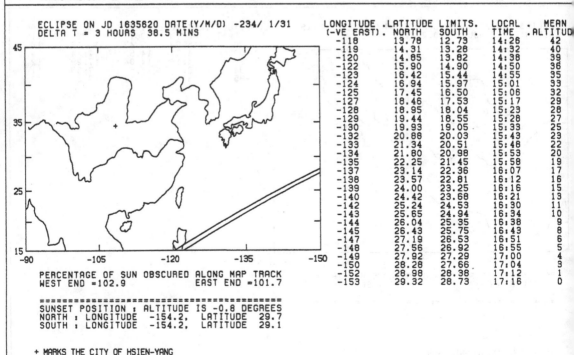

ECLIPSE ON JD 1635620 DATE(Y/M/D) -234/ 1/31
DELTA T = 3 HOURS 38.5 MINS

LONGITUDE (-VE EAST)	LATITUDE LIMITS. NORTH	SOUTH	LOCAL TIME	MEAN ALTITUDE
-118	13.78	12.73	14:26	42
-119	14.31	13.28	14:32	40
-120	14.85	13.82	14:38	39
-122	15.90	14.90	14:50	36
-123	16.42	15.44	14:55	35
-124	16.94	15.97	15:01	33
-125	17.45	16.50	15:06	32
-127	18.46	17.53	15:17	29
-128	18.95	18.04	15:23	28
-129	19.44	18.55	15:28	27
-130	19.93	19.05	15:33	25
-132	20.88	20.03	15:43	23
-133	21.34	20.51	15:48	22
-134	21.80	20.98	15:53	20
-135	22.25	21.45	15:58	19
-137	23.14	22.36	16:07	17
-138	23.57	22.81	16:12	16
-139	24.00	23.25	16:16	15
-140	24.42	23.68	16:21	13
-142	25.24	24.53	16:30	11
-143	25.65	24.94	16:34	10
-144	26.04	25.35	16:38	9
-145	26.43	25.75	16:43	8
-147	27.19	26.53	16:51	6
-148	27.56	26.92	16:55	5
-149	27.92	27.29	17:00	4
-150	28.28	27.66	17:04	3
-152	28.98	28.38	17:12	1
-153	29.32	28.73	17:16	0

PERCENTAGE OF SUN OBSCURED ALONG MAP TRACK
WEST END =102.9   EAST END =101.7

===
SUNSET POSITION : ALTITUDE IS -0.8 DEGREES
NORTH : LONGITUDE -154.2, LATITUDE 29.7
SOUTH : LONGITUDE -154.2, LATITUDE 29.1

+ MARKS THE CITY OF HSIEN-YANG

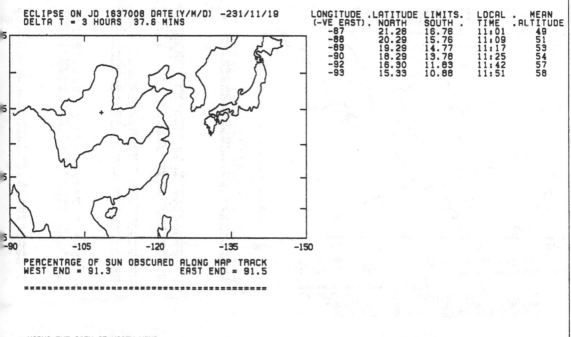

ECLIPSE ON JD 1637008 DATE(Y/M/D) -231/11/19
DELTA T = 3 HOURS 37.6 MINS

LONGITUDE (-VE EAST)	LATITUDE NORTH	LIMITS. SOUTH	LOCAL TIME	MEAN ALTITUDE
-87	21.28	16.78	11:01	49
-88	20.29	15.76	11:09	51
-89	19.29	14.77	11:17	53
-90	18.29	13.78	11:25	54
-92	16.30	11.83	11:42	57
-93	15.33	10.88	11:51	58

PERCENTAGE OF SUN OBSCURED ALONG MAP TRACK
WEST END = 91.3   EAST END = 91.5

===

+ MARKS THE CITY OF HSIEN-YANG

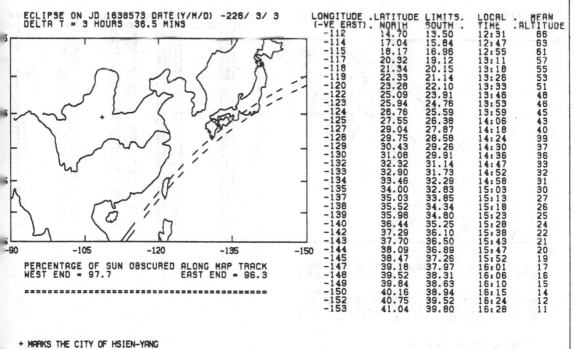

ECLIPSE ON JD 1638573 DATE(Y/M/D) -226/ 3/ 3
DELTA T = 3 HOURS 36.5 MINS

LONGITUDE (-VE EAST)	LATITUDE NORTH	LIMITS. SOUTH	LOCAL TIME	MEAN ALTITUDE
-112	14.70	13.50	12:31	66
-114	17.04	15.84	12:47	63
-115	18.17	16.96	12:55	61
-117	20.32	19.12	13:11	57
-118	21.34	20.15	13:18	55
-119	22.33	21.14	13:26	53
-120	23.28	22.10	13:33	51
-122	25.09	23.91	13:46	48
-123	25.94	24.76	13:53	46
-124	26.76	25.59	13:59	45
-125	27.55	26.38	14:06	43
-127	29.04	27.87	14:18	40
-128	29.75	28.58	14:24	39
-129	30.43	29.26	14:30	37
-130	31.08	29.91	14:36	36
-132	32.32	31.14	14:47	33
-133	32.90	31.73	14:52	32
-134	33.46	32.29	14:58	31
-135	34.00	32.83	15:03	30
-137	35.03	33.85	15:13	27
-138	35.52	34.34	15:18	26
-139	35.98	34.80	15:23	25
-140	36.44	35.25	15:28	24
-142	37.29	36.10	15:38	22
-143	37.70	36.50	15:43	21
-144	38.09	36.89	15:47	20
-145	38.47	37.26	15:52	19
-147	39.18	37.97	16:01	17
-148	39.52	38.31	16:06	16
-149	39.84	38.63	16:10	15
-150	40.16	38.94	16:15	14
-152	40.75	39.52	16:24	12
-153	41.04	39.80	16:28	11

PERCENTAGE OF SUN OBSCURED ALONG MAP TRACK
WEST END = 97.7   EAST END = 96.3

===

+ MARKS THE CITY OF HSIEN-YANG

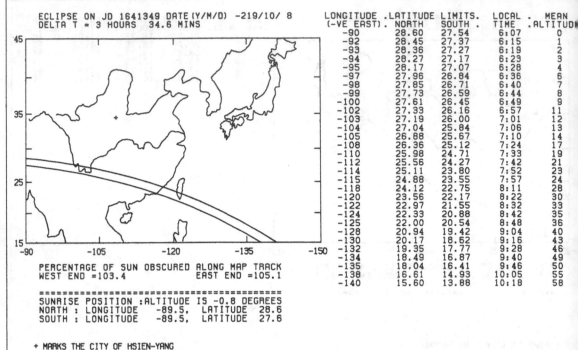

ECLIPSE ON JD 1641349 DATE(Y/M/D) -219/10/ 8
DELTA T = 3 HOURS 34.6 MINS

LONGITUDE (-VE EAST)	.LATITUDE LIMITS.		LOCAL	. MEAN
	NORTH	SOUTH .	TIME	.ALTITUDE
-90	28.60	27.54	6:07	0
-92	28.45	27.37	6:15	1
-93	28.36	27.27	6:19	2
-94	28.27	27.17	6:23	3
-95	28.17	27.07	6:28	4
-97	27.96	26.84	6:36	6
-98	27.85	26.71	6:40	7
-99	27.73	26.59	6:44	8
-100	27.61	26.45	6:49	9
-102	27.33	26.16	6:57	11
-103	27.19	26.00	7:01	12
-104	27.04	25.84	7:06	13
-105	26.88	25.67	7:10	14
-108	26.36	25.12	7:24	17
-110	25.98	24.71	7:33	19
-112	25.56	24.27	7:42	21
-114	25.11	23.80	7:52	23
-115	24.88	23.55	7:57	24
-118	24.12	22.75	8:11	28
-120	23.56	22.17	8:22	30
-122	22.97	21.55	8:32	33
-124	22.33	20.88	8:42	35
-125	22.00	20.54	8:48	36
-128	20.94	19.42	9:04	40
-130	20.17	18.62	9:16	43
-132	19.35	17.77	9:28	46
-134	18.49	16.87	9:40	49
-135	18.04	16.41	9:46	50
-138	16.61	14.93	10:05	55
-140	15.60	13.88	10:18	58

PERCENTAGE OF SUN OBSCURED ALONG MAP TRACK
WEST END =103.4   EAST END =105.1

===
SUNRISE POSITION :ALTITUDE IS -0.8 DEGREES
NORTH : LONGITUDE   -89.5,   LATITUDE   28.6
SOUTH : LONGITUDE   -89.5,   LATITUDE   27.6

+ MARKS THE CITY OF HSIEN-YANG

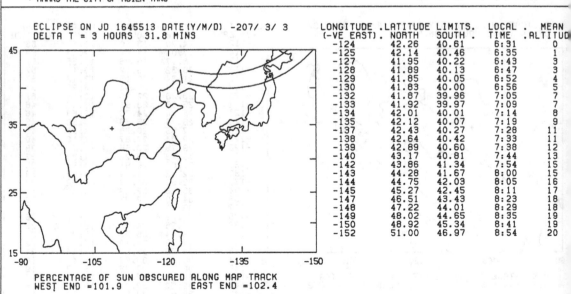

ECLIPSE ON JD 1645513 DATE(Y/M/D) -207/ 3/ 3
DELTA T = 3 HOURS 31.8 MINS

LONGITUDE (-VE EAST)	.LATITUDE LIMITS.		LOCAL	. MEAN
	NORTH	SOUTH .	TIME	.ALTITUDE
-124	42.26	40.61	6:31	0
-125	42.14	40.46	6:35	1
-127	41.95	40.22	6:43	3
-128	41.89	40.13	6:47	3
-129	41.85	40.05	6:52	4
-130	41.83	40.00	6:56	5
-132	41.87	39.96	7:05	7
-133	41.92	39.97	7:09	7
-134	42.01	40.01	7:14	8
-135	42.12	40.07	7:19	9
-137	42.43	40.27	7:28	11
-138	42.64	40.42	7:33	11
-139	42.89	40.60	7:38	12
-140	43.17	40.81	7:44	13
-142	43.86	41.34	7:54	15
-143	44.28	41.67	8:00	15
-144	44.75	42.03	8:05	16
-145	45.27	42.45	8:11	17
-147	46.51	43.43	8:23	18
-148	47.22	44.01	8:29	18
-149	48.02	44.65	8:35	19
-150	48.92	45.34	8:41	19
-152	51.00	46.97	8:54	20

PERCENTAGE OF SUN OBSCURED ALONG MAP TRACK
WEST END =101.9   EAST END =102.4

===
SUNRISE POSITION :ALTITUDE IS -0.8 DEGREES
NORTH : LONGITUDE -122.4,   LATITUDE   42.4
SOUTH : LONGITUDE -122.8,   LATITUDE   40.7

+ MARKS THE CITY OF HSIEN-YANG

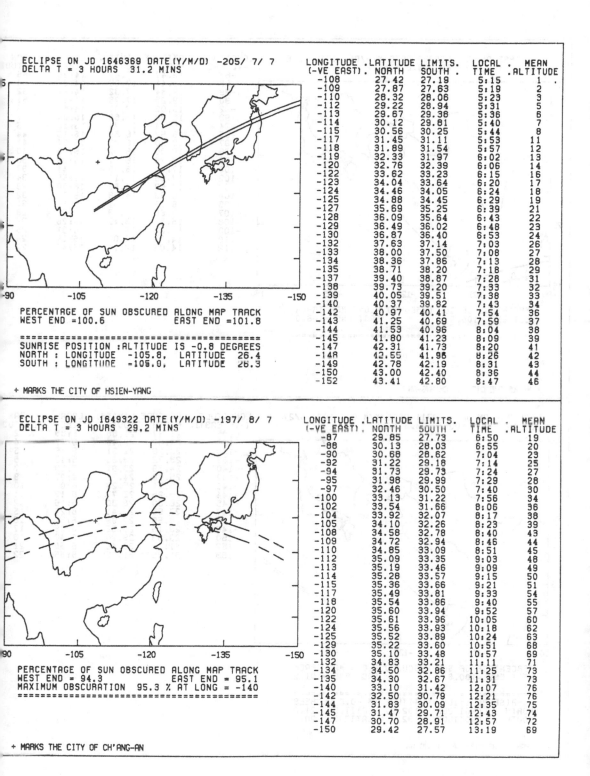

ECLIPSE ON JD 1646369 DATE(Y/M/D) -205/ 7/ 7
DELTA T = 3 HOURS 31.2 MINS

LONGITUDE	.LATITUDE	LIMITS.	LOCAL	. MEAN
(-VE EAST) .	NORTH	SOUTH .	TIME	.ALTITUDE
-108	27.42	27.19	5:15	1 .
-109	27.87	27.63	5:19	2
-110	28.32	28.06	5:23	3
-112	29.22	28.94	5:31	5
-113	29.67	29.38	5:36	6
-114	30.12	29.81	5:40	7
-115	30.56	30.25	5:44	8
-117	31.45	31.11	5:53	11
-118	31.89	31.54	5:57	12
-119	32.33	31.97	6:02	13
-120	32.76	32.39	6:06	14
-122	33.62	33.23	6:15	16
-123	34.04	33.64	6:20	17
-124	34.46	34.05	6:24	18
-125	34.88	34.45	6:29	19
-127	35.69	35.25	6:39	21
-128	36.09	35.64	6:43	22
-129	36.49	36.02	6:48	23
-130	36.87	36.40	6:53	24
-132	37.63	37.14	7:03	26
-133	38.00	37.50	7:08	27
-134	38.36	37.86	7:13	28
-135	38.71	38.20	7:18	29
-137	39.40	38.87	7:28	31
-138	39.73	39.20	7:33	32
-139	40.05	39.51	7:38	33
-140	40.37	39.82	7:43	34
-142	40.97	40.41	7:54	36
-143	41.25	40.69	7:59	37
-144	41.53	40.96	8:04	38
-145	41.80	41.23	8:09	39
-147	42.31	41.73	8:20	41
-148	42.55	41.96	8:26	42
-149	42.78	42.19	8:31	43
-150	43.00	42.40	8:36	44
-152	43.41	42.80	8:47	46

PERCENTAGE OF SUN OBSCURED ALONG MAP TRACK
WEST END =100.6   EAST END =101.8

===
SUNRISE POSITION :ALTITUDE IS -0.8 DEGREES
NORTH : LONGITUDE -105.8, LATITUDE 26.4
SOUTH : LONGITUDE -105.0, LATITUDE 26.3

+ MARKS THE CITY OF HSIEN-YANG

ECLIPSE ON JD 1649322 DATE(Y/M/D) -197/ 8/ 7
DELTA T = 3 HOURS 29.2 MINS

LONGITUDE	.LATITUDE	LIMITS.	LOCAL	. MEAN
(-VE EAST) .	NORTH	SOUTH .	TIME	.ALTITUDE
-87	29.85	27.73	6:50	19
-88	30.13	28.03	6:55	20
-90	30.68	28.62	7:04	23
-92	31.22	29.18	7:14	25
-94	31.73	29.73	7:24	27
-95	31.98	29.99	7:29	28
-97	32.46	30.50	7:40	30
-100	33.13	31.22	7:56	34
-102	33.54	31.66	8:06	36
-104	33.92	32.07	8:17	38
-105	34.10	32.26	8:23	39
-108	34.58	32.78	8:40	43
-109	34.72	32.94	8:46	44
-110	34.85	33.09	8:51	45
-112	35.09	33.35	9:03	48
-113	35.19	33.46	9:09	49
-114	35.28	33.57	9:15	50
-115	35.36	33.66	9:21	51
-117	35.49	33.81	9:33	54
-118	35.54	33.86	9:40	55
-120	35.60	33.94	9:52	57
-122	35.61	33.96	10:05	60
-124	35.56	33.93	10:18	62
-125	35.52	33.89	10:24	63
-129	35.22	33.60	10:51	68
-130	35.10	33.48	10:57	69
-132	34.83	33.21	11:11	71
-134	34.50	32.86	11:25	73
-135	34.30	32.67	11:31	73
-140	33.10	31.42	12:07	76
-142	32.50	30.79	12:21	76
-144	31.83	30.09	12:35	75
-145	31.47	29.71	12:43	74
-147	30.70	28.91	12:57	72
-150	29.42	27.57	13:19	69

PERCENTAGE OF SUN OBSCURED ALONG MAP TRACK
WEST END = 94.3   EAST END = 95.1
MAXIMUM OBSCURATION 95.3 % AT LONG = -140
===

+ MARKS THE CITY OF CH'ANG-AN

165

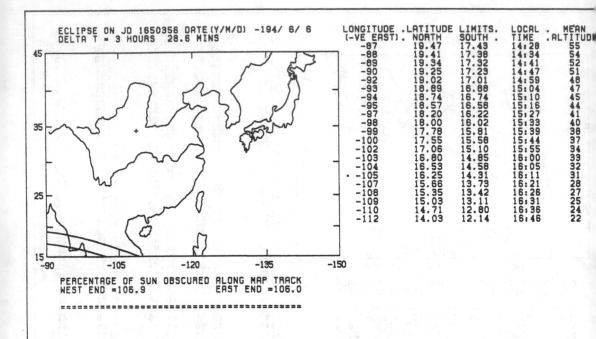

ECLIPSE ON JD 1650356 DATE(Y/M/D) -194/ 6/ 6
DELTA T = 3 HOURS 28.6 MINS

LONGITUDE (-VE EAST)	LATITUDE NORTH	LIMITS. SOUTH	LOCAL TIME	MEAN ALTITUDE
-87	19.47	17.43	14:28	55
-88	19.41	17.38	14:34	54
-89	19.34	17.32	14:41	52
-90	19.25	17.23	14:47	51
-92	19.02	17.01	14:59	48
-93	18.89	16.88	15:04	47
-94	18.74	16.74	15:10	45
-95	18.57	16.58	15:16	44
-97	18.20	16.22	15:27	41
-98	18.00	16.02	15:33	40
-99	17.78	15.81	15:39	38
-100	17.55	15.58	15:44	37
-102	17.06	15.10	15:55	34
-103	16.80	14.85	16:00	33
-104	16.53	14.58	16:05	32
-105	16.25	14.31	16:11	31
-107	15.66	13.73	16:21	28
-108	15.35	13.42	16:26	27
-109	15.03	13.11	16:31	25
-110	14.71	12.80	16:36	24
-112	14.03	12.14	16:46	22

PERCENTAGE OF SUN OBSCURED ALONG MAP TRACK
WEST END =106.9   EAST END =106.0

===

+ MARKS THE CITY OF CH'ANG-AN

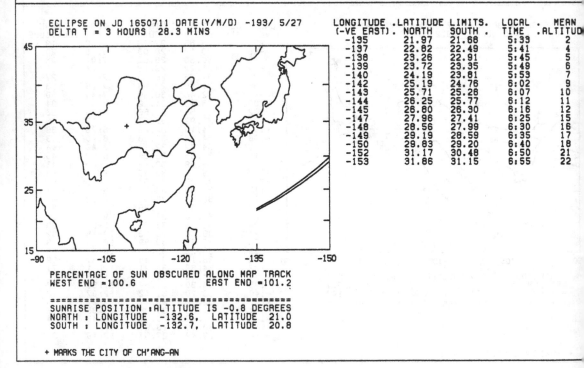

ECLIPSE ON JD 1650711 DATE(Y/M/D) -193/ 5/27
DELTA T = 3 HOURS 28.3 MINS

LONGITUDE (-VE EAST)	LATITUDE NORTH	LIMITS. SOUTH	LOCAL TIME	MEAN ALTITUDE
-135	21.97	21.68	5:39	2
-137	22.82	22.49	5:41	4
-138	23.26	22.91	5:45	5
-139	23.72	23.35	5:49	6
-140	24.19	23.81	5:53	7
-142	25.19	24.76	6:02	9
-143	25.71	25.26	6:07	10
-144	26.25	25.77	6:12	11
-145	26.80	26.30	6:16	12
-147	27.96	27.41	6:25	15
-148	28.56	27.99	6:30	16
-149	29.19	28.59	6:35	17
-150	29.83	29.20	6:40	18
-152	31.17	30.48	6:50	21
-153	31.86	31.15	6:55	22

PERCENTAGE OF SUN OBSCURED ALONG MAP TRACK
WEST END =100.6   EAST END =101.2

===
SUNRISE POSITION ;ALTITUDE IS -0.8 DEGREES
NORTH : LONGITUDE -132.6, LATITUDE 21.0
SOUTH : LONGITUDE -132.7, LATITUDE 20.8

+ MARKS THE CITY OF CH'ANG-AN

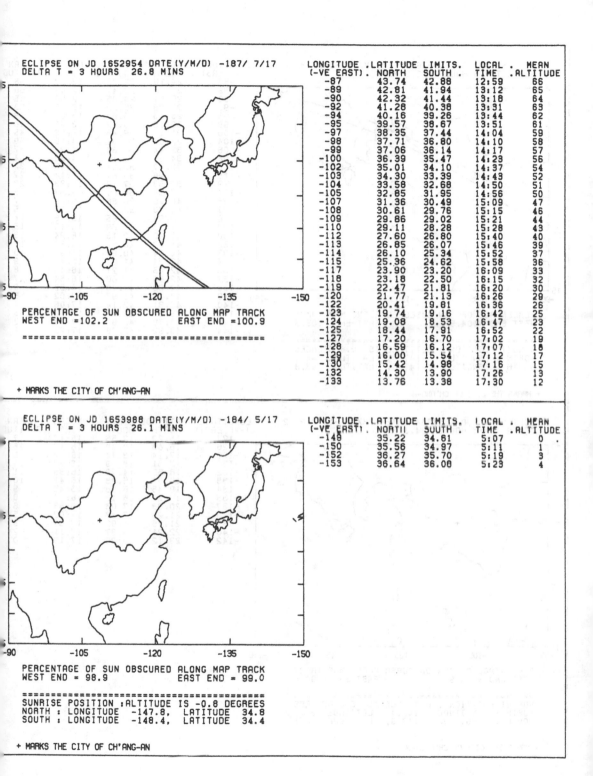

ECLIPSE ON JD 1652954 DATE(Y/M/D) -187/ 7/17
DELTA T = 3 HOURS 26.8 MINS

LONGITUDE (-VE EAST)	LATITUDE NORTH	LIMITS SOUTH	LOCAL TIME	MEAN ALTITUDE
-87	43.74	42.88	12:59	66
-89	42.81	41.94	13:12	65
-90	42.32	41.44	13:18	64
-92	41.28	40.38	13:31	63
-94	40.16	39.26	13:44	62
-95	39.57	38.67	13:51	61
-97	38.35	37.44	14:04	59
-98	37.71	36.80	14:10	58
-99	37.06	36.14	14:17	57
-100	36.39	35.47	14:23	56
-102	35.01	34.10	14:37	54
-103	34.30	33.39	14:43	52
-104	33.58	32.68	14:50	51
-105	32.85	31.95	14:56	50
-107	31.36	30.49	15:09	47
-108	30.61	29.76	15:15	46
-109	29.86	29.02	15:21	44
-110	29.11	28.28	15:28	43
-112	27.60	26.80	15:40	40
-113	26.85	26.07	15:46	39
-114	26.10	25.34	15:52	37
-115	25.36	24.62	15:58	36
-117	23.90	23.20	16:09	33
-118	23.18	22.50	16:15	32
-119	22.47	21.81	16:20	30
-120	21.77	21.13	16:26	29
-122	20.41	19.81	16:36	26
-123	19.74	19.16	16:42	25
-124	19.08	18.53	16:47	23
-125	18.44	17.91	16:52	22
-127	17.20	16.70	17:02	19
-128	16.59	16.12	17:07	18
-129	16.00	15.54	17:12	17
-130	15.42	14.98	17:16	15
-132	14.30	13.90	17:26	13
-133	13.76	13.38	17:30	12

PERCENTAGE OF SUN OBSCURED ALONG MAP TRACK
WEST END =102.2   EAST END =100.9

==

+ MARKS THE CITY OF CH'ANG-AN

ECLIPSE ON JD 1653988 DATE(Y/M/D) -184/ 5/17
DELTA T = 3 HOURS 26.1 MINS

LONGITUDE (-VE EAST)	LATITUDE NORTH	LIMITS SOUTH	LOCAL TIME	MEAN ALTITUDE
-148	35.22	34.61	5:07	0
-150	35.56	34.97	5:11	1
-152	36.27	35.70	5:19	3
-153	36.64	36.08	5:23	4

PERCENTAGE OF SUN OBSCURED ALONG MAP TRACK
WEST END = 98.9   EAST END = 99.0

==
SUNRISE POSITION ;ALTITUDE IS -0.8 DEGREES
NORTH : LONGITUDE -147.8, LATITUDE 34.8
SOUTH : LONGITUDE -148.4, LATITUDE 34.4

+ MARKS THE CITY OF CH'ANG-AN

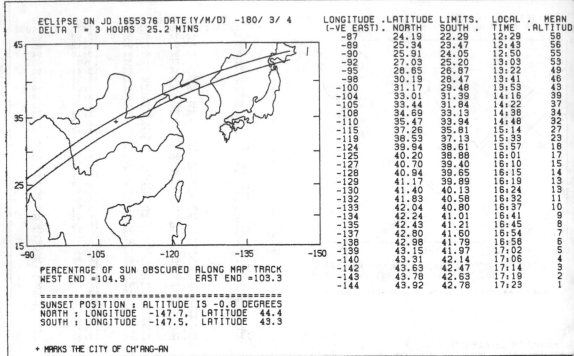

ECLIPSE ON JD 1655376 DATE(Y/M/D) -180/ 3/ 4
DELTA T = 3 HOURS 25.2 MINS

LONGITUDE (-VE EAST)	LATITUDE LIMITS. NORTH	SOUTH	LOCAL TIME	MEAN ALTITUD
-87	24.19	22.29	12:29	58
-89	25.34	23.47	12:43	56
-90	25.91	24.05	12:50	55
-92	27.03	25.20	13:03	53
-95	28.65	26.87	13:22	49
-98	30.19	28.47	13:41	46
-100	31.17	29.48	13:53	43
-104	33.01	31.39	14:16	39
-105	33.44	31.84	14:22	37
-108	34.69	33.13	14:38	34
-110	35.47	33.94	14:48	32
-115	37.26	35.81	15:14	27
-119	38.53	37.13	15:33	23
-124	39.94	38.61	15:57	18
-125	40.20	38.88	16:01	17
-127	40.70	39.40	16:10	15
-128	40.94	39.65	16:15	14
-129	41.17	39.89	16:19	13
-130	41.40	40.13	16:24	13
-132	41.83	40.58	16:32	11
-133	42.04	40.80	16:37	10
-134	42.24	41.01	16:41	9
-135	42.43	41.21	16:45	8
-137	42.80	41.60	16:54	7
-138	42.98	41.79	16:58	6
-139	43.15	41.97	17:02	5
-140	43.31	42.14	17:06	4
-142	43.63	42.47	17:14	3
-143	43.78	42.63	17:19	2
-144	43.92	42.78	17:23	1

PERCENTAGE OF SUN OBSCURED ALONG MAP TRACK
WEST END =104.9   EAST END =103.3

==
SUNSET POSITION : ALTITUDE IS -0.8 DEGREES
NORTH : LONGITUDE -147.7, LATITUDE 44.4
SOUTH : LONGITUDE -147.5, LATITUDE 43.3

+ MARKS THE CITY OF CH'ANG-AN

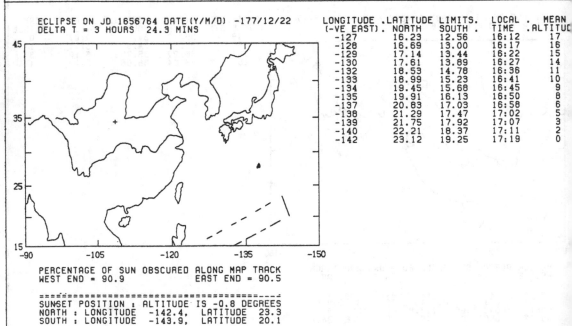

ECLIPSE ON JD 1656764 DATE(Y/M/D) -177/12/22
DELTA T = 3 HOURS 24.9 MINS

LONGITUDE (-VE EAST)	LATITUDE LIMITS. NORTH	SOUTH	LOCAL TIME	MEAN ALTITUD
-127	16.23	12.56	16:12	17
-128	16.69	13.00	16:17	16
-129	17.14	13.44	16:22	15
-130	17.61	13.89	16:27	14
-132	18.53	14.78	16:36	11
-133	18.99	15.23	16:41	10
-134	19.45	15.68	16:45	9
-135	19.91	16.13	16:50	8
-137	20.83	17.03	16:58	6
-138	21.29	17.47	17:02	5
-139	21.75	17.92	17:07	3
-140	22.21	18.37	17:11	2
-142	23.12	19.25	17:19	0

PERCENTAGE OF SUN OBSCURED ALONG MAP TRACK
WEST END = 90.9   EAST END = 90.5

SUNSET POSITION : ALTITUDE IS -0.8 DEGREES
NORTH : LONGITUDE -142.4, LATITUDE 23.3
SOUTH : LONGITUDE -143.9, LATITUDE 20.1

+ MARKS THE CITY OF CH'ANG-AN

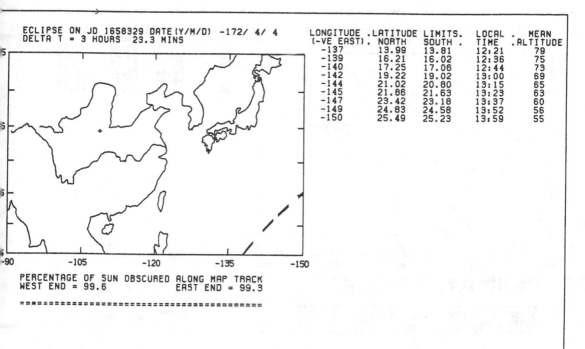

ECLIPSE ON JD 1658329 DATE(Y/M/D) -172/ 4/ 4
DELTA T = 3 HOURS 23.3 MINS

LONGITUDE (-VE EAST)	LATITUDE LIMITS. NORTH	SOUTH .	LOCAL TIME	MEAN .ALTITUDE
-137	13.99	13.81	12:21	79
-139	16.21	16.02	12:36	75
-140	17.25	17.06	12:44	73
-142	19.22	19.02	13:00	69
-144	21.02	20.80	13:15	65
-145	21.86	21.63	13:23	63
-147	23.42	23.18	13:37	60
-149	24.83	24.58	13:52	56
-150	25.49	25.23	13:59	55

PERCENTAGE OF SUN OBSCURED ALONG MAP TRACK
WEST END = 99.6   EAST END = 99.3

===

+ MARKS THE CITY OF CH'ANG-AN

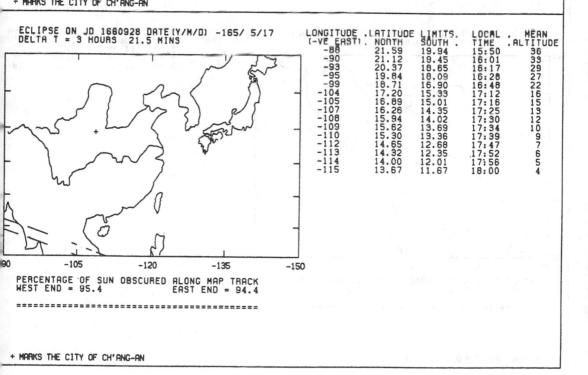

ECLIPSE ON JD 1680928 DATE(Y/M/D) -165/ 5/17
DELTA T = 3 HOURS 21.5 MINS

LONGITUDE (-VE EAST)	LATITUDE LIMITS. NORTH	SOUTH .	LOCAL TIME	MEAN .ALTITUDE
-88	21.59	19.94	15:50	36
-90	21.12	19.45	16:01	33
-93	20.37	18.65	16:17	29
-95	19.84	18.09	16:28	27
-99	18.71	16.90	16:48	22
-104	17.20	15.33	17:12	16
-105	16.89	15.01	17:16	15
-107	16.26	14.35	17:25	13
-108	15.94	14.02	17:30	12
-109	15.62	13.69	17:34	10
-110	15.30	13.36	17:39	9
-112	14.65	12.68	17:47	7
-113	14.32	12.35	17:52	6
-114	14.00	12.01	17:56	5
-115	13.67	11.67	18:00	4

PERCENTAGE OF SUN OBSCURED ALONG MAP TRACK
WEST END = 95.4   EAST END = 94.4

===

+ MARKS THE CITY OF CH'ANG-AN

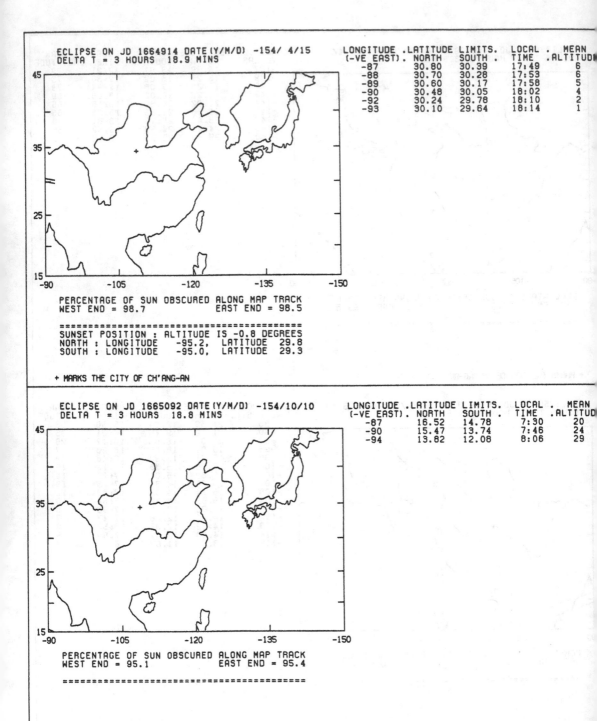

ECLIPSE ON JD 1664914 DATE(Y/M/D) -154/ 4/15
DELTA T = 3 HOURS 18.9 MINS

LONGITUDE (-VE EAST)	.LATITUDE . NORTH	LIMITS. SOUTH .	LOCAL . TIME	MEAN .ALTITUDE
-87	30.80	30.39	17:49	6
-88	30.70	30.28	17:53	6
-89	30.60	30.17	17:58	5
-90	30.48	30.05	18:02	4
-92	30.24	29.78	18:10	2
-93	30.10	29.64	18:14	1

PERCENTAGE OF SUN OBSCURED ALONG MAP TRACK
WEST END = 98.7   EAST END = 98.5

===
SUNSET POSITION : ALTITUDE IS -0.8 DEGREES
NORTH : LONGITUDE   -95.2, LATITUDE 29.8
SOUTH : LONGITUDE   -95.0, LATITUDE 29.3

+ MARKS THE CITY OF CH'ANG-AN

ECLIPSE ON JD 1665092 DATE(Y/M/D) -154/10/10
DELTA T = 3 HOURS 18.8 MINS

LONGITUDE (-VE EAST)	.LATITUDE . NORTH	LIMITS. SOUTH .	LOCAL . TIME	MEAN .ALTITUDE
-87	16.52	14.78	7:30	20
-90	15.47	13.74	7:46	24
-94	13.82	12.08	8:06	29

PERCENTAGE OF SUN OBSCURED ALONG MAP TRACK
WEST END = 95.1   EAST END = 95.4

===

+ MARKS THE CITY OF CH'ANG-AN

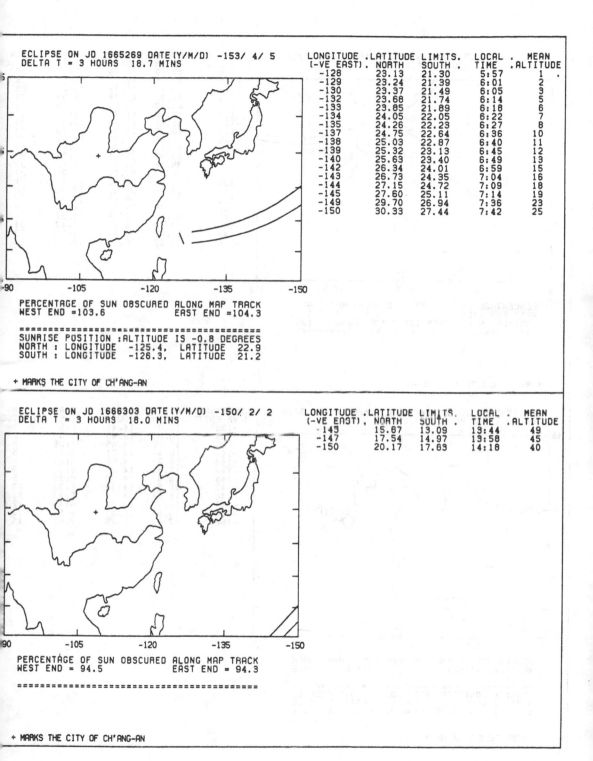

ECLIPSE ON JD 1665269 DATE(Y/M/D) -153/ 4/ 5
DELTA T = 3 HOURS 18.7 MINS

LONGITUDE (-VE EAST)	LATITUDE LIMITS		LOCAL TIME	MEAN ALTITUDE
	NORTH	SOUTH		
-128	23.13	21.30	5:57	1
-129	23.24	21.39	6:01	2
-130	23.37	21.49	6:05	3
-132	23.68	21.74	6:14	5
-133	23.85	21.89	6:18	6
-134	24.05	22.05	6:22	7
-135	24.26	22.23	6:27	8
-137	24.75	22.64	6:36	10
-138	25.03	22.87	6:40	11
-139	25.32	23.13	6:45	12
-140	25.63	23.40	6:49	13
-142	26.34	24.01	6:59	15
-143	26.73	24.35	7:04	16
-144	27.15	24.72	7:09	18
-145	27.60	25.11	7:14	19
-149	29.70	26.94	7:36	23
-150	30.33	27.44	7:42	25

PERCENTAGE OF SUN OBSCURED ALONG MAP TRACK
WEST END =103.6   EAST END =104.3

==
SUNRISE POSITION :ALTITUDE IS -0.8 DEGREES
NORTH : LONGITUDE -125.4, LATITUDE 22.9
SOUTH : LONGITUDE -126.3, LATITUDE 21.2

+ MARKS THE CITY OF CH'ANG-AN

-90   -105   -120   -135   -150

ECLIPSE ON JD 1666303 DATE(Y/M/D) -150/ 2/ 2
DELTA T = 3 HOURS 18.0 MINS

LONGITUDE (-VE EAST)	LATITUDE LIMITS		LOCAL TIME	MEAN ALTITUDE
	NORTH	SOUTH		
-145	15.87	13.09	13:44	49
-147	17.54	14.97	13:58	45
-150	20.17	17.63	14:18	40

PERCENTAGE OF SUN OBSCURED ALONG MAP TRACK
WEST END = 94.5   EAST END = 94.3

==

+ MARKS THE CITY OF CH'ANG-AN

-90   -105   -120   -135   -150

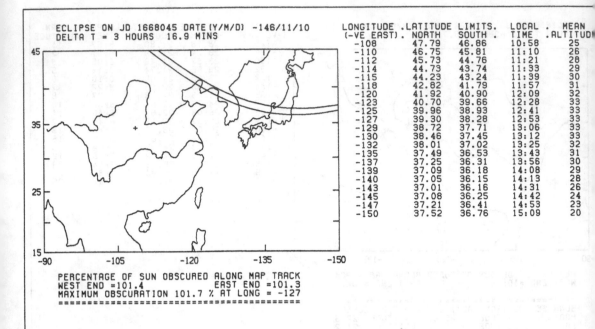

ECLIPSE ON JD 1668045 DATE(Y/M/D) -146/11/10
DELTA T = 3 HOURS 16.9 MINS

LONGITUDE (-VE EAST)	.LATITUDE NORTH	LIMITS. SOUTH	LOCAL TIME	. MEAN .ALTITUDE
-108	47.79	46.86	10:58	25
-110	46.75	45.81	11:10	26
-112	45.73	44.76	11:21	28
-114	44.73	43.74	11:33	29
-115	44.23	43.24	11:39	30
-118	42.82	41.79	11:57	31
-120	41.92	40.90	12:09	32
-123	40.70	39.66	12:28	33
-125	39.96	38.93	12:41	33
-127	39.30	38.28	12:53	33
-129	38.72	37.71	13:06	33
-130	38.46	37.45	13:12	33
-132	38.01	37.02	13:25	32
-135	37.49	36.53	13:43	31
-137	37.25	36.31	13:56	30
-139	37.09	36.18	14:08	29
-140	37.05	36.15	14:13	28
-143	37.01	36.16	14:31	26
-145	37.08	36.25	14:42	24
-147	37.21	36.41	14:53	23
-150	37.52	36.76	15:09	20

PERCENTAGE OF SUN OBSCURED ALONG MAP TRACK
WEST END =101.4   EAST END =101.3
MAXIMUM OBSCURATION 101.7 % AT LONG = -127
===

+ MARKS THE CITY OF CH'ANG-AN

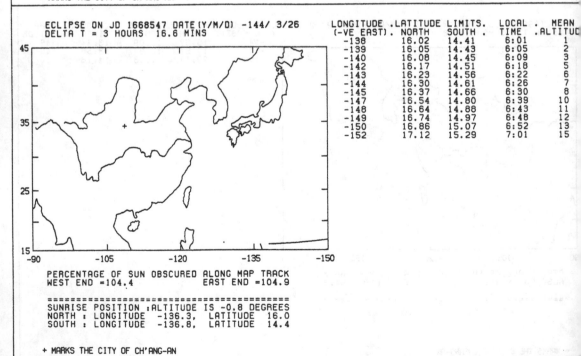

ECLIPSE ON JD 1668547 DATE(Y/M/D) -144/ 3/26
DELTA T = 3 HOURS 16.6 MINS

LONGITUDE (-VE EAST)	.LATITUDE NORTH	LIMITS. SOUTH	LOCAL TIME	. MEAN .ALTITUDE
-138	16.02	14.41	6:01	1
-139	16.05	14.43	6:05	2
-140	16.08	14.45	6:09	3
-142	16.17	14.51	6:18	5
-143	16.23	14.56	6:22	6
-144	16.30	14.61	6:26	7
-145	16.37	14.66	6:30	8
-147	16.54	14.80	6:39	10
-148	16.64	14.88	6:43	11
-149	16.74	14.97	6:48	12
-150	16.86	15.07	6:52	13
-152	17.12	15.29	7:01	15

PERCENTAGE OF SUN OBSCURED ALONG MAP TRACK
WEST END =104.4   EAST END =104.9

===
SUNRISE POSITION :ALTITUDE IS -0.8 DEGREES
NORTH : LONGITUDE -136.3, LATITUDE 16.0
SOUTH : LONGITUDE -136.8, LATITUDE 14.4

+ MARKS THE CITY OF CH'ANG-AN

ECLIPSE ON JD 1669078 DATE(Y/M/D) -143/ 9/ 8
DELTA T = 3 HOURS 16.2 MINS

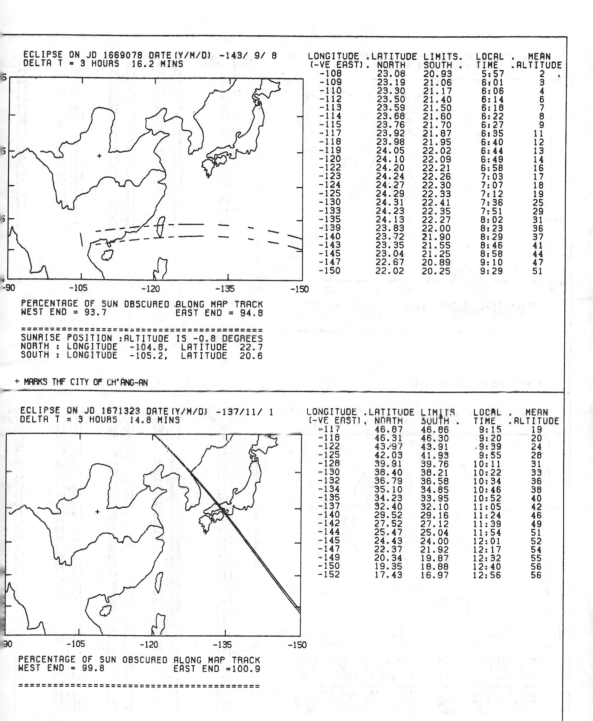

LONGITUDE (-VE EAST)	LATITUDE NORTH	LIMITS. SOUTH	LOCAL TIME	MEAN ALTITUDE
-108	23.08	20.93	5:57	2
-109	23.19	21.06	6:01	3
-110	23.30	21.17	6:06	4
-112	23.50	21.40	6:14	6
-113	23.59	21.50	6:18	7
-114	23.68	21.60	6:22	8
-115	23.76	21.70	6:27	9
-117	23.92	21.87	6:35	11
-118	23.98	21.95	6:40	12
-119	24.05	22.02	6:44	13
-120	24.10	22.09	6:49	14
-122	24.20	22.21	6:58	16
-123	24.24	22.26	7:03	17
-124	24.27	22.30	7:07	18
-125	24.29	22.33	7:12	19
-130	24.31	22.41	7:36	25
-133	24.23	22.35	7:51	29
-135	24.13	22.27	8:02	31
-139	23.83	22.00	8:23	36
-140	23.72	21.90	8:29	37
-143	23.35	21.55	8:46	41
-145	23.04	21.25	8:58	44
-147	22.67	20.89	9:10	47
-150	22.02	20.25	9:29	51

PERCENTAGE OF SUN OBSCURED ALONG MAP TRACK
WEST END = 93.7   EAST END = 94.8

===
SUNRISE POSITION :ALTITUDE IS -0.8 DEGREES
NORTH : LONGITUDE -104.8, LATITUDE 22.7
SOUTH : LONGITUDE -105.2, LATITUDE 20.6

+ MARKS THE CITY OF CH'ANG-AN

ECLIPSE ON JD 1671323 DATE(Y/M/D) -137/11/ 1
DELTA T = 3 HOURS 14.8 MINS

LONGITUDE (-VE EAST)	LATITUDE NORTH	LIMITS. SOUTH	LOCAL TIME	MEAN ALTITUDE
-117	46.87	46.86	9:15	19
-118	46.31	46.30	9:20	20
-122	43.97	43.91	9:39	24
-125	42.03	41.93	9:55	28
-128	39.91	39.76	10:11	31
-130	38.40	38.21	10:22	33
-132	36.79	36.58	10:34	36
-134	35.10	34.85	10:46	38
-135	34.23	33.95	10:52	40
-137	32.40	32.10	11:05	42
-140	29.52	29.16	11:24	46
-142	27.52	27.12	11:39	49
-144	25.47	25.04	11:54	51
-145	24.43	24.00	12:01	52
-147	22.37	21.92	12:17	54
-149	20.34	19.87	12:32	55
-150	19.35	18.88	12:40	56
-152	17.43	16.97	12:56	56

PERCENTAGE OF SUN OBSCURED ALONG MAP TRACK
WEST END = 99.8   EAST END =100.9

===

+ MARKS THE CITY OF CH'ANG-AN

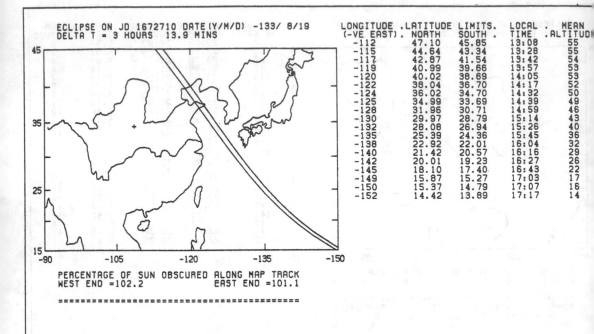

```
ECLIPSE ON JD 1672710 DATE(Y/M/D) -133/ 8/19
DELTA T = 3 HOURS  13.9 MINS
```

LONGITUDE (-VE EAST)	LATITUDE NORTH	LIMITS. SOUTH	LOCAL TIME	MEAN ALTITUD
-112	47.10	45.85	13:08	55
-115	44.64	43.34	13:28	55
-117	42.87	41.54	13:42	54
-119	40.99	39.66	13:57	53
-120	40.02	38.69	14:05	53
-122	38.04	36.70	14:17	52
-124	36.02	34.70	14:32	50
-125	34.99	33.69	14:39	49
-128	31.96	30.71	14:59	46
-130	29.97	28.79	15:14	43
-132	28.08	26.94	15:26	40
-135	25.39	24.36	15:45	36
-138	22.92	22.01	16:04	32
-140	21.42	20.57	16:16	29
-142	20.01	19.23	16:27	26
-145	18.10	17.40	16:43	22
-149	15.87	15.27	17:03	17
-150	15.37	14.79	17:07	16
-152	14.42	13.89	17:17	14

```
PERCENTAGE OF SUN OBSCURED ALONG MAP TRACK
WEST END =102.2              EAST END =101.1
=================================================
```

+ MARKS THE CITY OF CH'ANG-AN

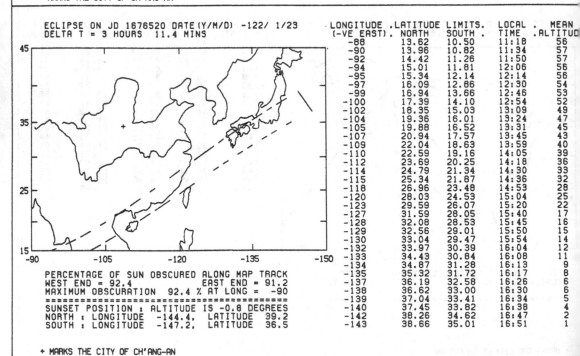

```
ECLIPSE ON JD 1676520 DATE(Y/M/D) -122/ 1/23
DELTA T = 3 HOURS  11.4 MINS
```

LONGITUDE (-VE EAST)	LATITUDE NORTH	LIMITS. SOUTH	LOCAL TIME	MEAN ALTITUD
-88	13.62	10.50	11:18	56
-90	13.96	10.82	11:34	57
-92	14.42	11.26	11:50	57
-94	15.01	11.81	12:06	56
-95	15.34	12.14	12:14	56
-97	16.09	12.86	12:30	54
-99	16.94	13.66	12:46	53
-100	17.39	14.10	12:54	52
-102	18.35	15.03	13:09	49
-104	19.36	16.01	13:24	47
-105	19.88	16.52	13:31	45
-107	20.94	17.57	13:45	43
-109	22.04	18.63	13:59	40
-110	22.59	19.16	14:05	39
-112	23.69	20.25	14:18	36
-114	24.79	21.34	14:30	33
-115	25.34	21.87	14:36	32
-118	26.96	23.48	14:53	28
-120	28.03	24.53	15:04	25
-123	29.59	26.07	15:20	22
-127	31.59	28.05	15:40	17
-128	32.08	28.53	15:45	16
-129	32.56	29.01	15:50	15
-130	33.04	29.47	15:54	14
-132	33.97	30.39	16:04	12
-133	34.43	30.84	16:08	11
-134	34.87	31.28	16:13	9
-135	35.32	31.72	16:17	8
-137	36.19	32.58	16:26	6
-138	36.62	33.00	16:30	6
-139	37.04	33.41	16:34	5
-140	37.45	33.82	16:38	4
-142	38.26	34.62	16:47	2
-143	38.66	35.01	16:51	1

```
PERCENTAGE OF SUN OBSCURED ALONG MAP TRACK
WEST END = 92.4              EAST END = 91.2
MAXIMUM OBSCURATION  92.4 % AT LONG =  -90
==============================================
SUNSET POSITION : ALTITUDE IS -0.8 DEGREES
NORTH : LONGITUDE  -144.4,  LATITUDE  39.2
SOUTH : LONGITUDE  -147.2,  LATITUDE  36.5
```

+ MARKS THE CITY OF CH'ANG-AN

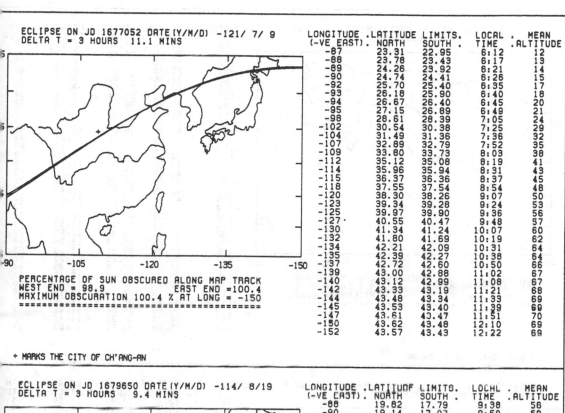

ECLIPSE ON JD 1677052 DATE(Y/M/D) -121/ 7/ 9
DELTA T = 3 HOURS 11.1 MINS

PERCENTAGE OF SUN OBSCURED ALONG MAP TRACK
WEST END = 98.9   EAST END =100.4
MAXIMUM OBSCURATION 100.4 % AT LONG = -150
==

+ MARKS THE CITY OF CH'ANG-AN

LONGITUDE (-VE EAST)	LATITUDE LIMITS. NORTH	SOUTH	LOCAL TIME	MEAN ALTITUDE
-87	23.31	22.95	6:12	12
-88	23.78	23.43	6:17	13
-89	24.26	23.92	6:21	14
-90	24.74	24.41	6:26	15
-92	25.70	25.40	6:35	17
-93	26.18	25.90	6:40	18
-94	26.67	26.40	6:45	20
-95	27.15	26.89	6:49	21
-98	28.61	28.39	7:05	24
-102	30.54	30.38	7:25	29
-104	31.49	31.36	7:36	32
-107	32.89	32.79	7:52	35
-109	33.80	33.73	8:03	38
-112	35.12	35.08	8:19	41
-114	35.96	35.94	8:31	43
-115	36.37	36.36	8:37	45
-118	37.55	37.54	8:54	48
-120	38.30	38.26	9:07	50
-123	39.34	39.28	9:24	53
-125	39.97	39.90	9:36	56
-127	40.55	40.47	9:48	57
-130	41.34	41.24	10:07	60
-132	41.80	41.69	10:19	62
-134	42.21	42.09	10:31	64
-135	42.39	42.27	10:38	64
-137	42.72	42.60	10:50	66
-139	43.00	42.88	11:02	67
-140	43.12	42.99	11:08	67
-142	43.33	43.19	11:21	68
-144	43.48	43.34	11:33	69
-145	43.53	43.40	11:39	69
-147	43.61	43.47	11:51	70
-150	43.62	43.48	12:10	69
-152	43.57	43.43	12:22	69

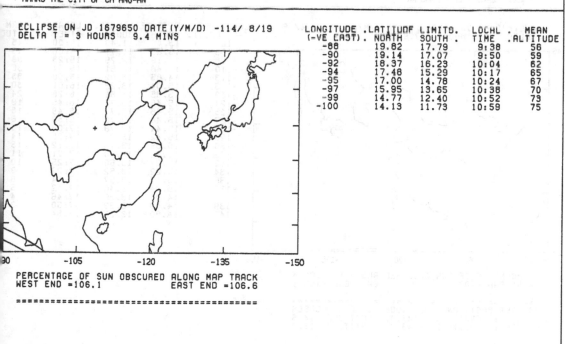

ECLIPSE ON JD 1679650 DATE(Y/M/D) -114/ 8/19
DELTA T = 3 HOURS   9.4 MINS

PERCENTAGE OF SUN OBSCURED ALONG MAP TRACK
WEST END =106.1   EAST END =106.6

==

+ MARKS THE CITY OF CH'ANG-AN

LONGITUDE (-VE EAST)	LATITUDE LIMITS. NORTH	SOUTH	LOCAL TIME	MEAN ALTITUDE
-88	19.82	17.79	9:38	56
-90	19.14	17.07	9:50	59
-92	18.37	16.23	10:04	62
-94	17.48	15.29	10:17	65
-95	17.00	14.78	10:24	67
-97	15.95	13.65	10:38	70
-99	14.77	12.40	10:52	73
-100	14.13	11.73	10:59	75

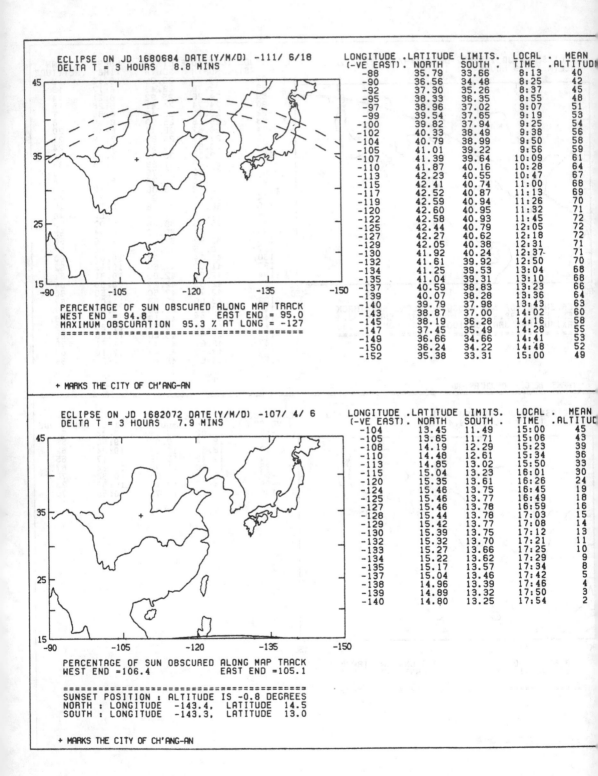

ECLIPSE ON JD 1680684 DATE(Y/M/D) -111/ 6/18
DELTA T = 3 HOURS   8.8 MINS

LONGITUDE (-VE EAST)	.LATITUDE LIMITS. NORTH	SOUTH .	LOCAL TIME	. MEAN .ALTITUDE
-88	35.79	33.66	8:13	40
-90	36.56	34.48	8:25	42
-92	37.30	35.26	8:37	45
-95	38.33	36.35	8:55	48
-97	38.96	37.02	9:07	51
-99	39.54	37.65	9:19	53
-100	39.82	37.94	9:25	54
-102	40.33	38.49	9:38	56
-104	40.79	38.99	9:50	58
-105	41.01	39.22	9:56	59
-107	41.39	39.64	10:09	61
-110	41.87	40.16	10:28	64
-113	42.23	40.55	10:47	67
-115	42.41	40.74	11:00	68
-117	42.52	40.87	11:13	69
-119	42.59	40.94	11:26	70
-120	42.60	40.95	11:32	71
-122	42.58	40.93	11:45	72
-125	42.44	40.79	12:05	72
-127	42.27	40.62	12:18	72
-129	42.05	40.38	12:31	71
-130	41.92	40.24	12:37	71
-132	41.61	39.92	12:50	70
-134	41.25	39.53	13:04	68
-135	41.04	39.31	13:10	68
-137	40.59	38.83	13:23	66
-139	40.07	38.28	13:36	64
-140	39.79	37.98	13:43	63
-143	38.87	37.00	14:02	60
-145	38.19	36.28	14:16	58
-147	37.45	35.49	14:28	55
-149	36.66	34.66	14:41	53
-150	36.24	34.22	14:48	52
-152	35.38	33.31	15:00	49

PERCENTAGE OF SUN OBSCURED ALONG MAP TRACK
WEST END = 94.8   EAST END = 95.0
MAXIMUM OBSCURATION 95.3 % AT LONG = -127
==

+ MARKS THE CITY OF CH'ANG-AN

ECLIPSE ON JD 1682072 DATE(Y/M/D) -107/ 4/ 6
DELTA T = 3 HOURS   7.9 MINS

LONGITUDE (-VE EAST)	.LATITUDE NORTH	LIMITS. SOUTH .	LOCAL TIME	. MEAN .ALTITUDE
-104	13.45	11.49	15:00	45
-105	13.65	11.71	15:06	43
-108	14.19	12.29	15:23	39
-110	14.48	12.61	15:34	36
-113	14.85	13.02	15:50	33
-115	15.04	13.23	16:01	30
-120	15.35	13.61	16:26	24
-124	15.46	13.75	16:45	19
-125	15.46	13.77	16:49	18
-127	15.46	13.78	16:59	16
-128	15.44	13.78	17:03	15
-129	15.42	13.77	17:08	14
-130	15.39	13.75	17:12	13
-132	15.32	13.70	17:21	11
-133	15.27	13.66	17:25	10
-134	15.22	13.62	17:29	9
-135	15.17	13.57	17:34	8
-137	15.04	13.46	17:42	5
-138	14.96	13.39	17:46	4
-139	14.89	13.32	17:50	3
-140	14.80	13.25	17:54	2

PERCENTAGE OF SUN OBSCURED ALONG MAP TRACK
WEST END =106.4   EAST END =105.1

==
SUNSET POSITION : ALTITUDE IS -0.8 DEGREES
NORTH : LONGITUDE -143.4, LATITUDE 14.5
SOUTH : LONGITUDE -143.3, LATITUDE 13.0

+ MARKS THE CITY OF CH'ANG-AN

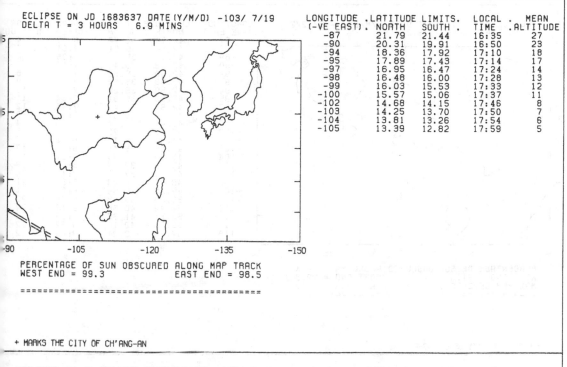

ECLIPSE ON JD 1683637 DATE(Y/M/D) -103/ 7/19
DELTA T = 3 HOURS   6.9 MINS

LONGITUDE (-VE EAST)	LATITUDE NORTH	LIMITS SOUTH	LOCAL TIME	MEAN ALTITUDE
-87	21.79	21.44	16:35	27
-90	20.31	19.91	16:50	23
-94	18.36	17.92	17:10	18
-95	17.89	17.43	17:14	17
-97	16.95	16.47	17:24	14
-98	16.48	16.00	17:28	13
-99	16.03	15.53	17:33	12
-100	15.57	15.06	17:37	11
-102	14.68	14.15	17:46	8
-103	14.25	13.70	17:50	7
-104	13.81	13.26	17:54	6
-105	13.39	12.82	17:59	5

-90   -105   -120   -135   -150

PERCENTAGE OF SUN OBSCURED ALONG MAP TRACK
WEST END = 99.3   EAST END = 98.5

==

+ MARKS THE CITY OF CH'ANG-AN

ECLIPSE ON JD 1684670 DATE(Y/M/D) -100/ 5/17
DELTA T = 3 HOURS   6.3 MINS

LONGITUDE (-VE EAST)	LATITUDE NORTH	LIMITS SOUTH	LOCAL TIME	MEAN ALTITUDE
-88	16.32	15.87	14:33	53
-90	16.56	16.12	14:45	50
-93	16.75	16.34	15:04	46
-95	16.79	16.40	15:16	43
-98	16.71	16.36	15:33	39
-100	16.58	16.25	15:44	37
-103	16.28	15.98	16:01	33
-105	16.02	15.74	16:11	30
-109	15.95	15.11	16:32	25
-113	14.51	14.31	16:51	20

-90   -105   -120   -135   -150

PERCENTAGE OF SUN OBSCURED ALONG MAP TRACK
WEST END =101.6   EAST END =100.3

==

+ MARKS THE CITY OF CH'ANG-AN

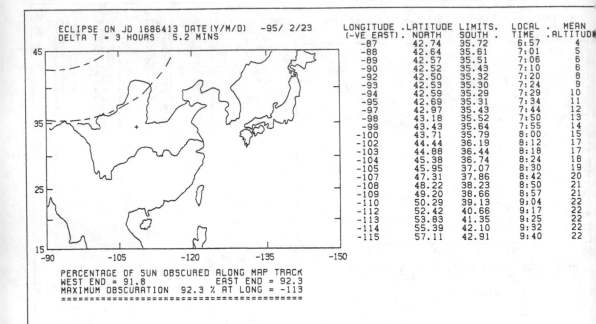

ECLIPSE ON JD 1686413 DATE(Y/M/D) -95/ 2/23
DELTA T = 3 HOURS   5.2 MINS

LONGITUDE (-VE EAST)	LATITUDE LIMITS. NORTH	SOUTH	LOCAL TIME	MEAN ALTITUDE
-87	42.74	35.72	6:57	4
-88	42.64	35.61	7:01	5
-89	42.57	35.51	7:06	6
-90	42.52	35.43	7:10	6
-92	42.50	35.32	7:20	8
-93	42.53	35.30	7:24	9
-94	42.59	35.29	7:29	10
-95	42.69	35.31	7:34	11
-97	42.97	35.43	7:44	12
-98	43.18	35.52	7:50	13
-99	43.43	35.64	7:55	14
-100	43.71	35.79	8:00	15
-102	44.44	36.19	8:12	17
-103	44.88	36.44	8:18	17
-104	45.38	36.74	8:24	18
-105	45.95	37.07	8:30	19
-107	47.31	37.86	8:42	20
-108	48.22	38.23	8:50	21
-109	49.20	38.66	8:57	21
-110	50.29	39.13	9:04	22
-112	52.42	40.66	9:17	22
-113	53.83	41.35	9:25	22
-114	55.39	42.10	9:32	22
-115	57.11	42.91	9:40	22

PERCENTAGE OF SUN OBSCURED ALONG MAP TRACK
WEST END = 91.8   EAST END = 92.3
MAXIMUM OBSCURATION 92.3 % AT LONG = -113
==

+ MARKS THE CITY OF CH'ANG-AN

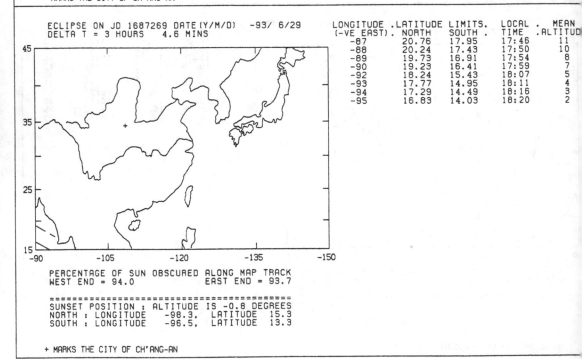

ECLIPSE ON JD 1687269 DATE(Y/M/D) -93/ 6/29
DELTA T = 3 HOURS   4.6 MINS

LONGITUDE (-VE EAST)	LATITUDE LIMITS. NORTH	SOUTH	LOCAL TIME	MEAN ALTITUDE
-87	20.76	17.95	17:46	11
-88	20.24	17.43	17:50	10
-89	19.73	16.91	17:54	8
-90	19.23	16.41	17:59	7
-92	18.24	15.43	18:07	5
-93	17.77	14.95	18:11	4
-94	17.29	14.49	18:16	3
-95	16.83	14.03	18:20	2

PERCENTAGE OF SUN OBSCURED ALONG MAP TRACK
WEST END = 94.0   EAST END = 93.7

==
SUNSET POSITION : ALTITUDE IS -0.8 DEGREES
NORTH : LONGITUDE -98.3, LATITUDE 15.3
SOUTH : LONGITUDE -96.5, LATITUDE 13.3

+ MARKS THE CITY OF CH'ANG-AN

ECLIPSE ON JD 1687801 DATE(Y/M/D) -92/12/12
DELTA T = 3 HOURS   4.3 MINS

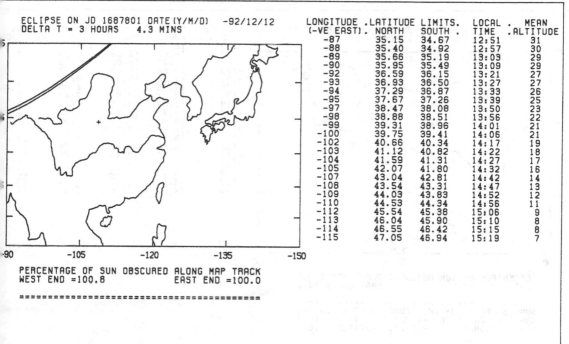

LONGITUDE (-VE EAST)	LATITUDE NORTH	LIMITS. SOUTH	LOCAL TIME	MEAN ALTITUDE
-87	35.15	34.67	12:51	31
-88	35.40	34.92	12:57	30
-89	35.66	35.19	13:03	29
-90	35.95	35.49	13:09	29
-92	36.59	36.15	13:21	27
-93	36.93	36.50	13:27	27
-94	37.29	36.87	13:33	26
-95	37.67	37.26	13:39	25
-97	38.47	38.08	13:50	23
-98	38.88	38.51	13:56	22
-99	39.31	38.96	14:01	21
-100	39.75	39.41	14:06	21
-102	40.66	40.34	14:17	19
-103	41.12	40.82	14:22	18
-104	41.59	41.31	14:27	17
-105	42.07	41.80	14:32	16
-107	43.04	42.81	14:42	14
-108	43.54	43.31	14:47	13
-109	44.03	43.83	14:52	12
-110	44.53	44.34	14:56	11
-112	45.54	45.38	15:06	9
-113	46.04	45.90	15:10	8
-114	46.55	46.42	15:15	8
-115	47.05	46.94	15:19	7

PERCENTAGE OF SUN OBSCURED ALONG MAP TRACK
WEST END =100.8   EAST END =100.0

===

+ MARKS THE CITY OF CH'ANG-AN

ECLIPSE ON JD 1688303 DATE(Y/M/D) -90/ 4/28
DELTA T = 3 HOURS   4.0 MINS

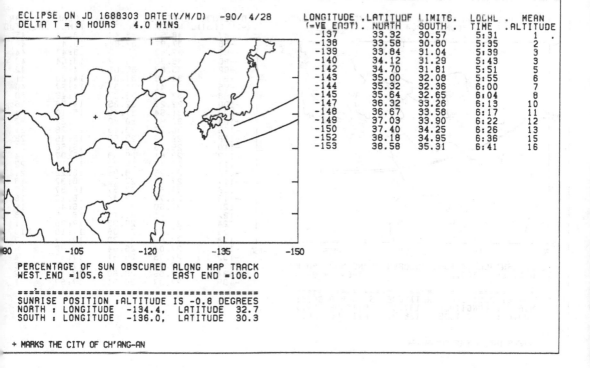

LONGITUDE (-VE EAST)	LATITUDE NORTH	LIMITS. SOUTH	LOCAL TIME	MEAN ALTITUDE
-137	33.32	30.57	5:31	1
-138	33.58	30.80	5:35	2
-139	33.84	31.04	5:39	3
-140	34.12	31.29	5:43	3
-142	34.70	31.81	5:51	5
-143	35.00	32.08	5:55	6
-144	35.32	32.36	6:00	7
-145	35.64	32.65	6:04	8
-147	36.32	33.26	6:13	10
-148	36.67	33.58	6:17	11
-149	37.03	33.90	6:22	12
-150	37.40	34.25	6:26	13
-152	38.18	34.95	6:36	15
-153	38.58	35.31	6:41	16

PERCENTAGE OF SUN OBSCURED ALONG MAP TRACK
WEST END =105.6   EAST END =106.0

===
SUNRISE POSITION :ALTITUDE IS -0.8 DEGREES
NORTH : LONGITUDE -194.4, LATITUDE 32.7
SOUTH : LONGITUDE -196.0, LATITUDE 30.3

+ MARKS THE CITY OF CH'ANG-AN

179

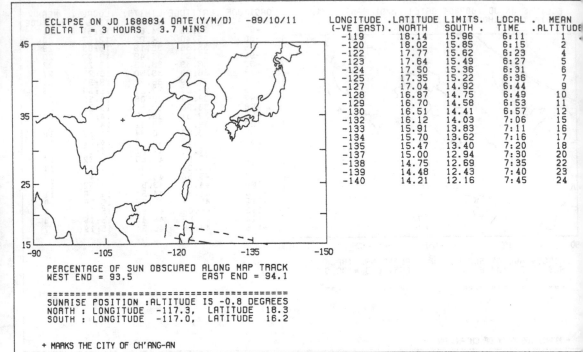

ECLIPSE ON JD 1688834 DATE(Y/M/D) -89/10/11
DELTA T = 3 HOURS   3.7 MINS

LONGITUDE	.LATITUDE	LIMITS.	LOCAL	. MEAN
(-VE EAST).	NORTH	SOUTH .	TIME	.ALTITUDE
-119	18.14	15.96	6:11	1
-120	18.02	15.85	6:15	2
-122	17.77	15.62	6:23	4
-123	17.64	15.49	6:27	5
-124	17.50	15.36	6:31	6
-125	17.35	15.22	6:36	7
-127	17.04	14.92	6:44	9
-128	16.87	14.75	6:49	10
-129	16.70	14.58	6:53	11
-130	16.51	14.41	6:57	12
-132	16.12	14.03	7:06	15
-133	15.91	13.83	7:11	16
-134	15.70	13.62	7:16	17
-135	15.47	13.40	7:20	18
-137	15.00	12.94	7:30	20
-138	14.75	12.69	7:35	22
-139	14.48	12.43	7:40	23
-140	14.21	12.16	7:45	24

PERCENTAGE OF SUN OBSCURED ALONG MAP TRACK
WEST END = 93.5   EAST END = 94.1

===
SUNRISE POSITION :ALTITUDE IS -0.8 DEGREES
NORTH : LONGITUDE -117.3,   LATITUDE   18.3
SOUTH : LONGITUDE -117.0,   LATITUDE   16.2

+ MARKS THE CITY OF CH'ANG-AN

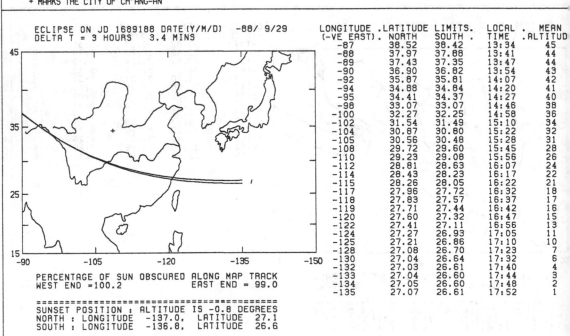

ECLIPSE ON JD 1689188 DATE(Y/M/D) -88/ 9/29
DELTA T = 3 HOURS   3.4 MINS

LONGITUDE	.LATITUDE	LIMITS.	LOCAL	. MEAN
(-VE EAST).	NORTH	SOUTH .	TIME	.ALTITUD
-87	38.52	38.42	13:34	45
-88	37.97	37.88	13:41	44
-89	37.43	37.35	13:47	44
-90	36.90	36.82	13:54	43
-92	35.87	35.81	14:07	42
-94	34.88	34.84	14:20	41
-95	34.41	34.37	14:27	40
-98	33.07	33.07	14:46	38
-100	32.27	32.25	14:58	36
-102	31.54	31.49	15:10	34
-104	30.87	30.80	15:22	32
-105	30.56	30.48	15:28	31
-108	29.72	29.60	15:45	28
-110	29.23	29.08	15:56	26
-112	28.81	28.63	16:07	24
-114	28.43	28.23	16:17	22
-115	28.26	28.05	16:22	21
-117	27.96	27.72	16:32	18
-118	27.83	27.57	16:37	17
-119	27.71	27.44	16:42	16
-120	27.60	27.32	16:47	15
-122	27.41	27.11	16:56	13
-124	27.27	26.93	17:05	11
-125	27.21	26.86	17:10	10
-128	27.08	26.70	17:23	7
-130	27.04	26.64	17:32	6
-132	27.03	26.61	17:40	4
-133	27.04	26.60	17:44	3
-134	27.05	26.60	17:48	2
-135	27.07	26.61	17:52	1

PERCENTAGE OF SUN OBSCURED ALONG MAP TRACK
WEST END =100.2   EAST END = 99.0

===
SUNSET POSITION : ALTITUDE IS -0.8 DEGREES
NORTH : LONGITUDE -137.0,   LATITUDE   27.1
SOUTH : LONGITUDE -136.8,   LATITUDE   26.6

+ MARKS THE CITY OF CH'ANG-AN

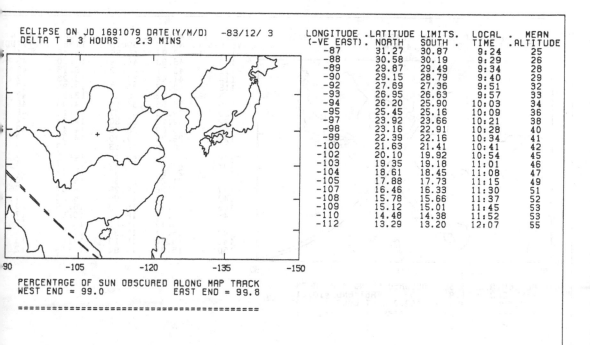

ECLIPSE ON JD 1691079 DATE(Y/M/D) -83/12/ 3
DELTA T = 3 HOURS   2.3 MINS

LONGITUDE (-VE EAST)	LATITUDE LIMITS. NORTH	SOUTH	LOCAL TIME	MEAN ALTITUDE
-87	31.27	30.87	9:24	25
-88	30.58	30.19	9:29	26
-89	29.87	29.49	9:34	28
-90	29.15	28.79	9:40	29
-92	27.69	27.36	9:51	32
-93	26.95	26.63	9:57	33
-94	26.20	25.90	10:03	34
-95	25.45	25.16	10:09	36
-97	23.92	23.66	10:21	38
-98	23.16	22.91	10:28	40
-99	22.39	22.16	10:34	41
-100	21.63	21.41	10:41	42
-102	20.10	19.92	10:54	45
-103	19.35	19.18	11:01	46
-104	18.61	18.45	11:08	47
-105	17.88	17.73	11:15	49
-107	16.46	16.33	11:30	51
-108	15.78	15.66	11:37	52
-109	15.12	15.01	11:45	53
-110	14.48	14.38	11:52	53
-112	13.29	13.20	12:07	55

PERCENTAGE OF SUN OBSCURED ALONG MAP TRACK
WEST END = 99.0   EAST END = 99.8

==

+ MARKS THE CITY OF CH'ANG-AN

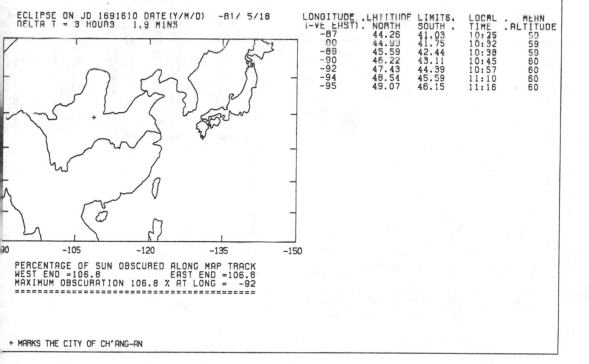

ECLIPSE ON JD 1691610 DATE(Y/M/D) -81/ 5/18
DELTA T = 3 HOURS   1.9 MINS

LONGITUDE (-VE EAST)	LATITUDE LIMITS. NORTH	SOUTH	LOCAL TIME	MEAN ALTITUDE
-87	44.26	41.03	10:25	59
-88	44.93	41.75	10:32	59
-89	45.59	42.44	10:38	59
-90	46.22	43.11	10:45	60
-92	47.43	44.39	10:57	60
-94	48.54	45.59	11:10	60
-95	49.07	46.15	11:16	60

PERCENTAGE OF SUN OBSCURED ALONG MAP TRACK
WEST END =106.8   EAST END =106.8
MAXIMUM OBSCURATION 106.8 % AT LONG = -92
==

+ MARKS THE CITY OF CH'ANG-AN

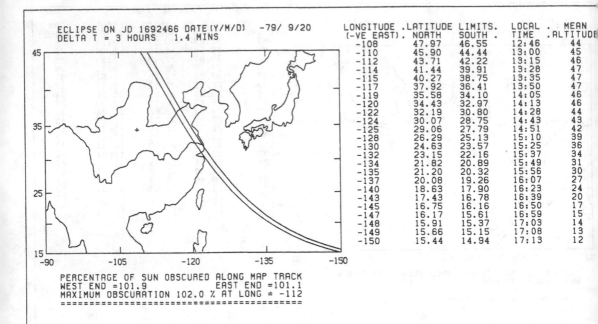

ECLIPSE ON JD 1692466 DATE(Y/M/D) -79/ 9/20
DELTA T = 3 HOURS   1.4 MINS

LONGITUDE	.LATITUDE	LIMITS.	LOCAL	. MEAN
(-VE EAST).	NORTH	SOUTH .	TIME	.ALTITUDE
-108	47.97	46.55	12:46	44
-110	45.90	44.44	13:00	45
-112	43.71	42.22	13:15	46
-114	41.44	39.91	13:28	47
-115	40.27	38.75	13:35	47
-117	37.92	36.41	13:50	47
-119	35.58	34.10	14:05	46
-120	34.43	32.97	14:13	46
-122	32.19	30.80	14:28	44
-124	30.07	28.75	14:43	43
-125	29.06	27.79	14:51	42
-128	26.29	25.13	15:10	39
-130	24.63	23.57	15:25	36
-132	23.15	22.16	15:37	34
-134	21.82	20.89	15:49	31
-135	21.20	20.32	15:56	30
-137	20.08	19.26	16:07	27
-140	18.63	17.90	16:23	24
-143	17.43	16.78	16:39	20
-145	16.75	16.16	16:50	17
-147	16.17	15.61	16:59	15
-148	15.91	15.37	17:03	14
-149	15.66	15.15	17:08	13
-150	15.44	14.94	17:13	12

PERCENTAGE OF SUN OBSCURED ALONG MAP TRACK
WEST END =101.9   EAST END =101.1
MAXIMUM OBSCURATION 102.0 % AT LONG = -112
===

+ MARKS THE CITY OF CH'ANG-AN

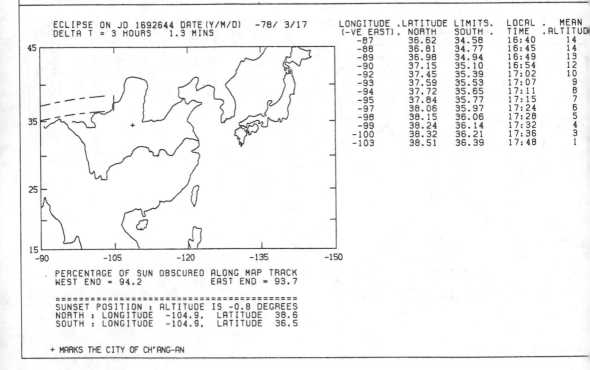

ECLIPSE ON JD 1692644 DATE(Y/M/D) -78/ 3/17
DELTA T = 3 HOURS   1.3 MINS

LONGITUDE	.LATITUDE	LIMITS.	LOCAL	. MEAN
(-VE EAST).	NORTH	SOUTH .	TIME	.ALTITUDE
-87	36.62	34.58	16:40	14
-88	36.81	34.77	16:45	14
-89	36.98	34.94	16:49	13
-90	37.15	35.10	16:54	12
-92	37.45	35.39	17:02	10
-93	37.59	35.53	17:07	9
-94	37.72	35.65	17:11	8
-95	37.84	35.77	17:15	7
-97	38.06	35.97	17:24	6
-98	38.15	36.06	17:28	5
-99	38.24	36.14	17:32	4
-100	38.32	36.21	17:36	3
-103	38.51	36.39	17:48	1

PERCENTAGE OF SUN OBSCURED ALONG MAP TRACK
WEST END = 94.2   EAST END = 93.7

===
SUNSET POSITION : ALTITUDE IS -0.8 DEGREES
NORTH : LONGITUDE -104.9, LATITUDE 38.6
SOUTH : LONGITUDE -104.9, LATITUDE 36.5

+ MARKS THE CITY OF CH'ANG-AN

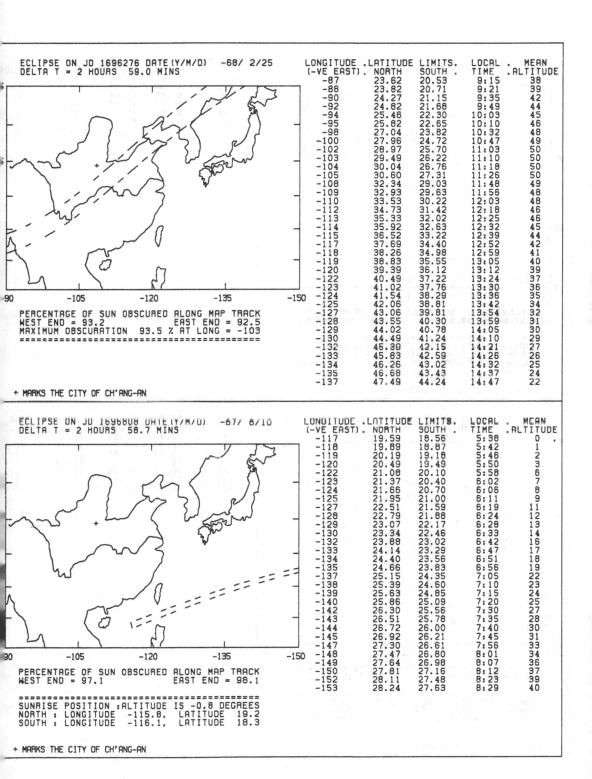

ECLIPSE ON JD 1696276 DATE(Y/M/D)   -68/ 2/25
DELTA T = 2 HOURS 59.0 MINS

LONGITUDE	.LATITUDE	LIMITS.	LOCAL	. MEAN
(-VE EAST).	NORTH	SOUTH .	TIME	.ALTITUDE
-87	23.62	20.53	9:15	38
-88	23.82	20.71	9:21	39
-90	24.27	21.15	9:35	42
-92	24.82	21.68	9:49	44
-94	25.46	22.30	10:03	45
-95	25.82	22.65	10:10	46
-98	27.04	23.82	10:32	48
-100	27.96	24.72	10:47	49
-102	28.97	25.70	11:03	50
-103	29.49	26.22	11:10	50
-104	30.04	26.76	11:18	50
-105	30.60	27.31	11:26	50
-108	32.34	29.03	11:48	49
-109	32.93	29.63	11:56	48
-110	33.53	30.22	12:03	48
-112	34.73	31.42	12:18	46
-113	35.33	32.02	12:25	46
-114	35.92	32.63	12:32	45
-115	36.52	33.22	12:39	44
-117	37.69	34.40	12:52	42
-118	38.26	34.98	12:59	41
-119	38.83	35.55	13:05	40
-120	39.39	36.12	13:12	39
-122	40.49	37.22	13:24	37
-123	41.02	37.76	13:30	36
-124	41.54	38.29	13:36	35
-125	42.06	38.81	13:42	34
-127	43.06	39.81	13:54	32
-128	43.55	40.30	13:59	31
-129	44.02	40.78	14:05	30
-130	44.49	41.24	14:10	29
-132	45.38	42.15	14:21	27
-133	45.83	42.59	14:26	26
-134	46.26	43.02	14:32	25
-135	46.68	43.43	14:37	24
-137	47.49	44.24	14:47	22

PERCENTAGE OF SUN OBSCURED ALONG MAP TRACK
WEST END = 93.2   EAST END = 92.5
MAXIMUM OBSCURATION 93.5 % AT LONG = -103
===

+ MARKS THE CITY OF CH'ANG-AN

ECLIPSE ON JD 1696808 DATE(Y/M/D)   -67/ 8/10
DELTA T = 2 HOURS 58.7 MINS

LONGITUDE	.LATITUDE	LIMITS.	LOCAL	. MEAN
(-VE EAST).	NORTH	SOUTH .	TIME	.ALTITUDE
-117	19.59	18.56	5:38	0 .
-118	19.89	18.87	5:42	1
-119	20.19	19.18	5:46	2
-120	20.49	19.49	5:50	3
-122	21.08	20.10	5:58	6
-123	21.37	20.40	6:02	7
-124	21.66	20.70	6:06	8
-125	21.95	21.00	6:11	9
-127	22.51	21.59	6:19	11
-128	22.79	21.88	6:24	12
-129	23.07	22.17	6:28	13
-130	23.34	22.46	6:33	14
-132	23.88	23.02	6:42	16
-133	24.14	23.29	6:47	17
-134	24.40	23.56	6:51	18
-135	24.66	23.83	6:56	19
-137	25.15	24.35	7:05	22
-138	25.39	24.60	7:10	23
-139	25.63	24.85	7:15	24
-140	25.86	25.09	7:20	25
-142	26.30	25.56	7:30	27
-143	26.51	25.78	7:35	28
-144	26.72	26.00	7:40	30
-145	26.92	26.21	7:45	31
-147	27.30	26.61	7:56	33
-148	27.47	26.80	8:01	34
-149	27.64	26.98	8:07	36
-150	27.81	27.16	8:12	37
-152	28.11	27.48	8:23	39
-153	28.24	27.63	8:29	40

PERCENTAGE OF SUN OBSCURED ALONG MAP TRACK
WEST END = 97.1   EAST END = 98.1

===
SUNRISE POSITION :ALTITUDE IS -0.8 DEGREES
NORTH : LONGITUDE -115.8, LATITUDE 19.2
SOUTH : LONGITUDE -116.1, LATITUDE 18.3

+ MARKS THE CITY OF CH'ANG-AN

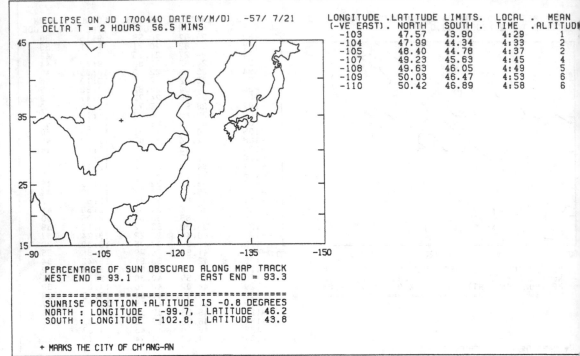

ECLIPSE ON JD 1700440 DATE(Y/M/D) -57/ 7/21
DELTA T = 2 HOURS 56.5 MINS

LONGITUDE	.LATITUDE	LIMITS.	LOCAL	. MEAN
(-VE EAST).	NORTH	SOUTH .	TIME	.ALTITUDE
-103	47.57	43.90	4:29	1
-104	47.99	44.34	4:33	2
-105	48.40	44.78	4:37	2
-107	49.23	45.63	4:45	4
-108	49.63	46.05	4:49	5
-109	50.03	46.47	4:53	6
-110	50.42	46.89	4:58	6

PERCENTAGE OF SUN OBSCURED ALONG MAP TRACK
WEST END = 93.1   EAST END = 93.3

==
SUNRISE POSITION :ALTITUDE IS -0.8 DEGREES
NORTH : LONGITUDE   -99.7, LATITUDE   46.2
SOUTH : LONGITUDE -102.8, LATITUDE   43.8

+ MARKS THE CITY OF CH'ANG-AN

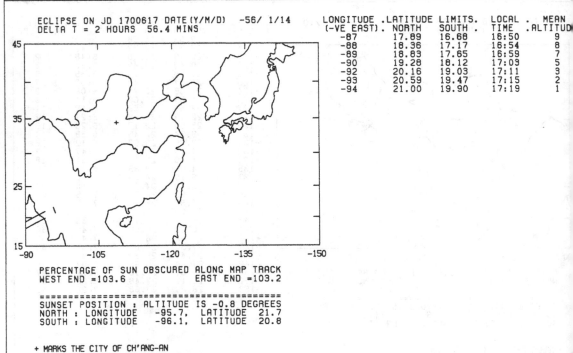

ECLIPSE ON JD 1700617 DATE(Y/M/D) -56/ 1/14
DELTA T = 2 HOURS 56.4 MINS

LONGITUDE	.LATITUDE	LIMITS.	LOCAL	. MEAN
(-VE EAST).	NORTH	SOUTH .	TIME	.ALTITUDE
-87	17.89	16.68	16:50	9
-88	18.36	17.17	16:54	8
-89	18.83	17.65	16:59	7
-90	19.28	18.12	17:03	5
-92	20.16	19.03	17:11	3
-93	20.59	19.47	17:15	2
-94	21.00	19.90	17:19	1

PERCENTAGE OF SUN OBSCURED ALONG MAP TRACK
WEST END =103.6   EAST END =103.2

==
SUNSET POSITION : ALTITUDE IS -0.8 DEGREES
NORTH : LONGITUDE   -95.7, LATITUDE   21.7
SOUTH : LONGITUDE   -96.1, LATITUDE   20.8

+ MARKS THE CITY OF CH'ANG-AN

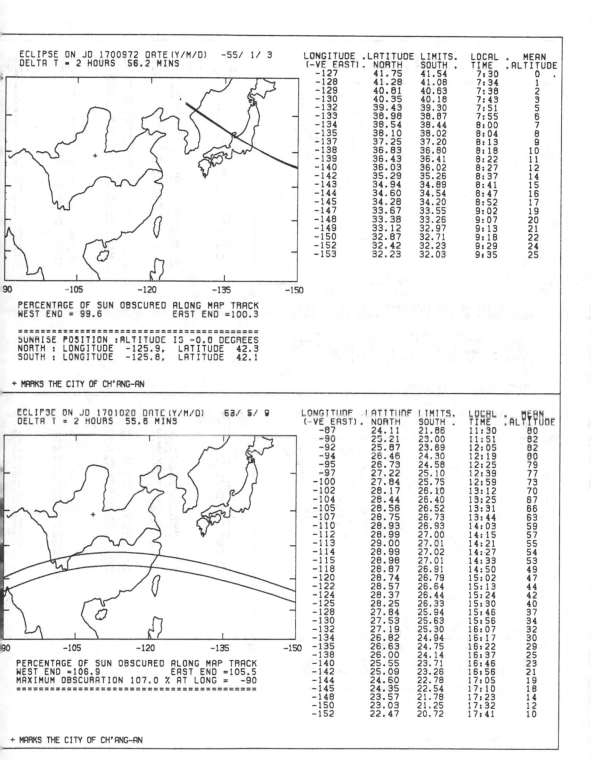

ECLIPSE ON JD 1700972 DATE(Y/M/D)   -55/ 1/ 3
DELTA T = 2 HOURS 56.2 MINS

LONGITUDE (-VE EAST) .	.LATITUDE LIMITS. NORTH SOUTH .	LOCAL TIME	MEAN .ALTITUDE .
-127	41.75	41.54	7:30	0
-128	41.28	41.08	7:34	1
-129	40.81	40.63	7:38	2
-130	40.35	40.18	7:43	3
-132	39.43	39.30	7:51	5
-133	38.98	38.87	7:55	6
-134	38.54	38.44	8:00	7
-135	38.10	38.02	8:04	8
-137	37.25	37.20	8:13	9
-138	36.83	36.80	8:18	10
-139	36.43	36.41	8:22	11
-140	36.03	36.02	8:27	12
-142	35.29	35.26	8:37	14
-143	34.94	34.89	8:41	15
-144	34.60	34.54	8:47	16
-145	34.28	34.20	8:52	17
-147	33.67	33.55	9:02	19
-148	33.38	33.26	9:07	20
-149	33.12	32.97	9:13	21
-150	32.87	32.71	9:18	22
-152	32.42	32.23	9:29	24
-153	32.23	32.03	9:35	25

PERCENTAGE OF SUN OBSCURED ALONG MAP TRACK
WEST END = 99.6   EAST END =100.3

==
SUNRISE POSITION :ALTITUDE IS -0.0 DEGREES
NORTH : LONGITUDE -125.9,   LATITUDE   42.3
SOUTH : LONGITUDE -125.8,   LATITUDE   42.1

+ MARKS THE CITY OF CH'ANG-AN

ECLIPSE ON JD 1701020 DATE(Y/M/D)   58/ 5/ 9
DELTA T = 2 HOURS 55.6 MINS

LONGITUDE (-VE EAST) .	.LATITUDE LIMITS. NORTH SOUTH .	LOCAL TIME	MEAN .ALTITUDE
-87	24.11	21.86	11:30	80
-90	25.21	23.00	11:51	82
-92	25.87	23.69	12:05	82
-94	26.46	24.30	12:19	80
-95	26.73	24.58	12:25	79
-97	27.22	25.10	12:39	77
-100	27.84	25.75	12:59	73
-102	28.17	26.10	13:12	70
-104	28.44	26.40	13:25	67
-105	28.56	26.52	13:31	66
-107	28.75	26.73	13:44	63
-110	28.93	26.93	14:03	59
-112	28.99	27.00	14:15	57
-113	29.00	27.01	14:21	55
-114	28.99	27.02	14:27	54
-115	28.98	27.01	14:33	53
-118	28.87	26.91	14:50	49
-120	28.74	26.79	15:02	47
-122	28.57	26.64	15:13	44
-124	28.37	26.44	15:24	42
-125	28.25	26.33	15:30	40
-128	27.84	25.94	15:46	37
-130	27.53	25.63	15:56	34
-132	27.19	25.30	16:07	32
-134	26.82	24.94	16:17	30
-135	26.63	24.75	16:22	29
-138	26.00	24.14	16:37	25
-140	25.55	23.71	16:46	23
-142	25.09	23.26	16:56	21
-144	24.60	22.78	17:05	19
-145	24.35	22.54	17:10	18
-148	23.57	21.78	17:23	14
-150	23.03	21.25	17:32	12
-152	22.47	20.72	17:41	10

PERCENTAGE OF SUN OBSCURED ALONG MAP TRACK
WEST END =106.9   EAST END =105.5
MAXIMUM OBSCURATION 107.0 % AT LONG =   -90
==

+ MARKS THE CITY OF CH'ANG-AN

185

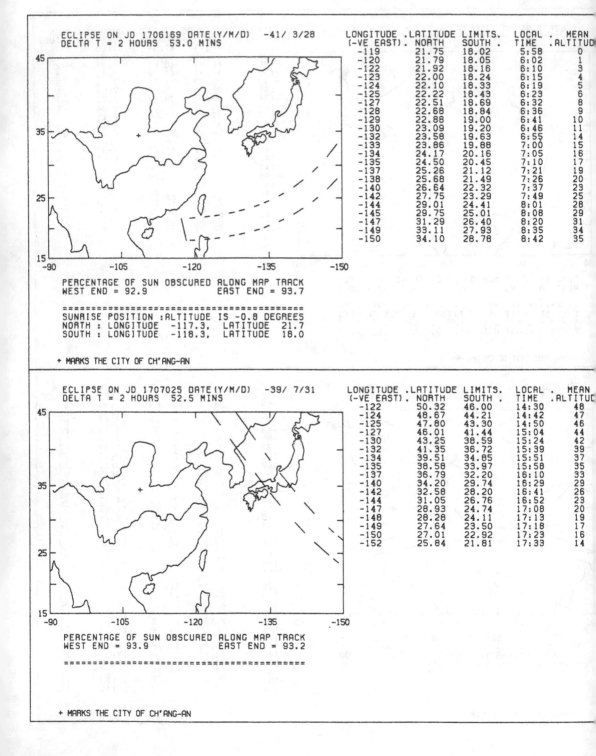

ECLIPSE ON JD 1706169 DATE(Y/M/D)   -41/ 3/28
DELTA T = 2 HOURS 53.0 MINS

LONGITUDE (-VE EAST)	.LATITUDE . NORTH	LIMITS. SOUTH .	LOCAL . TIME	MEAN .ALTITUDE
-119	21.75	18.02	5:58	0
-120	21.79	18.05	6:02	1
-122	21.92	18.16	6:10	3
-123	22.00	18.24	6:15	4
-124	22.10	18.33	6:19	5
-125	22.22	18.43	6:23	6
-127	22.51	18.69	6:32	8
-128	22.68	18.84	6:36	9
-129	22.88	19.00	6:41	10
-130	23.09	19.20	6:46	11
-132	23.58	19.63	6:55	14
-133	23.86	19.88	7:00	15
-134	24.17	20.16	7:05	16
-135	24.50	20.45	7:10	17
-137	25.26	21.12	7:21	19
-138	25.68	21.49	7:26	20
-140	26.64	22.32	7:37	23
-142	27.75	23.29	7:49	25
-144	29.01	24.41	8:01	28
-145	29.75	25.01	8:08	29
-147	31.29	26.40	8:20	31
-149	33.11	27.93	8:35	34
-150	34.10	28.78	8:42	35

PERCENTAGE OF SUN OBSCURED ALONG MAP TRACK
WEST END = 92.9   EAST END = 93.7

===
SUNRISE POSITION :ALTITUDE IS -0.8 DEGREES
NORTH : LONGITUDE -117.3, LATITUDE 21.7
SOUTH : LONGITUDE -118.3, LATITUDE 18.0

+ MARKS THE CITY OF CH'ANG-AN

ECLIPSE ON JD 1707025 DATE(Y/M/D)   -39/ 7/31
DELTA T = 2 HOURS 52.5 MINS

LONGITUDE (-VE EAST)	.LATITUDE . NORTH	LIMITS. SOUTH .	LOCAL . TIME	MEAN .ALTITUDE
-122	50.32	46.00	14:30	48
-124	48.67	44.21	14:42	47
-125	47.80	43.30	14:50	46
-127	46.01	41.44	15:04	44
-130	43.25	38.59	15:24	42
-132	41.35	36.72	15:39	39
-134	39.51	34.85	15:51	37
-135	38.58	33.97	15:58	35
-137	36.79	32.20	16:10	33
-140	34.20	29.74	16:29	29
-142	32.58	28.20	16:41	26
-144	31.05	26.76	16:52	23
-147	28.93	24.74	17:08	20
-148	28.28	24.11	17:13	19
-149	27.64	23.50	17:18	17
-150	27.01	22.92	17:23	16
-152	25.84	21.81	17:33	14

PERCENTAGE OF SUN OBSCURED ALONG MAP TRACK
WEST END = 93.9   EAST END = 93.2

===

+ MARKS THE CITY OF CH'ANG-AN

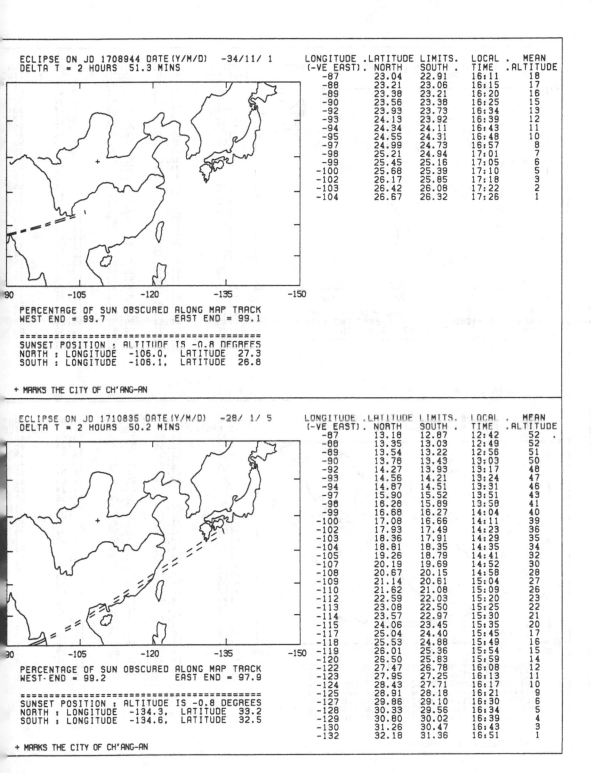

ECLIPSE ON JD 1708944 DATE(Y/M/D) -34/11/ 1
DELTA T = 2 HOURS 51.3 MINS

LONGITUDE	.LATITUDE	LIMITS.	LOCAL	. MEAN
(-VE EAST).	NORTH	SOUTH .	TIME	.ALTITUDE
-87	23.04	22.91	16:11	18
-88	23.21	23.06	16:15	17
-89	23.38	23.21	16:20	16
-90	23.56	23.38	16:25	15
-92	23.93	23.73	16:34	13
-93	24.13	23.92	16:39	12
-94	24.34	24.11	16:43	11
-95	24.55	24.31	16:48	10
-97	24.99	24.73	16:57	8
-98	25.21	24.94	17:01	7
-99	25.45	25.16	17:05	6
-100	25.68	25.39	17:10	5
-102	26.17	25.85	17:18	3
-103	26.42	26.08	17:22	2
-104	26.67	26.32	17:26	1

PERCENTAGE OF SUN OBSCURED ALONG MAP TRACK
WEST END = 99.7   EAST END = 99.1

==
SUNSET POSITION : ALTITUDE IS -0.8 DEGREES
NORTH : LONGITUDE -106.0, LATITUDE 27.3
SOUTH : LONGITUDE -106.1, LATITUDE 26.8

+ MARKS THE CITY OF CH'ANG-AN

ECLIPSE ON JD 1710835 DATE(Y/M/D) -28/ 1/ 5
DELTA T = 2 HOURS 50.2 MINS

LONGITUDE	.LATITUDE	LIMITS.	LOCAL	. MEAN
(-VE EAST).	NORTH	SOUTH .	TIME	.ALTITUDE
-87	13.18	12.87	12:42	52
-88	13.35	13.03	12:49	52
-89	13.54	13.22	12:56	51
-90	13.76	13.43	13:03	50
-92	14.27	13.93	13:17	48
-93	14.56	14.21	13:24	47
-94	14.87	14.51	13:31	46
-97	15.90	15.52	13:51	43
-98	16.28	15.89	13:58	41
-99	16.68	16.27	14:04	40
-100	17.08	16.66	14:11	39
-102	17.93	17.49	14:23	36
-103	18.36	17.91	14:29	35
-104	18.81	18.35	14:35	34
-105	19.26	18.79	14:41	32
-107	20.19	19.69	14:52	30
-108	20.67	20.15	14:58	28
-109	21.14	20.61	15:04	27
-110	21.62	21.08	15:09	26
-112	22.59	22.03	15:20	23
-113	23.08	22.50	15:25	22
-114	23.57	22.97	15:30	21
-115	24.06	23.45	15:35	20
-117	25.04	24.40	15:45	17
-118	25.53	24.88	15:49	16
-119	26.01	25.36	15:54	15
-120	26.50	25.83	15:59	14
-122	27.47	26.78	16:08	12
-123	27.95	27.25	16:13	11
-124	28.43	27.71	16:17	10
-125	28.91	28.18	16:21	9
-127	29.86	29.10	16:30	6
-128	30.33	29.56	16:34	5
-129	30.80	30.02	16:39	4
-130	31.26	30.47	16:43	3
-132	32.18	31.36	16:51	1

PERCENTAGE OF SUN OBSCURED ALONG MAP TRACK
WEST-END = 99.2   EAST END = 97.9

==
SUNSET POSITION : ALTITUDE IS -0.8 DEGREES
NORTH : LONGITUDE -134.3, LATITUDE 33.2
SOUTH : LONGITUDE -134.6, LATITUDE 32.5

+ MARKS THE CITY OF CH'ANG-AN

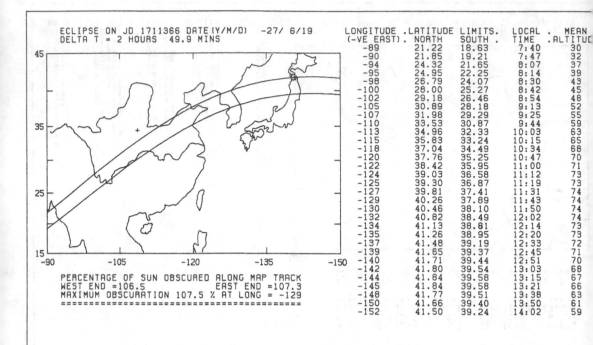

ECLIPSE ON JD 1711366 DATE(Y/M/D) -27/ 6/19
DELTA T = 2 HOURS 49.9 MINS

LONGITUDE (-VE EAST)	LATITUDE LIMITS. NORTH	LATITUDE LIMITS. SOUTH	LOCAL TIME	MEAN ALTITUDE
-89	21.22	18.63	7:40	30
-90	21.85	19.21	7:47	32
-94	24.32	21.65	8:07	37
-95	24.95	22.25	8:14	39
-98	26.79	24.07	8:30	43
-100	28.00	25.27	8:42	45
-102	29.18	26.46	8:54	48
-105	30.89	28.18	9:13	52
-107	31.98	29.29	9:25	55
-110	33.53	30.87	9:44	59
-113	34.96	32.33	10:03	63
-115	35.83	33.24	10:15	65
-118	37.04	34.49	10:34	68
-120	37.76	35.25	10:47	70
-122	38.42	35.95	11:00	71
-124	39.03	36.58	11:12	73
-125	39.30	36.87	11:19	73
-127	39.81	37.41	11:31	74
-129	40.26	37.89	11:43	74
-130	40.46	38.10	11:50	74
-132	40.82	38.49	12:02	74
-134	41.13	38.81	12:14	73
-135	41.26	38.95	12:20	73
-137	41.48	39.19	12:33	72
-139	41.65	39.37	12:45	71
-140	41.71	39.44	12:51	70
-142	41.80	39.54	13:03	68
-144	41.84	39.58	13:15	67
-145	41.84	39.58	13:21	66
-148	41.77	39.51	13:38	63
-150	41.66	39.40	13:50	61
-152	41.50	39.24	14:02	59

PERCENTAGE OF SUN OBSCURED ALONG MAP TRACK
WEST END =106.5   EAST END =107.3
MAXIMUM OBSCURATION 107.5 % AT LONG = -129
==

+ MARKS THE CITY OF CH'ANG-AN

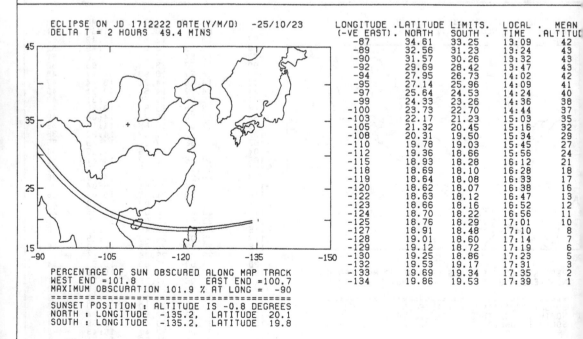

ECLIPSE ON JD 1712222 DATE(Y/M/D) -25/10/23
DELTA T = 2 HOURS 49.4 MINS

LONGITUDE (-VE EAST)	LATITUDE LIMITS. NORTH	LATITUDE LIMITS. SOUTH	LOCAL TIME	MEAN ALTITUDE
-87	34.61	33.25	13:09	42
-89	32.56	31.23	13:24	43
-90	31.57	30.26	13:32	43
-92	29.69	28.42	13:47	43
-94	27.95	26.73	14:02	42
-95	27.14	25.96	14:09	41
-97	25.64	24.53	14:24	40
-99	24.33	23.26	14:36	38
-100	23.73	22.70	14:44	37
-103	22.17	21.23	15:03	35
-105	21.32	20.45	15:16	32
-108	20.31	19.50	15:34	29
-110	19.78	19.03	15:45	27
-112	19.36	18.66	15:56	24
-115	18.93	18.28	16:12	21
-118	18.69	18.10	16:28	18
-119	18.64	18.08	16:33	17
-120	18.62	18.07	16:38	16
-122	18.63	18.12	16:47	13
-123	18.66	18.16	16:52	12
-124	18.70	18.22	16:56	11
-125	18.76	18.29	17:01	10
-127	18.91	18.48	17:10	8
-128	19.01	18.60	17:14	7
-129	19.12	18.72	17:19	6
-130	19.25	18.86	17:23	5
-132	19.53	19.17	17:31	3
-133	19.69	19.34	17:35	2
-134	19.86	19.53	17:39	1

PERCENTAGE OF SUN OBSCURED ALONG MAP TRACK
WEST END =101.8   EAST END =100.7
MAXIMUM OBSCURATION 101.9 % AT LONG = -90
==
SUNSET POSITION : ALTITUDE IS -0.8 DEGREES
NORTH : LONGITUDE -135.2, LATITUDE 20.1
SOUTH : LONGITUDE -135.2, LATITUDE 19.8

+ MARKS THE CITY OF CH'ANG-AN

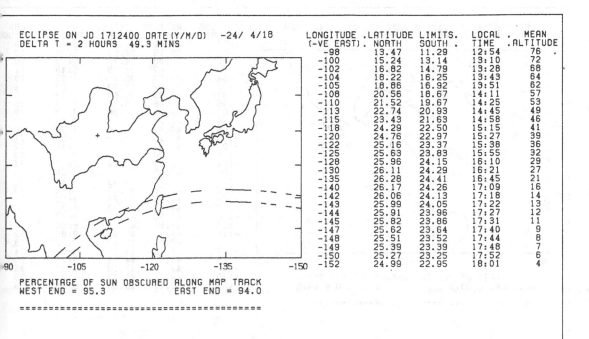

ECLIPSE ON JD 1712400 DATE(Y/M/D)   -24/ 4/18
DELTA T = 2 HOURS 49.3 MINS

LONGITUDE (-VE EAST)	LATITUDE NORTH	LIMITS SOUTH	LOCAL TIME	MEAN ALTITUDE
-98	13.47	11.29	12:54	76
-100	15.24	13.14	13:10	72
-102	16.82	14.79	13:28	68
-104	18.22	16.25	13:43	64
-105	18.86	16.92	13:51	62
-108	20.56	18.67	14:11	57
-110	21.52	19.67	14:25	53
-113	22.74	20.93	14:45	49
-115	23.43	21.63	14:58	46
-118	24.29	22.50	15:15	41
-120	24.76	22.97	15:27	39
-122	25.16	23.37	15:38	36
-125	25.63	23.83	15:55	32
-128	25.96	24.15	16:10	29
-130	26.11	24.29	16:21	27
-135	26.28	24.41	16:45	21
-140	26.17	24.26	17:09	16
-142	26.06	24.13	17:18	14
-143	25.99	24.05	17:22	13
-144	25.91	23.96	17:27	12
-145	25.82	23.86	17:31	11
-147	25.62	23.64	17:40	9
-148	25.51	23.52	17:44	8
-149	25.39	23.39	17:48	7
-150	25.27	23.25	17:52	6
-152	24.99	22.95	18:01	4

PERCENTAGE OF SUN OBSCURED ALONG MAP TRACK
WEST END = 95.3   EAST END = 94.0

==

+ MARKS THE CITY OF CH'ANG-AN

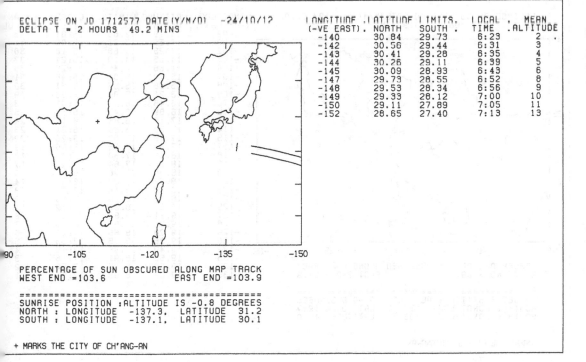

ECLIPSE ON JD 1712577 DATE(Y/M/D)   -24/10/12
DELTA T = 2 HOURS 49.2 MINS

LONGITUDE (-VE EAST)	LATITUDE NORTH	LIMITS SOUTH	LOCAL TIME	MEAN ALTITUDE
-140	30.84	29.73	6:23	2
-142	30.56	29.44	6:31	3
-143	30.41	29.28	6:35	4
-144	30.26	29.11	6:39	5
-145	30.09	28.93	6:43	6
-147	29.73	28.55	6:52	8
-148	29.53	28.34	6:56	9
-149	29.33	28.12	7:00	10
-150	29.11	27.89	7:05	11
-152	28.65	27.40	7:13	13

PERCENTAGE OF SUN OBSCURED ALONG MAP TRACK
WEST END =103.6   EAST END =103.9

==
SUNRISE POSITION ;ALTITUDE IS -0.8 DEGREES
NORTH : LONGITUDE -137.3, LATITUDE 31.2
SOUTH : LONGITUDE -137.1, LATITUDE 30.1

+ MARKS THE CITY OF CH'ANG-AN

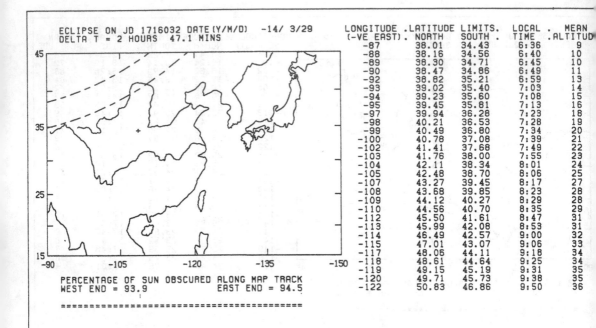

ECLIPSE ON JD 1716032 DATE(Y/M/D) -14/ 3/29
DELTA T = 2 HOURS 47.1 MINS

LONGITUDE (-VE EAST)	LATITUDE LIMITS. NORTH	SOUTH	LOCAL TIME	MEAN ALTITUDE
-87	38.01	34.43	6:36	9
-88	38.16	34.56	6:40	10
-89	38.30	34.71	6:45	10
-90	38.47	34.86	6:49	11
-92	38.82	35.21	6:59	13
-93	39.02	35.40	7:03	14
-94	39.23	35.60	7:08	15
-95	39.45	35.81	7:13	16
-97	39.94	36.28	7:23	18
-98	40.21	36.53	7:28	19
-99	40.49	36.80	7:34	20
-100	40.78	37.08	7:39	21
-102	41.41	37.68	7:49	22
-103	41.76	38.00	7:55	23
-104	42.11	38.34	8:01	24
-105	42.48	38.70	8:06	25
-107	43.27	39.45	8:17	27
-108	43.68	39.85	8:23	28
-109	44.12	40.27	8:29	28
-110	44.56	40.70	8:35	29
-112	45.50	41.61	8:47	31
-113	45.99	42.08	8:53	31
-114	46.49	42.57	9:00	32
-115	47.01	43.07	9:06	33
-117	48.06	44.11	9:18	34
-118	48.61	44.64	9:25	34
-119	49.15	45.19	9:31	35
-120	49.71	45.73	9:38	35
-122	50.83	46.86	9:50	36

PERCENTAGE OF SUN OBSCURED ALONG MAP TRACK
WEST END = 93.9   EAST END = 94.5

==

+ MARKS THE CITY OF CH'ANG-AN

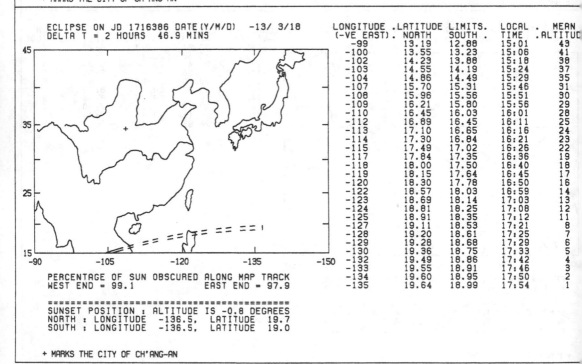

ECLIPSE ON JD 1716386 DATE(Y/M/D) -13/ 3/18
DELTA T = 2 HOURS 46.9 MINS

LONGITUDE (-VE EAST)	LATITUDE LIMITS. NORTH	SOUTH	LOCAL TIME	MEAN ALTITUDE
-99	13.19	12.88	15:01	43
-100	13.55	13.23	15:06	41
-102	14.23	13.88	15:18	38
-103	14.55	14.19	15:24	37
-104	14.86	14.49	15:29	35
-107	15.70	15.31	15:46	31
-108	15.96	15.56	15:51	30
-109	16.21	15.80	15:56	29
-110	16.45	16.03	16:01	28
-112	16.89	16.45	16:11	25
-113	17.10	16.65	16:16	24
-114	17.30	16.84	16:21	23
-115	17.49	17.02	16:26	22
-117	17.84	17.35	16:36	19
-118	18.00	17.50	16:40	18
-119	18.15	17.64	16:45	17
-120	18.30	17.78	16:50	16
-122	18.57	18.03	16:59	14
-123	18.69	18.14	17:03	13
-124	18.81	18.25	17:08	12
-125	18.91	18.35	17:12	11
-127	19.11	18.53	17:21	8
-128	19.20	18.61	17:25	7
-129	19.28	18.68	17:29	6
-130	19.36	18.75	17:33	5
-132	19.49	18.86	17:42	4
-133	19.55	18.91	17:46	3
-134	19.60	18.95	17:50	2
-135	19.64	18.99	17:54	1

PERCENTAGE OF SUN OBSCURED ALONG MAP TRACK
WEST END = 99.1   EAST END = 97.9

==
SUNSET POSITION : ALTITUDE IS -0.8 DEGREES
NORTH : LONGITUDE -136.5, LATITUDE 19.7
SOUTH : LONGITUDE -136.5, LATITUDE 19.0

+ MARKS THE CITY OF CH'ANG-AN

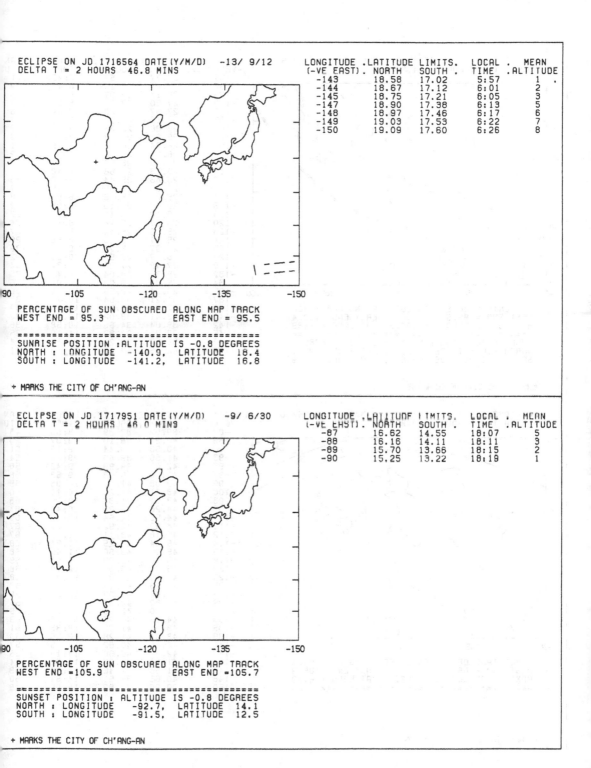

ECLIPSE ON JD 1716564 DATE(Y/M/D)   -13/ 9/12
DELTA T = 2 HOURS 46.8 MINS

LONGITUDE (-VE EAST).	LATITUDE LIMITS.		LOCAL TIME	MEAN ALTITUDE
	NORTH	SOUTH		
-143	18.58	17.02	5:57	1
-144	18.67	17.12	6:01	2
-145	18.75	17.21	6:05	3
-147	18.90	17.38	6:13	5
-148	18.97	17.46	6:17	6
-149	19.03	17.53	6:22	7
-150	19.09	17.60	6:26	8

90   -105   -120   -135   -150

PERCENTAGE OF SUN OBSCURED ALONG MAP TRACK
WEST END = 95.3   EAST END = 95.5

===
SUNRISE POSITION :ALTITUDE IS -0.8 DEGREES
NORTH : LONGITUDE -140.9, LATITUDE 18.4
SOUTH : LONGITUDE -141.2, LATITUDE 16.8

+ MARKS THE CITY OF CH'ANG-AN

ECLIPSE ON JD 1717951 DATE(Y/M/D)   -9/ 6/30
DELTA T = 2 HOURS 46.0 MINS

LONGITUDE (-VE EAST).	LATITUDE LIMITS.		LOCAL TIME	MEAN ALTITUDE
	NORTH	SOUTH		
-87	16.62	14.55	18:07	5
-88	16.16	14.11	18:11	3
-89	15.70	13.66	18:15	2
-90	15.25	13.22	18:19	1

90   -105   -120   -135   -150

PERCENTAGE OF SUN OBSCURED ALONG MAP TRACK
WEST END =105.9   EAST END =105.7

===
SUNSET POSITION : ALTITUDE IS -0.8 DEGREES
NORTH : LONGITUDE   -92.7, LATITUDE 14.1
SOUTH : LONGITUDE   -91.5, LATITUDE 12.5

+ MARKS THE CITY OF CH'ANG-AN

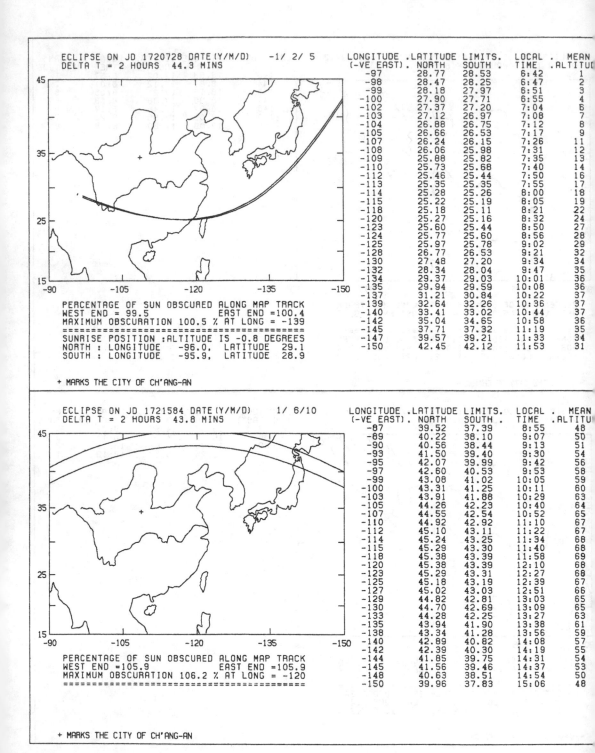

ECLIPSE ON JD 1720728 DATE(Y/M/D)   -1/ 2/ 5
DELTA T = 2 HOURS 44.3 MINS

LONGITUDE (-VE EAST)	LATITUDE NORTH	LIMITS. SOUTH	LOCAL TIME	MEAN ALTITUDE
-97	28.77	28.53	6:42	1
-98	28.47	28.25	6:47	2
-99	28.18	27.97	6:51	3
-100	27.90	27.71	6:55	4
-102	27.37	27.20	7:04	6
-103	27.12	26.97	7:08	7
-104	26.88	26.75	7:12	8
-105	26.66	26.53	7:17	9
-107	26.24	26.15	7:26	11
-108	26.06	25.98	7:31	12
-109	25.88	25.82	7:35	13
-110	25.73	25.68	7:40	14
-112	25.46	25.44	7:50	16
-113	25.35	25.35	7:55	17
-114	25.28	25.26	8:00	18
-115	25.22	25.19	8:05	19
-118	25.18	25.11	8:21	22
-120	25.27	25.16	8:32	24
-123	25.60	25.44	8:50	27
-124	25.77	25.60	8:56	28
-125	25.97	25.78	9:02	29
-128	26.77	26.53	9:21	32
-130	27.48	27.20	9:34	34
-132	28.34	28.04	9:47	35
-134	29.37	29.03	10:01	36
-135	29.94	29.59	10:08	36
-137	31.21	30.84	10:22	37
-139	32.64	32.26	10:36	37
-140	33.41	33.02	10:44	37
-142	35.04	34.65	10:58	36
-145	37.71	37.32	11:19	35
-147	39.57	39.21	11:33	34
-150	42.45	42.12	11:53	31

PERCENTAGE OF SUN OBSCURED ALONG MAP TRACK
WEST END = 99.5   EAST END =100.4
MAXIMUM OBSCURATION 100.5 % AT LONG = -139
==
SUNRISE POSITION :ALTITUDE IS -0.8 DEGREES
NORTH : LONGITUDE   -96.0, LATITUDE   29.1
SOUTH : LONGITUDE   -95.9, LATITUDE   28.9

+ MARKS THE CITY OF CH'ANG-AN

ECLIPSE ON JD 1721584 DATE(Y/M/D)   1/ 6/10
DELTA T = 2 HOURS 43.8 MINS

LONGITUDE (-VE EAST)	LATITUDE NORTH	LIMITS. SOUTH	LOCAL TIME	MEAN ALTITUDE
-87	39.52	37.39	8:55	48
-89	40.22	38.10	9:07	50
-90	40.56	38.44	9:13	51
-93	41.50	39.40	9:30	54
-95	42.07	39.99	9:42	56
-97	42.60	40.53	9:53	58
-99	43.08	41.02	10:05	59
-100	43.31	41.25	10:11	60
-103	43.91	41.88	10:29	63
-105	44.26	42.23	10:40	64
-107	44.55	42.54	10:52	65
-110	44.92	42.92	11:10	67
-112	45.10	43.11	11:22	67
-114	45.24	43.25	11:34	68
-115	45.29	43.30	11:40	68
-118	45.38	43.39	11:58	69
-120	45.38	43.39	12:10	68
-123	45.29	43.31	12:27	68
-125	45.18	43.19	12:39	67
-127	45.02	43.03	12:51	66
-129	44.82	42.81	13:03	65
-130	44.70	42.69	13:09	65
-133	44.28	42.25	13:27	63
-135	43.94	41.90	13:38	61
-138	43.34	41.28	13:56	59
-140	42.89	40.82	14:08	57
-142	42.39	40.30	14:19	55
-144	41.85	39.75	14:31	54
-145	41.56	39.46	14:37	53
-148	40.63	38.51	14:54	50
-150	39.96	37.83	15:06	48

PERCENTAGE OF SUN OBSCURED ALONG MAP TRACK
WEST END =105.9   EAST END =105.9
MAXIMUM OBSCURATION 106.2 % AT LONG = -120
==

+ MARKS THE CITY OF CH'ANG-AN

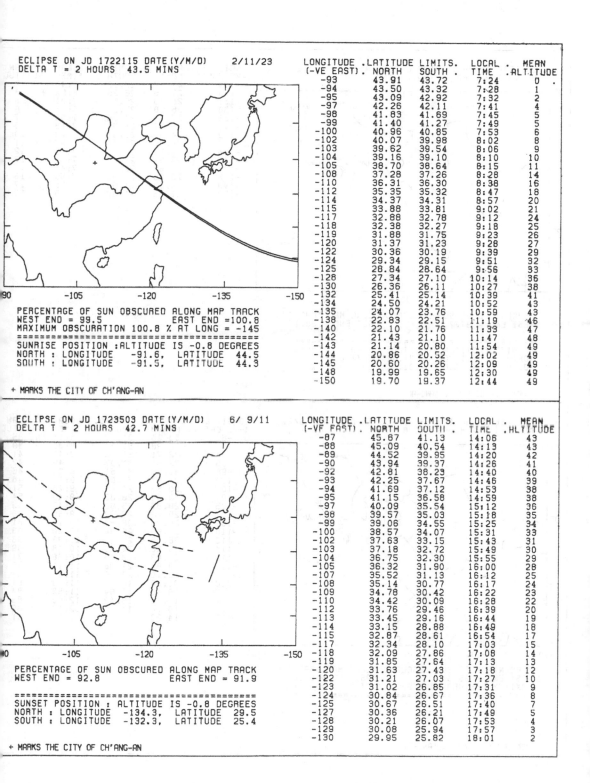

ECLIPSE ON JD 1722115 DATE(Y/M/D)   2/11/23
DELTA T = 2 HOURS 43.5 MINS

LONGITUDE (-VE EAST)	LATITUDE LIMITS. NORTH	SOUTH	LOCAL TIME	MEAN ALTITUDE
-93	43.91	43.72	7:24	0
-94	43.50	43.32	7:28	1
-95	43.09	42.92	7:32	2
-97	42.26	42.11	7:41	4
-98	41.83	41.69	7:45	5
-99	41.40	41.27	7:49	5
-100	40.96	40.85	7:53	6
-102	40.07	39.98	8:02	8
-103	39.62	39.54	8:06	9
-104	39.16	39.10	8:10	10
-105	38.70	38.64	8:15	11
-108	37.28	37.26	8:28	14
-110	36.31	36.30	8:38	16
-112	35.35	35.32	8:47	18
-114	34.37	34.31	8:57	20
-115	33.88	33.81	9:02	21
-117	32.88	32.78	9:12	24
-118	32.38	32.27	9:18	25
-119	31.88	31.75	9:23	26
-120	31.37	31.23	9:28	27
-122	30.36	30.19	9:39	29
-124	29.34	29.15	9:51	32
-125	28.84	28.64	9:56	33
-128	27.34	27.10	10:14	36
-130	26.36	26.11	10:27	38
-132	25.41	25.14	10:39	41
-134	24.50	24.21	10:52	43
-135	24.07	23.76	10:59	43
-138	22.83	22.51	11:19	46
-140	22.10	21.76	11:33	47
-142	21.43	21.10	11:47	48
-143	21.14	20.80	11:54	49
-144	20.86	20.52	12:02	49
-145	20.60	20.26	12:09	49
-148	19.99	19.65	12:30	49
-150	19.70	19.37	12:44	49

PERCENTAGE OF SUN OBSCURED ALONG MAP TRACK
WEST END = 99.5   EAST END =100.8
MAXIMUM OBSCURATION 100.8 % AT LONG = -145
==
SUNRISE POSITION :ALTITUDE IS -0.8 DEGREES
NORTH : LONGITUDE   -91.6, LATITUDE   44.5
SOUTH : LONGITUDE   -91.5, LATITUDE   44.3

+ MARKS THE CITY OF CH'ANG-AN

ECLIPSE ON JD 1723503 DATE(Y/M/D)   6/ 9/11
DELTA T = 2 HOURS 42.7 MINS

LONGITUDE (-VE EAST)	LATITUDE LIMITS. NORTH	SOUTH	LOCAL TIME	MEAN ALTITUDE
-87	45.87	41.13	14:06	43
-88	45.09	40.54	14:13	43
-89	44.52	39.95	14:20	42
-90	43.94	39.37	14:26	41
-92	42.81	38.29	14:40	40
-93	42.25	37.67	14:46	39
-94	41.69	37.12	14:53	38
-95	41.15	36.58	14:59	38
-97	40.09	35.54	15:12	36
-98	39.57	35.03	15:18	35
-99	39.06	34.55	15:25	34
-100	38.57	34.07	15:31	33
-102	37.63	33.15	15:43	31
-103	37.18	32.72	15:49	30
-104	36.75	32.30	15:55	29
-105	36.32	31.90	16:00	28
-107	35.52	31.13	16:12	25
-108	35.14	30.77	16:17	24
-109	34.78	30.42	16:22	23
-110	34.42	30.09	16:28	22
-112	33.76	29.46	16:39	20
-113	33.45	29.16	16:44	19
-114	33.15	28.88	16:49	18
-115	32.87	28.61	16:54	17
-117	32.34	28.10	17:03	15
-118	32.09	27.86	17:08	14
-119	31.85	27.64	17:13	13
-120	31.63	27.43	17:18	12
-122	31.21	27.03	17:27	10
-123	31.02	26.85	17:31	9
-124	30.84	26.67	17:36	8
-125	30.67	26.51	17:40	7
-127	30.36	26.21	17:49	5
-128	30.21	26.07	17:53	4
-129	30.08	25.94	17:57	3
-130	29.95	25.82	18:01	2

PERCENTAGE OF SUN OBSCURED ALONG MAP TRACK
WEST END = 92.8   EAST END = 91.9

==
SUNSET POSITION : ALTITUDE IS -0.8 DEGREES
NORTH : LONGITUDE   -134.3, LATITUDE   29.5
SOUTH : LONGITUDE   -132.3, LATITUDE   25.4

+ MARKS THE CITY OF CH'ANG-AN

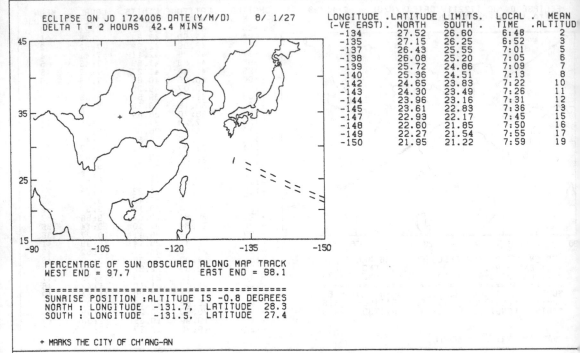

```
ECLIPSE ON JD 1724006 DATE(Y/M/D)    8/ 1/27
DELTA T = 2 HOURS  42.4 MINS
```

LONGITUDE (-VE EAST)	LATITUDE NORTH	LIMITS. SOUTH	LOCAL TIME	MEAN ALTITUD
-134	27.52	26.60	6:48	2
-135	27.15	26.25	6:52	3
-137	26.43	25.55	7:01	5
-138	26.08	25.20	7:05	6
-139	25.72	24.86	7:09	7
-140	25.36	24.51	7:13	8
-142	24.65	23.83	7:22	10
-143	24.30	23.49	7:26	11
-144	23.96	23.16	7:31	12
-145	23.61	22.83	7:36	13
-147	22.93	22.17	7:45	15
-148	22.60	21.85	7:50	16
-149	22.27	21.54	7:55	17
-150	21.95	21.22	7:59	19

```
PERCENTAGE OF SUN OBSCURED ALONG MAP TRACK
WEST END = 97.7              EAST END = 98.1

===========================================
SUNRISE POSITION :ALTITUDE IS -0.8 DEGREES
NORTH : LONGITUDE  -131.7,  LATITUDE  28.3
SOUTH : LONGITUDE  -131.5,  LATITUDE  27.4

+ MARKS THE CITY OF CH'ANG-AN
```

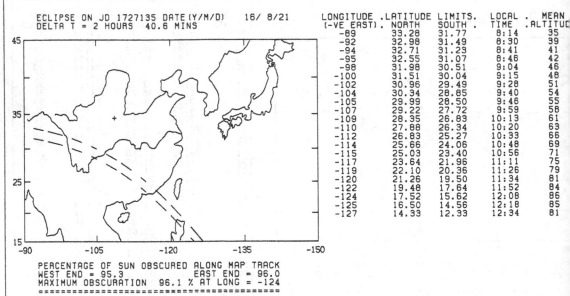

```
ECLIPSE ON JD 1727135 DATE(Y/M/D)   16/ 8/21
DELTA T = 2 HOURS  40.6 MINS
```

LONGITUDE (-VE EAST)	LATITUDE NORTH	LIMITS. SOUTH	LOCAL TIME	MEAN ALTITUD
-89	33.28	31.77	8:14	35
-92	32.98	31.49	8:30	39
-94	32.71	31.23	8:41	41
-95	32.55	31.07	8:46	42
-98	31.98	30.51	9:04	46
-100	31.51	30.04	9:15	48
-102	30.96	29.49	9:28	51
-104	30.34	28.85	9:40	54
-105	29.99	28.50	9:46	55
-107	29.22	27.72	9:59	58
-109	28.35	26.83	10:13	61
-110	27.88	26.34	10:20	63
-112	26.83	25.27	10:33	66
-114	25.66	24.06	10:48	69
-115	25.03	23.40	10:56	71
-117	23.64	21.96	11:11	75
-119	22.10	20.36	11:26	79
-120	21.26	19.50	11:34	81
-122	19.48	17.64	11:52	84
-124	17.52	15.62	12:08	86
-125	16.50	14.56	12:18	85
-127	14.33	12.33	12:34	81

```
PERCENTAGE OF SUN OBSCURED ALONG MAP TRACK
WEST END = 95.3              EAST END = 96.0
MAXIMUM OBSCURATION  96.1 % AT LONG = -124
===========================================

+ MARKS THE CITY OF CH'ANG-AN
```

ECLIPSE ON JD 1730591 DATE(Y/M/D)   26/ 2/ 6
DELTA T = 2 HOURS 38.6 MINS

LONGITUDE (-VE EAST)	LATITUDE LIMITS. NORTH	SOUTH	LOCAL TIME	MEAN ALTITUDE
-87	37.03	36.38	14:28	27
-88	37.53	36.85	14:34	25
-89	38.03	37.34	14:39	24
-90	38.53	37.82	14:44	23
-92	39.49	38.77	14:54	21
-93	39.96	39.23	14:59	20
-94	40.43	39.69	15:04	19
-95	40.89	40.13	15:09	18
-97	41.78	41.01	15:18	16
-98	42.22	41.43	15:23	15
-99	42.65	41.85	15:27	14
-100	43.07	42.27	15:32	14
-102	43.90	43.07	15:41	12
-103	44.30	43.47	15:45	11
-104	44.70	43.85	15:50	10
-105	45.08	44.23	15:54	9
-107	45.84	44.97	16:03	7
-108	46.21	45.33	16:07	7
-109	46.57	45.69	16:11	6
-110	46.93	46.03	16:15	5
-112	47.62	46.71	16:24	3

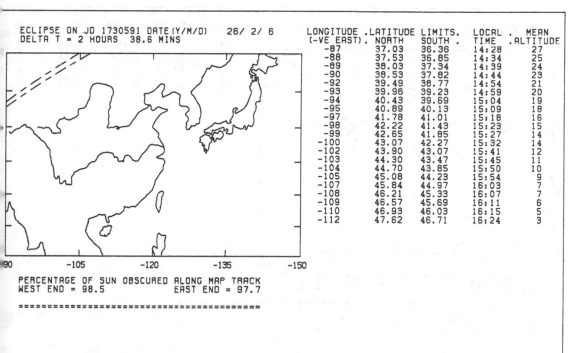

PERCENTAGE OF SUN OBSCURED ALONG MAP TRACK
WEST END = 98.5   EAST END = 97.7

==

+ MARKS THE CITY OF LO-YANG

ECLIPSE ON JD 1731122 DATE(Y/M/D)   27/ 7/22
DELTA T = 2 HOURS 38.3 MINS

LONGITUDE (-VE EAST)	LATITUDE LIMITS. NORTH	SOUTH	LOCAL TIME	MEAN ALTITUDE
-98	14.10	12.11	6:41	14
-99	14.51	12.50	6:45	15
-100	14.91	12.89	6:50	16
-102	15.71	13.68	6:59	19
-103	16.11	14.06	7:04	20
-104	16.51	14.45	7:08	21
-105	16.90	14.84	7:13	22
-107	17.68	15.60	7:23	25
-108	18.07	15.98	7:28	26
-109	18.45	16.35	7:33	27
-110	18.83	16.72	7:38	29
-112	19.57	17.45	7:48	31
-113	19.93	17.80	7:54	33
-114	20.29	18.16	7:59	34
-115	20.64	18.50	8:04	35
-117	21.32	19.18	8:15	38
-118	21.65	19.50	8:20	39
-119	21.98	19.82	8:26	40
-120	22.29	20.13	8:32	42
-122	22.90	20.74	8:43	44
-123	23.19	21.02	8:49	46
-124	23.47	21.30	8:54	47
-125	23.74	21.57	9:00	49
-127	24.25	22.08	9:12	51
-128	24.49	22.31	9:18	53
-129	24.71	22.54	9:24	54
-130	24.93	22.76	9:30	56
-132	25.32	23.15	9:42	58
-133	25.50	23.33	9:49	60
-135	25.82	23.64	10:01	63
-137	26.08	23.91	10:14	65
-139	26.28	24.11	10:27	68
-140	26.36	24.19	10:33	70
-142	26.48	24.30	10:46	73
-144	26.53	24.35	10:59	76
-145	26.54	24.35	11:06	77

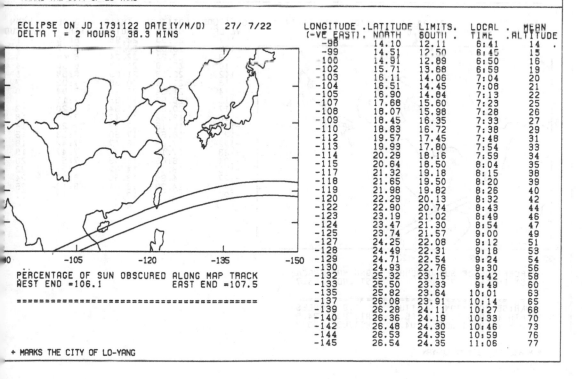

PERCENTAGE OF SUN OBSCURED ALONG MAP TRACK
WEST END =106.1   EAST END =107.5

==

+ MARKS THE CITY OF LO-YANG

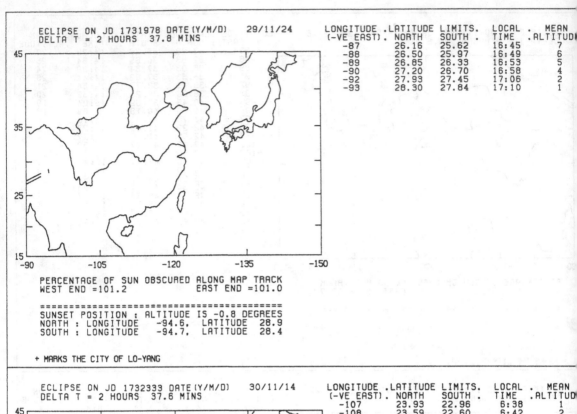

ECLIPSE ON JD 1731978 DATE(Y/M/D) 29/11/24
DELTA T = 2 HOURS 37.8 MINS

LONGITUDE (-VE EAST)	LATITUDE NORTH	LIMITS. SOUTH	LOCAL TIME	MEAN ALTITUDE
-87	26.16	25.62	16:45	7
-88	26.50	25.97	16:49	6
-89	26.85	26.33	16:53	5
-90	27.20	26.70	16:58	4
-92	27.93	27.45	17:06	2
-93	28.30	27.84	17:10	1

PERCENTAGE OF SUN OBSCURED ALONG MAP TRACK
WEST END =101.2   EAST END =101.0

==
SUNSET POSITION : ALTITUDE IS -0.8 DEGREES
NORTH : LONGITUDE -94.6, LATITUDE 28.9
SOUTH : LONGITUDE -94.7, LATITUDE 28.4

+ MARKS THE CITY OF LO-YANG

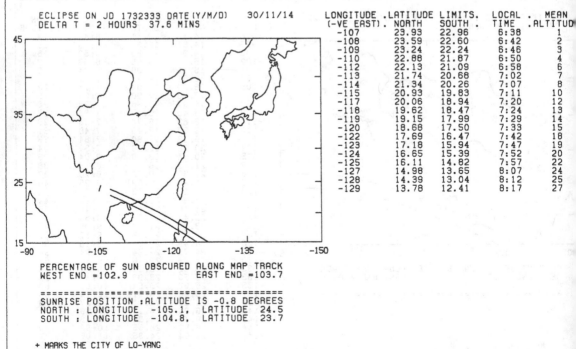

ECLIPSE ON JD 1732333 DATE(Y/M/D) 30/11/14
DELTA T = 2 HOURS 37.6 MINS

LONGITUDE (-VE EAST)	LATITUDE NORTH	LIMITS. SOUTH	LOCAL TIME	MEAN ALTITUDE
-107	23.93	22.96	6:38	1
-108	23.59	22.60	6:42	2
-109	23.24	22.24	6:46	3
-110	22.88	21.87	6:50	4
-112	22.13	21.09	6:58	6
-113	21.74	20.68	7:02	7
-114	21.34	20.26	7:07	8
-115	20.93	19.83	7:11	10
-117	20.06	18.94	7:20	12
-118	19.62	18.47	7:24	13
-119	19.15	17.99	7:29	14
-120	18.68	17.50	7:33	15
-122	17.69	16.47	7:42	18
-123	17.18	15.94	7:47	19
-124	16.65	15.39	7:52	20
-125	16.11	14.82	7:57	22
-127	14.98	13.65	8:07	24
-128	14.39	13.04	8:12	25
-129	13.78	12.41	8:17	27

PERCENTAGE OF SUN OBSCURED ALONG MAP TRACK
WEST END =102.9   EAST END =103.7

==
SUNRISE POSITION : ALTITUDE IS -0.8 DEGREES
NORTH : LONGITUDE -105.1, LATITUDE 24.5
SOUTH : LONGITUDE -104.8, LATITUDE 23.7

+ MARKS THE CITY OF LO-YANG

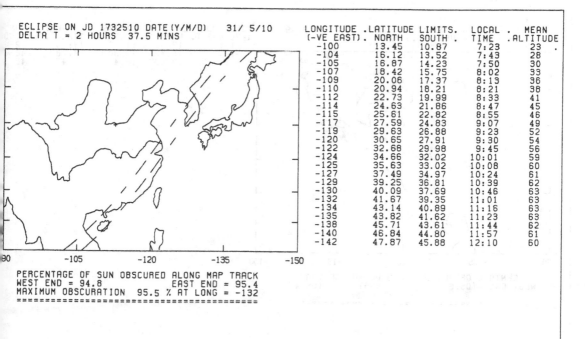

ECLIPSE ON JD 1732510 DATE(Y/M/D)   31/ 5/10
DELTA T = 2 HOURS 37.5 MINS

LONGITUDE (-VE EAST)	LATITUDE LIMITS. NORTH	SOUTH	LOCAL TIME	MEAN ALTITUDE
-100	13.45	10.87	7:23	23
-104	16.12	13.52	7:43	28
-105	16.87	14.23	7:50	30
-107	18.42	15.75	8:02	33
-109	20.06	17.37	8:13	36
-110	20.94	18.21	8:21	38
-112	22.73	19.99	8:33	41
-114	24.63	21.86	8:47	45
-115	25.61	22.82	8:55	46
-117	27.59	24.83	9:07	49
-119	29.63	26.88	9:23	52
-120	30.65	27.91	9:30	54
-122	32.68	29.98	9:45	56
-124	34.66	32.02	10:01	59
-125	35.63	33.02	10:08	60
-127	37.49	34.97	10:24	61
-129	39.25	36.81	10:39	62
-130	40.09	37.69	10:46	63
-132	41.67	39.35	11:01	63
-134	43.14	40.89	11:16	63
-135	43.82	41.62	11:23	63
-138	45.71	43.61	11:44	62
-140	46.84	44.80	11:57	61
-142	47.87	45.88	12:10	60

PERCENTAGE OF SUN OBSCURED ALONG MAP TRACK
WEST END = 94.8   EAST END = 95.4
MAXIMUM OBSCURATION 95.5 % AT LONG = -132
===

+ MARKS THE CITY OF LO-YANG

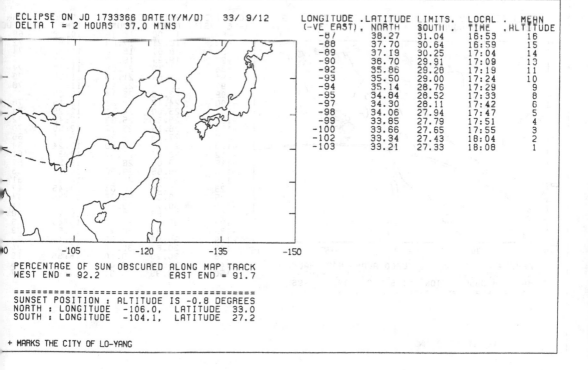

ECLIPSE ON JD 1733366 DATE(Y/M/D)   33/ 9/12
DELTA T = 2 HOURS 37.0 MINS

LONGITUDE (-VE EAST)	LATITUDE LIMITS. NORTH	SOUTH	LOCAL TIME	MEAN ALTITUDE
-87	38.27	31.04	16:53	16
-88	37.70	30.64	16:59	15
-89	37.19	30.25	17:04	14
-90	36.70	29.91	17:09	13
-92	35.86	29.28	17:19	11
-93	35.50	29.00	17:24	10
-94	35.14	28.76	17:29	9
-95	34.84	28.52	17:33	8
-97	34.30	28.11	17:42	6
-98	34.06	27.94	17:47	5
-99	33.85	27.79	17:51	4
-100	33.66	27.65	17:55	3
-102	33.34	27.43	18:04	2
-103	33.21	27.33	18:08	1

PERCENTAGE OF SUN OBSCURED ALONG MAP TRACK
WEST END = 92.2   EAST END = 91.7

===
SUNSET POSITION : ALTITUDE IS -0.8 DEGREES
NORTH : LONGITUDE -106.0,   LATITUDE 33.0
SOUTH : LONGITUDE -104.1,   LATITUDE 27.2

+ MARKS THE CITY OF LO-YANG

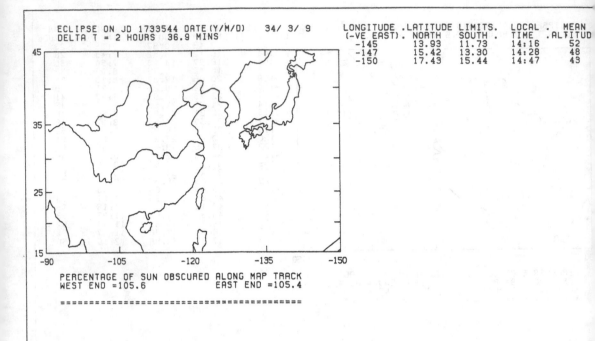

ECLIPSE ON JD 1733544 DATE(Y/M/D)   34/ 3/ 9
DELTA T = 2 HOURS 36.9 MINS

LONGITUDE (-VE EAST)	LATITUDE NORTH	LIMITS. SOUTH	LOCAL TIME	MEAN ALTITUD
-145	13.93	11.73	14:16	52
-147	15.42	13.30	14:28	48
-150	17.43	15.44	14:47	43

PERCENTAGE OF SUN OBSCURED ALONG MAP TRACK
WEST END =105.6   EAST END =105.4

===

+ MARKS THE CITY OF LO-YANG

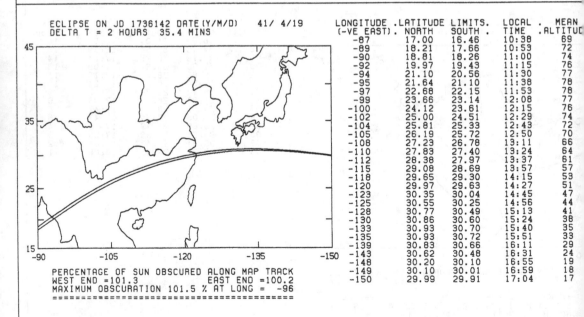

ECLIPSE ON JD 1736142 DATE(Y/M/D)   41/ 4/19
DELTA T = 2 HOURS 35.4 MINS

LONGITUDE (-VE EAST)	LATITUDE NORTH	LIMITS. SOUTH	LOCAL TIME	MEAN ALTITUD
-87	17.00	16.46	10:38	69
-89	18.21	17.66	10:53	72
-90	18.81	18.26	11:00	74
-92	19.97	19.43	11:15	76
-94	21.10	20.56	11:30	77
-95	21.64	21.10	11:38	78
-97	22.68	22.15	11:53	78
-99	23.66	23.14	12:08	77
-100	24.12	23.61	12:15	76
-102	25.00	24.51	12:29	74
-104	25.81	25.33	12:43	72
-105	26.19	25.72	12:50	70
-108	27.23	26.78	13:11	66
-110	27.83	27.40	13:24	64
-112	28.38	27.97	13:37	61
-115	29.08	28.69	13:57	57
-118	29.65	29.30	14:15	53
-120	29.97	29.63	14:27	51
-123	30.35	30.04	14:45	47
-125	30.55	30.25	14:56	44
-128	30.77	30.49	15:13	41
-130	30.86	30.60	15:24	38
-133	30.93	30.70	15:40	35
-135	30.93	30.72	15:51	33
-139	30.83	30.66	16:11	29
-143	30.62	30.48	16:31	24
-148	30.20	30.10	16:55	19
-149	30.10	30.01	16:59	18
-150	29.99	29.91	17:04	17

PERCENTAGE OF SUN OBSCURED ALONG MAP TRACK
WEST END =101.3   EAST END =100.2
MAXIMUM OBSCURATION 101.5 % AT LONG = -96

===

+ MARKS THE CITY OF LO-YANG

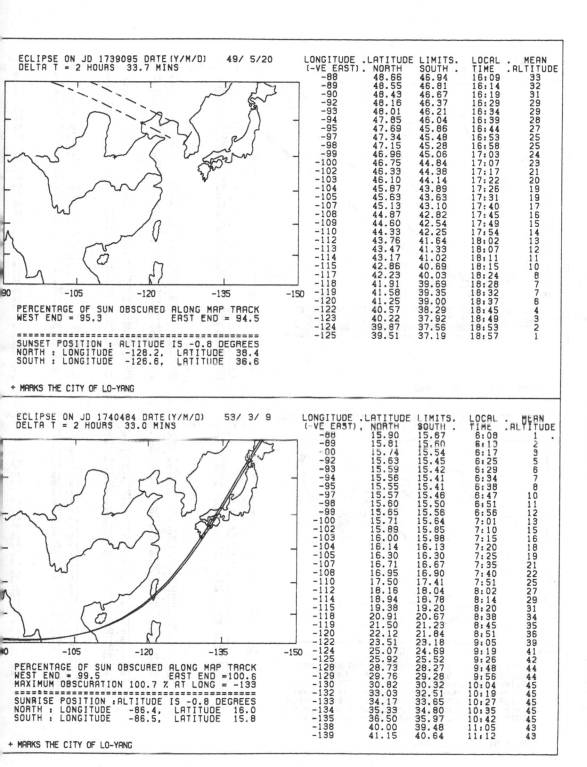

ECLIPSE ON JD 1739095 DATE(Y/M/D)   49/ 5/20
DELTA T = 2 HOURS 33.7 MINS

LONGITUDE (-VE EAST)	LATITUDE LIMITS. NORTH	SOUTH	LOCAL TIME	MEAN ALTITUDE
-88	48.66	46.94	16:09	33
-89	48.55	46.81	16:14	32
-90	48.43	46.67	16:19	31
-92	48.16	46.37	16:29	29
-93	48.01	46.21	16:34	29
-94	47.85	46.04	16:39	28
-95	47.69	45.86	16:44	27
-97	47.34	45.48	16:53	25
-98	47.15	45.28	16:58	25
-99	46.96	45.06	17:03	24
-100	46.75	44.84	17:07	23
-102	46.33	44.38	17:17	21
-103	46.10	44.14	17:22	20
-104	45.87	43.89	17:26	19
-105	45.63	43.63	17:31	19
-107	45.13	43.10	17:40	17
-108	44.87	42.82	17:45	16
-109	44.60	42.54	17:49	15
-110	44.33	42.25	17:54	14
-112	43.76	41.64	18:02	13
-113	43.47	41.33	18:07	12
-114	43.17	41.02	18:11	11
-115	42.86	40.69	18:15	10
-117	42.23	40.03	18:24	8
-118	41.91	39.69	18:28	7
-119	41.58	39.35	18:32	7
-120	41.25	39.00	18:37	6
-122	40.57	38.29	18:45	4
-123	40.22	37.92	18:49	3
-124	39.87	37.56	18:53	2
-125	39.51	37.19	18:57	1

PERCENTAGE OF SUN OBSCURED ALONG MAP TRACK
WEST END = 95.3   EAST END = 94.5
===
SUNSET POSITION : ALTITUDE IS -0.8 DEGREES
NORTH : LONGITUDE -128.2, LATITUDE   38.4
SOUTH : LONGITUDE -126.6, LATITUDE   36.6

+ MARKS THE CITY OF LO-YANG

ECLIPSE ON JD 1740484 DATE(Y/M/D)   53/ 3/ 9
DELTA T = 2 HOURS 33.0 MINS

LONGITUDE (-VE EAST)	LATITUDE LIMITS. NORTH	SOUTH	LOCAL TIME	MEAN ALTITUDE
-88	15.90	15.67	6:08	1
-89	15.81	15.60	6:13	2
-90	15.74	15.54	6:17	3
-92	15.63	15.45	6:25	5
-93	15.59	15.42	6:29	6
-94	15.56	15.41	6:34	7
-95	15.55	15.41	6:38	8
-97	15.57	15.46	6:47	10
-98	15.60	15.50	6:51	11
-99	15.65	15.56	6:56	12
-100	15.71	15.64	7:01	13
-102	15.89	15.85	7:10	15
-103	16.00	15.98	7:15	16
-104	16.14	16.13	7:20	18
-105	16.30	16.30	7:25	19
-107	16.71	16.67	7:35	21
-108	16.95	16.90	7:40	22
-110	17.50	17.41	7:51	25
-112	18.16	18.04	8:02	27
-114	18.94	18.78	8:14	29
-115	19.38	19.20	8:20	31
-118	20.91	20.67	8:38	34
-119	21.50	21.23	8:45	35
-120	22.12	21.84	8:51	36
-122	23.51	23.18	9:05	39
-124	25.07	24.69	9:19	41
-125	25.92	25.52	9:26	42
-128	28.73	28.27	9:48	44
-129	29.76	29.28	9:56	44
-130	30.82	30.32	10:04	45
-132	33.03	32.51	10:19	45
-133	34.17	33.65	10:27	45
-134	35.33	34.80	10:35	45
-135	36.50	35.97	10:42	45
-138	40.00	39.48	11:05	43
-139	41.15	40.64	11:12	43

PERCENTAGE OF SUN OBSCURED ALONG MAP TRACK
WEST END = 99.5   EAST END =100.6
MAXIMUM OBSCURATION 100.7 % AT LONG = -133
===
SUNRISE POSITION :ALTITUDE IS -0.8 DEGREES
NORTH : LONGITUDE   -86.4, LATITUDE   16.0
SOUTH : LONGITUDE   -86.5, LATITUDE   15.8

+ MARKS THE CITY OF LO-YANG

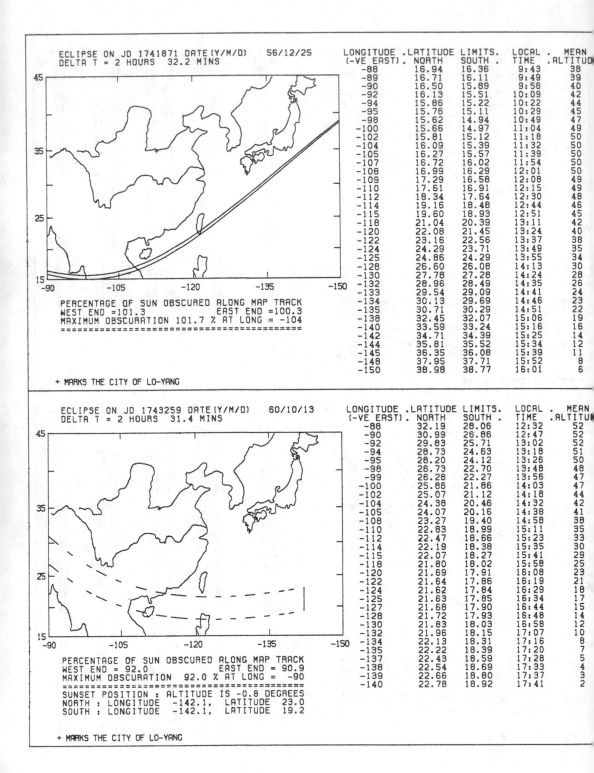

ECLIPSE ON JD 1741871 DATE(Y/M/D) 56/12/25
DELTA T = 2 HOURS 32.2 MINS

LONGITUDE (-VE EAST)	LATITUDE LIMITS. NORTH	SOUTH	LOCAL TIME	MEAN ALTITUDE
-88	16.94	16.36	9:43	38
-89	16.71	16.11	9:49	39
-90	16.50	15.89	9:56	40
-92	16.13	15.51	10:09	42
-94	15.86	15.22	10:22	44
-95	15.76	15.11	10:29	45
-98	15.62	14.94	10:49	47
-100	15.66	14.97	11:04	49
-102	15.81	15.12	11:18	50
-104	16.09	15.39	11:32	50
-105	16.27	15.57	11:39	50
-107	16.72	16.02	11:54	50
-108	16.99	16.29	12:01	50
-109	17.29	16.58	12:08	49
-110	17.61	16.91	12:15	49
-112	18.34	17.64	12:30	48
-114	19.16	18.48	12:44	46
-115	19.60	18.93	12:51	45
-118	21.04	20.39	13:11	42
-120	22.08	21.45	13:24	40
-122	23.16	22.56	13:37	38
-124	24.29	23.71	13:49	35
-125	24.86	24.29	13:55	34
-128	26.60	26.08	14:13	30
-130	27.78	27.28	14:24	28
-132	28.96	28.49	14:35	26
-133	29.54	29.09	14:41	24
-134	30.13	29.69	14:46	23
-135	30.71	30.29	14:51	22
-138	32.45	32.07	15:06	19
-140	33.59	33.24	15:16	16
-142	34.71	34.39	15:25	14
-144	35.81	35.52	15:34	12
-145	36.35	36.08	15:39	11
-148	37.95	37.71	15:52	8
-150	38.98	38.77	16:01	6

PERCENTAGE OF SUN OBSCURED ALONG MAP TRACK
WEST END =101.3   EAST END =100.3
MAXIMUM OBSCURATION 101.7 % AT LONG = -104
==

+ MARKS THE CITY OF LO-YANG

ECLIPSE ON JD 1743259 DATE(Y/M/D) 60/10/13
DELTA T = 2 HOURS 31.4 MINS

LONGITUDE (-VE EAST)	LATITUDE NORTH	SOUTH	LOCAL TIME	MEAN ALTITUDE
-88	32.19	28.06	12:32	52
-90	30.99	26.86	12:47	52
-92	29.83	25.71	13:02	52
-94	28.73	24.63	13:18	51
-95	28.20	24.12	13:26	50
-98	26.73	22.70	13:48	48
-99	26.28	22.27	13:56	47
-100	25.86	21.86	14:03	47
-102	25.07	21.12	14:18	44
-104	24.38	20.46	14:32	42
-105	24.07	20.16	14:38	41
-108	23.27	19.40	14:58	38
-110	22.83	18.99	15:11	35
-112	22.47	18.66	15:23	33
-114	22.19	18.38	15:35	30
-115	22.07	18.27	15:41	29
-118	21.80	18.02	15:58	25
-120	21.69	17.91	16:08	23
-122	21.64	17.86	16:19	21
-124	21.62	17.84	16:29	18
-125	21.63	17.85	16:34	17
-127	21.68	17.90	16:44	15
-128	21.72	17.93	16:48	14
-130	21.83	18.03	16:58	12
-132	21.96	18.15	17:07	10
-134	22.13	18.31	17:16	8
-135	22.22	18.39	17:20	7
-137	22.43	18.59	17:28	5
-138	22.54	18.69	17:33	4
-139	22.66	18.80	17:37	3
-140	22.78	18.92	17:41	2

PERCENTAGE OF SUN OBSCURED ALONG MAP TRACK
WEST END = 92.0   EAST END = 90.9
MAXIMUM OBSCURATION 92.0 % AT LONG = -90
==
SUNSET POSITION : ALTITUDE IS -0.8 DEGREES
NORTH : LONGITUDE -142.1, LATITUDE 23.0
SOUTH : LONGITUDE -142.1, LATITUDE 19.2

+ MARKS THE CITY OF LO-YANG

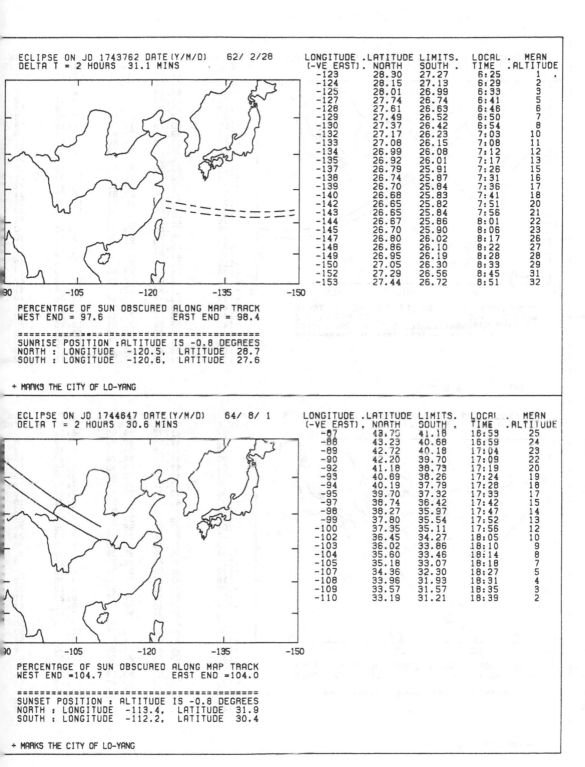

ECLIPSE ON JD 1743762 DATE(Y/M/D)   62/ 2/28
DELTA T = 2 HOURS 31.1 MINS

LONGITUDE (-VE EAST)	.LATITUDE NORTH	LIMITS. SOUTH	LOCAL TIME	. MEAN .ALTITUDE
-123	28.30	27.27	6:25	1 .
-124	28.15	27.13	6:29	2
-125	28.01	26.99	6:33	3
-127	27.74	26.74	6:41	5
-128	27.61	26.63	6:46	6
-129	27.49	26.52	6:50	7
-130	27.37	26.42	6:54	8
-132	27.17	26.23	7:03	10
-133	27.08	26.15	7:08	11
-134	26.99	26.08	7:12	12
-135	26.92	26.01	7:17	13
-137	26.79	25.91	7:26	15
-138	26.74	25.87	7:31	16
-139	26.70	25.84	7:36	17
-140	26.68	25.83	7:41	18
-142	26.65	25.82	7:51	20
-143	26.65	25.84	7:56	21
-144	26.67	25.86	8:01	22
-145	26.70	25.90	8:06	23
-147	26.80	26.02	8:17	26
-148	26.86	26.10	8:22	27
-149	26.95	26.19	8:28	28
-150	27.05	26.30	8:33	29
-152	27.29	26.56	8:45	31
-153	27.44	26.72	8:51	32

PERCENTAGE OF SUN OBSCURED ALONG MAP TRACK
WEST END = 97.6   EAST END = 98.4

===
SUNRISE POSITION :ALTITUDE IS -0.8 DEGREES
NORTH : LONGITUDE -120.5, LATITUDE 28.7
SOUTH : LONGITUDE -120.6, LATITUDE 27.6

+ MARKS THE CITY OF LO-YANG

ECLIPSE ON JD 1744647 DATE(Y/M/D)   64/ 8/ 1
DELTA T = 2 HOURS 30.6 MINS

LONGITUDE (-VE EAST)	.LATITUDE NORTH	LIMITS. SOUTH	LOCAL TIME	. MEAN .ALTITUDE
-87	43.75	41.18	16:59	25
-88	43.23	40.68	16:59	24
-89	42.72	40.18	17:04	23
-90	42.20	39.70	17:09	22
-92	41.18	38.73	17:19	20
-93	40.69	38.26	17:24	19
-94	40.19	37.79	17:28	18
-95	39.70	37.32	17:33	17
-97	38.74	36.42	17:42	15
-98	38.27	35.97	17:47	14
-99	37.80	35.54	17:52	13
-100	37.35	35.11	17:56	12
-102	36.45	34.27	18:05	10
-103	36.02	33.86	18:10	9
-104	35.60	33.46	18:14	8
-105	35.18	33.07	18:18	7
-107	34.36	32.30	18:27	5
-108	33.96	31.93	18:31	4
-109	33.57	31.57	18:35	3
-110	33.19	31.21	18:39	2

PERCENTAGE OF SUN OBSCURED ALONG MAP TRACK
WEST END =104.7   EAST END =104.0

===
SUNSET POSITION : ALTITUDE IS -0.8 DEGREES
NORTH : LONGITUDE -113.4, LATITUDE 31.9
SOUTH : LONGITUDE -112.2, LATITUDE 30.4

+ MARKS THE CITY OF LO-YANG

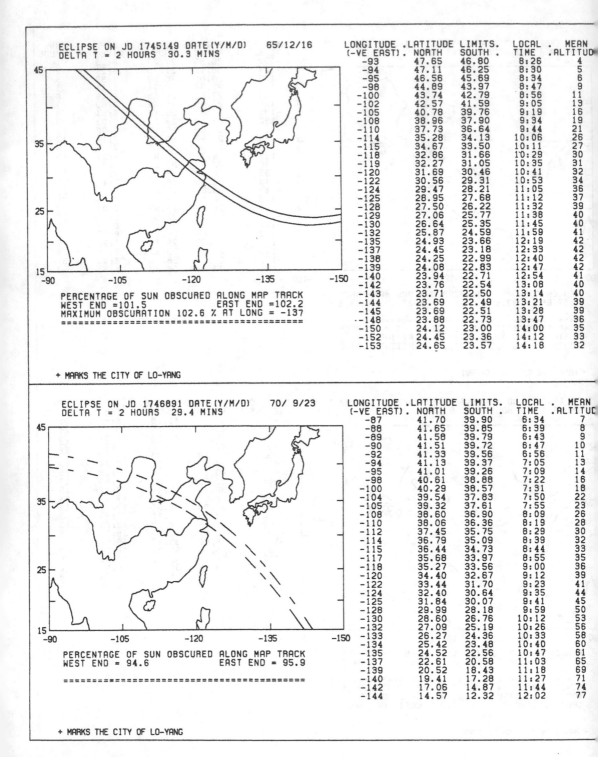

ECLIPSE ON JD 1745149 DATE (Y/M/D)   65/12/16
DELTA T = 2 HOURS   30.3 MINS

LONGITUDE (-VE EAST)	LATITUDE NORTH	LIMITS. SOUTH	LOCAL TIME	MEAN ALTITUD
-93	47.65	46.80	8:26	4
-94	47.11	46.25	8:30	5
-95	46.56	45.69	8:34	6
-98	44.89	43.97	8:47	9
-100	43.74	42.79	8:56	11
-102	42.57	41.59	9:05	13
-105	40.78	39.76	9:19	16
-108	38.96	37.90	9:34	19
-110	37.73	36.64	9:44	21
-114	35.28	34.13	10:06	26
-115	34.67	33.50	10:11	27
-118	32.86	31.66	10:29	30
-119	32.27	31.05	10:35	31
-120	31.69	30.46	10:41	32
-122	30.56	29.31	10:53	34
-124	29.47	28.21	11:05	36
-125	28.95	27.68	11:12	37
-128	27.50	26.22	11:32	39
-129	27.06	25.77	11:38	40
-130	26.64	25.35	11:45	40
-132	25.87	24.59	11:59	41
-135	24.93	23.66	12:19	42
-137	24.45	23.18	12:33	42
-138	24.25	22.99	12:40	42
-139	24.08	22.83	12:47	42
-140	23.94	22.71	12:54	41
-142	23.76	22.54	13:08	40
-143	23.71	22.50	13:14	40
-144	23.69	22.49	13:21	39
-145	23.69	22.51	13:28	39
-148	23.88	22.73	13:47	36
-150	24.12	23.00	14:00	35
-152	24.45	23.36	14:12	33
-153	24.65	23.57	14:18	32

PERCENTAGE OF SUN OBSCURED ALONG MAP TRACK
WEST END =101.5   EAST END =102.2
MAXIMUM OBSCURATION 102.6 % AT LONG = -137
===

+ MARKS THE CITY OF LO-YANG

ECLIPSE ON JD 1746891 DATE (Y/M/D)   70/ 9/23
DELTA T = 2 HOURS   29.4 MINS

LONGITUDE (-VE EAST)	LATITUDE NORTH	LIMITS. SOUTH	LOCAL TIME	MEAN ALTITUD
-87	41.70	39.90	6:34	7
-88	41.65	39.85	6:39	8
-89	41.58	39.79	6:43	9
-90	41.51	39.72	6:47	10
-92	41.33	39.56	6:56	11
-94	41.13	39.37	7:05	13
-95	41.01	39.26	7:09	14
-98	40.61	38.88	7:22	16
-100	40.29	38.57	7:31	18
-104	39.54	37.83	7:50	22
-105	39.32	37.61	7:55	23
-108	38.60	36.90	8:09	26
-110	38.06	36.36	8:19	28
-112	37.45	35.75	8:29	30
-114	36.79	35.09	8:39	32
-115	36.44	34.73	8:44	33
-117	35.68	33.97	8:55	35
-118	35.27	33.56	9:00	36
-120	34.40	32.67	9:12	39
-122	33.44	31.70	9:23	41
-124	32.40	30.64	9:35	44
-125	31.84	30.07	9:41	45
-128	29.99	28.18	9:59	50
-130	28.60	26.76	10:12	53
-132	27.09	25.19	10:26	56
-133	26.27	24.36	10:33	58
-134	25.42	23.48	10:40	60
-135	24.52	22.56	10:47	61
-137	22.61	20.58	11:03	65
-139	20.52	18.43	11:18	69
-140	19.41	17.28	11:27	71
-142	17.06	14.87	11:44	74
-144	14.57	12.32	12:02	77

PERCENTAGE OF SUN OBSCURED ALONG MAP TRACK
WEST END = 94.6   EAST END = 95.9

===

+ MARKS THE CITY OF LO-YANG

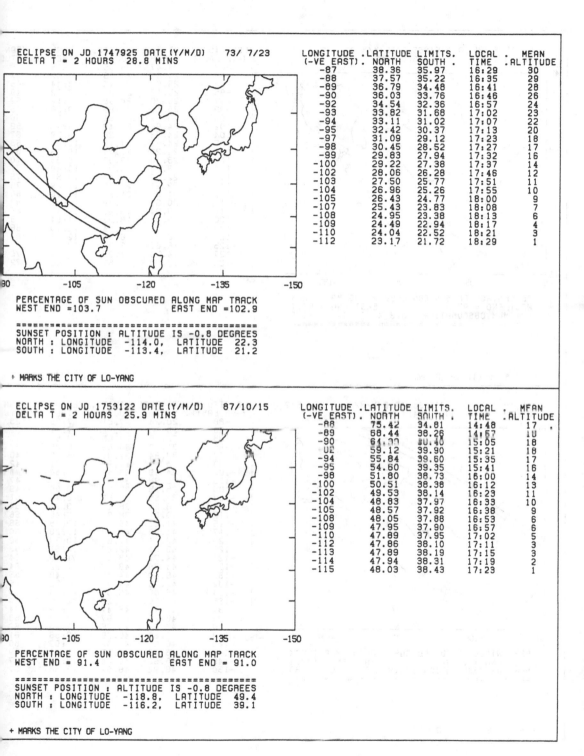

ECLIPSE ON JD 1747925 DATE (Y/M/D) 73/ 7/23
DELTA T = 2 HOURS 28.8 MINS

LONGITUDE (-VE EAST)	LATITUDE LIMITS NORTH	LATITUDE LIMITS SOUTH	LOCAL TIME	MEAN ALTITUDE
-87	38.36	35.97	16:29	30
-88	37.57	35.22	16:35	29
-89	36.79	34.48	16:41	28
-90	36.03	33.76	16:46	26
-92	34.54	32.36	16:57	24
-93	33.82	31.68	17:02	23
-94	33.11	31.02	17:07	22
-95	32.42	30.37	17:13	20
-97	31.09	29.12	17:23	18
-98	30.45	28.52	17:27	17
-99	29.83	27.94	17:32	16
-100	29.22	27.38	17:37	14
-102	28.06	26.28	17:46	12
-103	27.50	25.77	17:51	11
-104	26.96	25.26	17:55	10
-105	26.43	24.77	18:00	9
-107	25.43	23.83	18:08	7
-108	24.95	23.38	18:13	6
-109	24.49	22.94	18:17	4
-110	24.04	22.52	18:21	3
-112	23.17	21.72	18:29	1

PERCENTAGE OF SUN OBSCURED ALONG MAP TRACK
WEST END =103.7   EAST END =102.9

==
SUNSET POSITION : ALTITUDE IS -0.8 DEGREES
NORTH : LONGITUDE -114.0, LATITUDE 22.3
SOUTH : LONGITUDE -113.4, LATITUDE 21.2

+ MARKS THE CITY OF LO-YANG

ECLIPSE ON JD 1753122 DATE (Y/M/D) 87/10/15
DELTA T = 2 HOURS 25.9 MINS

LONGITUDE (-VE EAST)	LATITUDE LIMITS NORTH	LATITUDE LIMITS SOUTH	LOCAL TIME	MEAN ALTITUDE
-88	75.42	34.81	14:48	17
-89	68.44	38.26	14:57	18
-90	64.30	40.40	15:05	18
-92	59.12	39.90	15:21	18
-94	55.84	39.60	15:35	17
-95	54.60	39.35	15:41	16
-98	51.80	38.73	16:00	14
-100	50.51	38.38	16:12	13
-102	49.53	38.14	16:23	11
-104	48.83	97.97	16:33	10
-105	48.57	37.92	16:38	9
-108	48.05	37.88	16:53	6
-109	47.95	37.90	16:57	6
-110	47.89	37.95	17:02	5
-112	47.86	38.10	17:11	3
-113	47.89	38.19	17:15	3
-114	47.94	38.31	17:19	2
-115	48.03	38.43	17:23	1

PERCENTAGE OF SUN OBSCURED ALONG MAP TRACK
WEST END = 91.4   EAST END = 91.0

==
SUNSET POSITION : ALTITUDE IS -0.8 DEGREES
NORTH : LONGITUDE -118.8, LATITUDE 49.4
SOUTH : LONGITUDE -116.2, LATITUDE 39.1

+ MARKS THE CITY OF LO-YANG

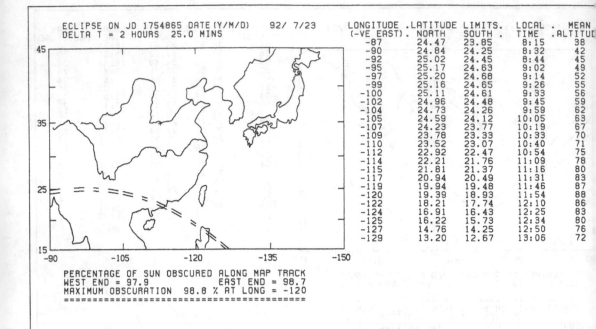

ECLIPSE ON JD 1754865 DATE (Y/M/D)   92/ 7/23
DELTA T = 2 HOURS 25.0 MINS

LONGITUDE (-VE EAST)	LATITUDE NORTH	LIMITS. SOUTH	LOCAL TIME	MEAN ALTITUDE
-87	24.47	23.85	8:15	38
-90	24.84	24.25	8:32	42
-92	25.02	24.45	8:44	45
-95	25.17	24.63	9:02	49
-97	25.20	24.68	9:14	52
-99	25.16	24.65	9:26	55
-100	25.11	24.61	9:33	56
-102	24.96	24.48	9:45	59
-104	24.73	24.26	9:59	62
-105	24.59	24.12	10:05	63
-107	24.23	23.77	10:19	67
-109	23.78	23.33	10:33	70
-110	23.52	23.07	10:40	71
-112	22.92	22.47	10:54	75
-114	22.21	21.76	11:09	78
-115	21.81	21.37	11:16	80
-117	20.94	20.49	11:31	83
-119	19.94	19.48	11:46	87
-120	19.39	18.93	11:54	88
-122	18.21	17.74	12:10	86
-124	16.91	16.43	12:25	83
-125	16.22	15.73	12:34	80
-127	14.76	14.25	12:50	76
-129	13.20	12.67	13:06	72

PERCENTAGE OF SUN OBSCURED ALONG MAP TRACK
WEST END = 97.9   EAST END = 98.7
MAXIMUM OBSCURATION 98.8 % AT LONG = -120
==

+ MARKS THE CITY OF LO-YANG

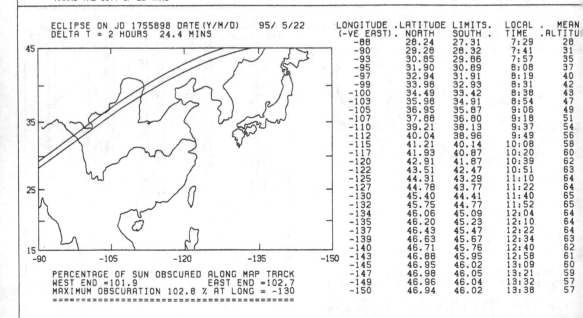

ECLIPSE ON JD 1755898 DATE (Y/M/D)   95/ 5/22
DELTA T = 2 HOURS 24.4 MINS

LONGITUDE (-VE EAST)	LATITUDE NORTH	LIMITS. SOUTH	LOCAL TIME	MEAN ALTITUDE
-88	28.24	27.31	7:29	28
-90	29.28	28.32	7:41	31
-93	30.85	29.86	7:57	35
-95	31.90	30.89	8:08	37
-97	32.94	31.91	8:19	40
-99	33.98	32.93	8:31	42
-100	34.49	33.42	8:38	43
-103	35.98	34.91	8:54	47
-105	36.95	35.87	9:06	49
-107	37.88	36.80	9:18	51
-110	39.21	38.13	9:37	54
-112	40.04	38.96	9:49	56
-115	41.21	40.14	10:08	58
-117	41.93	40.87	10:20	60
-120	42.91	41.87	10:39	62
-122	43.51	42.47	10:51	63
-125	44.31	43.29	11:10	64
-127	44.78	43.77	11:22	64
-130	45.40	44.41	11:40	65
-132	45.75	44.77	11:52	65
-134	46.06	45.09	12:04	64
-135	46.20	45.23	12:10	64
-137	46.43	45.47	12:22	64
-139	46.63	45.67	12:34	63
-140	46.71	45.76	12:40	62
-143	46.88	45.95	12:58	61
-145	46.95	46.02	13:09	60
-147	46.98	46.05	13:21	59
-149	46.96	46.04	13:32	57
-150	46.94	46.02	13:38	57

PERCENTAGE OF SUN OBSCURED ALONG MAP TRACK
WEST END =101.9   EAST END =102.7
MAXIMUM OBSCURATION 102.8 % AT LONG = -130
==

+ MARKS THE CITY OF LO-YANG

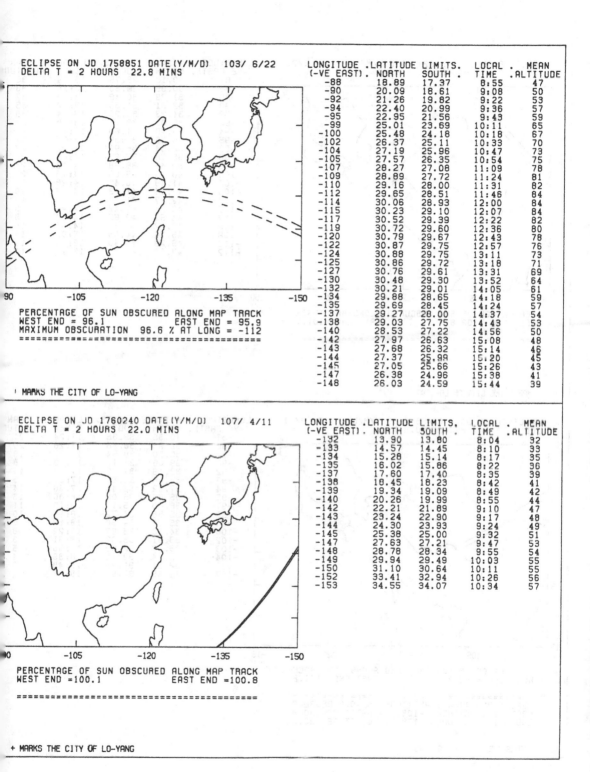

ECLIPSE ON JD 1758851 DATE(Y/M/D) 103/ 6/22
DELTA T = 2 HOURS 22.8 MINS

PERCENTAGE OF SUN OBSCURED ALONG MAP TRACK
WEST END = 96.1   EAST END = 95.9
MAXIMUM OBSCURATION 96.6 % AT LONG = -112
===

' MARKS THE CITY OF LO-YANG

LONGITUDE (-VE EAST)	LATITUDE NORTH	LIMITS SOUTH	LOCAL TIME	MEAN ALTITUDE
-88	18.89	17.37	8:55	47
-90	20.09	18.61	9:08	50
-92	21.26	19.82	9:22	53
-94	22.40	20.99	9:36	57
-95	22.95	21.56	9:43	59
-99	25.01	23.69	10:11	65
-100	25.48	24.18	10:18	67
-102	26.37	25.11	10:33	70
-104	27.19	25.96	10:47	73
-105	27.57	26.35	10:54	75
-107	28.27	27.08	11:09	78
-109	28.89	27.72	11:24	81
-110	29.16	28.00	11:31	82
-112	29.65	28.51	11:46	84
-114	30.06	28.93	12:00	84
-115	30.23	29.10	12:07	84
-117	30.52	29.39	12:22	82
-119	30.72	29.60	12:36	80
-120	30.79	29.67	12:43	78
-122	30.87	29.75	12:57	76
-124	30.88	29.75	13:11	73
-125	30.86	29.72	13:18	71
-127	30.76	29.61	13:31	69
-130	30.48	29.30	13:52	64
-132	30.21	29.01	14:05	61
-134	29.88	28.65	14:18	59
-135	29.69	28.45	14:24	57
-137	29.27	28.00	14:37	54
-138	29.03	27.75	14:43	53
-140	28.53	27.22	14:56	50
-142	27.97	26.63	15:08	48
-143	27.68	26.32	15:14	46
-144	27.37	25.99	15:20	45
-145	27.05	25.66	15:26	43
-147	26.38	24.96	15:38	41
-148	26.03	24.59	15:44	39

ECLIPSE ON JD 1760240 DATE(Y/M/D) 107/ 4/11
DELTA T = 2 HOURS 22.0 MINS

PERCENTAGE OF SUN OBSCURED ALONG MAP TRACK
WEST END =100.1   EAST END =100.8

===

+ MARKS THE CITY OF LO-YANG

LONGITUDE (-VE EAST)	LATITUDE NORTH	LIMITS SOUTH	LOCAL TIME	MEAN ALTITUDE
-132	13.90	13.80	8:04	32
-133	14.57	14.45	8:10	33
-134	15.28	15.14	8:17	35
-135	16.02	15.86	8:22	36
-137	17.60	17.40	8:35	39
-138	18.45	18.23	8:42	41
-139	19.34	19.09	8:49	42
-140	20.26	19.99	8:55	44
-142	22.21	21.89	9:10	47
-143	23.24	22.90	9:17	48
-144	24.30	23.93	9:24	49
-145	25.38	25.00	9:32	51
-147	27.63	27.21	9:47	53
-148	28.78	28.34	9:55	54
-149	29.94	29.49	10:03	55
-150	31.10	30.64	10:11	55
-152	33.41	32.94	10:26	56
-153	34.55	34.07	10:34	57

ECLIPSE ON JD 1761627 DATE(Y/M/D)   111/ 1/27
DELTA T = 2 HOURS 21.3 MINS

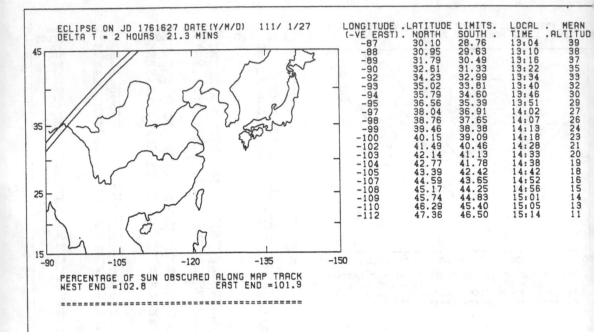

LONGITUDE (-VE EAST)	LATITUDE NORTH	LIMITS. SOUTH	LOCAL TIME	MEAN ALTITUD
-87	30.10	28.76	13:04	39
-88	30.95	29.63	13:10	38
-89	31.79	30.49	13:16	37
-90	32.61	31.33	13:22	35
-92	34.23	32.99	13:34	33
-93	35.02	33.81	13:40	32
-94	35.79	34.60	13:46	30
-95	36.56	35.39	13:51	29
-97	38.04	36.91	14:02	27
-98	38.76	37.65	14:07	26
-99	39.46	38.38	14:13	24
-100	40.15	39.09	14:18	23
-102	41.49	40.46	14:28	21
-103	42.14	41.13	14:33	20
-104	42.77	41.78	14:38	19
-105	43.39	42.42	14:42	18
-107	44.59	43.65	14:52	16
-108	45.17	44.25	14:56	15
-109	45.74	44.83	15:01	14
-110	46.29	45.40	15:05	13
-112	47.36	46.50	15:14	11

PERCENTAGE OF SUN OBSCURED ALONG MAP TRACK
WEST END =102.8   EAST END =101.9

+ MARKS THE CITY OF LO-YANG

ECLIPSE ON JD 1762483 DATE(Y/M/D)   113/ 6/ 1
DELTA T = 2 HOURS 20.8 MINS

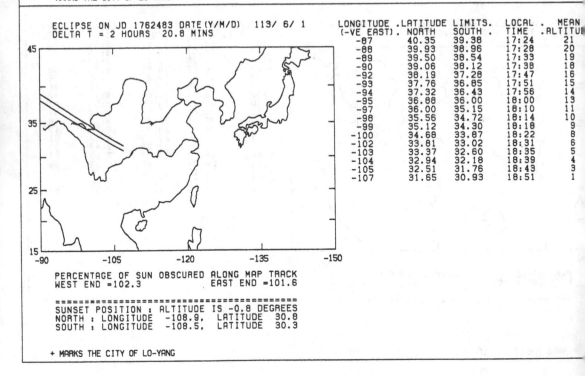

LONGITUDE (-VE EAST)	LATITUDE NORTH	LIMITS. SOUTH	LOCAL TIME	MEAN ALTITUD
-87	40.35	39.38	17:24	21
-88	39.93	38.96	17:28	20
-89	39.50	38.54	17:33	19
-90	39.06	38.12	17:38	18
-92	38.19	37.28	17:47	16
-93	37.76	36.85	17:51	15
-94	37.32	36.43	17:56	14
-95	36.88	36.00	18:00	13
-97	36.00	35.15	18:10	11
-98	35.56	34.72	18:14	10
-99	35.12	34.30	18:18	9
-100	34.68	33.87	18:22	8
-102	33.81	33.02	18:31	6
-103	33.37	32.60	18:35	5
-104	32.94	32.18	18:39	4
-105	32.51	31.76	18:43	3
-107	31.65	30.93	18:51	1

PERCENTAGE OF SUN OBSCURED ALONG MAP TRACK
WEST END =102.3   EAST END =101.6

SUNSET POSITION : ALTITUDE IS -0.8 DEGREES
NORTH : LONGITUDE -108.9, LATITUDE 30.8
SOUTH : LONGITUDE -108.5, LATITUDE 30.3

+ MARKS THE CITY OF LO-YANG

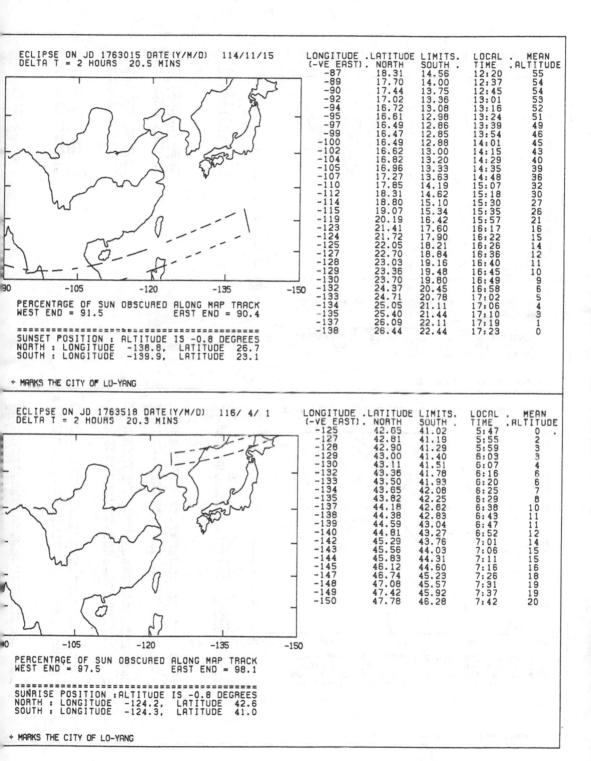

ECLIPSE ON JD 1763015 DATE (Y/M/D)   114/11/15
DELTA T = 2 HOURS   20.5 MINS

LONGITUDE (-VE EAST)	LATITUDE NORTH	LIMITS. SOUTH	LOCAL TIME	MEAN ALTITUDE
-87	18.31	14.56	12:20	55
-89	17.70	14.00	12:37	54
-90	17.44	13.75	12:45	54
-92	17.02	13.36	13:01	53
-94	16.72	13.08	13:16	52
-95	16.61	12.98	13:24	51
-97	16.49	12.86	13:39	49
-99	16.47	12.85	13:54	46
-100	16.49	12.88	14:01	45
-102	16.62	13.00	14:15	43
-104	16.82	13.20	14:29	40
-105	16.96	13.33	14:35	39
-107	17.27	13.63	14:48	36
-110	17.85	14.19	15:07	32
-112	18.31	14.62	15:18	30
-114	18.80	15.10	15:30	27
-115	19.07	15.34	15:35	26
-119	20.19	16.42	15:57	21
-123	21.41	17.60	16:17	16
-124	21.72	17.90	16:22	15
-125	22.05	18.21	16:26	14
-127	22.70	18.84	16:36	12
-128	23.03	19.16	16:40	11
-129	23.36	19.48	16:45	10
-130	23.70	19.80	16:49	9
-132	24.37	20.45	16:58	6
-133	24.71	20.78	17:02	5
-134	25.05	21.11	17:06	4
-135	25.40	21.44	17:10	3
-137	26.09	22.11	17:19	1
-138	26.44	22.44	17:23	0

PERCENTAGE OF SUN OBSCURED ALONG MAP TRACK
WEST END = 91.5   EAST END = 90.4

===
SUNSET POSITION : ALTITUDE IS -0.8 DEGREES
NORTH : LONGITUDE -138.8, LATITUDE 26.7
SOUTH : LONGITUDE -139.9, LATITUDE 23.1

+ MARKS THE CITY OF LO-YANG

ECLIPSE ON JD 1763518 DATE (Y/M/D)   116/ 4/ 1
DELTA T = 2 HOURS   20.3 MINS

LONGITUDE (-VE EAST)	LATITUDE NORTH	LIMITS. SOUTH	LOCAL TIME	MEAN ALTITUDE
-125	42.65	41.02	5:47	0
-127	42.81	41.19	5:55	2
-128	42.90	41.29	5:59	3
-129	43.00	41.40	6:03	3
-130	43.11	41.51	6:07	4
-132	43.36	41.78	6:16	6
-133	43.50	41.93	6:20	6
-134	43.65	42.08	6:25	7
-135	43.82	42.25	6:29	8
-137	44.18	42.62	6:38	10
-138	44.38	42.83	6:43	11
-139	44.59	43.04	6:47	11
-140	44.81	43.27	6:52	12
-142	45.29	43.76	7:01	14
-143	45.56	44.03	7:06	15
-144	45.83	44.31	7:11	15
-145	46.12	44.60	7:16	16
-147	46.74	45.23	7:26	18
-148	47.08	45.57	7:31	19
-149	47.42	45.92	7:37	19
-150	47.78	46.28	7:42	20

PERCENTAGE OF SUN OBSCURED ALONG MAP TRACK
WEST END = 97.5   EAST END = 98.1

===
SUNRISE POSITION : ALTITUDE IS -0.8 DEGREES
NORTH : LONGITUDE -124.2, LATITUDE 42.6
SOUTH : LONGITUDE -124.3, LATITUDE 41.0

+ MARKS THE CITY OF LO-YANG

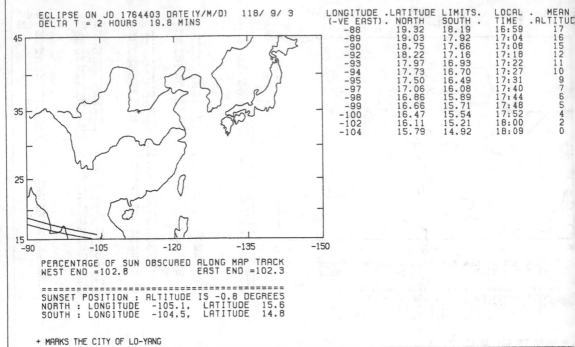

```
ECLIPSE ON JD 1764403 DATE(Y/M/D)   118/ 9/ 3
DELTA T = 2 HOURS  19.8 MINS
```

LONGITUDE (-VE EAST)	.LATITUDE NORTH	LIMITS. SOUTH	LOCAL TIME	MEAN .ALTITUDE
-88	19.32	18.19	16:59	17
-89	19.03	17.92	17:04	16
-90	18.75	17.66	17:08	15
-92	18.22	17.16	17:18	12
-93	17.97	16.93	17:22	11
-94	17.73	16.70	17:27	10
-95	17.50	16.49	17:31	9
-97	17.06	16.08	17:40	7
-98	16.86	15.89	17:44	6
-99	16.66	15.71	17:48	5
-100	16.47	15.54	17:52	4
-102	16.11	15.21	18:00	2
-104	15.79	14.92	18:09	0

```
PERCENTAGE OF SUN OBSCURED ALONG MAP TRACK
WEST END =102.8              EAST END =102.3

========================================
SUNSET POSITION : ALTITUDE IS -0.8 DEGREES
NORTH : LONGITUDE  -105.1,   LATITUDE  15.6
SOUTH : LONGITUDE  -104.5,   LATITUDE  14.8
```

+ MARKS THE CITY OF LO-YANG

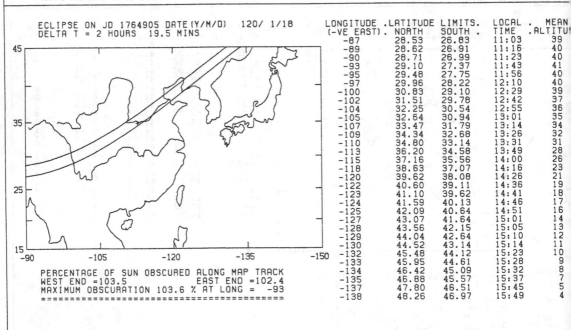

```
ECLIPSE ON JD 1764905 DATE(Y/M/D)   120/ 1/18
DELTA T = 2 HOURS  19.5 MINS
```

LONGITUDE (-VE EAST)	.LATITUDE NORTH	LIMITS. SOUTH	LOCAL TIME	MEAN .ALTITUDE
-87	28.53	26.83	11:03	39
-89	28.62	26.91	11:16	40
-90	28.71	26.99	11:23	40
-93	29.10	27.37	11:43	41
-95	29.48	27.75	11:56	40
-97	29.96	28.22	12:10	40
-100	30.83	29.10	12:29	39
-102	31.51	29.78	12:42	37
-104	32.25	30.54	12:55	36
-105	32.64	30.94	13:01	35
-107	33.47	31.79	13:14	34
-109	34.34	32.68	13:26	32
-110	34.80	33.14	13:31	31
-113	36.20	34.58	13:49	28
-115	37.16	35.56	14:00	26
-118	38.63	37.07	14:16	23
-120	39.62	38.08	14:26	21
-122	40.60	39.11	14:36	19
-123	41.10	39.62	14:41	18
-124	41.59	40.13	14:46	17
-125	42.09	40.64	14:51	16
-127	43.07	41.64	15:01	14
-128	43.56	42.15	15:05	13
-129	44.04	42.64	15:10	12
-130	44.52	43.14	15:14	11
-132	45.48	44.12	15:23	10
-133	45.95	44.61	15:28	9
-134	46.42	45.09	15:32	8
-135	46.88	45.57	15:37	7
-137	47.80	46.51	15:45	5
-138	48.26	46.97	15:49	4

```
PERCENTAGE OF SUN OBSCURED ALONG MAP TRACK
WEST END =103.5              EAST END =102.4
MAXIMUM OBSCURATION 103.6 % AT LONG =  -93
========================================
```

+ MARKS THE CITY OF LO-YANG

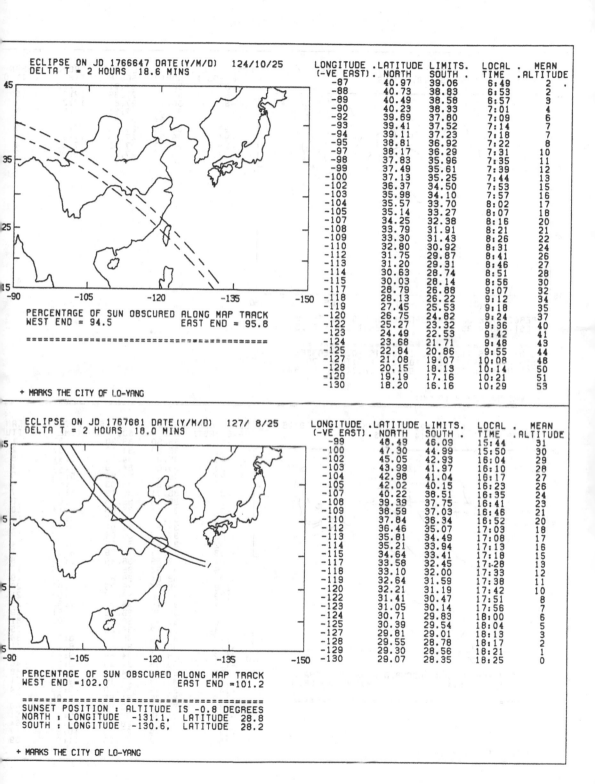

ECLIPSE ON JD 1766647 DATE(Y/M/D)   124/10/25
DELTA T = 2 HOURS 18.6 MINS

LONGITUDE (-VE EAST)	LATITUDE LIMITS. NORTH	SOUTH	LOCAL TIME	MEAN ALTITUDE
-87	40.97	39.06	6:49	2
-88	40.73	38.83	6:53	2
-89	40.49	38.58	6:57	3
-90	40.23	38.33	7:01	4
-92	39.69	37.80	7:09	6
-93	39.41	37.52	7:14	7
-94	39.11	37.23	7:18	7
-95	38.81	36.92	7:22	8
-97	38.17	36.29	7:31	10
-98	37.83	35.96	7:35	11
-99	37.49	35.61	7:39	12
-100	37.13	35.25	7:44	13
-102	36.37	34.50	7:53	15
-103	35.98	34.10	7:57	16
-104	35.57	33.70	8:02	17
-105	35.14	33.27	8:07	18
-107	34.25	32.38	8:16	20
-108	33.79	31.91	8:21	21
-109	33.30	31.43	8:26	22
-110	32.80	30.92	8:31	24
-112	31.75	29.87	8:41	26
-113	31.20	29.31	8:46	27
-114	30.63	28.74	8:51	28
-115	30.03	28.14	8:56	30
-117	28.79	26.88	9:07	32
-118	28.13	26.22	9:12	34
-119	27.45	25.53	9:18	35
-120	26.75	24.82	9:24	37
-122	25.27	23.32	9:36	40
-123	24.49	22.53	9:42	41
-124	23.68	21.71	9:48	43
-125	22.84	20.86	9:55	44
-127	21.08	19.07	10:08	48
-128	20.15	18.13	10:14	50
-120	19.19	17.16	10:21	51
-130	18.20	16.16	10:29	53

PERCENTAGE OF SUN OBSCURED ALONG MAP TRACK
WEST END = 94.5   EAST END = 95.8

===

+ MARKS THE CITY OF LO-YANG

ECLIPSE ON JD 1767681 DATE(Y/M/D)   127/ 8/25
DELTA T = 2 HOURS 18.0 MINS

LONGITUDE (-VE EAST)	LATITUDE LIMITS. NORTH	SOUTH	LOCAL TIME	MEAN ALTITUDE
-99	48.49	46.09	15:44	31
-100	47.30	44.99	15:50	30
-102	45.05	42.93	16:04	29
-103	43.99	41.97	16:10	28
-104	42.98	41.04	16:17	27
-105	42.02	40.15	16:23	26
-107	40.22	38.51	16:35	24
-108	39.39	37.75	16:41	23
-109	38.59	37.03	16:46	21
-110	37.84	36.34	16:52	20
-112	36.46	35.07	17:03	18
-113	35.81	34.49	17:08	17
-114	35.21	33.94	17:13	16
-115	34.64	33.41	17:18	15
-117	33.58	32.45	17:28	13
-118	33.10	32.00	17:33	12
-119	32.64	31.59	17:38	11
-120	32.21	31.19	17:42	10
-122	31.41	30.47	17:51	8
-123	31.05	30.14	17:56	7
-124	30.71	29.83	18:00	6
-125	30.39	29.54	18:04	5
-127	29.81	29.01	18:13	3
-128	29.55	28.78	18:17	2
-129	29.30	28.56	18:21	1
-130	29.07	28.35	18:25	0

PERCENTAGE OF SUN OBSCURED ALONG MAP TRACK
WEST END =102.0   EAST END =101.2

===
SUNSET POSITION : ALTITUDE IS -0.8 DEGREES
NORTH : LONGITUDE -131.1, LATITUDE 28.8
SOUTH : LONGITUDE -130.6, LATITUDE 28.2

+ MARKS THE CITY OF LO-YANG

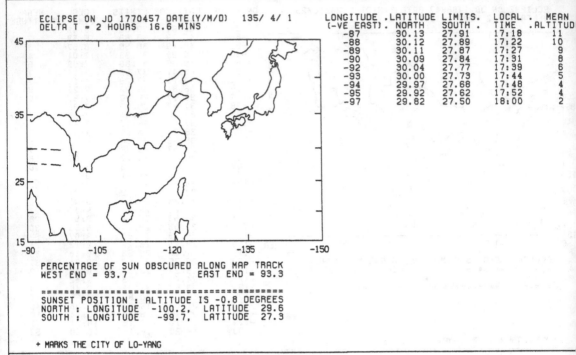

ECLIPSE ON JD 1770457 DATE(Y/M/D)   135/ 4/ 1
DELTA T = 2 HOURS 16.6 MINS

LONGITUDE (-VE EAST)	.LATITUDE . NORTH	LIMITS. SOUTH .	LOCAL TIME	. MEAN .ALTITUD
-87	30.13	27.91	17:18	11
-88	30.12	27.89	17:22	10
-89	30.11	27.87	17:27	9
-90	30.09	27.84	17:31	8
-92	30.04	27.77	17:39	6
-93	30.00	27.73	17:44	5
-94	29.97	27.68	17:48	4
-95	29.92	27.62	17:52	4
-97	29.82	27.50	18:00	2

PERCENTAGE OF SUN OBSCURED ALONG MAP TRACK
WEST END = 93.7   EAST END = 93.3

===
SUNSET POSITION : ALTITUDE IS -0.8 DEGREES
NORTH : LONGITUDE -100.2, LATITUDE 29.6
SOUTH : LONGITUDE -99.7, LATITUDE 27.3

+ MARKS THE CITY OF LO-YANG

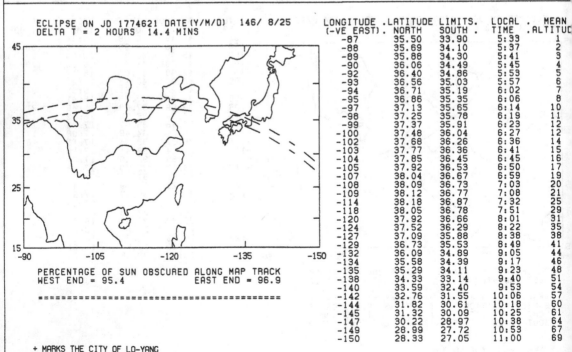

ECLIPSE ON JD 1774621 DATE(Y/M/D)   146/ 8/25
DELTA T = 2 HOURS 14.4 MINS

LONGITUDE (-VE EAST)	.LATITUDE . NORTH	LIMITS. SOUTH .	LOCAL TIME	. MEAN .ALTITUD
-87	35.50	33.90	5:33	1
-88	35.69	34.10	5:37	2
-89	35.88	34.30	5:41	3
-90	36.06	34.49	5:45	4
-92	36.40	34.86	5:53	5
-93	36.56	35.03	5:57	6
-94	36.71	35.19	6:02	7
-95	36.86	35.35	6:06	8
-97	37.13	35.65	6:14	10
-98	37.25	35.78	6:19	11
-99	37.37	35.91	6:23	12
-100	37.48	36.04	6:27	12
-102	37.68	36.26	6:36	14
-103	37.77	36.36	6:41	15
-104	37.85	36.45	6:45	16
-105	37.92	36.53	6:50	17
-107	38.04	36.67	6:59	19
-108	38.09	36.73	7:03	20
-109	38.12	36.77	7:08	21
-114	38.18	36.87	7:32	25
-118	38.05	36.78	7:51	29
-120	37.92	36.66	8:01	31
-124	37.52	36.29	8:22	35
-127	37.09	35.88	8:38	38
-129	36.73	35.53	8:49	41
-132	36.09	34.89	9:05	44
-134	35.58	34.39	9:17	46
-135	35.29	34.11	9:23	48
-138	34.33	33.14	9:40	51
-140	33.59	32.40	9:53	54
-142	32.76	31.55	10:06	57
-144	31.82	30.61	10:18	60
-145	31.32	30.09	10:25	61
-147	30.22	28.97	10:38	64
-149	28.99	27.72	10:53	67
-150	28.33	27.05	11:00	69

PERCENTAGE OF SUN OBSCURED ALONG MAP TRACK
WEST END = 95.4   EAST END = 96.9

===

+ MARKS THE CITY OF LO-YANG

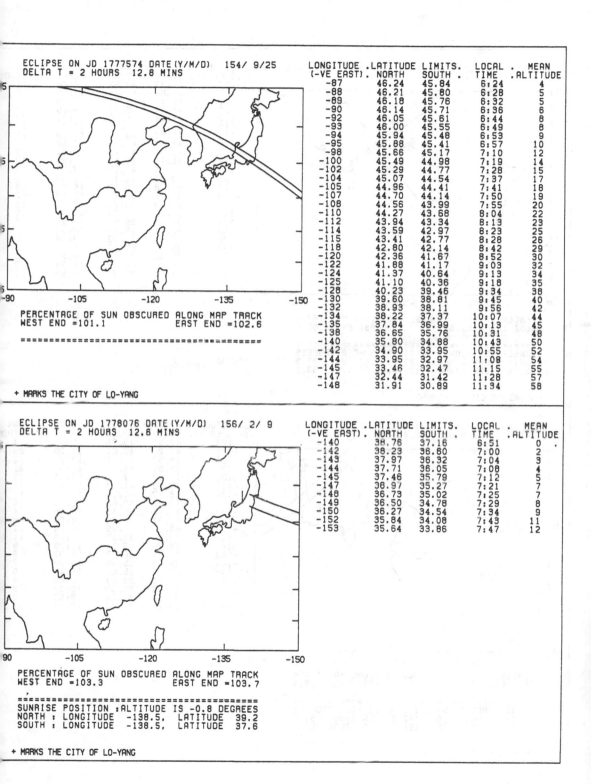

ECLIPSE ON JD 1777574 DATE(Y/M/D) 154/ 9/25
DELTA T = 2 HOURS 12.8 MINS

LONGITUDE (-VE EAST)	LATITUDE LIMITS. NORTH	SOUTH	LOCAL TIME	MEAN ALTITUDE
-87	46.24	45.84	6:24	4
-88	46.21	45.80	6:28	5
-89	46.18	45.76	6:32	5
-90	46.14	45.71	6:36	6
-92	46.05	45.61	6:44	8
-93	46.00	45.55	6:49	8
-94	45.94	45.48	6:53	9
-95	45.88	45.41	6:57	10
-98	45.66	45.17	7:10	12
-100	45.49	44.98	7:19	14
-102	45.29	44.77	7:28	15
-104	45.07	44.54	7:37	17
-105	44.96	44.41	7:41	18
-107	44.70	44.14	7:50	19
-108	44.56	43.99	7:55	20
-110	44.27	43.68	8:04	22
-112	43.94	43.34	8:13	23
-114	43.59	42.97	8:23	25
-115	43.41	42.77	8:28	26
-118	42.80	42.14	8:42	29
-120	42.36	41.67	8:52	30
-122	41.88	41.17	9:03	32
-124	41.37	40.64	9:13	34
-125	41.10	40.36	9:18	35
-128	40.23	39.46	9:34	38
-130	39.60	38.81	9:45	40
-132	38.93	38.11	9:56	42
-134	38.22	37.37	10:07	44
-135	37.84	36.99	10:13	45
-138	36.65	35.76	10:31	48
-140	35.80	34.88	10:43	50
-142	34.90	33.95	10:55	52
-144	33.95	32.97	11:08	54
-145	33.46	32.47	11:15	55
-147	32.44	31.42	11:28	57
-148	31.91	30.89	11:34	58

PERCENTAGE OF SUN OBSCURED ALONG MAP TRACK
WEST END =101.1   EAST END =102.6

==

+ MARKS THE CITY OF LO-YANG

ECLIPSE ON JD 1778076 DATE(Y/M/D) 156/ 2/ 9
DELTA T = 2 HOURS 12.6 MINS

LONGITUDE (-VE EAST)	LATITUDE LIMITS. NORTH	SOUTH	LOCAL TIME	MEAN ALTITUDE
-140	38.76	37.16	6:51	0
-142	38.23	36.60	7:00	2
-143	37.97	36.32	7:04	3
-144	37.71	36.05	7:08	4
-145	37.46	35.79	7:12	5
-147	36.97	35.27	7:21	7
-148	36.73	35.02	7:25	7
-149	36.50	34.78	7:29	8
-150	36.27	34.54	7:34	9
-152	35.84	34.08	7:43	11
-153	35.64	33.86	7:47	12

PERCENTAGE OF SUN OBSCURED ALONG MAP TRACK
WEST END =103.3   EAST END =103.7

==
SUNRISE POSITION :ALTITUDE IS -0.8 DEGREES
NORTH : LONGITUDE -138.5, LATITUDE 39.2
SOUTH : LONGITUDE -138.5, LATITUDE 37.6

+ MARKS THE CITY OF LO-YANG

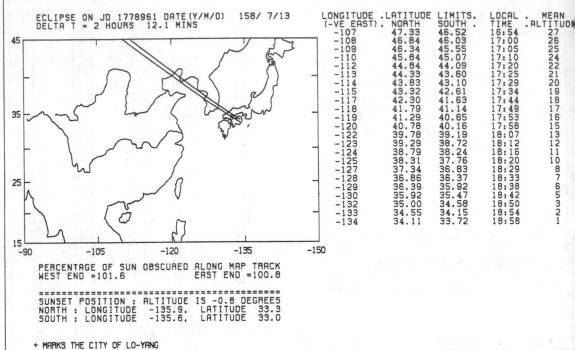

ECLIPSE ON JD 1778961 DATE(Y/M/D)   158/ 7/13
DELTA T = 2 HOURS 12.1 MINS

LONGITUDE (-VE EAST)	LATITUDE NORTH	LIMITS. SOUTH	LOCAL TIME	MEAN ALTITUDE
-107	47.33	46.52	16:54	27
-108	46.84	46.03	17:00	26
-109	46.34	45.55	17:05	25
-110	45.84	45.07	17:10	24
-112	44.84	44.09	17:20	22
-113	44.33	43.60	17:25	21
-114	43.83	43.10	17:29	20
-115	43.32	42.61	17:34	19
-117	42.30	41.63	17:44	18
-118	41.79	41.14	17:49	17
-119	41.29	40.65	17:53	16
-120	40.78	40.16	17:58	15
-122	39.78	39.19	18:07	13
-123	39.29	38.72	18:12	12
-124	38.79	38.24	18:16	11
-125	38.31	37.76	18:20	10
-127	37.34	36.83	18:29	8
-128	36.86	36.37	18:33	7
-129	36.39	35.92	18:38	6
-130	35.92	35.47	18:42	5
-132	35.00	34.58	18:50	3
-133	34.55	34.15	18:54	2
-134	34.11	33.72	18:58	1

PERCENTAGE OF SUN OBSCURED ALONG MAP TRACK
WEST END =101.6   EAST END =100.8

==
SUNSET POSITION : ALTITUDE IS -0.8 DEGREES
NORTH : LONGITUDE -135.9, LATITUDE   33.3
SOUTH : LONGITUDE -135.6, LATITUDE   33.0

+ MARKS THE CITY OF LO-YANG

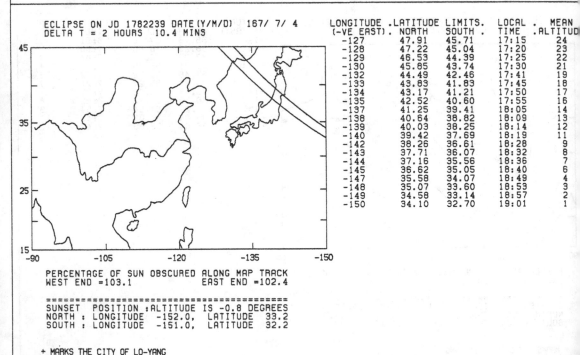

ECLIPSE ON JD 1782239 DATE(Y/M/D)   167/ 7/ 4
DELTA T = 2 HOURS 10.4 MINS

LONGITUDE (-VE EAST)	LATITUDE NORTH	LIMITS. SOUTH	LOCAL TIME	MEAN ALTITUDE
-127	47.91	45.71	17:15	24
-128	47.22	45.04	17:20	23
-129	46.53	44.39	17:25	22
-130	45.85	43.74	17:30	21
-132	44.49	42.46	17:41	19
-133	43.83	41.83	17:45	18
-134	43.17	41.21	17:50	17
-135	42.52	40.60	17:55	16
-137	41.25	39.41	18:05	14
-138	40.64	38.82	18:09	13
-139	40.03	38.25	18:14	12
-140	39.42	37.69	18:19	11
-142	38.26	36.61	18:28	9
-143	37.71	36.07	18:32	8
-144	37.16	35.56	18:36	7
-145	36.62	35.05	18:40	6
-147	35.58	34.07	18:49	4
-148	35.07	33.60	18:53	3
-149	34.58	33.14	18:57	2
-150	34.10	32.70	19:01	1

PERCENTAGE OF SUN OBSCURED ALONG MAP TRACK
WEST END =103.1   EAST END =102.4

==
SUNSET POSITION :ALTITUDE IS -0.8 DEGREES
NORTH : LONGITUDE -152.0, LATITUDE   33.2
SOUTH : LONGITUDE -151.0, LATITUDE   32.2

+ MARKS THE CITY OF LO-YANG

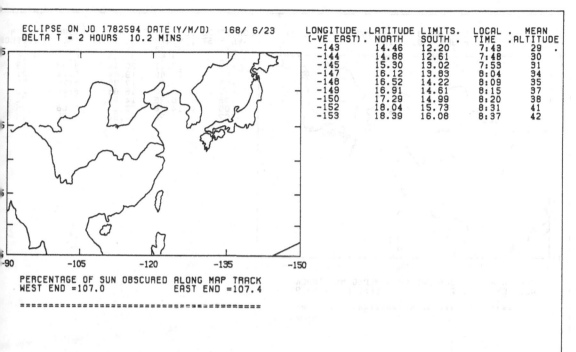

ECLIPSE ON JD 1782594 DATE(Y/M/D)   168/ 6/23
DELTA T = 2 HOURS 10.2 MINS

LONGITUDE (-VE EAST)	LATITUDE NORTH	LIMITS SOUTH	LOCAL TIME	MEAN ALTITUDE
-143	14.46	12.20	7:43	29
-144	14.88	12.61	7:48	30
-145	15.30	13.02	7:53	31
-147	16.12	13.83	8:04	34
-148	16.52	14.22	8:09	35
-149	16.91	14.61	8:15	37
-150	17.29	14.99	8:20	38
-152	18.04	15.73	8:31	41
-153	18.39	16.08	8:37	42

PERCENTAGE OF SUN OBSCURED ALONG MAP TRACK
WEST END =107.0   EAST END =107.4

=======================================

+ MARKS THE CITY OF LO-YANG

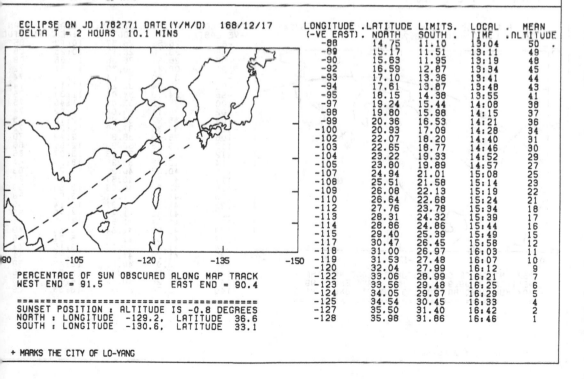

ECLIPSE ON JD 1782771 DATE(Y/M/D)   168/12/17
DELTA T = 2 HOURS 10.1 MINS

LONGITUDE (-VE EAST)	LATITUDE NORTH	LIMITS SOUTH	LOCAL TIME	MEAN ALTITUDE
-88	14.75	11.10	13:04	50
-89	15.17	11.51	13:11	49
-90	15.63	11.95	13:19	48
-92	16.59	12.87	13:34	45
-93	17.10	13.36	13:41	44
-94	17.61	13.87	13:48	43
-95	18.15	14.38	13:55	41
-97	19.24	15.44	14:08	38
-98	19.80	15.98	14:15	37
-99	20.36	16.53	14:21	36
-100	20.93	17.09	14:28	34
-102	22.07	18.20	14:40	31
-103	22.65	18.77	14:46	30
-104	23.22	19.33	14:52	29
-105	23.80	19.89	14:57	27
-107	24.94	21.01	15:08	25
-108	25.51	21.58	15:14	23
-109	26.08	22.13	15:19	22
-110	26.64	22.68	15:24	21
-112	27.76	23.78	15:34	18
-113	28.31	24.32	15:39	17
-114	28.86	24.86	15:44	16
-115	29.40	25.39	15:49	15
-117	30.47	26.45	15:58	12
-118	31.00	26.97	16:03	11
-119	31.53	27.48	16:07	10
-120	32.04	27.99	16:12	9
-122	33.06	28.99	16:21	7
-123	33.56	29.48	16:25	6
-124	34.05	29.97	16:29	5
-125	34.54	30.45	16:33	4
-127	35.50	31.40	16:42	2
-128	35.98	31.86	16:46	1

PERCENTAGE OF SUN OBSCURED ALONG MAP TRACK
WEST END = 91.5   EAST END = 90.4

===
SUNSET POSITION : ALTITUDE IS -0.8 DEGREES
NORTH : LONGITUDE -129.2, LATITUDE   36.6
SOUTH : LONGITUDE -130.6, LATITUDE   33.1

+ MARKS THE CITY OF LO-YANG

213

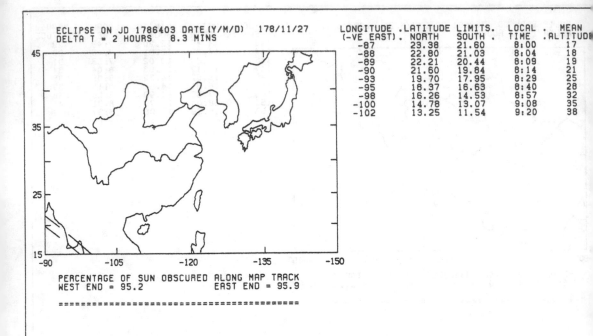

ECLIPSE ON JD 1786403 DATE (Y/M/D)   178/11/27
DELTA T = 2 HOURS   8.3 MINS

LONGITUDE (-VE EAST)	LATITUDE LIMITS. NORTH	SOUTH .	LOCAL TIME	MEAN .ALTITUDE
-87	23.38	21.60	8:00	17
-88	22.80	21.03	8:04	18
-89	22.21	20.44	8:09	19
-90	21.60	19.84	8:14	21
-93	19.70	17.95	8:29	25
-95	18.37	16.63	8:40	28
-98	16.26	14.53	8:57	32
-100	14.78	13.07	9:08	35
-102	13.25	11.54	9:20	38

PERCENTAGE OF SUN OBSCURED ALONG MAP TRACK
WEST END = 95.2   EAST END = 95.9

==

+ MARKS THE CITY OF LO-YANG

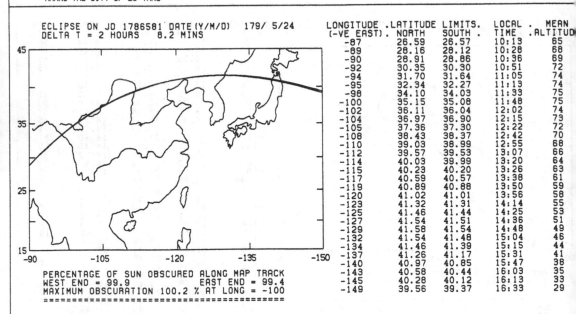

ECLIPSE ON JD 1786581 DATE (Y/M/D)   179/ 5/24
DELTA T = 2 HOURS   8.2 MINS

LONGITUDE (-VE EAST)	LATITUDE LIMITS. NORTH	SOUTH .	LOCAL TIME	MEAN .ALTITUDE
-87	26.59	26.57	10:13	65
-89	28.16	28.12	10:28	68
-90	28.91	28.86	10:36	69
-92	30.35	30.30	10:51	72
-94	31.70	31.64	11:05	74
-95	32.34	32.27	11:13	74
-98	34.10	34.03	11:33	75
-100	35.15	35.08	11:48	75
-102	36.11	36.04	12:02	74
-104	36.97	36.90	12:15	73
-105	37.36	37.30	12:22	72
-108	38.43	38.37	12:42	70
-110	39.03	38.99	12:55	68
-112	39.57	39.53	13:07	66
-114	40.03	39.99	13:20	64
-115	40.23	40.20	13:26	63
-117	40.59	40.57	13:38	61
-119	40.89	40.88	13:50	59
-120	41.02	41.01	13:56	58
-123	41.32	41.31	14:14	55
-125	41.46	41.44	14:25	53
-127	41.54	41.51	14:36	51
-129	41.58	41.54	14:48	49
-132	41.54	41.48	15:04	46
-134	41.46	41.39	15:15	44
-137	41.26	41.17	15:31	41
-140	40.97	40.85	15:47	38
-143	40.58	40.44	16:03	35
-145	40.28	40.12	16:13	33
-149	39.56	39.37	16:33	29

PERCENTAGE OF SUN OBSCURED ALONG MAP TRACK
WEST END = 99.9   EAST END = 99.4
MAXIMUM OBSCURATION 100.2 % AT LONG = -100

==

+ MARKS THE CITY OF LO-YANG

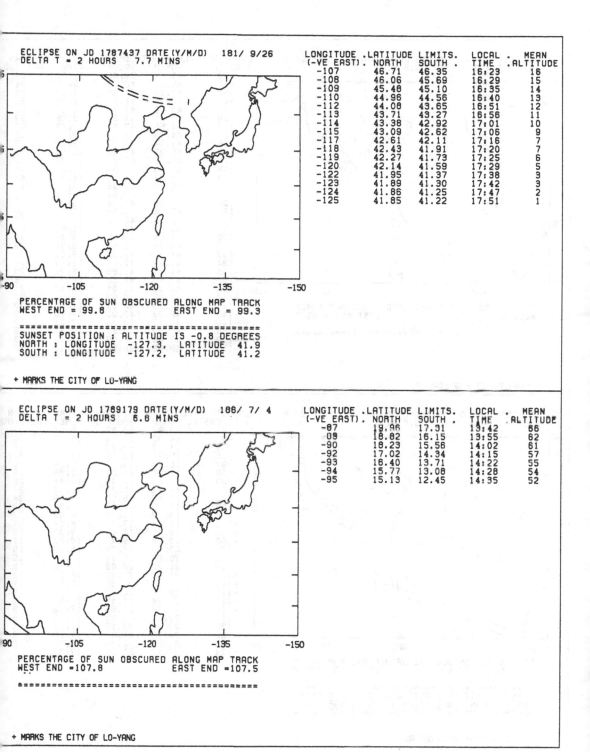

ECLIPSE ON JD 1787437 DATE(Y/M/D) 181/ 9/26
DELTA T = 2 HOURS   7.7 MINS

LONGITUDE (-VE EAST)	LATITUDE LIMITS. NORTH	SOUTH	LOCAL TIME	MEAN ALTITUDE
-107	46.71	46.35	16:23	18
-108	46.06	45.69	16:29	15
-109	45.48	45.10	16:35	14
-110	44.96	44.56	16:40	13
-112	44.08	43.65	16:51	12
-113	43.71	43.27	16:56	11
-114	43.38	42.92	17:01	10
-115	43.09	42.62	17:06	9
-117	42.61	42.11	17:16	7
-118	42.43	41.91	17:20	7
-119	42.27	41.73	17:25	6
-120	42.14	41.59	17:29	5
-122	41.95	41.37	17:38	3
-123	41.89	41.30	17:42	3
-124	41.86	41.25	17:47	2
-125	41.85	41.22	17:51	1

PERCENTAGE OF SUN OBSCURED ALONG MAP TRACK
WEST END = 99.8   EAST END = 99.3

===
SUNSET POSITION : ALTITUDE IS -0.8 DEGREES
NORTH : LONGITUDE -127.3, LATITUDE 41.9
SOUTH : LONGITUDE -127.2, LATITUDE 41.2

+ MARKS THE CITY OF LO-YANG

ECLIPSE ON JD 1789179 DATE(Y/M/D) 186/ 7/ 4
DELTA T = 2 HOURS   6.8 MINS

LONGITUDE (-VE EAST)	LATITUDE LIMITS. NORTH	SOUTH	LOCAL TIME	MEAN ALTITUDE
-87	19.86	17.91	13:42	66
-89	18.82	16.15	13:55	62
-90	18.23	15.56	14:02	61
-92	17.02	14.34	14:15	57
-93	16.40	13.71	14:22	55
-94	15.77	13.08	14:28	54
-95	15.13	12.45	14:35	52

PERCENTAGE OF SUN OBSCURED ALONG MAP TRACK
WEST END =107.8   EAST END =107.5

==

+ MARKS THE CITY OF LO-YANG

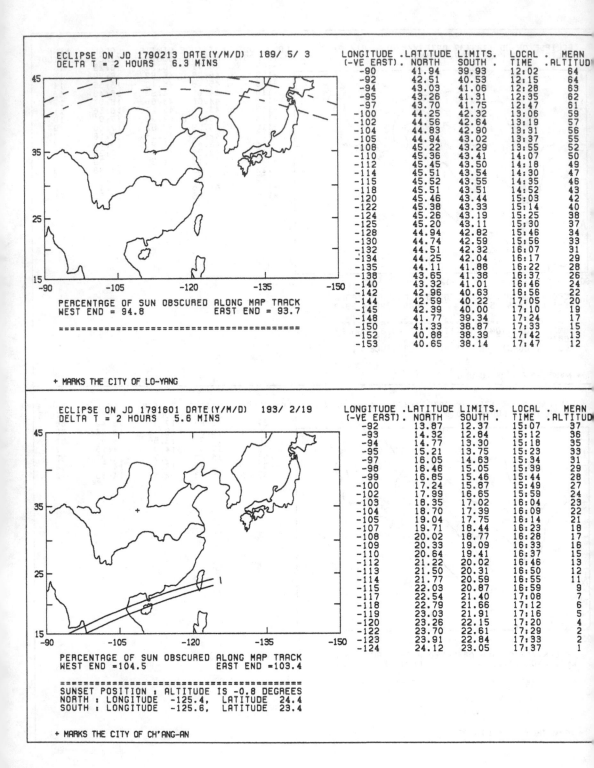

ECLIPSE ON JD 1790213 DATE(Y/M/D) 189/ 5/ 3
DELTA T = 2 HOURS   6.3 MINS

LONGITUDE (-VE EAST)	LATITUDE NORTH	LIMITS. SOUTH	LOCAL TIME	MEAN ALTITUDE
-90	41.94	39.93	12:02	64
-92	42.51	40.53	12:15	64
-94	43.03	41.06	12:28	63
-95	43.26	41.31	12:35	62
-97	43.70	41.75	12:47	61
-100	44.25	42.32	13:06	59
-102	44.56	42.64	13:19	57
-104	44.83	42.90	13:31	56
-105	44.94	43.02	13:37	55
-108	45.22	43.29	13:55	52
-110	45.36	43.41	14:07	50
-112	45.45	43.50	14:18	49
-114	45.51	43.54	14:30	47
-115	45.52	43.55	14:35	46
-118	45.51	43.51	14:52	43
-120	45.46	43.44	15:03	42
-122	45.38	43.33	15:14	40
-124	45.26	43.19	15:25	38
-125	45.20	43.11	15:30	37
-128	44.94	42.82	15:46	34
-130	44.74	42.59	15:56	33
-132	44.51	42.32	16:07	31
-134	44.25	42.04	16:17	29
-135	44.11	41.88	16:22	28
-138	43.65	41.38	16:37	26
-140	43.32	41.01	16:46	24
-142	42.96	40.63	16:56	22
-144	42.59	40.22	17:05	20
-145	42.39	40.00	17:10	19
-148	41.77	39.34	17:24	17
-150	41.33	38.87	17:33	15
-152	40.88	38.39	17:42	13
-153	40.65	38.14	17:47	12

PERCENTAGE OF SUN OBSCURED ALONG MAP TRACK
WEST END = 94.8   EAST END = 93.7

==

+ MARKS THE CITY OF LO-YANG

ECLIPSE ON JD 1791601 DATE(Y/M/D) 193/ 2/19
DELTA T = 2 HOURS   5.6 MINS

LONGITUDE (-VE EAST)	LATITUDE NORTH	LIMITS. SOUTH	LOCAL TIME	MEAN ALTITUDE
-92	13.87	12.37	15:07	37
-93	14.32	12.84	15:12	36
-94	14.77	13.30	15:18	35
-95	15.21	13.75	15:23	33
-97	16.05	14.63	15:34	31
-98	16.46	15.05	15:39	29
-99	16.85	15.46	15:44	28
-100	17.24	15.87	15:49	27
-102	17.99	16.65	15:59	24
-103	18.35	17.02	16:04	23
-104	18.70	17.39	16:09	22
-105	19.04	17.75	16:14	21
-107	19.71	18.44	16:23	18
-108	20.02	18.77	16:28	17
-109	20.33	19.09	16:33	16
-110	20.64	19.41	16:37	15
-112	21.22	20.02	16:46	13
-113	21.50	20.31	16:50	12
-114	21.77	20.59	16:55	11
-115	22.03	20.87	16:59	9
-117	22.54	21.40	17:08	7
-118	22.79	21.66	17:12	6
-119	23.03	21.91	17:16	5
-120	23.26	22.15	17:20	4
-122	23.70	22.61	17:29	2
-123	23.91	22.84	17:33	2
-124	24.12	23.05	17:37	1

PERCENTAGE OF SUN OBSCURED ALONG MAP TRACK
WEST END =104.5   EAST END =103.4

==
SUNSET POSITION : ALTITUDE IS -0.8 DEGREES
NORTH : LONGITUDE -125.4, LATITUDE 24.4
SOUTH : LONGITUDE -125.6, LATITUDE 23.4

+ MARKS THE CITY OF CH'ANG-AN

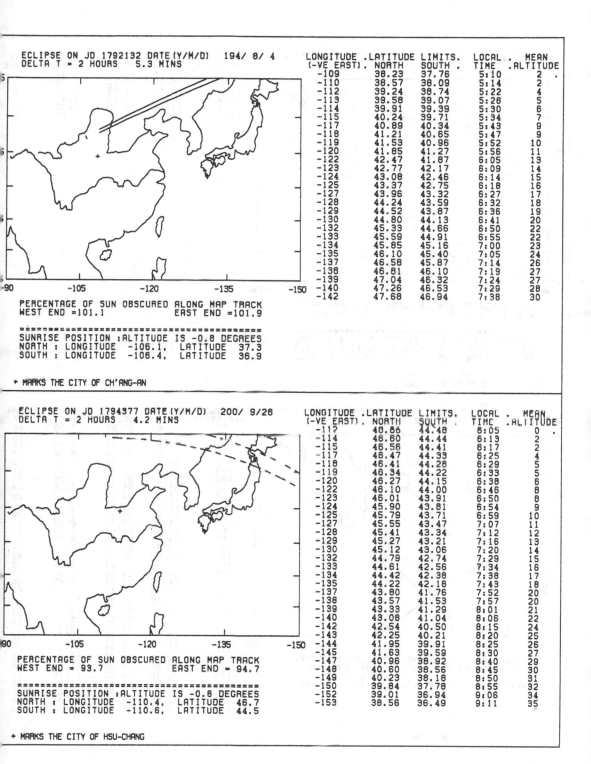

ECLIPSE ON JD 1792132 DATE(Y/M/D)   194/ 8/ 4
DELTA T = 2 HOURS   5.3 MINS

LONGITUDE (-VE EAST)	.LATITUDE NORTH	LIMITS. SOUTH	LOCAL TIME	. MEAN .ALTITUDE
-109	38.23	37.76	5:10	2
-110	38.57	38.09	5:14	2
-112	39.24	38.74	5:22	4
-113	39.58	39.07	5:26	5
-114	39.91	39.39	5:30	6
-115	40.24	39.71	5:34	7
-117	40.89	40.34	5:43	9
-118	41.21	40.65	5:47	9
-119	41.53	40.96	5:52	10
-120	41.85	41.27	5:56	11
-122	42.47	41.87	6:05	13
-123	42.77	42.17	6:09	14
-124	43.08	42.46	6:14	15
-125	43.37	42.75	6:18	16
-127	43.96	43.32	6:27	17
-128	44.24	43.59	6:32	18
-129	44.52	43.87	6:36	19
-130	44.80	44.13	6:41	20
-132	45.33	44.66	6:50	22
-133	45.59	44.91	6:55	22
-134	45.85	45.16	7:00	23
-135	46.10	45.40	7:05	24
-137	46.58	45.87	7:14	26
-138	46.81	46.10	7:19	27
-139	47.04	46.32	7:24	27
-140	47.26	46.53	7:29	28
-142	47.68	46.94	7:38	30

PERCENTAGE OF SUN OBSCURED ALONG MAP TRACK
WEST END =101.1   EAST END =101.9

===
SUNRISE POSITION :ALTITUDE IS -0.8 DEGREES
NORTH : LONGITUDE -106.1, LATITUDE 37.3
SOUTH : LONGITUDE -106.4, LATITUDE 36.9

+ MARKS THE CITY OF CH'ANG-AN

ECLIPSE ON JD 1794377 DATE(Y/M/D)   200/ 9/26
DELTA T = 2 HOURS   4.2 MINS

LONGITUDE (-VE EAST)	.LATITUDE NORTH	LIMITS. SOUTH	LOCAL TIME	. MEAN .ALTITUDE
-112	40.66	44.48	6:05	0
-114	46.60	44.44	6:13	2
-115	46.56	44.41	6:17	2
-117	46.47	44.39	6:25	4
-118	46.41	44.28	6:29	5
-119	46.34	44.22	6:33	5
-120	46.27	44.15	6:38	6
-122	46.10	44.00	6:46	8
-123	46.01	43.91	6:50	8
-124	45.90	43.81	6:54	9
-125	45.79	43.71	6:59	10
-127	45.55	43.47	7:07	11
-128	45.41	43.34	7:12	12
-129	45.27	43.21	7:16	13
-130	45.12	43.06	7:20	14
-132	44.79	42.74	7:29	15
-133	44.61	42.56	7:34	16
-134	44.42	42.38	7:38	17
-135	44.22	42.18	7:43	18
-137	43.80	41.76	7:52	20
-138	43.57	41.53	7:57	20
-139	43.33	41.29	8:01	21
-140	43.08	41.04	8:06	22
-142	42.54	40.50	8:15	24
-143	42.25	40.21	8:20	25
-144	41.95	39.91	8:25	26
-145	41.63	39.59	8:30	27
-147	40.96	38.92	8:40	29
-148	40.60	38.56	8:45	30
-149	40.23	38.18	8:50	31
-150	39.84	37.78	8:55	32
-152	39.01	36.94	9:06	34
-153	38.56	36.49	9:11	35

PERCENTAGE OF SUN OBSCURED ALONG MAP TRACK
WEST END = 93.7   EAST END = 94.7

===
SUNRISE POSITION :ALTITUDE IS -0.8 DEGREES
NORTH : LONGITUDE -110.4, LATITUDE 46.7
SOUTH : LONGITUDE -110.6, LATITUDE 44.5

+ MARKS THE CITY OF HSU-CHANG

217

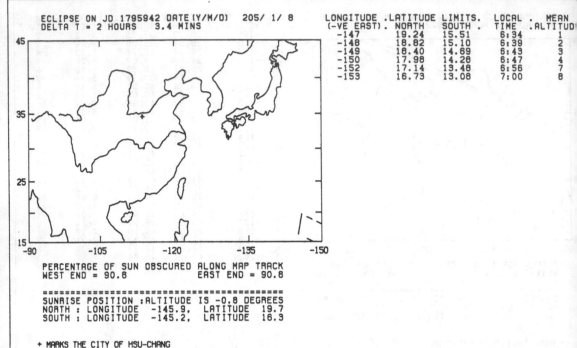

ECLIPSE ON JD 1795942 DATE(Y/M/D)   205/ 1/ 8
DELTA T = 2 HOURS   3.4 MINS

LONGITUDE (-VE EAST)	LATITUDE NORTH	LIMITS SOUTH	LOCAL TIME	MEAN ALTITUD
-147	19.24	15.51	6:34	1
-148	18.82	15.10	6:39	2
-149	18.40	14.69	6:43	3
-150	17.98	14.28	6:47	4
-152	17.14	13.48	6:56	7
-153	16.73	13.08	7:00	8

PERCENTAGE OF SUN OBSCURED ALONG MAP TRACK
WEST END = 90.8   EAST END = 90.8

SUNRISE POSITION :ALTITUDE IS -0.8 DEGREES
NORTH : LONGITUDE -145.9, LATITUDE 19.7
SOUTH : LONGITUDE -145.2, LATITUDE 16.3

+ MARKS THE CITY OF HSU-CHANG

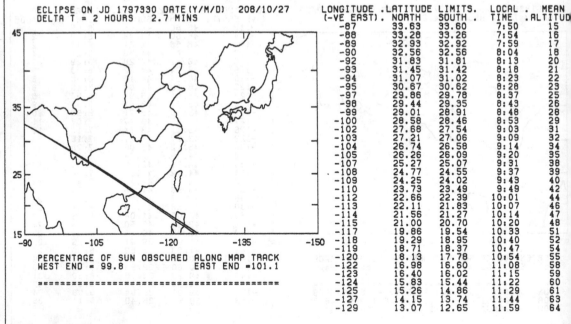

ECLIPSE ON JD 1797330 DATE(Y/M/D)   208/10/27
DELTA T = 2 HOURS   2.7 MINS

LONGITUDE (-VE EAST)	LATITUDE NORTH	LIMITS SOUTH	LOCAL TIME	MEAN ALTITUD
-87	33.63	33.60	7:50	15
-88	33.28	33.26	7:54	16
-89	32.93	32.92	7:59	17
-90	32.56	32.56	8:04	18
-92	31.83	31.81	8:13	20
-93	31.45	31.42	8:18	21
-94	31.07	31.02	8:23	22
-95	30.67	30.62	8:28	23
-97	29.86	29.78	8:37	25
-98	29.44	29.35	8:43	26
-99	29.01	28.91	8:48	28
-100	28.58	28.46	8:53	29
-102	27.68	27.54	9:03	31
-103	27.21	27.06	9:09	32
-104	26.74	26.58	9:14	34
-105	26.26	26.09	9:20	35
-107	25.27	25.07	9:31	38
-108	24.77	24.55	9:37	39
-109	24.25	24.02	9:43	40
-110	23.73	23.49	9:49	42
-112	22.66	22.39	10:01	44
-113	22.11	21.83	10:07	46
-114	21.56	21.27	10:14	47
-115	21.00	20.70	10:20	48
-117	19.86	19.54	10:33	51
-118	19.29	18.95	10:40	52
-119	18.71	18.37	10:47	54
-120	18.13	17.78	10:54	55
-122	16.98	16.60	11:08	58
-123	16.40	16.02	11:15	59
-124	15.83	15.44	11:22	60
-125	15.26	14.86	11:29	61
-127	14.15	13.74	11:44	63
-129	13.07	12.65	11:59	64

PERCENTAGE OF SUN OBSCURED ALONG MAP TRACK
WEST END = 99.8   EAST END =101.1

+ MARKS THE CITY OF HSU-CHANG

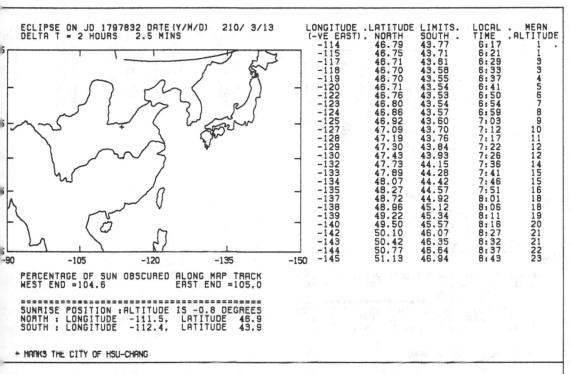

ECLIPSE ON JD 1797832 DATE(Y/M/D) 210/ 3/13
DELTA T = 2 HOURS   2.5 MINS

LONGITUDE (-VE EAST)	LATITUDE LIMITS NORTH	SOUTH	LOCAL TIME	MEAN ALTITUDE
-114	46.79	43.77	6:17	1
-115	46.75	43.71	6:21	1
-117	46.71	43.61	6:29	3
-118	46.70	43.58	6:33	3
-119	46.70	43.55	6:37	4
-120	46.71	43.54	6:41	5
-122	46.76	43.53	6:50	6
-123	46.80	43.54	6:54	7
-124	46.86	43.57	6:59	8
-125	46.92	43.60	7:03	9
-127	47.09	43.70	7:12	10
-128	47.19	43.76	7:17	11
-129	47.30	43.84	7:22	12
-130	47.43	43.93	7:26	12
-132	47.73	44.15	7:36	14
-133	47.89	44.28	7:41	15
-134	48.07	44.42	7:46	15
-135	48.27	44.57	7:51	16
-137	48.72	44.92	8:01	18
-138	48.96	45.12	8:06	18
-139	49.22	45.34	8:11	19
-140	49.50	45.57	8:16	20
-142	50.10	46.07	8:27	21
-143	50.42	46.35	8:32	21
-144	50.77	46.64	8:37	22
-145	51.13	46.94	8:43	23

PERCENTAGE OF SUN OBSCURED ALONG MAP TRACK
WEST END =104.6   EAST END =105.0

=======================================
SUNRISE POSITION :ALTITUDE IS -0.8 DEGREES
NORTH : LONGITUDE -111.5, LATITUDE 46.9
SOUTH : LONGITUDE -112.4, LATITUDE 43.9

+ MARKS THE CITY OF HSU-CHANG

ECLIPSE ON JD 1798717 DATE(Y/M/D) 212/ 8/14
DELTA T = 2 HOURS   2.0 MINS

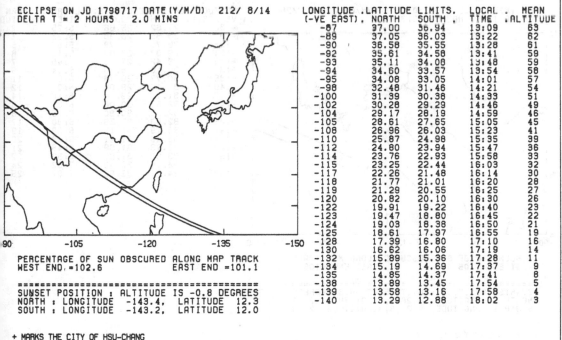

LONGITUDE (-VE EAST)	LATITUDE LIMITS NORTH	SOUTH	LOCAL TIME	MEAN ALTITUDE
-87	37.00	36.94	13:09	63
-89	37.05	36.03	13:22	62
-90	36.58	35.55	13:28	61
-92	35.61	34.58	13:41	59
-93	35.11	34.08	13:48	59
-94	34.60	33.57	13:54	58
-95	34.08	33.05	14:01	57
-98	32.48	31.46	14:21	54
-100	31.39	30.38	14:33	51
-102	30.28	29.29	14:46	49
-104	29.17	28.19	14:59	46
-105	28.61	27.65	15:05	45
-108	26.96	26.03	15:23	41
-110	25.87	24.98	15:35	39
-112	24.80	23.94	15:47	36
-114	23.76	22.93	15:58	33
-115	23.25	22.44	16:03	32
-117	22.26	21.48	16:14	30
-118	21.77	21.01	16:20	28
-119	21.29	20.55	16:25	27
-120	20.82	20.10	16:30	26
-122	19.91	19.22	16:40	23
-123	19.47	18.80	16:45	22
-124	19.03	18.38	16:50	21
-125	18.61	17.97	16:55	19
-128	17.39	16.80	17:10	16
-130	16.62	16.06	17:19	14
-132	15.89	15.36	17:28	11
-134	15.19	14.69	17:37	9
-135	14.85	14.37	17:41	8
-138	13.89	13.45	17:54	5
-139	13.58	13.16	17:58	4
-140	13.29	12.88	18:02	3

PERCENTAGE OF SUN OBSCURED ALONG MAP TRACK
WEST END =102.6   EAST END =101.1

=======================================
SUNSET POSITION : ALTITUDE IS -0.8 DEGREES
NORTH : LONGITUDE -143.4, LATITUDE 12.3
SOUTH : LONGITUDE -143.2, LATITUDE 12.0

+ MARKS THE CITY OF HSU-CHANG

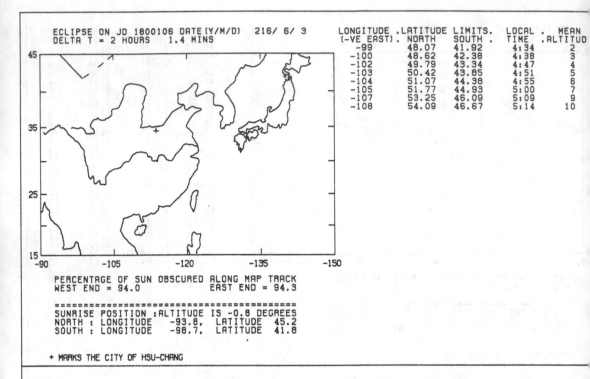

ECLIPSE ON JD 1800106 DATE(Y/M/D) 216/ 6/ 3
DELTA T = 2 HOURS   1.4 MINS

LONGITUDE (-VE EAST).	.LATITUDE LIMITS. NORTH	SOUTH .	LOCAL . TIME	MEAN .ALTITUD
-99	48.07	41.92	4:34	2
-100	48.62	42.38	4:38	3
-102	49.79	43.34	4:47	4
-103	50.42	43.85	4:51	5
-104	51.07	44.38	4:55	6
-105	51.77	44.93	5:00	7
-107	53.25	46.09	5:09	9
-108	54.09	46.67	5:14	10

PERCENTAGE OF SUN OBSCURED ALONG MAP TRACK
WEST END = 94.0   EAST END = 94.3

==
SUNRISE POSITION :ALTITUDE IS -0.8 DEGREES
NORTH : LONGITUDE -93.8, LATITUDE 45.2
SOUTH : LONGITUDE -98.7, LATITUDE 41.8

+ MARKS THE CITY OF HSU-CHANG

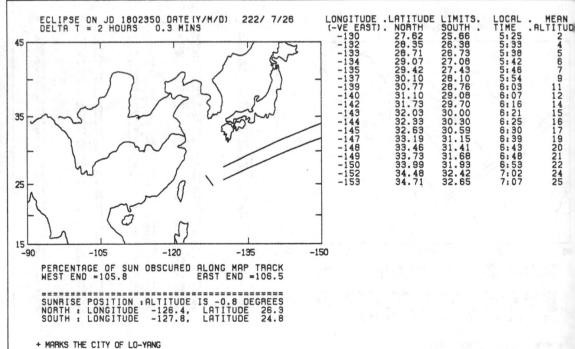

ECLIPSE ON JD 1802350 DATE(Y/M/D) 222/ 7/26
DELTA T = 2 HOURS   0.3 MINS

LONGITUDE (-VE EAST).	.LATITUDE LIMITS. NORTH	SOUTH .	LOCAL . TIME	MEAN .ALTITUD
-130	27.62	25.66	5:25	2
-132	28.35	26.38	5:33	4
-133	28.71	26.73	5:38	5
-134	29.07	27.08	5:42	6
-135	29.42	27.43	5:46	7
-137	30.10	28.10	5:54	9
-139	30.77	28.76	6:03	11
-140	31.10	29.08	6:07	12
-142	31.73	29.70	6:16	14
-143	32.03	30.00	6:21	15
-144	32.33	30.30	6:25	16
-145	32.63	30.59	6:30	17
-147	33.19	31.15	6:39	19
-148	33.46	31.41	6:43	20
-149	33.73	31.68	6:48	21
-150	33.99	31.93	6:53	22
-152	34.48	32.42	7:02	24
-153	34.71	32.65	7:07	25

PERCENTAGE OF SUN OBSCURED ALONG MAP TRACK
WEST END =105.8   EAST END =106.5

==
SUNRISE POSITION :ALTITUDE IS -0.8 DEGREES
NORTH : LONGITUDE -126.4, LATITUDE 26.3
SOUTH : LONGITUDE -127.8, LATITUDE 24.8

+ MARKS THE CITY OF LO-YANG

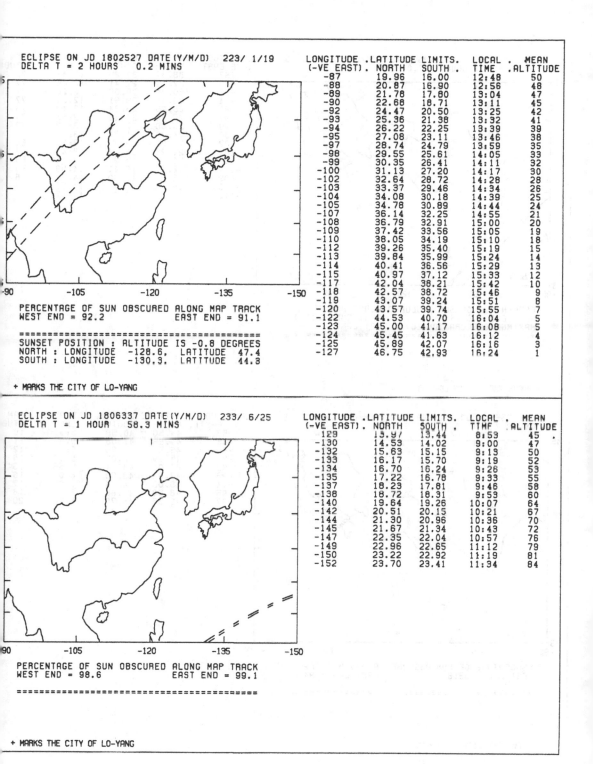

ECLIPSE ON JD 1802527 DATE(Y/M/D) 223/ 1/19
DELTA T = 2 HOURS   0.2 MINS

LONGITUDE (-VE EAST)	LATITUDE LIMITS. NORTH	SOUTH	LOCAL TIME	MEAN ALTITUDE
-87	19.96	16.00	12:48	50
-88	20.87	16.90	12:56	48
-89	21.78	17.80	13:04	47
-90	22.68	18.71	13:11	45
-92	24.47	20.50	13:25	42
-93	25.36	21.38	13:32	41
-94	26.22	22.25	13:39	39
-95	27.08	23.11	13:46	38
-97	28.74	24.79	13:59	35
-98	29.55	25.61	14:05	33
-99	30.35	26.41	14:11	32
-100	31.13	27.20	14:17	30
-102	32.64	28.72	14:28	28
-103	33.37	29.46	14:34	26
-104	34.08	30.18	14:39	25
-105	34.78	30.89	14:44	24
-107	36.14	32.25	14:55	21
-108	36.79	32.91	15:00	20
-109	37.42	33.56	15:05	19
-110	38.05	34.19	15:10	18
-112	39.26	35.40	15:19	15
-113	39.84	35.99	15:24	14
-114	40.41	36.56	15:29	13
-115	40.97	37.12	15:33	12
-117	42.04	38.21	15:42	10
-118	42.57	38.72	15:46	9
-119	43.07	39.24	15:51	8
-120	43.57	39.74	15:55	7
-122	44.53	40.70	16:04	5
-123	45.00	41.17	16:08	5
-124	45.45	41.63	16:12	4
-125	45.89	42.07	16:16	3
-127	46.75	42.93	16:24	1

PERCENTAGE OF SUN OBSCURED ALONG MAP TRACK
WEST END = 92.2   EAST END = 91.1

===

SUNSET POSITION : ALTITUDE IS -0.8 DEGREES
NORTH : LONGITUDE -128.6, LATITUDE 47.4
SOUTH : LONGITUDE -130.3, LATITUDE 44.3

+ MARKS THE CITY OF LO-YANG

ECLIPSE ON JD 1806337 DATE(Y/M/D) 233/ 6/25
DELTA T = 1 HOUR   58.3 MINS

LONGITUDE (-VE EAST)	LATITUDE LIMITS. NORTH	SOUTH	LOCAL TIME	MEAN ALTITUDE
-129	13.97	13.44	8:53	45
-130	14.53	14.02	9:00	47
-132	15.63	15.15	9:13	50
-133	16.17	15.70	9:19	52
-134	16.70	16.24	9:26	53
-135	17.22	16.78	9:33	55
-137	18.23	17.81	9:46	58
-138	18.72	18.31	9:53	60
-140	19.64	19.26	10:07	64
-142	20.51	20.15	10:21	67
-144	21.30	20.96	10:36	70
-145	21.67	21.34	10:43	72
-147	22.35	22.04	10:57	76
-149	22.96	22.65	11:12	79
-150	23.22	22.92	11:19	81
-152	23.70	23.41	11:34	84

PERCENTAGE OF SUN OBSCURED ALONG MAP TRACK
WEST END = 98.6   EAST END = 99.1

===

+ MARKS THE CITY OF LO-YANG

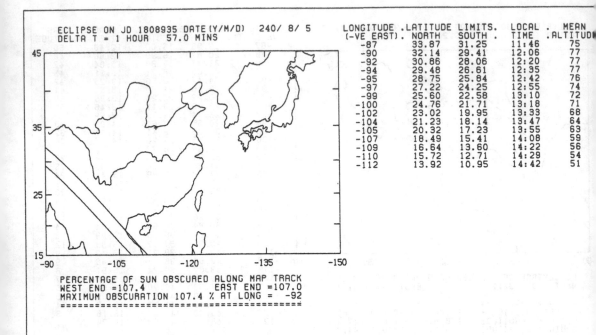

ECLIPSE ON JD 1808935 DATE(Y/M/D)   240/ 8/ 5
DELTA T = 1 HOUR   57.0 MINS

LONGITUDE (-VE EAST)	LATITUDE LIMITS. NORTH	SOUTH .	LOCAL TIME	MEAN ALTITUD
-87	33.87	31.25	11:46	75
-90	32.14	29.41	12:06	77
-92	30.86	28.06	12:20	77
-94	29.48	26.61	12:35	77
-95	28.75	25.84	12:42	76
-97	27.22	24.25	12:55	74
-99	25.60	22.58	13:10	72
-100	24.76	21.71	13:18	71
-102	23.02	19.95	13:33	68
-104	21.23	18.14	13:47	64
-105	20.32	17.23	13:55	63
-107	18.49	15.41	14:08	59
-109	16.64	13.60	14:22	56
-110	15.72	12.71	14:29	54
-112	13.92	10.95	14:42	51

PERCENTAGE OF SUN OBSCURED ALONG MAP TRACK
WEST END =107.4   EAST END =107.0
MAXIMUM OBSCURATION 107.4 % AT LONG = -92
===

+ MARKS THE CITY OF LO-YANG

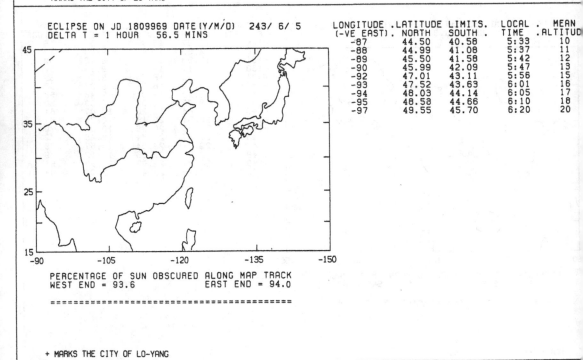

ECLIPSE ON JD 1809969 DATE(Y/M/D)   243/ 6/ 5
DELTA T = 1 HOUR   56.5 MINS

LONGITUDE (-VE EAST)	LATITUDE LIMITS. NORTH	SOUTH .	LOCAL TIME	MEAN ALTITUD
-87	44.50	40.58	5:33	10
-88	44.99	41.08	5:37	11
-89	45.50	41.58	5:42	12
-90	45.99	42.09	5:47	13
-92	47.01	43.11	5:56	15
-93	47.52	43.63	6:01	16
-94	48.03	44.14	6:05	17
-95	48.58	44.66	6:10	18
-97	49.55	45.70	6:20	20

PERCENTAGE OF SUN OBSCURED ALONG MAP TRACK
WEST END = 93.6   EAST END = 94.0

===

+ MARKS THE CITY OF LO-YANG

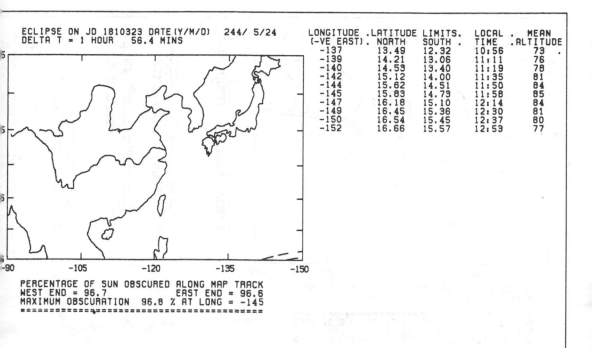

ECLIPSE ON JD 1810323 DATE(Y/M/D)   244/ 5/24
DELTA T = 1 HOUR   56.4 MINS

LONGITUDE (-VE EAST)	.LATITUDE NORTH	LIMITS. SOUTH .	LOCAL TIME	MEAN .ALTITUDE
-137	13.49	12.32	10:56	73
-139	14.21	13.06	11:11	76
-140	14.53	13.40	11:19	78
-142	15.12	14.00	11:35	81
-144	15.62	14.51	11:50	84
-145	15.83	14.73	11:58	85
-147	16.18	15.10	12:14	84
-149	16.45	15.36	12:30	81
-150	16.54	15.45	12:37	80
-152	16.66	15.57	12:53	77

PERCENTAGE OF SUN OBSCURED ALONG MAP TRACK
WEST END = 96.7   EAST END = 96.6
MAXIMUM OBSCURATION 96.8 % AT LONG = -145

+ MARKS THE CITY OF LO-YANG

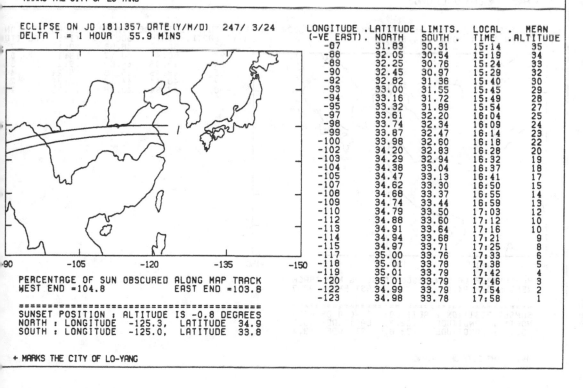

ECLIPSE ON JD 1811357 DATE(Y/M/D)   247/ 3/24
DELTA T = 1 HOUR   55.9 MINS

LONGITUDE (-VE EAST)	.LATITUDE NORTH	LIMITS. SOUTH .	LOCAL TIME	MEAN .ALTITUDE
-87	31.83	30.31	15:14	35
-88	32.05	30.54	15:19	34
-89	32.25	30.76	15:24	33
-90	32.45	30.97	15:29	32
-92	32.82	31.36	15:40	30
-93	33.00	31.55	15:45	29
-94	33.16	31.72	15:49	28
-95	33.32	31.89	15:54	27
-97	33.61	32.20	16:04	25
-98	33.74	32.34	16:09	24
-99	33.87	32.47	16:14	23
-100	33.98	32.60	16:18	22
-102	34.20	32.83	16:28	20
-103	34.29	32.94	16:32	19
-104	34.38	33.04	16:37	18
-105	34.47	33.13	16:41	17
-107	34.62	33.30	16:50	15
-108	34.68	33.37	16:55	14
-109	34.74	33.44	16:59	13
-110	34.79	33.50	17:03	12
-112	34.88	33.60	17:12	10
-113	34.91	33.64	17:16	10
-114	34.94	33.68	17:21	9
-115	34.97	33.71	17:25	8
-117	35.00	33.76	17:33	6
-118	35.01	33.78	17:38	5
-119	35.01	33.79	17:42	4
-120	35.01	33.79	17:46	3
-122	34.99	33.79	17:54	2
-123	34.98	33.78	17:58	1

PERCENTAGE OF SUN OBSCURED ALONG MAP TRACK
WEST END =104.8   EAST END =103.8

SUNSET POSITION : ALTITUDE IS -0.8 DEGREES
NORTH : LONGITUDE -125.3, LATITUDE 34.9
SOUTH : LONGITUDE -125.0, LATITUDE 33.8

+ MARKS THE CITY OF LO-YANG

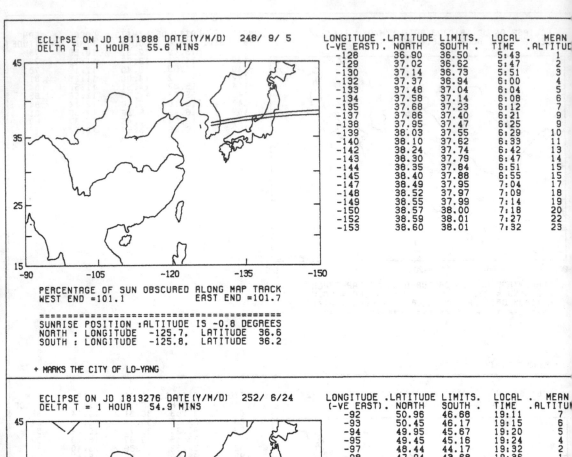

ECLIPSE ON JD 1811888 DATE(Y/M/D) 248/ 9/ 5
DELTA T = 1 HOUR   55.6 MINS

LONGITUDE (-VE EAST)	LATITUDE NORTH	LIMITS. SOUTH	LOCAL TIME	MEAN ALTITUDE
-128	36.90	36.50	5:43	1
-129	37.02	36.62	5:47	2
-130	37.14	36.73	5:51	3
-132	37.37	36.94	6:00	4
-133	37.48	37.04	6:04	5
-134	37.58	37.14	6:08	6
-135	37.68	37.23	6:12	7
-137	37.86	37.40	6:21	9
-138	37.95	37.47	6:25	9
-139	38.03	37.55	6:29	10
-140	38.10	37.62	6:33	11
-142	38.24	37.74	6:42	13
-143	38.30	37.79	6:47	14
-144	38.35	37.84	6:51	15
-145	38.40	37.88	6:55	15
-147	38.49	37.95	7:04	17
-148	38.52	37.97	7:09	18
-149	38.55	37.99	7:14	19
-150	38.57	38.00	7:18	20
-152	38.59	38.01	7:27	22
-153	38.60	38.01	7:32	23

PERCENTAGE OF SUN OBSCURED ALONG MAP TRACK
WEST END =101.1   EAST END =101.7

==
SUNRISE POSITION :ALTITUDE IS -0.8 DEGREES
NORTH : LONGITUDE -125.7, LATITUDE 36.6
SOUTH : LONGITUDE -125.8, LATITUDE 36.2

+ MARKS THE CITY OF LO-YANG

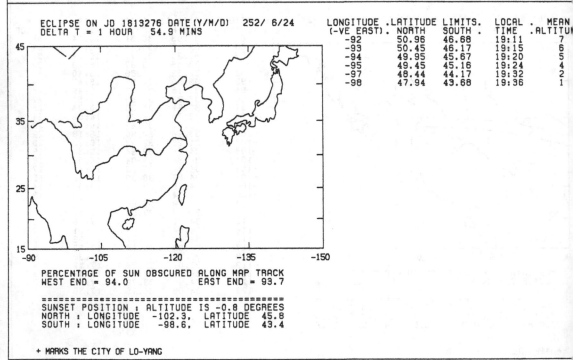

ECLIPSE ON JD 1813276 DATE(Y/M/D) 252/ 6/24
DELTA T = 1 HOUR   54.9 MINS

LONGITUDE (-VE EAST)	LATITUDE NORTH	LIMITS. SOUTH	LOCAL TIME	MEAN ALTITUDE
-92	50.96	46.68	19:11	7
-93	50.45	46.17	19:15	6
-94	49.95	45.67	19:20	5
-95	49.45	45.16	19:24	4
-97	48.44	44.17	19:32	2
-98	47.94	43.68	19:36	1

PERCENTAGE OF SUN OBSCURED ALONG MAP TRACK
WEST END = 94.0   EAST END = 93.7

==
SUNSET POSITION : ALTITUDE IS -0.8 DEGREES
NORTH : LONGITUDE -102.3, LATITUDE 45.8
SOUTH : LONGITUDE   -98.6, LATITUDE 43.4

+ MARKS THE CITY OF LO-YANG

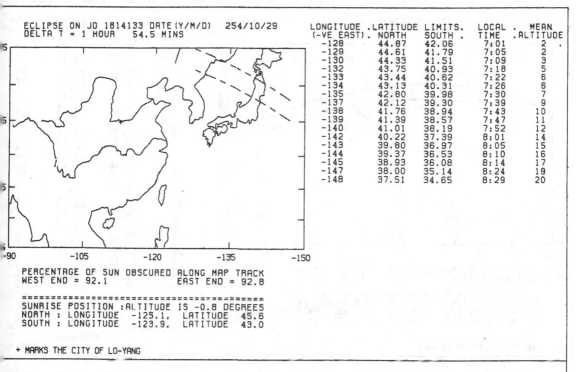

ECLIPSE ON JD 1814133 DATE(Y/M/D)   254/10/29
DELTA T = 1 HOUR   54.5 MINS

LONGITUDE (-VE EAST).	LATITUDE NORTH	LIMITS. SOUTH	LOCAL TIME	MEAN ALTITUDE
-128	44.87	42.06	7:01	2
-129	44.61	41.79	7:05	2
-130	44.33	41.51	7:09	3
-132	43.75	40.93	7:18	5
-133	43.44	40.62	7:22	6
-134	43.13	40.31	7:26	6
-135	42.80	39.98	7:30	7
-137	42.12	39.30	7:39	9
-138	41.76	38.94	7:43	10
-139	41.39	38.57	7:47	11
-140	41.01	38.19	7:52	12
-142	40.22	37.39	8:01	14
-143	39.80	36.97	8:05	15
-144	39.37	36.53	8:10	16
-145	38.93	36.08	8:14	17
-147	38.00	35.14	8:24	19
-148	37.51	34.65	8:29	20

PERCENTAGE OF SUN OBSCURED ALONG MAP TRACK
WEST END = 92.1   EAST END = 92.8

==
SUNRISE POSITION :ALTITUDE IS -0.8 DEGREES
NORTH : LONGITUDE -125.1, LATITUDE 45.6
SOUTH : LONGITUDE -123.9, LATITUDE 43.0

+ MARKS THE CITY OF LO-YANG

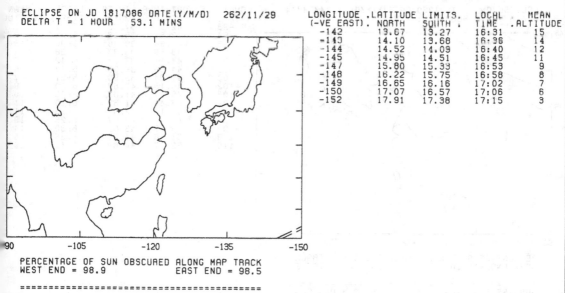

ECLIPSE ON JD 1817086 DATE(Y/M/D)   262/11/29
DELTA T = 1 HOUR   53.1 MINS

LONGITUDE (-VE EAST).	LATITUDE NORTH	LIMITS. SOUTH	LOCAL TIME	MEAN ALTITUDE
-142	13.67	13.27	16:31	15
-143	14.10	13.68	16:36	14
-144	14.52	14.09	16:40	12
-145	14.95	14.51	16:45	11
-147	15.80	15.33	16:53	9
-148	16.22	15.75	16:58	8
-149	16.65	16.16	17:02	7
-150	17.07	16.57	17:06	6
-152	17.91	17.38	17:15	3

PERCENTAGE OF SUN OBSCURED ALONG MAP TRACK
WEST END = 98.9   EAST END = 98.5

==

+ MARKS THE CITY OF LO-YANG

225

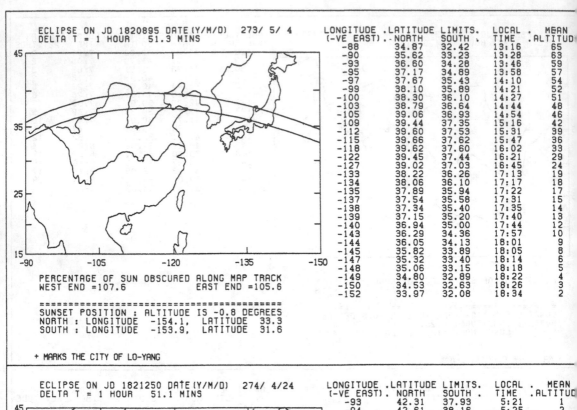

ECLIPSE ON JD 1820895 DATE(Y/M/D)   273/ 5/ 4
DELTA T = 1 HOUR   51.3 MINS

LONGITUDE (-VE EAST)	LATITUDE NORTH	LIMITS. SOUTH	LOCAL TIME	MEAN ALTITUD
-88	34.87	32.42	13:16	65
-90	35.62	33.23	13:28	63
-93	36.60	34.28	13:46	59
-95	37.17	34.89	13:58	57
-97	37.67	35.43	14:10	54
-99	38.10	35.89	14:21	52
-100	38.30	36.10	14:27	51
-103	38.79	36.64	14:44	48
-105	39.06	36.93	14:54	46
-109	39.44	37.35	15:16	42
-112	39.60	37.53	15:31	39
-115	39.66	37.62	15:47	36
-118	39.62	37.60	16:02	33
-122	39.45	37.44	16:21	29
-127	39.02	37.03	16:45	24
-133	38.22	36.26	17:13	19
-134	38.06	36.10	17:17	18
-135	37.89	35.94	17:22	17
-137	37.54	35.58	17:31	15
-138	37.34	35.40	17:35	14
-139	37.15	35.20	17:40	13
-140	36.94	35.00	17:44	12
-143	36.29	34.36	17:57	10
-144	36.05	34.13	18:01	9
-145	35.82	33.89	18:05	8
-147	35.32	33.40	18:14	6
-148	35.06	33.15	18:18	5
-149	34.80	32.89	18:22	4
-150	34.53	32.63	18:26	3
-152	33.97	32.08	18:34	2

PERCENTAGE OF SUN OBSCURED ALONG MAP TRACK
WEST END =107.6   EAST END =105.6

==
SUNSET POSITION : ALTITUDE IS -0.8 DEGREES
NORTH : LONGITUDE -154.1, LATITUDE   33.3
SOUTH : LONGITUDE -153.9, LATITUDE   31.6

+ MARKS THE CITY OF LO-YANG

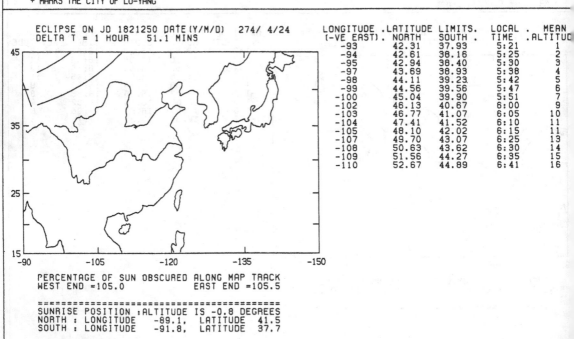

ECLIPSE ON JD 1821250 DATE(Y/M/D)   274/ 4/24
DELTA T = 1 HOUR   51.1 MINS

LONGITUDE (-VE EAST)	LATITUDE NORTH	LIMITS. SOUTH	LOCAL TIME	MEAN ALTITUD
-93	42.31	37.93	5:21	1
-94	42.61	38.16	5:25	2
-95	42.94	38.40	5:30	3
-97	43.69	38.93	5:38	4
-98	44.11	39.23	5:42	5
-99	44.56	39.56	5:47	6
-100	45.04	39.90	5:51	7
-102	46.13	40.67	6:00	9
-103	46.77	41.07	6:05	10
-104	47.41	41.52	6:10	11
-105	48.10	42.02	6:15	11
-107	49.70	43.07	6:25	13
-108	50.63	43.62	6:30	14
-109	51.56	44.27	6:35	15
-110	52.67	44.89	6:41	16

PERCENTAGE OF SUN OBSCURED ALONG MAP TRACK
WEST END =105.0   EAST END =105.5

==
SUNRISE POSITION : ALTITUDE IS -0.8 DEGREES
NORTH : LONGITUDE   -89.1, LATITUDE   41.5
SOUTH : LONGITUDE   -91.8, LATITUDE   37.7

+ MARKS THE CITY OF LO-YANG

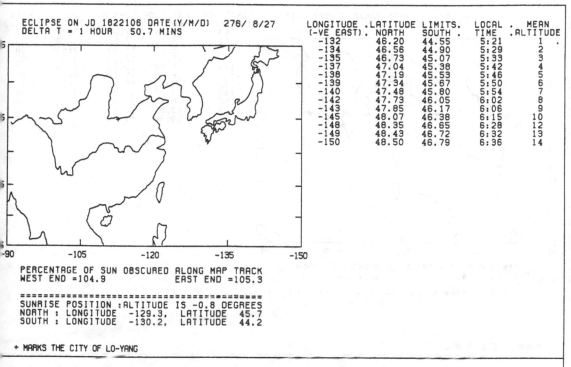

```
ECLIPSE ON JD 1822106 DATE(Y/M/D)   276/ 8/27
DELTA T = 1 HOUR   50.7 MINS
```

LONGITUDE (-VE EAST)	LATITUDE NORTH	LIMITS. SOUTH	LOCAL TIME	MEAN ALTITUDE
-132	46.20	44.55	5:21	1
-134	46.56	44.90	5:29	2
-135	46.73	45.07	5:33	3
-137	47.04	45.38	5:42	4
-138	47.19	45.53	5:46	5
-139	47.34	45.67	5:50	6
-140	47.48	45.80	5:54	7
-142	47.73	46.05	6:02	8
-143	47.85	46.17	6:06	9
-145	48.07	46.38	6:15	10
-148	48.35	46.65	6:28	12
-149	48.43	46.72	6:32	13
-150	48.50	46.79	6:36	14

```
PERCENTAGE OF SUN OBSCURED ALONG MAP TRACK
WEST END =104.9            EAST END =105.3

=========================================
SUNRISE POSITION :ALTITUDE IS -0.8 DEGREES
NORTH : LONGITUDE  -129.3,  LATITUDE  45.7
SOUTH : LONGITUDE  -130.2,  LATITUDE  44.2
```

+ MARKS THE CITY OF LO-YANG

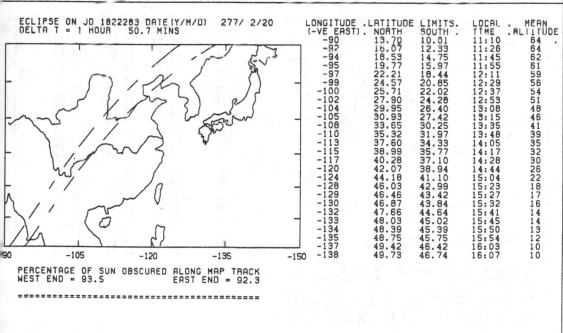

```
ECLIPSE ON JD 1822283 DATE(Y/M/D)   277/ 2/20
DELTA T = 1 HOUR   50.7 MINS
```

LONGITUDE (-VE EAST)	LATITUDE NORTH	LIMITS. SOUTH	LOCAL TIME	MEAN ALTITUDE
-90	13.70	10.01	11:10	64
-92	16.07	12.33	11:26	64
-94	18.53	14.75	11:45	62
-95	19.77	15.97	11:55	61
-97	22.21	18.44	12:11	59
-99	24.57	20.85	12:29	56
-100	25.71	22.02	12:37	54
-102	27.90	24.28	12:53	51
-104	29.95	26.40	13:08	48
-105	30.93	27.42	13:15	46
-108	33.65	30.25	13:35	41
-110	35.32	31.97	13:48	39
-113	37.60	34.33	14:05	35
-115	38.99	35.77	14:17	32
-117	40.28	37.10	14:28	30
-120	42.07	38.94	14:44	26
-124	44.18	41.10	15:04	22
-128	46.03	42.99	15:23	18
-129	46.46	43.42	15:27	17
-130	46.87	43.84	15:32	16
-132	47.66	44.64	15:41	14
-133	48.03	45.02	15:45	14
-134	48.39	45.39	15:50	13
-135	48.75	45.75	15:54	12
-137	49.42	46.42	16:03	10
-138	49.73	46.74	16:07	10

```
PERCENTAGE OF SUN OBSCURED ALONG MAP TRACK
WEST END = 93.5            EAST END = 92.3

=========================================
```

+ MARKS THE CITY OF LO-YANG

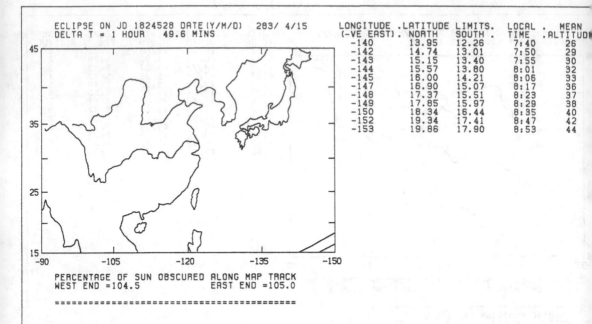

ECLIPSE ON JD 1824528 DATE(Y/M/D)   283/ 4/15
DELTA T = 1 HOUR   49.6 MINS

LONGITUDE (-VE EAST)	LATITUDE LIMITS. NORTH	SOUTH	LOCAL TIME	MEAN ALTITUD
-140	13.95	12.26	7:40	26
-142	14.74	13.01	7:50	29
-143	15.15	13.40	7:55	30
-144	15.57	13.80	8:01	32
-145	16.00	14.21	8:06	33
-147	16.90	15.07	8:17	36
-148	17.37	15.51	8:23	37
-149	17.85	15.97	8:29	38
-150	18.34	16.44	8:35	40
-152	19.34	17.41	8:47	42
-153	19.86	17.90	8:53	44

PERCENTAGE OF SUN OBSCURED ALONG MAP TRACK
WEST END =104.5   EAST END =105.0

===

+ MARKS THE CITY OF LO-YANG

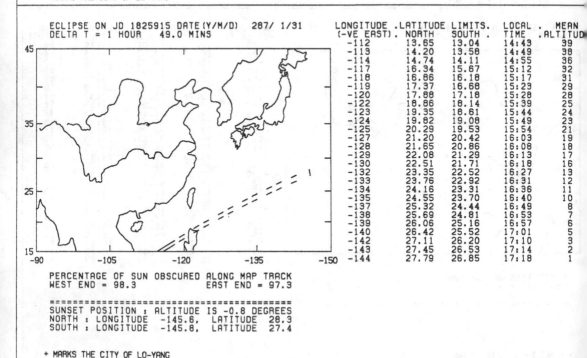

ECLIPSE ON JD 1825915 DATE(Y/M/D)   287/ 1/31
DELTA T = 1 HOUR   49.0 MINS

LONGITUDE (-VE EAST)	LATITUDE LIMITS. NORTH	SOUTH	LOCAL TIME	MEAN ALTITUD
-112	13.65	13.04	14:49	39
-113	14.20	13.58	14:49	38
-114	14.74	14.11	14:55	36
-117	16.34	15.67	15:12	32
-118	16.86	16.18	15:17	31
-119	17.37	16.68	15:23	29
-120	17.88	17.18	15:28	28
-122	18.86	18.14	15:39	25
-123	19.35	18.61	15:44	24
-124	19.82	19.08	15:49	23
-125	20.29	19.53	15:54	21
-127	21.20	20.42	16:03	19
-128	21.65	20.86	16:08	18
-129	22.08	21.29	16:13	17
-130	22.51	21.71	16:18	16
-132	23.35	22.52	16:27	13
-133	23.76	22.92	16:31	12
-134	24.16	23.31	16:36	11
-135	24.55	23.70	16:40	10
-137	25.32	24.44	16:49	8
-138	25.69	24.81	16:53	7
-139	26.06	25.16	16:57	6
-140	26.42	25.52	17:01	5
-142	27.11	26.20	17:10	3
-143	27.45	26.53	17:14	2
-144	27.79	26.85	17:18	1

PERCENTAGE OF SUN OBSCURED ALONG MAP TRACK
WEST END = 98.3   EAST END = 97.3

===
SUNSET POSITION : ALTITUDE IS -0.8 DEGREES
NORTH : LONGITUDE -145.6,   LATITUDE 28.3
SOUTH : LONGITUDE -145.8,   LATITUDE 27.4

+ MARKS THE CITY OF LO-YANG

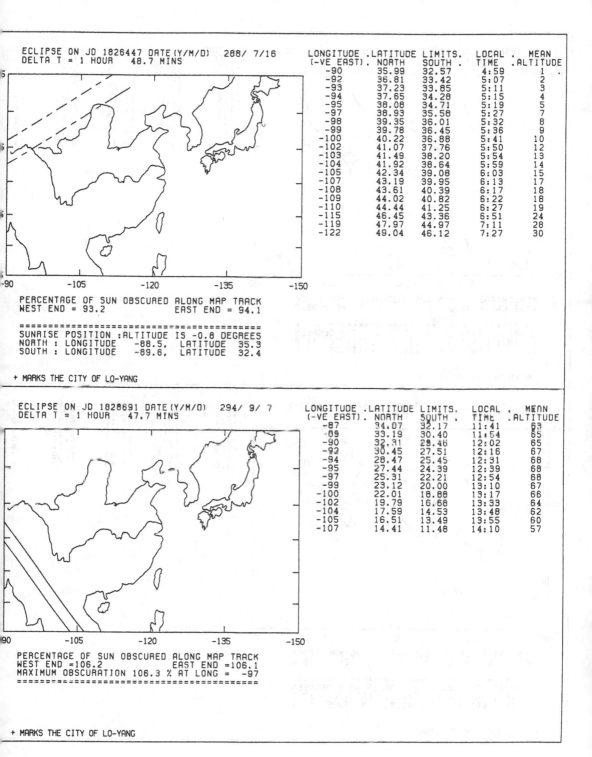

ECLIPSE ON JD 1826447 DATE(Y/M/D)   288/ 7/16
DELTA T = 1 HOUR   48.7 MINS

LONGITUDE	.LATITUDE	LIMITS.	LOCAL	. MEAN
(-VE EAST).	NORTH	SOUTH .	TIME	.ALTITUDE
-90	35.99	32.57	4:59	1
-92	36.81	33.42	5:07	2
-93	37.23	33.85	5:11	3
-94	37.65	34.28	5:15	4
-95	38.08	34.71	5:19	5
-97	38.93	35.58	5:27	7
-98	39.35	36.01	5:32	8
-99	39.78	36.45	5:36	9
-100	40.22	36.88	5:41	10
-102	41.07	37.76	5:50	12
-103	41.49	38.20	5:54	13
-104	41.92	38.64	5:59	14
-105	42.34	39.08	6:03	15
-107	43.19	39.95	6:13	17
-108	43.61	40.39	6:17	18
-109	44.02	40.82	6:22	18
-110	44.44	41.25	6:27	19
-115	46.45	43.36	6:51	24
-119	47.97	44.97	7:11	28
-122	49.04	46.12	7:27	30

PERCENTAGE OF SUN OBSCURED ALONG MAP TRACK
WEST END = 93.2   EAST END = 94.1

==
SUNRISE POSITION :ALTITUDE IS -0.8 DEGREES
NORTH : LONGITUDE   -88.5.   LATITUDE 35.3
SOUTH : LONGITUDE   -89.6.   LATITUDE 32.4

+ MARKS THE CITY OF LO-YANG

ECLIPSE ON JD 1828691 DATE(Y/M/D)   294/ 9/ 7
DELTA T = 1 HOUR   47.7 MINS

LONGITUDE	.LATITUDE	LIMITS.	LOCAL	. MEAN
(-VE EAST).	NORTH	SOUTH .	TIME	.ALTITUDE
-87	34.07	32.17	11:41	63
-89	33.19	30.40	11:54	65
-90	32.31	29.46	12:02	65
-92	30.45	27.51	12:16	67
-94	28.47	25.45	12:31	68
-95	27.44	24.39	12:39	68
-97	25.31	22.21	12:54	68
-99	23.12	20.00	13:10	67
-100	22.01	18.88	13:17	66
-102	19.79	16.68	13:33	64
-104	17.59	14.53	13:48	62
-105	16.51	13.49	13:55	60
-107	14.41	11.48	14:10	57

PERCENTAGE OF SUN OBSCURED ALONG MAP TRACK
WEST END =106.2   EAST END =106.1
MAXIMUM OBSCURATION 106.3 % AT LONG = -97
==

+ MARKS THE CITY OF LO-YANG

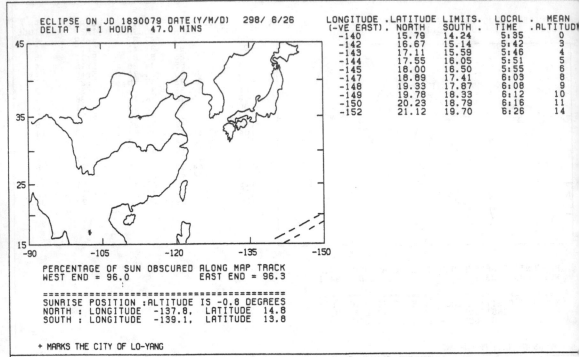

ECLIPSE ON JD 1830079 DATE(Y/M/D) 298/ 6/26
DELTA T = 1 HOUR   47.0 MINS

LONGITUDE (-VE EAST)	LATITUDE LIMITS. NORTH	SOUTH .	LOCAL TIME	MEAN . ALTITUDE
-140	15.79	14.24	5:35	0
-142	16.67	15.14	5:42	3
-143	17.11	15.59	5:46	4
-144	17.55	16.05	5:51	5
-145	18.00	16.50	5:55	6
-147	18.89	17.41	6:03	8
-148	19.33	17.87	6:08	9
-149	19.78	18.33	6:12	10
-150	20.23	18.79	6:16	11
-152	21.12	19.70	6:26	14

PERCENTAGE OF SUN OBSCURED ALONG MAP TRACK
WEST END = 96.0   EAST END = 96.3

==
SUNRISE POSITION :ALTITUDE IS -0.8 DEGREES
NORTH : LONGITUDE -137.8, LATITUDE 14.8
SOUTH : LONGITUDE -139.1, LATITUDE 13.8

+ MARKS THE CITY OF LO-YANG

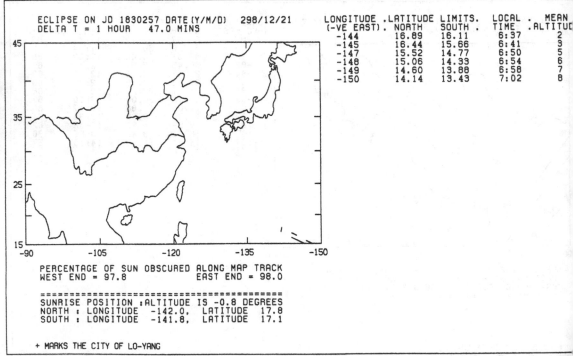

ECLIPSE ON JD 1830257 DATE(Y/M/D) 298/12/21
DELTA T = 1 HOUR   47.0 MINS

LONGITUDE (-VE EAST)	LATITUDE NORTH	LIMITS. SOUTH .	LOCAL TIME	MEAN . ALTITUDE
-144	16.89	16.11	6:37	2
-145	16.44	15.66	6:41	3
-147	15.52	14.77	6:50	5
-148	15.06	14.33	6:54	6
-149	14.60	13.88	6:58	7
-150	14.14	13.43	7:02	8

PERCENTAGE OF SUN OBSCURED ALONG MAP TRACK
WEST END = 97.8   EAST END = 98.0

==
SUNRISE POSITION :ALTITUDE IS -0.8 DEGREES
NORTH : LONGITUDE -142.0, LATITUDE 17.8
SOUTH : LONGITUDE -141.8, LATITUDE 17.1

+ MARKS THE CITY OF LO-YANG

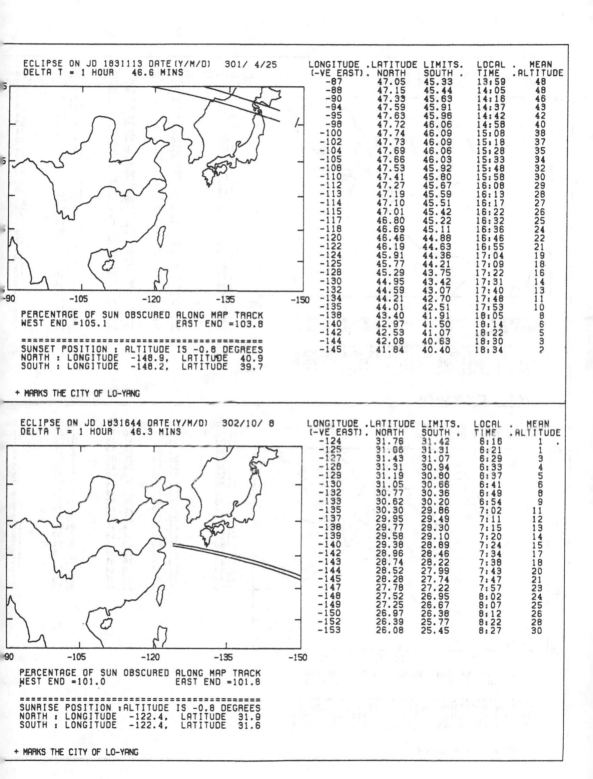

ECLIPSE ON JD 1831113 DATE(Y/M/D) 301/ 4/25
DELTA T = 1 HOUR   46.6 MINS

LONGITUDE (-VE EAST)	LATITUDE LIMITS. NORTH	SOUTH	LOCAL TIME	MEAN ALTITUDE
-87	47.05	45.33	13:59	48
-88	47.15	45.44	14:05	48
-90	47.33	45.63	14:16	46
-94	47.59	45.91	14:37	43
-95	47.63	45.96	14:42	42
-98	47.72	46.06	14:58	40
-100	47.74	46.09	15:08	38
-102	47.73	46.09	15:18	37
-104	47.69	46.06	15:28	35
-105	47.66	46.03	15:33	34
-108	47.53	45.92	15:48	32
-110	47.41	45.80	15:58	30
-112	47.27	45.67	16:08	29
-113	47.19	45.59	16:13	28
-114	47.10	45.51	16:17	27
-115	47.01	45.42	16:22	26
-117	46.80	45.22	16:32	25
-118	46.69	45.11	16:36	24
-120	46.46	44.88	16:46	22
-122	46.19	44.63	16:55	21
-124	45.91	44.36	17:04	19
-125	45.77	44.21	17:09	18
-128	45.29	43.75	17:22	16
-130	44.95	43.42	17:31	14
-132	44.59	43.07	17:40	13
-134	44.21	42.70	17:48	11
-135	44.01	42.51	17:53	10
-138	43.40	41.91	18:05	8
-140	42.97	41.50	18:14	6
-142	42.53	41.07	18:22	5
-144	42.08	40.63	18:30	3
-145	41.84	40.40	18:34	2

PERCENTAGE OF SUN OBSCURED ALONG MAP TRACK
WEST END =105.1   EAST END =103.8

===
SUNSET POSITION : ALTITUDE IS -0.8 DEGREES
NORTH : LONGITUDE -148.9, LATITUDE 40.9
SOUTH : LONGITUDE -148.2, LATITUDE 39.7

+ MARKS THE CITY OF LO-YANG

ECLIPSE ON JD 1831644 DATE(Y/M/D) 302/10/ 8
DELTA T = 1 HOUR   46.3 MINS

LONGITUDE (-VE EAST)	LATITUDE LIMITS. NORTH	SOUTH	LOCAL TIME	MEAN ALTITUDE
-124	31.76	31.42	6:16	1
-125	31.66	31.31	6:21	1
-127	31.43	31.07	6:29	3
-128	31.31	30.94	6:33	4
-129	31.19	30.80	6:37	5
-130	31.05	30.66	6:41	6
-132	30.77	30.36	6:49	8
-133	30.62	30.20	6:54	9
-135	30.30	29.86	7:02	11
-137	29.95	29.49	7:11	12
-138	29.77	29.30	7:15	13
-139	29.58	29.10	7:20	14
-140	29.38	28.89	7:24	15
-142	28.96	28.46	7:34	17
-143	28.74	28.22	7:38	18
-144	28.52	27.99	7:43	20
-145	28.28	27.74	7:47	21
-147	27.78	27.22	7:57	23
-148	27.52	26.95	8:02	24
-149	27.25	26.67	8:07	25
-150	26.97	26.38	8:12	26
-152	26.39	25.77	8:22	28
-153	26.08	25.45	8:27	30

PERCENTAGE OF SUN OBSCURED ALONG MAP TRACK
WEST END =101.0   EAST END =101.8

===
SUNRISE POSITION :ALTITUDE IS -0.8 DEGREES
NORTH : LONGITUDE -122.4, LATITUDE 31.9
SOUTH : LONGITUDE -122.4, LATITUDE 31.6

+ MARKS THE CITY OF LO-YANG

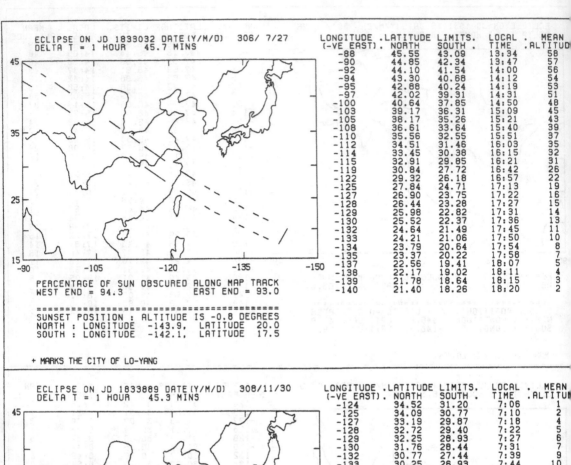

ECLIPSE ON JD 1833032 DATE(Y/M/D) 306/ 7/27
DELTA T = 1 HOUR   45.7 MINS

LONGITUDE (-VE EAST)	LATITUDE NORTH	LIMITS. SOUTH	LOCAL TIME	MEAN ALTITUD
-88	45.55	43.09	13:34	58
-90	44.85	42.34	13:47	57
-92	44.10	41.54	14:00	56
-94	43.30	40.68	14:12	54
-95	42.88	40.24	14:19	53
-97	42.02	39.31	14:31	51
-100	40.64	37.85	14:50	48
-103	39.17	36.31	15:09	45
-105	38.17	35.26	15:21	43
-108	36.61	33.64	15:40	39
-110	35.56	32.55	15:51	37
-112	34.51	31.46	16:03	35
-114	33.45	30.38	16:15	32
-115	32.91	29.85	16:21	31
-119	30.84	27.72	16:42	26
-122	29.32	26.18	16:57	22
-125	27.84	24.71	17:13	19
-127	26.90	23.75	17:22	16
-128	26.44	23.28	17:27	15
-129	25.98	22.82	17:31	14
-130	25.52	22.37	17:36	13
-132	24.64	21.49	17:45	11
-133	24.21	21.06	17:50	10
-134	23.79	20.64	17:54	8
-135	23.37	20.22	17:58	7
-137	22.56	19.41	18:07	5
-138	22.17	19.02	18:11	4
-139	21.78	18.64	18:15	3
-140	21.40	18.26	18:20	2

PERCENTAGE OF SUN OBSCURED ALONG MAP TRACK
WEST END = 94.3   EAST END = 93.0

===
SUNSET POSITION : ALTITUDE IS -0.8 DEGREES
NORTH : LONGITUDE -143.9, LATITUDE 20.0
SOUTH : LONGITUDE -142.1, LATITUDE 17.5

+ MARKS THE CITY OF LO-YANG

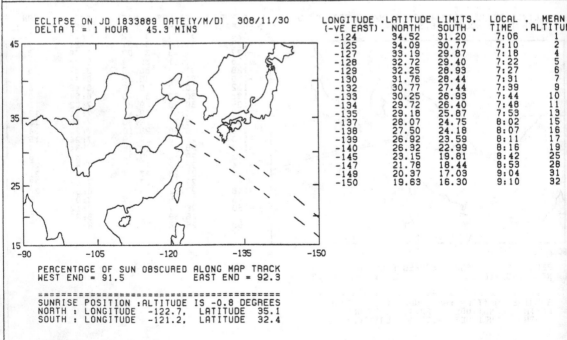

ECLIPSE ON JD 1833889 DATE(Y/M/D) 308/11/30
DELTA T = 1 HOUR   45.3 MINS

LONGITUDE (-VE EAST)	LATITUDE NORTH	LIMITS. SOUTH	LOCAL TIME	MEAN ALTITUD
-124	34.52	31.20	7:06	1
-125	34.09	30.77	7:10	2
-127	33.19	29.87	7:18	4
-128	32.72	29.40	7:22	5
-129	32.25	28.93	7:27	6
-130	31.76	28.44	7:31	7
-132	30.77	27.44	7:39	9
-133	30.25	26.93	7:44	10
-134	29.72	26.40	7:48	11
-135	29.18	25.87	7:53	13
-137	28.07	24.75	8:02	15
-138	27.50	24.18	8:07	16
-139	26.92	23.59	8:11	17
-140	26.32	22.99	8:16	19
-145	23.15	19.81	8:42	25
-147	21.78	18.44	8:53	28
-149	20.37	17.03	9:04	31
-150	19.63	16.30	9:10	32

PERCENTAGE OF SUN OBSCURED ALONG MAP TRACK
WEST END = 91.5   EAST END = 92.3

===
SUNRISE POSITION : ALTITUDE IS -0.8 DEGREES
NORTH : LONGITUDE -122.7, LATITUDE 35.1
SOUTH : LONGITUDE -121.2, LATITUDE 32.4

+ MARKS THE CITY OF LO-YANG

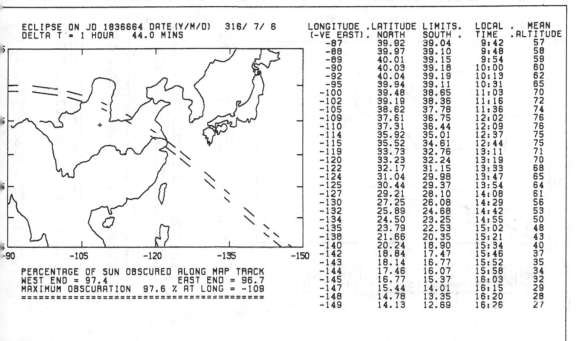

```
ECLIPSE ON JD 1836664 DATE(Y/M/D)   316/ 7/ 6
DELTA T = 1 HOUR    44.0 MINS
```

LONGITUDE (-VE EAST)	.LATITUDE NORTH	LIMITS. SOUTH .	LOCAL TIME	MEAN .ALTITUDE
-87	39.92	39.04	9:42	57
-88	39.97	39.10	9:48	58
-89	40.01	39.15	9:54	59
-90	40.03	39.18	10:00	60
-92	40.04	39.19	10:13	62
-95	39.94	39.11	10:31	65
-100	39.48	38.65	11:03	70
-102	39.19	38.36	11:16	72
-105	38.62	37.78	11:36	74
-109	37.61	36.75	12:02	76
-110	37.31	36.44	12:09	76
-114	35.92	35.01	12:37	75
-115	35.52	34.61	12:44	75
-119	33.73	32.76	13:11	71
-120	33.23	32.24	13:19	70
-122	32.17	31.15	13:33	68
-124	31.04	29.98	13:47	65
-125	30.44	29.37	13:54	64
-127	29.21	28.10	14:08	61
-130	27.25	26.08	14:29	56
-132	25.89	24.68	14:42	53
-134	24.50	23.25	14:55	50
-135	23.79	22.53	15:02	48
-138	21.66	20.35	15:21	43
-140	20.24	18.90	15:34	40
-142	18.84	17.47	15:46	37
-143	18.14	16.77	15:52	35
-144	17.46	16.07	15:58	34
-145	16.77	15.37	16:03	32
-147	15.44	14.01	16:15	29
-148	14.78	13.35	16:20	28
-149	14.13	12.69	16:26	27

```
PERCENTAGE OF SUN OBSCURED ALONG MAP TRACK
WEST END = 97.4              EAST END = 96.7
MAXIMUM OBSCURATION  97.6 % AT LONG = -109
=================================================
```

+ MARKS THE CITY OF CH'ANG-AN

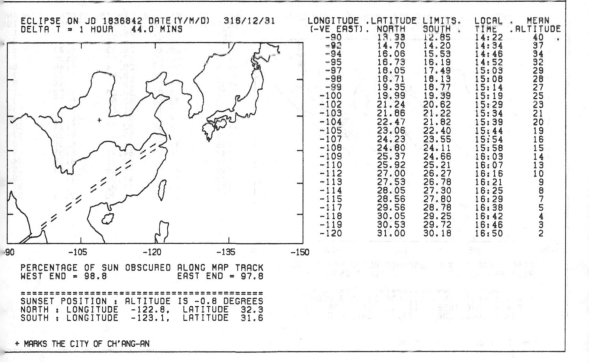

```
ECLIPSE ON JD 1836842 DATE(Y/M/D)   316/12/31
DELTA T = 1 HOUR    44.0 MINS
```

LONGITUDE (-VE EAST)	.LATITUDE NORTH	LIMITS. SOUTH .	LOCAL TIME	MEAN .ALTITUDE
-90	13.33	12.85	14:22	40 .
-92	14.70	14.20	14:34	37
-94	16.06	15.53	14:46	34
-95	16.73	16.19	14:52	32
-97	18.05	17.49	15:03	29
-98	18.71	18.13	15:08	28
-99	19.35	18.77	15:14	27
-100	19.99	19.39	15:19	25
-102	21.24	20.62	15:29	23
-103	21.86	21.22	15:34	21
-104	22.47	21.82	15:39	20
-105	23.06	22.40	15:44	19
-107	24.23	23.55	15:54	16
-108	24.80	24.11	15:58	15
-109	25.37	24.66	16:03	14
-110	25.92	25.21	16:07	13
-112	27.00	26.27	16:16	10
-113	27.53	26.78	16:21	9
-114	28.05	27.30	16:25	8
-115	28.56	27.80	16:29	7
-117	29.56	28.78	16:38	5
-118	30.05	29.25	16:42	4
-119	30.53	29.72	16:46	3
-120	31.00	30.18	16:50	2

```
PERCENTAGE OF SUN OBSCURED ALONG MAP TRACK
WEST END = 98.8              EAST END = 97.8

=================================================
SUNSET POSITION : ALTITUDE IS -0.8 DEGREES
NORTH : LONGITUDE  -122.8,  LATITUDE  32.3
SOUTH : LONGITUDE  -123.1,  LATITUDE  31.6
```

+ MARKS THE CITY OF CH'ANG-AN

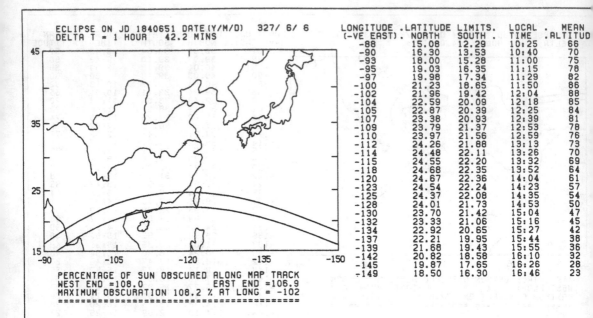

ECLIPSE ON JD 1840651 DATE(Y/M/D)   327/ 6/ 6
DELTA T = 1 HOUR   42.2 MINS

LONGITUDE (-VE EAST)	LATITUDE NORTH	LIMITS. SOUTH	LOCAL TIME	MEAN ALTITUD
-88	15.08	12.29	10:25	66
-90	16.30	13.53	10:40	70
-93	18.00	15.28	11:00	75
-95	19.03	16.35	11:15	78
-97	19.98	17.34	11:29	82
-100	21.23	18.65	11:50	86
-102	21.96	19.42	12:04	88
-104	22.59	20.09	12:18	85
-105	22.87	20.39	12:25	84
-107	23.38	20.93	12:39	81
-109	23.79	21.37	12:53	78
-110	23.97	21.56	12:59	76
-112	24.26	21.88	13:13	73
-114	24.48	22.11	13:26	70
-115	24.55	22.20	13:32	69
-118	24.68	22.35	13:52	64
-120	24.67	22.36	14:04	61
-123	24.54	22.24	14:23	57
-125	24.37	22.08	14:35	54
-128	24.01	21.73	14:53	50
-130	23.70	21.42	15:04	47
-132	23.33	21.06	15:16	45
-134	22.92	20.65	15:27	42
-137	22.21	19.95	15:44	38
-139	21.68	19.43	15:55	36
-142	20.82	18.58	16:10	32
-145	19.87	17.65	16:26	28
-149	18.50	16.30	16:46	23

PERCENTAGE OF SUN OBSCURED ALONG MAP TRACK
WEST END =108.0   EAST END =106.9
MAXIMUM OBSCURATION 108.2 % AT LONG = -102
===

+ MARKS THE CITY OF CHIEN-K'ANG

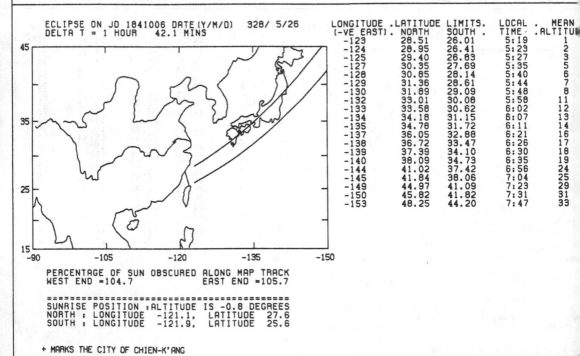

ECLIPSE ON JD 1841006 DATE(Y/M/D)   328/ 5/26
DELTA T = 1 HOUR   42.1 MINS

LONGITUDE (-VE EAST)	LATITUDE NORTH	LIMITS. SOUTH	LOCAL TIME	MEAN ALTITU
-123	28.51	26.01	5:19	1
-124	28.95	26.41	5:23	2
-125	29.40	26.83	5:27	3
-127	30.35	27.69	5:35	5
-128	30.85	28.14	5:40	6
-129	31.36	28.61	5:44	7
-130	31.89	29.09	5:48	8
-132	33.01	30.08	5:58	11
-133	33.58	30.62	6:02	12
-134	34.18	31.15	6:07	13
-135	34.78	31.72	6:11	14
-137	36.05	32.88	6:21	16
-138	36.72	33.47	6:26	17
-139	37.39	34.10	6:30	18
-140	38.09	34.73	6:35	19
-144	41.02	37.42	6:56	24
-145	41.84	38.06	7:04	25
-149	44.97	41.09	7:23	29
-150	45.82	41.82	7:31	31
-153	48.25	44.20	7:47	33

PERCENTAGE OF SUN OBSCURED ALONG MAP TRACK
WEST END =104.7   EAST END =105.7

===
SUNRISE POSITION :ALTITUDE IS -0.8 DEGREES
NORTH : LONGITUDE -121.1, LATITUDE 27.6
SOUTH : LONGITUDE -121.9, LATITUDE 25.6

+ MARKS THE CITY OF CHIEN-K'ANG

ECLIPSE ON JD 1842039 DATE(Y/M/D) 331/ 3/25
DELTA T = 1 HOUR   41.6 MINS

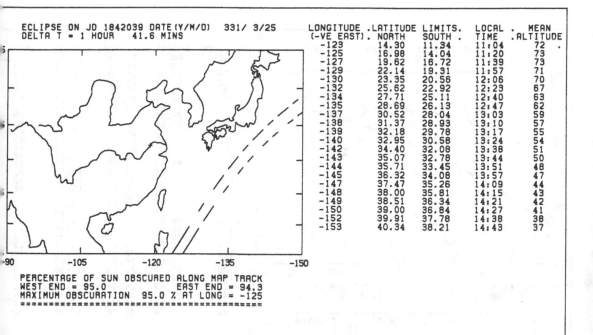

LONGITUDE (-VE EAST)	LATITUDE NORTH	LIMITS. SOUTH	LOCAL TIME	MEAN ALTITUDE
-123	14.30	11.34	11:04	72
-125	16.98	14.04	11:20	73
-127	19.62	16.72	11:39	73
-129	22.14	19.31	11:57	71
-130	23.35	20.56	12:06	70
-132	25.62	22.92	12:23	67
-134	27.71	25.11	12:40	63
-135	28.69	26.13	12:47	62
-137	30.52	28.04	13:03	59
-138	31.37	28.93	13:10	57
-139	32.18	29.78	13:17	55
-140	32.95	30.58	13:24	54
-142	34.40	32.08	13:38	51
-143	35.07	32.78	13:44	50
-144	35.71	33.45	13:51	48
-145	36.32	34.08	13:57	47
-147	37.47	35.26	14:09	44
-148	38.00	35.81	14:15	43
-149	38.51	36.34	14:21	42
-150	39.00	36.84	14:27	41
-152	39.91	37.78	14:38	38
-153	40.34	38.21	14:43	37

PERCENTAGE OF SUN OBSCURED ALONG MAP TRACK
WEST END = 95.0   EAST END = 94.3
MAXIMUM OBSCURATION 95.0 % AT LONG = -125
===

+ MARKS THE CITY OF CHIEN-K'ANG

ECLIPSE ON JD 1844283 DATE(Y/M/D) 337/ 5/17
DELTA T = 1 HOUR   40.6 MINS

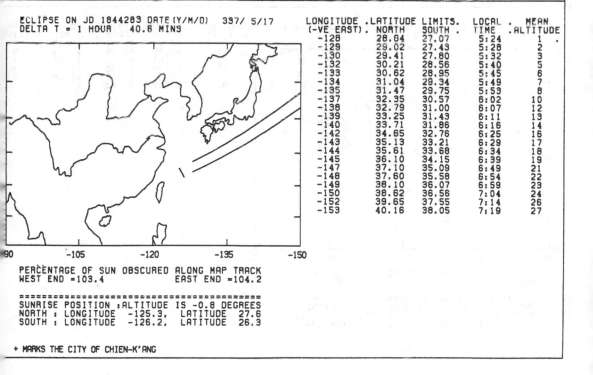

LONGITUDE (-VE EAST)	LATITUDE NORTH	LIMITS. SOUTH	LOCAL TIME	MEAN ALTITUDE
-128	28.64	27.07	5:24	1
-129	29.02	27.43	5:28	2
-130	29.41	27.80	5:32	3
-132	30.21	28.56	5:40	5
-133	30.62	28.95	5:45	6
-134	31.04	29.34	5:49	7
-135	31.47	29.75	5:53	8
-137	32.35	30.57	6:02	10
-138	32.79	31.00	6:07	12
-139	33.25	31.43	6:11	13
-140	33.71	31.86	6:16	14
-142	34.65	32.76	6:25	16
-143	35.13	33.21	6:29	17
-144	35.61	33.68	6:34	18
-145	36.10	34.15	6:39	19
-147	37.10	35.09	6:49	21
-148	37.60	35.58	6:54	22
-149	38.10	36.07	6:59	23
-150	38.62	36.56	7:04	24
-152	39.65	37.55	7:14	26
-153	40.16	38.05	7:19	27

PERCENTAGE OF SUN OBSCURED ALONG MAP TRACK
WEST END =103.4   EAST END =104.2

===
SUNRISE POSITION ;ALTITUDE IS -0.8 DEGREES
NORTH : LONGITUDE -125.3, LATITUDE 27.6
SOUTH : LONGITUDE -126.2, LATITUDE 26.3

+ MARKS THE CITY OF CHIEN-K'ANG

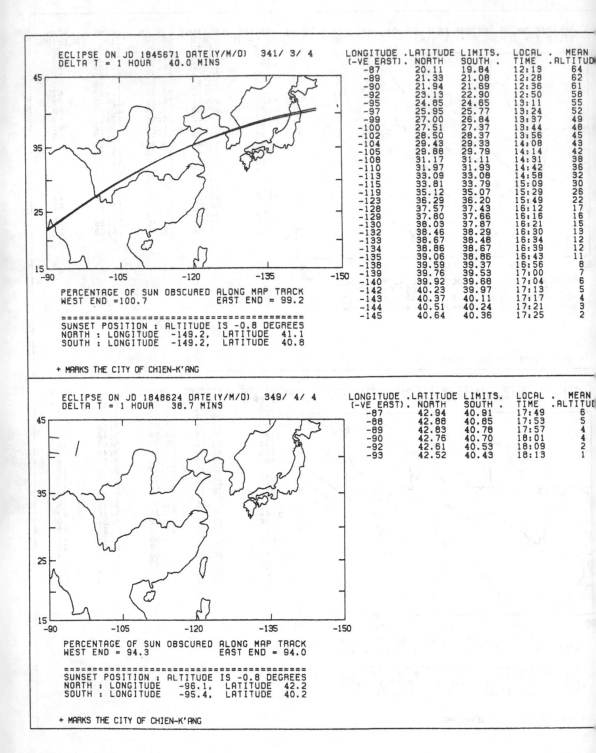

ECLIPSE ON JD 1845671 DATE(Y/M/D)   341/ 3/ 4
DELTA T = 1 HOUR   40.0 MINS

LONGITUDE (-VE EAST)	LATITUDE LIMITS. NORTH	SOUTH	LOCAL TIME	MEAN ALTITUDE
-87	20.11	19.84	12:19	64
-89	21.33	21.08	12:28	62
-90	21.94	21.69	12:36	61
-92	23.13	22.90	12:50	58
-95	24.85	24.65	13:11	55
-97	25.95	25.77	13:24	52
-99	27.00	26.84	13:37	49
-100	27.51	27.37	13:44	48
-102	28.50	28.37	13:56	45
-104	29.43	29.33	14:08	43
-105	29.88	29.79	14:14	42
-108	31.17	31.11	14:31	38
-110	31.97	31.93	14:42	36
-113	33.09	33.08	14:58	32
-115	33.81	33.79	15:09	30
-119	35.12	35.07	15:29	26
-123	36.29	36.20	15:49	22
-128	37.57	37.43	16:12	17
-129	37.80	37.66	16:16	16
-130	38.03	37.87	16:21	15
-132	38.46	38.29	16:30	13
-133	38.67	38.48	16:34	12
-134	38.86	38.67	16:39	12
-135	39.06	38.86	16:43	11
-138	39.59	39.37	16:56	8
-139	39.76	39.53	17:00	7
-140	39.92	39.68	17:04	6
-142	40.23	39.97	17:13	5
-143	40.37	40.11	17:17	4
-144	40.51	40.24	17:21	3
-145	40.64	40.36	17:25	2

PERCENTAGE OF SUN OBSCURED ALONG MAP TRACK
WEST END =100.7   EAST END = 99.2

==
SUNSET POSITION : ALTITUDE IS -0.8 DEGREES
NORTH : LONGITUDE -149.2, LATITUDE   41.1
SOUTH : LONGITUDE -149.2, LATITUDE   40.8

+ MARKS THE CITY OF CHIEN-K'ANG

ECLIPSE ON JD 1848624 DATE(Y/M/D)   349/ 4/ 4
DELTA T = 1 HOUR   38.7 MINS

LONGITUDE (-VE EAST)	LATITUDE LIMITS. NORTH	SOUTH	LOCAL TIME	MEAN ALTITUDE
-87	42.94	40.91	17:49	6
-88	42.88	40.85	17:53	5
-89	42.83	40.78	17:57	4
-90	42.76	40.70	18:01	4
-92	42.61	40.53	18:09	2
-93	42.52	40.43	18:13	1

PERCENTAGE OF SUN OBSCURED ALONG MAP TRACK
WEST END = 94.3   EAST END = 94.0

==
SUNSET POSITION : ALTITUDE IS -0.8 DEGREES
NORTH : LONGITUDE   -96.1, LATITUDE   42.2
SOUTH : LONGITUDE   -95.4, LATITUDE   40.2

+ MARKS THE CITY OF CHIEN-K'ANG

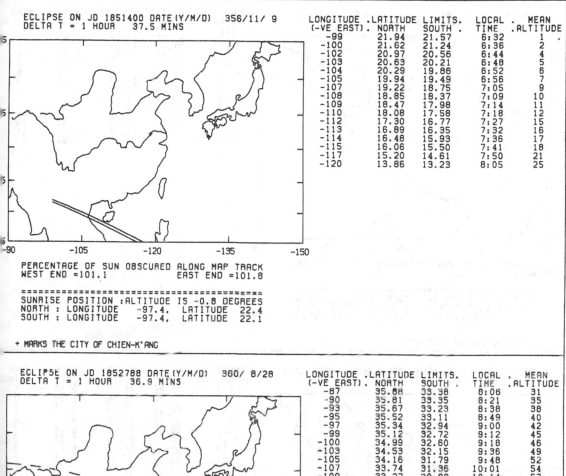

ECLIPSE ON JD 1851400 DATE(Y/M/D)   356/11/ 9
DELTA T = 1 HOUR   37.5 MINS

LONGITUDE (-VE EAST)	.LATITUDE . NORTH	LIMITS. SOUTH .	LOCAL TIME	. MEAN .ALTITUDE
-99	21.94	21.57	6:32	1 .
-100	21.62	21.24	6:36	2
-102	20.97	20.56	6:44	4
-103	20.63	20.21	6:48	5
-104	20.29	19.86	6:52	6
-105	19.94	19.49	6:56	7
-107	19.22	18.75	7:05	9
-108	18.85	18.37	7:09	10
-109	18.47	17.98	7:14	11
-110	18.08	17.58	7:18	12
-112	17.30	16.77	7:27	15
-113	16.89	16.35	7:32	16
-114	16.48	15.93	7:36	17
-115	16.06	15.50	7:41	18
-117	15.20	14.61	7:50	21
-120	13.86	13.23	8:05	25

PERCENTAGE OF SUN OBSCURED ALONG MAP TRACK
WEST END =101.1   EAST END =101.8

===
SUNRISE POSITION :ALTITUDE IS -0.8 DEGREES
NORTH : LONGITUDE   -97.4, LATITUDE 22.4
SOUTH : LONGITUDE   -97.4, LATITUDE 22.1

+ MARKS THE CITY OF CHIEN-K'ANG

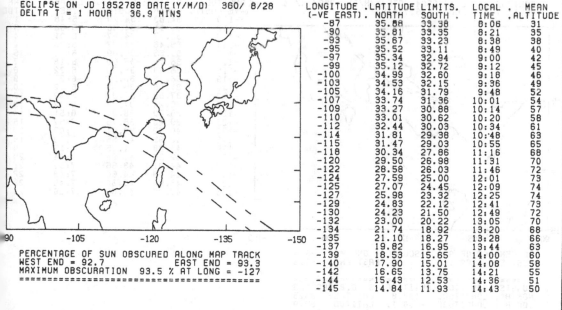

ECLIPSE ON JD 1852788 DATE(Y/M/D)   360/ 8/28
DELTA T = 1 HOUR   36.9 MINS

LONGITUDE (-VE EAST)	.LATITUDE . NORTH	LIMITS. SOUTH .	LOCAL TIME	. MEAN .ALTITUDE
-87	35.88	33.38	8:06	31
-90	35.81	33.35	8:21	35
-93	35.67	33.23	8:38	38
-95	35.52	33.11	8:49	40
-97	35.34	32.94	9:00	42
-99	35.12	32.72	9:12	45
-100	34.99	32.60	9:18	46
-103	34.53	32.15	9:36	49
-105	34.16	31.79	9:48	52
-107	33.74	31.36	10:01	54
-109	33.27	30.88	10:14	57
-110	33.01	30.62	10:20	58
-112	32.44	30.03	10:34	61
-114	31.81	29.38	10:48	63
-115	31.47	29.03	10:55	65
-118	30.34	27.86	11:16	68
-120	29.50	26.98	11:31	70
-122	28.58	26.03	11:46	72
-124	27.59	25.00	12:01	73
-125	27.07	24.45	12:09	74
-127	25.98	23.32	12:25	74
-129	24.83	22.12	12:41	73
-130	24.23	21.50	12:49	72
-132	23.00	20.22	13:05	70
-134	21.74	18.92	13:20	68
-135	21.10	18.27	13:28	66
-137	19.82	16.95	13:44	63
-139	18.53	15.65	14:00	60
-140	17.90	15.01	14:08	58
-142	16.65	13.75	14:21	55
-144	15.43	12.53	14:36	51
-145	14.84	11.93	14:43	50

PERCENTAGE OF SUN OBSCURED ALONG MAP TRACK
WEST END = 92.7   EAST END = 93.3
MAXIMUM OBSCURATION   93.5 % AT LONG = -127
===

+ MARKS THE CITY OF CHIEN-K'ANG

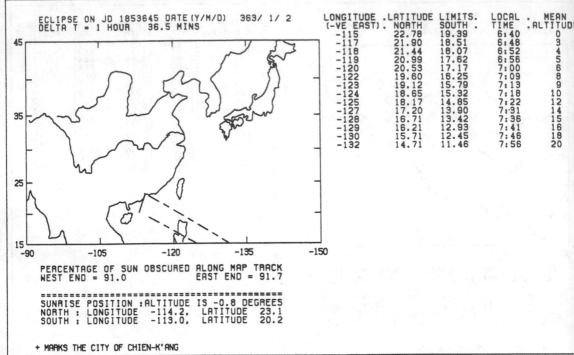

ECLIPSE ON JD 1853645 DATE(Y/M/D)   363/ 1/ 2
DELTA T = 1 HOUR   36.5 MINS

LONGITUDE (-VE EAST)	LATITUDE LIMITS. NORTH	SOUTH	LOCAL TIME	MEAN ALTITUDE
-115	22.78	19.39	6:40	0
-117	21.90	18.51	6:48	3
-118	21.44	18.07	6:52	4
-119	20.99	17.62	6:56	5
-120	20.53	17.17	7:00	6
-122	19.60	16.25	7:09	8
-123	19.12	15.79	7:13	9
-124	18.65	15.32	7:18	10
-125	18.17	14.85	7:22	12
-127	17.20	13.90	7:31	14
-128	16.71	13.42	7:36	15
-129	16.21	12.93	7:41	16
-130	15.71	12.45	7:46	18
-132	14.71	11.46	7:56	20

PERCENTAGE OF SUN OBSCURED ALONG MAP TRACK
WEST END = 91.0   EAST END = 91.7

==
SUNRISE POSITION :ALTITUDE IS -0.8 DEGREES
NORTH : LONGITUDE -114.2, LATITUDE 23.1
SOUTH : LONGITUDE -113.0, LATITUDE 20.2

+ MARKS THE CITY OF CHIEN-K'ANG

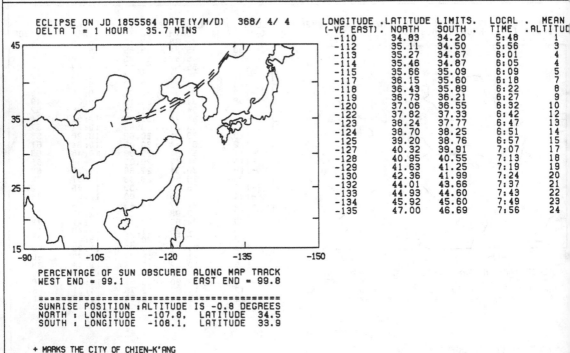

ECLIPSE ON JD 1855564 DATE(Y/M/D)   368/ 4/ 4
DELTA T = 1 HOUR   35.7 MINS

LONGITUDE (-VE EAST)	LATITUDE LIMITS. NORTH	SOUTH	LOCAL TIME	MEAN ALTITUDE
-110	34.83	34.20	5:48	1
-112	35.11	34.50	5:56	3
-113	35.27	34.67	6:01	4
-114	35.46	34.87	6:05	4
-115	35.66	35.09	6:09	5
-117	36.15	35.60	6:18	7
-118	36.43	35.89	6:22	8
-119	36.73	36.21	6:27	9
-120	37.06	36.55	6:32	10
-122	37.82	37.33	6:42	12
-123	38.24	37.77	6:47	13
-124	38.70	38.25	6:51	14
-125	39.20	38.76	6:57	15
-127	40.32	39.91	7:07	17
-128	40.95	40.55	7:13	18
-129	41.63	41.25	7:19	19
-130	42.36	41.99	7:24	20
-132	44.01	43.66	7:37	21
-133	44.93	44.60	7:43	22
-134	45.92	45.60	7:49	23
-135	47.00	46.69	7:56	24

PERCENTAGE OF SUN OBSCURED ALONG MAP TRACK
WEST END = 99.1   EAST END = 99.8

==
SUNRISE POSITION :ALTITUDE IS -0.8 DEGREES
NORTH : LONGITUDE -107.8, LATITUDE 34.5
SOUTH : LONGITUDE -108.1, LATITUDE 33.9

+ MARKS THE CITY OF CHIEN-K'ANG

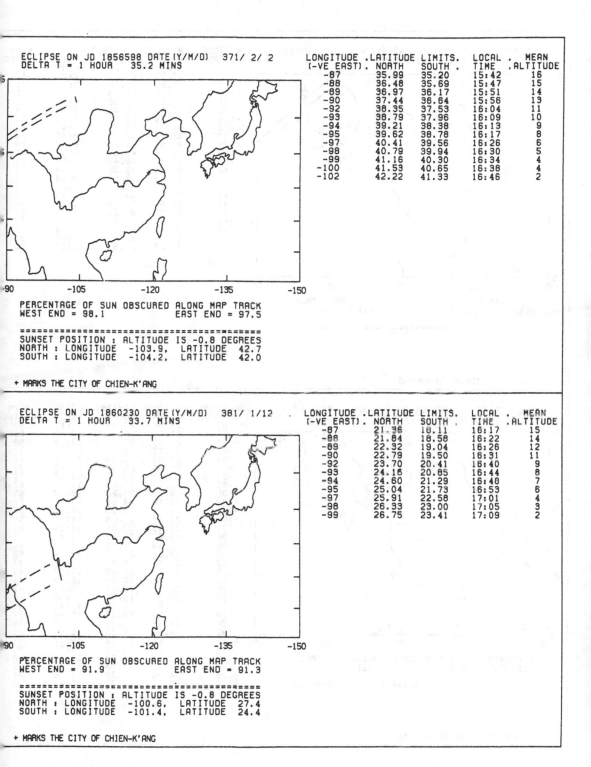

ECLIPSE ON JD 1856598 DATE(Y/M/D) 371/ 2/ 2
DELTA T = 1 HOUR   35.2 MINS

LONGITUDE (-VE EAST)	LATITUDE LIMITS. NORTH	LATITUDE LIMITS. SOUTH	LOCAL TIME	MEAN ALTITUDE
-87	35.99	35.20	15:42	16
-88	36.48	35.69	15:47	15
-89	36.97	36.17	15:51	14
-90	37.44	36.64	15:56	13
-92	38.35	37.53	16:04	11
-93	38.79	37.96	16:09	10
-94	39.21	38.38	16:13	9
-95	39.62	38.78	16:17	8
-97	40.41	39.56	16:26	6
-98	40.79	39.94	16:30	5
-99	41.16	40.30	16:34	4
-100	41.53	40.65	16:38	4
-102	42.22	41.33	16:46	2

PERCENTAGE OF SUN OBSCURED ALONG MAP TRACK
WEST END = 98.1   EAST END = 97.5

==
SUNSET POSITION : ALTITUDE IS -0.8 DEGREES
NORTH : LONGITUDE -103.9, LATITUDE 42.7
SOUTH : LONGITUDE -104.2, LATITUDE 42.0

+ MARKS THE CITY OF CHIEN-K'ANG

ECLIPSE ON JD 1860230 DATE(Y/M/D) 381/ 1/12
DELTA T = 1 HOUR   33.7 MINS

LONGITUDE (-VE EAST)	LATITUDE LIMITS. NORTH	LATITUDE LIMITS. SOUTH	LOCAL TIME	MEAN ALTITUDE
-87	21.36	18.11	16:17	15
-88	21.84	18.58	16:22	14
-89	22.32	19.04	16:26	12
-90	22.79	19.50	16:31	11
-92	23.70	20.41	16:40	9
-93	24.16	20.85	16:44	8
-94	24.60	21.29	16:48	7
-95	25.04	21.73	16:53	6
-97	25.91	22.58	17:01	4
-98	26.33	23.00	17:05	3
-99	26.75	23.41	17:09	2

PERCENTAGE OF SUN OBSCURED ALONG MAP TRACK
WEST END = 91.9   EAST END = 91.3

==
SUNSET POSITION : ALTITUDE IS -0.8 DEGREES
NORTH : LONGITUDE -100.6, LATITUDE 27.4
SOUTH : LONGITUDE -101.4, LATITUDE 24.4

+ MARKS THE CITY OF CHIEN-K'ANG

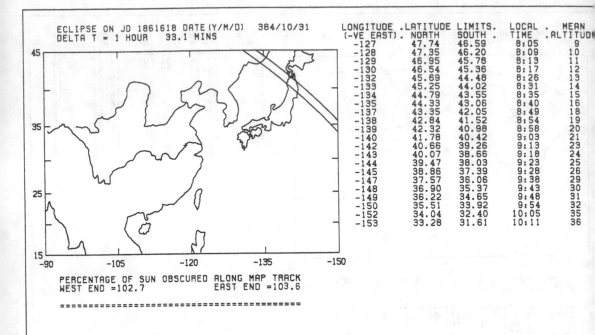

ECLIPSE ON JD 1861618 DATE(Y/M/D) 384/10/31
DELTA T = 1 HOUR 33.1 MINS

LONGITUDE (-VE EAST)	LATITUDE NORTH	LIMITS. SOUTH	LOCAL TIME	MEAN ALTITUDE
-127	47.74	46.59	8:05	9
-128	47.35	46.20	8:09	10
-129	46.95	45.78	8:13	11
-130	46.54	45.36	8:17	12
-132	45.69	44.48	8:26	13
-133	45.25	44.02	8:31	14
-134	44.79	43.55	8:35	15
-135	44.33	43.06	8:40	16
-137	43.35	42.05	8:49	18
-138	42.84	41.52	8:54	19
-139	42.32	40.98	8:58	20
-140	41.78	40.42	9:03	21
-142	40.66	39.26	9:13	23
-143	40.07	38.66	9:18	24
-144	39.47	38.03	9:23	25
-145	38.86	37.39	9:28	26
-147	37.57	36.06	9:38	29
-148	36.90	35.37	9:43	30
-149	36.22	34.65	9:48	31
-150	35.51	33.92	9:54	32
-152	34.04	32.40	10:05	35
-153	33.28	31.61	10:11	36

PERCENTAGE OF SUN OBSCURED ALONG MAP TRACK
WEST END =102.7   EAST END =103.6

==

+ MARKS THE CITY OF CHIEN-K'ANG

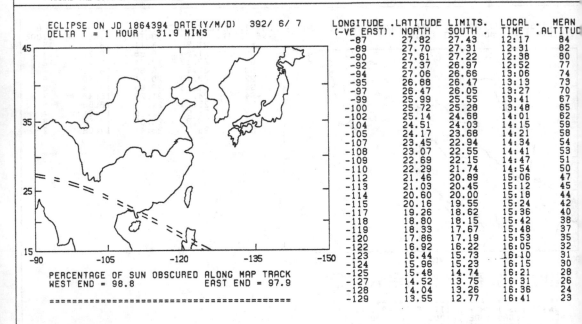

ECLIPSE ON JD 1864394 DATE(Y/M/D) 392/ 6/ 7
DELTA T = 1 HOUR 31.9 MINS

LONGITUDE (-VE EAST)	LATITUDE NORTH	LIMITS. SOUTH	LOCAL TIME	MEAN ALTITUDE
-87	27.82	27.43	12:17	84
-89	27.70	27.31	12:31	82
-90	27.61	27.22	12:38	80
-92	27.37	26.97	12:52	77
-94	27.06	26.66	13:06	74
-95	26.88	26.47	13:13	73
-97	26.47	26.05	13:27	70
-99	25.99	25.55	13:41	67
-100	25.72	25.28	13:48	65
-102	25.14	24.68	14:01	62
-104	24.51	24.03	14:15	59
-105	24.17	23.68	14:21	58
-107	23.45	22.94	14:34	54
-108	23.07	22.55	14:41	53
-109	22.69	22.15	14:47	51
-110	22.29	21.74	14:54	50
-112	21.46	20.89	15:06	47
-113	21.03	20.45	15:12	45
-114	20.60	20.00	15:18	44
-115	20.16	19.55	15:24	42
-117	19.26	18.62	15:36	40
-118	18.80	18.15	15:42	38
-119	18.33	17.67	15:48	37
-120	17.86	17.19	15:53	35
-122	16.92	16.22	16:05	32
-123	16.44	15.73	16:10	31
-124	15.96	15.23	16:15	30
-125	15.48	14.74	16:21	28
-127	14.52	13.75	16:31	26
-128	14.04	13.26	16:36	24
-129	13.55	12.77	16:41	23

PERCENTAGE OF SUN OBSCURED ALONG MAP TRACK
WEST END = 98.8   EAST END = 97.9

==

+ MARKS THE CITY OF CHIEN-K'ANG

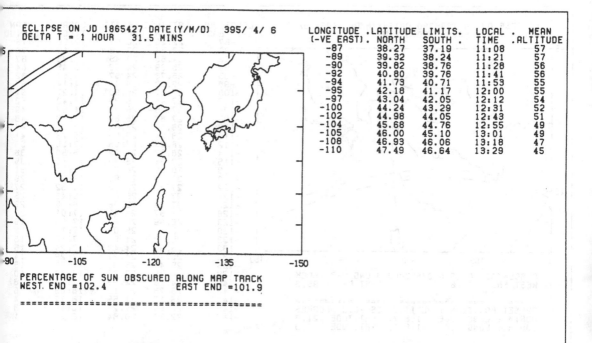

ECLIPSE ON JD 1865427 DATE(Y/M/D)   395/ 4/ 6
DELTA T = 1 HOUR   31.5 MINS

LONGITUDE (-VE EAST)	LATITUDE NORTH	LIMITS SOUTH	LOCAL TIME	MEAN ALTITUDE
-87	38.27	37.19	11:08	57
-89	39.32	38.24	11:21	57
-90	39.82	38.76	11:28	56
-92	40.80	39.76	11:41	56
-94	41.73	40.71	11:53	55
-95	42.18	41.17	12:00	55
-97	43.04	42.05	12:12	54
-100	44.24	43.29	12:31	52
-102	44.98	44.05	12:43	51
-104	45.68	44.76	12:55	49
-105	46.00	45.10	13:01	49
-108	46.93	46.06	13:18	47
-110	47.49	46.64	13:29	45

-90   -105   -120   -135   -150

PERCENTAGE OF SUN OBSCURED ALONG MAP TRACK
WEST END =102.4   EAST END =101.9

===

+ MARKS THE CITY OF CHIEN-K'ANG

ECLIPSE ON JD 1867347 DATE(Y/M/D)   400/ 7/ 8
DELTA T = 1 HOUR   30.7 MINS

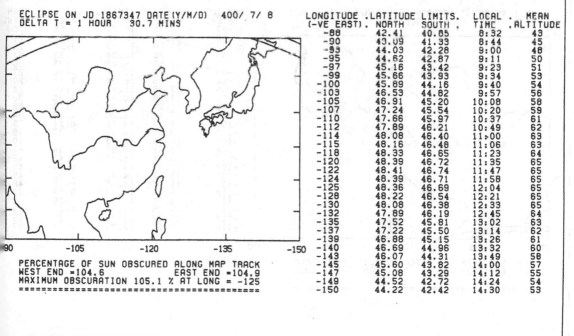

LONGITUDE (-VE EAST)	LATITUDE NORTH	LIMITS SOUTH	LOCAL TIME	MEAN ALTITUDE
-88	42.41	40.85	8:32	43
-90	43.09	41.33	8:44	45
-93	44.03	42.28	9:00	48
-95	44.62	42.87	9:11	50
-97	45.16	43.42	9:23	51
-99	45.66	43.93	9:34	53
-100	45.89	44.16	9:40	54
-103	46.53	44.82	9:57	56
-105	46.91	45.20	10:08	58
-107	47.24	45.54	10:20	59
-110	47.66	45.97	10:37	61
-112	47.89	46.21	10:49	62
-114	48.08	46.40	11:00	63
-115	48.16	46.48	11:06	63
-118	48.33	46.65	11:23	64
-120	48.39	46.72	11:35	65
-122	48.41	46.74	11:47	65
-124	48.39	46.71	11:58	65
-125	48.36	46.69	12:04	65
-128	48.22	46.54	12:21	65
-130	48.08	46.38	12:33	65
-132	47.89	46.19	12:45	64
-135	47.52	45.81	13:02	63
-137	47.22	45.50	13:14	62
-139	46.88	45.15	13:26	61
-140	46.69	44.96	13:32	60
-143	46.07	44.31	13:49	58
-145	45.60	43.82	14:00	57
-147	45.08	43.29	14:12	55
-149	44.52	42.72	14:24	54
-150	44.22	42.42	14:30	53

-90   -105   -120   -135   -150

PERCENTAGE OF SUN OBSCURED ALONG MAP TRACK
WEST END =104.6   EAST END =104.9
MAXIMUM OBSCURATION 105.1 % AT LONG = -125

===

+ MARKS THE CITY OF CHIEN-K'ANG

241

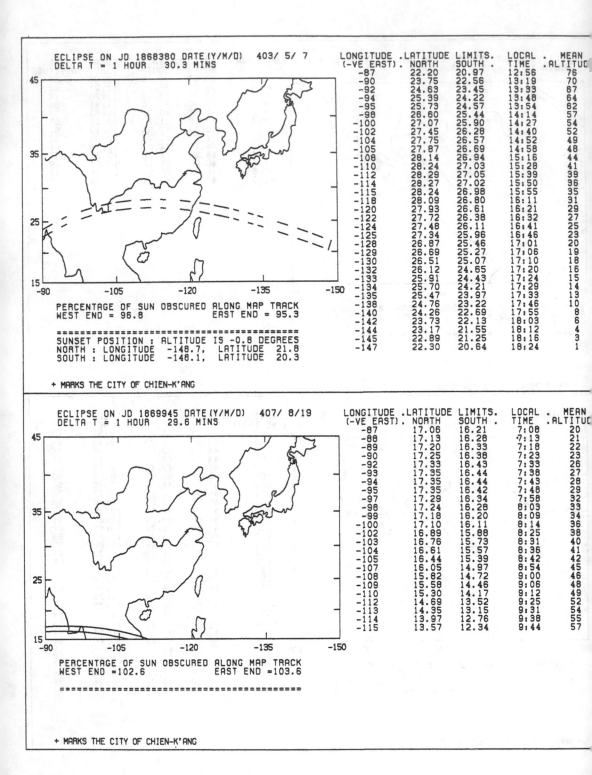

ECLIPSE ON JD 1868380 DATE(Y/M/D) 403/ 5/ 7
DELTA T = 1 HOUR   30.3 MINS

LONGITUDE (-VE EAST)	LATITUDE NORTH	LIMITS. SOUTH	LOCAL TIME	MEAN ALTITUDE
-87	22.20	20.97	12:56	76
-90	23.75	22.56	13:19	70
-92	24.63	23.45	13:33	67
-94	25.39	24.22	13:48	64
-95	25.73	24.57	13:54	62
-98	26.60	25.44	14:14	57
-100	27.07	25.90	14:27	54
-102	27.45	26.28	14:40	52
-104	27.75	26.57	14:52	49
-105	27.87	26.69	14:58	48
-108	28.14	26.94	15:16	44
-110	28.24	27.03	15:28	41
-112	28.29	27.05	15:39	39
-114	28.27	27.02	15:50	36
-115	28.24	26.98	15:55	35
-118	28.09	26.80	16:11	31
-120	27.93	26.61	16:21	29
-122	27.72	26.38	16:32	27
-124	27.48	26.11	16:41	25
-125	27.34	25.96	16:46	23
-128	26.87	25.46	17:01	20
-129	26.69	25.27	17:06	19
-130	26.51	25.07	17:10	18
-132	26.12	24.65	17:20	16
-133	25.91	24.43	17:24	15
-134	25.70	24.21	17:29	14
-135	25.47	23.97	17:33	13
-138	24.76	23.22	17:46	10
-140	24.26	22.69	17:55	8
-142	23.73	22.13	18:03	6
-144	23.17	21.55	18:12	4
-145	22.89	21.25	18:16	3
-147	22.30	20.64	18:24	1

PERCENTAGE OF SUN OBSCURED ALONG MAP TRACK
WEST END = 96.8   EAST END = 95.3

SUNSET POSITION : ALTITUDE IS -0.8 DEGREES
NORTH : LONGITUDE -148.7, LATITUDE 21.8
SOUTH : LONGITUDE -148.1, LATITUDE 20.3

+ MARKS THE CITY OF CHIEN-K'ANG

ECLIPSE ON JD 1869945 DATE(Y/M/D) 407/ 8/19
DELTA T = 1 HOUR   29.6 MINS

LONGITUDE (-VE EAST)	LATITUDE NORTH	LIMITS. SOUTH	LOCAL TIME	MEAN ALTITUDE
-87	17.06	16.21	7:08	20
-88	17.13	16.28	7:13	21
-89	17.20	16.33	7:18	22
-90	17.25	16.38	7:23	23
-92	17.33	16.43	7:33	26
-93	17.35	16.44	7:38	27
-94	17.35	16.44	7:43	28
-95	17.35	16.42	7:48	29
-97	17.29	16.34	7:58	32
-98	17.24	16.28	8:03	33
-99	17.18	16.20	8:09	34
-100	17.10	16.11	8:14	36
-102	16.89	15.88	8:25	38
-103	16.76	15.73	8:31	40
-104	16.61	15.57	8:36	41
-105	16.44	15.39	8:42	42
-107	16.05	14.97	8:54	45
-108	15.82	14.72	9:00	46
-109	15.58	14.46	9:06	48
-110	15.30	14.17	9:12	49
-112	14.69	13.52	9:25	52
-113	14.35	13.15	9:31	54
-114	13.97	12.76	9:38	55
-115	13.57	12.34	9:44	57

PERCENTAGE OF SUN OBSCURED ALONG MAP TRACK
WEST END =102.6   EAST END =103.6

+ MARKS THE CITY OF CHIEN-K'ANG

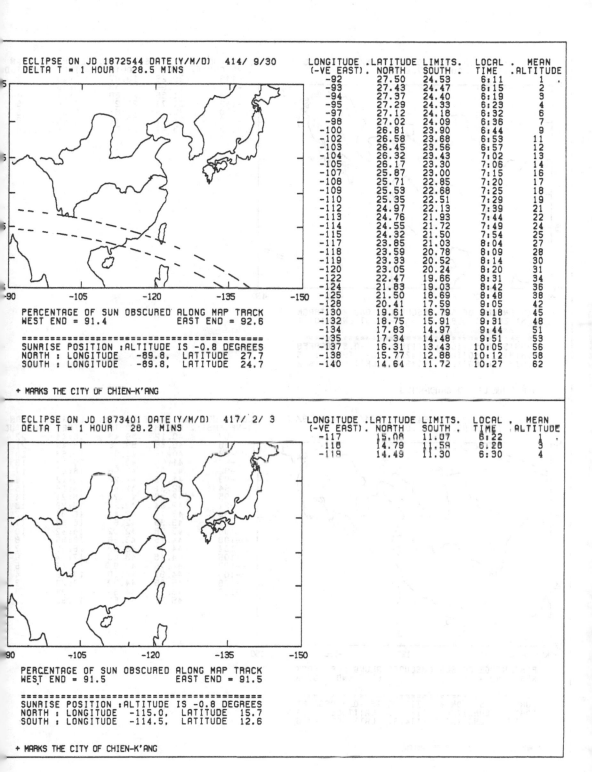

ECLIPSE ON JD 1872544 DATE(Y/M/D) 414/ 9/30
DELTA T = 1 HOUR   28.5 MINS

LONGITUDE (-VE EAST)	LATITUDE LIMITS. NORTH	SOUTH	LOCAL TIME	MEAN ALTITUDE
-92	27.50	24.53	6:11	1
-93	27.43	24.47	6:15	2
-94	27.37	24.40	6:19	3
-95	27.29	24.33	6:23	4
-97	27.12	24.18	6:32	6
-98	27.02	24.09	6:36	7
-100	26.81	23.90	6:44	9
-102	26.58	23.68	6:53	11
-103	26.45	23.56	6:57	12
-104	26.32	23.43	7:02	13
-105	26.17	23.30	7:06	14
-107	25.87	23.00	7:15	16
-108	25.71	22.85	7:20	17
-109	25.53	22.68	7:25	18
-110	25.35	22.51	7:29	19
-112	24.97	22.13	7:39	21
-113	24.76	21.93	7:44	22
-114	24.55	21.72	7:49	24
-115	24.32	21.50	7:54	25
-117	23.85	21.03	8:04	27
-118	23.59	20.78	8:09	28
-119	23.33	20.52	8:14	30
-120	23.05	20.24	8:20	31
-122	22.47	19.66	8:31	34
-124	21.83	19.03	8:42	36
-125	21.50	18.69	8:48	38
-128	20.41	17.59	9:05	42
-130	19.61	16.79	9:18	45
-132	18.75	15.91	9:31	48
-134	17.83	14.97	9:44	51
-135	17.34	14.48	9:51	53
-137	16.31	13.43	10:05	56
-138	15.77	12.88	10:12	58
-140	14.64	11.72	10:27	62

PERCENTAGE OF SUN OBSCURED ALONG MAP TRACK
WEST END = 91.4   EAST END = 92.6

SUNRISE POSITION :ALTITUDE IS -0.8 DEGREES
NORTH : LONGITUDE -89.8, LATITUDE 27.7
SOUTH : LONGITUDE -89.8, LATITUDE 24.7

+ MARKS THE CITY OF CHIEN-K'ANG

ECLIPSE ON JD 1873401 DATE(Y/M/D) 417/ 2/ 3
DELTA T = 1 HOUR   28.2 MINS

LONGITUDE (-VE EAST)	LATITUDE LIMITS. NORTH	SOUTH	LOCAL TIME	MEAN ALTITUDE
-117	15.08	11.07	6:22	1
-118	14.79	11.59	6:28	3
-119	14.49	11.30	6:30	4

PERCENTAGE OF SUN OBSCURED ALONG MAP TRACK
WEST END = 91.5   EAST END = 91.5

SUNRISE POSITION :ALTITUDE IS -0.8 DEGREES
NORTH : LONGITUDE -115.0, LATITUDE 15.7
SOUTH : LONGITUDE -114.5, LATITUDE 12.6

+ MARKS THE CITY OF CHIEN-K'ANG

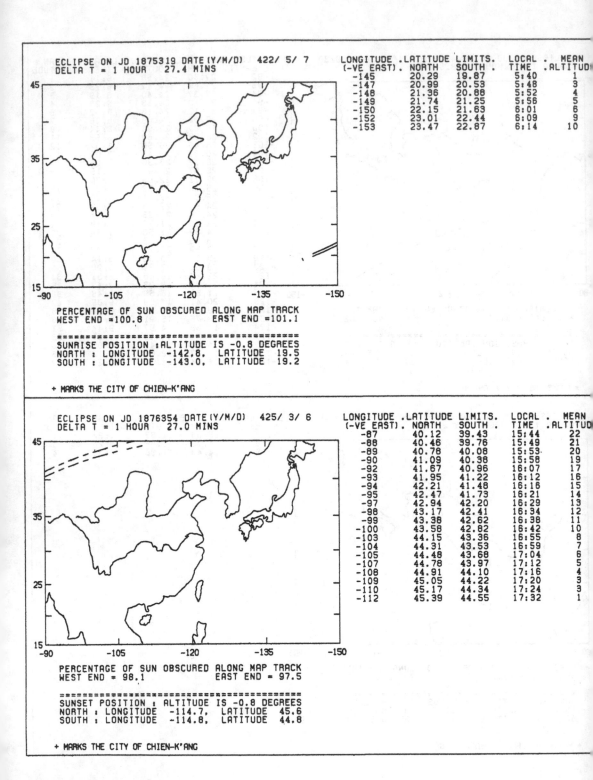

ECLIPSE ON JD 1875319 DATE(Y/M/D)   422/ 5/ 7
DELTA T = 1 HOUR   27.4 MINS

LONGITUDE (-VE EAST).	LATITUDE NORTH	LIMITS. SOUTH .	LOCAL TIME .	MEAN ALTITUDE
-145	20.29	19.87	5:40	1
-147	20.99	20.53	5:48	3
-148	21.36	20.88	5:52	4
-149	21.74	21.25	5:56	5
-150	22.15	21.63	6:01	6
-152	23.01	22.44	6:09	9
-153	23.47	22.87	6:14	10

PERCENTAGE OF SUN OBSCURED ALONG MAP TRACK
WEST END =100.8   EAST END =101.1

===
SUNRISE POSITION :ALTITUDE IS -0.8 DEGREES
NORTH : LONGITUDE -142.8. LATITUDE 19.5
SOUTH : LONGITUDE -143.0. LATITUDE 19.2

+ MARKS THE CITY OF CHIEN-K'ANG

ECLIPSE ON JD 1876354 DATE(Y/M/D)   425/ 3/ 6
DELTA T = 1 HOUR   27.0 MINS

LONGITUDE (-VE EAST).	LATITUDE NORTH	LIMITS. SOUTH	LOCAL TIME	MEAN ALTITUDE
-87	40.12	39.43	15:44	22
-88	40.46	39.76	15:49	21
-89	40.78	40.08	15:53	20
-90	41.09	40.38	15:58	19
-92	41.67	40.96	16:07	17
-93	41.95	41.22	16:12	16
-94	42.21	41.48	16:16	15
-95	42.47	41.73	16:21	14
-97	42.94	42.20	16:29	13
-98	43.17	42.41	16:34	12
-99	43.38	42.62	16:38	11
-100	43.58	42.82	16:42	10
-103	44.15	43.36	16:55	8
-104	44.31	43.53	16:59	7
-105	44.48	43.68	17:04	6
-107	44.78	43.97	17:12	5
-108	44.91	44.10	17:16	4
-109	45.05	44.22	17:20	3
-110	45.17	44.34	17:24	3
-112	45.39	44.55	17:32	1

PERCENTAGE OF SUN OBSCURED ALONG MAP TRACK
WEST END = 98.1   EAST END = 97.5

===
SUNSET POSITION : ALTITUDE IS -0.8 DEGREES
NORTH : LONGITUDE -114.7. LATITUDE 45.6
SOUTH : LONGITUDE -114.8. LATITUDE 44.8

+ MARKS THE CITY OF CHIEN-K'ANG

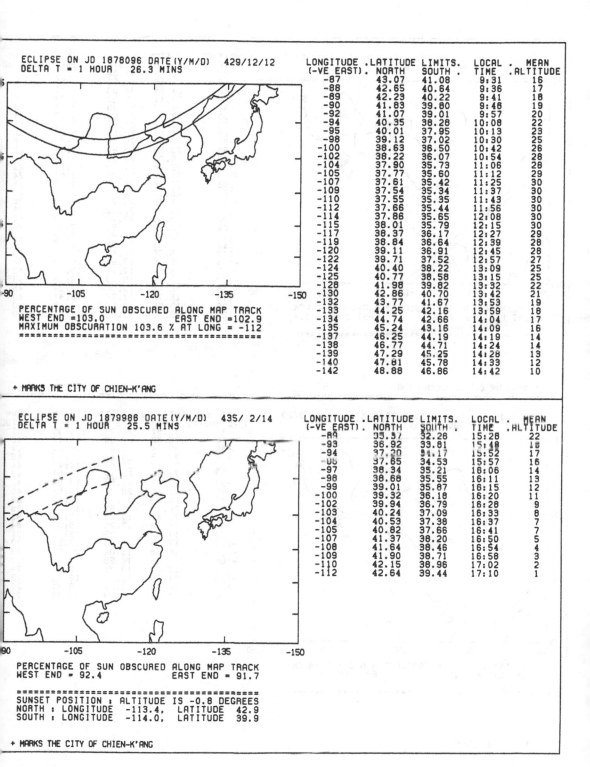

ECLIPSE ON JD 1878096 DATE(Y/M/D) 429/12/12
DELTA T = 1 HOUR   26.3 MINS

LONGITUDE (-VE EAST)	LATITUDE NORTH	LIMITS SOUTH	LOCAL TIME	MEAN ALTITUDE
-87	43.07	41.08	9:31	16
-88	42.65	40.64	9:36	17
-89	42.23	40.22	9:41	18
-90	41.83	39.80	9:48	19
-92	41.07	39.01	9:57	20
-94	40.35	38.28	10:08	22
-95	40.01	37.95	10:13	23
-98	39.12	37.02	10:30	25
-100	38.63	36.50	10:42	26
-102	38.22	36.07	10:54	28
-104	37.90	35.73	11:06	28
-105	37.77	35.60	11:12	29
-107	37.61	35.42	11:25	30
-109	37.54	35.34	11:37	30
-110	37.55	35.35	11:43	30
-112	37.66	35.44	11:56	30
-114	37.86	35.65	12:08	30
-115	38.01	35.79	12:15	30
-117	38.37	36.17	12:27	29
-119	38.84	36.64	12:39	28
-120	39.11	36.91	12:45	28
-122	39.71	37.52	12:57	27
-124	40.40	38.22	13:09	25
-125	40.77	38.58	13:15	25
-128	41.98	39.82	13:32	22
-130	42.86	40.70	13:42	21
-132	43.77	41.67	13:53	19
-133	44.25	42.16	13:59	18
-134	44.74	42.66	14:04	17
-135	45.24	43.16	14:09	16
-137	46.25	44.19	14:19	14
-138	46.77	44.71	14:24	14
-139	47.29	45.25	14:28	13
-140	47.81	45.78	14:33	12
-142	48.88	46.86	14:42	10

PERCENTAGE OF SUN OBSCURED ALONG MAP TRACK
WEST END =103.0   EAST END =102.9
MAXIMUM OBSCURATION 103.6 % AT LONG = -112
===

+ MARKS THE CITY OF CHIEN-K'ANG

ECLIPSE ON JD 1879988 DATE(Y/M/D) 435/ 2/14
DELTA T = 1 HOUR   25.5 MINS

LONGITUDE (-VE EAST)	LATITUDE NORTH	LIMITS SOUTH	LOCAL TIME	MEAN ALTITUDE
-89	35.57	32.26	15:28	22
-93	36.92	33.81	15:48	18
-94	37.20	34.17	15:52	17
-96	37.65	34.53	15:57	16
-97	38.34	35.21	16:06	14
-98	38.68	35.55	16:11	13
-99	39.01	35.87	16:15	12
-100	39.32	36.18	16:20	11
-102	39.94	36.79	16:28	9
-103	40.24	37.09	16:33	8
-104	40.53	37.38	16:37	7
-105	40.82	37.66	16:41	7
-107	41.37	38.20	16:50	5
-108	41.64	38.46	16:54	4
-109	41.90	38.71	16:58	3
-110	42.15	38.96	17:02	2
-112	42.64	39.44	17:10	1

PERCENTAGE OF SUN OBSCURED ALONG MAP TRACK
WEST END = 92.4   EAST END = 91.7

==
SUNSET POSITION : ALTITUDE IS -0.8 DEGREES
NORTH : LONGITUDE -113.4, LATITUDE   42.9
SOUTH : LONGITUDE -114.0, LATITUDE   39.9

+ MARKS THE CITY OF CHIEN-K'ANG

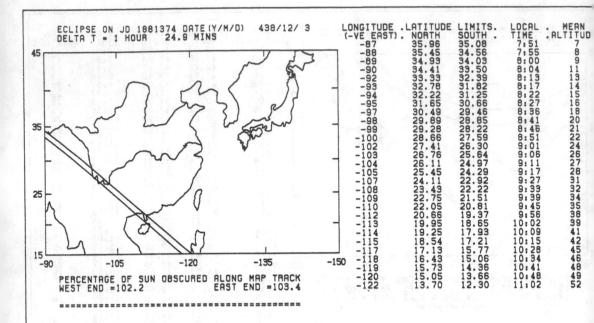

ECLIPSE ON JD 1881374 DATE(Y/M/D)   438/12/ 3
DELTA T = 1 HOUR   24.9 MINS

LONGITUDE (-VE EAST)	.LATITUDE NORTH	LIMITS. SOUTH .	LOCAL TIME	. MEAN .ALTITUD
-87	35.96	35.08	7:51	7
-88	35.45	34.56	7:55	8
-89	34.93	34.03	8:00	9
-90	34.41	33.50	8:04	11
-92	33.93	32.39	8:13	13
-93	32.78	31.82	8:17	14
-94	32.22	31.25	8:22	15
-95	31.65	30.66	8:27	16
-97	30.49	29.46	8:36	18
-98	29.89	28.85	8:41	20
-99	29.28	28.22	8:46	21
-100	28.66	27.59	8:51	22
-102	27.41	26.30	9:01	24
-103	26.76	25.64	9:06	26
-104	26.11	24.97	9:11	27
-105	25.45	24.29	9:17	28
-107	24.11	22.92	9:27	31
-108	23.43	22.22	9:33	32
-109	22.75	21.51	9:39	34
-110	22.05	20.81	9:45	35
-112	20.66	19.37	9:56	38
-113	19.95	18.65	10:02	39
-114	19.25	17.93	10:09	41
-115	18.54	17.21	10:15	42
-117	17.13	15.77	10:28	45
-118	16.43	15.06	10:34	46
-119	15.73	14.36	10:41	48
-120	15.05	13.66	10:48	49
-122	13.70	12.30	11:02	52

PERCENTAGE OF SUN OBSCURED ALONG MAP TRACK
WEST END =102.2   EAST END =103.4

===

+ MARKS THE CITY OF CHIEN-K'ANG

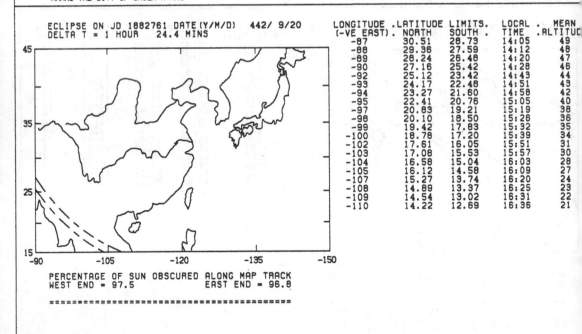

ECLIPSE ON JD 1882761 DATE(Y/M/D)   442/ 9/20
DELTA T = 1 HOUR   24.4 MINS

LONGITUDE (-VE EAST)	.LATITUDE NORTH	LIMITS. SOUTH .	LOCAL TIME	. MEAN .ALTITUD
-87	30.51	28.73	14:05	49
-88	29.36	27.59	14:12	48
-89	28.24	26.48	14:20	47
-90	27.16	25.42	14:28	46
-92	25.12	23.42	14:43	44
-93	24.17	22.48	14:51	43
-94	23.27	21.60	14:58	42
-95	22.41	20.76	15:05	40
-97	20.83	19.21	15:19	38
-98	20.10	18.50	15:26	36
-99	19.42	17.83	15:32	35
-100	18.78	17.20	15:39	34
-102	17.61	16.05	15:51	31
-103	17.08	15.53	15:57	30
-104	16.58	15.04	16:03	28
-105	16.12	14.58	16:09	27
-107	15.27	13.74	16:20	24
-108	14.89	13.37	16:25	23
-109	14.54	13.02	16:31	22
-110	14.22	12.69	16:36	21

PERCENTAGE OF SUN OBSCURED ALONG MAP TRACK
WEST END = 97.5   EAST END = 96.8

===

+ MARKS THE CITY OF CHIEN-K'ANG

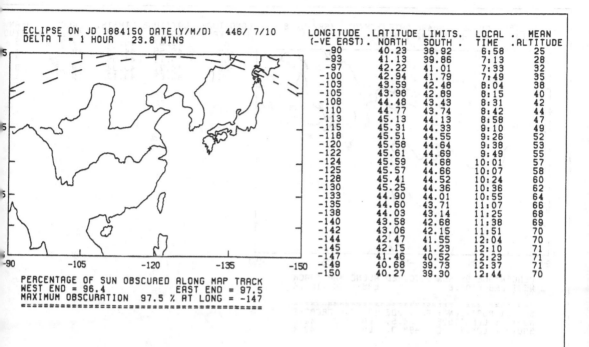

```
ECLIPSE ON JD 1884150 DATE (Y/M/D)   446/ 7/10
DELTA T = 1 HOUR   23.8 MINS
```

LONGITUDE (-VE EAST)	LATITUDE LIMITS. NORTH	SOUTH	LOCAL TIME	MEAN ALTITUDE
-90	40.23	38.92	6:58	25
-93	41.13	39.86	7:13	28
-97	42.22	41.01	7:33	32
-100	42.94	41.79	7:49	35
-103	43.59	42.48	8:04	38
-105	43.98	42.89	8:15	40
-108	44.48	43.43	8:31	42
-110	44.77	43.74	8:42	44
-113	45.13	44.13	8:58	47
-115	45.31	44.33	9:10	49
-118	45.51	44.55	9:26	52
-120	45.58	44.64	9:38	53
-122	45.61	44.69	9:49	55
-124	45.59	44.68	10:01	57
-125	45.57	44.66	10:07	58
-128	45.41	44.52	10:24	60
-130	45.25	44.36	10:36	62
-133	44.90	44.01	10:55	64
-135	44.60	43.71	11:07	66
-138	44.03	43.14	11:25	68
-140	43.58	42.68	11:38	69
-142	43.06	42.15	11:51	70
-144	42.47	41.55	12:04	70
-145	42.15	41.23	12:10	71
-147	41.46	40.52	12:23	71
-149	40.68	39.73	12:37	71
-150	40.27	39.30	12:44	70

```
PERCENTAGE OF SUN OBSCURED ALONG MAP TRACK
WEST END = 96.4           EAST END = 97.5
MAXIMUM OBSCURATION  97.5 % AT LONG = -147
==========================================
```

+ MARKS THE CITY OF CHIEN-K'ANG

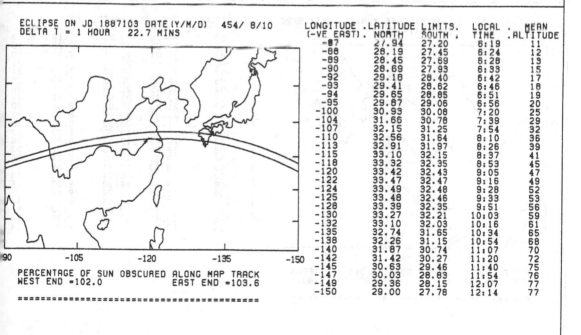

```
ECLIPSE ON JD 1887103 DATE (Y/M/D)   454/ 8/10
DELTA T = 1 HOUR   22.7 MINS
```

LONGITUDE (-VE EAST)	LATITUDE LIMITS. NORTH	SOUTH	LOCAL TIME	MEAN ALTITUDE
-87	27.94	27.20	6:19	11
-88	28.19	27.45	6:24	12
-89	28.45	27.69	6:28	13
-90	28.69	27.93	6:33	15
-92	29.18	28.40	6:42	17
-93	29.41	28.62	6:46	18
-94	29.65	28.85	6:51	19
-95	29.87	29.06	6:56	20
-100	30.93	30.08	7:20	25
-104	31.66	30.78	7:39	29
-107	32.15	31.25	7:54	32
-110	32.56	31.64	8:10	36
-113	32.91	31.97	8:26	39
-115	33.10	32.15	8:37	41
-118	33.32	32.35	8:53	45
-120	33.42	32.43	9:05	47
-122	33.47	32.47	9:16	49
-124	33.49	32.48	9:28	52
-125	33.48	32.46	9:33	53
-128	33.39	32.35	9:51	56
-130	33.27	32.21	10:03	59
-132	33.10	32.03	10:16	61
-135	32.74	31.65	10:34	65
-138	32.26	31.15	10:54	68
-140	31.87	30.74	11:07	70
-142	31.42	30.27	11:20	72
-145	30.63	29.46	11:40	75
-147	30.03	28.83	11:54	76
-149	29.36	28.15	12:07	77
-150	29.00	27.78	12:14	77

```
PERCENTAGE OF SUN OBSCURED ALONG MAP TRACK
WEST END =102.0           EAST END =103.6

==========================================
```

+ MARKS THE CITY OF CHIEN-K'ANG

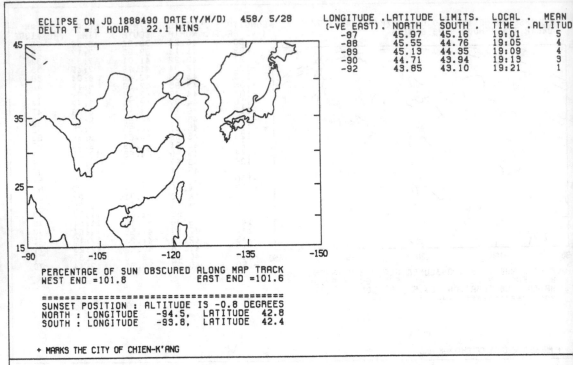

ECLIPSE ON JD 1888490 DATE(Y/M/D) 458/ 5/28
DELTA T = 1 HOUR   22.1 MINS

LONGITUDE (-VE EAST)	.LATITUDE NORTH	LIMITS. SOUTH .	LOCAL TIME	. MEAN .ALTITUD
-87	45.97	45.16	19:01	5
-88	45.55	44.76	19:05	4
-89	45.13	44.35	19:09	4
-90	44.71	43.94	19:13	3
-92	43.85	43.10	19:21	1

PERCENTAGE OF SUN OBSCURED ALONG MAP TRACK
WEST END =101.8   EAST END =101.6
===
SUNSET POSITION : ALTITUDE IS -0.8 DEGREES
NORTH : LONGITUDE   -94.5, LATITUDE   42.8
SOUTH : LONGITUDE   -93.8, LATITUDE   42.4

+ MARKS THE CITY OF CHIEN-K'ANG

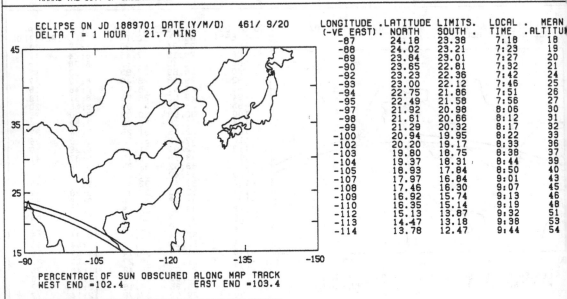

ECLIPSE ON JD 1889701 DATE(Y/M/D) 461/ 9/20
DELTA T = 1 HOUR   21.7 MINS

LONGITUDE (-VE EAST)	.LATITUDE NORTH	LIMITS. SOUTH .	LOCAL TIME	. MEAN .ALTITU
-87	24.18	23.38	7:18	18
-88	24.02	23.21	7:23	19
-89	23.84	23.01	7:27	20
-90	23.65	22.81	7:32	21
-92	23.23	22.36	7:42	24
-93	23.00	22.12	7:46	25
-94	22.75	21.86	7:51	26
-95	22.49	21.58	7:56	27
-97	21.92	20.98	8:06	30
-98	21.61	20.66	8:12	31
-99	21.29	20.32	8:17	32
-100	20.94	19.95	8:22	33
-102	20.20	19.17	8:33	36
-103	19.80	18.75	8:38	37
-104	19.37	18.31	8:44	39
-105	18.93	17.84	8:50	40
-107	17.97	16.84	9:01	43
-108	17.46	16.30	9:07	45
-109	16.92	15.74	9:13	46
-110	16.35	15.14	9:19	48
-112	15.13	13.87	9:32	51
-113	14.47	13.18	9:38	53
-114	13.78	12.47	9:44	54

PERCENTAGE OF SUN OBSCURED ALONG MAP TRACK
WEST END =102.4   EAST END =103.4
===

+ MARKS THE CITY OF CHIEN-K'ANG

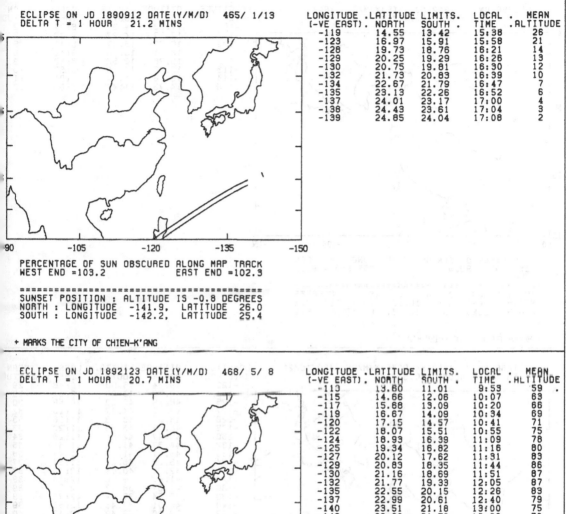

ECLIPSE ON JD 1890912 DATE(Y/M/D)   465/ 1/13
DELTA T = 1 HOUR   21.2 MINS

LONGITUDE (-VE EAST)	LATITUDE NORTH	LIMITS. SOUTH	LOCAL TIME	MEAN ALTITUDE
-119	14.55	13.42	15:38	26
-123	16.97	15.91	15:58	21
-128	19.73	18.76	16:21	14
-129	20.25	19.29	16:26	13
-130	20.75	19.81	16:30	12
-132	21.73	20.83	16:39	10
-134	22.67	21.79	16:47	7
-135	23.13	22.26	16:52	6
-137	24.01	23.17	17:00	4
-138	24.43	23.61	17:04	3
-139	24.85	24.04	17:08	2

PERCENTAGE OF SUN OBSCURED ALONG MAP TRACK
WEST END =103.2   EAST END =102.3

===
SUNSET POSITION : ALTITUDE IS -0.8 DEGREES
NORTH : LONGITUDE -141.9, LATITUDE 26.0
SOUTH : LONGITUDE -142.2, LATITUDE 25.4

+ MARKS THE CITY OF CHIEN-K'ANG

ECLIPSE ON JD 1892123 DATE(Y/M/D)   468/ 5/ 8
DELTA T = 1 HOUR   20.7 MINS

LONGITUDE (-VE EAST)	LATITUDE NORTH	LIMITS. SOUTH	LOCAL TIME	MEAN ALTITUDE
-113	13.60	11.01	9:53	59
-115	14.66	12.06	10:07	63
-117	15.68	13.09	10:20	66
-119	16.67	14.09	10:34	69
-120	17.15	14.57	10:41	71
-122	18.07	15.51	10:55	75
-124	18.93	16.39	11:09	78
-125	19.34	16.82	11:16	80
-127	20.12	17.62	11:31	83
-129	20.83	18.35	11:44	86
-130	21.16	18.69	11:51	87
-132	21.77	19.33	12:05	87
-135	22.55	20.15	12:26	83
-137	22.99	20.61	12:40	79
-140	23.51	21.18	13:00	75
-143	23.90	21.59	13:20	70
-145	24.07	21.78	13:33	67
-147	24.19	21.91	13:46	64
-150	24.25	22.00	14:05	60
-152	24.23	21.99	14:17	57

PERCENTAGE OF SUN OBSCURED ALONG MAP TRACK
WEST END =107.7   EAST END =107.7
MAXIMUM OBSCURATION 108.0 % AT LONG = -130
===

+ MARKS THE CITY OF CHIEN-K'ANG

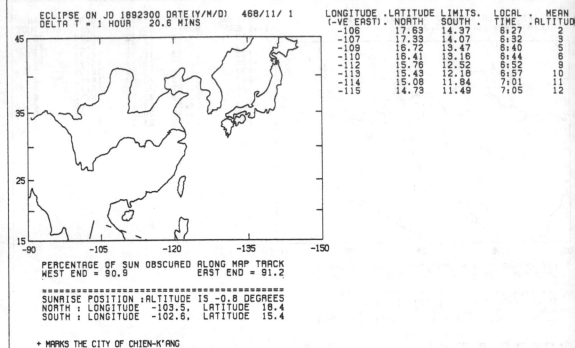

ECLIPSE ON JD 1892300 DATE(Y/M/D)   468/11/ 1
DELTA T = 1 HOUR   20.6 MINS

LONGITUDE (-VE EAST)	.LATITUDE LIMITS. NORTH	SOUTH .	LOCAL TIME	. MEAN .ALTITU
-106	17.63	14.37	6:27	2
-107	17.33	14.07	6:32	3
-109	16.72	13.47	6:40	5
-110	16.41	13.16	6:44	6
-112	15.76	12.52	6:52	9
-113	15.43	12.18	6:57	10
-114	15.08	11.84	7:01	11
-115	14.73	11.49	7:05	12

PERCENTAGE OF SUN OBSCURED ALONG MAP TRACK
WEST END = 90.9   EAST END = 91.2

==
SUNRISE POSITION :ALTITUDE IS -0.8 DEGREES
NORTH : LONGITUDE -103.5,   LATITUDE 18.4
SOUTH : LONGITUDE -102.6,   LATITUDE 15.4

+ MARKS THE CITY OF CHIEN-K'ANG

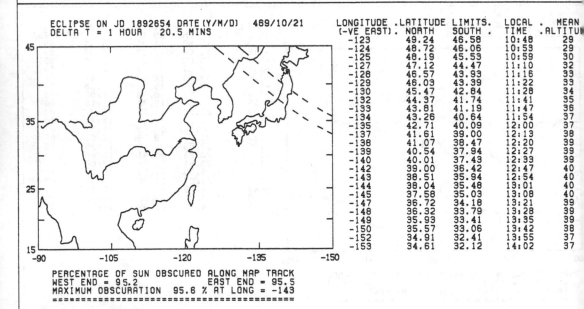

ECLIPSE ON JD 1892654 DATE(Y/M/D)   469/10/21
DELTA T = 1 HOUR   20.5 MINS

LONGITUDE (-VE EAST)	.LATITUDE LIMITS. NORTH	SOUTH .	LOCAL TIME	. MEAN .ALTITU
-123	49.24	46.58	10:48	29
-124	48.72	46.06	10:53	29
-125	48.19	45.53	10:59	30
-127	47.12	44.47	11:10	32
-128	46.57	43.93	11:16	33
-129	46.03	43.39	11:22	33
-130	45.47	42.84	11:28	34
-132	44.37	41.74	11:41	35
-133	43.81	41.19	11:47	36
-134	43.26	40.64	11:54	37
-135	42.71	40.09	12:00	37
-137	41.61	39.00	12:13	38
-138	41.07	38.47	12:20	39
-139	40.54	37.94	12:27	39
-140	40.01	37.43	12:33	39
-142	39.00	36.42	12:47	40
-143	38.51	35.94	12:54	40
-144	38.04	35.48	13:01	40
-145	37.58	35.03	13:08	40
-147	36.72	34.18	13:21	39
-148	36.32	33.79	13:28	39
-149	35.93	33.41	13:35	39
-150	35.57	33.06	13:42	38
-152	34.91	32.41	13:55	37
-153	34.61	32.12	14:02	37

PERCENTAGE OF SUN OBSCURED ALONG MAP TRACK
WEST END = 95.2   EAST END = 95.5
MAXIMUM OBSCURATION 95.6 % AT LONG = -143
==

+ MARKS THE CITY OF CHIEN-K'ANG

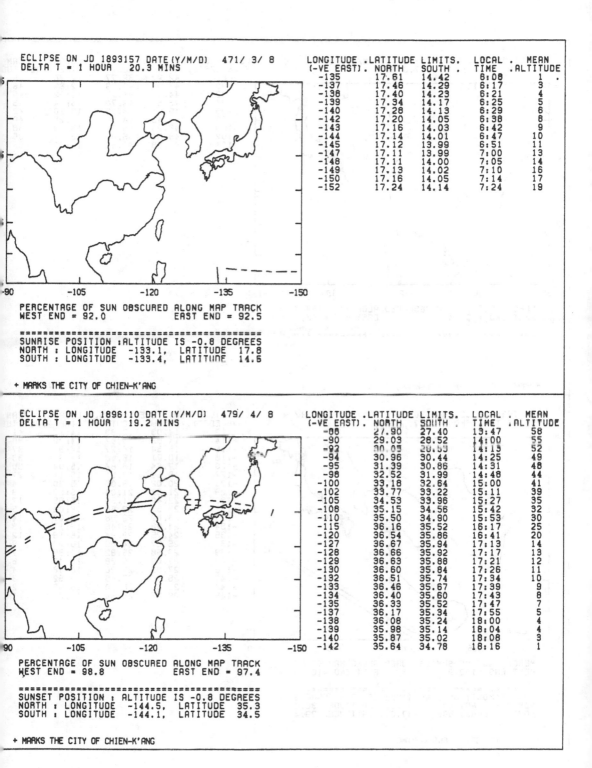

ECLIPSE ON JD 1893157 DATE(Y/M/D) 471/ 3/ 8
DELTA T = 1 HOUR   20.3 MINS

LONGITUDE (-VE EAST)	.LATITUDE LIMITS. NORTH	SOUTH .	LOCAL TIME	MEAN .ALTITUDE
-135	17.61	14.42	6:08	1
-137	17.46	14.29	6:17	3
-138	17.40	14.23	6:21	4
-139	17.34	14.17	6:25	5
-140	17.28	14.13	6:29	6
-142	17.20	14.05	6:38	8
-143	17.16	14.03	6:42	9
-144	17.14	14.01	6:47	10
-145	17.12	13.99	6:51	11
-147	17.11	13.99	7:00	13
-148	17.11	14.00	7:05	14
-149	17.13	14.02	7:10	16
-150	17.16	14.05	7:14	17
-152	17.24	14.14	7:24	19

-90   -105   -120   -135   -150

PERCENTAGE OF SUN OBSCURED ALONG MAP TRACK
WEST END = 92.0   EAST END = 92.5

==
SUNRISE POSITION :ALTITUDE IS -0.8 DEGREES
NORTH : LONGITUDE -133.1, LATITUDE 17.8
SOUTH : LONGITUDE -133.4, LATITUDE 14.6

+ MARKS THE CITY OF CHIEN-K'ANG

ECLIPSE ON JD 1896110 DATE(Y/M/D) 479/ 4/ 8
DELTA T = 1 HOUR   19.2 MINS

LONGITUDE (-VE EAST)	.LATITUDE LIMITS. NORTH	SOUTH .	LOCAL TIME	MEAN .ALTITUDE
-88	27.90	27.40	13:47	58
-90	29.03	28.52	14:00	55
-92	30.05	29.55	14:15	52
-94	30.96	30.44	14:25	49
-95	31.39	30.86	14:31	48
-98	32.52	31.99	14:48	44
-100	33.18	32.64	15:00	41
-102	33.77	33.22	15:11	39
-105	34.53	33.96	15:27	35
-108	35.15	34.56	15:42	32
-110	35.50	34.90	15:53	30
-115	36.16	35.52	16:17	25
-120	36.54	35.86	16:41	20
-127	36.67	35.94	17:13	14
-128	36.66	35.92	17:17	13
-129	36.63	35.88	17:21	12
-130	36.60	35.84	17:26	11
-132	36.51	35.74	17:34	10
-133	36.46	35.67	17:39	9
-134	36.40	35.60	17:43	8
-135	36.33	35.52	17:47	7
-137	36.17	35.34	17:55	5
-138	36.08	35.24	18:00	4
-139	35.98	35.14	18:04	4
-140	35.87	35.02	18:08	3
-142	35.64	34.78	18:16	1

-90   -105   -120   -135   -150

PERCENTAGE OF SUN OBSCURED ALONG MAP TRACK
WEST END = 98.8   EAST END = 97.4

==
SUNSET POSITION : ALTITUDE IS -0.8 DEGREES
NORTH : LONGITUDE -144.5, LATITUDE 35.3
SOUTH : LONGITUDE -144.1, LATITUDE 34.5

+ MARKS THE CITY OF CHIEN-K'ANG

251

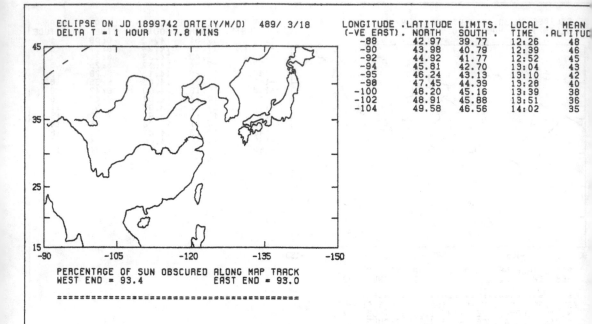

ECLIPSE ON JD 1899742 DATE(Y/M/D) 489/ 3/18
DELTA T = 1 HOUR 17.8 MINS

LONGITUDE (-VE EAST)	LATITUDE NORTH	LIMITS. SOUTH	LOCAL TIME	MEAN ALTITUDE
-88	42.97	39.77	12:26	48
-90	43.98	40.79	12:39	46
-92	44.92	41.77	12:52	45
-94	45.81	42.70	13:04	43
-95	46.24	43.13	13:10	42
-98	47.45	44.39	13:28	40
-100	48.20	45.16	13:39	38
-102	48.91	45.88	13:51	36
-104	49.58	46.56	14:02	35

PERCENTAGE OF SUN OBSCURED ALONG MAP TRACK
WEST END = 93.4   EAST END = 93.0

==

+ MARKS THE CITY OF CHIEN-K'ANG

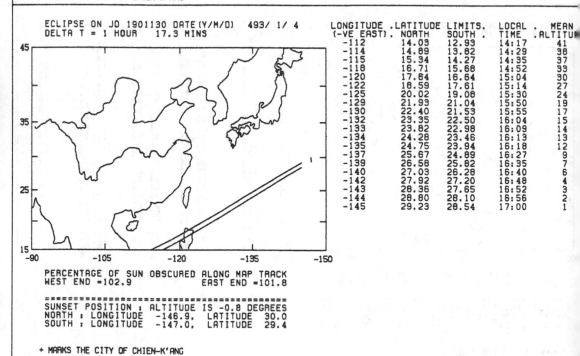

ECLIPSE ON JD 1901130 DATE(Y/M/D) 493/ 1/ 4
DELTA T = 1 HOUR 17.3 MINS

LONGITUDE (-VE EAST)	LATITUDE NORTH	LIMITS. SOUTH	LOCAL TIME	MEAN ALTITUDE
-112	14.03	12.93	14:17	41
-114	14.89	13.82	14:29	38
-115	15.34	14.27	14:35	37
-118	16.71	15.68	14:52	33
-120	17.64	16.64	15:04	30
-122	18.59	17.61	15:14	27
-125	20.02	19.08	15:30	24
-129	21.93	21.04	15:50	19
-130	22.40	21.53	15:55	17
-132	23.35	22.50	16:04	15
-133	23.82	22.98	16:09	14
-134	24.28	23.46	16:13	13
-135	24.75	23.94	16:18	12
-137	25.67	24.89	16:27	9
-139	26.58	25.82	16:35	7
-140	27.03	26.28	16:40	6
-142	27.92	27.20	16:48	4
-143	28.36	27.65	16:52	3
-144	28.80	28.10	16:56	2
-145	29.23	28.54	17:00	1

PERCENTAGE OF SUN OBSCURED ALONG MAP TRACK
WEST END =102.9   EAST END =101.8

==
SUNSET POSITION : ALTITUDE IS -0.8 DEGREES
NORTH : LONGITUDE -146.9, LATITUDE 30.0
SOUTH : LONGITUDE -147.0, LATITUDE 29.4

+ MARKS THE CITY OF CHIEN-K'ANG

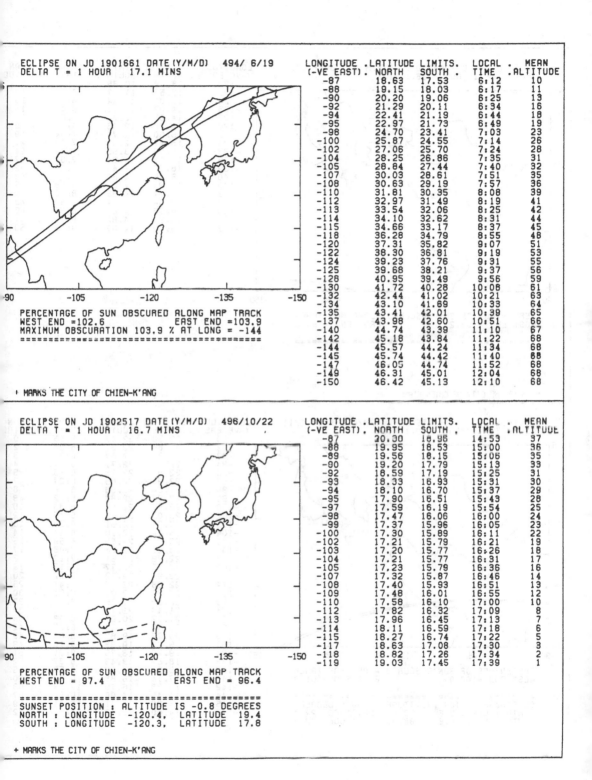

ECLIPSE ON JD 1901661 DATE(Y/M/D)   494/ 6/19
DELTA T = 1 HOUR   17.1 MINS

LONGITUDE (-VE EAST)	LATITUDE NORTH	LIMITS. SOUTH	LOCAL TIME	MEAN ALTITUDE
-87	18.63	17.53	6:12	10
-88	19.15	18.03	6:17	11
-90	20.20	19.06	6:25	13
-92	21.29	20.11	6:34	16
-94	22.41	21.19	6:44	18
-95	22.97	21.73	6:49	19
-98	24.70	23.41	7:03	23
-100	25.87	24.55	7:14	26
-102	27.06	25.70	7:24	28
-104	28.25	26.86	7:35	31
-105	28.84	27.44	7:40	32
-107	30.03	28.61	7:51	35
-108	30.63	29.19	7:57	36
-110	31.81	30.35	8:08	39
-112	32.97	31.49	8:19	41
-113	33.54	32.06	8:25	42
-114	34.10	32.62	8:31	44
-115	34.66	33.17	8:37	45
-118	36.28	34.79	8:55	48
-120	37.31	35.82	9:07	51
-122	38.30	36.81	9:19	53
-124	39.23	37.76	9:31	55
-125	39.68	38.21	9:37	56
-128	40.95	39.49	9:56	59
-130	41.72	40.28	10:08	61
-132	42.44	41.02	10:21	63
-134	43.10	41.69	10:33	64
-135	43.41	42.01	10:39	65
-137	43.98	42.60	10:51	66
-140	44.74	43.39	11:10	67
-142	45.18	43.84	11:22	68
-144	45.57	44.24	11:34	68
-145	45.74	44.42	11:40	68
-147	46.05	44.74	11:52	68
-149	46.31	45.01	12:04	68
-150	46.42	45.13	12:10	68

PERCENTAGE OF SUN OBSCURED ALONG MAP TRACK
WEST END =102.6   EAST END =103.9
MAXIMUM OBSCURATION 103.9 % AT LONG = -144
===

' MARKS THE CITY OF CHIEN-K'ANG

ECLIPSE ON JD 1902517 DATE(Y/M/D)   496/10/22
DELTA T = 1 HOUR   16.7 MINS

LONGITUDE (-VE EAST)	LATITUDE NORTH	LIMITS. SOUTH	LOCAL TIME	MEAN ALTITUDE
-87	20.30	18.96	14:53	37
-88	19.95	18.53	15:00	36
-89	19.56	18.15	15:06	35
-90	19.20	17.79	15:13	33
-92	18.59	17.19	15:25	31
-93	18.33	16.93	15:31	30
-94	18.10	16.70	15:37	29
-95	17.90	16.51	15:43	28
-97	17.59	16.19	15:54	25
-98	17.47	16.06	16:00	24
-99	17.37	15.96	16:05	23
-100	17.30	15.89	16:11	22
-102	17.21	15.79	16:21	19
-103	17.20	15.77	16:26	18
-104	17.21	15.77	16:31	17
-105	17.23	15.79	16:36	16
-107	17.32	15.87	16:46	14
-108	17.40	15.93	16:51	13
-109	17.48	16.01	16:55	12
-110	17.58	16.10	17:00	10
-112	17.82	16.32	17:09	8
-113	17.96	16.45	17:13	7
-114	18.11	16.59	17:18	6
-115	18.27	16.74	17:22	5
-117	18.63	17.08	17:30	3
-118	18.82	17.26	17:34	2
-119	19.03	17.45	17:39	1

PERCENTAGE OF SUN OBSCURED ALONG MAP TRACK
WEST END = 97.4   EAST END = 96.4

===
SUNSET POSITION : ALTITUDE IS -0.8 DEGREES
NORTH : LONGITUDE -120.4, LATITUDE 19.4
SOUTH : LONGITUDE -120.3, LATITUDE 17.8

+ MARKS THE CITY OF CHIEN-K'ANG

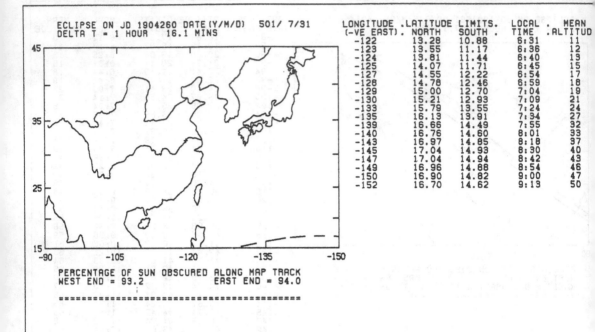

ECLIPSE ON JD 1904260 DATE(Y/M/D) 501/ 7/31
DELTA T = 1 HOUR 16.1 MINS

LONGITUDE (-VE EAST)	LATITUDE NORTH	LIMITS SOUTH	LOCAL TIME	MEAN ALTITUD
-122	13.28	10.88	6:31	11
-123	13.55	11.17	6:36	12
-124	13.81	11.44	6:40	13
-125	14.07	11.71	6:45	15
-127	14.55	12.22	6:54	17
-128	14.78	12.46	6:59	18
-129	15.00	12.70	7:04	19
-130	15.21	12.93	7:09	21
-133	15.79	13.55	7:24	24
-135	16.13	13.91	7:34	27
-139	16.66	14.49	7:55	32
-140	16.76	14.60	8:01	33
-143	16.97	14.85	8:18	37
-145	17.04	14.93	8:30	40
-147	17.04	14.94	8:42	43
-149	16.96	14.88	8:54	46
-150	16.90	14.82	9:00	47
-152	16.70	14.62	9:13	50

PERCENTAGE OF SUN OBSCURED ALONG MAP TRACK
WEST END = 93.2   EAST END = 94.0

==

+ MARKS THE CITY OF CHIEN-K'ANG

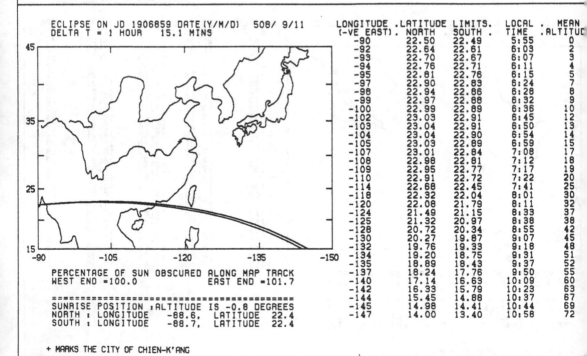

ECLIPSE ON JD 1906859 DATE(Y/M/D) 508/ 9/11
DELTA T = 1 HOUR 15.1 MINS

LONGITUDE (-VE EAST)	LATITUDE NORTH	LIMITS SOUTH	LOCAL TIME	MEAN ALTITUD
-90	22.50	22.49	5:55	0
-92	22.64	22.61	6:03	2
-93	22.70	22.67	6:07	3
-94	22.76	22.71	6:11	4
-95	22.81	22.76	6:15	5
-97	22.90	22.83	6:24	7
-98	22.94	22.86	6:28	8
-99	22.97	22.88	6:32	9
-100	22.99	22.89	6:36	10
-102	23.03	22.91	6:45	12
-103	23.04	22.91	6:50	13
-104	23.04	22.90	6:54	14
-105	23.03	22.89	6:59	15
-107	23.01	22.84	7:08	17
-108	22.98	22.81	7:12	18
-109	22.95	22.77	7:17	19
-110	22.91	22.72	7:22	20
-114	22.68	22.45	7:41	25
-118	22.32	22.04	8:01	30
-120	22.08	21.79	8:11	32
-124	21.49	21.15	8:33	37
-125	21.32	20.97	8:38	38
-128	20.72	20.34	8:55	42
-130	20.27	19.87	9:07	45
-132	19.76	19.33	9:18	48
-134	19.20	18.75	9:31	51
-135	18.89	18.43	9:37	52
-137	18.24	17.76	9:50	55
-140	17.14	16.63	10:09	60
-142	16.33	15.79	10:23	63
-144	15.45	14.88	10:37	67
-145	14.98	14.41	10:44	69
-147	14.00	13.40	10:58	72

PERCENTAGE OF SUN OBSCURED ALONG MAP TRACK
WEST END =100.0   EAST END =101.7

==
SUNRISE POSITION ,ALTITUDE IS -0.8 DEGREES
NORTH : LONGITUDE   -88.6, LATITUDE 22.4
SOUTH : LONGITUDE   -88.7, LATITUDE 22.4

+ MARKS THE CITY OF CHIEN-K'ANG

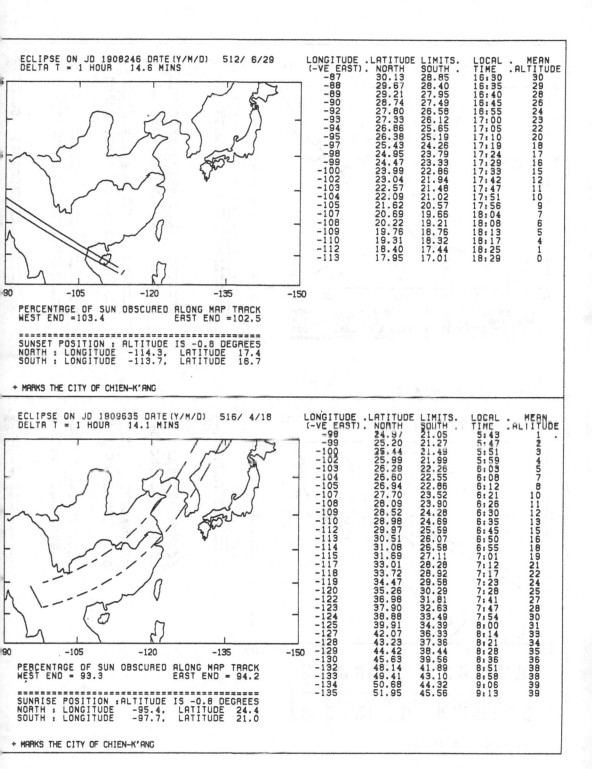

ECLIPSE ON JD 1908246 DATE(Y/M/D)   512/ 6/29
DELTA T = 1 HOUR   14.6 MINS

LONGITUDE	.LATITUDE	LIMITS.	LOCAL	. MEAN
(-VE EAST).	NORTH	SOUTH .	TIME	.ALTITUDE
-87	30.13	28.85	16:30	30
-88	29.67	28.40	16:35	29
-89	29.21	27.95	16:40	28
-90	28.74	27.49	16:45	26
-92	27.80	26.58	16:55	24
-93	27.33	26.12	17:00	23
-94	26.86	25.65	17:05	22
-95	26.38	25.19	17:10	20
-97	25.43	24.26	17:19	18
-98	24.95	23.79	17:24	17
-99	24.47	23.33	17:29	16
-100	23.99	22.86	17:33	15
-102	23.04	21.94	17:42	12
-103	22.57	21.48	17:47	11
-104	22.09	21.02	17:51	10
-105	21.62	20.57	17:56	9
-107	20.69	19.66	18:04	7
-108	20.22	19.21	18:08	6
-109	19.76	18.76	18:13	5
-110	19.31	18.32	18:17	4
-112	18.40	17.44	18:25	1
-113	17.95	17.01	18:29	0

PERCENTAGE OF SUN OBSCURED ALONG MAP TRACK
WEST END =103.4   EAST END =102.5

===
SUNSET POSITION : ALTITUDE IS -0.8 DEGREES
NORTH : LONGITUDE -114.3, LATITUDE 17.4
SOUTH : LONGITUDE -113.7, LATITUDE 16.7

+ MARKS THE CITY OF CHIEN-K'ANG

ECLIPSE ON JD 1909635 DATE(Y/M/D)   516/ 4/18
DELTA T = 1 HOUR   14.1 MINS

LONGITUDE	.LATITUDE	LIMITS.	LOCAL	. MEAN
(-VE EAST).	NORTH	SOUTH .	TIME	.ALTITUDE
-98	24.97	21.05	5:43	1
-99	25.20	21.27	5:47	2
-100	25.44	21.49	5:51	3
-102	25.99	21.99	5:59	4
-103	26.29	22.26	6:03	5
-104	26.60	22.55	6:08	7
-105	26.94	22.86	6:12	8
-107	27.70	23.52	6:21	10
-108	28.09	23.90	6:26	11
-109	28.52	24.28	6:30	12
-110	28.98	24.69	6:35	13
-112	29.97	25.59	6:45	15
-113	30.51	26.07	6:50	16
-114	31.08	26.58	6:55	18
-115	31.69	27.11	7:01	19
-117	33.01	28.28	7:12	21
-118	33.72	28.92	7:17	22
-119	34.47	29.58	7:23	24
-120	35.26	30.29	7:28	25
-122	36.98	31.81	7:41	27
-123	37.90	32.63	7:47	28
-124	38.88	33.49	7:54	30
-125	39.91	34.39	8:00	31
-127	42.07	36.33	8:14	33
-128	43.23	37.36	8:21	34
-129	44.42	38.44	8:28	35
-130	45.63	39.56	8:36	36
-132	48.14	41.89	8:51	38
-133	49.41	43.10	8:58	38
-134	50.68	44.32	9:06	39
-135	51.95	45.56	9:13	39

PERCENTAGE OF SUN OBSCURED ALONG MAP TRACK
WEST END = 93.3   EAST END = 94.2

===
SUNRISE POSITION :ALTITUDE IS -0.8 DEGREES
NORTH : LONGITUDE   -95.4, LATITUDE 24.4
SOUTH : LONGITUDE   -97.7, LATITUDE 21.0

+ MARKS THE CITY OF CHIEN-K'ANG

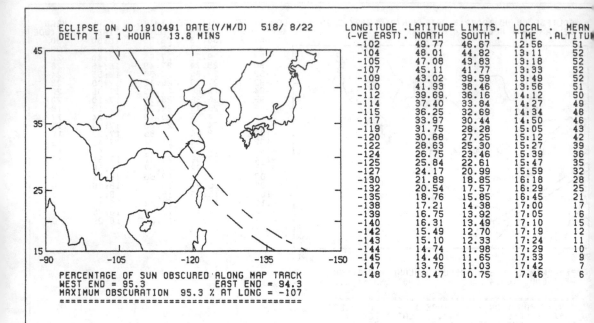

ECLIPSE ON JD 1910491 DATE(Y/M/D)   518/ 8/22
DELTA T = 1 HOUR   13.8 MINS

LONGITUDE (-VE EAST)	LATITUDE NORTH	LIMITS. SOUTH	LOCAL TIME	MEAN ALTITU
-102	49.77	46.67	12:56	51
-104	48.01	44.82	13:11	52
-105	47.08	43.83	13:18	52
-107	45.11	41.77	13:33	52
-109	43.02	39.59	13:49	52
-110	41.93	38.46	13:56	51
-112	39.69	36.16	14:12	50
-114	37.40	33.84	14:27	49
-115	36.25	32.69	14:34	48
-117	33.97	30.44	14:50	46
-119	31.75	28.28	15:05	43
-120	30.68	27.25	15:12	42
-122	28.63	25.30	15:27	39
-124	26.75	23.46	15:39	36
-125	25.84	22.61	15:47	35
-127	24.17	20.99	15:59	32
-130	21.89	18.85	16:18	28
-132	20.54	17.57	16:29	25
-135	18.76	15.85	16:45	21
-138	17.21	14.38	17:00	17
-139	16.75	13.92	17:05	16
-140	16.31	13.49	17:10	15
-142	15.49	12.70	17:19	12
-143	15.10	12.33	17:24	11
-144	14.74	11.98	17:29	10
-145	14.40	11.65	17:33	9
-147	13.76	11.03	17:42	7
-148	13.47	10.75	17:46	6

PERCENTAGE OF SUN OBSCURED ALONG MAP TRACK
WEST END = 95.3   EAST END = 94.3
MAXIMUM OBSCURATION 95.3 % AT LONG = -107
==

+ MARKS THE CITY OF CHIEN-K'ANG

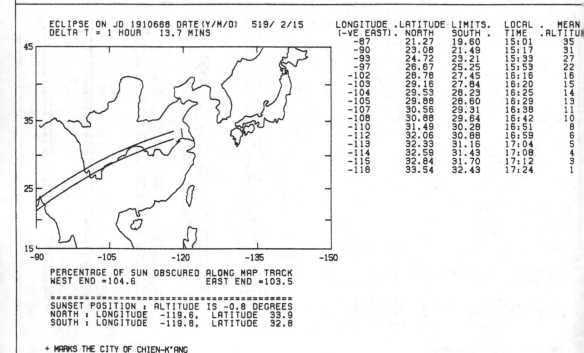

ECLIPSE ON JD 1910668 DATE(Y/M/D)   519/ 2/15
DELTA T = 1 HOUR   13.7 MINS

LONGITUDE (-VE EAST)	LATITUDE NORTH	LIMITS. SOUTH	LOCAL TIME	MEAN ALTITU
-87	21.27	19.60	15:01	35
-90	23.08	21.49	15:17	31
-93	24.72	23.21	15:33	27
-97	26.67	25.25	15:53	22
-102	28.78	27.45	16:16	16
-103	29.16	27.84	16:20	15
-104	29.53	28.23	16:25	14
-105	29.88	28.60	16:29	13
-107	30.56	29.31	16:38	11
-108	30.88	29.64	16:42	10
-110	31.49	30.28	16:51	8
-112	32.06	30.88	16:59	6
-113	32.33	31.16	17:04	5
-114	32.59	31.43	17:08	4
-115	32.84	31.70	17:12	3
-118	33.54	32.43	17:24	1

PERCENTAGE OF SUN OBSCURED ALONG MAP TRACK
WEST END =104.6   EAST END =103.5

==
SUNSET POSITION : ALTITUDE IS -0.8 DEGREES
NORTH : LONGITUDE -119.6, LATITUDE 33.9
SOUTH : LONGITUDE -119.8, LATITUDE 32.8

+ MARKS THE CITY OF CHIEN-K'ANG

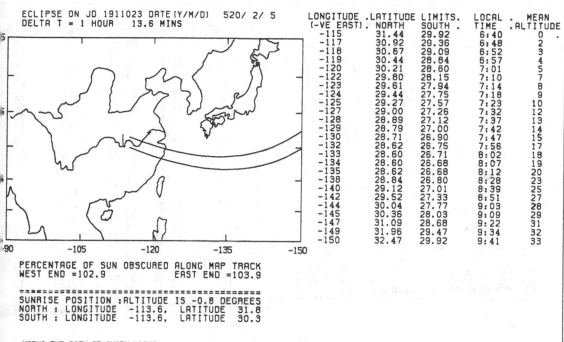

```
ECLIPSE ON JD 1911023 DATE(Y/M/D)   520/ 2/ 5
DELTA T = 1 HOUR   13.6 MINS
```

LONGITUDE (-VE EAST)	LATITUDE NORTH	LIMITS. SOUTH	LOCAL TIME	MEAN ALTITUDE
-115	31.44	29.92	6:40	0
-117	30.92	29.36	6:48	2
-118	30.67	29.09	6:52	3
-119	30.44	28.84	6:57	4
-120	30.21	28.60	7:01	5
-122	29.80	28.15	7:10	7
-123	29.61	27.94	7:14	8
-124	29.44	27.75	7:18	9
-125	29.27	27.57	7:23	10
-127	29.00	27.26	7:32	12
-128	28.89	27.12	7:37	13
-129	28.79	27.00	7:42	14
-130	28.71	26.90	7:47	15
-132	28.62	26.75	7:56	17
-133	28.60	26.71	8:02	18
-134	28.60	26.68	8:07	19
-135	28.62	26.68	8:12	20
-138	28.84	26.80	8:28	23
-140	29.12	27.01	8:39	25
-142	29.52	27.33	8:51	27
-144	30.04	27.77	9:03	28
-145	30.36	28.03	9:09	29
-147	31.09	28.68	9:22	31
-149	31.96	29.47	9:34	32
-150	32.47	29.92	9:41	33

PERCENTAGE OF SUN OBSCURED ALONG MAP TRACK
WEST END =102.9   EAST END =103.9

==
SUNRISE POSITION :ALTITUDE IS -0.8 DEGREES
NORTH : LONGITUDE -113.6, LATITUDE 31.8
SOUTH : LONGITUDE -113.6, LATITUDE 30.3

+ MARKS THE CITY OF CHIEN-K'ANG

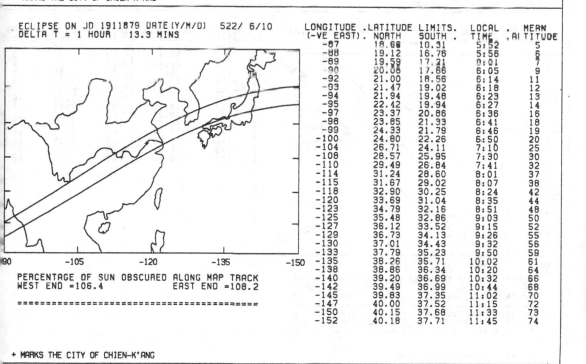

```
ECLIPSE ON JD 1911879 DATE(Y/M/D)   522/ 6/10
DELTA T = 1 HOUR   13.3 MINS
```

LONGITUDE (-VE EAST)	LATITUDE NORTH	LIMITS. SOUTH	LOCAL TIME	MEAN ALTITUDE
-87	18.66	16.31	5:52	5
-88	19.12	16.76	5:56	6
-89	19.59	17.21	6:01	7
-90	20.06	17.66	6:05	9
-92	21.00	18.56	6:14	11
-93	21.47	19.02	6:18	12
-94	21.94	19.48	6:23	13
-95	22.42	19.94	6:27	14
-97	23.37	20.86	6:36	16
-98	23.85	21.33	6:41	18
-99	24.33	21.79	6:46	19
-100	24.80	22.26	6:50	20
-104	26.71	24.11	7:10	25
-108	28.57	25.95	7:30	30
-110	29.49	26.84	7:41	32
-114	31.24	28.60	8:01	37
-115	31.67	29.02	8:07	38
-118	32.90	30.25	8:24	42
-120	33.69	31.04	8:35	44
-123	34.79	32.16	8:51	48
-125	35.48	32.86	9:03	50
-127	36.12	33.52	9:15	52
-129	36.73	34.13	9:26	55
-130	37.01	34.43	9:32	56
-133	37.79	35.23	9:50	59
-135	38.26	35.71	10:02	61
-138	38.86	36.34	10:20	64
-140	39.20	36.69	10:32	66
-142	39.49	36.99	10:44	68
-145	39.83	37.35	11:02	70
-147	40.00	37.52	11:15	72
-150	40.15	37.68	11:33	73
-152	40.18	37.71	11:45	74

PERCENTAGE OF SUN OBSCURED ALONG MAP TRACK
WEST END =106.4   EAST END =108.2

==

+ MARKS THE CITY OF CHIEN-K'ANG
```
```

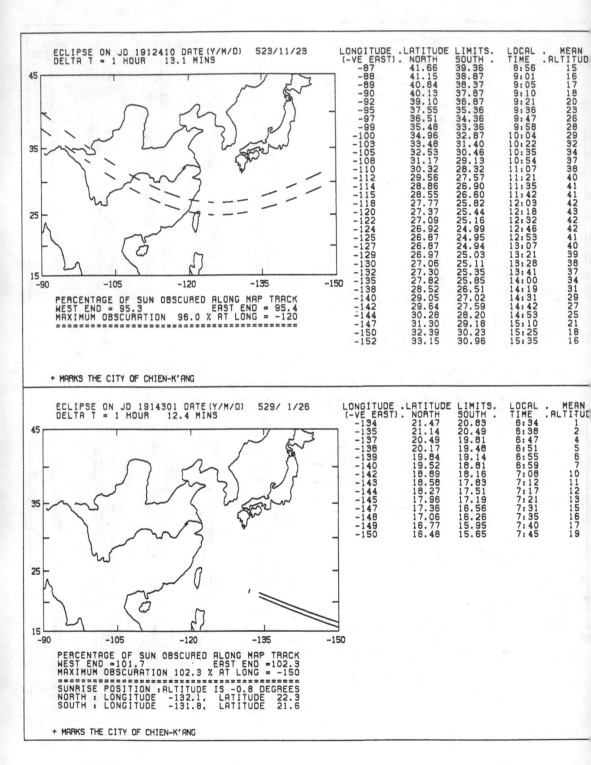

ECLIPSE ON JD 1912410 DATE(Y/M/D) 523/11/23
DELTA T = 1 HOUR   13.1 MINS

LONGITUDE (-VE EAST)	.LATITUDE NORTH	LIMITS. SOUTH .	LOCAL TIME	. MEAN .ALTITUD
-87	41.66	39.36	8:56	15
-88	41.15	38.87	9:01	16
-89	40.64	38.37	9:05	17
-90	40.13	37.87	9:10	18
-92	39.10	36.87	9:21	20
-95	37.55	35.36	9:36	23
-97	36.51	34.36	9:47	26
-99	35.48	33.36	9:58	28
-100	34.96	32.87	10:04	29
-103	33.48	31.40	10:22	32
-105	32.53	30.46	10:35	34
-108	31.17	29.13	10:54	37
-110	30.32	28.32	11:07	38
-112	29.56	27.57	11:21	40
-114	28.86	26.90	11:35	41
-115	28.55	26.60	11:42	41
-118	27.77	25.82	12:03	42
-120	27.37	25.44	12:18	43
-122	27.09	25.16	12:32	42
-124	26.92	24.99	12:46	42
-125	26.87	24.95	12:53	41
-127	26.87	24.94	13:07	40
-129	26.97	25.03	13:21	39
-130	27.06	25.11	13:28	38
-132	27.30	25.35	13:41	37
-135	27.82	25.85	14:00	34
-138	28.52	26.51	14:19	31
-140	29.05	27.02	14:31	29
-142	29.64	27.59	14:42	27
-144	30.28	28.20	14:53	25
-147	31.30	29.18	15:10	21
-150	32.39	30.23	15:25	18
-152	33.15	30.96	15:35	16

PERCENTAGE OF SUN OBSCURED ALONG MAP TRACK
WEST END = 95.3   EAST END = 95.4
MAXIMUM OBSCURATION 96.0 % AT LONG = -120
==

+ MARKS THE CITY OF CHIEN-K'ANG

ECLIPSE ON JD 1914301 DATE(Y/M/D) 529/ 1/26
DELTA T = 1 HOUR   12.4 MINS

LONGITUDE (-VE EAST)	.LATITUDE NORTH	LIMITS. SOUTH .	LOCAL TIME	. MEAN .ALTITUD
-134	21.47	20.83	6:34	1
-135	21.14	20.49	6:38	2
-137	20.49	19.81	6:47	4
-138	20.17	19.48	6:51	5
-139	19.84	19.14	6:55	6
-140	19.52	18.81	6:59	7
-142	18.89	18.16	7:08	10
-143	18.58	17.83	7:12	11
-144	18.27	17.51	7:17	12
-145	17.96	17.19	7:21	13
-147	17.36	16.56	7:31	15
-148	17.06	16.26	7:35	16
-149	16.77	15.95	7:40	17
-150	16.48	15.65	7:45	19

PERCENTAGE OF SUN OBSCURED ALONG MAP TRACK
WEST END =101.7   EAST END =102.3
MAXIMUM OBSCURATION 102.3 % AT LONG = -150
==
SUNRISE POSITION :ALTITUDE IS -0.8 DEGREES
NORTH : LONGITUDE -132.1, LATITUDE   22.3
SOUTH : LONGITUDE -131.8, LATITUDE   21.6

+ MARKS THE CITY OF CHIEN-K'ANG

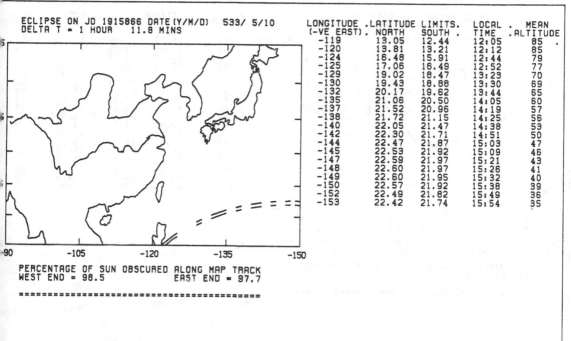

ECLIPSE ON JD 1915866 DATE(Y/M/D) 533/ 5/10
DELTA T = 1 HOUR   11.8 MINS

LONGITUDE (-VE EAST)	LATITUDE NORTH	LIMITS. SOUTH	LOCAL TIME	MEAN ALTITUDE
-119	13.05	12.44	12:05	85
-120	13.81	13.21	12:12	85
-124	16.48	15.91	12:44	79
-125	17.06	16.49	12:52	77
-129	19.02	18.47	13:23	70
-130	19.43	18.88	13:30	69
-132	20.17	19.62	13:44	65
-135	21.06	20.50	14:05	60
-137	21.52	20.96	14:19	57
-138	21.72	21.15	14:25	56
-140	22.05	21.47	14:38	53
-142	22.30	21.71	14:51	50
-144	22.47	21.87	15:03	47
-145	22.53	21.92	15:09	46
-147	22.59	21.97	15:21	43
-148	22.60	21.97	15:26	41
-149	22.60	21.95	15:32	40
-150	22.57	21.92	15:38	39
-152	22.49	21.82	15:49	36
-153	22.42	21.74	15:54	35

PERCENTAGE OF SUN OBSCURED ALONG MAP TRACK
WEST END = 98.5   EAST END = 97.7

==

+ MARKS THE CITY OF CHIEN-K'ANG

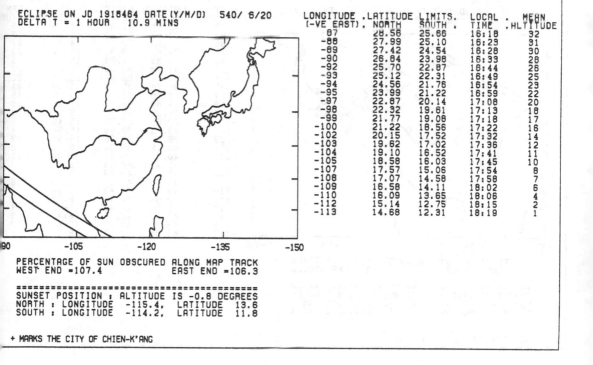

ECLIPSE ON JD 1918464 DATE(Y/M/D) 540/ 6/20
DELTA T = 1 HOUR   10.9 MINS

LONGITUDE (-VE EAST)	LATITUDE NORTH	LIMITS. SOUTH	LOCAL TIME	MEAN ALTITUDE
87	28.56	25.66	16:18	32
-88	27.99	25.10	16:23	31
-89	27.42	24.54	16:28	30
-90	26.84	23.98	16:33	28
-92	25.70	22.87	16:44	26
-93	25.12	22.31	16:49	25
-94	24.56	21.76	16:54	23
-95	23.99	21.22	16:59	22
-97	22.87	20.14	17:08	20
-98	22.32	19.61	17:13	18
-99	21.77	19.08	17:18	17
-100	21.22	18.56	17:22	16
-102	20.15	17.52	17:32	14
-103	19.62	17.02	17:36	12
-104	19.10	16.52	17:41	11
-105	18.58	16.03	17:45	10
-107	17.57	15.06	17:54	8
-108	17.07	14.58	17:58	7
-109	16.58	14.11	18:02	6
-110	16.09	13.65	18:06	4
-112	15.14	12.75	18:15	2
-113	14.68	12.31	18:19	1

PERCENTAGE OF SUN OBSCURED ALONG MAP TRACK
WEST END =107.4   EAST END =106.3

==
SUNSET POSITION : ALTITUDE IS -0.8 DEGREES
NORTH : LONGITUDE -115.4, LATITUDE 13.6
SOUTH : LONGITUDE -114.2, LATITUDE 11.8

+ MARKS THE CITY OF CHIEN-K'ANG

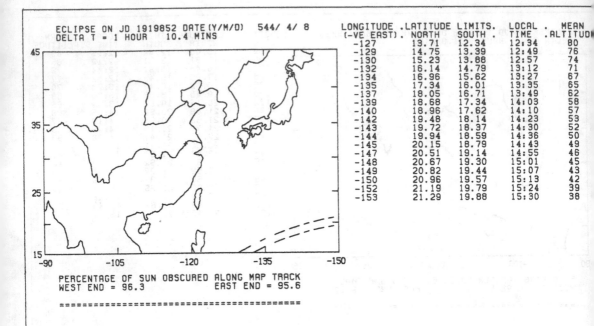

ECLIPSE ON JD 1919852 DATE(Y/M/D)   544/ 4/ 8
DELTA T = 1 HOUR   10.4 MINS

LONGITUDE (-VE EAST)	.LATITUDE NORTH	LIMITS. SOUTH	LOCAL TIME	. MEAN .ALTITUDE
-127	13.71	12.34	12:34	80
-129	14.75	13.39	12:49	76
-130	15.23	13.88	12:57	74
-132	16.14	14.79	13:12	71
-134	16.96	15.62	13:27	67
-135	17.34	16.01	13:35	65
-137	18.05	16.71	13:49	62
-139	18.68	17.34	14:03	58
-140	18.96	17.62	14:10	57
-142	19.48	18.14	14:23	53
-143	19.72	18.37	14:30	52
-144	19.94	18.59	14:36	50
-145	20.15	18.79	14:43	49
-147	20.51	19.14	14:55	46
-148	20.67	19.30	15:01	45
-149	20.82	19.44	15:07	43
-150	20.96	19.57	15:13	42
-152	21.19	19.79	15:24	39
-153	21.29	19.88	15:30	38

PERCENTAGE OF SUN OBSCURED ALONG MAP TRACK
WEST END = 96.3   EAST END = 95.6

==

+ MARKS THE CITY OF CHIEN-K'ANG

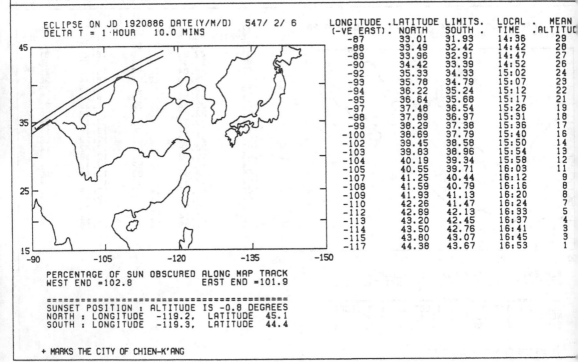

ECLIPSE ON JD 1920886 DATE(Y/M/D)   547/ 2/ 6
DELTA T = 1 HOUR   10.0 MINS

LONGITUDE (-VE EAST)	.LATITUDE NORTH	LIMITS. SOUTH .	LOCAL TIME	. MEAN .ALTITUDE
-87	33.01	31.93	14:36	29
-88	33.49	32.42	14:42	28
-89	33.96	32.91	14:47	27
-90	34.42	33.39	14:52	26
-92	35.33	34.33	15:02	24
-93	35.78	34.79	15:07	23
-94	36.22	35.24	15:12	22
-95	36.64	35.68	15:17	21
-97	37.48	36.54	15:26	19
-98	37.89	36.97	15:31	18
-99	38.29	37.38	15:36	17
-100	38.69	37.79	15:40	16
-102	39.45	38.58	15:50	14
-103	39.83	38.96	15:54	13
-104	40.19	39.34	15:58	12
-105	40.55	39.71	16:03	11
-107	41.25	40.44	16:12	9
-108	41.59	40.79	16:16	8
-109	41.93	41.13	16:20	8
-110	42.26	41.47	16:24	7
-112	42.89	42.13	16:33	5
-113	43.20	42.45	16:37	4
-114	43.50	42.76	16:41	3
-115	43.80	43.07	16:45	3
-117	44.38	43.67	16:53	1

PERCENTAGE OF SUN OBSCURED ALONG MAP TRACK
WEST END =102.8   EAST END =101.9

==
SUNSET POSITION : ALTITUDE IS -0.8 DEGREES
NORTH : LONGITUDE -119.2, LATITUDE 45.1
SOUTH : LONGITUDE -119.3, LATITUDE 44.4

+ MARKS THE CITY OF CHIEN-K'ANG

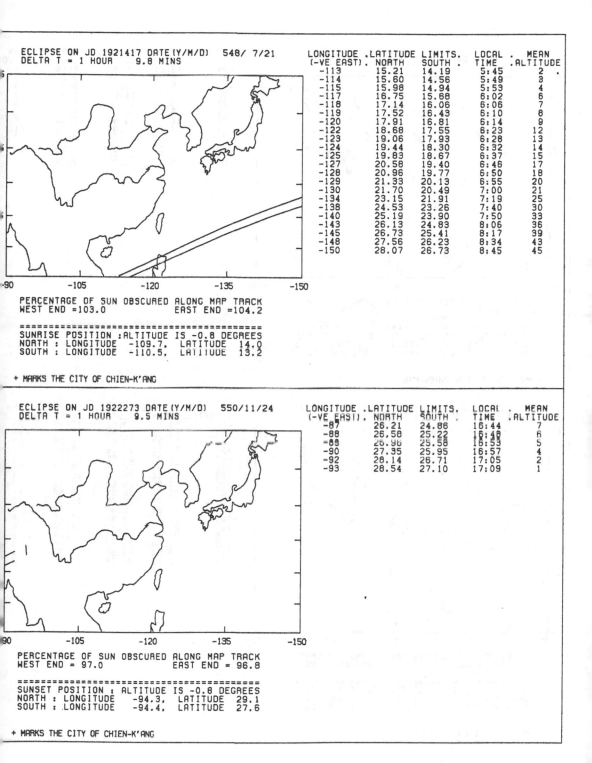

ECLIPSE ON JD 1921417 DATE (Y/M/D) 548/ 7/21
DELTA T = 1 HOUR   9.8 MINS

LONGITUDE (-VE EAST)	LATITUDE NORTH	LIMITS SOUTH	LOCAL TIME	MEAN ALTITUDE
-113	15.21	14.19	5:45	2
-114	15.60	14.56	5:49	3
-115	15.98	14.94	5:53	4
-117	16.75	15.68	6:02	6
-118	17.14	16.06	6:06	7
-119	17.52	16.43	6:10	8
-120	17.91	16.81	6:14	9
-122	18.68	17.55	6:23	12
-123	19.06	17.93	6:28	13
-124	19.44	18.30	6:32	14
-125	19.83	18.67	6:37	15
-127	20.58	19.40	6:46	17
-128	20.96	19.77	6:50	18
-129	21.33	20.13	6:55	20
-130	21.70	20.49	7:00	21
-134	23.15	21.91	7:19	25
-138	24.53	23.26	7:40	30
-140	25.19	23.90	7:50	33
-143	26.13	24.83	8:06	36
-145	26.73	25.41	8:17	39
-148	27.56	26.23	8:34	43
-150	28.07	26.73	8:45	45

PERCENTAGE OF SUN OBSCURED ALONG MAP TRACK
WEST END =103.0   EAST END =104.2

===
SUNRISE POSITION :ALTITUDE IS -0.8 DEGREES
NORTH : LONGITUDE -109.7, LATITUDE 14.0
SOUTH : LONGITUDE -110.5, LATITUDE 13.2

+ MARKS THE CITY OF CHIEN-K'ANG

ECLIPSE ON JD 1922273 DATE (Y/M/D) 550/11/24
DELTA T = 1 HOUR   9.5 MINS

LONGITUDE (-VE EAST)	LATITUDE NORTH	LIMITS SOUTH	LOCAL TIME	MEAN ALTITUDE
-87	26.21	24.86	16:44	7
-88	26.58	25.22	16:48	6
-89	26.96	25.58	16:53	5
-90	27.35	25.95	16:57	4
-92	28.14	26.71	17:05	2
-93	28.54	27.10	17:09	1

PERCENTAGE OF SUN OBSCURED ALONG MAP TRACK
WEST END = 97.0   EAST END = 96.8

===
SUNSET POSITION : ALTITUDE IS -0.8 DEGREES
NORTH : LONGITUDE -94.3, LATITUDE 29.1
SOUTH : LONGITUDE -94.4, LATITUDE 27.6

+ MARKS THE CITY OF CHIEN-K'ANG

ECLIPSE ON JD 1926615 DATE(Y/M/D) 562/10/14
DELTA T = 1 HOUR   8.0 MINS

LONGITUDE (-VE EAST)	LATITUDE NORTH	LIMITS. SOUTH	LOCAL TIME	MEAN ALTITUD
-88	16.21	15.65	6:16	1
-89	16.04	15.49	6:20	2
-90	15.86	15.32	6:24	3
-92	15.48	14.95	6:32	5
-93	15.28	14.76	6:36	6
-94	15.07	14.56	6:41	8
-95	14.86	14.36	6:45	9
-97	14.40	13.93	6:54	11
-98	14.17	13.70	6:58	12
-99	13.92	13.46	7:02	13
-100	13.67	13.22	7:07	14
-102	13.13	12.71	7:16	16

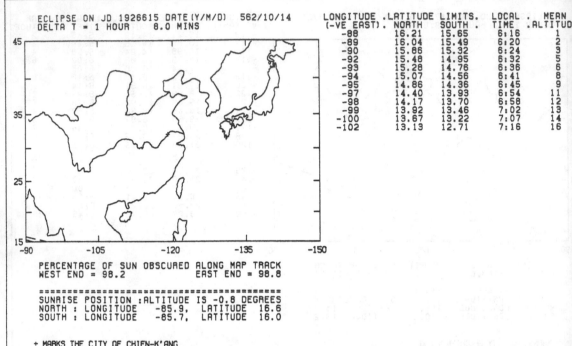

PERCENTAGE OF SUN OBSCURED ALONG MAP TRACK
WEST END = 98.2   EAST END = 98.8

===
SUNRISE POSITION :ALTITUDE IS -0.8 DEGREES
NORTH : LONGITUDE -85.9, LATITUDE 16.6
SOUTH : LONGITUDE -85.7, LATITUDE 16.0

+ MARKS THE CITY OF CHIEN-K'ANG

ECLIPSE ON JD 1926969 DATE(Y/M/D) 563/10/ 3
DELTA T = 1 HOUR   7.9 MINS

LONGITUDE (-VE EAST)	LATITUDE NORTH	LIMITS. SOUTH	LOCAL TIME	MEAN ALTITU
-88	33.02	29.84	15:48	25
-90	32.69	29.53	15:59	23
-94	32.20	29.05	16:20	19
-95	32.11	28.96	16:25	18
-97	31.97	28.82	16:34	16
-98	31.92	28.77	16:39	15
-99	31.88	28.73	16:44	14
-100	31.85	28.69	16:49	13
-102	31.81	28.66	16:58	11
-103	31.81	28.65	17:03	10
-104	31.82	28.66	17:07	9
-105	31.84	28.67	17:11	8
-107	31.89	28.72	17:20	6
-108	31.94	28.76	17:24	5
-109	31.99	28.80	17:29	4
-110	32.04	28.86	17:33	3
-112	32.18	28.98	17:41	2

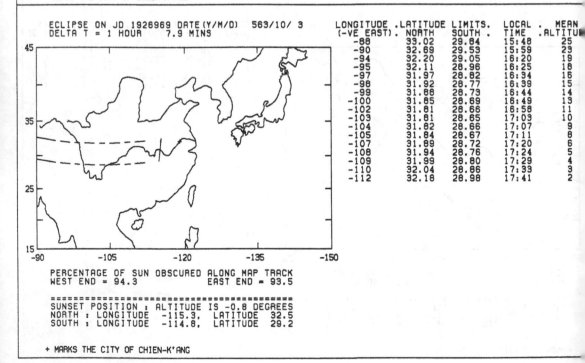

PERCENTAGE OF SUN OBSCURED ALONG MAP TRACK
WEST END = 94.3   EAST END = 93.5

===
SUNSET POSITION : ALTITUDE IS -0.8 DEGREES
NORTH : LONGITUDE -115.3, LATITUDE 32.5
SOUTH : LONGITUDE -114.8, LATITUDE 29.2

+ MARKS THE CITY OF CHIEN-K'ANG

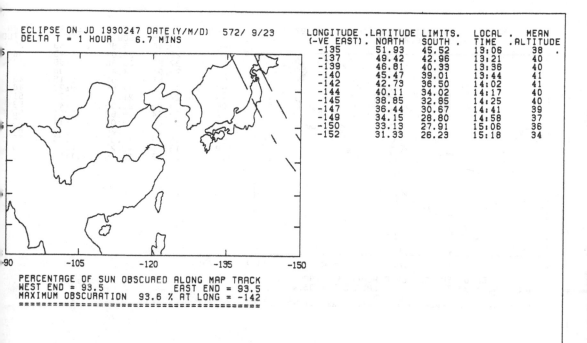

ECLIPSE ON JD 1930247 DATE(Y/M/D)   572/ 9/23
DELTA T = 1 HOUR   6.7 MINS

LONGITUDE	.LATITUDE	LIMITS.	LOCAL	. MEAN
(-VE EAST).	NORTH	SOUTH .	TIME	.ALTITUDE
-135	51.93	45.52	13:06	38
-137	49.42	42.96	13:21	40
-139	46.81	40.39	13:36	40
-140	45.47	39.01	13:44	41
-142	42.73	36.50	14:02	41
-144	40.11	34.02	14:17	40
-145	38.85	32.85	14:25	40
-147	36.44	30.67	14:41	39
-149	34.15	28.80	14:58	37
-150	33.13	27.91	15:06	36
-152	31.33	26.23	15:18	34

PERCENTAGE OF SUN OBSCURED ALONG MAP TRACK
WEST END = 93.5   EAST END = 93.5
MAXIMUM OBSCURATION 93.6 % AT LONG = -142
===

+ MARKS THE CITY OF CHIEN-K'ANG

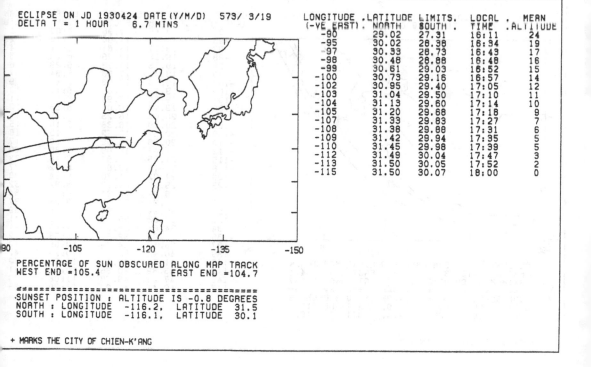

ECLIPSE ON JD 1930424 DATE(Y/M/D)   573/ 3/19
DELTA T = 1 HOUR   6.7 MINS

LONGITUDE	.LATITUDE	LIMITS.	LOCAL	. MEAN
(-VE EAST).	NORTH	SOUTH .	TIME	.ALTITUDE
-90	29.02	27.31	16:11	24
-95	30.02	28.38	16:34	19
-97	30.33	28.73	16:43	17
-98	30.48	28.88	16:48	16
-99	30.61	29.03	16:52	15
-100	30.73	29.16	16:57	14
-102	30.95	29.40	17:05	12
-103	31.04	29.50	17:10	11
-104	31.13	29.60	17:14	10
-105	31.20	29.68	17:18	9
-107	31.33	29.83	17:27	7
-108	31.38	29.88	17:31	6
-109	31.42	29.94	17:35	5
-110	31.45	29.98	17:39	5
-112	31.49	30.04	17:47	3
-113	31.50	30.05	17:52	2
-115	31.50	30.07	18:00	0

PERCENTAGE OF SUN OBSCURED ALONG MAP TRACK
WEST END =105.4   EAST END =104.7

===
SUNSET POSITION : ALTITUDE IS -0.8 DEGREES
NORTH : LONGITUDE -116.2, LATITUDE 31.5
SOUTH : LONGITUDE -116.1, LATITUDE 30.1

+ MARKS THE CITY OF CHIEN-K'ANG

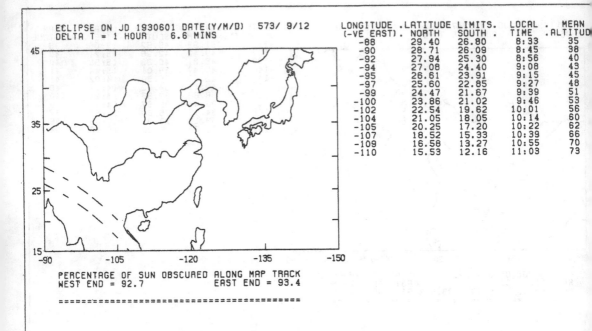

ECLIPSE ON JD 1930601 DATE(Y/M/D) 573/ 9/12
DELTA T = 1 HOUR   6.6 MINS

LONGITUDE (-VE EAST)	LATITUDE NORTH	LIMITS. SOUTH	LOCAL TIME	MEAN ALTITU
-88	29.40	26.80	8:33	35
-90	28.71	26.09	8:45	38
-92	27.94	25.30	8:56	40
-94	27.08	24.40	9:08	43
-95	26.61	23.91	9:15	45
-97	25.60	22.85	9:27	48
-99	24.47	21.67	9:39	51
-100	23.86	21.02	9:46	53
-102	22.54	19.62	10:01	56
-104	21.05	18.05	10:14	60
-105	20.25	17.20	10:22	62
-107	18.52	15.33	10:39	66
-109	16.58	13.27	10:55	70
-110	15.53	12.16	11:03	73

PERCENTAGE OF SUN OBSCURED ALONG MAP TRACK
WEST END = 92.7   EAST END = 93.4

==

+ MARKS THE CITY OF CHIEN-K'ANG

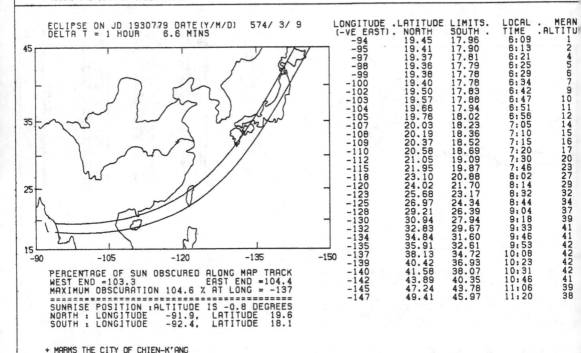

ECLIPSE ON JD 1930779 DATE(Y/M/D) 574/ 3/ 9
DELTA T = 1 HOUR   6.6 MINS

LONGITUDE (-VE EAST)	LATITUDE NORTH	LIMITS. SOUTH	LOCAL TIME	MEAN ALTITU
-94	19.45	17.96	6:09	1
-95	19.41	17.90	6:13	2
-97	19.37	17.81	6:21	4
-98	19.36	17.79	6:25	5
-99	19.38	17.78	6:29	6
-100	19.40	17.78	6:34	7
-102	19.50	17.83	6:42	9
-103	19.57	17.88	6:47	10
-104	19.66	17.94	6:51	11
-105	19.76	18.02	6:56	12
-107	20.03	18.23	7:05	14
-108	20.19	18.36	7:10	15
-109	20.37	18.52	7:15	16
-110	20.58	18.69	7:20	17
-112	21.05	19.09	7:30	20
-115	21.95	19.87	7:46	23
-118	23.10	20.88	8:02	27
-120	24.02	21.70	8:14	29
-123	25.68	23.17	8:32	32
-125	26.97	24.34	8:44	34
-128	29.21	26.39	9:04	37
-130	30.94	27.94	9:18	39
-132	32.83	29.67	9:33	41
-134	34.84	31.60	9:46	41
-135	35.91	32.61	9:53	42
-137	38.13	34.72	10:08	42
-139	40.42	36.93	10:23	42
-140	41.58	38.07	10:31	42
-142	43.89	40.35	10:46	41
-145	47.24	43.78	11:06	39
-147	49.41	45.97	11:20	38

PERCENTAGE OF SUN OBSCURED ALONG MAP TRACK
WEST END =103.3   EAST END =104.4
MAXIMUM OBSCURATION 104.6 % AT LONG = -137
===
SUNRISE POSITION :ALTITUDE IS -0.8 DEGREES
NORTH : LONGITUDE   -91.9, LATITUDE 19.6
SOUTH : LONGITUDE   -92.4, LATITUDE 18.1

+ MARKS THE CITY OF CHIEN-K'ANG

ECLIPSE ON JD 1931635 DATE(Y/M/D)   576/ 7/12
DELTA T = 1 HOUR   6.3 MINS

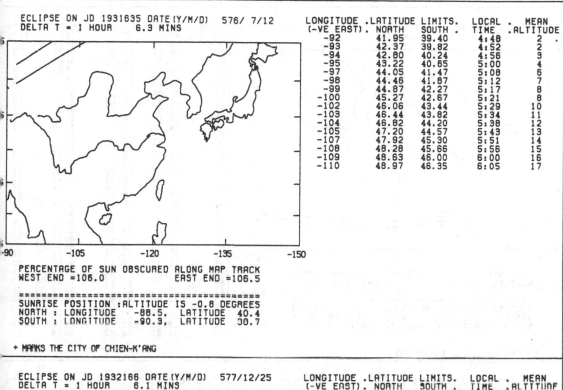

LONGITUDE	.LATITUDE	LIMITS.	LOCAL	. MEAN
(-VE EAST)	. NORTH	SOUTH .	TIME	.ALTITUDE
-92	41.95	39.40	4:48	2
-93	42.37	39.82	4:52	2
-94	42.80	40.24	4:56	3
-95	43.22	40.65	5:00	4
-97	44.05	41.47	5:08	6
-98	44.46	41.87	5:12	7
-99	44.87	42.27	5:17	8
-100	45.27	42.67	5:21	8
-102	46.06	43.44	5:29	10
-103	46.44	43.82	5:34	11
-104	46.82	44.20	5:38	12
-105	47.20	44.57	5:43	13
-107	47.92	45.30	5:51	14
-108	48.28	45.66	5:56	15
-109	48.63	46.00	6:00	16
-110	48.97	46.35	6:05	17

PERCENTAGE OF SUN OBSCURED ALONG MAP TRACK
WEST END =106.0   EAST END =106.5

==
SUNRISE POSITION :ALTITUDE IS -0.8 DEGREES
NORTH : LONGITUDE   -88.5. LATITUDE   40.4
SOUTH : LONGITUDE   -90.3. LATITUDE   30.7

+ MARKS THE CITY OF CHIEN-K'ANG

ECLIPSE ON JD 1932166 DATE(Y/M/D)   577/12/25
DELTA T = 1 HOUR   6.1 MINS

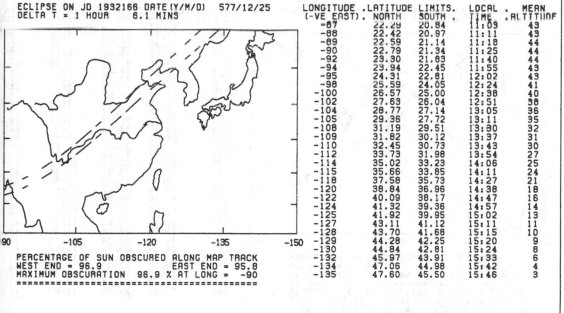

LONGITUDE	.LATITUDE	LIMITS.	LOCAL	. MEAN
(-VE EAST)	. NORTH	SOUTH .	TIME	.ALTITUDE
-87	22.29	20.84	11:09	43
-88	22.42	20.97	11:11	43
-89	22.59	21.14	11:18	44
-90	22.79	21.34	11:25	44
-92	23.30	21.83	11:40	44
-94	23.94	22.45	11:55	43
-95	24.31	22.81	12:02	43
-98	25.59	24.05	12:24	41
-100	26.57	25.00	12:38	40
-102	27.63	26.04	12:51	38
-104	28.77	27.14	13:05	36
-105	29.36	27.72	13:11	35
-108	31.19	29.51	13:30	32
-109	31.82	30.12	13:37	31
-110	32.45	30.73	13:43	30
-112	33.73	31.98	13:54	27
-114	35.02	33.23	14:06	25
-115	35.66	33.85	14:11	24
-118	37.58	35.73	14:27	21
-120	38.84	36.96	14:38	18
-122	40.09	38.17	14:47	16
-124	41.32	39.36	14:57	14
-125	41.92	39.95	15:02	13
-127	43.11	41.12	15:11	11
-128	43.70	41.68	15:15	10
-129	44.28	42.25	15:20	9
-130	44.84	42.81	15:24	8
-132	45.97	43.91	15:33	6
-134	47.06	44.98	15:42	4
-135	47.60	45.50	15:46	3

PERCENTAGE OF SUN OBSCURED ALONG MAP TRACK
WEST END = 96.9   EAST END = 95.8
MAXIMUM OBSCURATION 96.9 % AT LONG = -90
==

+ MARKS THE CITY OF CHIEN-K'ANG

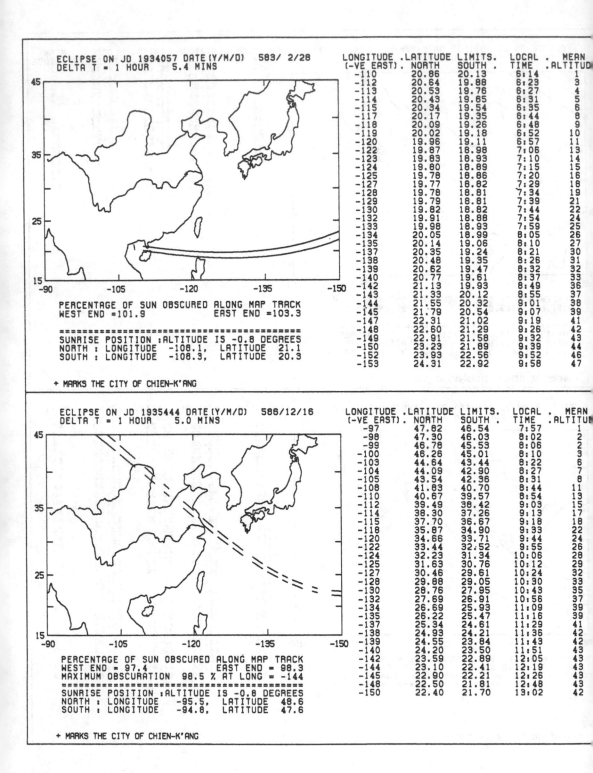

ECLIPSE ON JD 1934057 DATE(Y/M/D)   583/ 2/28
DELTA T = 1 HOUR   5.4 MINS

LONGITUDE (-VE EAST)	LATITUDE NORTH	LIMITS. SOUTH	LOCAL TIME	MEAN ALTITUDE
-110	20.86	20.13	6:14	1
-112	20.64	19.88	6:23	3
-113	20.53	19.76	6:27	4
-114	20.43	19.65	6:31	5
-115	20.34	19.54	6:35	6
-117	20.17	19.35	6:44	8
-118	20.09	19.26	6:48	9
-119	20.02	19.18	6:52	10
-120	19.96	19.11	6:57	11
-122	19.87	18.98	7:06	13
-123	19.83	18.93	7:10	14
-124	19.80	18.89	7:15	15
-125	19.78	18.86	7:20	16
-127	19.77	18.82	7:29	18
-128	19.78	18.81	7:34	19
-129	19.79	18.81	7:39	21
-130	19.82	18.82	7:44	22
-132	19.91	18.88	7:54	24
-133	19.98	18.93	7:59	25
-134	20.05	18.99	8:05	26
-135	20.14	19.06	8:10	27
-137	20.35	19.24	8:21	30
-138	20.48	19.35	8:26	31
-139	20.62	19.47	8:32	32
-140	20.77	19.61	8:37	33
-142	21.13	19.93	8:49	36
-143	21.33	20.12	8:55	37
-144	21.55	20.32	9:01	38
-145	21.79	20.54	9:07	39
-147	22.31	21.02	9:19	41
-148	22.60	21.29	9:26	42
-149	22.91	21.58	9:32	43
-150	23.23	21.89	9:39	44
-152	23.93	22.56	9:52	46
-153	24.31	22.92	9:58	47

PERCENTAGE OF SUN OBSCURED ALONG MAP TRACK
WEST END =101.9   EAST END =103.3

===
SUNRISE POSITION :ALTITUDE IS -0.8 DEGREES
NORTH : LONGITUDE -108.1, LATITUDE 21.1
SOUTH : LONGITUDE -108.3, LATITUDE 20.3

+ MARKS THE CITY OF CHIEN-K'ANG

ECLIPSE ON JD 1935444 DATE(Y/M/D)   586/12/16
DELTA T = 1 HOUR   5.0 MINS

LONGITUDE (-VE EAST)	LATITUDE NORTH	LIMITS. SOUTH	LOCAL TIME	MEAN ALTITUDE
-97	47.82	46.54	7:57	1
-98	47.30	46.03	8:02	2
-99	46.78	45.53	8:06	2
-100	46.26	45.01	8:10	3
-103	44.64	43.44	8:22	6
-104	44.09	42.90	8:27	7
-105	43.54	42.36	8:31	8
-108	41.83	40.70	8:44	11
-110	40.67	39.57	8:54	13
-112	39.49	38.42	9:03	15
-114	38.30	37.26	9:13	17
-115	37.70	36.67	9:18	18
-118	35.87	34.90	9:33	22
-120	34.66	33.71	9:44	24
-122	33.44	32.52	9:55	26
-124	32.23	31.34	10:06	28
-125	31.63	30.76	10:12	29
-127	30.46	29.61	10:24	32
-128	29.88	29.05	10:30	33
-130	28.76	27.95	10:43	35
-132	27.69	26.91	10:56	37
-134	26.69	25.93	11:09	39
-135	26.22	25.47	11:16	39
-137	25.34	24.61	11:29	41
-138	24.93	24.21	11:36	42
-139	24.55	23.84	11:43	42
-140	24.20	23.50	11:51	43
-142	23.59	22.89	12:05	43
-144	23.10	22.41	12:19	43
-145	22.90	22.21	12:26	43
-148	22.50	21.81	12:48	43
-150	22.40	21.70	13:02	42

PERCENTAGE OF SUN OBSCURED ALONG MAP TRACK
WEST END = 97.4   EAST END = 98.3
MAXIMUM OBSCURATION 98.5 % AT LONG = -144
===
SUNRISE POSITION :ALTITUDE IS -0.8 DEGREES
NORTH : LONGITUDE -95.5, LATITUDE 48.6
SOUTH : LONGITUDE -94.8, LATITUDE 47.6

+ MARKS THE CITY OF CHIEN-K'ANG

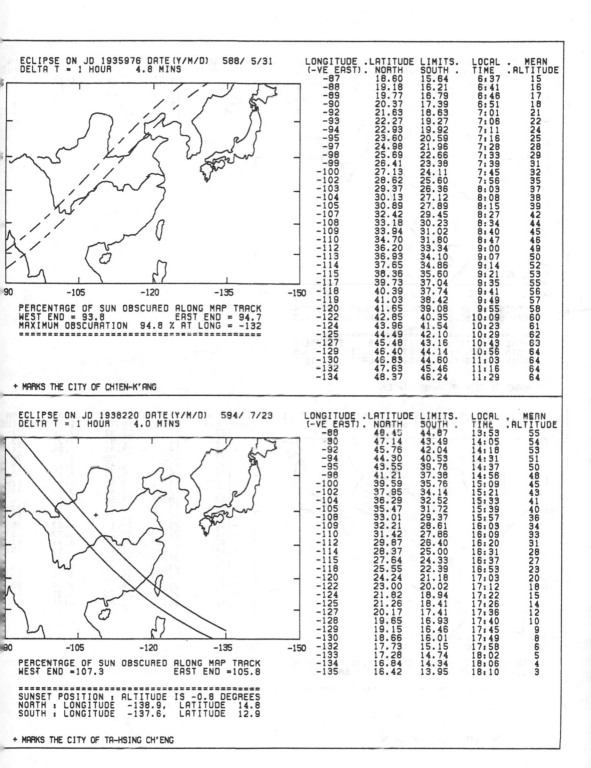

ECLIPSE ON JD 1935976 DATE(Y/M/D) 588/ 5/31
DELTA T = 1 HOUR   4.8 MINS

LONGITUDE (-VE EAST)	LATITUDE NORTH	LIMITS. SOUTH	LOCAL TIME	MEAN ALTITUDE
-87	18.60	15.64	6:37	15
-88	19.18	16.21	6:41	16
-89	19.77	16.79	6:46	17
-90	20.37	17.39	6:51	18
-92	21.63	18.63	7:01	21
-93	22.27	19.27	7:06	22
-94	22.93	19.92	7:11	24
-95	23.60	20.59	7:16	25
-97	24.98	21.96	7:28	28
-98	25.69	22.66	7:33	29
-99	26.41	23.38	7:39	31
-100	27.13	24.11	7:45	32
-102	28.62	25.60	7:56	35
-103	29.37	26.36	8:03	37
-104	30.13	27.12	8:08	38
-105	30.89	27.89	8:15	39
-107	32.42	29.45	8:27	42
-108	33.18	30.23	8:34	44
-109	33.94	31.02	8:40	45
-110	34.70	31.80	8:47	46
-112	36.20	33.34	9:00	49
-113	36.93	34.10	9:07	50
-114	37.65	34.86	9:14	52
-115	38.36	35.60	9:21	53
-117	39.73	37.04	9:35	55
-118	40.39	37.74	9:41	56
-119	41.03	38.42	9:49	57
-120	41.65	39.08	9:55	58
-122	42.85	40.35	10:09	60
-124	43.96	41.54	10:23	61
-125	44.49	42.10	10:29	62
-127	45.48	43.16	10:43	63
-129	46.40	44.14	10:56	64
-130	46.83	44.60	11:03	64
-132	47.63	45.46	11:16	64
-134	48.37	46.24	11:29	64

PERCENTAGE OF SUN OBSCURED ALONG MAP TRACK
WEST END = 93.8   EAST END = 94.7
MAXIMUM OBSCURATION 94.8 % AT LONG = -132
==

+ MARKS THE CITY OF CHIEN-K'ANG

ECLIPSE ON JD 1938220 DATE(Y/M/D) 594/ 7/23
DELTA T = 1 HOUR   4.0 MINS

LONGITUDE (-VE EAST)	LATITUDE NORTH	LIMITS. SOUTH	LOCAL TIME	MEAN ALTITUDE
-88	48.45	44.87	13:53	55
-90	47.14	43.49	14:05	54
-92	45.76	42.04	14:18	53
-94	44.30	40.53	14:31	51
-95	43.55	39.76	14:37	50
-98	41.21	37.38	14:56	48
-100	39.59	35.76	15:09	45
-102	37.95	34.14	15:21	43
-104	36.29	32.52	15:33	41
-105	35.47	31.72	15:39	40
-108	33.01	29.37	15:57	36
-109	32.21	28.61	16:03	34
-110	31.42	27.86	16:09	33
-112	29.87	26.40	16:20	31
-114	28.37	25.00	16:31	28
-115	27.64	24.33	16:37	27
-118	25.55	22.39	16:53	23
-120	24.24	21.18	17:03	20
-122	23.00	20.02	17:12	18
-124	21.82	18.94	17:22	15
-125	21.26	18.41	17:26	14
-127	20.17	17.41	17:36	12
-128	19.65	16.93	17:40	10
-129	19.15	16.46	17:45	9
-130	18.66	16.01	17:49	8
-132	17.73	15.15	17:58	6
-133	17.28	14.74	18:02	5
-134	16.84	14.34	18:06	4
-135	16.42	13.95	18:10	3

PERCENTAGE OF SUN OBSCURED ALONG MAP TRACK
WEST END =107.3   EAST END =105.8

==
SUNSET POSITION : ALTITUDE IS -0.8 DEGREES
NORTH : LONGITUDE -138.9, LATITUDE 14.8
SOUTH : LONGITUDE -137.6, LATITUDE 12.9

+ MARKS THE CITY OF TA-HSING CH'ENG

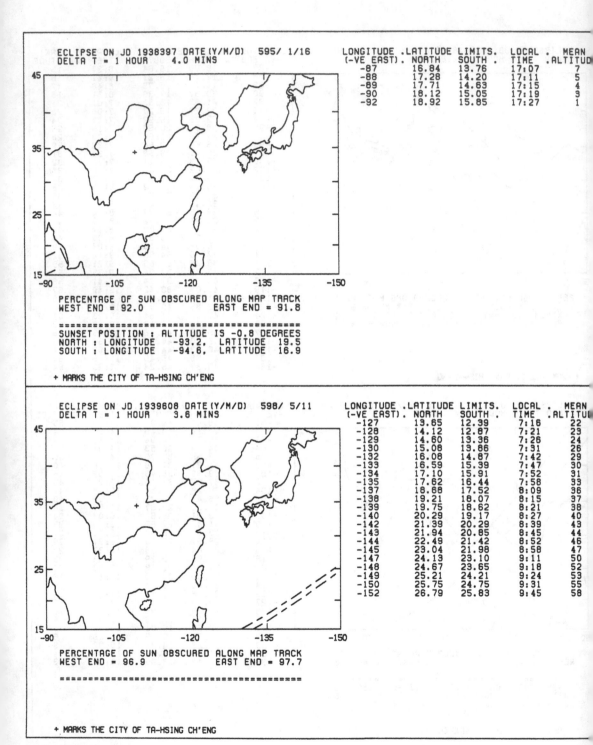

ECLIPSE ON JD 1938397 DATE(Y/M/D)   595/ 1/16
DELTA T = 1 HOUR   4.0 MINS

LONGITUDE (-VE EAST)	LATITUDE NORTH	LIMITS. SOUTH	LOCAL TIME	MEAN ALTITUDE
-87	16.84	13.76	17:07	7
-88	17.28	14.20	17:11	5
-89	17.71	14.63	17:15	4
-90	18.12	15.05	17:19	3
-92	18.92	15.85	17:27	1

PERCENTAGE OF SUN OBSCURED ALONG MAP TRACK
WEST END = 92.0   EAST END = 91.8

==
SUNSET POSITION : ALTITUDE IS -0.8 DEGREES
NORTH : LONGITUDE -93.2, LATITUDE 19.5
SOUTH : LONGITUDE -94.6, LATITUDE 16.9

+ MARKS THE CITY OF TA-HSING CH'ENG

ECLIPSE ON JD 1939608 DATE(Y/M/D)   598/ 5/11
DELTA T = 1 HOUR   3.8 MINS

LONGITUDE (-VE EAST)	LATITUDE NORTH	LIMITS. SOUTH	LOCAL TIME	MEAN ALTITUDE
-127	13.65	12.39	7:16	22
-128	14.12	12.87	7:21	23
-129	14.60	13.36	7:26	24
-130	15.08	13.86	7:31	26
-132	16.08	14.87	7:42	29
-133	16.59	15.39	7:47	30
-134	17.10	15.91	7:52	31
-135	17.62	16.44	7:58	33
-137	18.68	17.52	8:09	36
-138	19.21	18.07	8:15	37
-139	19.75	18.62	8:21	38
-140	20.29	19.17	8:27	40
-142	21.39	20.29	8:39	43
-143	21.94	20.85	8:45	44
-144	22.49	21.42	8:52	46
-145	23.04	21.98	8:58	47
-147	24.13	23.10	9:11	50
-148	24.67	23.65	9:18	52
-149	25.21	24.21	9:24	53
-150	25.75	24.75	9:31	55
-152	26.79	25.83	9:45	58

PERCENTAGE OF SUN OBSCURED ALONG MAP TRACK
WEST END = 96.9   EAST END = 97.7

==

+ MARKS THE CITY OF TA-HSING CH'ENG

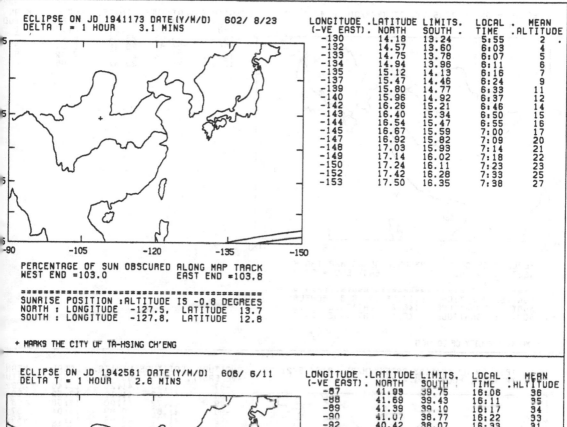

ECLIPSE ON JD 1941173 DATE(Y/M/D) 602/ 8/23
DELTA T = 1 HOUR   3.1 MINS

LONGITUDE (-VE EAST)	LATITUDE NORTH	LIMITS. SOUTH	LOCAL TIME	MEAN ALTITUDE
-130	14.18	13.24	5:55	2
-132	14.57	13.60	6:03	4
-133	14.75	13.78	6:07	5
-134	14.94	13.96	6:11	6
-135	15.12	14.13	6:16	7
-137	15.47	14.46	6:24	9
-139	15.80	14.77	6:33	11
-140	15.96	14.92	6:37	12
-142	16.26	15.21	6:46	14
-143	16.40	15.34	6:50	15
-144	16.54	15.47	6:55	16
-145	16.67	15.59	7:00	17
-147	16.92	15.82	7:09	20
-148	17.03	15.93	7:14	21
-149	17.14	16.02	7:18	22
-150	17.24	16.11	7:23	23
-152	17.42	16.28	7:33	25
-153	17.50	16.35	7:38	27

PERCENTAGE OF SUN OBSCURED ALONG MAP TRACK
WEST END =103.0   EAST END =103.8

==
SUNRISE POSITION :ALTITUDE IS -0.8 DEGREES
NORTH : LONGITUDE -127.5, LATITUDE 13.7
SOUTH : LONGITUDE -127.8, LATITUDE 12.8

+ MARKS THE CITY OF TA-HSING CH'ENG

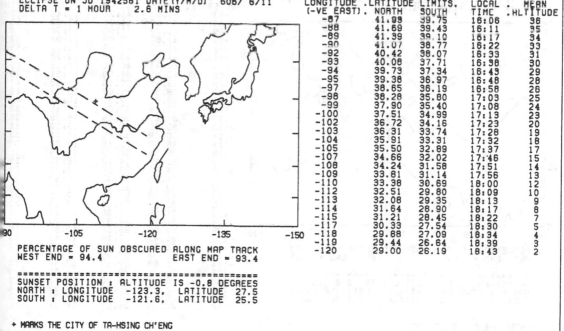

ECLIPSE ON JD 1942561 DATE(Y/M/D) 606/ 6/11
DELTA T = 1 HOUR   2.6 MINS

LONGITUDE (-VE EAST)	LATITUDE NORTH	LIMITS. SOUTH	LOCAL TIME	MEAN ALTITUDE
-87	41.99	39.75	16:06	36
-88	41.69	39.43	16:11	35
-89	41.39	39.10	16:17	34
-90	41.07	38.77	16:22	33
-92	40.42	38.07	16:33	31
-93	40.08	37.71	16:38	30
-94	39.73	37.34	16:43	29
-95	39.38	36.97	16:48	28
-97	38.65	36.19	16:58	26
-98	38.28	35.80	17:03	25
-99	37.90	35.40	17:08	24
-100	37.51	34.99	17:13	23
-102	36.72	34.16	17:23	20
-103	36.31	33.74	17:28	19
-104	35.91	33.31	17:32	18
-105	35.50	32.89	17:37	17
-107	34.66	32.02	17:46	15
-108	34.24	31.58	17:51	14
-109	33.81	31.14	17:56	13
-110	33.38	30.69	18:00	12
-112	32.51	29.80	18:09	10
-113	32.08	29.35	18:13	9
-114	31.64	28.90	18:17	8
-115	31.21	28.45	18:22	7
-117	30.33	27.54	18:30	5
-118	29.88	27.09	18:34	4
-119	29.44	26.64	18:39	3
-120	29.00	26.19	18:43	2

PERCENTAGE OF SUN OBSCURED ALONG MAP TRACK
WEST END = 94.4   EAST END = 93.4

==
SUNSET POSITION : ALTITUDE IS -0.8 DEGREES
NORTH : LONGITUDE -123.3, LATITUDE 27.5
SOUTH : LONGITUDE -121.6, LATITUDE 25.5

+ MARKS THE CITY OF TA-HSING CH'ENG

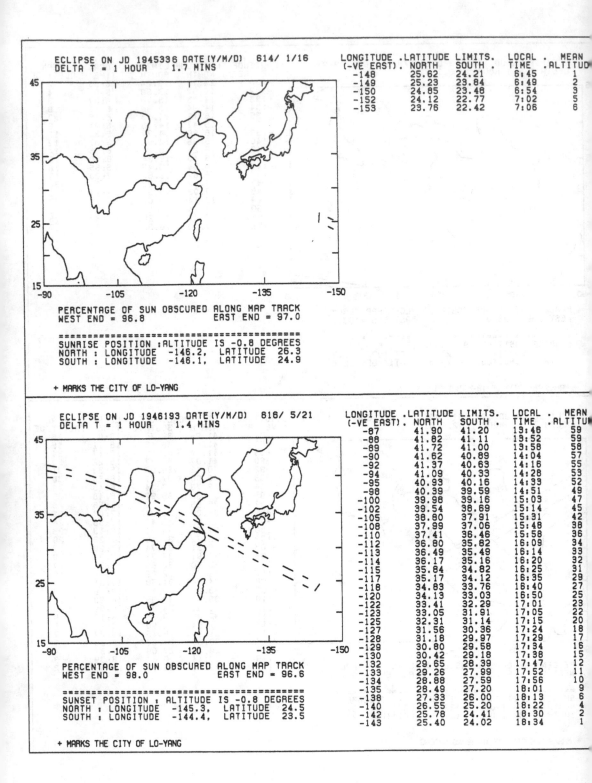

ECLIPSE ON JD 1945336 DATE(Y/M/D)   614/ 1/16
DELTA T = 1 HOUR   1.7 MINS

LONGITUDE (-VE EAST)	LATITUDE NORTH	LIMITS SOUTH	LOCAL TIME	MEAN ALTITU
-148	25.62	24.21	6:45	1
-149	25.23	23.84	6:49	2
-150	24.85	23.48	6:54	3
-152	24.12	22.77	7:02	5
-153	23.76	22.42	7:06	6

PERCENTAGE OF SUN OBSCURED ALONG MAP TRACK
WEST END = 96.8   EAST END = 97.0

===
SUNRISE POSITION :ALTITUDE IS -0.8 DEGREES
NORTH : LONGITUDE -146.2, LATITUDE 26.3
SOUTH : LONGITUDE -146.1, LATITUDE 24.9

+ MARKS THE CITY OF LO-YANG

ECLIPSE ON JD 1946193 DATE(Y/M/D)   616/ 5/21
DELTA T = 1 HOUR   1.4 MINS

LONGITUDE (-VE EAST)	LATITUDE NORTH	LIMITS SOUTH	LOCAL TIME	MEAN ALTITU
-87	41.90	41.20	13:46	59
-88	41.82	41.11	13:52	59
-89	41.72	41.00	13:58	58
-90	41.62	40.89	14:04	57
-92	41.37	40.63	14:16	55
-94	41.09	40.33	14:28	53
-95	40.93	40.16	14:39	52
-98	40.39	39.59	14:51	49
-100	39.98	39.16	15:03	47
-102	39.54	38.69	15:14	45
-105	38.80	37.91	15:31	42
-108	37.99	37.06	15:48	38
-110	37.41	36.46	15:58	36
-112	36.80	35.82	16:09	34
-113	36.49	35.49	16:14	33
-114	36.17	35.16	16:20	32
-115	35.84	34.82	16:25	31
-117	35.17	34.12	16:35	29
-118	34.83	33.76	16:40	27
-120	34.13	33.03	16:50	25
-122	33.41	32.29	17:01	23
-123	33.05	31.91	17:05	22
-125	32.31	31.14	17:15	20
-127	31.56	30.36	17:24	18
-128	31.18	29.97	17:29	17
-129	30.80	29.58	17:34	16
-130	30.42	29.18	17:38	15
-132	29.65	28.39	17:47	12
-133	29.26	27.99	17:52	11
-134	28.88	27.59	17:56	10
-135	28.49	27.20	18:01	9
-138	27.33	26.00	18:13	6
-140	26.55	25.20	18:22	4
-142	25.78	24.41	18:30	2
-143	25.40	24.02	18:34	1

PERCENTAGE OF SUN OBSCURED ALONG MAP TRACK
WEST END = 98.0   EAST END = 96.6

===
SUNSET POSITION : ALTITUDE IS -0.8 DEGREES
NORTH : LONGITUDE -145.3, LATITUDE 24.5
SOUTH : LONGITUDE -144.4, LATITUDE 23.5

+ MARKS THE CITY OF LO-YANG

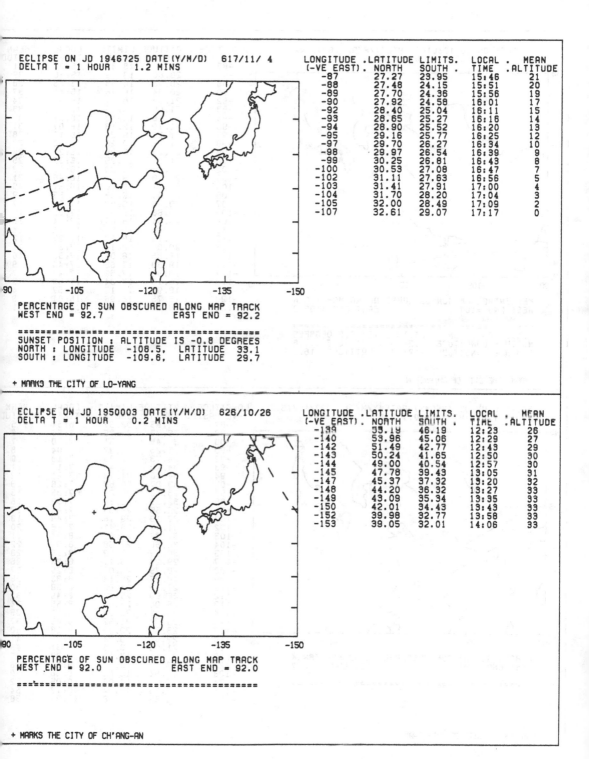

ECLIPSE ON JD 1946725 DATE(Y/M/D) 617/11/ 4
DELTA T = 1 HOUR   1.2 MINS

LONGITUDE (-VE EAST)	LATITUDE NORTH	LIMITS. SOUTH	LOCAL TIME	MEAN ALTITUDE
-87	27.27	29.95	15:46	21
-88	27.48	24.15	15:51	20
-89	27.70	24.36	15:56	19
-90	27.92	24.58	16:01	17
-92	28.40	25.04	16:11	15
-93	28.65	25.27	16:16	14
-94	28.90	25.52	16:20	13
-95	29.16	25.77	16:25	12
-97	29.70	26.27	16:34	10
-98	29.97	26.54	16:39	9
-99	30.25	26.81	16:43	8
-100	30.53	27.08	16:47	7
-102	31.11	27.63	16:56	5
-103	31.41	27.91	17:00	4
-104	31.70	28.20	17:04	3
-105	32.00	28.49	17:09	2
-107	32.61	29.07	17:17	0

PERCENTAGE OF SUN OBSCURED ALONG MAP TRACK
WEST END = 92.7   EAST END = 92.2

==
SUNSET POSITION : ALTITUDE IS -0.8 DEGREES
NORTH : LONGITUDE -108.5, LATITUDE 33.1
SOUTH : LONGITUDE -109.6, LATITUDE 29.7

+ MARKS THE CITY OF LO-YANG

ECLIPSE ON JD 1950003 DATE(Y/M/D) 626/10/26
DELTA T = 1 HOUR   0.2 MINS

LONGITUDE (-VE EAST)	LATITUDE NORTH	LIMITS. SOUTH	LOCAL TIME	MEAN ALTITUDE
-139	55.19	46.19	12:23	26
-140	53.96	45.06	12:29	27
-142	51.49	42.77	12:43	29
-143	50.24	41.65	12:50	30
-144	49.00	40.54	12:57	30
-145	47.78	39.43	13:05	31
-147	45.37	37.32	13:20	32
-148	44.20	36.32	13:27	33
-149	43.09	35.34	13:35	33
-150	42.01	34.43	13:43	33
-152	39.98	32.77	13:58	33
-153	39.05	32.01	14:06	33

PERCENTAGE OF SUN OBSCURED ALONG MAP TRACK
WEST END = 92.0   EAST END = 92.0

==

+ MARKS THE CITY OF CH'ANG-AN

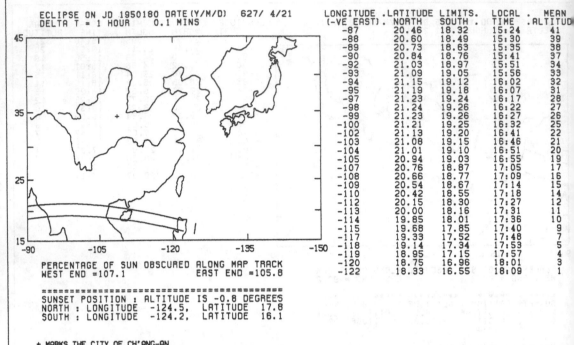

ECLIPSE ON JD 1950180 DATE(Y/M/D) 627/ 4/21
DELTA T = 1 HOUR   0.1 MINS

LONGITUDE (-VE EAST)	LATITUDE NORTH	LIMITS. SOUTH	LOCAL TIME	MEAN ALTITUD
-87	20.46	18.32	15:24	41
-88	20.60	18.49	15:30	39
-89	20.73	18.63	15:35	38
-90	20.84	18.76	15:41	37
-92	21.03	18.97	15:51	34
-93	21.09	19.05	15:56	33
-94	21.15	19.12	16:02	32
-95	21.19	19.18	16:07	31
-97	21.23	19.24	16:17	28
-98	21.24	19.26	16:22	27
-99	21.23	19.26	16:27	26
-100	21.21	19.25	16:32	25
-102	21.13	19.20	16:41	22
-103	21.08	19.15	16:46	21
-104	21.01	19.10	16:51	20
-105	20.94	19.03	16:55	19
-107	20.76	18.87	17:05	17
-108	20.66	18.77	17:09	16
-109	20.54	18.67	17:14	15
-110	20.42	18.55	17:18	14
-112	20.15	18.30	17:27	12
-113	20.00	18.16	17:31	11
-114	19.85	18.01	17:36	10
-115	19.68	17.85	17:40	9
-117	19.33	17.52	17:48	7
-118	19.14	17.34	17:53	5
-119	18.95	17.15	17:57	4
-120	18.75	16.96	18:01	3
-122	-18.33	16.55	18:09	1

PERCENTAGE OF SUN OBSCURED ALONG MAP TRACK
WEST END =107.1   EAST END =105.8

SUNSET POSITION : ALTITUDE IS -0.8 DEGREES
NORTH : LONGITUDE -124.5, LATITUDE 17.8
SOUTH : LONGITUDE -124.2, LATITUDE 16.1

+ MARKS THE CITY OF CH'ANG-AN

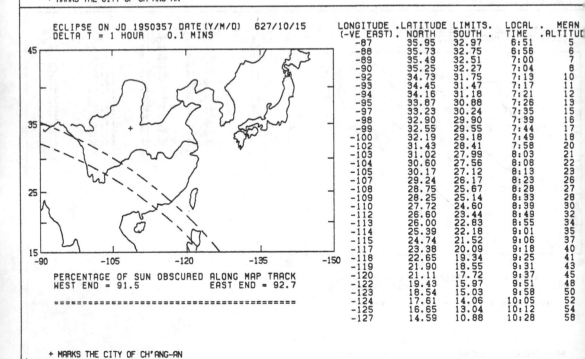

ECLIPSE ON JD 1950357 DATE(Y/M/D) 627/10/15
DELTA T = 1 HOUR   0.1 MINS

LONGITUDE (-VE EAST)	LATITUDE NORTH	LIMITS. SOUTH	LOCAL TIME	MEAN ALTITUD
-87	35.95	32.97	6:51	5
-88	35.73	32.75	6:56	6
-89	35.49	32.51	7:00	7
-90	35.25	32.27	7:04	8
-92	34.73	31.75	7:13	10
-93	34.45	31.47	7:17	11
-94	34.16	31.18	7:21	12
-95	33.87	30.88	7:26	13
-97	33.23	30.24	7:35	15
-98	32.90	29.90	7:39	16
-99	32.55	29.55	7:44	17
-100	32.19	29.18	7:49	18
-102	31.43	28.41	7:58	20
-103	31.02	27.99	8:03	21
-104	30.60	27.56	8:08	22
-105	30.17	27.12	8:13	23
-107	29.24	26.17	8:23	26
-108	28.75	25.67	8:28	27
-109	28.25	25.14	8:33	28
-110	27.72	24.60	8:39	30
-112	26.60	23.44	8:49	32
-113	26.00	22.83	8:55	34
-114	25.39	22.18	9:01	35
-115	24.74	21.52	9:06	37
-117	23.38	20.09	9:18	40
-118	22.65	19.34	9:25	41
-119	21.90	18.55	9:31	43
-120	21.11	17.72	9:37	45
-122	19.43	15.97	9:51	48
-123	18.54	15.03	9:58	50
-124	17.61	14.06	10:05	52
-125	16.65	13.04	10:12	54
-127	14.59	10.88	10:28	58

PERCENTAGE OF SUN OBSCURED ALONG MAP TRACK
WEST END = 91.5   EAST END = 92.7

+ MARKS THE CITY OF CH'ANG-AN

ECLIPSE ON JD 1950535 DATE(Y/M/D)   628/ 4/10
DELTA T = 1 HOUR   0.0 MINS

LONGITUDE (-VE EAST)	LATITUDE LIMITS. NORTH	SOUTH	LOCAL TIME	MEAN ALTITUDE
-110	13.11	11.35	6:55	15
-112	13.92	12.09	7:04	18
-113	14.35	12.49	7:09	19
-114	14.81	12.91	7:14	20
-117	16.32	14.30	7:30	24
-120	18.06	15.91	7:45	28
-122	19.37	17.10	7:57	31
-124	20.79	18.42	8:09	33
-125	21.56	19.10	8:16	35
-127	23.17	20.61	8:28	38
-129	24.90	22.24	8:40	40
-130	25.83	23.09	8:47	42
-133	28.75	25.86	9:06	45
-135	30.84	27.83	9:21	48
-137	32.99	29.90	9:36	50
-139	35.17	32.03	9:51	51
-140	36.26	33.11	9:59	52
-142	38.41	35.28	10:11	53
-144	40.51	37.41	10:26	53
-145	41.53	38.46	10:34	53
-147	43.50	40.49	10:48	53
-149	45.35	42.43	11:01	53
-150	46.24	43.35	11:08	52

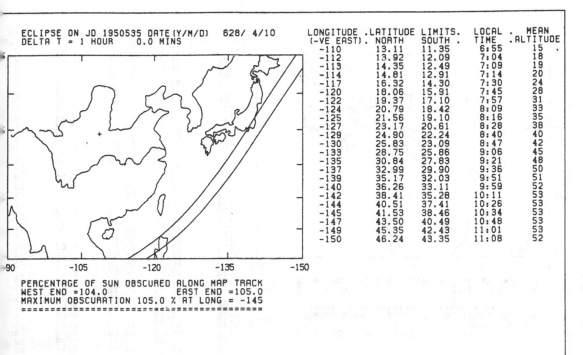

PERCENTAGE OF SUN OBSCURED ALONG MAP TRACK
WEST END =104.0   EAST END =105.0
MAXIMUM OBSCURATION 105.0 % AT LONG = -145
==

+ MARKS THE CITY OF CH'ANG-AN

ECLIPSE ON JD 1951922 DATE(Y/M/D)   632/ 1/27
DELTA T =   59.6 MINS

LONGITUDE (-VE EAST)	LATITUDE LIMITS. NORTH	SOUTH	LOCAL TIME	MEAN ALTITUDE
-87	40.97	39.89	13:16	29
-89	42.63	41.46	13:28	27
-90	43.44	42.22	13:34	26
-94	46.49	45.25	13:54	21

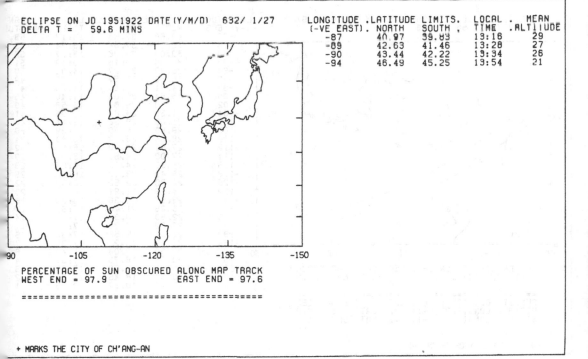

PERCENTAGE OF SUN OBSCURED ALONG MAP TRACK
WEST END = 97.9   EAST END = 97.6

==

+ MARKS THE CITY OF CH'ANG-AN

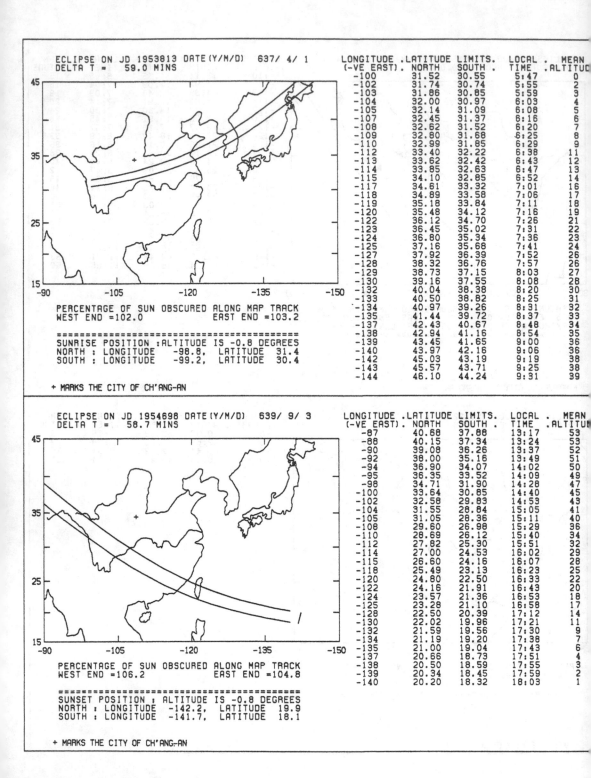

ECLIPSE ON JD 1953813 DATE(Y/M/D) 637/ 4/ 1
DELTA T =   59.0 MINS

LONGITUDE (-VE EAST)	.LATITUDE LIMITS. NORTH	SOUTH .	LOCAL TIME	MEAN .ALTITUD
-100	31.52	30.55	5:47	0
-102	31.74	30.74	5:55	2
-103	31.86	30.85	5:59	3
-104	32.00	30.97	6:03	4
-105	32.14	31.09	6:08	5
-107	32.45	31.37	6:16	6
-108	32.62	31.52	6:20	7
-109	32.80	31.68	6:25	8
-110	32.99	31.85	6:29	9
-112	33.40	32.22	6:38	11
-113	33.62	32.42	6:43	12
-114	33.85	32.63	6:47	13
-115	34.10	32.85	6:52	14
-117	34.61	33.32	7:01	16
-118	34.89	33.58	7:06	17
-119	35.18	33.84	7:11	18
-120	35.48	34.12	7:16	19
-122	36.12	34.70	7:26	21
-123	36.45	35.02	7:31	22
-124	36.80	35.34	7:36	23
-125	37.16	35.68	7:41	24
-127	37.92	36.39	7:52	26
-128	38.32	36.76	7:57	26
-129	38.73	37.15	8:03	27
-130	39.16	37.55	8:08	28
-132	40.04	38.38	8:20	30
-133	40.50	38.82	8:25	31
-134	40.97	39.26	8:31	32
-135	41.44	39.72	8:37	33
-137	42.43	40.67	8:48	34
-138	42.94	41.16	8:54	35
-139	43.45	41.65	9:00	36
-140	43.97	42.16	9:06	36
-142	45.03	43.19	9:19	38
-143	45.57	43.71	9:25	38
-144	46.10	44.24	9:31	39

PERCENTAGE OF SUN OBSCURED ALONG MAP TRACK
WEST END =102.0   EAST END =103.2

==
SUNRISE POSITION :ALTITUDE IS -0.8 DEGREES
NORTH : LONGITUDE   -98.8, LATITUDE   31.4
SOUTH : LONGITUDE   -99.2, LATITUDE   30.4

+ MARKS THE CITY OF CH'ANG-AN

ECLIPSE ON JD 1954698 DATE(Y/M/D) 639/ 9/ 3
DELTA T =   58.7 MINS

LONGITUDE (-VE EAST)	.LATITUDE LIMITS. NORTH	SOUTH .	LOCAL TIME	MEAN .ALTITUD
-87	40.68	37.88	13:17	53
-88	40.15	37.34	13:24	53
-90	39.08	36.26	13:37	52
-92	38.00	35.16	13:49	51
-94	36.90	34.07	14:02	50
-95	36.35	33.52	14:09	49
-98	34.71	31.90	14:28	47
-100	33.64	30.85	14:40	45
-102	32.58	29.83	14:53	43
-104	31.55	28.84	15:05	41
-105	31.05	28.36	15:11	40
-108	29.60	26.98	15:29	36
-110	28.69	26.12	15:40	34
-112	27.82	25.30	15:51	32
-114	27.00	24.53	16:02	29
-115	26.60	24.16	16:07	28
-118	25.49	23.13	16:23	25
-120	24.80	22.50	16:33	22
-122	24.16	21.91	16:43	20
-124	23.57	21.36	16:53	18
-125	23.28	21.10	16:58	17
-128	22.50	20.39	17:12	14
-130	22.02	19.96	17:21	11
-132	21.59	19.56	17:30	9
-134	21.19	19.20	17:38	7
-135	21.00	19.04	17:43	6
-137	20.66	18.73	17:51	4
-138	20.50	18.59	17:55	3
-139	20.34	18.45	17:59	2
-140	20.20	18.32	18:03	1

PERCENTAGE OF SUN OBSCURED ALONG MAP TRACK
WEST END =106.2   EAST END =104.8

==
SUNSET POSITION : ALTITUDE IS -0.8 DEGREES
NORTH : LONGITUDE -142.2, LATITUDE   19.9
SOUTH : LONGITUDE -141.7, LATITUDE   18.1

+ MARKS THE CITY OF CH'ANG-AN

274

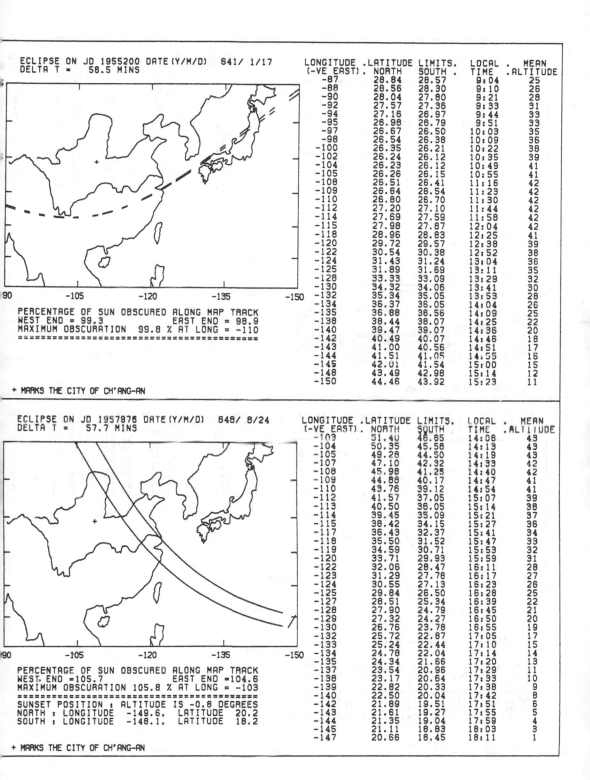

ECLIPSE ON JD 1955200 DATE(Y/M/D) 641/ 1/17
DELTA T =   58.5 MINS

LONGITUDE (-VE EAST)	LATITUDE LIMITS NORTH	SOUTH	LOCAL TIME	MEAN ALTITUDE
-87	28.84	28.57	9:04	25
-88	28.56	28.30	9:10	26
-90	28.04	27.80	9:21	28
-92	27.57	27.36	9:33	31
-94	27.16	26.97	9:44	33
-95	26.98	26.79	9:51	33
-97	26.67	26.50	10:03	35
-98	26.54	26.38	10:09	36
-100	26.35	26.21	10:22	38
-102	26.24	26.12	10:35	39
-104	26.23	26.12	10:49	41
-105	26.26	26.15	10:55	41
-108	26.51	26.41	11:16	42
-109	26.64	26.54	11:23	42
-110	26.80	26.70	11:30	42
-112	27.20	27.10	11:44	42
-114	27.69	27.59	11:58	42
-115	27.98	27.87	12:04	42
-118	28.96	28.83	12:25	41
-120	29.72	29.57	12:38	39
-122	30.54	30.38	12:52	38
-124	31.43	31.24	13:04	36
-125	31.89	31.69	13:11	35
-128	33.33	33.09	13:29	32
-130	34.32	34.06	13:41	30
-132	35.34	35.05	13:53	28
-134	36.37	36.05	14:04	26
-135	36.88	36.56	14:09	25
-138	38.44	38.07	14:25	22
-140	39.47	39.07	14:36	20
-142	40.49	40.07	14:46	18
-143	41.00	40.56	14:51	17
-144	41.51	41.05	14:55	16
-145	42.01	41.54	15:00	15
-148	43.49	42.98	15:14	12
-150	44.46	43.92	15:23	11

PERCENTAGE OF SUN OBSCURED ALONG MAP TRACK
WEST END = 99.3   EAST END = 98.9
MAXIMUM OBSCURATION 99.8 % AT LONG = -110
==

+ MARKS THE CITY OF CH'ANG-AN

ECLIPSE ON JD 1957976 DATE(Y/M/D) 648/ 8/24
DELTA T =   57.7 MINS

LONGITUDE (-VE EAST)	LATITUDE LIMITS NORTH	SOUTH	LOCAL TIME	MEAN ALTITUDE
-103	51.40	46.65	14:06	43
-104	50.35	45.58	14:13	43
-105	49.28	44.50	14:19	43
-107	47.10	42.32	14:33	42
-108	45.98	41.25	14:40	42
-109	44.88	40.17	14:47	41
-110	43.76	39.12	14:54	41
-112	41.57	37.05	15:07	39
-113	40.50	36.05	15:14	38
-114	39.45	35.09	15:21	37
-115	38.42	34.15	15:27	36
-117	36.43	32.37	15:41	34
-118	35.50	31.52	15:47	33
-119	34.59	30.71	15:53	32
-120	33.71	29.93	15:59	31
-122	32.06	28.47	16:11	28
-123	31.29	27.78	16:17	27
-124	30.55	27.13	16:23	26
-125	29.84	26.50	16:28	25
-127	28.51	25.34	16:39	22
-128	27.90	24.79	16:45	21
-129	27.32	24.27	16:50	20
-130	26.76	23.78	16:55	19
-132	25.72	22.87	17:05	17
-133	25.24	22.44	17:10	15
-134	24.78	22.04	17:14	14
-135	24.34	21.66	17:20	13
-137	23.54	20.96	17:29	11
-138	23.17	20.64	17:33	10
-139	22.82	20.33	17:38	9
-140	22.50	20.04	17:42	8
-142	21.89	19.51	17:51	6
-143	21.61	19.27	17:55	5
-144	21.35	19.04	17:59	4
-145	21.11	18.83	18:03	3
-147	20.66	18.45	18:11	1

PERCENTAGE OF SUN OBSCURED ALONG MAP TRACK
WEST END =105.7   EAST END =104.6
MAXIMUM OBSCURATION 105.8 % AT LONG = -103
==
SUNSET POSITION : ALTITUDE IS -0.8 DEGREES
NORTH : LONGITUDE -149.6, LATITUDE 20.2
SOUTH : LONGITUDE -148.1, LATITUDE 18.2

+ MARKS THE CITY OF CH'ANG-AN

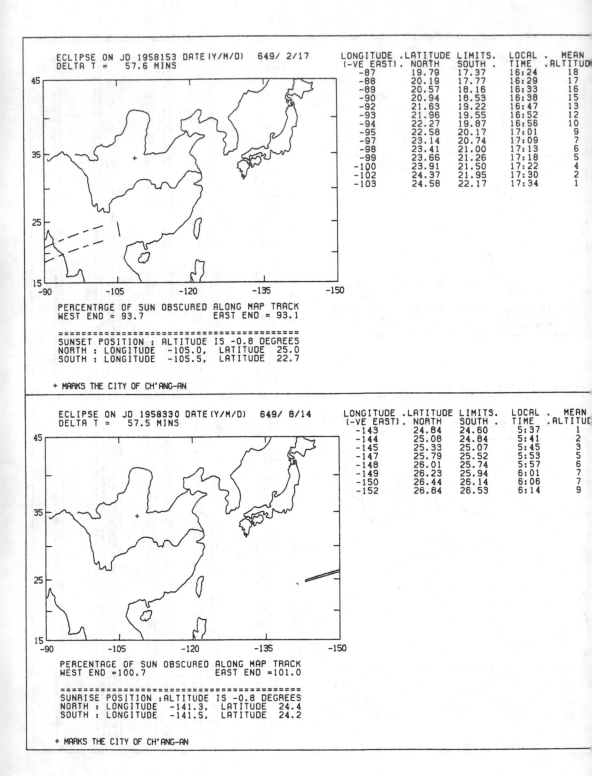

ECLIPSE ON JD 1958153 DATE(Y/M/D)   649/ 2/17
DELTA T =   57.6 MINS

LONGITUDE (-VE EAST)	LATITUDE NORTH	LIMITS. SOUTH	LOCAL TIME	MEAN ALTITUDE
-87	19.79	17.37	16:24	18
-88	20.19	17.77	16:29	17
-89	20.57	18.16	16:33	16
-90	20.94	18.53	16:38	15
-92	21.63	19.22	16:47	13
-93	21.96	19.55	16:52	12
-94	22.27	19.87	16:56	10
-95	22.58	20.17	17:01	9
-97	23.14	20.74	17:09	7
-98	23.41	21.00	17:13	6
-99	23.66	21.26	17:18	5
-100	23.91	21.50	17:22	4
-102	24.37	21.95	17:30	2
-103	24.58	22.17	17:34	1

PERCENTAGE OF SUN OBSCURED ALONG MAP TRACK
WEST END = 93.7   EAST END = 93.1

===
SUNSET POSITION : ALTITUDE IS -0.8 DEGREES
NORTH : LONGITUDE -105.0, LATITUDE 25.0
SOUTH : LONGITUDE -105.5, LATITUDE 22.7

+ MARKS THE CITY OF CH'ANG-AN

ECLIPSE ON JD 1958330 DATE(Y/M/D)   649/ 8/14
DELTA T =   57.5 MINS

LONGITUDE (-VE EAST)	LATITUDE NORTH	LIMITS. SOUTH	LOCAL TIME	MEAN ALTITUDE
-143	24.84	24.60	5:37	1
-144	25.08	24.84	5:41	2
-145	25.33	25.07	5:45	3
-147	25.79	25.52	5:53	5
-148	26.01	25.74	5:57	6
-149	26.23	25.94	6:01	7
-150	26.44	26.14	6:06	7
-152	26.84	26.53	6:14	9

PERCENTAGE OF SUN OBSCURED ALONG MAP TRACK
WEST END =100.7   EAST END =101.0

===
SUNRISE POSITION :ALTITUDE IS -0.8 DEGREES
NORTH : LONGITUDE -141.3, LATITUDE 24.4
SOUTH : LONGITUDE -141.5, LATITUDE 24.2

+ MARKS THE CITY OF CH'ANG-AN

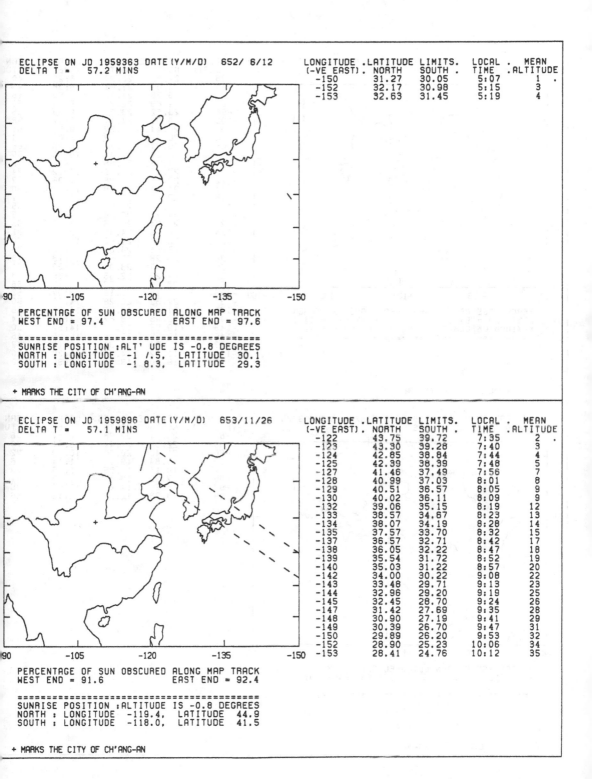

ECLIPSE ON JD 1959363 DATE(Y/M/D)   652/ 6/12
DELTA T =   57.2 MINS

LONGITUDE (-VE EAST)	LATITUDE NORTH	LIMITS. SOUTH	LOCAL TIME	MEAN ALTITUDE
-150	31.27	30.05	5:07	1 .
-152	32.17	30.98	5:15	3
-153	32.63	31.45	5:19	4

PERCENTAGE OF SUN OBSCURED ALONG MAP TRACK
WEST END = 97.4   EAST END = 97.6

SUNRISE POSITION :ALT' UDE IS -0.8 DEGREES
NORTH : LONGITUDE -1 /.5, LATITUDE 30.1
SOUTH : LONGITUDE -1 8.3, LATITUDE 29.3

+ MARKS THE CITY OF CH'ANG-AN

ECLIPSE ON JD 1959896 DATE(Y/M/D)   653/11/26
DELTA T =   57.1 MINS

LONGITUDE (-VE EAST)	LATITUDE NORTH	LIMITS. SOUTH	LOCAL TIME	MEAN ALTITUDE
-122	43.75	39.72	7:35	2 .
-123	43.30	39.28	7:40	3
-124	42.85	38.84	7:44	4
-125	42.39	38.39	7:48	5
-127	41.46	37.49	7:56	7
-128	40.99	37.03	8:01	8
-129	40.51	36.57	8:05	9
-130	40.02	36.11	8:09	9
-132	39.06	35.15	8:19	12
-133	38.57	34.67	8:23	13
-134	38.07	34.19	8:28	14
-135	37.57	33.70	8:32	15
-137	36.57	32.71	8:42	17
-138	36.05	32.22	8:47	18
-139	35.54	31.72	8:52	19
-140	35.03	31.22	8:57	20
-142	34.00	30.22	9:08	22
-143	33.48	29.71	9:13	23
-144	32.96	29.20	9:19	25
-145	32.45	28.70	9:24	26
-147	31.42	27.69	9:35	28
-148	30.90	27.19	9:41	29
-149	30.39	26.70	9:47	31
-150	29.89	26.20	9:53	32
-152	28.90	25.23	10:06	34
-153	28.41	24.76	10:12	35

PERCENTAGE OF SUN OBSCURED ALONG MAP TRACK
WEST END = 91.6   EAST END = 92.4

SUNRISE POSITION :ALTITUDE IS -0.8 DEGREES
NORTH : LONGITUDE -119.4, LATITUDE 44.9
SOUTH : LONGITUDE -118.0, LATITUDE 41.5

+ MARKS THE CITY OF CH'ANG-AN

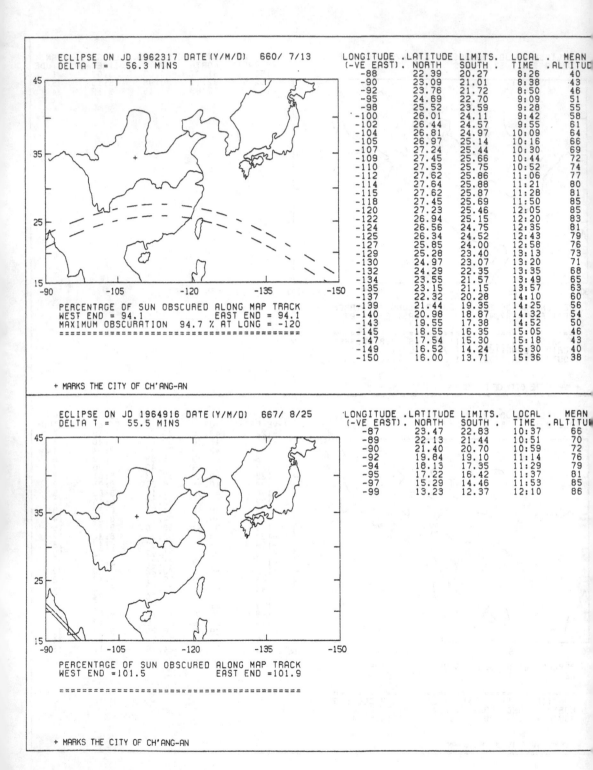

ECLIPSE ON JD 1962317 DATE(Y/M/D) 660/ 7/13
DELTA T =   56.3 MINS

LONGITUDE (-VE EAST)	LATITUDE NORTH	LIMITS. SOUTH	LOCAL TIME	MEAN ALTITUDE
-88	22.39	20.27	8:26	40
-90	23.09	21.01	8:38	43
-92	23.76	21.72	8:50	46
-95	24.69	22.70	9:09	51
-98	25.52	23.59	9:28	55
-100	26.01	24.11	9:42	58
-102	26.44	24.57	9:55	61
-104	26.81	24.97	10:09	64
-105	26.97	25.14	10:16	66
-107	27.24	25.44	10:30	69
-109	27.45	25.66	10:44	72
-110	27.53	25.75	10:52	74
-112	27.62	25.86	11:06	77
-114	27.64	25.88	11:21	80
-115	27.62	25.87	11:28	81
-118	27.45	25.69	11:50	85
-120	27.23	25.46	12:05	85
-122	26.94	25.15	12:20	83
-124	26.56	24.75	12:35	81
-125	26.34	24.52	12:43	79
-127	25.85	24.00	12:58	76
-129	25.28	23.40	13:13	73
-130	24.97	23.07	13:20	71
-132	24.29	22.35	13:35	68
-134	23.55	21.57	13:49	65
-135	23.15	21.15	13:57	63
-137	22.32	20.28	14:10	60
-139	21.44	19.35	14:25	56
-140	20.98	18.87	14:32	54
-143	19.55	17.38	14:52	50
-145	18.55	16.35	15:05	46
-147	17.54	15.30	15:18	43
-149	16.52	14.24	15:30	40
-150	16.00	13.71	15:36	38

PERCENTAGE OF SUN OBSCURED ALONG MAP TRACK
WEST END = 94.1   EAST END = 94.1
MAXIMUM OBSCURATION 94.7 % AT LONG = -120
===

+ MARKS THE CITY OF CH'ANG-AN

ECLIPSE ON JD 1964916 DATE(Y/M/D) 667/ 8/25
DELTA T =   55.5 MINS

LONGITUDE (-VE EAST)	LATITUDE NORTH	LIMITS. SOUTH	LOCAL TIME	MEAN ALTITUDE
-87	23.47	22.83	10:37	66
-89	22.13	21.44	10:51	70
-90	21.40	20.70	10:59	72
-92	19.84	19.10	11:14	76
-94	18.13	17.35	11:29	79
-95	17.22	16.42	11:37	81
-97	15.29	14.46	11:53	85
-99	13.23	12.37	12:10	86

PERCENTAGE OF SUN OBSCURED ALONG MAP TRACK
WEST END =101.5   EAST END =101.9

===

+ MARKS THE CITY OF CH'ANG-AN

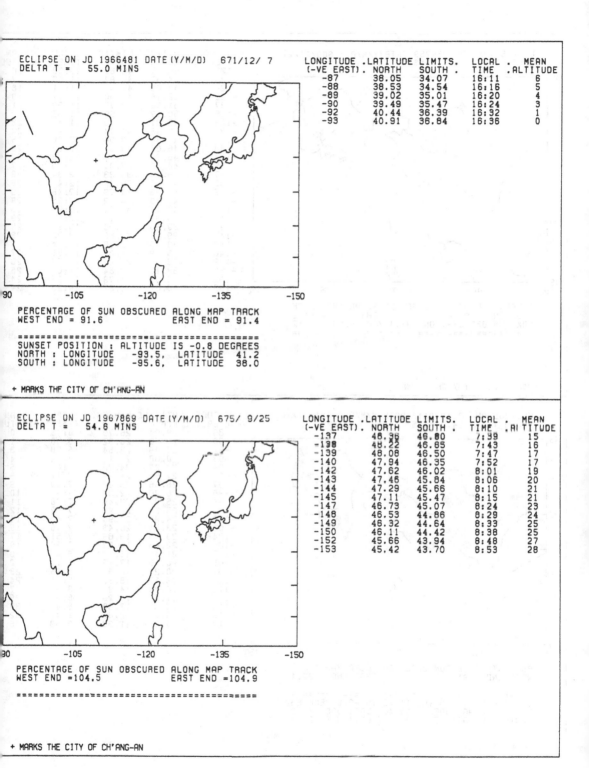

ECLIPSE ON JD 1966481 DATE(Y/M/D) 671/12/ 7
DELTA T =   55.0 MINS

LONGITUDE (-VE EAST)	.LATITUDE LIMITS. NORTH	SOUTH .	LOCAL TIME	. MEAN .ALTITUDE
-87	38.05	34.07	16:11	6
-88	38.53	34.54	16:16	5
-89	39.02	35.01	16:20	4
-90	39.49	35.47	16:24	3
-92	40.44	36.39	16:32	1
-93	40.91	36.84	16:36	0

PERCENTAGE OF SUN OBSCURED ALONG MAP TRACK
WEST END = 91.6   EAST END = 91.4

==
SUNSET POSITION : ALTITUDE IS -0.8 DEGREES
NORTH : LONGITUDE   -93.5, LATITUDE   41.2
SOUTH : LONGITUDE   -95.6, LATITUDE   38.0

+ MARKS THE CITY OF CH'ANG-AN

ECLIPSE ON JD 1967869 DATE(Y/M/D) 675/ 9/25
DELTA T =   54.6 MINS

LONGITUDE (-VE EAST)	.LATITUDE LIMITS. NORTH	SOUTH .	LOCAL TIME	. MEAN .ALTITUDE
-137	48.36	46.80	7:39	15
-138	48.22	46.65	7:43	16
-139	48.08	46.50	7:47	17
-140	47.94	46.35	7:52	17
-142	47.62	46.02	8:01	19
-143	47.46	45.84	8:06	20
-144	47.29	45.66	8:10	21
-145	47.11	45.47	8:15	21
-147	46.73	45.07	8:24	23
-148	46.53	44.86	8:29	24
-149	46.32	44.64	8:33	25
-150	46.11	44.42	8:38	25
-152	45.66	43.94	8:48	27
-153	45.42	43.70	8:53	28

PERCENTAGE OF SUN OBSCURED ALONG MAP TRACK
WEST END =104.5   EAST END =104.9

==

+ MARKS THE CITY OF CH'ANG-AN

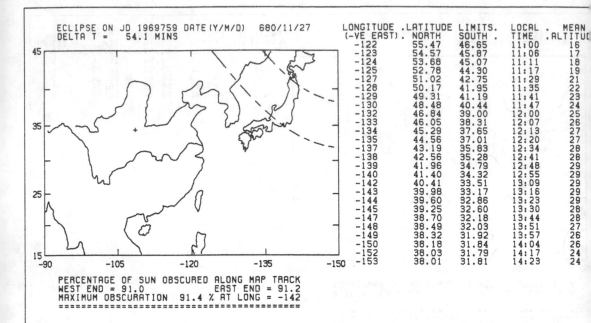

ECLIPSE ON JD 1969759 DATE (Y/M/D) 680/11/27
DELTA T = 54.1 MINS

LONGITUDE (-VE EAST)	LATITUDE NORTH	LIMITS SOUTH	LOCAL TIME	MEAN ALTITU
-122	55.47	46.65	11:00	16
-123	54.57	45.87	11:06	17
-124	53.68	45.07	11:11	18
-125	52.78	44.30	11:17	19
-127	51.02	42.75	11:29	21
-128	50.17	41.95	11:35	22
-129	49.31	41.19	11:41	23
-130	48.48	40.44	11:47	24
-132	46.84	39.00	12:00	25
-133	46.05	38.31	12:07	26
-134	45.29	37.65	12:13	27
-135	44.56	37.01	12:20	27
-137	43.19	35.83	12:34	28
-138	42.56	35.28	12:41	28
-139	41.96	34.79	12:48	29
-140	41.40	34.32	12:55	29
-142	40.41	33.51	13:09	29
-143	39.98	33.17	13:16	29
-144	39.60	32.86	13:23	29
-145	39.25	32.60	13:30	28
-147	38.70	32.18	13:44	28
-148	38.49	32.03	13:51	27
-149	38.32	31.92	13:57	26
-150	38.18	31.84	14:04	26
-152	38.03	31.79	14:17	24
-153	38.01	31.81	14:23	24

PERCENTAGE OF SUN OBSCURED ALONG MAP TRACK
WEST END = 91.0   EAST END = 91.2
MAXIMUM OBSCURATION 91.4 % AT LONG = -142
===

+ MARKS THE CITY OF CH'ANG-AN

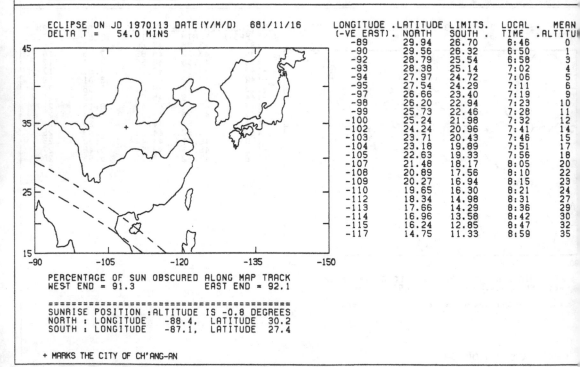

ECLIPSE ON JD 1970113 DATE (Y/M/D) 681/11/16
DELTA T = 54.0 MINS

LONGITUDE (-VE EAST)	LATITUDE NORTH	LIMITS SOUTH	LOCAL TIME	MEAN ALTITU
-89	29.94	26.70	6:46	0
-90	29.56	26.32	6:50	1
-92	28.79	25.54	6:58	3
-93	28.38	25.14	7:02	4
-94	27.97	24.72	7:06	5
-95	27.54	24.29	7:11	6
-97	26.66	23.40	7:19	9
-98	26.20	22.94	7:23	10
-99	25.73	22.46	7:28	11
-100	25.24	21.98	7:32	12
-102	24.24	20.96	7:41	14
-103	23.71	20.43	7:46	15
-104	23.18	19.89	7:51	17
-105	22.63	19.33	7:56	18
-107	21.48	18.17	8:05	20
-108	20.89	17.56	8:10	22
-109	20.27	16.94	8:15	23
-110	19.65	16.30	8:21	24
-112	18.34	14.98	8:31	27
-113	17.66	14.29	8:36	29
-114	16.96	13.58	8:42	30
-115	16.24	12.85	8:47	32
-117	14.75	11.33	8:59	35

PERCENTAGE OF SUN OBSCURED ALONG MAP TRACK
WEST END = 91.3   EAST END = 92.1

===
SUNRISE POSITION :ALTITUDE IS -0.8 DEGREES
NORTH : LONGITUDE -88.4, LATITUDE 30.2
SOUTH : LONGITUDE -87.1, LATITUDE 27.4

+ MARKS THE CITY OF CH'ANG-AN

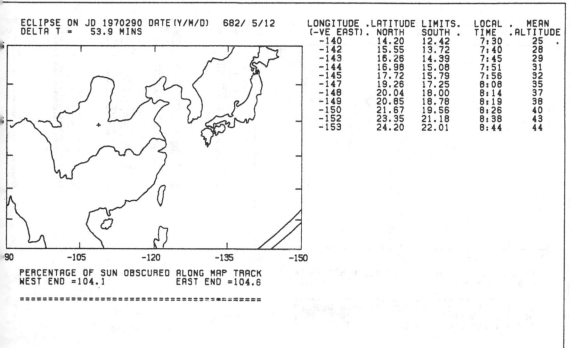

ECLIPSE ON JD 1970290 DATE(Y/M/D)   682/ 5/12
DELTA T =   53.9 MINS

LONGITUDE	.LATITUDE	LIMITS.	LOCAL	. MEAN
(-VE EAST).	NORTH	SOUTH .	TIME	.ALTITUDE
-140	14.20	12.42	7:30	25 .
-142	15.55	13.72	7:40	28
-143	16.26	14.39	7:45	29
-144	16.98	15.08	7:51	31
-145	17.72	15.79	7:56	32
-147	19.26	17.25	8:08	35
-148	20.04	18.00	8:14	37
-149	20.85	18.78	8:19	38
-150	21.67	19.56	8:26	40
-152	23.35	21.18	8:38	43
-153	24.20	22.01	8:44	44

-90   -105   -120   -135   -150

PERCENTAGE OF SUN OBSCURED ALONG MAP TRACK
WEST END =104.1   EAST END =104.6

===■=======

+ MARKS THE CITY OF CH'ANG-AN

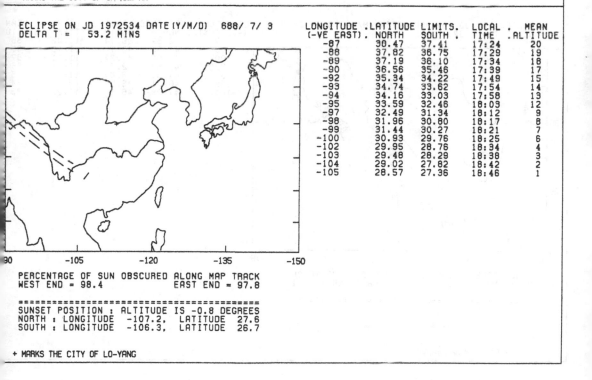

ECLIPSE ON JD 1972534 DATE(Y/M/D)   688/ 7/ 3
DELTA T =   53.2 MINS

LONGITUDE	.LATITUDE	LIMITS.	LOCAL	. MEAN
(-VE EAST).	NORTH	SOUTH .	TIME	.ALTITUDE
-87	30.47	37.41	17:24	20
-88	37.82	36.75	17:29	19
-89	37.19	36.10	17:34	18
-90	36.56	35.46	17:39	17
-92	35.34	34.22	17:49	15
-93	34.74	33.62	17:54	14
-94	34.16	33.03	17:58	13
-95	33.59	32.46	18:03	12
-97	32.49	31.34	18:12	9
-98	31.96	30.80	18:17	8
-99	31.44	30.27	18:21	7
-100	30.93	29.76	18:25	6
-102	29.95	28.76	18:34	4
-103	29.48	28.29	18:38	3
-104	29.02	27.82	18:42	2
-105	28.57	27.36	18:46	1

-30   -105   -120   -135   -150

PERCENTAGE OF SUN OBSCURED ALONG MAP TRACK
WEST END = 98.4   EAST END = 97.8

==
SUNSET POSITION : ALTITUDE IS -0.8 DEGREES
NORTH : LONGITUDE -107.2, LATITUDE 27.6
SOUTH : LONGITUDE -106.3, LATITUDE 26.7

+ MARKS THE CITY OF LO-YANG

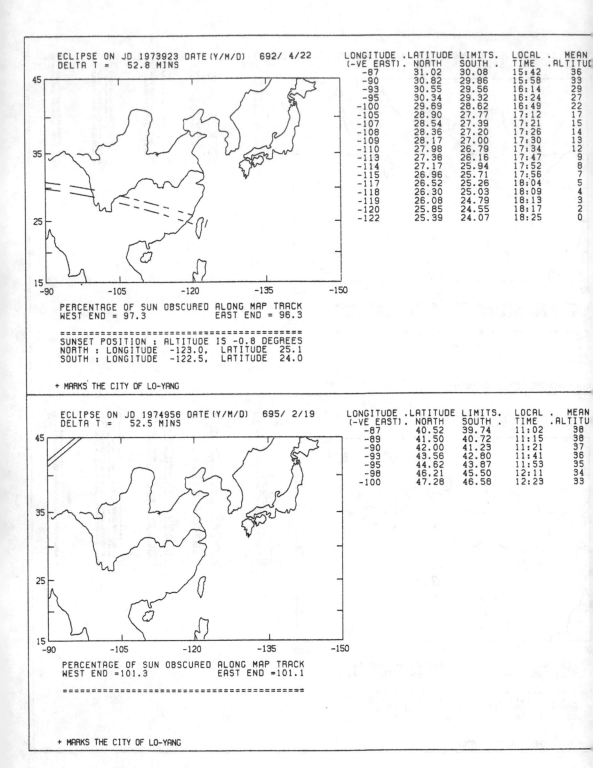

ECLIPSE ON JD 1973923 DATE(Y/M/D)   692/ 4/22
DELTA T =   52.8 MINS

LONGITUDE (-VE EAST)	LATITUDE NORTH	LIMITS. SOUTH	LOCAL TIME	MEAN ALTITUDE
-87	31.02	30.08	15:42	36
-90	30.82	29.86	15:58	33
-93	30.55	29.56	16:14	29
-95	30.34	29.32	16:24	27
-100	29.69	28.62	16:49	22
-105	28.90	27.77	17:12	17
-107	28.54	27.39	17:21	15
-108	28.36	27.20	17:26	14
-109	28.17	27.00	17:30	13
-110	27.98	26.79	17:34	12
-113	27.38	26.16	17:47	9
-114	27.17	25.94	17:52	8
-115	26.96	25.71	17:56	7
-117	26.52	25.26	18:04	5
-118	26.30	25.03	18:09	4
-119	26.08	24.79	18:13	3
-120	25.85	24.55	18:17	2
-122	25.39	24.07	18:25	0

PERCENTAGE OF SUN OBSCURED ALONG MAP TRACK
WEST END = 97.3   EAST END = 96.3

===
SUNSET POSITION : ALTITUDE IS -0.8 DEGREES
NORTH : LONGITUDE -123.0, LATITUDE 25.1
SOUTH : LONGITUDE -122.5, LATITUDE 24.0

+ MARKS THE CITY OF LO-YANG

ECLIPSE ON JD 1974956 DATE(Y/M/D)   695/ 2/19
DELTA T =   52.5 MINS

LONGITUDE (-VE EAST)	LATITUDE NORTH	LIMITS. SOUTH	LOCAL TIME	MEAN ALTITUDE
-87	40.52	39.74	11:02	38
-89	41.50	40.72	11:15	38
-90	42.00	41.23	11:21	37
-93	43.56	42.80	11:41	36
-95	44.62	43.87	11:53	35
-98	46.21	45.50	12:11	34
-100	47.28	46.58	12:23	33

PERCENTAGE OF SUN OBSCURED ALONG MAP TRACK
WEST END =101.3   EAST END =101.1

===

+ MARKS THE CITY OF LO-YANG

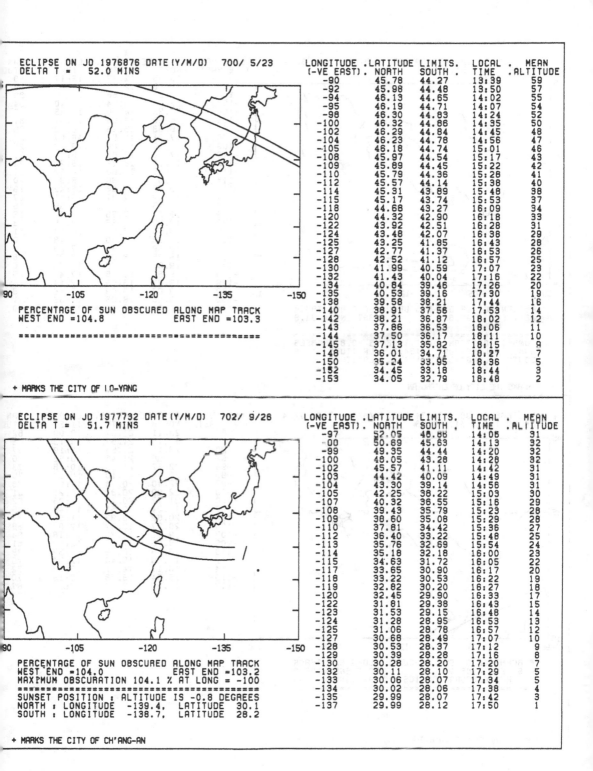

ECLIPSE ON JD 1976876 DATE(Y/M/D) 700/ 5/23
DELTA T =   52.0 MINS

LONGITUDE (-VE EAST)	LATITUDE LIMITS. NORTH	SOUTH	LOCAL TIME	MEAN ALTITUDE
-90	45.78	44.27	13:39	59
-92	45.98	44.48	13:50	57
-94	46.13	44.65	14:02	55
-95	46.19	44.71	14:07	54
-98	46.30	44.83	14:24	52
-100	46.32	44.86	14:35	50
-102	46.29	44.84	14:45	48
-104	46.23	44.78	14:56	47
-105	46.18	44.74	15:01	46
-108	45.97	44.54	15:17	43
-109	45.89	44.45	15:22	42
-110	45.79	44.36	15:28	41
-112	45.57	44.14	15:38	40
-114	45.31	43.89	15:48	38
-115	45.17	43.74	15:53	37
-118	44.68	43.27	16:09	34
-120	44.32	42.90	16:18	33
-122	43.92	42.51	16:28	31
-124	43.48	42.07	16:38	29
-125	43.25	41.85	16:43	28
-127	42.77	41.37	16:53	26
-128	42.52	41.12	16:57	25
-130	41.99	40.59	17:07	23
-132	41.43	40.04	17:16	22
-134	40.84	39.46	17:26	20
-135	40.53	39.16	17:30	19
-138	39.58	38.21	17:44	16
-140	38.91	37.56	17:53	14
-142	38.21	36.87	18:02	12
-143	37.86	36.53	18:06	11
-144	37.50	36.17	18:11	10
-145	37.13	35.82	18:15	9
-148	36.01	34.71	18:27	7
-150	35.24	33.95	18:36	5
-152	34.45	33.18	18:44	3
-153	34.05	32.79	18:48	2

PERCENTAGE OF SUN OBSCURED ALONG MAP TRACK
WEST END =104.8   EAST END =103.3

===

+ MARKS THE CITY OF LO-YANG

ECLIPSE ON JD 1977732 DATE(Y/M/D) 702/ 9/26
DELTA T =   51.7 MINS

LONGITUDE (-VE EAST)	LATITUDE LIMITS. NORTH	SOUTH	LOCAL TIME	MEAN ALTITUDE
-97	52.05	46.86	14:06	31
-00	50.69	45.63	14:13	32
-99	49.35	44.44	14:20	32
-100	48.05	43.28	14:28	32
-102	45.57	41.11	14:42	31
-103	44.42	40.09	14:49	31
-104	43.30	39.14	14:56	31
-105	42.25	38.22	15:03	30
-107	40.32	36.55	15:16	29
-108	39.43	35.79	15:23	28
-109	38.60	35.08	15:29	28
-110	37.81	34.42	15:36	27
-112	36.40	33.22	15:48	25
-113	35.76	32.69	15:54	24
-114	35.18	32.18	16:00	23
-115	34.63	31.72	16:05	22
-117	33.65	30.90	16:17	20
-118	33.22	30.53	16:22	19
-119	32.82	30.20	16:27	18
-120	32.45	29.90	16:33	17
-122	31.81	29.38	16:43	15
-123	31.53	29.15	16:48	14
-124	31.28	28.95	16:53	13
-125	31.06	28.78	16:57	12
-127	30.68	28.49	17:07	10
-128	30.53	28.37	17:12	9
-129	30.39	28.28	17:16	8
-130	30.28	28.20	17:20	7
-132	30.11	28.10	17:29	5
-133	30.06	28.07	17:34	5
-134	30.02	28.06	17:38	4
-135	29.99	28.07	17:42	3
-137	29.99	28.12	17:50	1

PERCENTAGE OF SUN OBSCURED ALONG MAP TRACK
WEST END =104.0   EAST END =103.2
MAXIMUM OBSCURATION 104.1 % AT LONG = -100
===
SUNSET POSITION : ALTITUDE IS -0.8 DEGREES
NORTH : LONGITUDE -139.4, LATITUDE   30.1
SOUTH : LONGITUDE -138.7, LATITUDE   28.2

+ MARKS THE CITY OF CH'ANG-AN

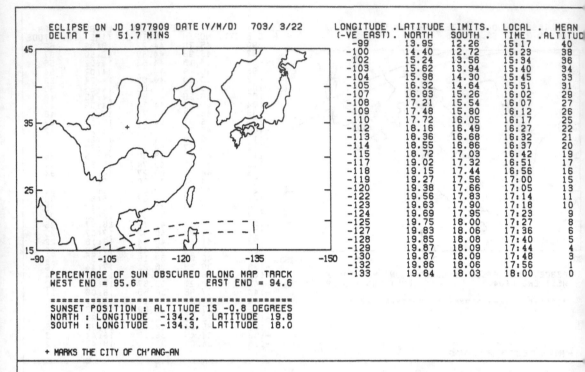

ECLIPSE ON JD 1977909 DATE (Y/M/D)   703/ 3/22
DELTA T =   51.7 MINS

LONGITUDE (-VE EAST)	LATITUDE NORTH	LIMITS. SOUTH	LOCAL TIME	MEAN ALTITUDE
-99	13.95	12.26	15:17	40
-100	14.40	12.72	15:23	38
-102	15.24	13.56	15:34	36
-103	15.62	13.94	15:40	34
-104	15.98	14.30	15:45	33
-105	16.32	14.64	15:51	31
-107	16.93	15.26	16:02	29
-108	17.21	15.54	16:07	27
-109	17.48	15.80	16:12	26
-110	17.72	16.05	16:17	25
-112	18.16	16.49	16:27	22
-113	18.36	16.68	16:32	21
-114	18.55	16.86	16:37	20
-115	18.72	17.03	16:42	19
-117	19.02	17.32	16:51	17
-118	19.15	17.44	16:56	16
-119	19.27	17.56	17:00	15
-120	19.38	17.66	17:05	13
-122	19.56	17.83	17:14	11
-123	19.63	17.90	17:18	10
-124	19.69	17.95	17:23	9
-125	19.75	18.00	17:27	8
-127	19.83	18.06	17:36	6
-128	19.85	18.08	17:40	5
-129	19.87	18.09	17:44	4
-130	19.87	18.09	17:48	3
-132	19.86	18.06	17:56	1
-133	19.84	18.03	18:00	0

PERCENTAGE OF SUN OBSCURED ALONG MAP TRACK
WEST END = 95.6   EAST END = 94.6

===
SUNSET POSITION : ALTITUDE IS -0.8 DEGREES
NORTH : LONGITUDE -134.2, LATITUDE 19.8
SOUTH : LONGITUDE -134.3, LATITUDE 18.0

+ MARKS THE CITY OF CH'ANG-AN

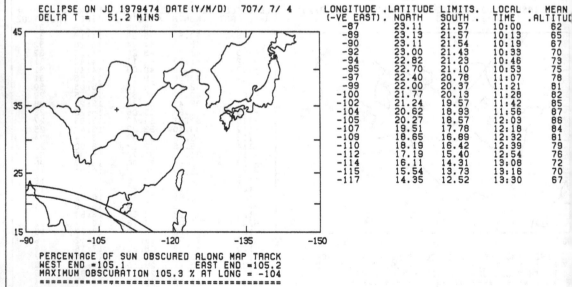

ECLIPSE ON JD 1979474 DATE (Y/M/D)   707/ 7/ 4
DELTA T =   51.2 MINS

LONGITUDE (-VE EAST)	LATITUDE NORTH	LIMITS. SOUTH	LOCAL TIME	MEAN ALTITUDE
-87	23.11	21.57	10:00	62
-89	23.13	21.57	10:13	65
-90	23.11	21.54	10:19	67
-92	23.00	21.43	10:33	70
-94	22.82	21.23	10:46	73
-95	22.70	21.10	10:53	75
-97	22.40	20.78	11:07	78
-99	22.00	20.37	11:21	81
-100	21.77	20.13	11:28	82
-102	21.24	19.57	11:42	85
-104	20.62	18.93	11:56	87
-105	20.27	18.57	12:03	86
-107	19.51	17.78	12:18	84
-109	18.65	16.89	12:32	81
-110	18.19	16.42	12:39	79
-112	17.19	15.40	12:54	76
-114	16.11	14.31	13:08	72
-115	15.54	13.73	13:16	70
-117	14.35	12.52	13:30	67

PERCENTAGE OF SUN OBSCURED ALONG MAP TRACK
WEST END =105.1   EAST END =105.2
MAXIMUM OBSCURATION 105.3 % AT LONG = -104
===

+ MARKS THE CITY OF CH'ANG-AN

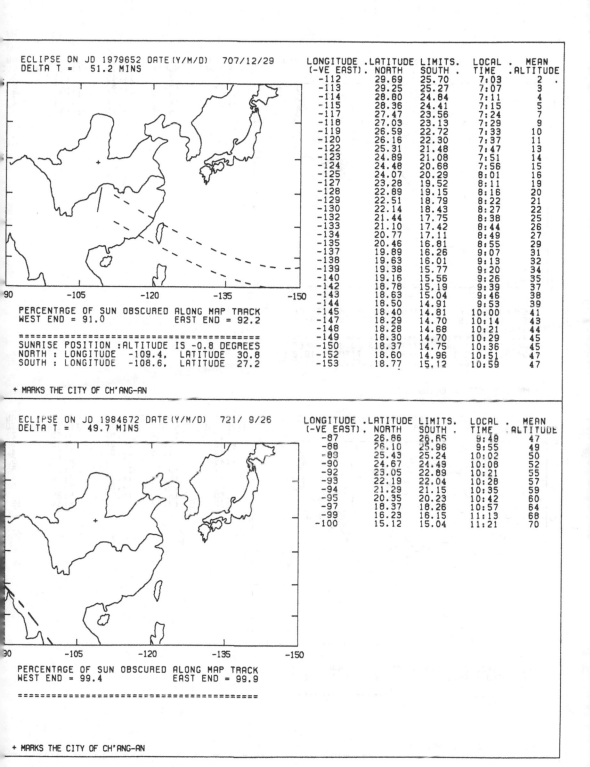

ECLIPSE ON JD 1979652 DATE(Y/M/D) 707/12/29
DELTA T =   51.2 MINS

LONGITUDE (-VE EAST)	LATITUDE LIMITS. NORTH	SOUTH	LOCAL TIME	MEAN ALTITUDE
-112	29.69	25.70	7:03	2
-113	29.25	25.27	7:07	3
-114	28.80	24.84	7:11	4
-115	28.36	24.41	7:15	5
-117	27.47	23.56	7:24	7
-118	27.03	23.13	7:29	9
-119	26.59	22.72	7:33	10
-120	26.16	22.30	7:37	11
-122	25.31	21.48	7:47	13
-123	24.89	21.08	7:51	14
-124	24.48	20.68	7:56	15
-125	24.07	20.29	8:01	16
-127	23.28	19.52	8:11	19
-128	22.89	19.15	8:16	20
-129	22.51	18.79	8:22	21
-130	22.14	18.43	8:27	22
-132	21.44	17.75	8:38	25
-133	21.10	17.42	8:44	26
-134	20.77	17.11	8:49	27
-135	20.46	16.81	8:55	29
-137	19.89	16.26	9:07	31
-138	19.63	16.01	9:13	32
-139	19.38	15.77	9:20	34
-140	19.16	15.56	9:26	35
-142	18.78	15.19	9:39	37
-143	18.63	15.04	9:46	38
-144	18.50	14.91	9:53	39
-145	18.40	14.81	10:00	41
-147	18.29	14.70	10:14	43
-148	18.28	14.68	10:21	44
-149	18.30	14.70	10:29	45
-150	18.37	14.75	10:36	45
-152	18.60	14.96	10:51	47
-153	18.77	15.12	10:59	47

PERCENTAGE OF SUN OBSCURED ALONG MAP TRACK
WEST END = 91.0   EAST END = 92.2

===

SUNRISE POSITION :ALTITUDE IS -0.8 DEGREES
NORTH : LONGITUDE -109.4, LATITUDE 30.8
SOUTH : LONGITUDE -108.6, LATITUDE 27.2

+ MARKS THE CITY OF CH'ANG-AN

ECLIPSE ON JD 1984672 DATE(Y/M/D) 721/ 9/26
DELTA T =   49.7 MINS

LONGITUDE (-VE EAST)	LATITUDE NORTH	LIMITS. SOUTH	LOCAL TIME	MEAN ALTITUDE
-87	26.86	26.65	9:49	47
-88	26.10	25.96	9:55	49
-89	25.43	25.24	10:02	50
-90	24.67	24.49	10:08	52
-92	23.05	22.89	10:21	55
-93	22.19	22.04	10:28	57
-94	21.29	21.15	10:35	59
-95	20.35	20.23	10:42	60
-97	18.37	18.26	10:57	64
-99	16.23	16.15	11:13	68
-100	15.12	15.04	11:21	70

PERCENTAGE OF SUN OBSCURED ALONG MAP TRACK
WEST END = 99.4   EAST END = 99.9

===

+ MARKS THE CITY OF CH'ANG-AN

285

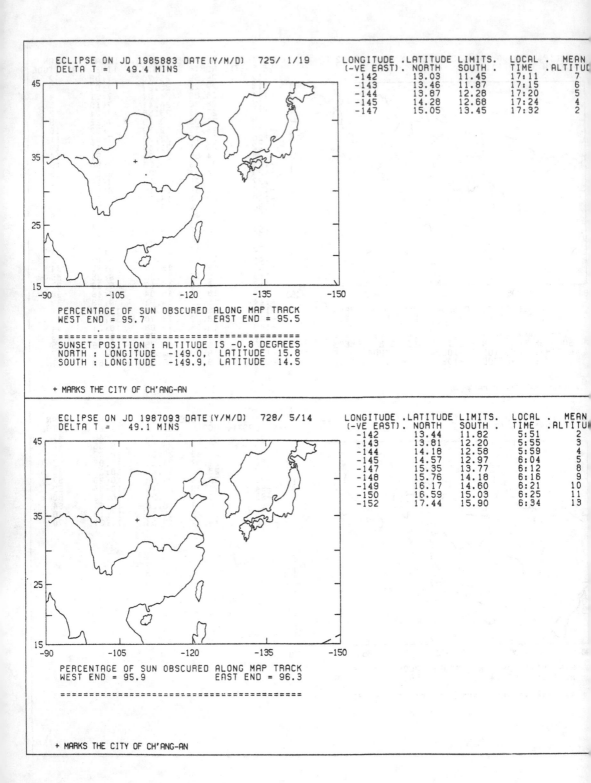

ECLIPSE ON JD 1985883 DATE(Y/M/D)   725/ 1/19
DELTA T =   49.4 MINS

LONGITUDE (-VE EAST).	LATITUDE NORTH	LIMITS. SOUTH .	LOCAL TIME	MEAN ALTITU...
-142	13.03	11.45	17:11	7
-143	13.46	11.87	17:15	6
-144	13.87	12.28	17:20	5
-145	14.28	12.68	17:24	4
-147	15.05	13.45	17:32	2

PERCENTAGE OF SUN OBSCURED ALONG MAP TRACK
WEST END = 95.7   EAST END = 95.5

==
SUNSET POSITION : ALTITUDE IS -0.8 DEGREES
NORTH : LONGITUDE -149.0, LATITUDE 15.8
SOUTH : LONGITUDE -149.9, LATITUDE 14.5

+ MARKS THE CITY OF CH'ANG-AN

ECLIPSE ON JD 1987093 DATE(Y/M/D)   728/ 5/14
DELTA T =   49.1 MINS

LONGITUDE (-VE EAST).	LATITUDE NORTH	LIMITS. SOUTH .	LOCAL TIME	MEAN ALTITU...
-142	13.44	11.82	5:51	2
-143	13.81	12.20	5:55	3
-144	14.18	12.58	5:59	4
-145	14.57	12.97	6:04	5
-147	15.35	13.77	6:12	8
-148	15.76	14.18	6:16	9
-149	16.17	14.60	6:21	10
-150	16.59	15.03	6:25	11
-152	17.44	15.90	6:34	13

PERCENTAGE OF SUN OBSCURED ALONG MAP TRACK
WEST END = 95.9   EAST END = 96.3

==

+ MARKS THE CITY OF CH'ANG-AN

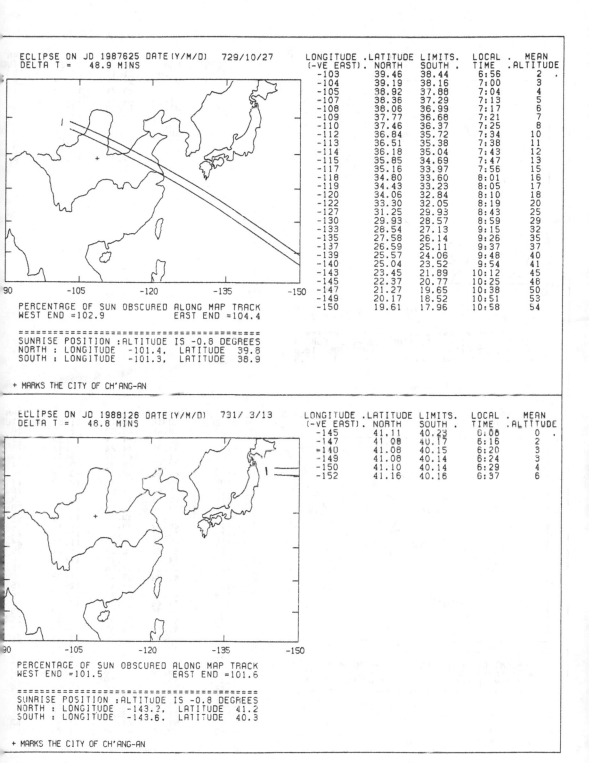

ECLIPSE ON JD 1987625 DATE(Y/M/D) 729/10/27
DELTA T = 48.9 MINS

LONGITUDE (-VE EAST)	LATITUDE NORTH	LIMITS. SOUTH	LOCAL TIME	MEAN ALTITUDE
-103	39.46	38.44	6:56	2
-104	39.19	38.16	7:00	3
-105	38.92	37.88	7:04	4
-107	38.36	37.29	7:13	5
-108	38.06	36.99	7:17	6
-109	37.77	36.68	7:21	7
-110	37.46	36.37	7:25	8
-112	36.84	35.72	7:34	10
-113	36.51	35.38	7:38	11
-114	36.18	35.04	7:43	12
-115	35.85	34.69	7:47	13
-117	35.16	33.97	7:56	15
-118	34.80	33.60	8:01	16
-119	34.43	33.23	8:05	17
-120	34.06	32.84	8:10	18
-122	33.30	32.05	8:19	20
-127	31.25	29.93	8:43	25
-130	29.93	28.57	8:59	29
-133	28.54	27.13	9:15	32
-135	27.58	26.14	9:26	35
-137	26.59	25.11	9:37	37
-139	25.57	24.06	9:48	40
-140	25.04	23.52	9:54	41
-143	23.45	21.89	10:12	45
-145	22.37	20.77	10:25	48
-147	21.27	19.65	10:38	50
-149	20.17	18.52	10:51	53
-150	19.61	17.96	10:58	54

PERCENTAGE OF SUN OBSCURED ALONG MAP TRACK
WEST END =102.9   EAST END =104.4

===
SUNRISE POSITION :ALTITUDE IS -0.8 DEGREES
NORTH : LONGITUDE -101.4, LATITUDE 39.8
SOUTH : LONGITUDE -101.3, LATITUDE 38.9

+ MARKS THE CITY OF CH'ANG-AN

ECLIPSE ON JD 1988126 DATE(Y/M/D) 731/ 3/13
DELTA T = 48.8 MINS

LONGITUDE (-VE EAST)	LATITUDE NORTH	LIMITS. SOUTH	LOCAL TIME	MEAN ALTITUDE
-145	41.11	40.23	6:08	0
-147	41.08	40.17	6:16	2
-148	41.08	40.15	6:20	3
-149	41.08	40.14	6:24	3
-150	41.10	40.14	6:29	4
-152	41.16	40.16	6:37	6

PERCENTAGE OF SUN OBSCURED ALONG MAP TRACK
WEST END =101.5   EAST END =101.6

===
SUNRISE POSITION :ALTITUDE IS -0.8 DEGREES
NORTH : LONGITUDE -143.2, LATITUDE 41.2
SOUTH : LONGITUDE -143.6, LATITUDE 40.3

+ MARKS THE CITY OF CH'ANG-AN

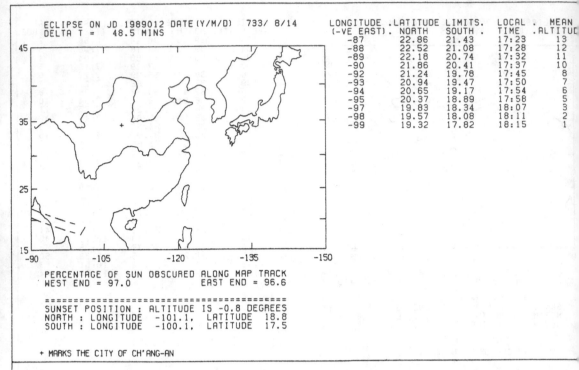

ECLIPSE ON JD 1989012 DATE(Y/M/D) 733/ 8/14
DELTA T = 48.5 MINS

LONGITUDE	.LATITUDE	LIMITS.	LOCAL .	MEAN
(-VE EAST).	NORTH	SOUTH .	TIME	.ALTITUD
-87	22.86	21.43	17:23	13
-88	22.52	21.08	17:28	12
-89	22.18	20.74	17:32	11
-90	21.86	20.41	17:37	10
-92	21.24	19.78	17:45	8
-93	20.94	19.47	17:50	7
-94	20.65	19.17	17:54	6
-95	20.37	18.89	17:58	5
-97	19.83	18.34	18:07	3
-98	19.57	18.08	18:11	2
-99	19.32	17.82	18:15	1

PERCENTAGE OF SUN OBSCURED ALONG MAP TRACK
WEST END = 97.0   EAST END = 96.6

===
SUNSET POSITION : ALTITUDE IS -0.8 DEGREES
NORTH : LONGITUDE -101.1, LATITUDE   18.8
SOUTH : LONGITUDE -100.1, LATITUDE   17.5

+ MARKS THE CITY OF CH'ANG-AN

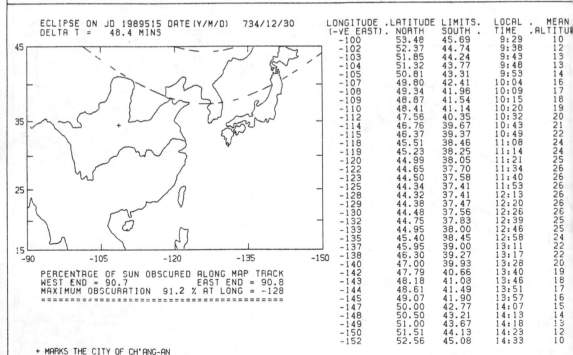

ECLIPSE ON JD 1989515 DATE(Y/M/D) 734/12/30
DELTA T = 48.4 MINS

LONGITUDE	.LATITUDE	LIMITS.	LOCAL	MEAN
(-VE EAST).	NORTH	SOUTH .	TIME	.ALTITU
-100	53.48	45.69	9:29	10
-102	52.37	44.74	9:38	12
-103	51.85	44.24	9:43	13
-104	51.32	43.77	9:48	13
-105	50.81	43.31	9:53	14
-107	49.80	42.41	10:04	16
-108	49.34	41.96	10:09	17
-109	48.87	41.54	10:15	18
-110	48.41	41.14	10:20	19
-112	47.56	40.35	10:32	20
-114	46.76	39.67	10:43	21
-115	46.37	39.37	10:49	22
-118	45.51	38.46	11:08	24
-119	45.23	38.25	11:14	24
-120	44.99	38.05	11:21	25
-122	44.65	37.70	11:34	26
-123	44.50	37.58	11:40	26
-125	44.34	37.41	11:53	26
-128	44.32	37.41	12:13	26
-129	44.38	37.47	12:20	26
-130	44.48	37.56	12:26	26
-132	44.75	37.83	12:39	25
-133	44.95	38.00	12:46	25
-135	45.40	38.45	12:58	24
-137	45.95	39.00	13:11	22
-138	46.30	39.27	13:17	22
-140	47.00	39.93	13:28	20
-142	47.79	40.66	13:40	19
-143	48.18	41.03	13:46	18
-144	48.61	41.49	13:51	17
-145	49.07	41.90	13:57	16
-147	50.00	42.77	14:07	15
-148	50.50	43.21	14:13	14
-149	51.00	43.67	14:18	13
-150	51.51	44.13	14:23	12
-152	52.56	45.08	14:33	10

PERCENTAGE OF SUN OBSCURED ALONG MAP TRACK
WEST END = 90.7   EAST END = 90.8
MAXIMUM OBSCURATION 91.2 % AT LONG = -128
===

+ MARKS THE CITY OF CH'ANG-AN

ECLIPSE ON JD 1992290 DATE(Y/M/D) 742/ 8/ 5
DELTA T = 47.6 MINS

LONGITUDE (-VE EAST)	LATITUDE NORTH	LIMITS. SOUTH	LOCAL TIME	MEAN ALTITUDE
-135	47.75	46.23	17:16	19
-137	46.03	44.56	17:27	17
-138	45.24	43.79	17:33	16
-139	44.50	43.05	17:38	15
-140	43.79	42.36	17:43	14
-142	42.49	41.07	17:53	12
-143	41.88	40.48	17:58	11
-144	41.32	39.92	18:03	10
-145	40.78	39.38	18:07	9
-147	39.79	38.40	18:16	7
-148	39.33	37.95	18:21	6
-149	38.90	37.52	18:25	5
-150	38.49	37.11	18:30	4
-152	37.74	36.36	18:38	2
-153	37.40	36.02	18:42	1

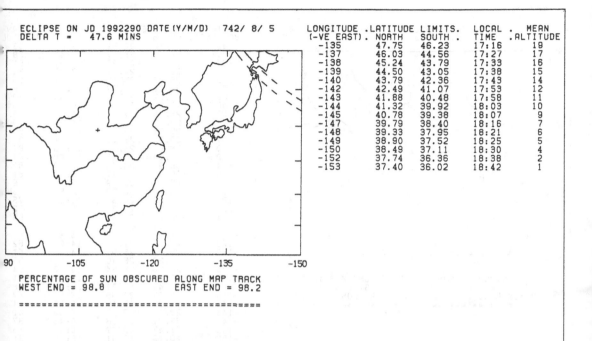

PERCENTAGE OF SUN OBSCURED ALONG MAP TRACK
WEST END = 98.8   EAST END = 98.2

==

+ MARKS THE CITY OF CH'ANG-AN

ECLIPSE ON JD 1993679 DATE(Y/M/D) 746/ 5/25
DELTA T = 47.3 MINS

LONGITUDE (-VE EAST)	LATITUDE NORTH	LIMITS. SOUTH	LOCAL TIME	MEAN ALTITUDE
-87	44.48	43.32	10:03	57
-88	44.78	43.64	10:09	58
-89	45.07	43.94	10:10	59
-90	45.35	44.22	10:22	59
-92	45.86	44.76	10:34	60
-94	46.33	45.25	10:46	62
-95	46.55	45.48	10:53	62
-97	46.94	45.89	11:05	63
-99	47.29	46.25	11:17	63
-100	47.45	46.41	11:23	64
-102	47.73	46.70	11:36	64
-104	47.96	46.94	11:48	64
-123	47.99	46.94	13:42	57
-124	47.88	46.83	13:47	56
-125	47.77	46.71	13:53	56
-127	47.51	46.43	14:05	54
-128	47.37	46.28	14:11	53
-129	47.22	46.12	14:16	53
-130	47.05	45.94	14:22	52
-132	46.70	45.56	14:34	50
-133	46.50	45.36	14:40	49
-134	46.30	45.14	14:45	49
-135	46.09	44.92	14:51	48
-137	45.64	44.44	15:02	46
-138	45.39	44.19	15:08	45
-139	45.15	43.93	15:14	44
-140	44.89	43.65	15:19	43
-142	44.34	43.08	15:30	41
-143	44.06	42.78	15:36	40
-144	43.77	42.47	15:41	39
-145	43.47	42.16	15:47	39
-147	42.84	41.50	15:58	37
-148	42.52	41.16	16:03	36
-149	42.19	40.82	16:08	35
-150	41.85	40.46	16:14	34

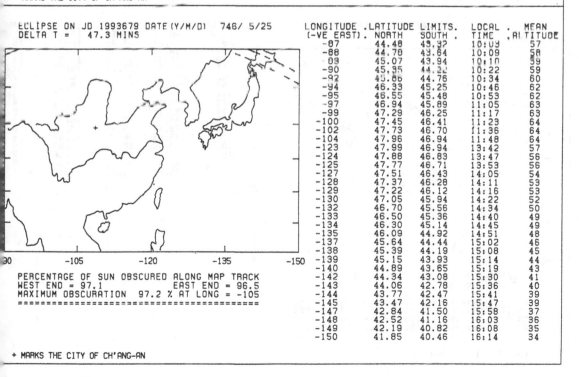

PERCENTAGE OF SUN OBSCURED ALONG MAP TRACK
WEST END = 97.1   EAST END = 96.5
MAXIMUM OBSCURATION 97.2 % AT LONG = -105

==

+ MARKS THE CITY OF CH'ANG-AN

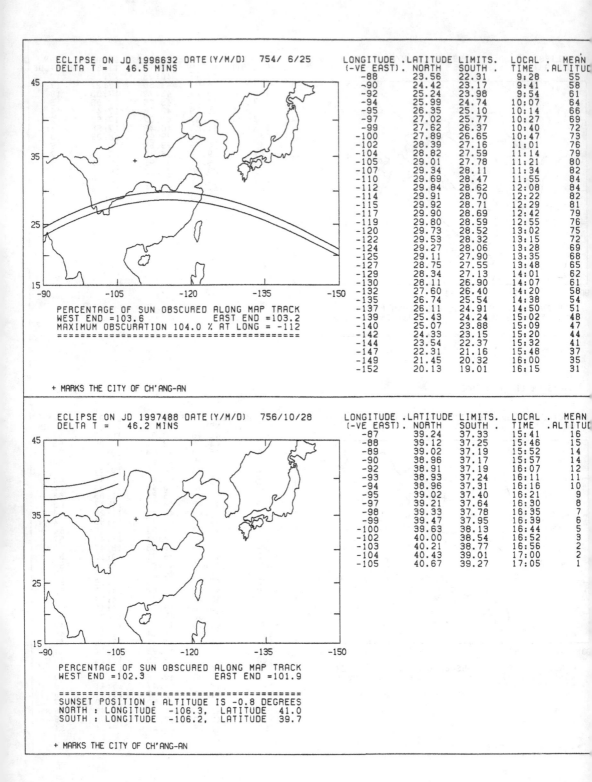

ECLIPSE ON JD 1996632 DATE (Y/M/D)   754/ 6/25
DELTA T =   46.5 MINS

LONGITUDE (-VE EAST)	LATITUDE LIMITS. NORTH	LATITUDE LIMITS. SOUTH	LOCAL TIME	MEAN ALTITUDE
-88	23.56	22.31	9:28	55
-90	24.42	23.17	9:41	58
-92	25.24	23.98	9:54	61
-94	25.99	24.74	10:07	64
-95	26.35	25.10	10:14	66
-97	27.02	25.77	10:27	69
-99	27.62	26.37	10:40	72
-100	27.89	26.65	10:47	73
-102	28.39	27.16	11:01	76
-104	28.82	27.59	11:14	79
-105	29.01	27.78	11:21	80
-107	29.34	28.11	11:34	82
-110	29.69	28.47	11:55	84
-112	29.84	28.62	12:08	84
-114	29.91	28.70	12:22	82
-115	29.92	28.71	12:29	81
-117	29.90	28.69	12:42	79
-119	29.80	28.59	12:55	76
-120	29.73	28.52	13:02	75
-122	29.53	28.32	13:15	72
-124	29.27	28.06	13:28	69
-125	29.11	27.90	13:35	68
-127	28.75	27.55	13:48	65
-129	28.34	27.13	14:01	62
-130	28.11	26.90	14:07	61
-132	27.60	26.40	14:20	58
-135	26.74	25.54	14:38	54
-137	26.11	24.91	14:50	51
-139	25.43	24.24	15:02	48
-140	25.07	23.88	15:09	47
-142	24.33	23.15	15:20	44
-144	23.54	22.37	15:32	41
-147	22.31	21.16	15:48	37
-149	21.45	20.32	16:00	35
-152	20.13	19.01	16:15	31

PERCENTAGE OF SUN OBSCURED ALONG MAP TRACK
WEST END =103.6   EAST END =103.2
MAXIMUM OBSCURATION 104.0 % AT LONG = -112
==

+ MARKS THE CITY OF CH'ANG-AN

ECLIPSE ON JD 1997488 DATE (Y/M/D)   756/10/28
DELTA T =   46.2 MINS

LONGITUDE (-VE EAST)	LATITUDE LIMITS. NORTH	LATITUDE LIMITS. SOUTH	LOCAL TIME	MEAN ALTITUDE
-87	39.24	37.33	15:41	16
-88	39.12	37.25	15:46	15
-89	39.02	37.19	15:52	14
-90	38.96	37.17	15:57	14
-92	38.91	37.19	16:07	12
-93	38.93	37.24	16:11	11
-94	38.96	37.31	16:16	10
-95	39.02	37.40	16:21	9
-97	39.21	37.64	16:30	8
-98	39.33	37.78	16:35	7
-99	39.47	37.95	16:39	6
-100	39.63	38.13	16:44	5
-102	40.00	38.54	16:52	3
-103	40.21	38.77	16:56	2
-104	40.43	39.01	17:00	2
-105	40.67	39.27	17:05	1

PERCENTAGE OF SUN OBSCURED ALONG MAP TRACK
WEST END =102.3   EAST END =101.9

==
SUNSET POSITION : ALTITUDE IS -0.8 DEGREES
NORTH : LONGITUDE -106.3, LATITUDE 41.0
SOUTH : LONGITUDE -106.2, LATITUDE 39.7

+ MARKS THE CITY OF CH'ANG-AN

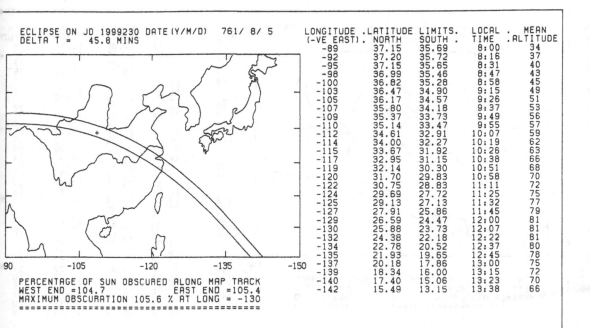

ECLIPSE ON JD 1999230 DATE(Y/M/D)   761/ 8/ 5
DELTA T =   45.8 MINS

LONGITUDE	.LATITUDE	LIMITS.	LOCAL	MEAN
(-VE EAST).	NORTH	SOUTH .	TIME	.ALTITUDE
-89	37.15	35.69	8:00	34
-92	37.20	35.72	8:16	37
-95	37.15	35.65	8:31	40
-98	36.99	35.46	8:47	43
-100	36.82	35.28	8:58	45
-103	36.47	34.90	9:15	49
-105	36.17	34.57	9:26	51
-107	35.80	34.18	9:37	53
-109	35.37	33.73	9:49	56
-110	35.14	33.47	9:55	57
-112	34.61	32.91	10:07	59
-114	34.00	32.27	10:19	62
-115	33.67	31.92	10:26	63
-117	32.95	31.15	10:38	66
-119	32.14	30.30	10:51	68
-120	31.70	29.83	10:58	70
-122	30.75	28.83	11:11	72
-124	29.69	27.72	11:25	75
-125	29.13	27.13	11:32	77
-127	27.91	25.86	11:45	79
-129	26.59	24.47	12:00	81
-130	25.88	23.73	12:07	81
-132	24.38	22.18	12:22	81
-134	22.78	20.52	12:37	80
-135	21.93	19.65	12:45	78
-137	20.18	17.86	13:00	75
-139	18.34	16.00	13:15	72
-140	17.40	15.06	13:23	70
-142	15.49	13.15	13:38	66

PERCENTAGE OF SUN OBSCURED ALONG MAP TRACK
WEST END =104.7   EAST END =105.4
MAXIMUM OBSCURATION 105.6 % AT LONG = -130
===

+ MARKS THE CITY OF CH'ANG-AN

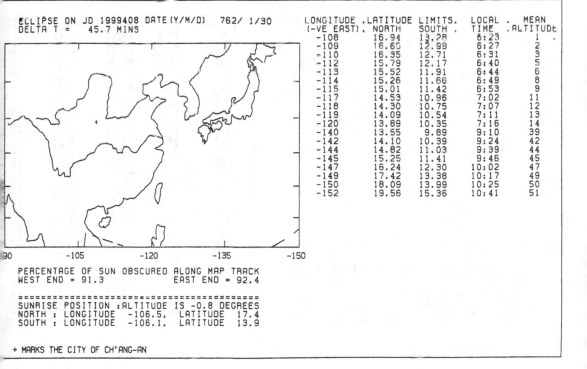

ECLIPSE ON JD 1999408 DATE(Y/M/D)   762/ 1/30
DELTA T =   45.7 MINS

LONGITUDE	.LATITUDE	LIMITS.	LOCAL	MEAN
(-VE EAST).	NORTH	SOUTH .	TIME	.ALTITUDE
-108	16.94	13.28	6:23	1
-109	16.65	12.99	6:27	2
-110	16.35	12.71	6:31	3
-112	15.79	12.17	6:40	5
-113	15.52	11.91	6:44	6
-114	15.26	11.66	6:49	8
-115	15.01	11.42	6:53	9
-117	14.53	10.96	7:02	11
-118	14.30	10.75	7:07	12
-119	14.09	10.54	7:11	13
-120	13.89	10.35	7:16	14
-140	13.55	9.89	9:10	39
-142	14.10	10.39	9:24	42
-144	14.82	11.03	9:39	44
-145	15.25	11.41	9:46	45
-147	16.24	12.30	10:02	47
-149	17.42	13.38	10:17	49
-150	18.09	13.99	10:25	50
-152	19.56	15.36	10:41	51

PERCENTAGE OF SUN OBSCURED ALONG MAP TRACK
WEST END = 91.3   EAST END = 92.4

===
SUNRISE POSITION :ALTITUDE IS -0.8 DEGREES
NORTH : LONGITUDE -106.5, LATITUDE 17.4
SOUTH : LONGITUDE -106.1, LATITUDE 13.9

+ MARKS THE CITY OF CH'ANG-AN

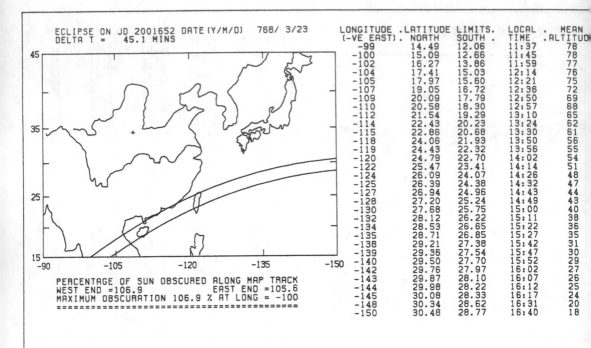

ECLIPSE ON JD 2001652 DATE(Y/M/D) 768/ 3/23
DELTA T =   45.1 MINS

PERCENTAGE OF SUN OBSCURED ALONG MAP TRACK
WEST END =106.9   EAST END =105.6
MAXIMUM OBSCURATION 106.9 % AT LONG = -100
===

+ MARKS THE CITY OF CH'ANG-AN

LONGITUDE (-VE EAST)	LATITUDE NORTH	LIMITS. SOUTH	LOCAL TIME	MEAN ALTITUD
-99	14.49	12.06	11:37	78
-100	15.09	12.66	11:45	78
-102	16.27	13.86	11:59	77
-104	17.41	15.03	12:14	76
-105	17.97	15.60	12:21	75
-107	19.05	16.72	12:36	72
-109	20.09	17.79	12:50	69
-110	20.58	18.30	12:57	68
-112	21.54	19.29	13:10	65
-114	22.43	20.23	13:24	62
-115	22.86	20.68	13:30	61
-118	24.06	21.93	13:50	56
-119	24.43	22.32	13:56	55
-120	24.79	22.70	14:02	54
-122	25.47	23.41	14:14	51
-124	26.09	24.07	14:26	48
-125	26.39	24.38	14:32	47
-127	26.94	24.96	14:43	44
-128	27.20	25.24	14:49	43
-130	27.68	25.75	15:00	40
-132	28.12	26.22	15:11	38
-134	28.53	26.65	15:22	36
-135	28.71	26.85	15:27	35
-138	29.21	27.38	15:42	31
-139	29.36	27.54	15:47	30
-140	29.50	27.70	15:52	29
-142	29.76	27.97	16:02	27
-143	29.87	28.10	16:07	26
-144	29.98	28.22	16:12	25
-145	30.08	28.33	16:17	24
-148	30.34	28.62	16:31	20
-150	30.48	28.77	16:40	18

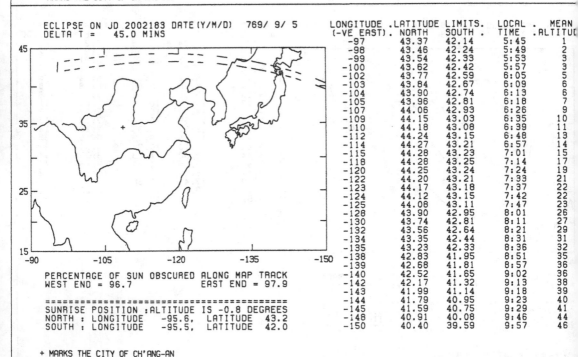

ECLIPSE ON JD 2002183 DATE(Y/M/D) 769/ 9/ 5
DELTA T =   45.0 MINS

PERCENTAGE OF SUN OBSCURED ALONG MAP TRACK
WEST END = 96.7   EAST END = 97.9

===
SUNRISE POSITION :ALTITUDE IS -0.8 DEGREES
NORTH : LONGITUDE -95.6, LATITUDE   43.2
SOUTH : LONGITUDE -95.5, LATITUDE   42.0

+ MARKS THE CITY OF CH'ANG-AN

LONGITUDE (-VE EAST)	LATITUDE NORTH	LIMITS. SOUTH	LOCAL TIME	MEAN ALTITUD
-97	43.37	42.14	5:45	1
-98	43.46	42.24	5:49	2
-99	43.54	42.33	5:53	3
-100	43.62	42.42	5:57	3
-102	43.77	42.59	6:05	5
-103	43.84	42.67	6:09	6
-104	43.90	42.74	6:13	6
-105	43.96	42.81	6:18	7
-107	44.06	42.93	6:26	9
-109	44.15	43.03	6:35	10
-110	44.18	43.08	6:39	11
-112	44.24	43.15	6:48	13
-114	44.27	43.21	6:57	14
-115	44.28	43.23	7:01	15
-118	44.28	43.25	7:14	17
-120	44.25	43.24	7:24	19
-122	44.20	43.21	7:33	21
-123	44.17	43.18	7:37	22
-124	44.12	43.15	7:42	22
-125	44.08	43.11	7:47	23
-128	43.90	42.95	8:01	26
-130	43.74	42.81	8:11	27
-132	43.56	42.64	8:21	29
-134	43.35	42.44	8:31	31
-135	43.23	42.33	8:36	32
-138	42.83	41.95	8:51	35
-139	42.68	41.81	8:57	36
-140	42.52	41.65	9:02	36
-142	42.17	41.32	9:13	38
-143	41.99	41.14	9:18	39
-144	41.79	40.95	9:23	40
-145	41.59	40.75	9:29	41
-148	40.91	40.08	9:46	44
-150	40.40	39.59	9:57	46

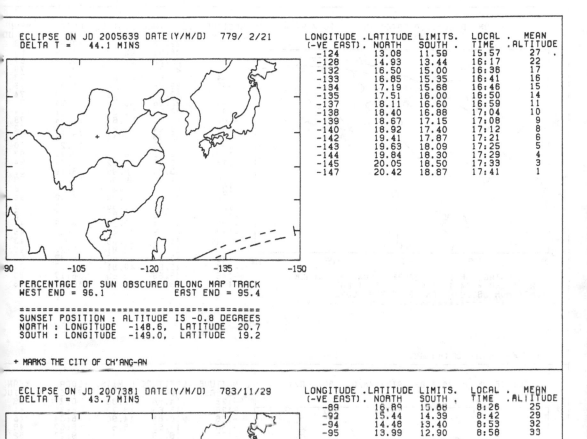

ECLIPSE ON JD 2005639 DATE(Y/M/D) 779/ 2/21
DELTA T =   44.1 MINS

LONGITUDE (-VE EAST)	.LATITUDE NORTH	LIMITS. SOUTH	LOCAL TIME	MEAN .ALTITUDE
-124	13.08	11.59	15:57	27
-128	14.93	13.44	16:17	22
-132	16.50	15.00	16:36	17
-133	16.85	15.35	16:41	16
-134	17.19	15.68	16:46	15
-135	17.51	16.00	16:50	14
-137	18.11	16.60	16:59	11
-138	18.40	16.88	17:04	10
-139	18.67	17.15	17:08	9
-140	18.92	17.40	17:12	8
-142	19.41	17.87	17:21	6
-143	19.63	18.09	17:25	5
-144	19.84	18.30	17:29	4
-145	20.05	18.50	17:33	3
-147	20.42	18.87	17:41	1

PERCENTAGE OF SUN OBSCURED ALONG MAP TRACK
WEST END = 96.1   EAST END = 95.4

===
SUNSET POSITION : ALTITUDE IS -0.8 DEGREES
NORTH : LONGITUDE -148.6, LATITUDE 20.7
SOUTH : LONGITUDE -149.0, LATITUDE 19.2

+ MARKS THE CITY OF CH'ANG-AN

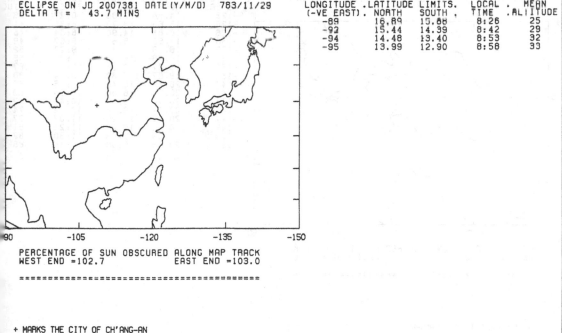

ECLIPSE ON JD 2007381 DATE(Y/M/D) 783/11/29
DELTA T =   43.7 MINS

LONGITUDE (-VE EAST)	.LATITUDE NORTH	LIMITS. SOUTH	LOCAL TIME	MEAN .ALTITUDE
-89	16.89	15.88	8:26	25
-92	15.44	14.39	8:42	29
-94	14.48	13.40	8:53	32
-95	13.99	12.90	8:58	33

PERCENTAGE OF SUN OBSCURED ALONG MAP TRACK
WEST END =102.7   EAST END =103.0

===

+ MARKS THE CITY OF CH'ANG-AN

293

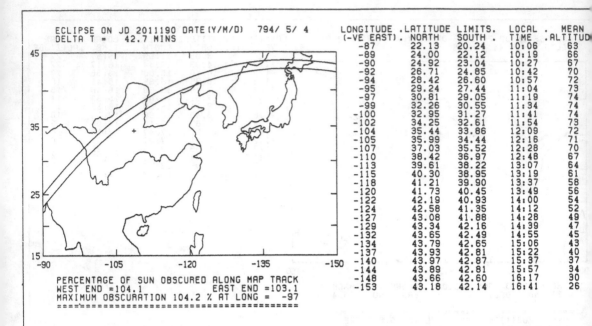

ECLIPSE ON JD 2011190 DATE(Y/M/D) 794/ 5/ 4
DELTA T = 42.7 MINS

LONGITUDE (-VE EAST)	LATITUDE NORTH	LIMITS. SOUTH	LOCAL TIME	MEAN ALTITUDE
-87	22.13	20.24	10:06	63
-89	24.00	22.12	10:19	66
-90	24.92	23.04	10:27	67
-92	26.71	24.85	10:42	70
-94	28.42	26.60	10:57	72
-95	29.24	27.44	11:04	73
-97	30.81	29.05	11:19	74
-99	32.26	30.55	11:34	74
-100	32.95	31.27	11:41	74
-102	34.25	32.61	11:54	73
-104	35.44	33.86	12:09	72
-105	35.99	34.44	12:16	71
-107	37.03	35.52	12:28	70
-110	38.42	36.97	12:48	67
-113	39.61	38.22	13:07	64
-115	40.30	38.95	13:19	61
-118	41.21	39.90	13:37	58
-120	41.73	40.45	13:49	56
-122	42.19	40.93	14:00	54
-124	42.58	41.35	14:12	52
-127	43.08	41.88	14:28	49
-129	43.34	42.16	14:39	47
-132	43.65	42.49	14:55	45
-134	43.79	42.65	15:06	43
-137	43.93	42.81	15:22	40
-140	43.97	42.87	15:37	37
-144	43.89	42.81	15:57	34
-148	43.66	42.60	16:17	30
-153	43.18	42.14	16:41	26

PERCENTAGE OF SUN OBSCURED ALONG MAP TRACK
WEST END =104.1   EAST END =103.1
MAXIMUM OBSCURATION 104.2 % AT LONG = -97
===

+ MARKS THE CITY OF CH'ANG-AN

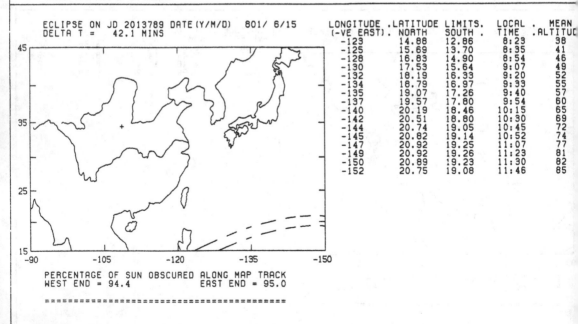

ECLIPSE ON JD 2013789 DATE(Y/M/D) 801/ 6/15
DELTA T = 42.1 MINS

LONGITUDE (-VE EAST)	LATITUDE NORTH	LIMITS SOUTH	LOCAL TIME	MEAN ALTITUDE
-123	14.88	12.86	8:23	38
-125	15.69	13.70	8:35	41
-128	16.83	14.90	8:54	46
-130	17.53	15.64	9:07	49
-132	18.19	16.33	9:20	52
-134	18.79	16.97	9:33	55
-135	19.07	17.26	9:40	57
-137	19.57	17.80	9:54	60
-140	20.19	18.46	10:15	65
-142	20.51	18.80	10:30	69
-144	20.74	19.05	10:45	72
-145	20.82	19.14	10:52	74
-147	20.92	19.25	11:07	77
-149	20.92	19.26	11:23	81
-150	20.89	19.23	11:30	82
-152	20.75	19.08	11:46	85

PERCENTAGE OF SUN OBSCURED ALONG MAP TRACK
WEST END = 94.4   EAST END = 95.0

===

+ MARKS THE CITY OF CH'ANG-AN

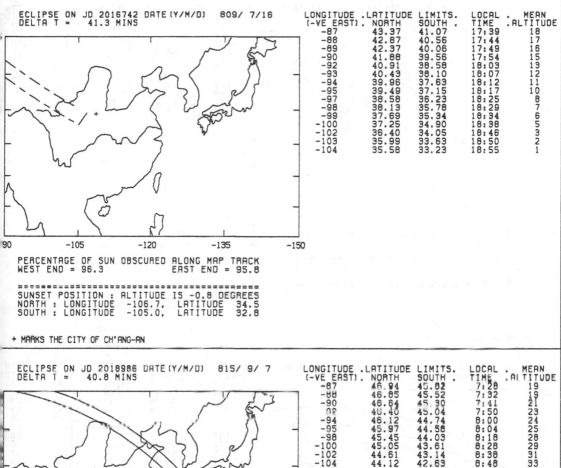

ECLIPSE ON JD 2016742 DATE(Y/M/D)   809/ 7/16
DELTA T =   41.3 MINS

LONGITUDE (-VE EAST)	LATITUDE LIMITS. NORTH	SOUTH	LOCAL TIME	MEAN ALTITUDE
-87	43.37	41.07	17:39	18
-88	42.87	40.56	17:44	17
-89	42.37	40.06	17:49	16
-90	41.88	39.56	17:54	15
-92	40.91	38.58	18:03	13
-93	40.43	38.10	18:07	12
-94	39.96	37.63	18:12	11
-95	39.49	37.15	18:17	10
-97	38.58	36.23	18:25	8
-98	38.13	35.78	18:29	7
-99	37.69	35.34	18:34	6
-100	37.25	34.90	18:38	5
-102	36.40	34.05	18:46	3
-103	35.99	33.63	18:50	2
-104	35.58	33.23	18:55	1

PERCENTAGE OF SUN OBSCURED ALONG MAP TRACK
WEST END = 96.3   EAST END = 95.8

==
SUNSET POSITION : ALTITUDE IS -0.8 DEGREES
NORTH : LONGITUDE -106.7, LATITUDE   34.5
SOUTH : LONGITUDE -105.0, LATITUDE   32.8

+ MARKS THE CITY OF CH'ANG-AN

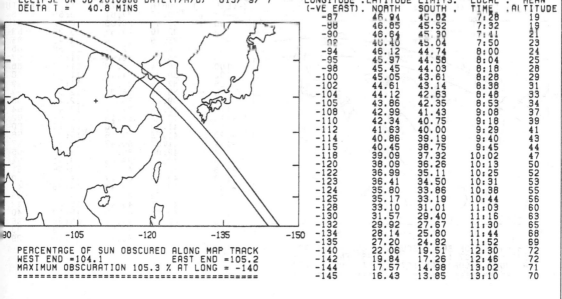

ECLIPSE ON JD 2018986 DATE(Y/M/D)   815/ 9/ 7
DELTA T =   40.8 MINS

LONGITUDE (-VE EAST)	LATITUDE LIMITS. NORTH	SOUTH	LOCAL TIME	MEAN ALTITUDE
-87	46.94	45.02	7:28	19
-88	46.85	45.52	7:32	19
-90	46.64	45.30	7:41	21
-92	46.40	45.04	7:50	23
-94	46.12	44.74	8:00	24
-95	45.97	44.58	8:04	25
-98	45.45	44.03	8:18	28
-100	45.05	43.61	8:28	29
-102	44.61	43.14	8:38	31
-104	44.12	42.63	8:48	33
-105	43.86	42.35	8:53	34
-108	42.99	41.43	9:08	37
-110	42.34	40.75	9:18	39
-112	41.63	40.00	9:29	41
-114	40.86	39.19	9:40	43
-115	40.45	38.75	9:45	44
-118	39.09	37.32	10:02	47
-120	38.09	36.26	10:13	50
-122	36.99	35.11	10:25	52
-123	36.41	34.50	10:31	53
-124	35.80	33.86	10:38	55
-125	35.17	33.19	10:44	56
-128	33.10	31.01	11:03	60
-130	31.57	29.40	11:16	63
-132	29.92	27.67	11:30	65
-134	28.14	25.80	11:44	68
-135	27.20	24.82	11:52	69
-140	22.06	19.51	12:30	72
-142	19.84	17.26	12:46	72
-144	17.57	14.98	13:02	71
-145	16.43	13.85	13:10	70

PERCENTAGE OF SUN OBSCURED ALONG MAP TRACK
WEST END =104.1   EAST END =105.2
MAXIMUM OBSCURATION 105.3 % AT LONG = -140
==

+ MARKS THE CITY OF CH'ANG-AN

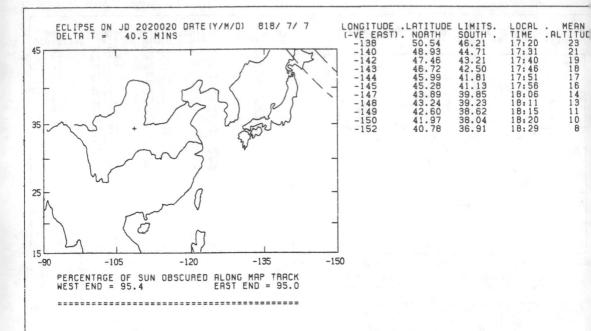

```
ECLIPSE ON JD 2020020 DATE(Y/M/D)  818/ 7/ 7
DELTA T =   40.5 MINS
```

LONGITUDE (-VE EAST)	.LATITUDE LIMITS. NORTH	SOUTH .	LOCAL . TIME	MEAN .ALTITUD
-138	50.54	46.21	17:20	23
-140	48.93	44.71	17:31	21
-142	47.46	43.21	17:40	19
-143	46.72	42.50	17:46	18
-144	45.99	41.81	17:51	17
-145	45.28	41.13	17:56	16
-147	43.89	39.95	18:06	14
-148	43.24	39.23	18:11	13
-149	42.60	38.62	18:15	11
-150	41.97	38.04	18:20	10
-152	40.78	36.91	18:29	8

```
PERCENTAGE OF SUN OBSCURED ALONG MAP TRACK
WEST END = 95.4          EAST END = 95.0

========================================

+ MARKS THE CITY OF CH'ANG-AN
```

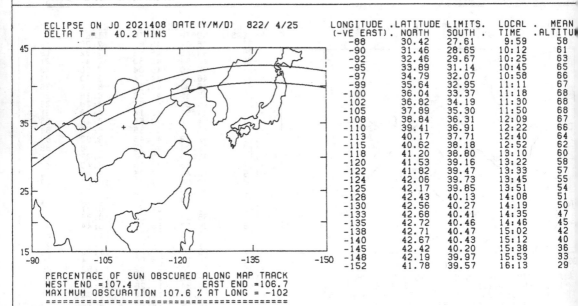

```
ECLIPSE ON JD 2021408 DATE(Y/M/D)  822/ 4/25
DELTA T =   40.2 MINS
```

LONGITUDE (-VE EAST)	.LATITUDE LIMITS. NORTH	SOUTH .	LOCAL TIME	MEAN .ALTITU
-88	30.42	27.71	9:59	58
-90	31.46	28.65	10:12	61
-92	32.46	29.67	10:25	63
-95	33.89	31.14	10:45	65
-97	34.79	32.07	10:58	66
-99	35.64	32.95	11:11	67
-100	36.04	33.37	11:18	68
-102	36.82	34.19	11:30	68
-105	37.89	35.30	11:50	68
-108	38.84	36.31	12:09	67
-110	39.41	36.91	12:22	66
-113	40.17	37.71	12:40	64
-115	40.62	38.18	12:52	62
-118	41.20	38.80	13:10	60
-120	41.53	39.16	13:22	58
-122	41.82	39.47	13:33	57
-124	42.06	39.73	13:45	55
-125	42.17	39.85	13:51	54
-128	42.43	40.13	14:08	51
-130	42.56	40.27	14:19	50
-133	42.68	40.41	14:35	47
-135	42.72	40.46	14:46	45
-138	42.71	40.47	15:02	42
-140	42.67	40.43	15:12	40
-145	42.42	40.20	15:38	36
-148	42.19	39.97	15:53	33
-152	41.78	39.57	16:13	29

```
PERCENTAGE OF SUN OBSCURED ALONG MAP TRACK
WEST END =107.4          EAST END =106.7
MAXIMUM OBSCURATION 107.6 % AT LONG = -102

========================================

+ MARKS THE CITY OF CH'ANG-AN
```

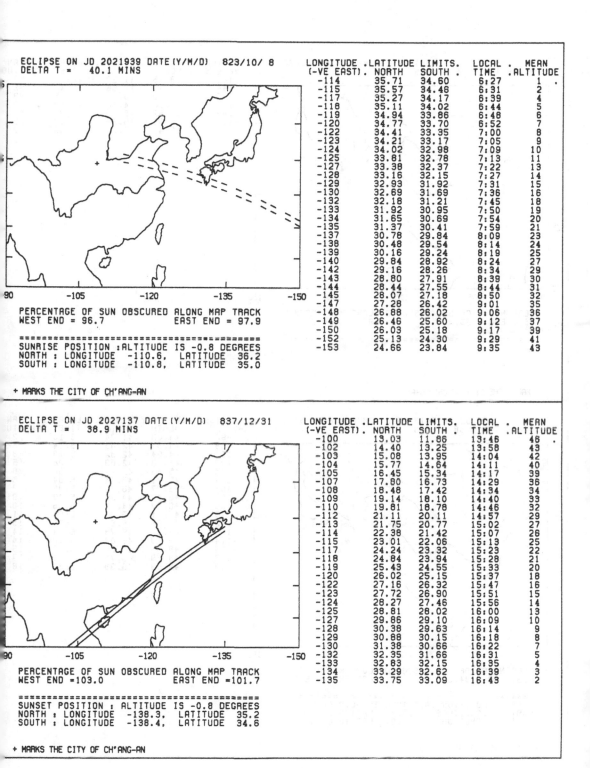

ECLIPSE ON JD 2021939 DATE(Y/M/D) 823/10/ 8
DELTA T =   40.1 MINS

LONGITUDE (-VE EAST)	LATITUDE LIMITS. NORTH	SOUTH	LOCAL TIME	MEAN ALTITUDE
-114	35.71	34.60	6:27	1
-115	35.57	34.48	6:31	2
-117	35.27	34.17	6:39	4
-118	35.11	34.02	6:44	5
-119	34.94	33.86	6:48	6
-120	34.77	33.70	6:52	7
-122	34.41	33.35	7:00	8
-123	34.21	33.17	7:05	9
-124	34.02	32.98	7:09	10
-125	33.81	32.78	7:13	11
-127	33.38	32.37	7:22	13
-128	33.16	32.15	7:27	14
-129	32.93	31.92	7:31	15
-130	32.69	31.69	7:36	16
-132	32.18	31.21	7:45	18
-133	31.92	30.95	7:50	19
-134	31.65	30.69	7:54	20
-135	31.37	30.41	7:59	21
-137	30.78	29.84	8:09	23
-138	30.48	29.54	8:14	24
-139	30.16	29.24	8:19	25
-140	29.84	28.92	8:24	27
-142	29.16	28.26	8:34	29
-143	28.80	27.91	8:39	30
-144	28.44	27.55	8:44	31
-145	28.07	27.18	8:50	32
-147	27.28	26.42	9:01	35
-148	26.88	26.02	9:06	36
-149	26.46	25.60	9:12	37
-150	26.03	25.18	9:17	39
-152	25.13	24.30	9:29	41
-153	24.66	23.84	9:35	43

PERCENTAGE OF SUN OBSCURED ALONG MAP TRACK
WEST END = 96.7   EAST END = 97.9

=====================================
SUNRISE POSITION :ALTITUDE IS -0.8 DEGREES
NORTH : LONGITUDE -110.6, LATITUDE 36.2
SOUTH : LONGITUDE -110.8, LATITUDE 35.0

+ MARKS THE CITY OF CH'ANG-AN

ECLIPSE ON JD 2027137 DATE(Y/M/D) 837/12/31
DELTA T =   38.9 MINS

LONGITUDE (-VE EAST)	LATITUDE LIMITS. NORTH	SOUTH	LOCAL TIME	MEAN ALTITUDE
-100	13.03	11.86	13:46	46
-102	14.40	13.25	13:58	43
-103	15.08	13.95	14:04	42
-104	15.77	14.64	14:11	40
-105	16.45	15.34	14:17	39
-107	17.80	16.73	14:29	36
-108	18.48	17.42	14:34	34
-109	19.14	18.10	14:40	33
-110	19.81	18.78	14:46	32
-112	21.11	20.11	14:57	29
-113	21.75	20.77	15:02	27
-114	22.38	21.42	15:07	26
-115	23.01	22.06	15:13	25
-117	24.24	23.32	15:23	22
-118	24.84	23.94	15:28	21
-119	25.43	24.55	15:33	20
-120	26.02	25.15	15:37	18
-122	27.16	26.32	15:47	16
-123	27.72	26.90	15:51	15
-124	28.27	27.46	15:56	14
-125	28.81	28.02	16:00	13
-127	29.86	29.10	16:09	10
-128	30.38	29.63	16:14	9
-129	30.88	30.15	16:18	8
-130	31.38	30.66	16:22	7
-132	32.35	31.66	16:31	5
-133	32.83	32.15	16:35	4
-134	33.29	32.62	16:39	3
-135	33.75	33.09	16:43	2

PERCENTAGE OF SUN OBSCURED ALONG MAP TRACK
WEST END =103.0   EAST END =101.7

=====================================
SUNSET POSITION : ALTITUDE IS -0.8 DEGREES
NORTH : LONGITUDE -138.3, LATITUDE 35.2
SOUTH : LONGITUDE -138.4, LATITUDE 34.6

+ MARKS THE CITY OF CH'ANG-AN

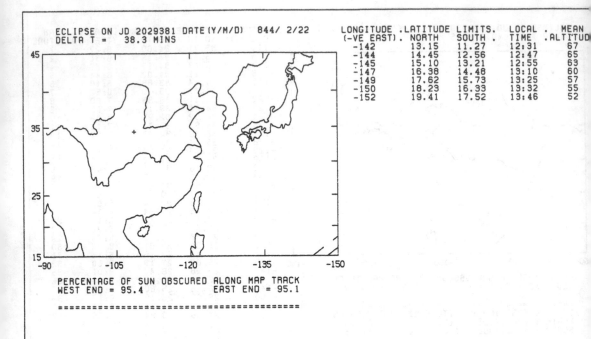

ECLIPSE ON JD 2029381 DATE(Y/M/D) 844/ 2/22
DELTA T =   38.3 MINS

LONGITUDE (-VE EAST)	LATITUDE LIMITS NORTH	SOUTH	LOCAL TIME	MEAN ALTITUDE
-142	13.15	11.27	12:31	67
-144	14.45	12.56	12:47	65
-145	15.10	13.21	12:55	63
-147	16.38	14.48	13:10	60
-149	17.62	15.73	13:25	57
-150	18.23	16.33	13:32	55
-152	19.41	17.52	13:46	52

PERCENTAGE OF SUN OBSCURED ALONG MAP TRACK
WEST END = 95.4   EAST END = 95.1

+ MARKS THE CITY OF CH'ANG-AN

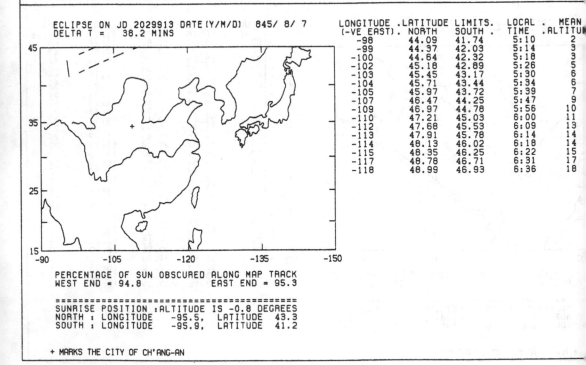

ECLIPSE ON JD 2029913 DATE(Y/M/D) 845/ 8/ 7
DELTA T =   38.2 MINS

LONGITUDE (-VE EAST)	LATITUDE LIMITS NORTH	SOUTH	LOCAL TIME	MEAN ALTITUDE
-98	44.09	41.74	5:10	2
-99	44.37	42.03	5:14	3
-100	44.64	42.32	5:18	3
-102	45.18	42.89	5:26	5
-103	45.45	43.17	5:30	6
-104	45.71	43.44	5:34	6
-105	45.97	43.72	5:39	7
-107	46.47	44.25	5:47	9
-109	46.97	44.78	5:56	10
-110	47.21	45.03	6:00	11
-112	47.68	45.53	6:09	13
-113	47.91	45.78	6:14	14
-114	48.13	46.02	6:18	14
-115	48.35	46.25	6:22	15
-117	48.78	46.71	6:31	17
-118	48.99	46.93	6:36	18

PERCENTAGE OF SUN OBSCURED ALONG MAP TRACK
WEST END = 94.8   EAST END = 95.3

SUNRISE POSITION :ALTITUDE IS -0.8 DEGREES
NORTH : LONGITUDE   -95.5, LATITUDE   43.3
SOUTH : LONGITUDE   -95.9, LATITUDE   41.2

+ MARKS THE CITY OF CH'ANG-AN

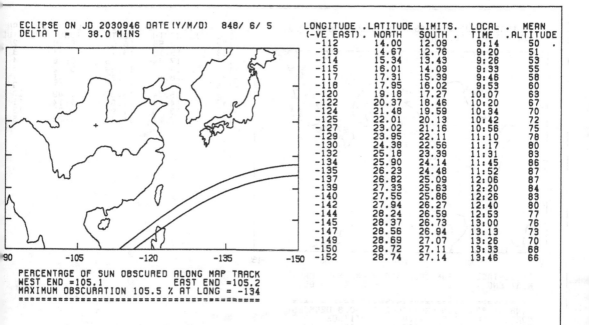

```
ECLIPSE ON JD 2030946 DATE(Y/M/D)  848/ 6/ 5
DELTA T =   38.0 MINS
```

LONGITUDE (-VE EAST)	.LATITUDE. NORTH	LIMITS. SOUTH	LOCAL TIME	MEAN ALTITUDE
-112	14.00	12.09	9:14	50
-113	14.67	12.76	9:20	51
-114	15.34	13.43	9:26	53
-115	16.01	14.09	9:39	55
-117	17.31	15.39	9:46	58
-118	17.95	16.02	9:53	60
-120	19.18	17.27	10:07	63
-122	20.37	18.46	10:20	67
-124	21.48	19.59	10:34	70
-125	22.01	20.13	10:42	72
-127	23.02	21.16	10:56	75
-129	23.95	22.11	11:10	78
-130	24.38	22.56	11:17	80
-132	25.18	23.39	11:31	83
-134	25.90	24.14	11:45	86
-135	26.23	24.48	11:52	87
-137	26.82	25.09	12:06	87
-139	27.33	25.63	12:20	84
-140	27.55	25.86	12:26	83
-142	27.94	26.27	12:40	80
-144	28.24	26.59	12:53	77
-145	28.37	26.73	13:00	76
-147	28.56	26.94	13:13	73
-149	28.69	27.07	13:26	70
-150	28.72	27.11	13:33	68
-152	28.74	27.14	13:46	66

```
PERCENTAGE OF SUN OBSCURED ALONG MAP TRACK
WEST END =105.1              EAST END =105.2
MAXIMUM OBSCURATION 105.5 % AT LONG = -134
==========================================
```

+ MARKS THE CITY OF CH'ANG-AN

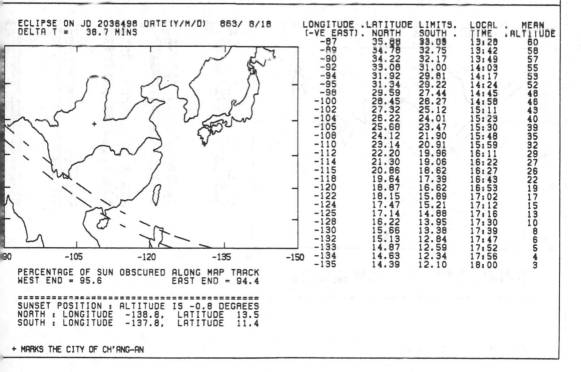

```
ECLIPSE ON JD 2036498 DATE(Y/M/D)  863/ 8/18
DELTA T =   38.7 MINS
```

LONGITUDE (-VE EAST)	.LATITUDE. NORTH	LIMITS. SOUTH	LOCAL TIME	MEAN ALTITUDE
-87	35.88	33.09	13:28	60
-89	34.78	32.75	13:42	58
-90	34.22	32.17	13:49	57
-92	33.08	31.00	14:03	55
-94	31.92	29.81	14:17	53
-95	31.34	29.22	14:24	52
-98	29.59	27.44	14:45	48
-100	28.45	26.27	14:58	46
-102	27.32	25.12	15:11	43
-104	26.22	24.01	15:23	40
-105	25.68	23.47	15:30	39
-108	24.12	21.90	15:48	35
-110	23.14	20.91	15:59	32
-112	22.20	19.96	16:11	29
-114	21.30	19.06	16:22	27
-115	20.86	18.62	16:27	26
-118	19.64	17.39	16:43	22
-120	18.87	16.62	16:53	19
-122	18.15	15.89	17:02	17
-124	17.47	15.21	17:12	15
-125	17.14	14.88	17:16	13
-128	16.22	13.95	17:30	10
-130	15.66	13.38	17:39	8
-132	15.13	12.84	17:47	6
-133	14.87	12.59	17:52	5
-134	14.63	12.34	17:56	4
-135	14.39	12.10	18:00	3

```
PERCENTAGE OF SUN OBSCURED ALONG MAP TRACK
WEST END = 95.6             EAST END = 94.4

=================================================
SUNSET POSITION : ALTITUDE IS -0.8 DEGREES
NORTH : LONGITUDE  -138.8,  LATITUDE   13.5
SOUTH : LONGITUDE  -137.8,  LATITUDE   11.4
```

+ MARKS THE CITY OF CH'ANG-AN

299

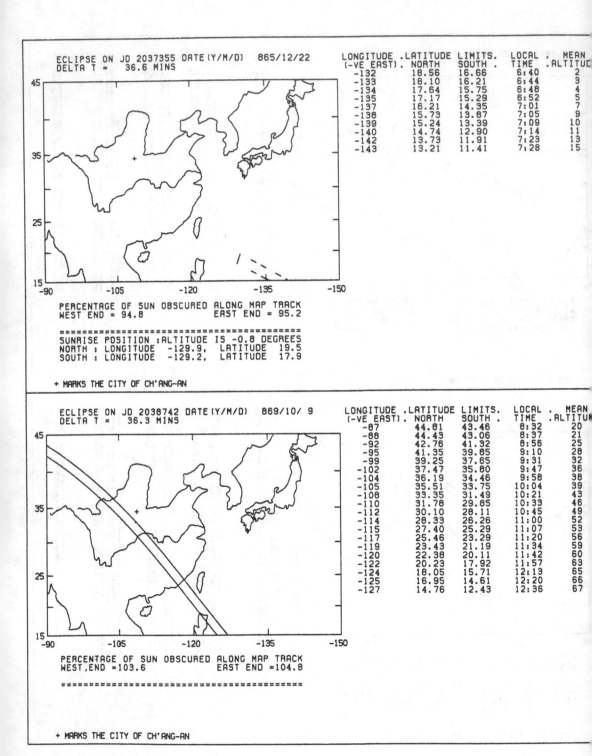

ECLIPSE ON JD 2037355 DATE(Y/M/D) 865/12/22
DELTA T = 36.6 MINS

LONGITUDE (-VE EAST)	LATITUDE NORTH	LIMITS SOUTH	LOCAL TIME	MEAN ALTITUDE
-132	18.56	16.66	6:40	2
-133	18.10	16.21	6:44	3
-134	17.64	15.75	6:48	4
-135	17.17	15.29	6:52	5
-137	16.21	14.35	7:01	7
-138	15.73	13.87	7:05	9
-139	15.24	13.39	7:09	10
-140	14.74	12.90	7:14	11
-142	13.73	11.91	7:23	13
-143	13.21	11.41	7:28	15

PERCENTAGE OF SUN OBSCURED ALONG MAP TRACK
WEST END = 94.8   EAST END = 95.2

==
SUNRISE POSITION :ALTITUDE IS -0.8 DEGREES
NORTH : LONGITUDE -129.9, LATITUDE 19.5
SOUTH : LONGITUDE -129.2, LATITUDE 17.9

+ MARKS THE CITY OF CH'ANG-AN

ECLIPSE ON JD 2038742 DATE(Y/M/D) 869/10/ 9
DELTA T = 36.3 MINS

LONGITUDE (-VE EAST)	LATITUDE NORTH	LIMITS SOUTH	LOCAL TIME	MEAN ALTITUDE
-87	44.81	43.46	8:32	20
-88	44.43	43.06	8:37	21
-92	42.76	41.32	8:56	25
-95	41.35	39.85	9:10	28
-99	39.25	37.65	9:31	32
-102	37.47	35.80	9:47	36
-104	36.19	34.46	9:58	38
-105	35.51	33.75	10:04	39
-108	33.35	31.49	10:21	43
-110	31.78	29.85	10:33	46
-112	30.10	28.11	10:45	49
-114	28.33	26.26	11:00	52
-115	27.40	25.29	11:07	53
-117	25.46	23.29	11:20	56
-119	23.43	21.19	11:34	59
-120	22.38	20.11	11:42	60
-122	20.23	17.92	11:57	63
-124	18.05	15.71	12:13	65
-125	16.95	14.61	12:20	66
-127	14.76	12.43	12:36	67

PERCENTAGE OF SUN OBSCURED ALONG MAP TRACK
WEST END =103.6   EAST END =104.8

==

+ MARKS THE CITY OF CH'ANG-AN

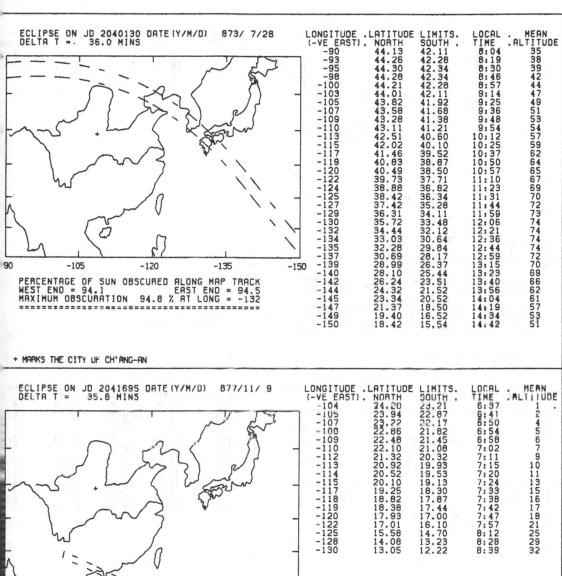

ECLIPSE ON JD 2040130 DATE (Y/M/D) 873/ 7/28
DELTA T =. 36.0 MINS

LONGITUDE (-VE EAST)	LATITUDE NORTH	LIMITS. SOUTH	LOCAL TIME	MEAN ALTITUDE
-90	44.13	42.11	8:04	35
-93	44.26	42.28	8:19	38
-95	44.30	42.34	8:30	39
-98	44.28	42.34	8:46	42
-100	44.21	42.28	8:57	44
-103	44.01	42.11	9:14	47
-105	43.82	41.92	9:25	49
-107	43.58	41.68	9:36	51
-109	43.28	41.38	9:48	53
-110	43.11	41.21	9:54	54
-113	42.51	40.60	10:12	57
-115	42.02	40.10	10:25	59
-117	41.46	39.52	10:37	62
-119	40.83	38.87	10:50	64
-120	40.49	38.50	10:57	65
-122	39.73	37.71	11:10	67
-124	38.88	36.82	11:23	69
-125	38.42	36.34	11:31	70
-127	37.42	35.28	11:44	72
-129	36.31	34.11	11:59	73
-130	35.72	33.48	12:06	74
-132	34.44	32.12	12:21	74
-134	33.03	30.64	12:36	74
-135	32.28	29.84	12:44	74
-137	30.69	28.17	12:59	72
-139	28.99	26.37	13:15	70
-140	28.10	25.44	13:23	69
-142	26.24	23.51	13:40	66
-144	24.32	21.52	13:56	62
-145	23.34	20.52	14:04	61
-147	21.37	18.50	14:19	57
-149	19.40	16.52	14:34	53
-150	18.42	15.54	14:42	51

PERCENTAGE OF SUN OBSCURED ALONG MAP TRACK
WEST END = 94.1   EAST END = 94.5
MAXIMUM OBSCURATION 94.8 % AT LONG = -132
===

+ MARKS THE CITY OF CH'ANG-AN

ECLIPSE ON JD 2041695 DATE (Y/M/D) 877/11/ 9
DELTA T = 35.8 MINS

LONGITUDE (-VE EAST)	LATITUDE NORTH	LIMITS. SOUTH	LOCAL TIME	MEAN ALTITUDE
-104	24.20	23.21	6:37	1
-105	23.94	22.87	6:41	2
-107	23.22	22.17	6:50	4
-100	22.86	21.82	6:54	5
-109	22.48	21.45	6:58	6
-110	22.10	21.08	7:02	7
-112	21.32	20.32	7:11	9
-113	20.92	19.93	7:15	10
-114	20.52	19.53	7:20	11
-115	20.10	19.13	7:24	13
-117	19.25	18.30	7:33	15
-118	18.82	17.87	7:38	16
-119	18.38	17.44	7:42	17
-120	17.93	17.00	7:47	18
-122	17.01	16.10	7:57	21
-125	15.58	14.70	8:12	25
-128	14.08	13.23	8:28	29
-130	13.05	12.22	8:39	32

PERCENTAGE OF SUN OBSCURED ALONG MAP TRACK
WEST END = 96.9   EAST END = 97.7

===
SUNRISE POSITION :ALTITUDE IS -0.8 DEGREES
NORTH : LONGITUDE -102.3, LATITUDE 24.9
SOUTH : LONGITUDE -102.0, LATITUDE 23.9

+ MARKS THE CITY OF CH'ANG-AN

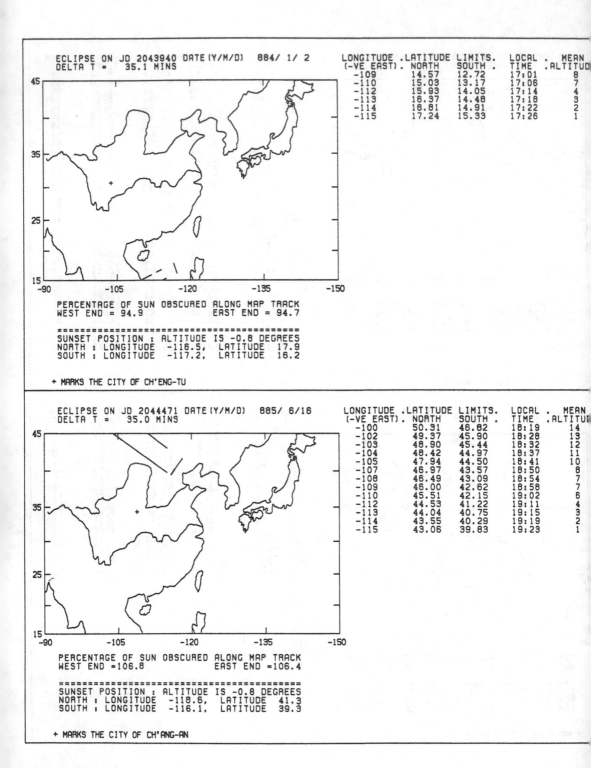

ECLIPSE ON JD 2043940 DATE (Y/M/D) 884/ 1/ 2
DELTA T =   35.1 MINS

LONGITUDE (-VE EAST)	.LATITUDE LIMITS. . NORTH	SOUTH .	LOCAL TIME	. MEAN .ALTITUD
-109	14.57	12.72	17:01	8
-110	15.03	13.17	17:06	7
-112	15.93	14.05	17:14	4
-113	16.37	14.48	17:18	3
-114	16.81	14.91	17:22	2
-115	17.24	15.33	17:26	1

PERCENTAGE OF SUN OBSCURED ALONG MAP TRACK
WEST END = 94.9   EAST END = 94.7

===
SUNSET POSITION : ALTITUDE IS -0.8 DEGREES
NORTH : LONGITUDE -116.5, LATITUDE 17.9
SOUTH : LONGITUDE -117.2, LATITUDE 16.2

+ MARKS THE CITY OF CH'ENG-TU

ECLIPSE ON JD 2044471 DATE (Y/M/D) 885/ 6/16
DELTA T =   35.0 MINS

LONGITUDE (-VE EAST)	.LATITUDE LIMITS. . NORTH	SOUTH .	LOCAL TIME	. MEAN .ALTITU
-100	50.31	46.82	18:19	14
-102	49.37	45.90	18:28	13
-103	48.90	45.44	18:32	12
-104	48.42	44.97	18:37	11
-105	47.94	44.50	18:41	10
-107	46.97	43.57	18:50	8
-108	46.49	43.09	18:54	7
-109	46.00	42.62	18:58	7
-110	45.51	42.15	19:02	6
-112	44.53	41.22	19:11	4
-113	44.04	40.75	19:15	3
-114	43.55	40.29	19:19	2
-115	43.06	39.83	19:23	1

PERCENTAGE OF SUN OBSCURED ALONG MAP TRACK
WEST END =106.8   EAST END =106.4

===
SUNSET POSITION : ALTITUDE IS -0.8 DEGREES
NORTH : LONGITUDE -118.6, LATITUDE 41.3
SOUTH : LONGITUDE -116.1, LATITUDE 39.3

+ MARKS THE CITY OF CH'ANG-AN

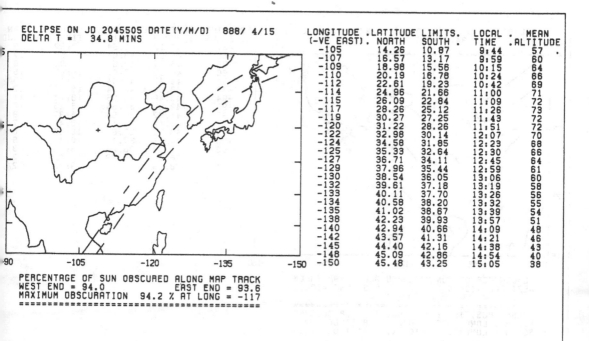

ECLIPSE ON JD 2045505 DATE(Y/M/D) 888/ 4/15
DELTA T = 34.8 MINS

LONGITUDE (-VE EAST)	LATITUDE LIMITS. NORTH	SOUTH	LOCAL TIME	MEAN ALTITUDE
-105	14.26	10.87	9:44	57
-107	16.57	13.17	9:59	60
-109	18.98	15.56	10:15	64
-110	20.19	16.78	10:24	66
-112	22.61	19.23	10:42	69
-114	24.96	21.66	11:00	71
-115	26.09	22.84	11:09	72
-117	28.26	25.12	11:26	73
-119	30.27	27.25	11:43	72
-120	31.22	28.26	11:51	72
-122	32.98	30.14	12:07	70
-124	34.58	31.85	12:23	68
-125	35.33	32.64	12:30	66
-127	36.71	34.11	12:45	64
-129	37.96	35.44	12:59	61
-130	38.54	36.05	13:06	60
-132	39.61	37.18	13:19	58
-133	40.11	37.70	13:26	56
-134	40.58	38.20	13:32	55
-135	41.02	38.67	13:39	54
-138	42.23	39.93	13:57	51
-140	42.94	40.66	14:09	48
-142	43.57	41.31	14:21	46
-145	44.40	42.16	14:38	43
-148	45.09	42.86	14:54	40
-150	45.48	43.25	15:05	38

PERCENTAGE OF SUN OBSCURED ALONG MAP TRACK
WEST END = 94.0   EAST END = 93.6
MAXIMUM OBSCURATION 94.2 % AT LONG = -117

+ MARKS THE CITY OF CH'ANG-AN

ECLIPSE ON JD 2046893 DATE(Y/M/D) 892/ 2/ 2
DELTA T = 34.5 MINS

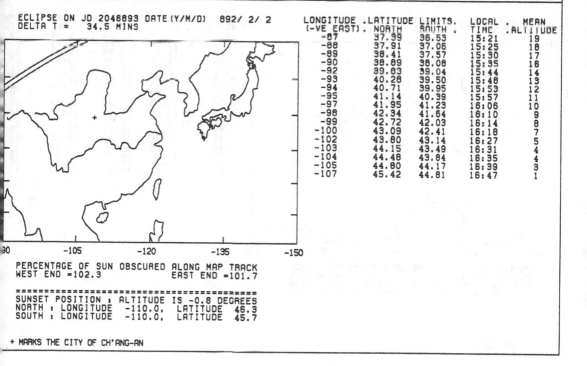

LONGITUDE (-VE EAST)	LATITUDE LIMITS. NORTH	SOUTH	LOCAL TIME	MEAN ALTITUDE
-87	37.99	36.53	15:21	19
-88	37.91	37.06	15:25	18
-89	38.41	37.57	15:30	17
-90	38.89	38.08	15:35	16
-92	39.83	39.04	15:44	14
-93	40.28	39.50	15:48	13
-94	40.71	39.95	15:53	12
-95	41.14	40.39	15:57	11
-97	41.95	41.23	16:06	10
-98	42.34	41.64	16:10	9
-99	42.72	42.03	16:14	8
-100	43.09	42.41	16:18	7
-102	43.80	43.14	16:27	5
-103	44.15	43.49	16:31	4
-104	44.48	43.84	16:35	4
-105	44.80	44.17	16:39	3
-107	45.42	44.81	16:47	1

PERCENTAGE OF SUN OBSCURED ALONG MAP TRACK
WEST END =102.3   EAST END =101.7

SUNSET POSITION : ALTITUDE IS -0.8 DEGREES
NORTH : LONGITUDE -110.0, LATITUDE 46.3
SOUTH : LONGITUDE -110.0, LATITUDE 45.7

+ MARKS THE CITY OF CH'ANG-AN

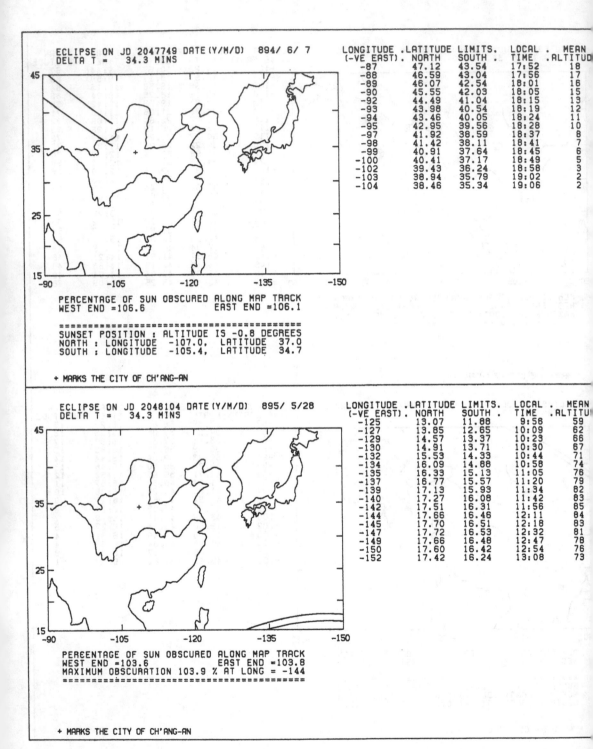

ECLIPSE ON JD 2047749 DATE(Y/M/D) 894/ 6/ 7
DELTA T =   34.3 MINS

LONGITUDE (-VE EAST)	LATITUDE NORTH	LIMITS. SOUTH	LOCAL TIME	MEAN ALTITUD
-87	47.12	43.54	17:52	18
-88	46.59	43.04	17:56	17
-89	46.07	42.54	18:01	16
-90	45.55	42.03	18:05	15
-92	44.49	41.04	18:15	13
-93	43.98	40.54	18:19	12
-94	43.46	40.05	18:24	11
-95	42.95	39.56	18:28	10
-97	41.92	38.59	18:37	8
-98	41.42	38.11	18:41	7
-99	40.91	37.64	18:45	6
-100	40.41	37.17	18:49	5
-102	39.43	36.24	18:58	3
-103	38.94	35.79	19:02	2
-104	38.46	35.34	19:06	2

PERCENTAGE OF SUN OBSCURED ALONG MAP TRACK
WEST END =106.6   EAST END =106.1

==
SUNSET POSITION : ALTITUDE IS -0.8 DEGREES
NORTH : LONGITUDE -107.0, LATITUDE 37.0
SOUTH : LONGITUDE -105.4, LATITUDE 34.7

+ MARKS THE CITY OF CH'ANG-AN

ECLIPSE ON JD 2048104 DATE(Y/M/D) 895/ 5/28
DELTA T =   34.3 MINS

LONGITUDE (-VE EAST)	LATITUDE NORTH	LIMITS. SOUTH	LOCAL TIME	MEAN ALTITU
-125	13.07	11.88	9:56	59
-127	13.85	12.65	10:09	62
-129	14.57	13.37	10:23	66
-130	14.91	13.71	10:30	67
-132	15.53	14.33	10:44	71
-134	16.09	14.88	10:58	74
-135	16.33	15.13	11:05	76
-137	16.77	15.57	11:20	79
-139	17.13	15.93	11:34	82
-140	17.27	16.08	11:42	83
-142	17.51	16.31	11:56	85
-144	17.66	16.46	12:11	84
-145	17.70	16.51	12:18	83
-147	17.72	16.53	12:32	81
-149	17.66	16.48	12:47	78
-150	17.60	16.42	12:54	76
-152	17.42	16.24	13:08	73

PERCENTAGE OF SUN OBSCURED ALONG MAP TRACK
WEST END =103.6   EAST END =103.8
MAXIMUM OBSCURATION 103.9 % AT LONG = -144
==

+ MARKS THE CITY OF CH'ANG-AN

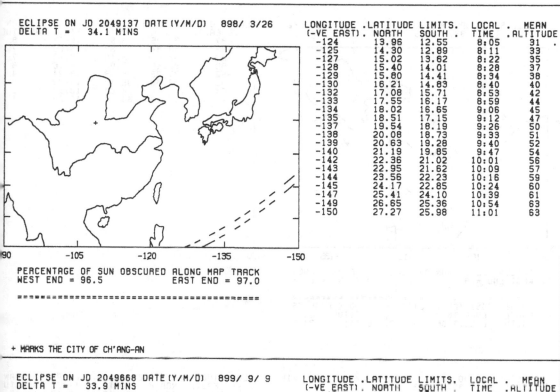

ECLIPSE ON JD 2049137 DATE(Y/M/D) 898/ 3/26
DELTA T =   34.1 MINS

LONGITUDE (-VE EAST)	LATITUDE NORTH	LIMITS. SOUTH	LOCAL TIME	MEAN ALTITUDE
-124	13.96	12.55	8:05	31
-125	14.30	12.89	8:11	33
-127	15.02	13.62	8:22	35
-128	15.40	14.01	8:28	37
-129	15.80	14.41	8:34	38
-130	16.21	14.83	8:40	40
-132	17.08	15.71	8:53	42
-133	17.55	16.17	8:59	44
-134	18.02	16.65	9:06	45
-135	18.51	17.15	9:12	47
-137	19.54	18.19	9:26	50
-138	20.08	18.73	9:33	51
-139	20.63	19.28	9:40	52
-140	21.19	19.85	9:47	54
-142	22.36	21.02	10:01	56
-143	22.95	21.62	10:09	57
-144	23.56	22.23	10:16	59
-145	24.17	22.85	10:24	60
-147	25.41	24.10	10:39	61
-149	26.65	25.36	10:54	63
-150	27.27	25.98	11:01	63

PERCENTAGE OF SUN OBSCURED ALONG MAP TRACK
WEST END = 96.5   EAST END = 97.0

==

+ MARKS THE CITY OF CH'ANG-AN

ECLIPSE ON JD 2049668 DATE(Y/M/D) 899/ 9/ 9
DELTA T =   33.9 MINS

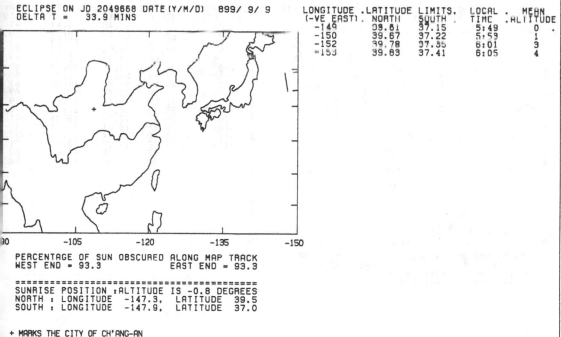

LONGITUDE (-VE EAST)	LATITUDE NORTH	LIMITS. SOUTH	LOCAL TIME	MEAN ALTITUDE
-149	39.61	37.15	5:49	0
-150	39.67	37.22	5:53	1
-152	39.78	37.35	6:01	3
-153	39.83	37.41	6:05	4

PERCENTAGE OF SUN OBSCURED ALONG MAP TRACK
WEST END = 93.3   EAST END = 93.3

==
SUNRISE POSITION : ALTITUDE IS -0.8 DEGREES
NORTH : LONGITUDE -147.3, LATITUDE 39.5
SOUTH : LONGITUDE -147.9, LATITUDE 37.0

+ MARKS THE CITY OF CH'ANG-AN

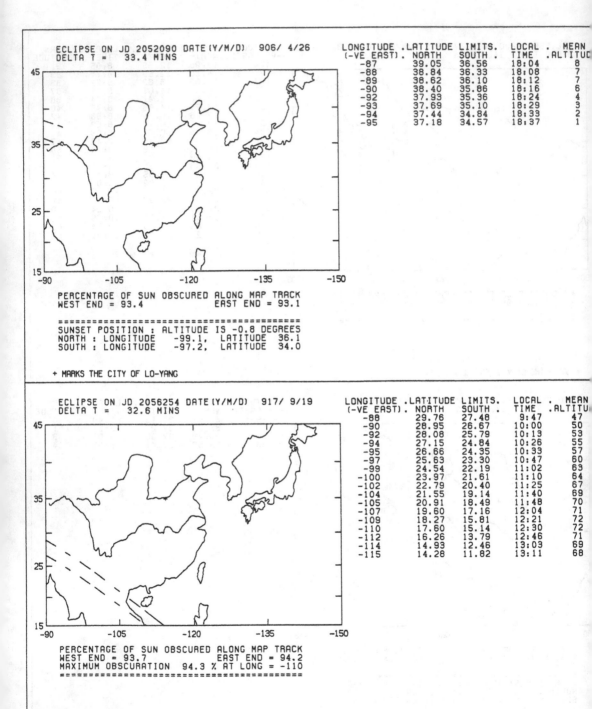

ECLIPSE ON JD 2052090 DATE(Y/M/D)   906/ 4/26
DELTA T =   33.4 MINS

LONGITUDE (-VE EAST)	.LATITUDE LIMITS. NORTH	SOUTH .	LOCAL TIME	. MEAN .ALTITU
-87	39.05	36.56	18:04	8
-88	38.84	36.33	18:08	7
-89	38.62	36.10	18:12	7
-90	38.40	35.86	18:16	6
-92	37.93	35.36	18:24	4
-93	37.69	35.10	18:29	3
-94	37.44	34.84	18:33	2
-95	37.18	34.57	18:37	1

PERCENTAGE OF SUN OBSCURED ALONG MAP TRACK
WEST END = 93.4   EAST END = 93.1

===
SUNSET POSITION : ALTITUDE IS -0.8 DEGREES
NORTH : LONGITUDE   -99.1, LATITUDE 36.1
SOUTH : LONGITUDE   -97.2, LATITUDE 34.0

+ MARKS THE CITY OF LO-YANG

ECLIPSE ON JD 2056254 DATE(Y/M/D)   917/ 9/19
DELTA T =   32.6 MINS

LONGITUDE (-VE EAST)	.LATITUDE LIMITS. NORTH	SOUTH .	LOCAL TIME	. MEAN .ALTITU
-88	29.76	27.48	9:47	47
-90	28.95	26.67	10:00	50
-92	28.08	25.79	10:13	53
-94	27.15	24.84	10:26	55
-95	26.66	24.35	10:33	57
-97	25.63	23.30	10:47	60
-99	24.54	22.19	11:02	63
-100	23.97	21.61	11:10	64
-102	22.79	20.40	11:25	67
-104	21.55	19.14	11:40	69
-105	20.91	18.49	11:48	70
-107	19.60	17.16	12:04	71
-109	18.27	15.81	12:21	72
-110	17.60	15.14	12:30	72
-112	16.26	13.79	12:46	71
-114	14.93	12.46	13:03	69
-115	14.28	11.82	13:11	68

PERCENTAGE OF SUN OBSCURED ALONG MAP TRACK
WEST END = 93.7   EAST END = 94.2
MAXIMUM OBSCURATION 94.3 % AT LONG = -110
===

+ MARKS THE CITY OF K'AI-FENG

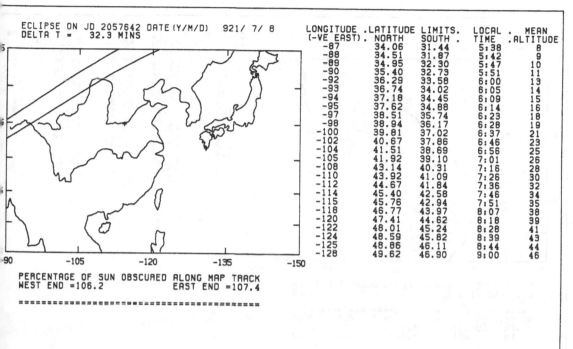

ECLIPSE ON JD 2057642 DATE(Y/M/D)   921/ 7/ 8
DELTA T =   32.3 MINS

LONGITUDE (-VE EAST)	LATITUDE LIMITS. NORTH	SOUTH	LOCAL TIME	MEAN ALTITUDE
-87	34.06	31.44	5:38	8
-88	34.51	31.87	5:42	9
-89	34.95	32.30	5:47	10
-90	35.40	32.73	5:51	11
-92	36.29	33.58	6:00	13
-93	36.74	34.02	6:05	14
-94	37.18	34.45	6:09	15
-95	37.62	34.88	6:14	16
-97	38.51	35.74	6:23	18
-98	38.94	36.17	6:28	19
-100	39.81	37.02	6:37	21
-102	40.67	37.86	6:46	23
-104	41.51	38.69	6:56	25
-105	41.92	39.10	7:01	26
-108	43.14	40.31	7:16	28
-110	43.92	41.09	7:26	30
-112	44.67	41.84	7:36	32
-114	45.40	42.58	7:46	34
-115	45.76	42.94	7:51	35
-118	46.77	43.97	8:07	38
-120	47.41	44.62	8:18	39
-122	48.01	45.24	8:28	41
-124	48.59	45.82	8:39	43
-125	48.86	46.11	8:44	44
-128	49.62	46.90	9:00	46

PERCENTAGE OF SUN OBSCURED ALONG MAP TRACK
WEST END =106.2   EAST END =107.4

===

+ MARKS THE CITY OF K'AI-FENG

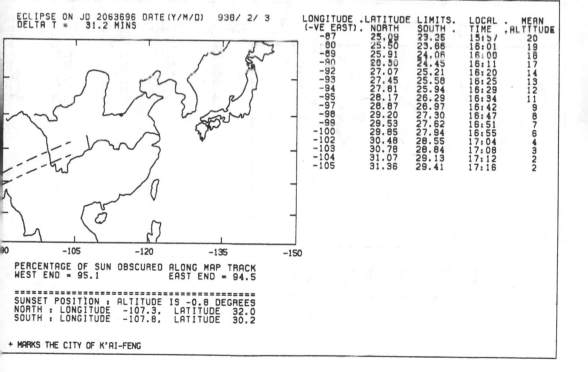

ECLIPSE ON JD 2063696 DATE(Y/M/D)   938/ 2/ 3
DELTA T =   31.2 MINS

LONGITUDE (-VE EAST)	LATITUDE LIMITS. NORTH	SOUTH	LOCAL TIME	MEAN ALTITUDE
-87	25.09	23.25	15:57	20
-80	25.50	23.66	16:01	19
-89	25.91	24.06	16:00	18
-90	26.50	24.45	16:11	17
-92	27.07	25.21	16:20	14
-93	27.45	25.58	16:25	13
-94	27.81	25.94	16:29	12
-95	28.17	26.29	16:34	11
-97	28.87	26.97	16:42	9
-98	29.20	27.30	16:47	8
-99	29.53	27.62	16:51	7
-100	29.85	27.94	16:55	6
-102	30.48	28.55	17:04	4
-103	30.78	28.84	17:08	3
-104	31.07	29.13	17:12	2
-105	31.36	29.41	17:16	2

PERCENTAGE OF SUN OBSCURED ALONG MAP TRACK
WEST END = 95.1   EAST END = 94.5

===
SUNSET POSITION : ALTITUDE IS -0.8 DEGREES
NORTH : LONGITUDE -107.3, LATITUDE 32.0
SOUTH : LONGITUDE -107.8, LATITUDE 30.2

+ MARKS THE CITY OF K'AI-FENG

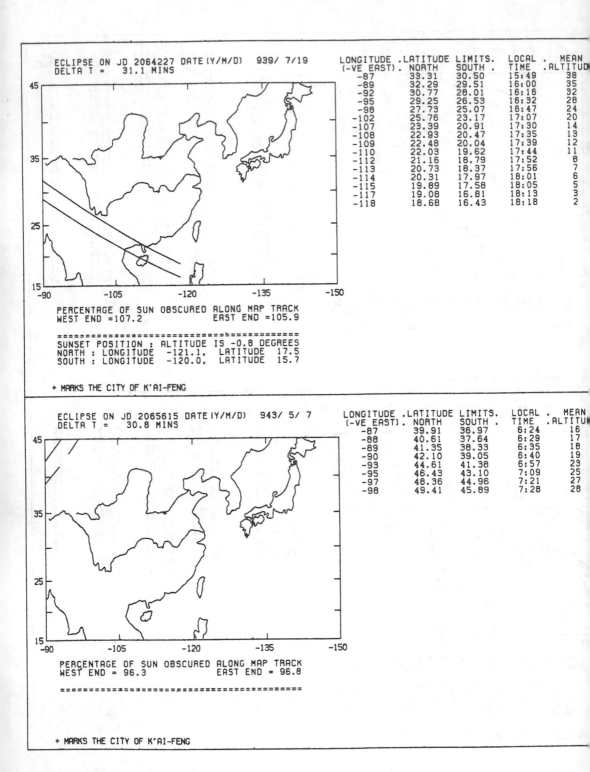

ECLIPSE ON JD 2064227 DATE(Y/M/D) 939/ 7/19
DELTA T =   31.1 MINS

LONGITUDE (-VE EAST)	LATITUDE LIMITS. NORTH	SOUTH	LOCAL TIME	MEAN ALTITUDE
-87	33.31	30.50	15:49	38
-89	32.29	29.51	16:00	35
-92	30.77	28.01	16:16	32
-95	29.25	26.53	16:32	28
-98	27.73	25.07	16:47	24
-102	25.76	23.17	17:07	20
-107	23.39	20.91	17:30	14
-108	22.93	20.47	17:35	13
-109	22.48	20.04	17:39	12
-110	22.03	19.62	17:44	11
-112	21.16	18.79	17:52	8
-113	20.73	18.37	17:56	7
-114	20.31	17.97	18:01	6
-115	19.89	17.58	18:05	5
-117	19.08	16.81	18:13	3
-118	18.68	16.43	18:18	2

PERCENTAGE OF SUN OBSCURED ALONG MAP TRACK
WEST END =107.2   EAST END =105.9

==
SUNSET POSITION : ALTITUDE IS -0.8 DEGREES
NORTH : LONGITUDE -121.1, LATITUDE 17.5
SOUTH : LONGITUDE -120.0, LATITUDE 15.7

+ MARKS THE CITY OF K'AI-FENG

ECLIPSE ON JD 2065615 DATE(Y/M/D) 943/ 5/ 7
DELTA T =   30.8 MINS

LONGITUDE (-VE EAST)	LATITUDE LIMITS. NORTH	SOUTH	LOCAL TIME	MEAN ALTITUDE
-87	39.91	36.97	6:24	16
-88	40.61	37.64	6:29	17
-89	41.35	38.33	6:35	18
-90	42.10	39.05	6:40	19
-93	44.61	41.38	6:57	23
-95	46.43	43.10	7:09	25
-97	48.36	44.96	7:21	27
-98	49.41	45.89	7:28	28

PERCENTAGE OF SUN OBSCURED ALONG MAP TRACK
WEST END = 96.3   EAST END = 96.8

==

+ MARKS THE CITY OF K'AI-FENG

ECLIPSE ON JD 2066471 DATE(Y/M/D) 945/ 9/ 9
DELTA T =   30.7 MINS

LONGITUDE (-VE EAST)	LATITUDE NORTH	LIMITS. SOUTH	LOCAL TIME	MEAN ALTITUDE
-87	29.35	25.27	13:30	58
-88	28.07	23.99	13:39	57
-89	26.79	22.73	13:48	56
-90	25.54	21.50	13:56	55
-92	23.10	19.15	14:13	53
-93	21.94	18.03	14:22	52
-94	20.81	16.97	14:30	50
-95	19.74	15.95	14:38	49
-97	17.73	14.07	14:53	46
-98	16.80	13.20	15:01	44
-99	15.91	12.38	15:08	42
-100	15.08	11.59	15:15	41

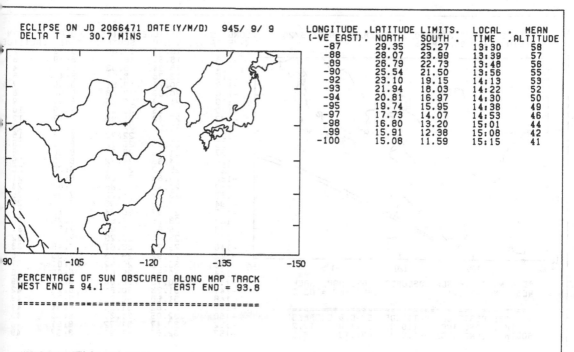

PERCENTAGE OF SUN OBSCURED ALONG MAP TRACK
WEST END = 94.1   EAST END = 93.8

==

+ MARKS THE CITY OF K'AI-FENG

ECLIPSE ON JD 2067505 DATE(Y/M/D) 948/ 7/ 9
DELTA T =   30.5 MINS

LONGITUDE (-VE EAST)	LATITUDE NORTH	LIMITS. SOUTH	LOCAL TIME	MEAN ALTITUDE
-130	52.71	46.45	18:35	11
-132	51.16	45.30	18:45	9
-133	50.48	44.71	18:49	8
-134	49.79	44.18	18:53	7
-135	49.12	43.68	18:58	6
-137	47.87	42.72	19:06	4
-138	47.29	42.26	19:11	3
-139	46.73	41.82	19:15	3
-140	46.19	41.41	19:19	2

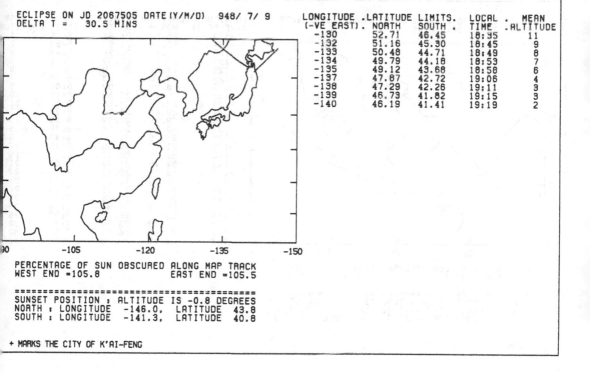

PERCENTAGE OF SUN OBSCURED ALONG MAP TRACK
WEST END =105.8   EAST END =105.5

==
SUNSET POSITION : ALTITUDE IS -0.8 DEGREES
NORTH : LONGITUDE -146.0, LATITUDE 43.8
SOUTH : LONGITUDE -141.3, LATITUDE 40.8

+ MARKS THE CITY OF K'AI-FENG

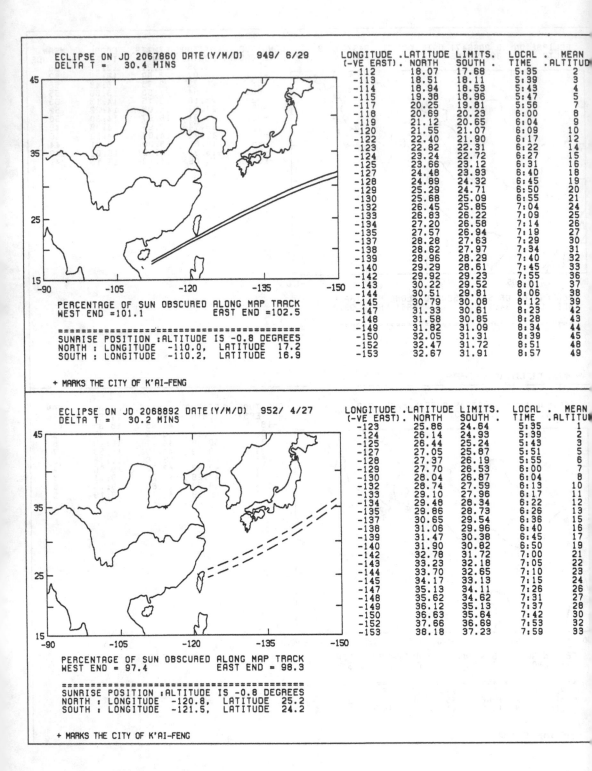

ECLIPSE ON JD 2067860 DATE(Y/M/D) 949/ 6/29
DELTA T =   30.4 MINS

LONGITUDE (-VE EAST)	LATITUDE NORTH	LIMITS. SOUTH	LOCAL TIME	MEAN ALTITUDE
-112	18.07	17.68	5:35	2
-113	18.51	18.11	5:39	3
-114	18.94	18.53	5:43	4
-115	19.38	18.96	5:47	5
-117	20.25	19.81	5:56	7
-118	20.69	20.23	6:00	8
-119	21.12	20.65	6:04	9
-120	21.55	21.07	6:09	10
-122	22.40	21.90	6:17	12
-123	22.82	22.31	6:22	14
-124	23.24	22.72	6:27	15
-125	23.66	23.12	6:31	16
-127	24.48	23.93	6:40	18
-128	24.89	24.32	6:45	19
-129	25.29	24.71	6:50	20
-130	25.68	25.09	6:55	21
-132	26.45	25.85	7:04	24
-133	26.83	26.22	7:09	25
-134	27.20	26.58	7:14	26
-135	27.57	26.94	7:19	27
-137	28.28	27.63	7:29	30
-138	28.62	27.97	7:34	31
-139	28.96	28.29	7:40	32
-140	29.29	28.61	7:45	33
-142	29.92	29.23	7:55	36
-143	30.22	29.52	8:01	37
-144	30.51	29.81	8:06	38
-145	30.79	30.08	8:12	39
-147	31.33	30.61	8:23	42
-148	31.58	30.85	8:28	43
-149	31.82	31.09	8:34	44
-150	32.05	31.31	8:39	45
-152	32.47	31.72	8:51	48
-153	32.67	31.91	8:57	49

PERCENTAGE OF SUN OBSCURED ALONG MAP TRACK
WEST END =101.1   EAST END =102.5

==
SUNRISE POSITION :ALTITUDE IS -0.8 DEGREES
NORTH : LONGITUDE -110.0, LATITUDE 17.2
SOUTH : LONGITUDE -110.2, LATITUDE 16.9

+ MARKS THE CITY OF K'AI-FENG

ECLIPSE ON JD 2068892 DATE(Y/M/D) 952/ 4/27
DELTA T =   30.2 MINS

LONGITUDE (-VE EAST)	LATITUDE NORTH	LIMITS. SOUTH	LOCAL TIME	MEAN ALTITUDE
-123	25.86	24.64	5:35	1
-124	26.14	24.93	5:39	2
-125	26.44	25.24	5:43	3
-127	27.05	25.87	5:51	5
-128	27.37	26.19	5:55	6
-129	27.70	26.53	6:00	7
-130	28.04	26.87	6:04	8
-132	28.74	27.59	6:13	10
-133	29.10	27.96	6:17	11
-134	29.48	28.34	6:22	12
-135	29.86	28.73	6:26	13
-137	30.65	29.54	6:36	15
-138	31.06	29.96	6:40	16
-139	31.47	30.38	6:45	17
-140	31.90	30.82	6:50	19
-142	32.78	31.72	7:00	21
-143	33.23	32.18	7:05	22
-144	33.70	32.65	7:10	23
-145	34.17	33.13	7:15	24
-147	35.13	34.11	7:26	26
-148	35.62	34.62	7:31	27
-149	36.12	35.13	7:37	28
-150	36.63	35.64	7:42	30
-152	37.66	36.69	7:53	32
-153	38.18	37.23	7:59	33

PERCENTAGE OF SUN OBSCURED ALONG MAP TRACK
WEST END = 97.4   EAST END = 98.3

==
SUNRISE POSITION :ALTITUDE IS -0.8 DEGREES
NORTH : LONGITUDE -120.8, LATITUDE 25.2
SOUTH : LONGITUDE -121.5, LATITUDE 24.2

+ MARKS THE CITY OF K'AI-FENG

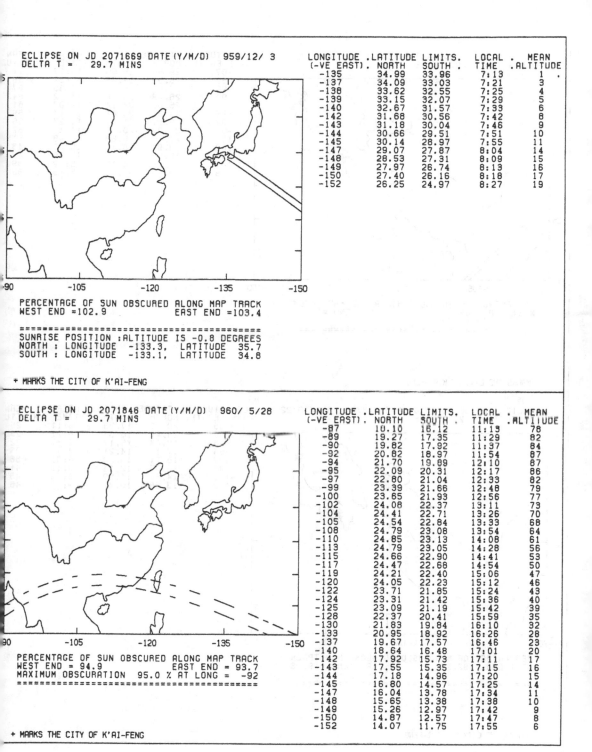

ECLIPSE ON JD 2071669 DATE(Y/M/D) 959/12/ 3
DELTA T =   29.7 MINS

LONGITUDE (-VE EAST)	LATITUDE NORTH	LIMITS. SOUTH	LOCAL TIME	MEAN ALTITUDE
-135	34.99	33.96	7:13	1
-137	34.09	33.03	7:21	3
-138	33.62	32.55	7:25	4
-139	33.15	32.07	7:29	5
-140	32.67	31.57	7:33	6
-142	31.68	30.56	7:42	8
-143	31.18	30.04	7:46	9
-144	30.66	29.51	7:51	10
-145	30.14	28.97	7:55	11
-147	29.07	27.87	8:04	14
-148	28.53	27.31	8:09	15
-149	27.97	26.74	8:13	16
-150	27.40	26.16	8:18	17
-152	26.25	24.97	8:27	19

PERCENTAGE OF SUN OBSCURED ALONG MAP TRACK
WEST END =102.9   EAST END =103.4

==
SUNRISE POSITION :ALTITUDE IS -0.8 DEGREES
NORTH : LONGITUDE -133.3, LATITUDE 35.7
SOUTH : LONGITUDE -133.1, LATITUDE 34.8

+ MARKS THE CITY OF K'AI-FENG

ECLIPSE ON JD 2071846 DATE(Y/M/D) 960/ 5/28
DELTA T =   29.7 MINS

LONGITUDE (-VE EAST)	LATITUDE NORTH	LIMITS. SOUTH	LOCAL TIME	MEAN ALTITUDE
-87	18.10	16.12	11:19	78
-89	19.27	17.35	11:29	82
-90	19.82	17.92	11:37	84
-92	20.82	18.97	11:54	87
-94	21.70	19.89	12:10	87
-95	22.09	20.31	12:17	86
-97	22.80	21.04	12:33	82
-99	23.39	21.66	12:48	79
-100	23.65	21.93	12:56	77
-102	24.08	22.37	13:11	73
-104	24.41	22.71	13:26	70
-105	24.54	22.84	13:33	68
-108	24.79	23.08	13:54	64
-110	24.85	23.13	14:08	61
-113	24.79	23.05	14:28	56
-115	24.66	22.90	14:41	53
-117	24.47	22.68	14:54	50
-119	24.21	22.40	15:06	47
-120	24.05	22.23	15:12	46
-122	23.71	21.85	15:24	43
-124	23.31	21.42	15:36	40
-125	23.09	21.19	15:42	39
-128	22.37	20.41	15:59	35
-130	21.83	19.84	16:10	32
-133	20.95	18.92	16:26	28
-137	19.67	17.57	16:46	23
-140	18.64	16.48	17:01	20
-142	17.92	15.73	17:11	17
-143	17.55	15.35	17:15	16
-144	17.18	14.96	17:20	15
-145	16.80	14.57	17:25	14
-147	16.04	13.78	17:34	11
-148	15.65	13.38	17:38	10
-149	15.26	12.97	17:42	9
-150	14.87	12.57	17:47	8
-152	14.07	11.75	17:55	6

PERCENTAGE OF SUN OBSCURED ALONG MAP TRACK
WEST END = 94.9   EAST END = 93.7
MAXIMUM OBSCURATION   95.0 % AT LONG = -92
==

+ MARKS THE CITY OF K'AI-FENG

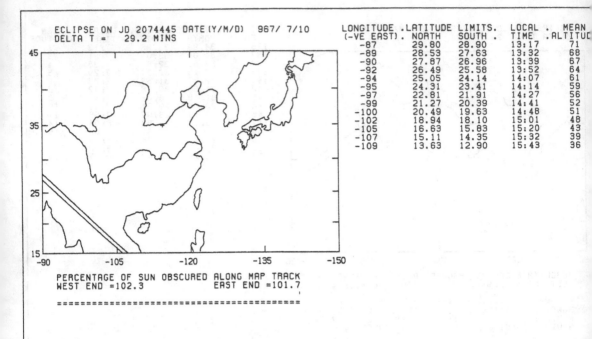

ECLIPSE ON JD 2074445 DATE(Y/M/D) 967/ 7/10
DELTA T =   29.2 MINS

LONGITUDE (-VE EAST)	.LATITUDE . NORTH	LIMITS. SOUTH .	LOCAL . TIME	MEAN .ALTITUDE
-87	29.80	28.90	13:17	71
-89	28.53	27.63	13:32	68
-90	27.87	26.96	13:39	67
-92	26.49	25.58	13:52	64
-94	25.05	24.14	14:07	61
-95	24.31	23.41	14:14	59
-97	22.81	21.91	14:27	56
-99	21.27	20.39	14:41	52
-100	20.49	19.63	14:48	51
-102	18.94	18.10	15:01	48
-105	16.63	15.83	15:20	43
-107	15.11	14.35	15:32	39
-109	13.63	12.90	15:43	36

PERCENTAGE OF SUN OBSCURED ALONG MAP TRACK
WEST END =102.3   EAST END =101.7

+ MARKS THE CITY OF K'AI-FENG

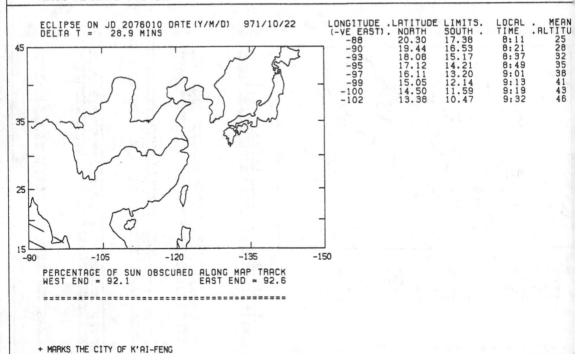

ECLIPSE ON JD 2076010 DATE(Y/M/D) 971/10/22
DELTA T =   28.9 MINS

LONGITUDE (-VE EAST)	.LATITUDE . NORTH	LIMITS. SOUTH .	LOCAL . TIME	MEAN .ALTITUDE
-88	20.30	17.38	8:11	25
-90	19.44	16.53	8:21	28
-93	18.08	15.17	8:37	32
-95	17.12	14.21	8:49	35
-97	16.11	13.20	9:01	38
-99	15.05	12.14	9:13	41
-100	14.50	11.59	9:19	43
-102	13.38	10.47	9:32	46

PERCENTAGE OF SUN OBSCURED ALONG MAP TRACK
WEST END = 92.1   EAST END = 92.6

+ MARKS THE CITY OF K'AI-FENG

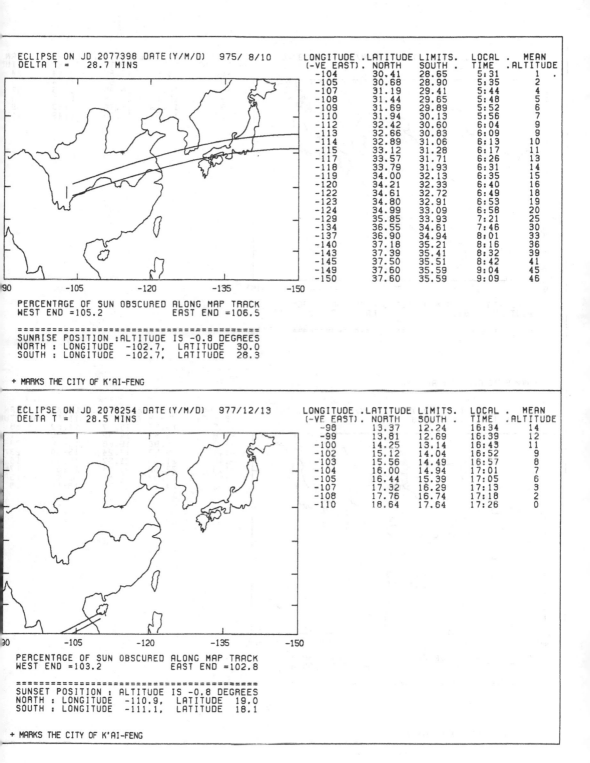

ECLIPSE ON JD 2077398 DATE(Y/M/D)   975/ 8/10
DELTA T =   28.7 MINS

LONGITUDE	.LATITUDE	LIMITS.	LOCAL	. MEAN
(-VE EAST).	NORTH	SOUTH .	TIME	.ALTITUDE
-104	30.41	28.65	5:31	1
-105	30.68	28.90	5:35	2
-107	31.19	29.41	5:44	4
-108	31.44	29.65	5:48	5
-109	31.69	29.89	5:52	6
-110	31.94	30.13	5:56	7
-112	32.42	30.60	6:04	9
-113	32.66	30.83	6:09	9
-114	32.89	31.06	6:13	10
-115	33.12	31.28	6:17	11
-117	33.57	31.71	6:26	13
-118	33.79	31.93	6:31	14
-119	34.00	32.13	6:35	15
-120	34.21	32.33	6:40	16
-122	34.61	32.72	6:49	18
-123	34.80	32.91	6:53	19
-124	34.99	33.09	6:58	20
-129	35.85	33.93	7:21	25
-134	36.55	34.61	7:46	30
-137	36.90	34.94	8:01	33
-140	37.18	35.21	8:16	36
-143	37.39	35.41	8:32	39
-145	37.50	35.51	8:42	41
-149	37.60	35.59	9:04	45
-150	37.60	35.59	9:09	46

PERCENTAGE OF SUN OBSCURED ALONG MAP TRACK
WEST END =105.2   EAST END =106.5

==
SUNRISE POSITION :ALTITUDE IS -0.8 DEGREES
NORTH : LONGITUDE -102.7, LATITUDE 30.0
SOUTH : LONGITUDE -102.7, LATITUDE 28.3

+ MARKS THE CITY OF K'AI-FENG

ECLIPSE ON JD 2078254 DATE(Y/M/D)   977/12/13
DELTA T =   28.5 MINS

LONGITUDE	.LATITUDE	LIMITS.	LOCAL	. MEAN
(-VE EAST).	NORTH	SOUTH .	TIME	.ALTITUDE
-98	13.37	12.24	16:34	14
-99	13.81	12.69	16:39	12
-100	14.25	13.14	16:43	11
-102	15.12	14.04	16:52	9
-103	15.56	14.49	16:57	8
-104	16.00	14.94	17:01	7
-105	16.44	15.39	17:05	6
-107	17.32	16.29	17:13	3
-108	17.76	16.74	17:18	2
-110	18.64	17.64	17:26	0

PERCENTAGE OF SUN OBSCURED ALONG MAP TRACK
WEST END =103.2   EAST END =102.8

==
SUNSET POSITION : ALTITUDE IS -0.8 DEGREES
NORTH : LONGITUDE -110.9, LATITUDE 19.0
SOUTH : LONGITUDE -111.1, LATITUDE 18.1

+ MARKS THE CITY OF K'AI-FENG

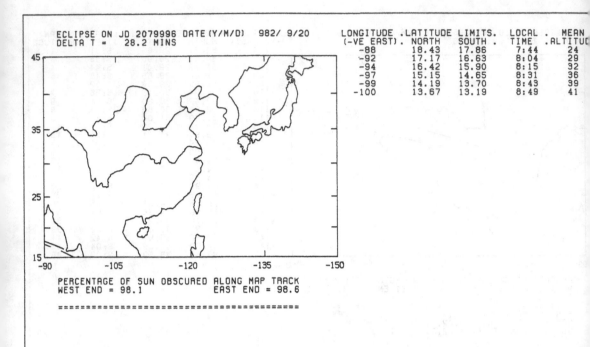

ECLIPSE ON JD 2079996 DATE(Y/M/D)   982/ 9/20
DELTA T =   28.2 MINS

LONGITUDE (-VE EAST)	LATITUDE NORTH	LIMITS. SOUTH	LOCAL TIME	MEAN ALTITUD
-88	18.43	17.86	7:44	24
-92	17.17	16.63	8:04	29
-94	16.42	15.90	8:15	32
-97	15.15	14.65	8:31	36
-99	14.19	13.70	8:43	39
-100	13.67	13.19	8:49	41

PERCENTAGE OF SUN OBSCURED ALONG MAP TRACK
WEST END = 98.1   EAST END = 98.6

===

+ MARKS THE CITY OF K'AI-FENG

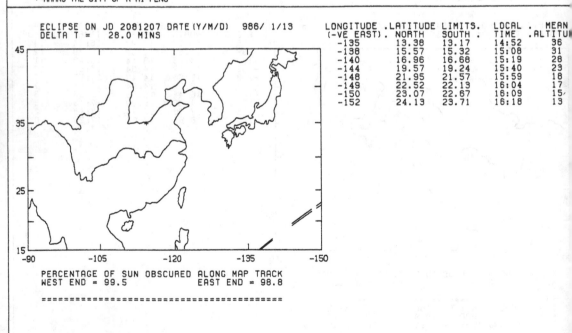

ECLIPSE ON JD 2081207 DATE(Y/M/D)   986/ 1/13
DELTA T =   28.0 MINS

LONGITUDE (-VE EAST)	LATITUDE NORTH	LIMITS. SOUTH	LOCAL TIME	MEAN ALTITUD
-135	13.38	13.17	14:52	36
-138	15.57	15.32	15:08	31
-140	16.96	16.68	15:19	28
-144	19.57	19.24	15:40	23
-148	21.95	21.57	15:59	18
-149	22.52	22.13	16:04	17
-150	23.07	22.67	16:09	15
-152	24.13	23.71	16:18	13

PERCENTAGE OF SUN OBSCURED ALONG MAP TRACK
WEST END = 99.5   EAST END = 98.8

===

+ MARKS THE CITY OF K'AI-FENG

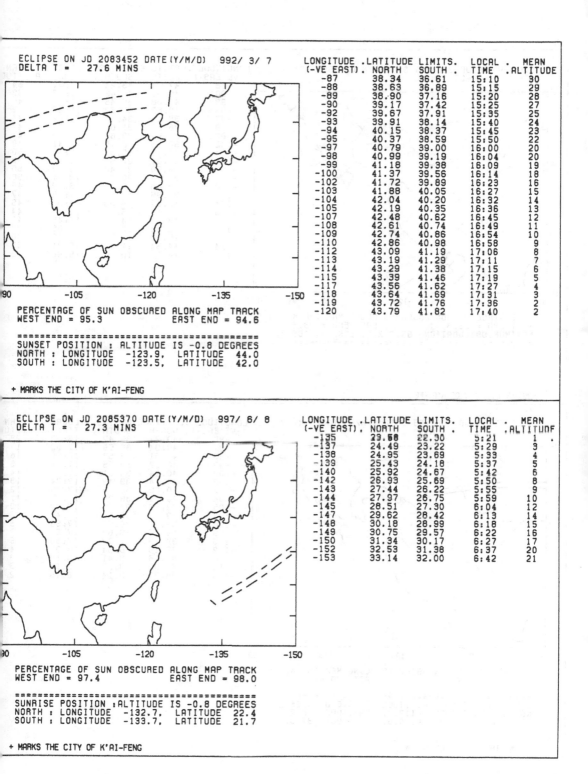

ECLIPSE ON JD 2083452 DATE(Y/M/D) 992/ 3/ 7
DELTA T = 27.6 MINS

LONGITUDE (-VE EAST).	LATITUDE NORTH	LIMITS. SOUTH .	LOCAL TIME	MEAN .ALTITUDE
-87	38.34	36.61	15:10	30
-88	38.63	36.89	15:15	29
-89	38.90	37.16	15:20	28
-90	39.17	37.42	15:25	27
-92	39.67	37.91	15:35	25
-93	39.91	38.14	15:40	24
-94	40.15	38.37	15:45	23
-95	40.37	38.59	15:50	22
-97	40.79	39.00	16:00	20
-98	40.99	39.19	16:04	20
-99	41.18	39.38	16:09	19
-100	41.37	39.56	16:14	18
-102	41.72	39.89	16:23	16
-103	41.88	40.05	16:27	15
-104	42.04	40.20	16:32	14
-105	42.19	40.35	16:36	13
-107	42.48	40.62	16:45	12
-108	42.61	40.74	16:49	11
-109	42.74	40.86	16:54	10
-110	42.86	40.98	16:58	9
-112	43.09	41.19	17:06	8
-113	43.19	41.29	17:11	7
-114	43.29	41.38	17:15	6
-115	43.39	41.46	17:19	5
-117	43.56	41.62	17:27	4
-118	43.64	41.69	17:31	3
-119	43.72	41.76	17:36	2
-120	43.79	41.82	17:40	2

PERCENTAGE OF SUN OBSCURED ALONG MAP TRACK
WEST END = 95.3   EAST END = 94.6

==
SUNSET POSITION : ALTITUDE IS -0.8 DEGREES
NORTH : LONGITUDE -123.9, LATITUDE 44.0
SOUTH : LONGITUDE -123.5, LATITUDE 42.0

+ MARKS THE CITY OF K'AI-FENG

ECLIPSE ON JD 2085370 DATE(Y/M/D) 997/ 6/ 8
DELTA T = 27.3 MINS

LONGITUDE (-VE EAST).	LATITUDE NORTH	LIMITS. SOUTH .	LOCAL TIME	MEAN .ALTITUDE
-135	23.58	22.30	5:21	1
-137	24.49	23.22	5:29	3
-138	24.95	23.69	5:33	4
-139	25.43	24.18	5:37	5
-140	25.92	24.67	5:42	6
-142	26.93	25.69	5:50	8
-143	27.44	26.22	5:55	9
-144	27.97	26.75	5:59	10
-145	28.51	27.30	6:04	12
-147	29.62	28.42	6:13	14
-148	30.18	28.99	6:18	15
-149	30.75	29.57	6:22	16
-150	31.34	30.17	6:27	17
-152	32.53	31.38	6:37	20
-153	33.14	32.00	6:42	21

PERCENTAGE OF SUN OBSCURED ALONG MAP TRACK
WEST END = 97.4   EAST END = 98.0

==
SUNRISE POSITION :ALTITUDE IS -0.8 DEGREES
NORTH : LONGITUDE -132.7, LATITUDE 22.4
SOUTH : LONGITUDE -133.7, LATITUDE 21.7

+ MARKS THE CITY OF K'AI-FENG

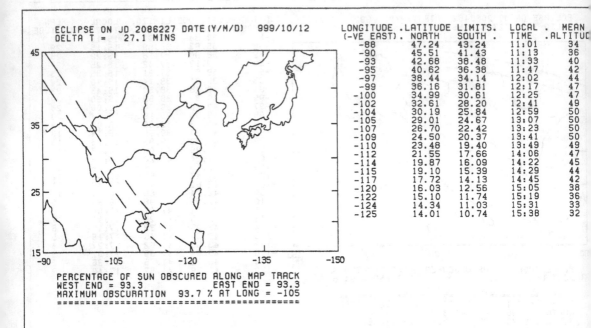

ECLIPSE ON JD 2086227 DATE(Y/M/D) 999/10/12
DELTA T = 27.1 MINS

LONGITUDE (-VE EAST)	LATITUDE NORTH	LIMITS SOUTH	LOCAL TIME	MEAN ALTITUDE
-88	47.24	43.24	11:01	34
-90	45.51	41.43	11:13	36
-93	42.68	38.48	11:33	40
-95	40.62	36.38	11:47	42
-97	38.44	34.14	12:02	44
-99	36.16	31.81	12:17	47
-100	34.99	30.61	12:25	47
-102	32.61	28.20	12:41	49
-104	30.19	25.84	12:59	50
-105	29.01	24.67	13:07	50
-107	26.70	22.42	13:23	50
-109	24.50	20.37	13:41	50
-110	23.48	19.40	13:49	49
-112	21.55	17.66	14:06	47
-114	19.87	16.09	14:22	45
-115	19.10	15.39	14:29	44
-117	17.72	14.13	14:45	42
-120	16.03	12.56	15:05	38
-122	15.10	11.74	15:19	36
-124	14.34	11.03	15:31	33
-125	14.01	10.74	15:38	32

PERCENTAGE OF SUN OBSCURED ALONG MAP TRACK
WEST END = 93.3   EAST END = 93.3
MAXIMUM OBSCURATION 93.7 % AT LONG = -105
==

+ MARKS THE CITY OF K'AI-FENG

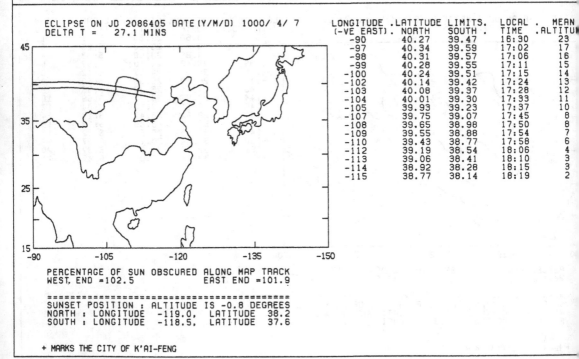

ECLIPSE ON JD 2086405 DATE(Y/M/D) 1000/ 4/ 7
DELTA T = 27.1 MINS

LONGITUDE (-VE EAST)	LATITUDE NORTH	LIMITS SOUTH	LOCAL TIME	MEAN ALTITUDE
-90	40.27	39.47	16:30	23
-97	40.34	39.59	17:02	17
-98	40.31	39.57	17:06	16
-99	40.28	39.55	17:11	15
-100	40.24	39.51	17:15	14
-102	40.14	39.42	17:24	13
-103	40.08	39.37	17:28	12
-104	40.01	39.30	17:33	11
-105	39.93	39.23	17:37	10
-107	39.75	39.07	17:45	8
-108	39.65	38.98	17:50	8
-109	39.55	38.88	17:54	7
-110	39.43	38.77	17:58	6
-112	39.19	38.54	18:06	4
-113	39.06	38.41	18:10	3
-114	38.92	38.28	18:15	3
-115	38.77	38.14	18:19	2

PERCENTAGE OF SUN OBSCURED ALONG MAP TRACK
WEST END =102.5   EAST END =101.9

==
SUNSET POSITION : ALTITUDE IS -0.8 DEGREES
NORTH : LONGITUDE -119.0, LATITUDE 38.2
SOUTH : LONGITUDE -118.5, LATITUDE 37.6

+ MARKS THE CITY OF K'AI-FENG

ECLIPSE ON JD 2087615 DATE(Y/M/D) 1003/ 8/ 1
DELTA T =   26.9 MINS

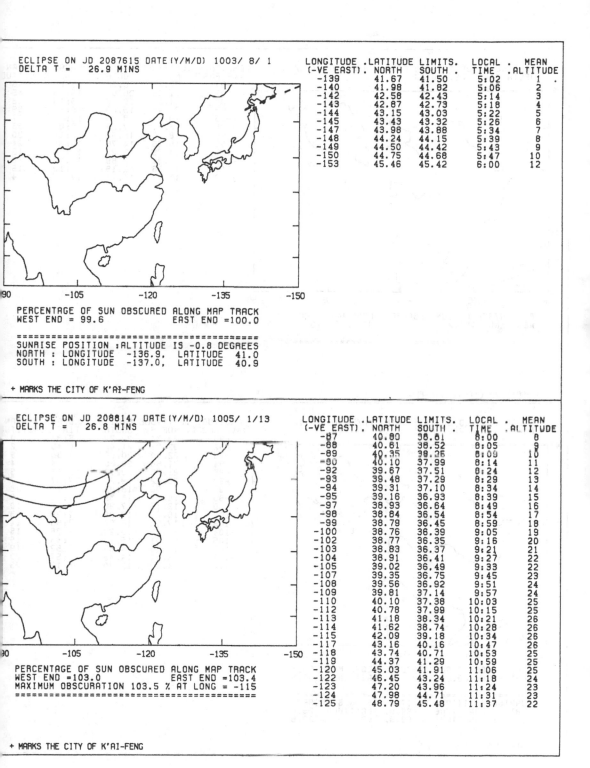

LONGITUDE .LATITUDE LIMITS.			LOCAL .	MEAN
(-VE EAST).	NORTH	SOUTH .	TIME	.ALTITUDE .
-139	41.67	41.50	5:02	1
-140	41.98	41.82	5:06	2
-142	42.58	42.43	5:14	3
-143	42.87	42.73	5:18	4
-144	43.15	43.03	5:22	5
-145	43.43	43.32	5:26	6
-147	43.98	43.88	5:34	7
-148	44.24	44.15	5:39	8
-149	44.50	44.42	5:43	9
-150	44.75	44.68	5:47	10
-153	45.46	45.42	6:00	12

PERCENTAGE OF SUN OBSCURED ALONG MAP TRACK
WEST END = 99.6   EAST END =100.0

==
SUNRISE POSITION :ALTITUDE IS -0.8 DEGREES
NORTH : LONGITUDE -136.9, LATITUDE   41.0
SOUTH : LONGITUDE -137.0, LATITUDE   40.9

+ MARKS THE CITY OF K'AI-FENG

ECLIPSE ON JD 2088147 DATE(Y/M/D) 1005/ 1/13
DELTA T =   26.8 MINS

LONGITUDE .LATITUDE LIMITS.			LOCAL .	MEAN
(-VE EAST).	NORTH	SOUTH .	TIME	.ALTITUDE
-87	40.80	38.61	8:00	8
-88	40.61	38.52	8:05	9
-89	40.35	38.26	8:09	10
-90	40.10	37.99	8:14	11
-92	39.67	37.51	8:24	12
-93	39.48	37.29	8:29	13
-94	39.31	37.10	8:34	14
-95	39.16	36.93	8:39	15
-97	38.93	36.64	8:49	16
-98	38.84	36.54	8:54	17
-99	38.79	36.45	8:59	18
-100	38.76	36.39	9:05	19
-102	38.77	36.35	9:16	20
-103	38.83	36.37	9:21	21
-104	38.91	36.41	9:27	22
-105	39.02	36.49	9:33	22
-107	39.35	36.75	9:45	23
-108	39.56	36.92	9:51	24
-109	39.81	37.14	9:57	24
-110	40.10	37.38	10:03	25
-112	40.78	37.99	10:15	25
-113	41.18	38.34	10:21	26
-114	41.62	38.74	10:28	26
-115	42.09	39.18	10:34	26
-117	43.16	40.16	10:47	26
-118	43.74	40.71	10:53	25
-119	44.37	41.29	10:59	25
-120	45.03	41.91	11:06	25
-122	46.45	43.24	11:18	24
-123	47.20	43.96	11:24	23
-124	47.98	44.71	11:31	23
-125	48.79	45.48	11:37	22

PERCENTAGE OF SUN OBSCURED ALONG MAP TRACK
WEST END =103.0   EAST END =103.4
MAXIMUM OBSCURATION 103.5 % AT LONG = -115
==

+ MARKS THE CITY OF K'AI-FENG

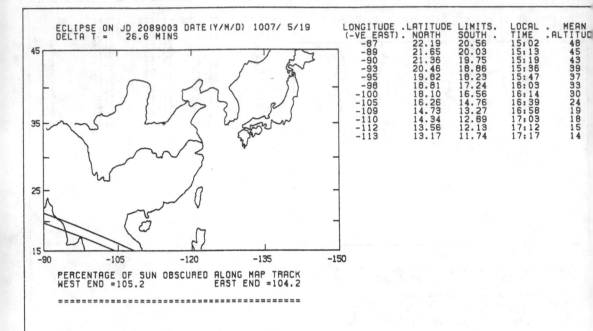

ECLIPSE ON JD 2089003 DATE(Y/M/D) 1007/ 5/19
DELTA T =   26.6 MINS

LONGITUDE (-VE EAST)	.LATITUDE NORTH	LIMITS. SOUTH .	LOCAL TIME	. MEAN .ALTITUD
-87	22.19	20.56	15:02	48
-89	21.65	20.03	15:13	45
-90	21.36	19.75	15:19	43
-93	20.46	18.86	15:36	39
-95	19.82	18.23	15:47	37
-98	18.81	17.24	16:03	33
-100	18.10	16.56	16:14	30
-105	16.26	14.76	16:39	24
-109	14.73	13.27	16:58	19
-110	14.34	12.89	17:03	18
-112	13.56	12.13	17:12	15
-113	13.17	11.74	17:17	14

PERCENTAGE OF SUN OBSCURED ALONG MAP TRACK
WEST END =105.2   EAST END =104.2

===

+ MARKS THE CITY OF K'AI-FENG

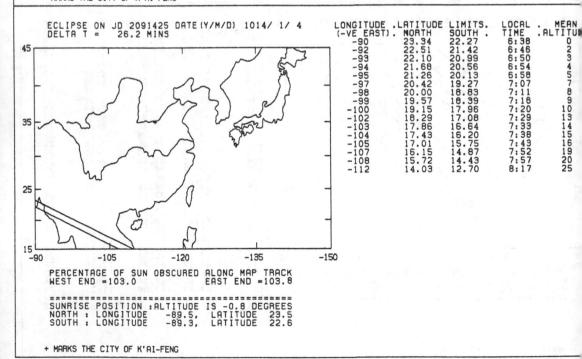

ECLIPSE ON JD 2091425 DATE(Y/M/D) 1014/ 1/ 4
DELTA T =   26.2 MINS

LONGITUDE (-VE EAST)	.LATITUDE NORTH	LIMITS. SOUTH .	LOCAL TIME	. MEAN .ALTITU
-90	23.34	22.27	6:38	0
-92	22.51	21.42	6:46	2
-93	22.10	20.99	6:50	3
-94	21.68	20.56	6:54	4
-95	21.26	20.13	6:58	5
-97	20.42	19.27	7:07	7
-98	20.00	18.83	7:11	8
-99	19.57	18.39	7:16	9
-100	19.15	17.96	7:20	10
-102	18.29	17.08	7:29	13
-103	17.86	16.64	7:33	14
-104	17.43	16.20	7:38	15
-105	17.01	15.75	7:43	16
-107	16.15	14.87	7:52	19
-108	15.72	14.43	7:57	20
-112	14.03	12.70	8:17	25

PERCENTAGE OF SUN OBSCURED ALONG MAP TRACK
WEST END =103.0   EAST END =103.8

===
SUNRISE POSITION :ALTITUDE IS -0.8 DEGREES
NORTH : LONGITUDE -89.5, LATITUDE 23.5
SOUTH : LONGITUDE -89.3, LATITUDE 22.6

+ MARKS THE CITY OF K'AI-FENG

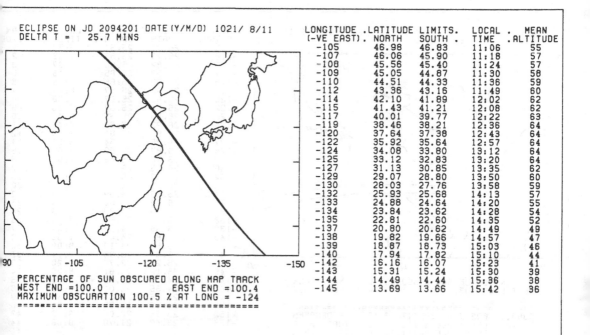

ECLIPSE ON JD 2094201 DATE(Y/M/D) 1021/ 8/11
DELTA T =   25.7 MINS

LONGITUDE	.LATITUDE LIMITS.		LOCAL .	MEAN
(-VE EAST).	NORTH	SOUTH .	TIME	.ALTITUDE
-105	46.98	46.83	11:06	55
-107	46.06	45.90	11:18	57
-108	45.56	45.40	11:24	57
-109	45.05	44.87	11:30	58
-110	44.51	44.33	11:36	59
-112	43.36	43.16	11:49	60
-114	42.10	41.89	12:02	62
-115	41.43	41.21	12:08	62
-117	40.01	39.77	12:22	63
-119	38.46	38.21	12:36	64
-120	37.64	37.38	12:43	64
-122	35.92	35.64	12:57	64
-124	34.08	33.80	13:12	64
-125	33.12	32.83	13:20	64
-127	31.13	30.85	13:35	62
-129	29.07	28.80	13:50	60
-130	28.03	27.76	13:58	59
-132	25.93	25.68	14:13	57
-133	24.88	24.64	14:20	55
-134	23.84	23.62	14:28	54
-135	22.81	22.60	14:35	52
-137	20.80	20.62	14:49	49
-138	19.82	19.66	14:57	47
-139	18.87	18.73	15:03	46
-140	17.94	17.82	15:10	44
-142	16.16	16.07	15:23	41
-143	15.31	15.24	15:30	39
-144	14.49	14.44	15:36	38
-145	13.69	13.66	15:42	36

PERCENTAGE OF SUN OBSCURED ALONG MAP TRACK
WEST END =100.0   EAST END =100.4
MAXIMUM OBSCURATION 100.5 % AT LONG = -124
===

+ MARKS THE CITY OF K'AI-FENG

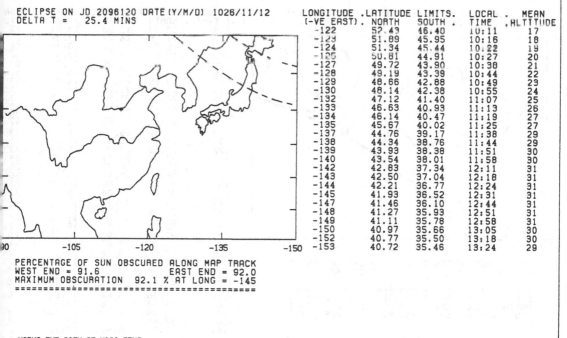

ECLIPSE ON JD 2096120 DATE(Y/M/D) 1026/11/12
DELTA T =   25.4 MINS

LONGITUDE	.LATITUDE LIMITS.		LOCAL .	MEAN
(-VE EAST).	NORTH	SOUTH .	TIME	.ALTITUDE
-122	52.43	46.40	10:11	17
-123	51.89	45.95	10:16	18
-124	51.34	45.44	10:22	19
-125	50.81	44.91	10:27	20
-127	49.72	43.90	10:38	21
-128	49.19	43.39	10:44	22
-129	48.66	42.88	10:49	23
-130	48.14	42.38	10:55	24
-132	47.12	41.40	11:07	25
-133	46.63	40.93	11:13	26
-134	46.14	40.47	11:19	27
-135	45.67	40.02	11:25	27
-137	44.76	39.17	11:38	29
-138	44.34	38.76	11:44	29
-139	43.93	38.38	11:51	30
-140	43.54	38.01	11:58	30
-142	42.83	37.34	12:11	31
-143	42.50	37.04	12:18	31
-144	42.21	36.77	12:24	31
-145	41.93	36.52	12:31	31
-147	41.46	36.10	12:44	31
-148	41.27	35.93	12:51	31
-149	41.11	35.78	12:58	31
-150	40.97	35.66	13:05	30
-152	40.77	35.50	13:18	30
-153	40.72	35.46	13:24	29

PERCENTAGE OF SUN OBSCURED ALONG MAP TRACK
WEST END = 91.6   EAST END = 92.0
MAXIMUM OBSCURATION 92.1 % AT LONG = -145
===

+ MARKS THE CITY OF K'AI-FENG

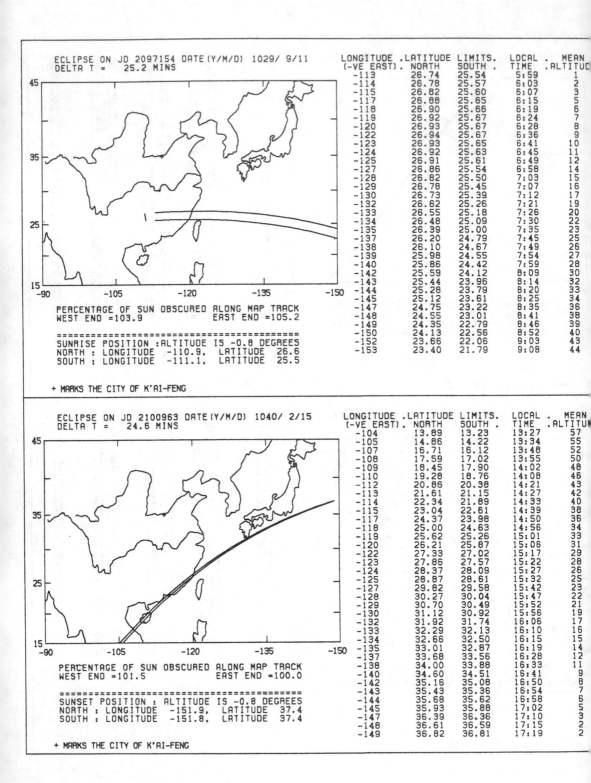

ECLIPSE ON JD 2097154 DATE(Y/M/D) 1029/ 9/11
DELTA T = 25.2 MINS

PERCENTAGE OF SUN OBSCURED ALONG MAP TRACK
WEST END =103.9   EAST END =105.2

===
SUNRISE POSITION :ALTITUDE IS -0.8 DEGREES
NORTH : LONGITUDE -110.9, LATITUDE 26.6
SOUTH : LONGITUDE -111.1, LATITUDE 25.5

+ MARKS THE CITY OF K'AI-FENG

LONGITUDE (-VE EAST)	LATITUDE LIMITS. NORTH	SOUTH	LOCAL TIME	MEAN ALTITUDE
-113	26.74	25.54	5:59	1
-114	26.78	25.57	6:03	2
-115	26.82	25.60	6:07	3
-117	26.88	25.65	6:15	5
-118	26.90	25.66	6:19	6
-119	26.92	25.67	6:24	7
-120	26.93	25.67	6:28	8
-122	26.94	25.67	6:36	9
-123	26.93	25.65	6:41	10
-124	26.92	25.63	6:45	11
-125	26.91	25.61	6:49	12
-127	26.86	25.54	6:58	14
-128	26.82	25.50	7:03	15
-129	26.78	25.45	7:07	16
-130	26.73	25.39	7:12	17
-132	26.62	25.26	7:21	19
-133	26.55	25.18	7:26	20
-134	26.48	25.09	7:30	22
-135	26.39	25.00	7:35	23
-137	26.20	24.79	7:45	25
-138	26.10	24.67	7:49	26
-139	25.98	24.55	7:54	27
-140	25.86	24.42	7:59	28
-142	25.59	24.12	8:09	30
-143	25.44	23.96	8:14	32
-144	25.28	23.79	8:20	33
-145	25.12	23.61	8:25	34
-147	24.75	23.22	8:35	36
-148	24.55	23.01	8:41	38
-149	24.35	22.79	8:46	39
-150	24.13	22.56	8:52	40
-152	23.66	22.06	9:03	43
-153	23.40	21.79	9:08	44

ECLIPSE ON JD 2100963 DATE(Y/M/D) 1040/ 2/15
DELTA T = 24.6 MINS

PERCENTAGE OF SUN OBSCURED ALONG MAP TRACK
WEST END =101.5   EAST END =100.0

===
SUNSET POSITION : ALTITUDE IS -0.8 DEGREES
NORTH : LONGITUDE -151.9, LATITUDE 37.4
SOUTH : LONGITUDE -151.8, LATITUDE 37.4

+ MARKS THE CITY OF K'AI-FENG

LONGITUDE (-VE EAST)	LATITUDE LIMITS. NORTH	SOUTH	LOCAL TIME	MEAN ALTITUDE
-104	13.89	13.23	13:27	57
-105	14.86	14.22	13:34	55
-107	16.71	16.12	13:48	52
-108	17.59	17.02	13:55	50
-109	18.45	17.90	14:02	48
-110	19.28	18.76	14:08	46
-112	20.86	20.38	14:21	43
-113	21.61	21.15	14:27	42
-114	22.34	21.89	14:33	40
-115	23.04	22.61	14:39	38
-117	24.37	23.98	14:50	36
-118	25.00	24.63	14:56	34
-119	25.62	25.26	15:01	33
-120	26.21	25.87	15:06	31
-122	27.33	27.02	15:17	29
-123	27.86	27.57	15:22	28
-124	28.37	28.09	15:27	26
-125	28.87	28.61	15:32	25
-127	29.82	29.58	15:42	23
-128	30.27	30.04	15:47	22
-129	30.70	30.49	15:52	21
-130	31.12	30.92	15:56	19
-132	31.92	31.74	16:06	17
-133	32.29	32.13	16:10	16
-134	32.66	32.50	16:15	15
-135	33.01	32.87	16:19	14
-137	33.68	33.56	16:28	12
-138	34.00	33.88	16:33	11
-140	34.60	34.51	16:41	9
-142	35.16	35.08	16:50	8
-143	35.43	35.36	16:54	7
-144	35.68	35.62	16:58	6
-145	35.93	35.88	17:02	5
-147	36.39	36.36	17:10	3
-148	36.61	36.59	17:15	2
-149	36.82	36.81	17:19	2

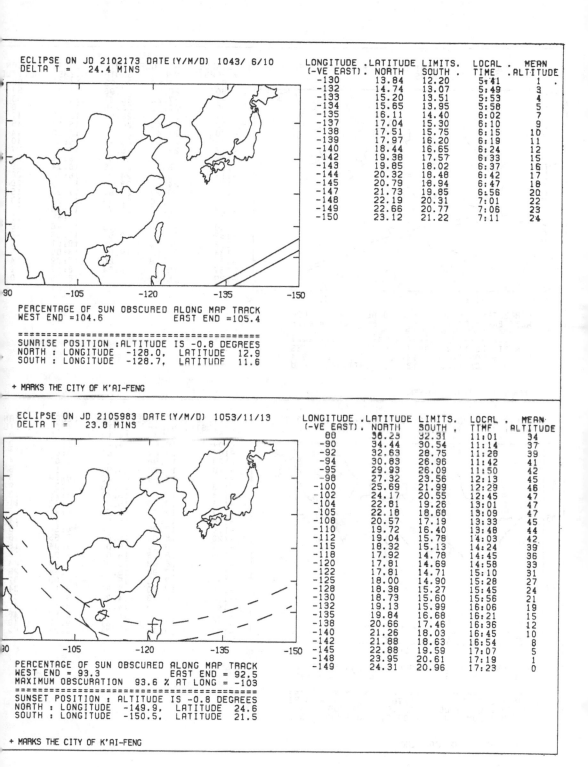

ECLIPSE ON JD 2102173 DATE (Y/M/D) 1043/ 6/10
DELTA T =   24.4 MINS

LONGITUDE	.LATITUDE LIMITS.		LOCAL	. MEAN
(-VE EAST).	NORTH	SOUTH .	TIME	.ALTITUDE .
-130	13.84	12.20	5:41	1
-132	14.74	13.07	5:49	3
-133	15.20	13.51	5:53	4
-134	15.65	13.95	5:58	5
-135	16.11	14.40	6:02	7
-137	17.04	15.30	6:10	9
-138	17.51	15.75	6:15	10
-139	17.97	16.20	6:19	11
-140	18.44	16.65	6:24	12
-142	19.38	17.57	6:33	15
-143	19.85	18.02	6:37	16
-144	20.32	18.48	6:42	17
-145	20.79	18.94	6:47	18
-147	21.73	19.85	6:56	20
-148	22.19	20.31	7:01	22
-149	22.66	20.77	7:06	23
-150	23.12	21.22	7:11	24

PERCENTAGE OF SUN OBSCURED ALONG MAP TRACK
WEST END =104.6   EAST END =105.4

==
SUNRISE POSITION :ALTITUDE IS -0.8 DEGREES
NORTH : LONGITUDE -128.0, LATITUDE 12.9
SOUTH : LONGITUDE -128.7, LATITUDE 11.6

+ MARKS THE CITY OF K'AI-FENG

ECLIPSE ON JD 2105983 DATE (Y/M/D) 1053/11/13
DELTA T =   23.8 MINS

LONGITUDE	.LATITUDE LIMITS.		LOCAL	. MEAN
(-VE EAST).	NORTH	SOUTH .	TIME	ALTITUDE
80	36.25	32.31	11:01	34
-90	34.44	30.54	11:14	37
-92	32.63	28.75	11:28	39
-94	30.83	26.96	11:42	41
-95	29.93	26.09	11:50	42
-98	27.32	23.56	12:13	45
-100	25.69	21.99	12:29	46
-102	24.17	20.55	12:45	47
-104	22.81	19.26	13:01	47
-105	22.18	18.68	13:09	47
-108	20.57	17.19	13:33	45
-110	19.72	16.40	13:48	44
-112	19.04	15.78	14:03	42
-115	18.32	15.13	14:24	39
-118	17.92	14.78	14:45	36
-120	17.81	14.69	14:58	33
-122	17.81	14.71	15:10	31
-125	18.00	14.90	15:28	27
-128	18.38	15.27	15:45	24
-130	18.73	15.60	15:56	21
-132	19.13	15.99	16:06	19
-135	19.84	16.68	16:21	15
-138	20.66	17.46	16:36	12
-140	21.26	18.03	16:45	10
-142	21.88	18.63	16:54	8
-145	22.88	19.59	17:07	5
-148	23.95	20.61	17:19	1
-149	24.31	20.96	17:23	0

PERCENTAGE OF SUN OBSCURED ALONG MAP TRACK
WEST END = 93.3   EAST END = 92.5
MAXIMUM OBSCURATION 93.6 % AT LONG = -103
==
SUNSET POSITION : ALTITUDE IS -0.8 DEGREES
NORTH : LONGITUDE -149.9, LATITUDE 24.6
SOUTH : LONGITUDE -150.5, LATITUDE 21.5

+ MARKS THE CITY OF K'AI-FENG

321

ECLIPSE ON JD 2106161 DATE(Y/M/D) 1054/ 5/10
DELTA T =   23.7 MINS

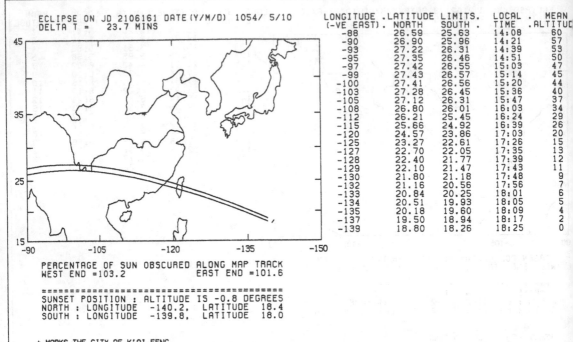

LONGITUDE (-VE EAST)	.LATITUDE LIMITS. NORTH	SOUTH .	LOCAL . TIME	MEAN .ALTITU
-88	26.59	25.63	14:08	60
-90	26.90	25.96	14:21	57
-93	27.22	26.31	14:39	53
-95	27.35	26.46	14:51	50
-97	27.42	26.55	15:03	47
-99	27.43	26.57	15:14	45
-100	27.41	26.56	15:20	44
-103	27.28	26.45	15:36	40
-105	27.12	26.31	15:47	37
-108	26.80	26.01	16:03	34
-112	26.21	25.45	16:24	29
-115	25.66	24.92	16:39	26
-120	24.57	23.86	17:03	20
-125	23.27	22.61	17:26	15
-127	22.70	22.05	17:35	13
-128	22.40	21.77	17:39	12
-129	22.10	21.47	17:43	11
-130	21.80	21.18	17:48	9
-132	21.16	20.56	17:56	7
-133	20.84	20.25	18:01	6
-134	20.51	19.93	18:05	5
-135	20.18	19.60	18:09	4
-137	19.50	18.94	18:17	2
-139	18.80	18.26	18:25	0

PERCENTAGE OF SUN OBSCURED ALONG MAP TRACK
WEST END =103.2   EAST END =101.6

==
SUNSET POSITION : ALTITUDE IS -0.8 DEGREES
NORTH : LONGITUDE -140.2, LATITUDE 18.4
SOUTH : LONGITUDE -139.8, LATITUDE 18.0

+ MARKS THE CITY OF K'AI-FENG

ECLIPSE ON JD 2107726 DATE(Y/M/D) 1058/ 8/22
DELTA T =   23.5 MINS

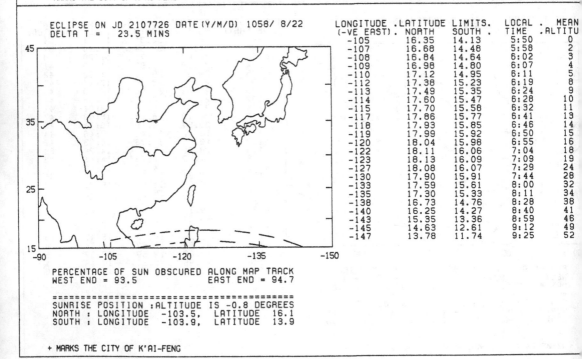

LONGITUDE (-VE EAST)	.LATITUDE LIMITS. NORTH	SOUTH .	LOCAL . TIME	MEAN .ALTITU
-105	16.35	14.13	5:50	0
-107	16.68	14.48	5:58	2
-108	16.84	14.64	6:02	3
-109	16.98	14.80	6:07	4
-110	17.12	14.95	6:11	5
-112	17.38	15.23	6:19	8
-113	17.49	15.35	6:24	9
-114	17.60	15.47	6:28	10
-115	17.70	15.58	6:32	11
-117	17.86	15.77	6:41	13
-118	17.93	15.85	6:46	14
-119	17.99	15.92	6:50	15
-120	18.04	15.98	6:55	16
-122	18.11	16.06	7:04	18
-123	18.13	16.09	7:09	19
-127	18.08	16.07	7:29	24
-130	17.90	15.91	7:44	28
-133	17.59	15.61	8:00	32
-135	17.30	15.33	8:11	34
-138	16.73	14.76	8:28	38
-140	16.25	14.27	8:40	41
-143	15.35	13.36	8:59	46
-145	14.63	12.61	9:12	49
-147	13.78	11.74	9:25	52

PERCENTAGE OF SUN OBSCURED ALONG MAP TRACK
WEST END = 93.5   EAST END = 94.7

==
SUNRISE POSITION :ALTITUDE IS -0.8 DEGREES
NORTH : LONGITUDE -103.5, LATITUDE 16.1
SOUTH : LONGITUDE -103.9, LATITUDE 13.9

+ MARKS THE CITY OF K'AI-FENG

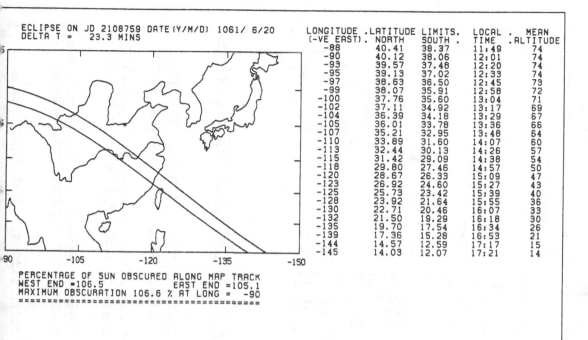

```
ECLIPSE ON JD 2108759 DATE(Y/M/D) 1061/ 6/20
DELTA T =    23.3 MINS
```

LONGITUDE (-VE EAST)	LATITUDE NORTH	LIMITS SOUTH	LOCAL TIME	MEAN ALTITUDE
-88	40.41	38.37	11:49	74
-90	40.12	38.06	12:01	74
-93	39.57	37.48	12:20	74
-95	39.13	37.02	12:33	74
-97	38.63	36.50	12:45	73
-99	38.07	35.91	12:58	72
-100	37.76	35.60	13:04	71
-102	37.11	34.92	13:17	69
-104	36.39	34.18	13:29	67
-105	36.01	33.78	13:36	66
-107	35.21	32.95	13:48	64
-110	33.89	31.60	14:07	60
-113	32.44	30.13	14:26	57
-115	31.42	29.09	14:38	54
-118	29.80	27.46	14:57	50
-120	28.67	26.33	15:09	47
-123	26.92	24.60	15:27	43
-125	25.73	23.42	15:39	40
-128	23.92	21.64	15:55	36
-130	22.71	20.46	16:07	33
-132	21.50	19.29	16:18	30
-135	19.70	17.54	16:34	26
-139	17.36	15.28	16:53	21
-144	14.57	12.59	17:17	15
-145	14.03	12.07	17:21	14

```
PERCENTAGE OF SUN OBSCURED ALONG MAP TRACK
WEST END =106.5              EAST END =105.1
MAXIMUM OBSCURATION 106.6 % AT LONG =   -90
==============================================
```

+ MARKS THE CITY OF K'AI-FENG

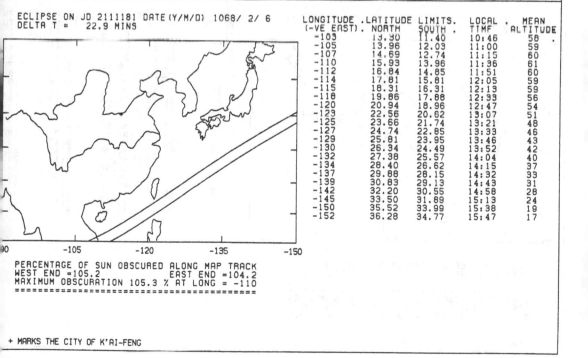

```
ECLIPSE ON JD 2111181 DATE(Y/M/D) 1068/ 2/ 6
DELTA T =    22.9 MINS
```

LONGITUDE (-VE EAST)	LATITUDE NORTH	LIMITS SOUTH	LOCAL TIME	MEAN ALTITUDE
-103	13.30	11.40	10:46	58
-105	13.96	12.03	11:00	59
-107	14.69	12.74	11:15	60
-110	15.93	13.96	11:36	61
-112	16.84	14.85	11:51	60
-114	17.81	15.81	12:05	59
-115	18.31	16.31	12:13	59
-118	19.86	17.88	12:33	56
-120	20.94	18.96	12:47	54
-123	22.56	20.62	13:07	51
-125	23.66	21.74	13:21	48
-127	24.74	22.85	13:33	46
-129	25.81	23.95	13:46	43
-130	26.34	24.49	13:52	42
-132	27.38	25.57	14:04	40
-134	28.40	26.62	14:15	37
-137	29.88	28.15	14:32	33
-139	30.83	29.13	14:43	31
-142	32.20	30.55	14:58	28
-145	33.50	31.89	15:13	24
-150	35.52	33.99	15:38	19
-152	36.28	34.77	15:47	17

```
PERCENTAGE OF SUN OBSCURED ALONG MAP TRACK
WEST END =105.2              EAST END =104.2
MAXIMUM OBSCURATION 105.3 % AT LONG = -110
==============================================
```

+ MARKS THE CITY OF K'AI-FENG

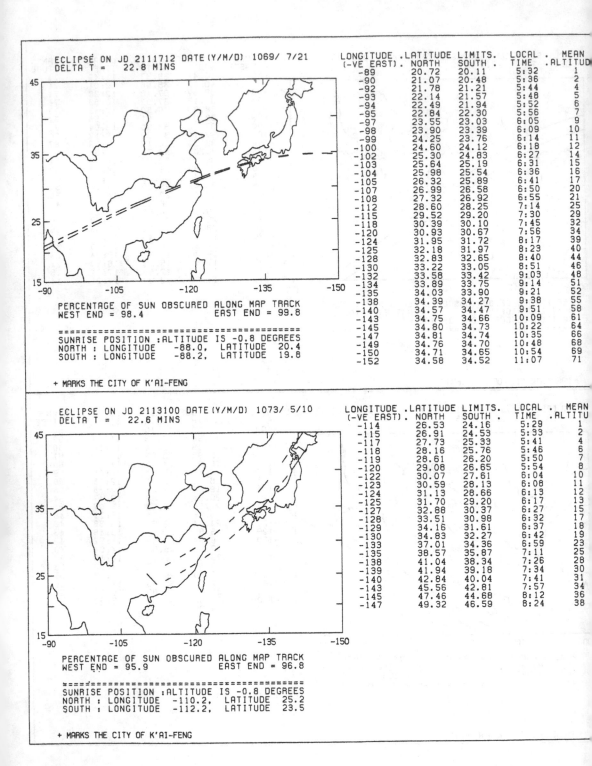

ECLIPSE ON JD 2111712 DATE(Y/M/D) 1069/ 7/21
DELTA T =   22.8 MINS

45

35

25

15
-90   -105   -120   -135   -150

PERCENTAGE OF SUN OBSCURED ALONG MAP TRACK
WEST END = 98.4   EAST END = 99.8

===
SUNRISE POSITION :ALTITUDE IS -0.8 DEGREES
NORTH : LONGITUDE -88.0, LATITUDE 20.4
SOUTH : LONGITUDE -88.2, LATITUDE 19.8

+ MARKS THE CITY OF K'AI-FENG

LONGITUDE (-VE EAST)	LATITUDE NORTH	LIMITS. SOUTH	LOCAL TIME	MEAN ALTITUDE
-89	20.72	20.11	5:32	1
-90	21.07	20.48	5:36	2
-92	21.78	21.21	5:44	4
-93	22.14	21.57	5:48	5
-94	22.49	21.94	5:52	6
-95	22.84	22.30	5:56	7
-97	23.55	23.03	6:05	9
-98	23.90	23.39	6:09	10
-99	24.25	23.76	6:14	11
-100	24.60	24.12	6:18	12
-102	25.30	24.83	6:27	14
-103	25.64	25.19	6:31	15
-104	25.98	25.54	6:36	16
-105	26.32	25.89	6:41	17
-107	26.99	26.58	6:50	20
-108	27.32	26.92	6:55	21
-112	28.60	28.25	7:14	25
-115	29.52	29.20	7:30	29
-118	30.39	30.10	7:45	32
-120	30.93	30.67	7:56	34
-124	31.95	31.72	8:17	39
-125	32.18	31.97	8:23	40
-128	32.83	32.65	8:40	44
-130	33.22	33.05	8:51	46
-132	33.58	33.42	9:03	48
-134	33.89	33.75	9:14	51
-135	34.03	33.90	9:21	52
-138	34.39	34.27	9:38	55
-140	34.57	34.47	9:51	58
-143	34.75	34.66	10:09	61
-145	34.80	34.73	10:22	64
-147	34.81	34.74	10:35	66
-149	34.76	34.70	10:48	68
-150	34.71	34.65	10:54	69
-152	34.58	34.52	11:07	71

ECLIPSE ON JD 2113100 DATE(Y/M/D) 1073/ 5/10
DELTA T =   22.6 MINS

45

35

25

15
-90   -105   -120   -135   -150

PERCENTAGE OF SUN OBSCURED ALONG MAP TRACK
WEST END = 95.9   EAST END = 96.8

===
SUNRISE POSITION :ALTITUDE IS -0.8 DEGREES
NORTH : LONGITUDE -110.2, LATITUDE 25.2
SOUTH : LONGITUDE -112.2, LATITUDE 23.5

+ MARKS THE CITY OF K'AI-FENG

LONGITUDE (-VE EAST)	LATITUDE NORTH	LIMITS. SOUTH	LOCAL TIME	MEAN ALTITUDE
-114	26.53	24.16	5:29	1
-115	26.91	24.53	5:33	2
-117	27.73	25.33	5:41	4
-118	28.16	25.76	5:46	6
-119	28.61	26.20	5:50	7
-120	29.08	26.65	5:54	8
-122	30.07	27.61	6:04	10
-123	30.59	28.13	6:08	11
-124	31.13	28.66	6:13	12
-125	31.70	29.20	6:17	13
-127	32.88	30.37	6:27	15
-128	33.51	30.98	6:32	17
-129	34.16	31.61	6:37	18
-130	34.83	32.27	6:42	19
-133	37.01	34.36	6:59	23
-135	38.57	35.87	7:11	25
-138	41.04	38.34	7:26	28
-139	41.94	39.18	7:34	30
-140	42.84	40.04	7:41	31
-143	45.56	42.81	7:57	34
-145	47.46	44.68	8:12	36
-147	49.32	46.59	8:24	38

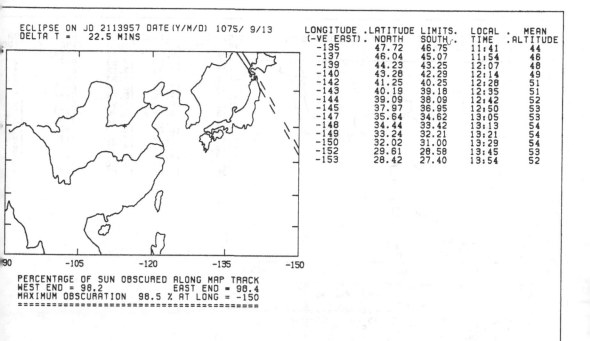

ECLIPSE ON JD 2113957 DATE(Y/M/D) 1075/ 9/13
DELTA T = 22.5 MINS

LONGITUDE (-VE EAST)	LATITUDE NORTH	LIMITS. SOUTH	LOCAL TIME	MEAN ALTITUDE
-135	47.72	46.75	11:41	44
-137	46.04	45.07	11:54	46
-139	44.23	43.25	12:07	48
-140	43.28	42.29	12:14	49
-142	41.25	40.25	12:28	51
-143	40.19	39.18	12:35	51
-144	39.09	38.09	12:42	52
-145	37.97	36.95	12:50	53
-147	35.64	34.62	13:05	53
-148	34.44	33.42	13:13	54
-149	33.24	32.21	13:21	54
-150	32.02	31.00	13:29	54
-152	29.61	28.58	13:45	53
-153	28.42	27.40	13:54	52

PERCENTAGE OF SUN OBSCURED ALONG MAP TRACK
WEST END = 98.2   EAST END = 98.4
MAXIMUM OBSCURATION 98.5 % AT LONG = -150
==

+ MARKS THE CITY OF K'AI-FENG

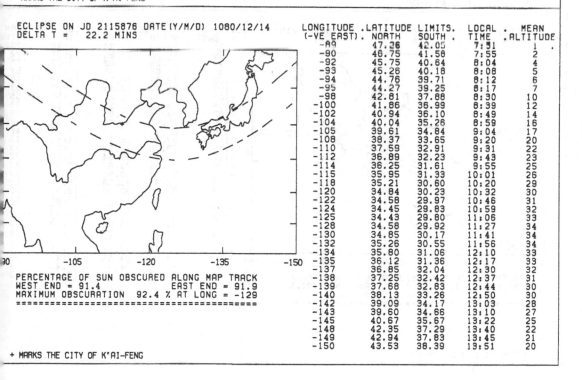

ECLIPSE ON JD 2115876 DATE(Y/M/D) 1080/12/14
DELTA T = 22.2 MINS

LONGITUDE (-VE EAST)	LATITUDE NORTH	LIMITS. SOUTH	LOCAL TIME	MEAN ALTITUDE
-89	47.26	42.05	7:51	1
-90	46.75	41.58	7:55	2
-92	45.75	40.64	8:04	4
-93	45.26	40.18	8:08	5
-94	44.76	39.71	8:12	6
-95	44.27	39.25	8:17	7
-98	42.81	37.88	8:30	10
-100	41.86	36.99	8:39	12
-102	40.94	36.10	8:49	14
-104	40.04	35.26	8:59	16
-105	39.61	34.84	9:04	17
-108	38.37	33.65	9:20	20
-110	37.59	32.91	9:31	22
-112	36.89	32.23	9:43	23
-114	36.25	31.61	9:55	25
-115	35.95	31.33	10:01	26
-118	35.21	30.60	10:20	29
-120	34.84	30.23	10:32	30
-122	34.58	29.97	10:46	31
-124	34.45	29.83	10:59	32
-125	34.43	29.80	11:06	33
-128	34.58	29.92	11:27	34
-130	34.85	30.17	11:41	34
-132	35.26	30.55	11:56	34
-134	35.80	31.06	12:10	33
-135	36.12	31.36	12:17	33
-137	36.85	32.04	12:30	32
-138	37.25	32.42	12:37	31
-139	37.68	32.83	12:44	30
-140	38.13	33.26	12:50	30
-142	39.09	34.17	13:03	28
-143	39.60	34.66	13:10	27
-145	40.67	35.67	13:22	25
-148	42.35	37.29	13:40	22
-149	42.94	37.83	13:45	21
-150	43.53	38.39	13:51	20

PERCENTAGE OF SUN OBSCURED ALONG MAP TRACK
WEST END = 91.4   EAST END = 91.9
MAXIMUM OBSCURATION 92.4 % AT LONG = -129
==

+ MARKS THE CITY OF K'AI-FENG

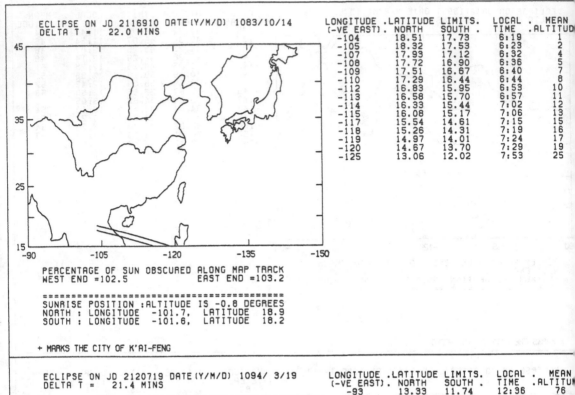

```
ECLIPSE ON JD 2116910 DATE(Y/M/D) 1083/10/14
DELTA T =   22.0 MINS
```

LONGITUDE (-VE EAST)	.LATITUDE NORTH	LIMITS. SOUTH .	LOCAL TIME	. MEAN .ALTITUD
-104	18.51	17.73	6:19	1
-105	18.32	17.53	6:23	2
-107	17.93	17.12	6:32	4
-108	17.72	16.90	6:36	5
-109	17.51	16.67	6:40	7
-110	17.29	16.44	6:44	8
-112	16.83	15.95	6:53	10
-113	16.58	15.70	6:57	11
-114	16.33	15.44	7:02	12
-115	16.08	15.17	7:06	13
-117	15.54	14.61	7:15	15
-118	15.26	14.31	7:19	16
-119	14.97	14.01	7:24	17
-120	14.67	13.70	7:29	19
-125	13.06	12.02	7:53	25

```
PERCENTAGE OF SUN OBSCURED ALONG MAP TRACK
WEST END =102.5          EAST END =103.2

==========================================
SUNRISE POSITION :ALTITUDE IS -0.8 DEGREES
NORTH : LONGITUDE  -101.7,  LATITUDE  18.9
SOUTH : LONGITUDE  -101.6,  LATITUDE  18.2

+ MARKS THE CITY OF K'AI-FENG
```

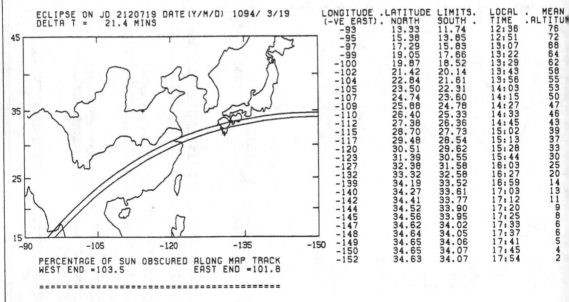

```
ECLIPSE ON JD 2120719 DATE(Y/M/D) 1094/ 3/19
DELTA T =   21.4 MINS
```

LONGITUDE (-VE EAST)	.LATITUDE NORTH	LIMITS. SOUTH .	LOCAL TIME	. MEAN .ALTITU
-93	13.33	11.74	12:36	76
-95	15.38	13.85	12:51	72
-97	17.29	15.83	13:07	68
-99	19.05	17.66	13:22	64
-100	19.87	18.52	13:29	62
-102	21.42	20.14	13:43	58
-104	22.84	21.61	13:56	55
-105	23.50	22.31	14:03	53
-107	24.74	23.60	14:15	50
-109	25.88	24.78	14:27	47
-110	26.40	25.33	14:33	46
-112	27.38	26.36	14:45	43
-115	28.70	27.73	15:02	39
-117	29.48	28.54	15:13	37
-120	30.51	29.62	15:28	33
-123	31.39	30.55	15:44	30
-127	32.38	31.58	16:03	25
-132	33.32	32.58	16:27	20
-139	34.19	33.52	16:59	14
-140	34.27	33.61	17:03	13
-142	34.41	33.77	17:12	11
-144	34.52	33.90	17:20	9
-145	34.56	33.95	17:25	8
-147	34.62	34.02	17:33	6
-148	34.64	34.05	17:37	6
-149	34.65	34.06	17:41	5
-150	34.65	34.07	17:45	4
-152	34.63	34.07	17:54	2

```
PERCENTAGE OF SUN OBSCURED ALONG MAP TRACK
WEST END =103.5          EAST END =101.8

==========================================

+ MARKS THE CITY OF K'AI-FENG
```

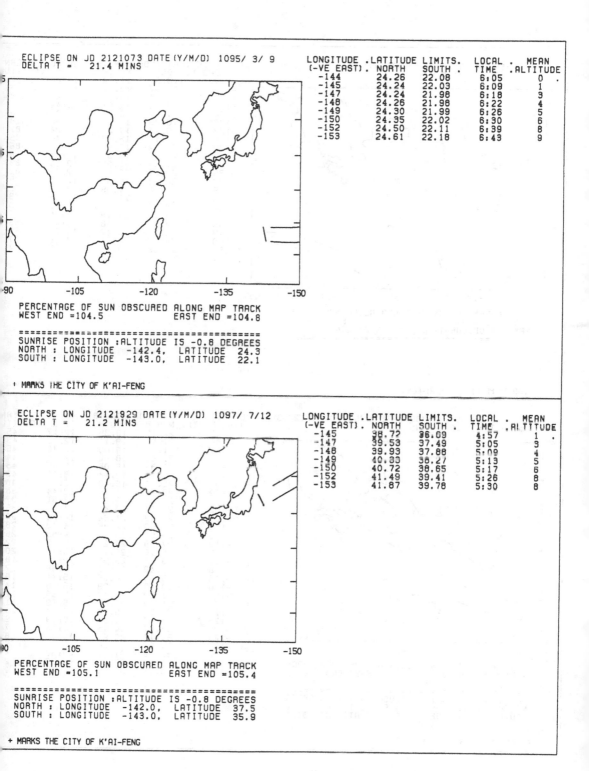

ECLIPSE ON JD 2121073 DATE(Y/M/D) 1095/ 3/ 9
DELTA T =   21.4 MINS

LONGITUDE (-VE EAST)	.LATITUDE NORTH	LIMITS. SOUTH .	LOCAL TIME	. MEAN .ALTITUDE
-144	24.26	22.08	6:05	0 .
-145	24.24	22.03	6:09	1
-147	24.24	21.98	6:18	3
-148	24.26	21.98	6:22	4
-149	24.30	21.99	6:26	5
-150	24.35	22.02	6:30	6
-152	24.50	22.11	6:39	8
-153	24.61	22.18	6:43	9

PERCENTAGE OF SUN OBSCURED ALONG MAP TRACK
WEST END =104.5   EAST END =104.8

===
SUNRISE POSITION :ALTITUDE IS -0.8 DEGREES
NORTH : LONGITUDE -142.4, LATITUDE 24.3
SOUTH : LONGITUDE -143.0, LATITUDE 22.1

+ MARKS THE CITY OF K'AI-FENG

ECLIPSE ON JD 2121929 DATE(Y/M/D) 1097/ 7/12
DELTA T =   21.2 MINS

LONGITUDE (-VE EAST)	.LATITUDE NORTH	LIMITS. SOUTH .	LOCAL TIME	. MEAN .ALTITUDE
-145	38.72	36.09	4:57	1 .
-147	39.53	37.49	5:05	3
-148	39.93	37.88	5:09	4
-149	40.30	38.27	5:13	5
-150	40.72	38.65	5:17	6
-152	41.49	39.41	5:26	8
-153	41.87	39.78	5:30	8

PERCENTAGE OF SUN OBSCURED ALONG MAP TRACK
WEST END =105.1   EAST END =105.4

===
SUNRISE POSITION :ALTITUDE IS -0.8 DEGREES
NORTH : LONGITUDE -142.0, LATITUDE 37.5
SOUTH : LONGITUDE -143.0, LATITUDE 35.9

+ MARKS THE CITY OF K'AI-FENG

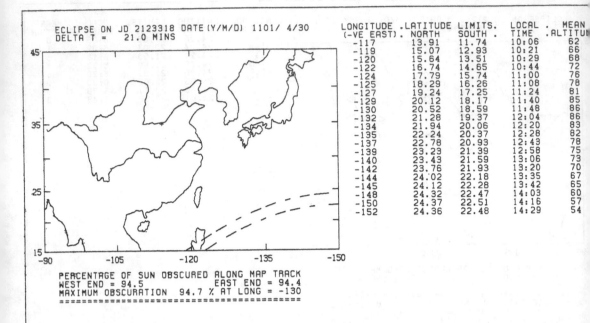

ECLIPSE ON JD 2123318 DATE(Y/M/D) 1101/ 4/30
DELTA T = 21.0 MINS

LONGITUDE (-VE EAST)	.LATITUDE . NORTH	LIMITS. SOUTH .	LOCAL TIME	. MEAN .ALTITU
-117	13.91	11.74	10:06	62
-119	15.07	12.93	10:21	66
-120	15.64	13.51	10:29	68
-122	16.74	14.65	10:44	72
-124	17.79	15.74	11:00	76
-125	18.29	16.26	11:08	78
-127	19.24	17.25	11:24	81
-129	20.12	18.17	11:40	85
-130	20.52	18.59	11:48	86
-132	21.28	19.37	12:04	86
-134	21.94	20.06	12:20	83
-135	22.24	20.37	12:28	82
-137	22.78	20.93	12:43	78
-139	23.23	21.39	12:58	75
-140	23.43	21.59	13:06	73
-142	23.76	21.93	13:20	70
-144	24.02	22.18	13:35	67
-145	24.12	22.28	13:42	65
-148	24.32	22.47	14:03	60
-150	24.37	22.51	14:16	57
-152	24.36	22.48	14:29	54

PERCENTAGE OF SUN OBSCURED ALONG MAP TRACK
WEST END = 94.5   EAST END = 94.4
MAXIMUM OBSCURATION 94.7 % AT LONG = -130
==

+ MARKS THE CITY OF K'AI-FENG

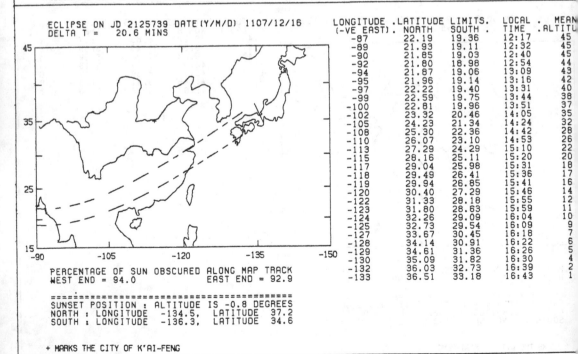

ECLIPSE ON JD 2125739 DATE(Y/M/D) 1107/12/16
DELTA T = 20.6 MINS

LONGITUDE (-VE EAST)	.LATITUDE . NORTH	LIMITS. SOUTH .	LOCAL TIME	. MEAN .ALTITU
-87	22.19	19.36	12:17	45
-89	21.93	19.11	12:32	45
-90	21.85	19.03	12:40	45
-92	21.80	18.98	12:54	44
-94	21.87	19.06	13:09	43
-95	21.96	19.14	13:16	42
-97	22.22	19.40	13:31	40
-99	22.59	19.75	13:44	38
-100	22.81	19.96	13:51	37
-102	23.32	20.46	14:05	35
-105	24.23	21.34	14:24	32
-108	25.30	22.36	14:42	28
-110	26.07	23.10	14:53	26
-113	27.29	24.29	15:10	22
-115	28.16	25.11	15:20	20
-117	29.04	25.98	15:31	18
-118	29.49	26.41	15:36	17
-119	29.94	26.85	15:41	16
-120	30.40	27.29	15:46	14
-122	31.33	28.18	15:55	12
-123	31.80	28.63	15:59	11
-124	32.26	29.09	16:04	10
-125	32.73	29.54	16:09	9
-127	33.67	30.45	16:18	7
-128	34.14	30.91	16:22	6
-129	34.61	31.36	16:26	5
-130	35.09	31.82	16:30	4
-132	36.03	32.73	16:39	2
-133	36.51	33.18	16:43	1

PERCENTAGE OF SUN OBSCURED ALONG MAP TRACK
WEST END = 94.0   EAST END = 92.9

==
SUNSET POSITION : ALTITUDE IS -0.8 DEGREES
NORTH : LONGITUDE -134.5, LATITUDE 37.2
SOUTH : LONGITUDE -136.3, LATITUDE 34.6

+ MARKS THE CITY OF K'AI-FENG

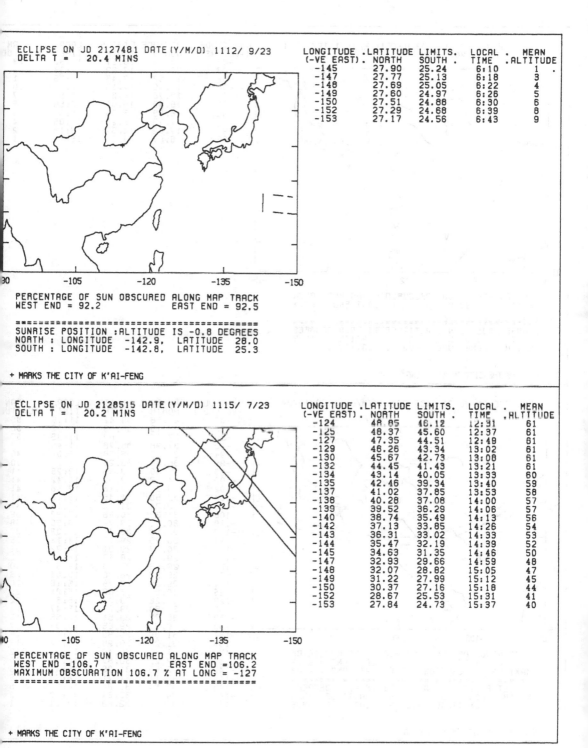

ECLIPSE ON JD 2127481 DATE (Y/M/D) 1112/ 9/23
DELTA T =   20.4 MINS

LONGITUDE (-VE EAST)	LATITUDE NORTH	LIMITS. SOUTH	LOCAL TIME	MEAN ALTITUDE
-145	27.90	25.24	6:10	1
-147	27.77	25.13	6:18	3
-148	27.69	25.05	6:22	4
-149	27.60	24.97	6:26	5
-150	27.51	24.88	6:30	6
-152	27.29	24.68	6:39	8
-153	27.17	24.56	6:43	9

PERCENTAGE OF SUN OBSCURED ALONG MAP TRACK
WEST END = 92.2   EAST END = 92.5

SUNRISE POSITION :ALTITUDE IS -0.8 DEGREES
NORTH : LONGITUDE -142.9, LATITUDE 28.0
SOUTH : LONGITUDE -142.8, LATITUDE 25.3

+ MARKS THE CITY OF K'AI-FENG

ECLIPSE ON JD 2128515 DATE (Y/M/D) 1115/ 7/23
DELTA T =   20.2 MINS

LONGITUDE (-VE EAST)	LATITUDE NORTH	LIMITS. SOUTH	LOCAL TIME	MEAN ALTITUDE
-124	48.85	46.12	12:31	61
-125	48.37	45.60	12:37	61
-127	47.35	44.51	12:49	61
-129	46.26	43.34	13:02	61
-130	45.67	42.73	13:08	61
-132	44.45	41.43	13:21	61
-134	43.14	40.05	13:33	60
-135	42.46	39.34	13:40	59
-137	41.02	37.85	13:53	58
-138	40.28	37.08	14:00	57
-139	39.52	36.29	14:06	57
-140	38.74	35.49	14:13	56
-142	37.13	33.85	14:26	54
-143	36.31	33.02	14:33	53
-144	35.47	32.19	14:39	52
-145	34.63	31.35	14:46	50
-147	32.93	29.66	14:59	48
-148	32.07	28.82	15:05	47
-149	31.22	27.99	15:12	45
-150	30.37	27.16	15:18	44
-152	28.67	25.53	15:31	41
-153	27.84	24.73	15:37	40

PERCENTAGE OF SUN OBSCURED ALONG MAP TRACK
WEST END =106.7   EAST END =106.2
MAXIMUM OBSCURATION 106.7 % AT LONG = -127

+ MARKS THE CITY OF K'AI-FENG

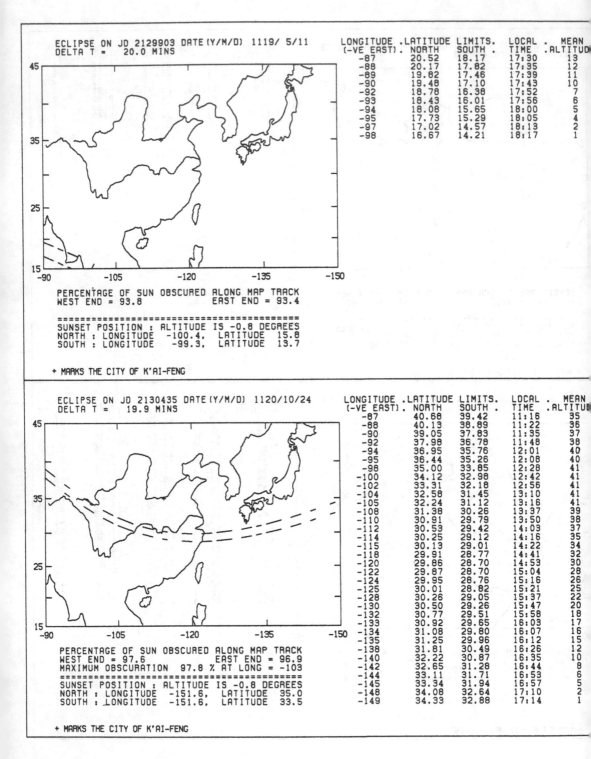

ECLIPSE ON JD 2129903 DATE(Y/M/D) 1119/ 5/11
DELTA T =   20.0 MINS

LONGITUDE (-VE EAST)	LATITUDE NORTH	LIMITS. SOUTH	LOCAL TIME	MEAN ALTITUDE
-87	20.52	18.17	17:30	13
-88	20.17	17.82	17:35	12
-89	19.82	17.46	17:39	11
-90	19.48	17.10	17:43	10
-92	18.78	16.38	17:52	7
-93	18.43	16.01	17:56	6
-94	18.08	15.65	18:00	5
-95	17.73	15.29	18:05	4
-97	17.02	14.57	18:13	2
-98	16.67	14.21	18:17	1

PERCENTAGE OF SUN OBSCURED ALONG MAP TRACK
WEST END = 93.8   EAST END = 93.4

==
SUNSET POSITION : ALTITUDE IS -0.8 DEGREES
NORTH : LONGITUDE -100.4, LATITUDE 15.8
SOUTH : LONGITUDE   -99.3, LATITUDE 13.7

+ MARKS THE CITY OF K'AI-FENG

ECLIPSE ON JD 2130435 DATE(Y/M/D) 1120/10/24
DELTA T =   19.9 MINS

LONGITUDE (-VE EAST)	LATITUDE NORTH	LIMITS. SOUTH	LOCAL TIME	MEAN ALTITUDE
-87	40.68	39.42	11:16	35
-88	40.13	38.89	11:22	36
-90	39.05	37.83	11:35	37
-92	37.98	36.78	11:48	38
-94	36.95	35.76	12:01	40
-95	36.44	35.26	12:08	40
-98	35.00	33.85	12:28	41
-100	34.12	32.98	12:42	41
-102	33.31	32.18	12:56	41
-104	32.58	31.45	13:10	41
-105	32.24	31.12	13:16	41
-108	31.38	30.26	13:37	39
-110	30.91	29.79	13:50	38
-112	30.53	29.42	14:03	37
-114	30.25	29.12	14:16	35
-115	30.13	29.01	14:22	34
-118	29.91	28.77	14:41	32
-120	29.86	28.70	14:53	30
-122	29.87	28.70	15:04	28
-124	29.95	28.76	15:16	26
-125	30.01	28.82	15:21	25
-128	30.26	29.05	15:37	22
-130	30.50	29.26	15:47	20
-132	30.77	29.51	15:58	18
-133	30.92	29.65	16:03	17
-134	31.08	29.80	16:07	16
-135	31.25	29.96	16:12	15
-138	31.81	30.49	16:26	12
-140	32.22	30.87	16:35	10
-142	32.65	31.28	16:44	8
-144	33.11	31.71	16:53	5
-145	33.34	31.94	16:57	5
-148	34.08	32.64	17:10	2
-149	34.33	32.88	17:14	1

PERCENTAGE OF SUN OBSCURED ALONG MAP TRACK
WEST END = 97.6   EAST END = 96.9
MAXIMUM OBSCURATION 97.8 % AT LONG = -103
==
SUNSET POSITION : ALTITUDE IS -0.8 DEGREES
NORTH : LONGITUDE -151.6, LATITUDE 35.0
SOUTH : LONGITUDE -151.6, LATITUDE 33.5

+ MARKS THE CITY OF K'AI-FENG

ECLIPSE ON JD 2130937 DATE(Y/M/D) 1122/ 3/10
DELTA T =   19.9 MINS

LONGITUDE (-VE EAST)	LATITUDE LIMITS. NORTH	SOUTH	LOCAL TIME	MEAN ALTITUDE
-87	34.04	31.49	11:25	55
-88	34.59	32.05	11:32	54
-89	35.14	32.61	11:39	54
-90	35.69	33.17	11:45	54
-92	36.76	34.27	11:58	53
-93	37.29	34.81	12:05	52
-94	37.81	35.35	12:11	52
-95	38.32	35.88	12:18	51
-97	39.32	36.91	12:30	50
-98	39.81	37.42	12:36	49
-99	40.28	37.91	12:43	48
-100	40.75	38.40	12:49	47
-102	41.65	39.34	13:01	46
-103	42.08	39.79	13:07	45
-104	42.51	40.24	13:12	44
-105	42.93	40.67	13:18	43
-107	43.73	41.51	13:29	41
-108	44.11	41.91	13:35	40
-109	44.49	42.31	13:41	40
-110	44.86	42.69	13:46	39
-112	45.56	43.43	13:57	37
-113	45.90	43.78	14:02	36
-114	46.22	44.12	14:08	35
-115	46.54	44.46	14:13	34
-117	47.16	45.10	14:23	33
-118	47.45	45.41	14:28	32
-119	47.73	45.71	14:33	31
-120	48.01	46.00	14:38	30
-122	48.54	46.56	14:48	29
-123	48.79	46.82	14:53	28

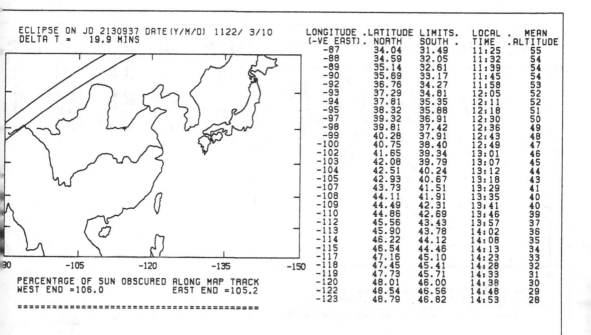

PERCENTAGE OF SUN OBSCURED ALONG MAP TRACK
WEST END =106.0   EAST END =105.2

===

+ MARKS THE CITY OF K'AI-FENG

ECLIPSE ON JD 2131467 DATE(Y/M/D) 1123/ 8/23
DELTA T =   19.8 MINS

LONGITUDE (-VE EAST)	LATITUDE LIMITS. NORTH	SOUTH	LOCAL TIME	MEAN ALTITUDE
-125	18.13	17.68	5:40	0
-127	18.46	18.04	5:54	2
-128	18.62	18.21	5:58	3
-129	18.78	18.37	6:02	4
-130	18.93	18.54	6:06	5
-132	19.22	18.85	6:14	7
-133	19.36	19.00	6:18	8
-134	19.50	19.14	6:23	9
-135	19.63	19.28	6:27	10
-137	19.87	19.55	6:36	12
-138	19.99	19.67	6:40	13
-139	20.10	19.79	6:45	14
-140	20.20	19.91	6:49	15
-142	20.40	20.12	6:58	17
-143	20.49	20.22	7:03	18
-144	20.57	20.31	7:08	19
-145	20.64	20.39	7:12	20
-147	20.77	20.54	7:22	23
-148	20.83	20.61	7:27	24
-149	20.88	20.67	7:32	25
-150	20.92	20.72	7:37	26
-152	20.97	20.79	7:47	28
-153	20.99	20.82	7:52	30

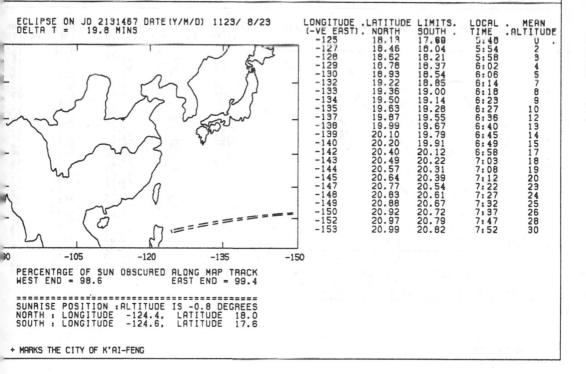

PERCENTAGE OF SUN OBSCURED ALONG MAP TRACK
WEST END = 98.6   EAST END = 99.4

===
SUNRISE POSITION :ALTITUDE IS -0.8 DEGREES
NORTH : LONGITUDE -124.4, LATITUDE 18.0
SOUTH : LONGITUDE -124.6, LATITUDE 17.6

+ MARKS THE CITY OF K'AI-FENG

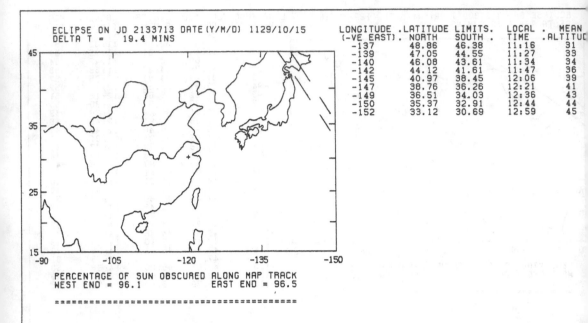

ECLIPSE ON JD 2133713 DATE(Y/M/D) 1129/10/15
DELTA T =   19.4 MINS

LONGITUDE (-VE EAST)	.LATITUDE LIMITS. NORTH	SOUTH .	LOCAL . TIME	MEAN .ALTITUDE
-137	48.86	46.38	11:16	31
-139	47.05	44.55	11:27	33
-140	46.08	43.61	11:34	34
-142	44.12	41.61	11:47	36
-145	40.97	38.45	12:06	39
-147	38.76	36.26	12:21	41
-149	36.51	34.03	12:36	43
-150	35.37	32.91	12:44	44
-152	33.12	30.69	12:59	45

PERCENTAGE OF SUN OBSCURED ALONG MAP TRACK
WEST END = 96.1   EAST END = 96.5

===

+ MARKS THE CITY OF LIN-AN

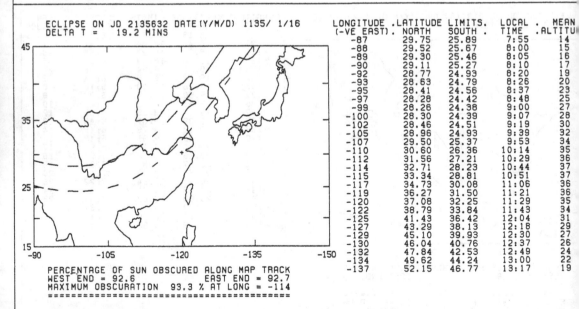

ECLIPSE ON JD 2135632 DATE(Y/M/D) 1135/ 1/16
DELTA T =   19.2 MINS

LONGITUDE (-VE EAST)	.LATITUDE LIMITS. NORTH	SOUTH .	LOCAL . TIME	MEAN .ALTITUDE
-87	29.75	25.89	7:55	14
-88	29.52	25.67	8:00	15
-89	29.30	25.46	8:05	16
-90	29.11	25.27	8:10	17
-92	28.77	24.93	8:20	19
-93	28.63	24.79	8:26	20
-95	28.41	24.56	8:37	23
-97	28.28	24.42	8:48	25
-99	28.26	24.38	9:00	27
-100	28.30	24.39	9:07	28
-102	28.46	24.51	9:19	30
-105	28.96	24.93	9:39	32
-107	29.50	25.37	9:53	34
-110	30.60	26.36	10:14	35
-112	31.56	27.21	10:29	36
-114	32.71	28.23	10:44	37
-115	33.34	28.81	10:51	37
-117	34.73	30.08	11:06	36
-119	36.27	31.50	11:21	36
-120	37.08	32.25	11:29	35
-122	38.79	33.84	11:43	34
-125	41.43	36.42	12:04	31
-127	43.29	38.13	12:18	29
-129	45.10	39.93	12:30	27
-130	46.04	40.76	12:37	26
-132	47.84	42.53	12:49	24
-134	49.62	44.24	13:00	22
-137	52.15	46.77	13:17	19

PERCENTAGE OF SUN OBSCURED ALONG MAP TRACK
WEST END = 92.6   EAST END = 92.7
MAXIMUM OBSCURATION 93.3 % AT LONG = -114

===

+ MARKS THE CITY OF LIN-AN

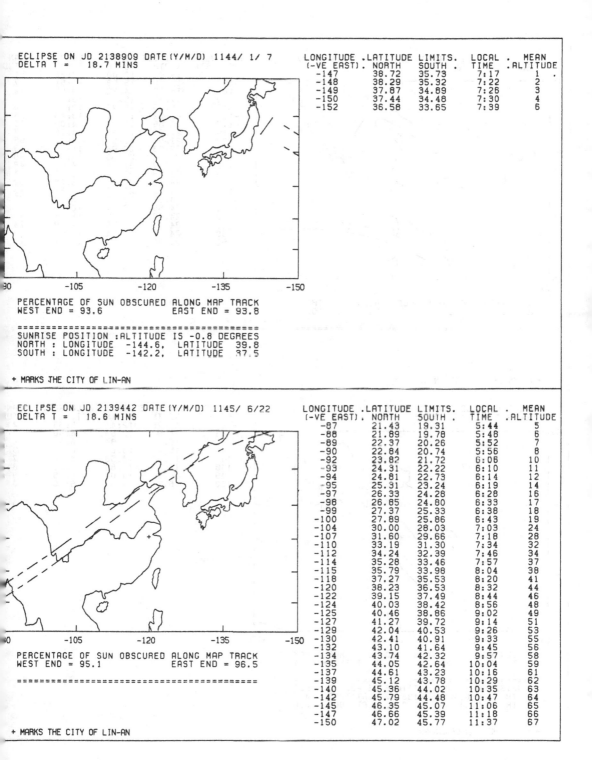

ECLIPSE ON JD 2138909 DATE(Y/M/D) 1144/ 1/ 7
DELTA T =   18.7 MINS

LONGITUDE (-VE EAST) .	.LATITUDE NORTH	LIMITS. SOUTH .	LOCAL TIME	. MEAN .ALTITUDE
-147	38.72	35.73	7:17	1 .
-148	38.29	35.32	7:22	2
-149	37.87	34.89	7:26	3
-150	37.44	34.48	7:30	4
-152	36.58	33.65	7:39	6

-30   -105   -120   -135   -150

PERCENTAGE OF SUN OBSCURED ALONG MAP TRACK
WEST END = 93.6   EAST END = 93.8

==
SUNRISE POSITION :ALTITUDE IS -0.8 DEGREES
NORTH : LONGITUDE -144.6, LATITUDE 39.8
SOUTH : LONGITUDE -142.2, LATITUDE 37.5

+ MARKS THE CITY OF LIN-AN

ECLIPSE ON JD 2139442 DATE(Y/M/D) 1145/ 6/22
DELTA T =   18.6 MINS

LONGITUDE (-VE EAST) .	.LATITUDE NORTH	LIMITS. SOUTH .	LOCAL TIME	. MEAN .ALTITUDE
-87	21.43	19.31	5:44	5
-88	21.89	19.78	5:48	6
-89	22.37	20.26	5:52	7
-90	22.84	20.74	5:56	8
-92	23.82	21.72	6:06	10
-93	24.31	22.22	6:10	11
-94	24.81	22.73	6:14	12
-95	25.31	23.24	6:19	14
-97	26.33	24.28	6:28	16
-98	26.85	24.80	6:33	17
-99	27.37	25.33	6:38	18
-100	27.89	25.86	6:43	19
-104	30.00	28.03	7:03	24
-107	31.60	29.66	7:18	28
-110	33.19	31.30	7:34	32
-112	34.24	32.39	7:46	34
-114	35.28	33.46	7:57	37
-115	35.79	33.98	8:04	38
-118	37.27	35.53	8:20	41
-120	38.23	36.53	8:32	44
-122	39.15	37.49	8:44	46
-124	40.03	38.42	8:56	48
-125	40.46	38.86	9:02	49
-127	41.27	39.72	9:14	51
-129	42.04	40.53	9:26	53
-130	42.41	40.91	9:33	55
-132	43.10	41.64	9:45	56
-134	43.74	42.32	9:57	58
-135	44.05	42.64	10:04	59
-137	44.61	43.23	10:16	61
-139	45.12	43.78	10:29	62
-140	45.36	44.02	10:35	63
-142	45.79	44.48	10:47	64
-145	46.35	45.07	11:06	65
-147	46.66	45.39	11:18	66
-150	47.02	45.77	11:37	67

-30   -105   -120   -135   -150

PERCENTAGE OF SUN OBSCURED ALONG MAP TRACK
WEST END = 95.1   EAST END = 96.5

==

+ MARKS THE CITY OF LIN-AN

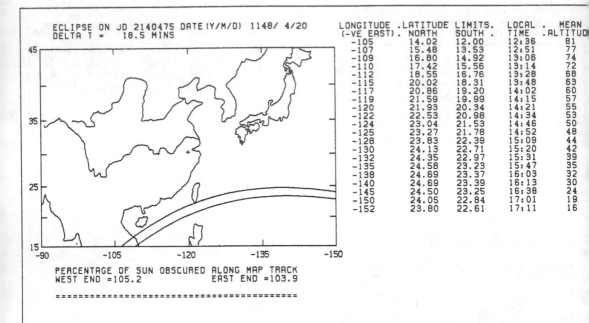

ECLIPSE ON JD 2140475 DATE(Y/M/D) 1148/ 4/20
DELTA T =   18.5 MINS

LONGITUDE (-VE EAST)	LATITUDE LIMITS. NORTH	SOUTH	LOCAL TIME	MEAN ALTITUD
-105	14.02	12.00	12:36	81
-107	15.48	13.53	12:51	77
-109	16.80	14.92	13:06	74
-110	17.42	15.56	13:14	72
-112	18.55	16.76	13:28	68
-115	20.02	18.31	13:48	63
-117	20.86	19.20	14:02	60
-119	21.59	19.99	14:15	57
-120	21.93	20.34	14:21	55
-122	22.53	20.98	14:34	53
-124	23.04	21.53	14:46	50
-125	23.27	21.78	14:52	48
-128	23.83	22.39	15:09	44
-130	24.13	22.71	15:20	42
-132	24.35	22.97	15:31	39
-135	24.58	23.23	15:47	35
-138	24.69	23.37	16:03	32
-140	24.69	23.39	16:13	30
-145	24.50	23.25	16:38	24
-150	24.05	22.84	17:01	19
-152	23.80	22.61	17:11	16

PERCENTAGE OF SUN OBSCURED ALONG MAP TRACK
WEST END =105.2   EAST END =103.9

==

+ MARKS THE CITY OF LIN-AN

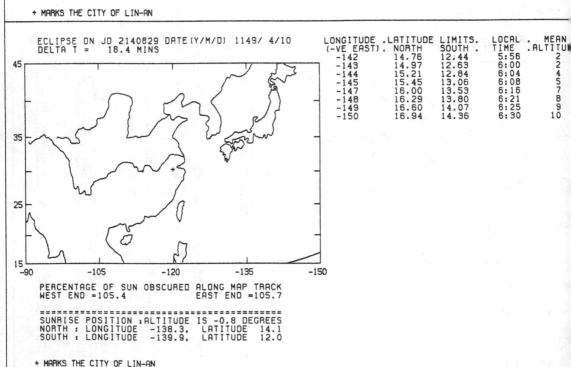

ECLIPSE ON JD 2140829 DATE(Y/M/D) 1149/ 4/10
DELTA T =   18.4 MINS

LONGITUDE (-VE EAST)	LATITUDE LIMITS. NORTH	SOUTH	LOCAL TIME	MEAN ALTITU
-142	14.76	12.44	5:56	2
-143	14.97	12.63	6:00	2
-144	15.21	12.84	6:04	4
-145	15.45	13.06	6:08	5
-147	16.00	13.53	6:16	7
-148	16.29	13.80	6:21	8
-149	16.60	14.07	6:25	9
-150	16.94	14.36	6:30	10

PERCENTAGE OF SUN OBSCURED ALONG MAP TRACK
WEST END =105.4   EAST END =105.7

==
SUNRISE POSITION :ALTITUDE IS -0.8 DEGREES
NORTH : LONGITUDE -138.3, LATITUDE   14.1
SOUTH : LONGITUDE -139.9, LATITUDE   12.0

+ MARKS THE CITY OF LIN-AN

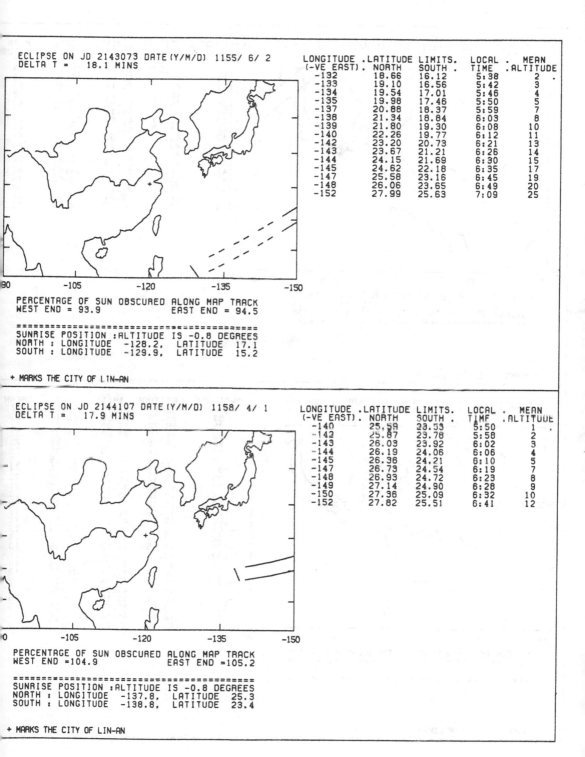

ECLIPSE ON JD 2143073 DATE(Y/M/D) 1155/ 6/ 2
DELTA T =   18.1 MINS

LONGITUDE (-VE EAST) .	LATITUDE NORTH	LIMITS. SOUTH .	LOCAL TIME	MEAN ALTITUDE
-132	18.66	16.12	5:38	2
-133	19.10	16.56	5:42	3
-134	19.54	17.01	5:46	4
-135	19.98	17.46	5:50	5
-137	20.88	18.37	5:59	7
-138	21.34	18.84	6:03	8
-139	21.80	19.30	6:08	10
-140	22.26	19.77	6:12	11
-142	23.20	20.73	6:21	13
-143	23.67	21.21	6:26	14
-144	24.15	21.69	6:30	15
-145	24.62	22.18	6:35	17
-147	25.58	23.16	6:45	19
-148	26.06	23.65	6:49	20
-152	27.99	25.63	7:09	25

PERCENTAGE OF SUN OBSCURED ALONG MAP TRACK
WEST END = 93.9   EAST END = 94.5

===
SUNRISE POSITION :ALTITUDE IS -0.8 DEGREES
NORTH : LONGITUDE -128.2, LATITUDE 17.1
SOUTH : LONGITUDE -129.9, LATITUDE 15.2

+ MARKS THE CITY OF LIN-AN

ECLIPSE ON JD 2144107 DATE(Y/M/D) 1158/ 4/ 1
DELTA T =   17.9 MINS

LONGITUDE (-VE EAST) .	LATITUDE NORTH	LIMITS. SOUTH .	LOCAL TIME	MEAN ALTITUDE
-140	25.59	23.53	5:50	1
-142	25.87	23.78	5:58	2
-143	26.03	23.92	6:02	3
-144	26.19	24.06	6:06	4
-145	26.36	24.21	6:10	5
-147	26.73	24.54	6:19	7
-148	26.93	24.72	6:23	8
-149	27.14	24.90	6:28	9
-150	27.36	25.09	6:32	10
-152	27.82	25.51	6:41	12

PERCENTAGE OF SUN OBSCURED ALONG MAP TRACK
WEST END =104.9   EAST END =105.2

===
SUNRISE POSITION :ALTITUDE IS -0.8 DEGREES
NORTH : LONGITUDE -137.8, LATITUDE 25.3
SOUTH : LONGITUDE -138.8, LATITUDE 23.4

+ MARKS THE CITY OF LIN-AN

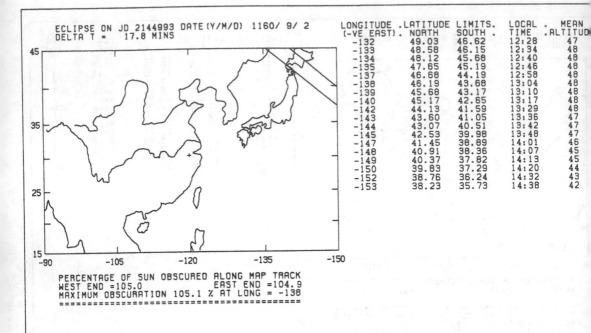

ECLIPSE ON JD 2144993 DATE(Y/M/D) 1160/ 9/ 2
DELTA T =   17.8 MINS

LONGITUDE (-VE EAST)	.LATITUDE NORTH	LIMITS. SOUTH .	LOCAL TIME	. MEAN .ALTITUDE
-132	49.03	46.62	12:28	47
-133	48.58	46.15	12:34	48
-134	48.12	45.68	12:40	48
-135	47.65	45.19	12:46	48
-137	46.68	44.19	12:58	48
-138	46.19	43.68	13:04	48
-139	45.68	43.17	13:10	48
-140	45.17	42.65	13:17	48
-142	44.13	41.59	13:29	48
-143	43.60	41.05	13:36	47
-144	43.07	40.51	13:42	47
-145	42.53	39.98	13:48	47
-147	41.45	38.89	14:01	46
-148	40.91	38.36	14:07	45
-149	40.37	37.82	14:13	45
-150	39.83	37.29	14:20	44
-152	38.76	36.24	14:32	43
-153	38.23	35.73	14:38	42

PERCENTAGE OF SUN OBSCURED ALONG MAP TRACK
WEST END =105.0   EAST END =104.9
MAXIMUM OBSCURATION 105.1 % AT LONG = -138
===

+ MARKS THE CITY OF LIN-AN

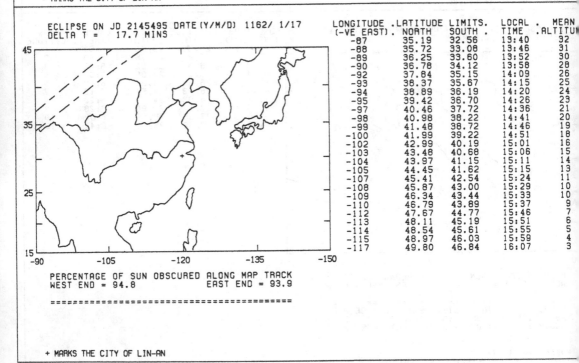

ECLIPSE ON JD 2145495 DATE(Y/M/D) 1162/ 1/17
DELTA T =   17.7 MINS

LONGITUDE (-VE EAST)	.LATITUDE NORTH	LIMITS. SOUTH .	LOCAL TIME	. MEAN .ALTITUDE
-87	35.19	32.56	13:40	32
-88	35.72	33.08	13:46	31
-89	36.25	33.60	13:52	30
-90	36.78	34.12	13:58	28
-92	37.84	35.15	14:09	26
-93	38.37	35.67	14:15	25
-94	38.89	36.19	14:20	24
-95	39.42	36.70	14:26	23
-97	40.46	37.72	14:36	21
-98	40.98	38.22	14:41	20
-99	41.48	38.72	14:46	19
-100	41.99	39.22	14:51	18
-102	42.99	40.19	15:01	16
-103	43.48	40.68	15:06	15
-104	43.97	41.15	15:11	14
-105	44.45	41.62	15:15	13
-107	45.41	42.54	15:24	11
-108	45.87	43.00	15:29	10
-109	46.34	43.44	15:33	10
-110	46.79	43.89	15:37	9
-112	47.67	44.77	15:46	7
-113	48.11	45.19	15:51	6
-114	48.54	45.61	15:55	5
-115	48.97	46.03	15:59	4
-117	49.80	46.84	16:07	3

PERCENTAGE OF SUN OBSCURED ALONG MAP TRACK
WEST END = 94.8   EAST END = 93.9

===

+ MARKS THE CITY OF LIN-AN

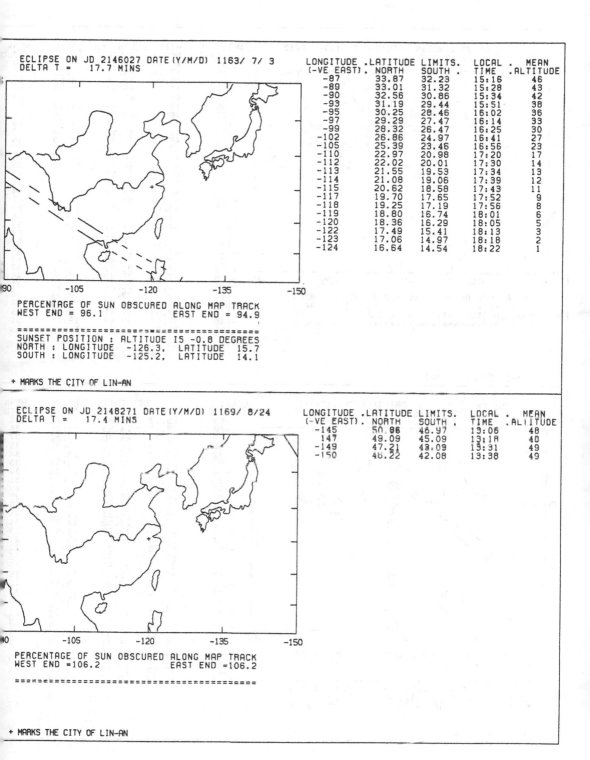

ECLIPSE ON JD 2146027 DATE (Y/M/D) 1163/ 7/ 3
DELTA T =   17.7 MINS

LONGITUDE (-VE EAST)	LATITUDE LIMITS. NORTH	SOUTH	LOCAL TIME	MEAN ALTITUDE
-87	33.87	32.23	15:16	46
-89	33.01	31.32	15:28	43
-90	32.56	30.86	15:34	42
-93	31.19	29.44	15:51	38
-95	30.25	28.46	16:02	36
-97	29.29	27.47	16:14	33
-99	28.32	26.47	16:25	30
-102	26.86	24.97	16:41	27
-105	25.39	23.46	16:56	23
-110	22.97	20.98	17:20	17
-112	22.02	20.01	17:30	14
-113	21.55	19.53	17:34	13
-114	21.08	19.06	17:39	12
-115	20.62	18.58	17:43	11
-117	19.70	17.65	17:52	9
-118	19.25	17.19	17:56	8
-119	18.80	16.74	18:01	6
-120	18.36	16.29	18:05	5
-122	17.49	15.41	18:13	3
-123	17.06	14.97	18:18	2
-124	16.64	14.54	18:22	1

-90   -105   -120   -135   -150

PERCENTAGE OF SUN OBSCURED ALONG MAP TRACK
WEST END = 96.1   EAST END = 94.9

===

SUNSET POSITION : ALTITUDE IS -0.8 DEGREES
NORTH : LONGITUDE -126.3,   LATITUDE   15.7
SOUTH : LONGITUDE -125.2,   LATITUDE   14.1

+ MARKS THE CITY OF LIN-AN

ECLIPSE ON JD 2148271 DATE (Y/M/D) 1169/ 8/24
DELTA T =   17.4 MINS

LONGITUDE (-VE EAST)	LATITUDE LIMITS. NORTH	SOUTH	LOCAL TIME	MEAN ALTITUDE
-145	50.86	46.97	13:06	48
147	49.09	45.09	13:18	40
-149	47.21	43.09	13:31	49
-150	46.22	42.08	13:38	49

-90   -105   -120   -135   -150

PERCENTAGE OF SUN OBSCURED ALONG MAP TRACK
WEST END =106.2   EAST END =106.2

===

+ MARKS THE CITY OF LIN-AN

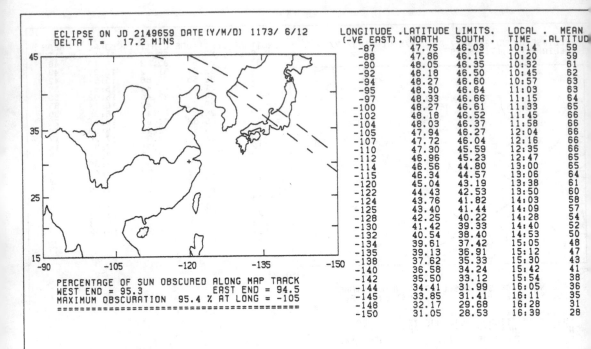

ECLIPSE ON JD 2149659 DATE(Y/M/D) 1173/ 6/12
DELTA T =   17.2 MINS

LONGITUDE (-VE EAST)	LATITUDE NORTH	LIMITS SOUTH	LOCAL TIME	MEAN ALTITUDE
-87	47.75	46.03	10:14	59
-88	47.86	46.15	10:20	59
-90	48.05	46.35	10:32	61
-92	48.18	46.50	10:45	62
-94	48.27	46.60	10:57	63
-95	48.30	46.64	11:03	63
-97	48.33	46.66	11:15	64
-100	48.27	46.61	11:33	65
-102	48.18	46.52	11:45	66
-104	48.03	46.37	11:58	66
-105	47.94	46.27	12:04	66
-107	47.72	46.04	12:16	66
-110	47.30	45.59	12:35	66
-112	46.96	45.23	12:47	65
-114	46.56	44.80	13:00	65
-115	46.34	44.57	13:06	64
-120	45.04	43.19	13:38	61
-122	44.43	42.53	13:50	60
-124	43.76	41.82	14:03	58
-125	43.40	41.44	14:09	57
-128	42.25	40.22	14:28	54
-130	41.42	39.33	14:40	52
-132	40.54	38.40	14:53	50
-134	39.61	37.42	15:05	48
-135	39.13	36.91	15:12	47
-138	37.62	35.33	15:30	43
-140	36.58	34.24	15:42	41
-142	35.50	33.12	15:54	38
-144	34.41	31.99	16:05	36
-145	33.85	31.41	16:11	35
-148	32.17	29.68	16:28	31
-150	31.05	28.53	16:39	28

PERCENTAGE OF SUN OBSCURED ALONG MAP TRACK
WEST END = 95.3   EAST END = 94.5
MAXIMUM OBSCURATION 95.4 % AT LONG = -105
==

+ MARKS THE CITY OF LIN-AN

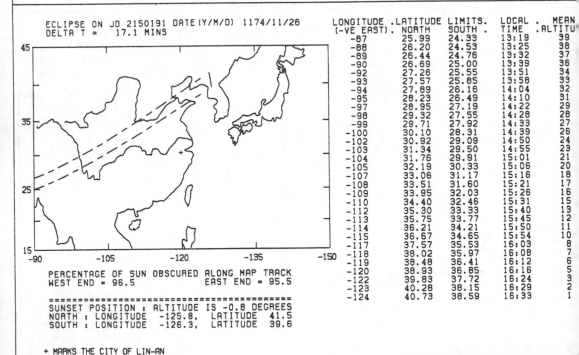

ECLIPSE ON JD 2150191 DATE(Y/M/D) 1174/11/26
DELTA T =   17.1 MINS

LONGITUDE (-VE EAST)	LATITUDE NORTH	LIMITS SOUTH	LOCAL TIME	MEAN ALTITUDE
-87	25.99	24.33	13:19	39
-88	26.20	24.53	13:25	38
-89	26.44	24.76	13:32	37
-90	26.69	25.00	13:39	36
-92	27.26	25.55	13:51	34
-93	27.57	25.85	13:58	33
-94	27.89	26.16	14:04	32
-95	28.23	26.49	14:10	31
-97	28.95	27.19	14:22	29
-98	29.32	27.55	14:28	28
-99	29.71	27.92	14:33	27
-100	30.10	28.31	14:39	26
-102	30.92	29.09	14:50	24
-103	31.34	29.50	14:55	23
-104	31.76	29.91	15:01	21
-105	32.19	30.33	15:06	20
-107	33.06	31.17	15:16	18
-108	33.51	31.60	15:21	17
-109	33.95	32.03	15:26	16
-110	34.40	32.46	15:31	15
-112	35.30	33.33	15:40	13
-113	35.75	33.77	15:45	12
-114	36.21	34.21	15:50	11
-115	36.67	34.65	15:54	10
-117	37.57	35.53	16:03	8
-118	38.02	35.97	16:08	7
-119	38.48	36.41	16:12	6
-120	38.93	36.85	16:16	5
-122	39.83	37.72	16:24	3
-123	40.28	38.15	16:29	2
-124	40.73	38.59	16:33	1

PERCENTAGE OF SUN OBSCURED ALONG MAP TRACK
WEST END = 96.5   EAST END = 95.5

==
SUNSET POSITION : ALTITUDE IS -0.8 DEGREES
NORTH : LONGITUDE -125.8, LATITUDE 41.5
SOUTH : LONGITUDE -126.3, LATITUDE 39.6

+ MARKS THE CITY OF LIN-AN

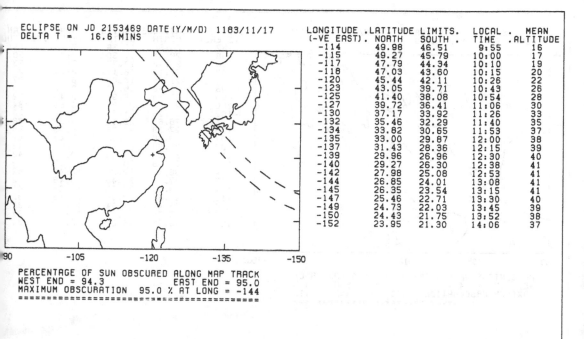

```
ECLIPSE ON JD 2153469 DATE(Y/M/D) 1183/11/17
DELTA T =    16.6 MINS
```

LONGITUDE (-VE EAST)	LATITUDE NORTH	LIMITS. SOUTH	LOCAL TIME	MEAN ALTITUDE
-114	49.98	46.51	9:55	16
-115	49.27	45.79	10:00	17
-117	47.79	44.34	10:10	19
-118	47.03	43.60	10:15	20
-120	45.44	42.11	10:26	22
-123	43.05	39.71	10:43	26
-125	41.40	38.08	10:54	28
-127	39.72	36.41	11:06	30
-130	37.17	33.92	11:26	33
-132	35.46	32.29	11:40	35
-134	33.82	30.65	11:53	37
-135	33.00	29.87	12:00	38
-137	31.43	28.36	12:15	39
-139	29.96	26.96	12:30	40
-140	29.27	26.30	12:38	41
-142	27.98	25.08	12:53	41
-144	26.85	24.01	13:08	41
-145	26.35	23.54	13:15	41
-147	25.46	22.71	13:30	40
-149	24.73	22.03	13:45	39
-150	24.43	21.75	13:52	38
-152	23.95	21.30	14:06	37

```
PERCENTAGE OF SUN OBSCURED ALONG MAP TRACK
WEST END = 94.3              EAST END = 95.0
MAXIMUM OBSCURATION  95.0 % AT LONG = -144
=================================================
```

+ MARKS THE CITY OF LIN-AN

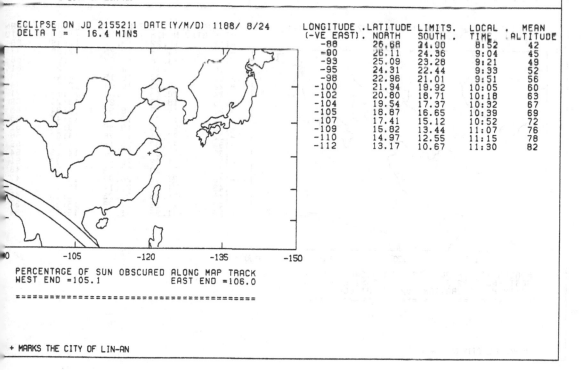

```
ECLIPSE ON JD 2155211 DATE(Y/M/D) 1188/ 8/24
DELTA T =    16.4 MINS
```

LONGITUDE (-VE EAST)	LATITUDE NORTH	LIMITS. SOUTH	LOCAL TIME	MEAN ALTITUDE
-88	26.68	24.00	8:52	42
-90	26.11	24.36	9:04	45
-93	25.09	23.28	9:21	49
-95	24.31	22.44	9:33	52
-98	22.96	21.01	9:51	56
-100	21.94	19.92	10:05	60
-102	20.80	18.71	10:18	63
-104	19.54	17.37	10:32	67
-105	18.87	16.65	10:39	69
-107	17.41	15.12	10:52	72
-109	15.82	13.44	11:07	76
-110	14.97	12.55	11:15	78
-112	13.17	10.67	11:30	82

```
PERCENTAGE OF SUN OBSCURED ALONG MAP TRACK
WEST END =105.1              EAST END =106.0

=================================================
```

+ MARKS THE CITY OF LIN-AN

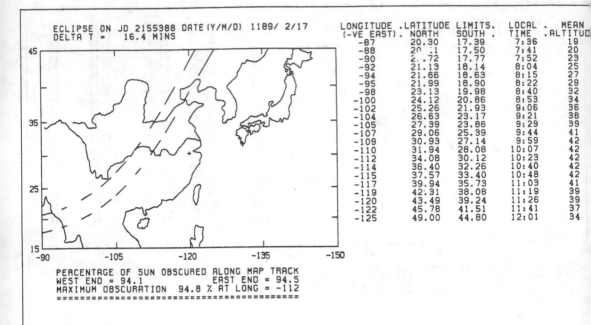

ECLIPSE ON JD 2155388 DATE(Y/M/D) 1189/ 2/17
DELTA T =   16.4 MINS

LONGITUDE (-VE EAST)	LATITUDE NORTH	LIMITS. SOUTH	LOCAL TIME	MEAN ALTITUD
-87	20.30	17.39	7:36	19
-88	20.51	17.50	7:41	20
-90	20.72	17.77	7:52	23
-92	21.13	18.14	8:04	25
-94	21.66	18.63	8:15	27
-95	21.99	18.90	8:22	29
-98	23.13	19.98	8:40	32
-100	24.12	20.86	8:53	34
-102	25.26	21.93	9:06	36
-104	26.63	23.17	9:21	38
-105	27.39	23.86	9:29	39
-107	29.06	25.39	9:44	41
-109	30.93	27.14	9:59	42
-110	31.94	28.08	10:07	42
-112	34.08	30.12	10:23	42
-114	36.40	32.26	10:40	42
-115	37.57	33.40	10:48	42
-117	39.94	35.73	11:03	41
-119	42.31	38.08	11:19	39
-120	43.49	39.24	11:26	39
-122	45.78	41.51	11:41	37
-125	49.00	44.80	12:01	34

PERCENTAGE OF SUN OBSCURED ALONG MAP TRACK
WEST END = 94.1   EAST END = 94.5
MAXIMUM OBSCURATION 94.8 % AT LONG = -112
==

+ MARKS THE CITY OF LIN-AN

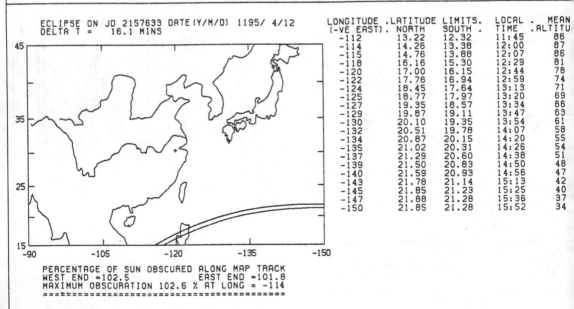

ECLIPSE ON JD 2157633 DATE(Y/M/D) 1195/ 4/12
DELTA T =   16.1 MINS

LONGITUDE (-VE EAST)	LATITUDE NORTH	LIMITS. SOUTH	LOCAL TIME	MEAN ALTITU
-112	13.22	12.32	11:45	86
-114	14.26	13.38	12:00	87
-115	14.76	13.88	12:07	86
-118	16.16	15.30	12:29	81
-120	17.00	16.15	12:44	78
-122	17.76	16.94	12:59	74
-124	18.45	17.64	13:13	71
-125	18.77	17.97	13:20	69
-127	19.35	18.57	13:34	66
-129	19.87	19.11	13:47	63
-130	20.10	19.35	13:54	61
-132	20.51	19.78	14:07	58
-134	20.87	20.15	14:20	55
-135	21.02	20.31	14:26	54
-137	21.29	20.60	14:38	51
-139	21.50	20.83	14:50	48
-140	21.59	20.93	14:56	47
-143	21.78	21.14	15:13	42
-145	21.85	21.23	15:25	40
-147	21.88	21.28	15:36	37
-150	21.85	21.28	15:52	34

PERCENTAGE OF SUN OBSCURED ALONG MAP TRACK
WEST END =102.5   EAST END =101.8
MAXIMUM OBSCURATION 102.6 % AT LONG = -114
==

+ MARKS THE CITY OF LIN-AN

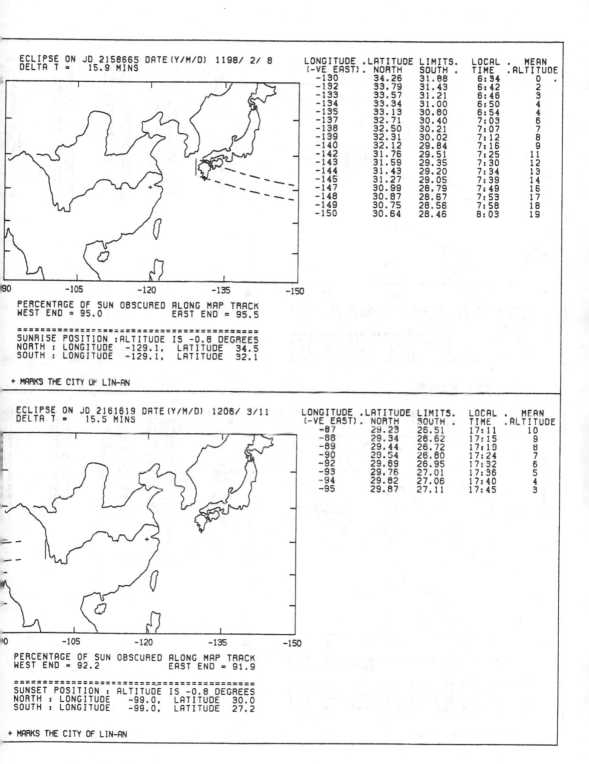

ECLIPSE ON JD 2158665 DATE(Y/M/D) 1198/ 2/ 8
DELTA T =   15.9 MINS

LONGITUDE (-VE EAST)	LATITUDE NORTH	LIMITS. SOUTH	LOCAL TIME	MEAN ALTITUDE
-130	34.26	31.88	6:34	0
-132	33.79	31.43	6:42	2
-133	33.57	31.21	6:46	3
-134	33.34	31.00	6:50	4
-135	33.13	30.80	6:54	4
-137	32.71	30.40	7:03	6
-138	32.50	30.21	7:07	7
-139	32.31	30.02	7:12	8
-140	32.12	29.84	7:16	9
-142	31.76	29.51	7:25	11
-143	31.59	29.35	7:30	12
-144	31.43	29.20	7:34	13
-145	31.27	29.05	7:39	14
-147	30.99	28.79	7:49	16
-148	30.87	28.67	7:53	17
-149	30.75	28.56	7:58	18
-150	30.64	28.46	8:03	19

90   -105   -120   -135   -150

PERCENTAGE OF SUN OBSCURED ALONG MAP TRACK
WEST END = 95.0   EAST END = 95.5

==
SUNRISE POSITION :ALTITUDE IS -0.8 DEGREES
NORTH : LONGITUDE -129.1,   LATITUDE   34.5
SOUTH : LONGITUDE -129.1,   LATITUDE   32.1

+ MARKS THE CITY OF LIN-AN

ECLIPSE ON JD 2161619 DATE(Y/M/D) 1206/ 3/11
DELTA T =   15.5 MINS

LONGITUDE (-VE EAST)	LATITUDE NORTH	LIMITS. SOUTH	LOCAL TIME	MEAN ALTITUDE
-87	29.23	26.51	17:11	10
-88	29.34	26.62	17:15	9
-89	29.44	26.72	17:19	8
-90	29.54	26.80	17:24	7
-92	29.69	26.95	17:32	6
-93	29.76	27.01	17:36	5
-94	29.82	27.06	17:40	4
-95	29.87	27.11	17:45	3

0   -105   -120   -135   -150

PERCENTAGE OF SUN OBSCURED ALONG MAP TRACK
WEST END = 92.2   EAST END = 91.9

==
SUNSET POSITION : ALTITUDE IS -0.8 DEGREES
NORTH : LONGITUDE   -99.0,   LATITUDE   30.0
SOUTH : LONGITUDE   -99.0,   LATITUDE   27.2

+ MARKS THE CITY OF LIN-AN

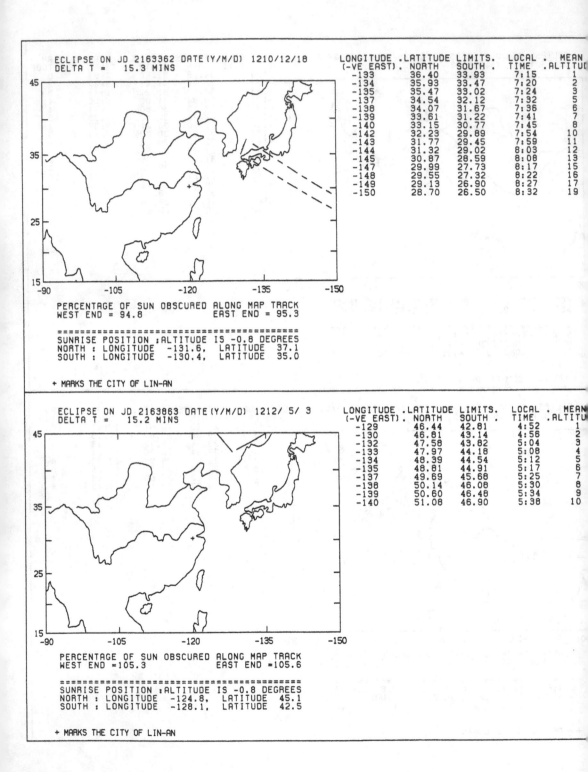

ECLIPSE ON JD 2163362 DATE(Y/M/D) 1210/12/18
DELTA T = 15.3 MINS

LONGITUDE (-VE EAST)	LATITUDE NORTH	LIMITS. SOUTH	LOCAL TIME	MEAN ALTITU
-133	36.40	33.93	7:15	1
-134	35.93	33.47	7:20	2
-135	35.47	33.02	7:24	3
-137	34.54	32.12	7:32	5
-138	34.07	31.67	7:36	6
-139	33.61	31.22	7:41	7
-140	33.15	30.77	7:45	8
-142	32.23	29.89	7:54	10
-143	31.77	29.45	7:59	11
-144	31.32	29.02	8:03	12
-145	30.87	28.59	8:08	13
-147	29.99	27.73	8:17	15
-148	29.55	27.32	8:22	16
-149	29.13	26.90	8:27	17
-150	28.70	26.50	8:32	19

PERCENTAGE OF SUN OBSCURED ALONG MAP TRACK
WEST END = 94.8   EAST END = 95.3

==
SUNRISE POSITION :ALTITUDE IS -0.8 DEGREES
NORTH : LONGITUDE -131.6,   LATITUDE 37.1
SOUTH : LONGITUDE -130.4,   LATITUDE 35.0

+ MARKS THE CITY OF LIN-AN

ECLIPSE ON JD 2163863 DATE(Y/M/D) 1212/ 5/ 3
DELTA T = 15.2 MINS

LONGITUDE (-VE EAST)	LATITUDE NORTH	LIMITS. SOUTH	LOCAL TIME	MEAN ALTITU
-129	46.44	42.81	4:52	1
-130	46.81	43.14	4:56	2
-132	47.58	43.82	5:04	3
-133	47.97	44.18	5:08	4
-134	48.39	44.54	5:12	5
-135	48.81	44.91	5:17	6
-137	49.69	45.68	5:25	7
-138	50.14	46.08	5:30	8
-139	50.60	46.48	5:34	9
-140	51.08	46.90	5:38	10

PERCENTAGE OF SUN OBSCURED ALONG MAP TRACK
WEST END =105.3   EAST END =105.6

==
SUNRISE POSITION :ALTITUDE IS -0.8 DEGREES
NORTH : LONGITUDE -124.8,   LATITUDE 45.1
SOUTH : LONGITUDE -128.1,   LATITUDE 42.5

+ MARKS THE CITY OF LIN-AN

ECLIPSE ON JD 2164749 DATE(Y/M/D) 1214/10/ 5
DELTA T =   15.1 MINS

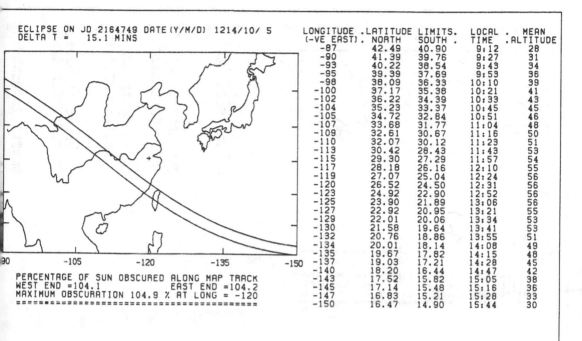

LONGITUDE	.LATITUDE	LIMITS.	LOCAL	. MEAN
(-VE EAST).	NORTH	SOUTH .	TIME	.ALTITUDE
-87	42.49	40.90	9:12	28
-90	41.39	39.76	9:27	31
-93	40.22	38.54	9:43	34
-95	39.39	37.69	9:53	36
-98	38.09	36.33	10:10	39
-100	37.17	35.38	10:21	41
-102	36.22	34.39	10:33	43
-104	35.23	33.37	10:45	45
-105	34.72	32.84	10:51	46
-107	33.68	31.77	11:04	48
-109	32.61	30.67	11:16	50
-110	32.07	30.12	11:23	51
-113	30.42	28.43	11:43	53
-115	29.30	27.29	11:57	54
-117	28.18	26.16	12:10	55
-119	27.07	25.04	12:24	56
-120	26.52	24.50	12:31	56
-123	24.92	22.90	12:52	56
-125	23.90	21.89	13:06	56
-127	22.92	20.95	13:21	55
-129	22.01	20.06	13:34	53
-130	21.58	19.64	13:41	53
-132	20.76	18.86	13:55	51
-134	20.01	18.14	14:08	49
-135	19.67	17.82	14:15	48
-137	19.03	17.21	14:28	45
-140	18.20	16.44	14:47	42
-143	17.52	15.82	15:05	38
-145	17.14	15.48	15:16	36
-147	16.83	15.21	15:28	33
-150	16.47	14.90	15:44	30

PERCENTAGE OF SUN OBSCURED ALONG MAP TRACK
WEST END =104.1   EAST END =104.2
MAXIMUM OBSCURATION 104.9 % AT LONG = -120
===

+ MARKS THE CITY OF LIN-AN

ECLIPSE ON JD 2165783 DATE(Y/M/D) 1217/ 8/ 4
DELTA T =   15.0 MINS

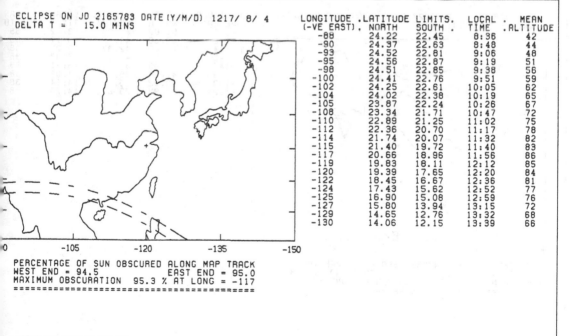

LONGITUDE	.LATITUDE	LIMITS.	LOCAL	. MEAN
(-VE EAST).	NORTH	SOUTH .	TIME	.ALTITUDE
-88	24.22	22.45	8:36	42
-90	24.37	22.63	8:48	44
-93	24.52	22.81	9:06	48
-95	24.56	22.87	9:19	51
-98	24.51	22.85	9:38	56
-100	24.41	22.76	9:51	59
-102	24.25	22.61	10:05	62
-104	24.02	22.38	10:19	65
-105	23.87	22.24	10:26	67
-108	23.34	21.71	10:47	72
-110	22.89	21.25	11:02	75
-112	22.36	20.70	11:17	78
-114	21.74	20.07	11:32	82
-115	21.40	19.72	11:40	83
-117	20.66	18.96	11:56	86
-119	19.83	18.11	12:12	85
-120	19.39	17.65	12:20	84
-122	18.45	16.67	12:36	81
-124	17.43	15.62	12:52	77
-125	16.90	15.08	12:59	76
-127	15.80	13.94	13:15	72
-129	14.65	12.76	13:32	68
-130	14.06	12.15	13:39	66

PERCENTAGE OF SUN OBSCURED ALONG MAP TRACK
WEST END = 94.5   EAST END = 95.0
MAXIMUM OBSCURATION 95.3 % AT LONG = -117
===

+ MARKS THE CITY OF LIN-AN

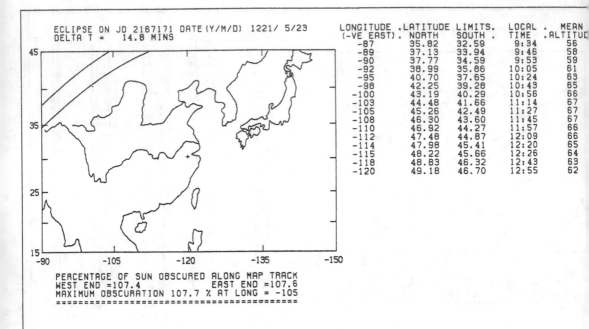

ECLIPSE ON JD 2167171 DATE(Y/M/D) 1221/ 5/23
DELTA T = 14.8 MINS

LONGITUDE (-VE EAST)	LATITUDE LIMITS. NORTH	SOUTH	LOCAL TIME	MEAN ALTITUDE
-87	35.82	32.59	9:34	56
-89	37.13	33.94	9:46	58
-90	37.77	34.59	9:53	59
-92	38.99	35.86	10:05	61
-95	40.70	37.65	10:24	63
-98	42.25	39.28	10:43	65
-100	43.19	40.29	10:56	66
-103	44.48	41.66	11:14	67
-105	45.26	42.49	11:27	67
-108	46.30	43.60	11:45	67
-110	46.92	44.27	11:57	66
-112	47.48	44.87	12:09	66
-114	47.98	45.41	12:20	65
-115	48.22	45.66	12:26	64
-118	48.83	46.32	12:43	63
-120	49.18	46.70	12:55	62

PERCENTAGE OF SUN OBSCURED ALONG MAP TRACK
WEST END =107.4   EAST END =107.6
MAXIMUM OBSCURATION 107.7 % AT LONG = -105
===

+ MARKS THE CITY OF LIN-AN

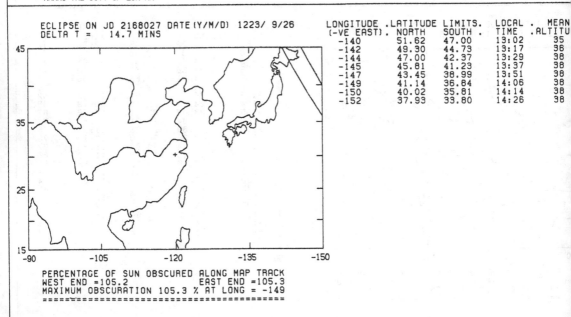

ECLIPSE ON JD 2168027 DATE(Y/M/D) 1223/ 9/26
DELTA T = 14.7 MINS

LONGITUDE (-VE EAST)	LATITUDE LIMITS. NORTH	SOUTH	LOCAL TIME	MEAN ALTITUDE
-140	51.62	47.00	13:02	35
-142	49.30	44.73	13:17	36
-144	47.00	42.37	13:29	38
-145	45.81	41.23	13:37	38
-147	43.45	38.99	13:51	38
-149	41.14	36.84	14:06	38
-150	40.02	35.81	14:14	38
-152	37.93	33.80	14:26	38

PERCENTAGE OF SUN OBSCURED ALONG MAP TRACK
WEST END =105.2   EAST END =105.3
MAXIMUM OBSCURATION 105.3 % AT LONG = -149
===

+ MARKS THE CITY OF LIN-AN

ECLIPSE ON JD 2172722 DATE(Y/M/D) 1236/ 8/ 3
DELTA T = 14.1 MINS

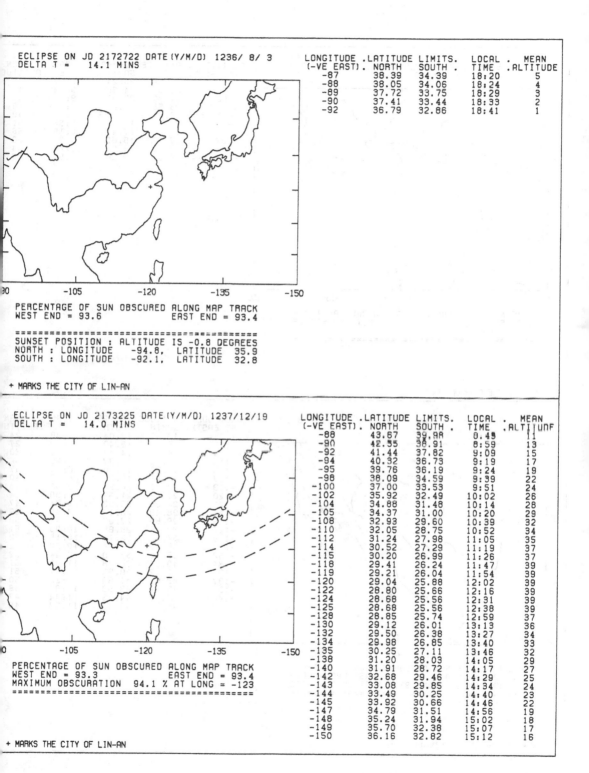

LONGITUDE (-VE EAST)	LATITUDE NORTH	LIMITS. SOUTH	LOCAL TIME	MEAN ALTITUDE
-87	38.39	34.39	18:20	5
-88	38.05	34.06	18:24	4
-89	37.72	33.75	18:29	3
-90	37.41	33.44	18:33	2
-92	36.79	32.86	18:41	1

```
          -105          -120          -135          -150
```

PERCENTAGE OF SUN OBSCURED ALONG MAP TRACK
WEST END = 93.6   EAST END = 93.4

```
==============================================
```

SUNSET POSITION : ALTITUDE IS -0.8 DEGREES
NORTH : LONGITUDE   -94.8, LATITUDE   35.9
SOUTH : LONGITUDE   -92.1, LATITUDE   32.8

+ MARKS THE CITY OF LIN-AN

ECLIPSE ON JD 2173225 DATE(Y/M/D) 1237/12/19
DELTA T = 14.0 MINS

LONGITUDE (-VE EAST)	LATITUDE NORTH	LIMITS. SOUTH	LOCAL TIME	MEAN ALTITUDE
-88	43.67	39.98	8:49	11
-90	42.55	38.91	8:59	13
-92	41.44	37.82	9:09	15
-94	40.32	36.73	9:19	17
-95	39.76	36.19	9:24	19
-98	38.09	34.59	9:39	22
-100	37.00	33.53	9:51	24
-102	35.92	32.49	10:02	26
-104	34.88	31.48	10:14	28
-105	34.37	31.00	10:20	29
-108	32.93	29.60	10:39	32
-110	32.05	28.75	10:52	34
-112	31.24	27.98	11:05	35
-114	30.52	27.29	11:19	37
-115	30.20	26.99	11:26	37
-118	29.41	26.24	11:47	39
-119	29.21	26.04	11:54	39
-120	29.04	25.88	12:02	39
-122	28.80	25.66	12:16	39
-124	28.68	25.56	12:31	39
-125	28.68	25.56	12:38	39
-128	28.85	25.74	12:59	37
-130	29.12	26.01	13:13	36
-132	29.50	26.38	13:27	34
-134	29.98	26.85	13:40	33
-135	30.25	27.11	13:46	32
-138	31.20	28.03	14:05	29
-140	31.91	28.72	14:17	27
-142	32.68	29.46	14:29	25
-143	33.08	29.85	14:34	24
-144	33.49	30.25	14:40	23
-145	33.92	30.66	14:46	22
-147	34.79	31.51	14:56	19
-148	35.24	31.94	15:02	18
-149	35.70	32.38	15:07	17
-150	36.16	32.82	15:12	16

```
          -105          -120          -135          -150
```

PERCENTAGE OF SUN OBSCURED ALONG MAP TRACK
WEST END = 93.3   EAST END = 93.4
MAXIMUM OBSCURATION 94.1 % AT LONG = -123

```
==============================================
```

+ MARKS THE CITY OF LIN-AN

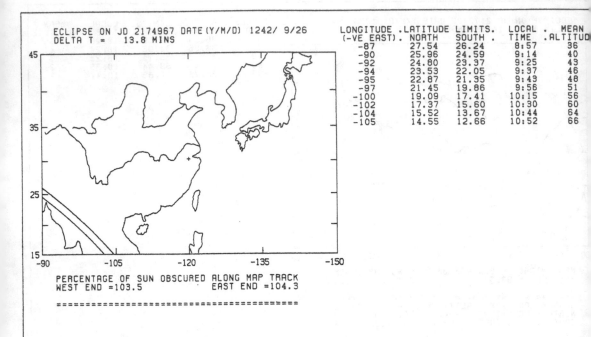

ECLIPSE ON JD 2174967 DATE (Y/M/D) 1242/ 9/26
DELTA T =   13.8 MINS

LONGITUDE (-VE EAST)	LATITUDE NORTH	LIMITS. SOUTH	LOCAL TIME	MEAN ALTITUD
-87	27.54	26.24	8:57	36
-90	25.96	24.59	9:14	40
-92	24.80	23.37	9:25	43
-94	23.53	22.05	9:37	46
-95	22.87	21.35	9:49	48
-97	21.45	19.86	9:56	51
-100	19.09	17.41	10:15	56
-102	17.37	15.60	10:30	60
-104	15.52	13.67	10:44	64
-105	14.55	12.66	10:52	66

PERCENTAGE OF SUN OBSCURED ALONG MAP TRACK
WEST END =103.5   EAST END =104.3

==

+ MARKS THE CITY OF LIN-AN

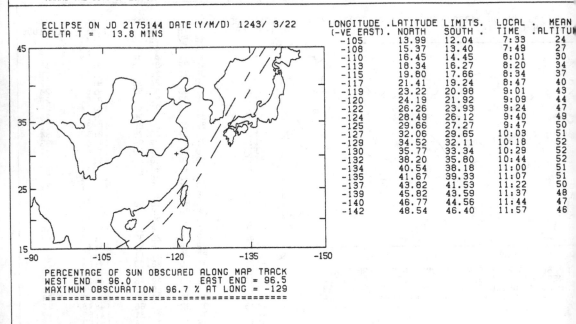

ECLIPSE ON JD 2175144 DATE (Y/M/D) 1243/ 3/22
DELTA T =   13.8 MINS

LONGITUDE (-VE EAST)	LATITUDE NORTH	LIMITS. SOUTH	LOCAL TIME	MEAN ALTITU
-105	13.99	12.04	7:39	24
-108	15.37	13.40	7:49	27
-110	16.45	14.45	8:01	30
-113	18.34	16.27	8:20	34
-115	19.80	17.66	8:34	37
-117	21.41	19.24	8:47	40
-119	23.22	20.98	9:01	43
-120	24.19	21.92	9:09	44
-122	26.26	23.93	9:24	47
-124	28.49	26.12	9:40	49
-125	29.66	27.27	9:47	50
-127	32.06	29.65	10:03	51
-129	34.52	32.11	10:18	52
-130	35.77	33.34	10:29	52
-132	38.20	35.80	10:44	52
-134	40.54	38.18	11:00	51
-135	41.67	39.33	11:07	51
-137	43.82	41.53	11:22	50
-139	45.82	43.59	11:37	48
-140	46.77	44.56	11:44	47
-142	48.54	46.40	11:57	46

PERCENTAGE OF SUN OBSCURED ALONG MAP TRACK
WEST END = 96.0   EAST END = 96.5
MAXIMUM OBSCURATION 96.7 % AT LONG = -129

==

+ MARKS THE CITY OF LIN-AN

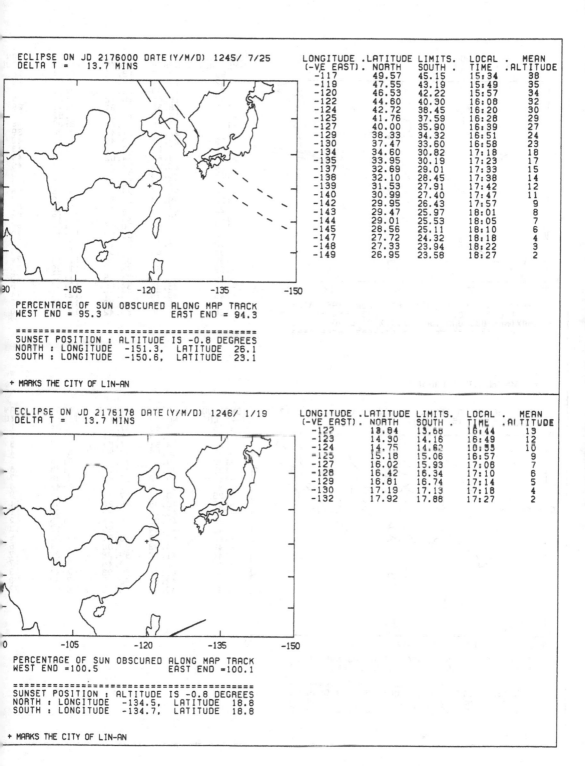

ECLIPSE ON JD 2176000 DATE (Y/M/D) 1245/ 7/25
DELTA T =   13.7 MINS

LONGITUDE (-VE EAST)	LATITUDE LIMITS. NORTH	SOUTH	LOCAL TIME	MEAN ALTITUDE
-117	49.57	45.15	15:34	38
-119	47.55	43.19	15:49	35
-120	46.53	42.22	15:57	34
-122	44.60	40.30	16:08	32
-124	42.72	38.45	16:20	30
-125	41.76	37.59	16:28	29
-127	40.00	35.90	16:39	27
-129	38.33	34.32	16:51	24
-130	37.47	33.60	16:58	23
-134	34.60	30.82	17:18	18
-135	33.95	30.19	17:23	17
-137	32.69	29.01	17:33	15
-138	32.10	28.45	17:38	14
-139	31.53	27.91	17:42	12
-140	30.99	27.40	17:47	11
-142	29.95	26.43	17:57	9
-143	29.47	25.97	18:01	8
-144	29.01	25.53	18:05	7
-145	28.56	25.11	18:10	6
-147	27.72	24.32	18:18	4
-148	27.33	23.94	18:22	3
-149	26.95	23.58	18:27	2

PERCENTAGE OF SUN OBSCURED ALONG MAP TRACK
WEST END = 95.3   EAST END = 94.3

==
SUNSET POSITION : ALTITUDE IS -0.8 DEGREES
NORTH : LONGITUDE -151.3, LATITUDE 26.1
SOUTH : LONGITUDE -150.6, LATITUDE 23.1

+ MARKS THE CITY OF LIN-AN

ECLIPSE ON JD 2176178 DATE (Y/M/D) 1246/ 1/19
DELTA T =   13.7 MINS

LONGITUDE (-VE EAST)	LATITUDE LIMITS. NORTH	SOUTH	LOCAL TIME	MEAN ALTITUDE
-122	13.84	13.68	16:44	13
-123	14.30	14.16	16:49	12
-124	14.75	14.62	16:53	10
-125	15.18	15.06	16:57	9
-127	16.02	15.93	17:08	7
-128	16.42	16.34	17:10	6
-129	16.81	16.74	17:14	5
-130	17.19	17.13	17:18	4
-132	17.92	17.88	17:27	2

PERCENTAGE OF SUN OBSCURED ALONG MAP TRACK
WEST END =100.5   EAST END =100.1

==
SUNSET POSITION : ALTITUDE IS -0.8 DEGREES
NORTH : LONGITUDE -134.5, LATITUDE 18.8
SOUTH : LONGITUDE -134.7, LATITUDE 18.8

+ MARKS THE CITY OF LIN-AN

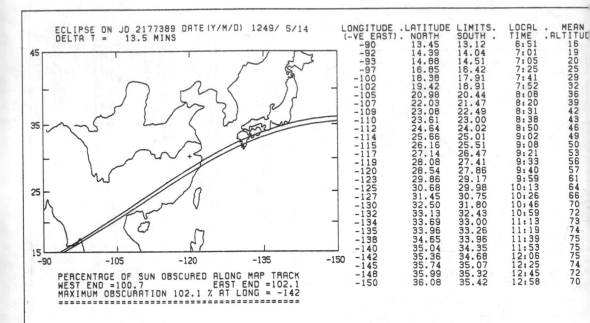

ECLIPSE ON JD 2177389 DATE(Y/M/D) 1249/ 5/14
DELTA T =   13.5 MINS

LONGITUDE (-VE EAST)	LATITUDE NORTH	LIMITS SOUTH	LOCAL TIME	MEAN ALTITU
-90	13.45	13.12	6:51	16
-92	14.39	14.04	7:01	19
-93	14.88	14.51	7:05	20
-97	16.85	16.42	7:25	25
-100	18.38	17.91	7:41	29
-102	19.42	18.91	7:52	32
-105	20.98	20.44	8:08	36
-107	22.03	21.47	8:20	39
-109	23.08	22.49	8:31	42
-110	23.61	23.00	8:38	43
-112	24.64	24.02	8:50	46
-114	25.66	25.01	9:02	49
-115	26.16	25.51	9:08	50
-117	27.14	26.47	9:21	53
-119	28.08	27.41	9:33	56
-120	28.54	27.86	9:40	57
-123	29.86	29.17	9:59	61
-125	30.68	29.98	10:13	64
-127	31.45	30.75	10:26	66
-130	32.50	31.80	10:46	70
-132	33.13	32.43	10:59	72
-134	33.69	33.00	11:13	73
-135	33.96	33.26	11:19	74
-138	34.65	33.96	11:39	75
-140	35.04	34.35	11:53	75
-142	35.36	34.68	12:06	75
-145	35.74	35.07	12:25	74
-148	35.99	35.32	12:45	72
-150	36.08	35.42	12:58	70

PERCENTAGE OF SUN OBSCURED ALONG MAP TRACK
WEST END =100.7   EAST END =102.1
MAXIMUM OBSCURATION 102.1 % AT LONG = -142
===

+ MARKS THE CITY OF LIN-AN

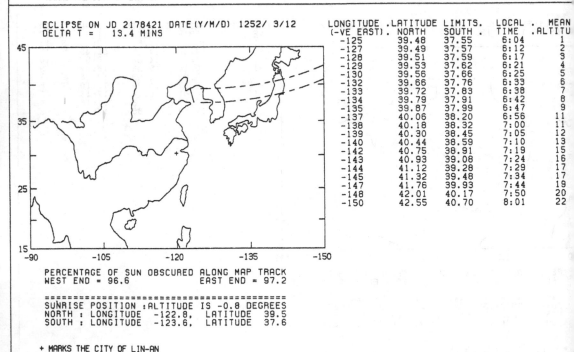

ECLIPSE ON JD 2178421 DATE(Y/M/D) 1252/ 3/12
DELTA T =   13.4 MINS

LONGITUDE (-VE EAST)	LATITUDE NORTH	LIMITS SOUTH	LOCAL TIME	MEAN ALTITU
-125	39.48	37.55	6:04	1
-127	39.49	37.57	6:12	2
-128	39.51	37.59	6:17	3
-129	39.53	37.62	6:21	4
-130	39.56	37.66	6:25	5
-132	39.66	37.76	6:33	6
-133	39.72	37.83	6:38	7
-134	39.79	37.91	6:42	8
-135	39.87	37.99	6:47	9
-137	40.06	38.20	6:56	11
-138	40.18	38.32	7:00	11
-139	40.30	38.45	7:05	12
-140	40.44	38.59	7:10	13
-142	40.75	38.91	7:19	15
-143	40.93	39.08	7:24	16
-144	41.12	39.28	7:29	17
-145	41.32	39.48	7:34	17
-147	41.76	39.93	7:44	19
-148	42.01	40.17	7:50	20
-150	42.55	40.70	8:01	22

PERCENTAGE OF SUN OBSCURED ALONG MAP TRACK
WEST END = 96.6   EAST END = 97.2

===
SUNRISE POSITION :ALTITUDE IS -0.8 DEGREES
NORTH : LONGITUDE -122.8, LATITUDE   39.5
SOUTH : LONGITUDE -123.6, LATITUDE   37.6

+ MARKS THE CITY OF LIN-AN

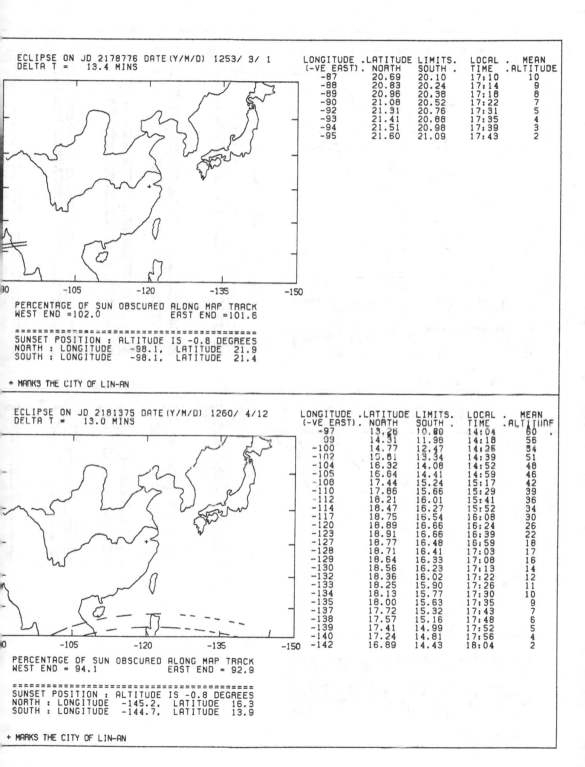

ECLIPSE ON JD 2178776 DATE(Y/M/D) 1253/ 3/ 1
DELTA T =   13.4 MINS

LONGITUDE (-VE EAST)	.LATITUDE LIMITS.		LOCAL TIME	. MEAN ALTITUDE
	NORTH	SOUTH		
-87	20.69	20.10	17:10	10
-88	20.83	20.24	17:14	9
-89	20.96	20.38	17:18	8
-90	21.08	20.52	17:22	7
-92	21.31	20.76	17:31	5
-93	21.41	20.88	17:35	4
-94	21.51	20.98	17:39	3
-95	21.60	21.09	17:43	2

PERCENTAGE OF SUN OBSCURED ALONG MAP TRACK
WEST END =102.0   EAST END =101.6

===
SUNSET POSITION : ALTITUDE IS -0.8 DEGREES
NORTH : LONGITUDE   -98.1, LATITUDE   21.9
SOUTH : LONGITUDE   -98.1, LATITUDE   21.4

+ MARKS THE CITY OF LIN-AN

ECLIPSE ON JD 2181375 DATE(Y/M/D) 1260/ 4/12
DELTA T =   13.0 MINS

LONGITUDE (-VE EAST)	.LATITUDE LIMITS.		LOCAL TIME	. MEAN ALTITUDE
	NORTH	SOUTH		
-97	13.26	10.80	14:04	60
-99	14.31	11.98	14:18	56
-100	14.77	12.47	14:26	54
-102	15.61	13.34	14:39	51
-104	16.32	14.08	14:52	48
-105	16.64	14.41	14:59	46
-108	17.44	15.24	15:17	42
-110	17.86	15.66	15:29	39
-112	18.21	16.01	15:41	36
-114	18.47	16.27	15:52	34
-117	18.75	16.54	16:08	30
-120	18.89	16.66	16:24	26
-123	18.91	16.66	16:39	22
-127	18.77	16.48	16:59	18
-128	18.71	16.41	17:03	17
-129	18.64	16.33	17:08	16
-130	18.56	16.23	17:13	14
-132	18.36	16.02	17:22	12
-133	18.25	15.90	17:26	11
-134	18.13	15.77	17:30	10
-135	18.00	15.63	17:35	9
-137	17.72	15.32	17:43	7
-138	17.57	15.16	17:48	6
-139	17.41	14.99	17:52	5
-140	17.24	14.81	17:56	4
-142	16.89	14.43	18:04	2

PERCENTAGE OF SUN OBSCURED ALONG MAP TRACK
WEST END = 94.1   EAST END = 92.9

===
SUNSET POSITION : ALTITUDE IS -0.8 DEGREES
NORTH : LONGITUDE -145.2, LATITUDE   16.3
SOUTH : LONGITUDE -144.7, LATITUDE   13.9

+ MARKS THE CITY OF LIN-AN

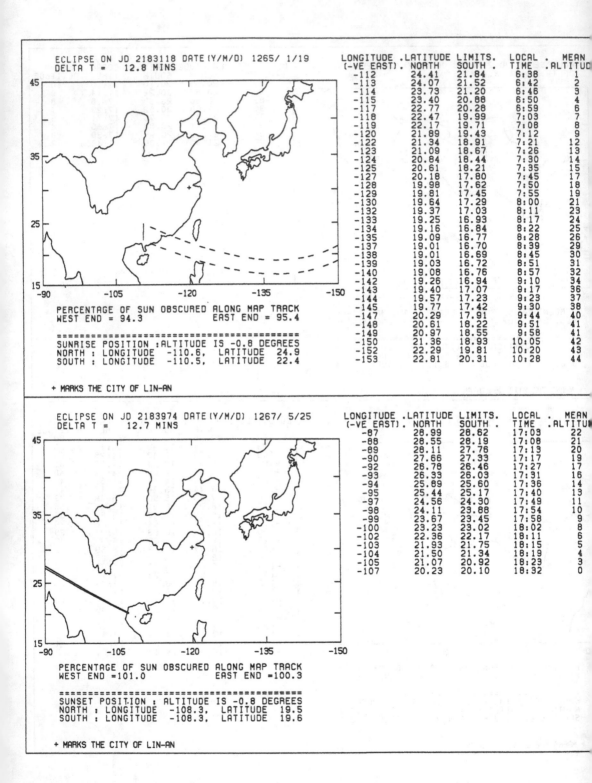

ECLIPSE ON JD 2183118 DATE(Y/M/D) 1265/ 1/19
DELTA T = 12.8 MINS

LONGITUDE (-VE EAST)	LATITUDE LIMITS. NORTH	SOUTH	LOCAL TIME	MEAN ALTITUD
-112	24.41	21.84	6:38	1
-113	24.07	21.52	6:42	2
-114	23.73	21.20	6:46	3
-115	23.40	20.88	6:50	4
-117	22.77	20.28	6:59	6
-118	22.47	19.99	7:03	7
-119	22.17	19.71	7:08	8
-120	21.89	19.43	7:12	9
-122	21.34	18.91	7:21	12
-123	21.09	18.67	7:26	13
-124	20.84	18.44	7:30	14
-125	20.61	18.21	7:35	15
-127	20.18	17.80	7:45	17
-128	19.98	17.62	7:50	18
-129	19.81	17.45	7:55	19
-130	19.64	17.29	8:00	21
-132	19.37	17.03	8:11	23
-133	19.25	16.93	8:17	24
-134	19.16	16.84	8:22	25
-135	19.09	16.77	8:28	26
-137	19.01	16.70	8:39	29
-138	19.01	16.69	8:45	30
-139	19.03	16.72	8:51	31
-140	19.08	16.76	8:57	32
-142	19.26	16.94	9:10	34
-143	19.40	17.07	9:17	36
-144	19.57	17.23	9:23	37
-145	19.77	17.42	9:30	38
-147	20.29	17.91	9:44	40
-148	20.61	18.22	9:51	41
-149	20.97	18.55	9:58	41
-150	21.36	18.93	10:05	42
-152	22.29	19.81	10:20	43
-153	22.81	20.31	10:28	44

PERCENTAGE OF SUN OBSCURED ALONG MAP TRACK
WEST END = 94.3   EAST END = 95.4

===
SUNRISE POSITION :ALTITUDE IS -0.8 DEGREES
NORTH : LONGITUDE -110.6, LATITUDE 24.9
SOUTH : LONGITUDE -110.5, LATITUDE 22.4

+ MARKS THE CITY OF LIN-AN

ECLIPSE ON JD 2183974 DATE(Y/M/D) 1267/ 5/25
DELTA T = 12.7 MINS

LONGITUDE (-VE EAST)	LATITUDE LIMITS. NORTH	SOUTH	LOCAL TIME	MEAN ALTITU
-87	28.99	28.62	17:03	22
-88	28.55	28.19	17:08	21
-89	28.11	27.76	17:13	20
-90	27.66	27.33	17:17	19
-92	26.78	26.46	17:27	17
-93	26.33	26.03	17:31	16
-94	25.89	25.60	17:36	14
-95	25.44	25.17	17:40	13
-97	24.56	24.30	17:49	11
-98	24.11	23.88	17:54	10
-99	23.67	23.45	17:58	9
-100	23.23	23.02	18:02	8
-102	22.36	22.17	18:11	6
-103	21.93	21.75	18:15	5
-104	21.50	21.34	18:19	4
-105	21.07	20.92	18:23	3
-107	20.23	20.10	18:32	0

PERCENTAGE OF SUN OBSCURED ALONG MAP TRACK
WEST END =101.0   EAST END =100.3

===
SUNSET POSITION : ALTITUDE IS -0.8 DEGREES
NORTH : LONGITUDE -108.3, LATITUDE 19.5
SOUTH : LONGITUDE -108.3, LATITUDE 19.6

+ MARKS THE CITY OF LIN-AN

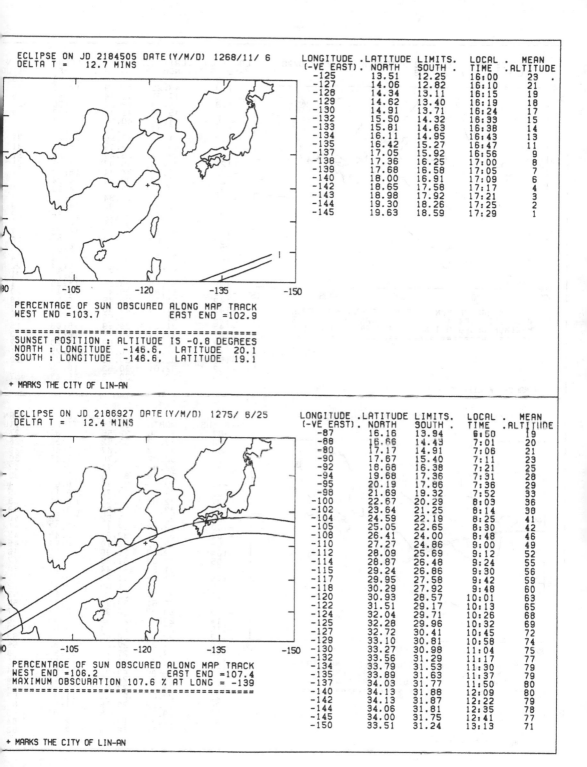

ECLIPSE ON JD 2184505 DATE(Y/M/D) 1268/11/ 6
DELTA T =   12.7 MINS

LONGITUDE (-VE EAST)	.LATITUDE NORTH	LIMITS. SOUTH .	LOCAL . TIME	MEAN .ALTITUDE
-125	13.51	12.25	16:00	23
-127	14.06	12.82	16:10	21
-128	14.34	13.11	16:15	19
-129	14.62	13.40	16:19	18
-130	14.91	13.71	16:24	17
-132	15.50	14.32	16:33	15
-133	15.81	14.63	16:38	14
-134	16.11	14.95	16:43	13
-135	16.42	15.27	16:47	11
-137	17.05	15.92	16:56	9
-138	17.36	16.25	17:00	8
-139	17.68	16.58	17:05	7
-140	18.00	16.91	17:09	6
-142	18.65	17.58	17:17	4
-143	18.98	17.92	17:21	3
-144	19.30	18.26	17:25	2
-145	19.63	18.59	17:29	1

PERCENTAGE OF SUN OBSCURED ALONG MAP TRACK
WEST END =103.7   EAST END =102.9

===
SUNSET POSITION : ALTITUDE IS -0.8 DEGREES
NORTH : LONGITUDE -146.6,   LATITUDE   20.1
SOUTH : LONGITUDE -146.6,   LATITUDE   19.1

+ MARKS THE CITY OF LIN-AN

ECLIPSE ON JD 2186927 DATE(Y/M/D) 1275/ 6/25
DELTA T =   12.4 MINS

LONGITUDE (-VE EAST)	.LATITUDE NORTH	LIMITS. SOUTH .	LOCAL TIME	MEAN .ALTITUDE
-87	16.16	13.94	6:50	19
-88	16.66	14.43	7:01	20
-80	17.17	14.91	7:06	21
-90	17.67	15.40	7:11	23
-92	18.68	16.38	7:21	25
-94	19.68	17.36	7:31	28
-95	20.19	17.86	7:36	29
-98	21.69	19.32	7:52	33
-100	22.67	20.29	8:09	36
-102	23.64	21.25	8:14	38
-104	24.59	22.19	8:25	41
-105	25.05	22.65	8:30	42
-108	26.41	24.00	8:48	46
-110	27.27	24.86	9:00	49
-112	28.09	25.69	9:12	52
-114	28.87	26.48	9:24	55
-115	29.24	26.86	9:30	56
-117	29.95	27.58	9:42	59
-118	30.29	27.92	9:48	60
-120	30.93	28.57	10:01	63
-122	31.51	29.17	10:13	65
-124	32.04	29.71	10:26	68
-125	32.28	29.96	10:32	69
-127	32.72	30.41	10:45	72
-129	33.10	30.81	10:58	74
-130	33.27	30.98	11:04	75
-132	33.56	31.29	11:17	77
-134	33.79	31.53	11:30	79
-135	33.89	31.63	11:37	79
-137	34.03	31.77	11:50	80
-140	34.13	31.88	12:09	80
-142	34.13	31.87	12:22	79
-144	34.06	31.81	12:35	78
-145	34.00	31.75	12:41	77
-150	33.51	31.24	13:13	71

PERCENTAGE OF SUN OBSCURED ALONG MAP TRACK
WEST END =106.2   EAST END =107.4
MAXIMUM OBSCURATION 107.6 % AT LONG = -139
===

+ MARKS THE CITY OF LIN-AN

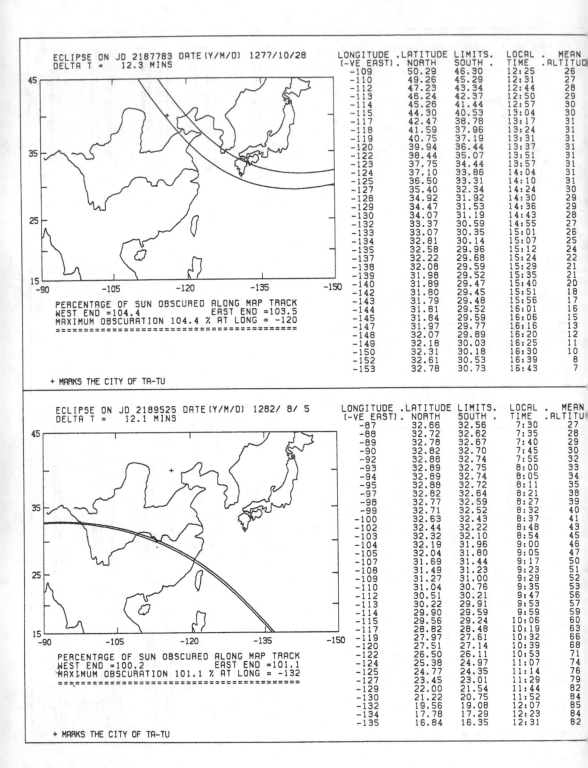

ECLIPSE ON JD 2187783 DATE(Y/M/D) 1277/10/28
DELTA T =   12.3 MINS

LONGITUDE (-VE EAST)	.LATITUDE LIMITS.		LOCAL TIME	MEAN .ALTITUDE
	NORTH	SOUTH		
-109	50.29	46.30	12:25	26
-110	49.26	45.29	12:31	27
-112	47.23	43.34	12:44	28
-113	46.24	42.37	12:50	29
-114	45.26	41.44	12:57	30
-115	44.30	40.53	13:04	30
-117	42.47	38.78	13:17	31
-118	41.59	37.96	13:24	31
-119	40.75	37.19	13:31	31
-120	39.94	36.44	13:37	31
-122	38.44	35.07	13:51	31
-123	37.75	34.44	13:57	31
-124	37.10	33.86	14:04	31
-125	36.50	33.31	14:10	31
-127	35.40	32.34	14:24	30
-128	34.92	31.92	14:30	29
-129	34.47	31.53	14:36	29
-130	34.07	31.19	14:43	28
-132	33.37	30.59	14:55	27
-133	33.07	30.35	15:01	26
-134	32.81	30.14	15:07	25
-135	32.58	29.96	15:12	24
-137	32.22	29.68	15:24	22
-138	32.08	29.59	15:29	21
-139	31.98	29.52	15:35	21
-140	31.89	29.47	15:40	20
-142	31.80	29.45	15:51	18
-143	31.79	29.48	15:56	17
-144	31.81	29.52	16:01	16
-145	31.84	29.59	16:06	15
-147	31.97	29.77	16:16	13
-148	32.07	29.89	16:20	12
-149	32.18	30.03	16:25	11
-150	32.31	30.18	16:30	10
-152	32.61	30.53	16:39	8
-153	32.78	30.73	16:43	7

PERCENTAGE OF SUN OBSCURED ALONG MAP TRACK
WEST END =104.4   EAST END =103.5
MAXIMUM OBSCURATION 104.4 % AT LONG = -120
==

+ MARKS THE CITY OF TA-TU

ECLIPSE ON JD 2189525 DATE(Y/M/D) 1282/ 8/ 5
DELTA T =   12.1 MINS

LONGITUDE (-VE EAST)	.LATITUDE LIMITS.		LOCAL TIME	MEAN .ALTITUDE
	NORTH	SOUTH		
-87	32.66	32.56	7:30	27
-88	32.72	32.62	7:35	28
-89	32.78	32.67	7:40	29
-90	32.82	32.70	7:45	30
-92	32.88	32.74	7:55	32
-93	32.89	32.75	8:00	33
-94	32.89	32.74	8:05	34
-95	32.88	32.72	8:11	35
-97	32.82	32.64	8:21	38
-98	32.77	32.59	8:27	39
-99	32.71	32.52	8:32	40
-100	32.63	32.43	8:37	41
-102	32.44	32.22	8:48	43
-103	32.32	32.10	8:54	45
-104	32.19	31.96	9:00	46
-105	32.04	31.80	9:05	47
-107	31.69	31.44	9:17	50
-108	31.49	31.23	9:23	51
-109	31.27	31.00	9:29	52
-110	31.04	30.76	9:35	53
-112	30.51	30.21	9:47	56
-113	30.22	29.91	9:53	57
-114	29.90	29.59	9:59	59
-115	29.56	29.24	10:06	60
-117	28.82	28.48	10:19	63
-119	27.97	27.61	10:32	66
-120	27.51	27.14	10:39	68
-122	26.50	26.11	10:53	71
-124	25.38	24.97	11:07	74
-125	24.77	24.35	11:14	76
-127	23.45	23.01	11:29	79
-129	22.00	21.54	11:44	82
-130	21.22	20.75	11:52	84
-132	19.56	19.08	12:07	85
-134	17.78	17.29	12:23	84
-135	16.84	16.35	12:31	82

PERCENTAGE OF SUN OBSCURED ALONG MAP TRACK
WEST END =100.2   EAST END =101.1
MAXIMUM OBSCURATION 101.1 % AT LONG = -132
==

+ MARKS THE CITY OF TA-TU

ECLIPSE ON JD 2192478 DATE(Y/M/D) 1290/ 9/ 5
DELTA T =   11.7 MINS

LONGITUDE (-VE EAST)	LATITUDE LIMITS NORTH	SOUTH	LOCAL TIME	MEAN ALTITUDE
-88	26.38	23.45	15:00	42
-90	25.42	22.52	15:14	39
-92	24.53	21.63	15:26	37
-94	23.70	20.81	15:38	34
-95	23.30	20.42	15:45	33
-98	22.22	19.34	16:01	29
-100	21.57	18.70	16:12	26
-103	20.70	17.84	16:28	23
-105	20.17	17.33	16:39	20
-107	19.71	16.85	16:49	18
-108	19.49	16.64	16:53	17
-109	19.29	16.43	16:58	16
-110	19.09	16.24	17:03	15
-112	18.73	15.87	17:12	12
-113	18.57	15.71	17:17	11
-114	18.41	15.55	17:21	10
-115	18.27	15.40	17:26	9
-117	18.00	15.13	17:35	7
-118	17.88	15.01	17:39	6
-119	17.77	14.90	17:43	5
-120	17.67	14.79	17:47	4
-122	17.48	14.60	17:56	2

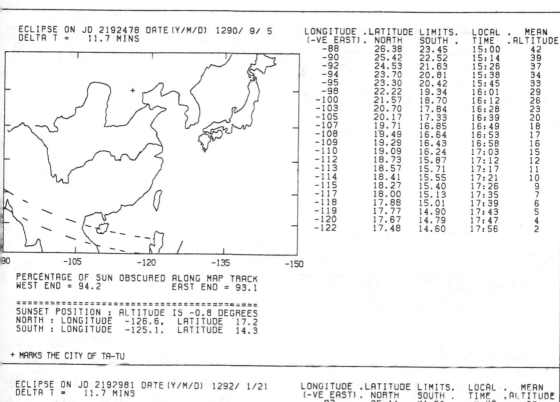

PERCENTAGE OF SUN OBSCURED ALONG MAP TRACK
WEST END = 94.2   EAST END = 93.1

===
SUNSET POSITION : ALTITUDE IS -0.8 DEGREES
NORTH : LONGITUDE -126.6, LATITUDE 17.2
SOUTH : LONGITUDE -125.1, LATITUDE 14.3

+ MARKS THE CITY OF TA-TU

ECLIPSE ON JD 2192981 DATE(Y/M/D) 1292/ 1/21
DELTA T =   11.7 MINS

LONGITUDE (-VE EAST)	LATITUDE LIMITS NORTH	SOUTH	LOCAL TIME	MEAN ALTITUDE
-87	35.11	31.50	9:38	28
-89	34.96	31.34	9:51	30
-90	34.91	31.29	9:57	31
-92	34.89	31.20	10:10	32
-95	35.02	31.38	10:29	34
-97	35.24	31.57	10:43	35
-100	35.75	32.07	11:03	36
-102	36.24	32.52	11:17	36
-105	37.13	33.39	11:38	36
-107	37.86	34.07	11:52	36
-110	39.08	35.27	12:12	35
-112	40.00	36.13	12:26	33
-114	40.95	37.08	12:39	32
-115	41.45	37.55	12:45	31
-117	42.46	38.55	12:58	30
-119	43.49	39.58	13:10	28
-120	44.05	40.07	13:16	27
-122	45.11	41.11	13:27	25
-125	46.69	42.69	13:44	23
-127	47.78	43.71	13:55	21
-130	49.36	45.25	14:10	18
-132	50.37	46.28	14:21	16
-133	50.88	46.78	14:25	16

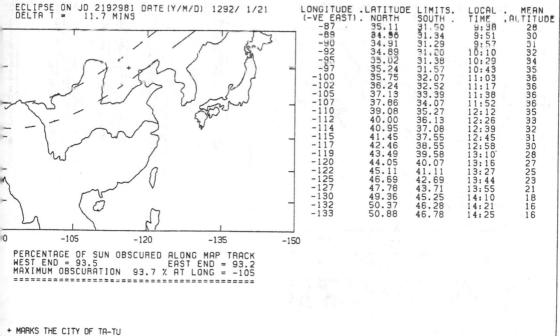

PERCENTAGE OF SUN OBSCURED ALONG MAP TRACK
WEST END = 93.5   EAST END = 93.2
MAXIMUM OBSCURATION 93.7 % AT LONG = -105
===

+ MARKS THE CITY OF TA-TU

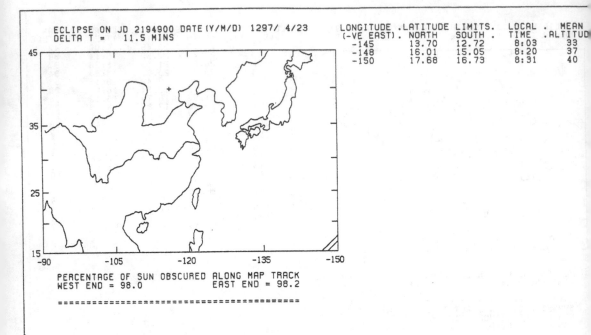

ECLIPSE ON JD 2194900 DATE(Y/M/D) 1297/ 4/23
DELTA T =   11.5 MINS

LONGITUDE (-VE EAST)	.LATITUDE LIMITS. NORTH	SOUTH .	LOCAL . TIME	MEAN .ALTITUD
-145	13.70	12.72	8:03	33
-148	16.01	15.05	8:20	37
-150	17.68	16.73	8:31	40

PERCENTAGE OF SUN OBSCURED ALONG MAP TRACK
WEST END = 98.0   EAST END = 98.2

===

+ MARKS THE CITY OF TA-TU

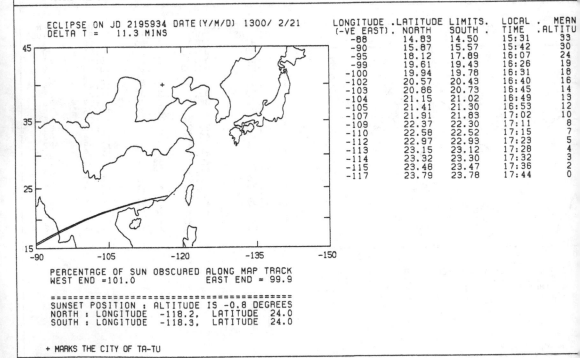

ECLIPSE ON JD 2195934 DATE(Y/M/D) 1300/ 2/21
DELTA T =   11.3 MINS

LONGITUDE (-VE EAST)	.LATITUDE NORTH	LIMITS. SOUTH .	LOCAL . TIME	MEAN .ALTITU
-88	14.83	14.50	15:31	33
-90	15.87	15.57	15:42	30
-95	18.12	17.89	16:07	24
-99	19.61	19.43	16:26	19
-100	19.94	19.78	16:31	18
-102	20.57	20.43	16:40	16
-103	20.86	20.73	16:45	14
-104	21.15	21.02	16:49	13
-105	21.41	21.30	16:53	12
-107	21.91	21.83	17:02	10
-109	22.37	22.30	17:11	8
-110	22.58	22.52	17:15	7
-112	22.97	22.93	17:23	5
-113	23.15	23.12	17:28	4
-114	23.32	23.30	17:32	3
-115	23.48	23.47	17:36	2
-117	23.79	23.78	17:44	0

PERCENTAGE OF SUN OBSCURED ALONG MAP TRACK
WEST END =101.0   EAST END = 99.9

===
SUNSET POSITION : ALTITUDE IS -0.8 DEGREES
NORTH : LONGITUDE -118.2, LATITUDE 24.0
SOUTH : LONGITUDE -118.3, LATITUDE 24.0

+ MARKS THE CITY OF TA-TU

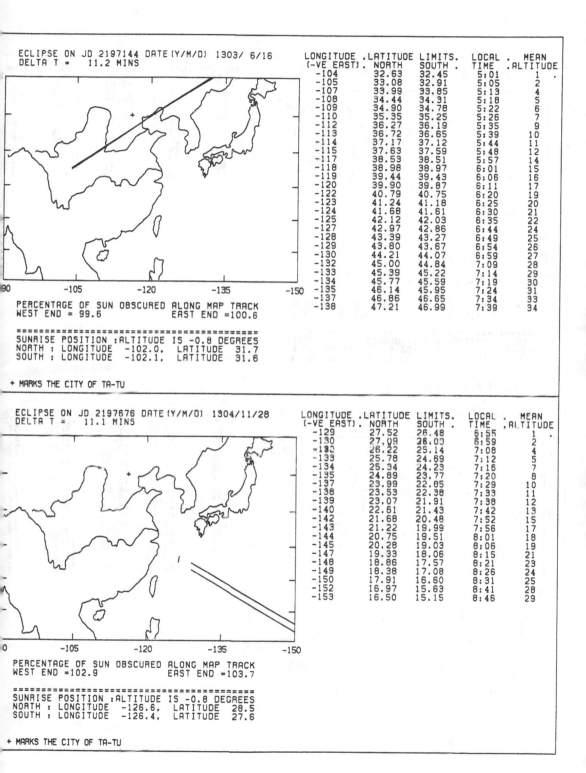

ECLIPSE ON JD 2197144 DATE (Y/M/D) 1303/ 6/16
DELTA T =   11.2 MINS

LONGITUDE (-VE EAST)	LATITUDE LIMITS. NORTH	SOUTH	LOCAL TIME	MEAN ALTITUDE
-104	32.63	32.45	5:01	1
-105	33.08	32.91	5:05	2
-107	33.99	33.85	5:13	4
-108	34.44	34.31	5:18	5
-109	34.90	34.78	5:22	6
-110	35.35	35.25	5:26	7
-112	36.27	36.19	5:35	9
-113	36.72	36.65	5:39	10
-114	37.17	37.12	5:44	11
-115	37.63	37.59	5:48	12
-117	38.53	38.51	5:57	14
-118	38.98	38.97	6:01	15
-119	39.44	39.43	6:06	16
-120	39.90	39.87	6:11	17
-122	40.79	40.75	6:20	19
-123	41.24	41.18	6:25	20
-124	41.68	41.61	6:30	21
-125	42.12	42.03	6:35	22
-127	42.97	42.86	6:44	24
-128	43.39	43.27	6:49	25
-129	43.80	43.67	6:54	26
-130	44.21	44.07	6:59	27
-132	45.00	44.84	7:09	28
-133	45.39	45.22	7:14	29
-134	45.77	45.59	7:19	30
-135	46.14	45.95	7:24	31
-137	46.86	46.65	7:34	33
-138	47.21	46.99	7:39	34

PERCENTAGE OF SUN OBSCURED ALONG MAP TRACK
WEST END = 99.6   EAST END =100.6

===
SUNRISE POSITION :ALTITUDE IS -0.8 DEGREES
NORTH : LONGITUDE -102.0, LATITUDE   31.7
SOUTH : LONGITUDE -102.1, LATITUDE   31.6

+ MARKS THE CITY OF TA-TU

ECLIPSE ON JD 2197676 DATE (Y/M/D) 1304/11/28
DELTA T =   11.1 MINS

LONGITUDE (-VE EAST)	LATITUDE LIMITS. NORTH	SOUTH	LOCAL TIME	MEAN ALTITUDE
-129	27.52	26.48	6:55	1
-130	27.09	26.03	6:59	2
-132	26.22	25.14	7:08	4
-133	25.78	24.69	7:12	5
-134	25.34	24.23	7:16	7
-135	24.89	23.77	7:20	8
-137	23.99	22.85	7:29	10
-138	23.53	22.38	7:33	11
-139	23.07	21.91	7:38	12
-140	22.61	21.43	7:42	13
-142	21.68	20.48	7:52	15
-143	21.22	19.99	7:56	17
-144	20.75	19.51	8:01	18
-145	20.28	19.03	8:06	19
-147	19.33	18.06	8:15	21
-148	18.86	17.57	8:21	23
-149	18.38	17.08	8:26	24
-150	17.91	16.60	8:31	25
-152	16.97	15.63	8:41	28
-153	16.50	15.15	8:46	29

PERCENTAGE OF SUN OBSCURED ALONG MAP TRACK
WEST END =102.9   EAST END =103.7

===
SUNRISE POSITION :ALTITUDE IS -0.8 DEGREES
NORTH : LONGITUDE -126.6, LATITUDE   28.5
SOUTH : LONGITUDE -126.4, LATITUDE   27.6

+ MARKS THE CITY OF TA-TU

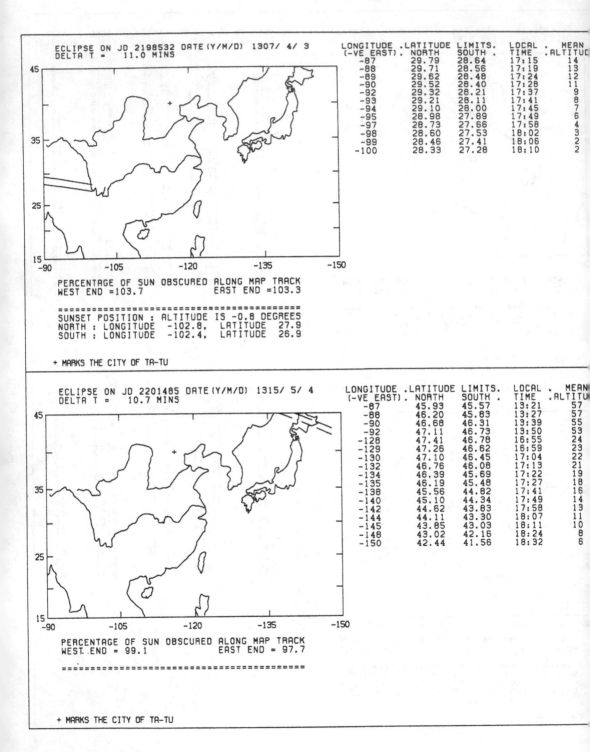

ECLIPSE ON JD 2198532 DATE(Y/M/D) 1307/ 4/ 3
DELTA T =   11.0 MINS

LONGITUDE (-VE EAST)	.LATITUDE LIMITS. NORTH	SOUTH	LOCAL TIME	. MEAN .ALTITUDE
-87	29.79	28.64	17:15	14
-88	29.71	28.56	17:19	13
-89	29.62	28.48	17:24	12
-90	29.52	28.40	17:28	11
-92	29.32	28.21	17:37	9
-93	29.21	28.11	17:41	8
-94	29.10	28.00	17:45	7
-95	28.98	27.89	17:49	6
-97	28.73	27.66	17:58	4
-98	28.60	27.53	18:02	3
-99	28.46	27.41	18:06	2
-100	28.33	27.28	18:10	2

PERCENTAGE OF SUN OBSCURED ALONG MAP TRACK
WEST END =103.7   EAST END =103.3

===
SUNSET POSITION : ALTITUDE IS -0.8 DEGREES
NORTH : LONGITUDE -102.8, LATITUDE 27.9
SOUTH : LONGITUDE -102.4, LATITUDE 26.9

+ MARKS THE CITY OF TA-TU

ECLIPSE ON JD 2201485 DATE(Y/M/D) 1315/ 5/ 4
DELTA T =   10.7 MINS

LONGITUDE (-VE EAST)	.LATITUDE LIMITS. NORTH	SOUTH	LOCAL TIME	. MEAN .ALTITUD
-87	45.93	45.57	13:21	57
-88	46.20	45.83	13:27	57
-90	46.68	46.31	13:39	55
-92	47.11	46.73	13:50	53
-128	47.41	46.78	16:55	24
-129	47.26	46.62	16:59	23
-130	47.10	46.45	17:04	22
-132	46.76	46.08	17:13	21
-134	46.39	45.69	17:22	19
-135	46.19	45.48	17:27	18
-138	45.56	44.82	17:41	16
-140	45.10	44.34	17:49	14
-142	44.62	43.83	17:58	13
-144	44.11	43.30	18:07	11
-145	43.85	43.03	18:11	10
-148	43.02	42.16	18:24	8
-150	42.44	41.56	18:32	6

PERCENTAGE OF SUN OBSCURED ALONG MAP TRACK
WEST END = 99.1   EAST END = 97.7

===

+ MARKS THE CITY OF TA-TU

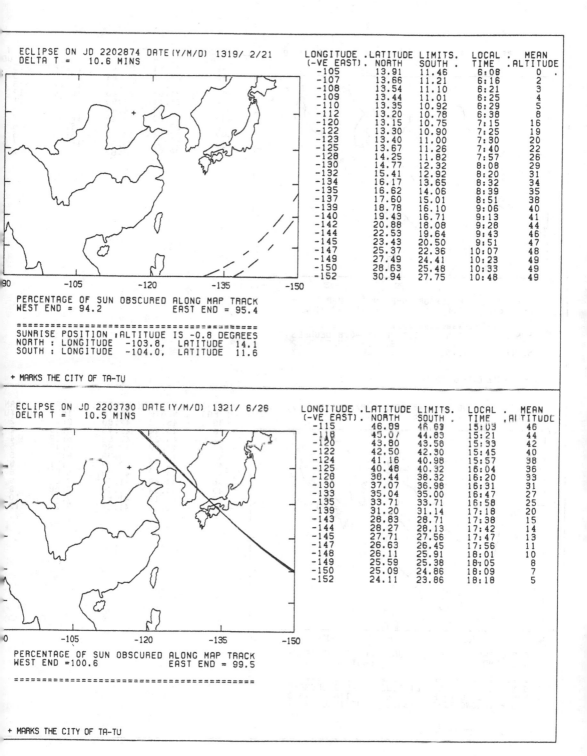

ECLIPSE ON JD 2202874 DATE(Y/M/D) 1319/ 2/21
DELTA T =   10.6 MINS

LONGITUDE (-VE EAST)	LATITUDE NORTH	LIMITS. SOUTH	LOCAL TIME	MEAN ALTITUDE
-105	13.91	11.46	6:08	0
-107	13.66	11.21	6:16	2
-108	13.54	11.10	6:21	3
-109	13.44	11.01	6:25	4
-110	13.35	10.92	6:29	5
-112	13.20	10.78	6:38	8
-120	13.15	10.75	7:15	16
-122	13.30	10.90	7:25	19
-123	13.40	11.00	7:30	20
-125	13.67	11.26	7:40	22
-128	14.25	11.82	7:57	26
-130	14.77	12.32	8:08	29
-132	15.41	12.92	8:20	31
-134	16.17	13.65	8:32	34
-135	16.62	14.06	8:39	35
-137	17.60	15.01	8:51	38
-139	18.78	16.10	9:06	40
-140	19.43	16.71	9:13	41
-142	20.88	18.08	9:28	44
-144	22.53	19.64	9:43	46
-145	23.43	20.50	9:51	47
-147	25.37	22.36	10:07	48
-149	27.49	24.41	10:23	49
-150	28.63	25.48	10:33	49
-152	30.94	27.75	10:48	49

PERCENTAGE OF SUN OBSCURED ALONG MAP TRACK
WEST END = 94.2   EAST END = 95.4

===
SUNRISE POSITION ,ALTITUDE IS -0.8 DEGREES
NORTH : LONGITUDE -103.8, LATITUDE 14.1
SOUTH : LONGITUDE -104.0, LATITUDE 11.6

+ MARKS THE CITY OF TA-TU

ECLIPSE ON JD 2203730 DATE(Y/M/D) 1321/ 6/26
DELTA T =   10.5 MINS

LONGITUDE (-VE EAST)	LATITUDE NORTH	LIMITS. SOUTH	LOCAL TIME	MEAN ALTITUDE
-115	46.09	46.69	15:03	46
-118	45.07	44.83	15:21	44
-120	43.80	43.58	15:33	42
-122	42.50	42.30	15:45	40
-124	41.16	40.98	15:57	38
-125	40.48	40.32	16:04	36
-128	38.44	38.32	16:20	33
-130	37.07	36.98	16:31	31
-133	35.04	35.00	16:47	27
-135	33.71	33.71	16:58	25
-139	31.20	31.14	17:18	20
-143	28.83	28.71	17:38	15
-144	28.27	28.13	17:42	14
-145	27.71	27.56	17:47	13
-147	26.63	26.45	17:56	11
-148	26.11	25.91	18:01	10
-149	25.59	25.38	18:05	8
-150	25.09	24.86	18:09	7
-152	24.11	23.86	18:18	5

PERCENTAGE OF SUN OBSCURED ALONG MAP TRACK
WEST END =100.6   EAST END = 99.5

===

+ MARKS THE CITY OF TA-TU

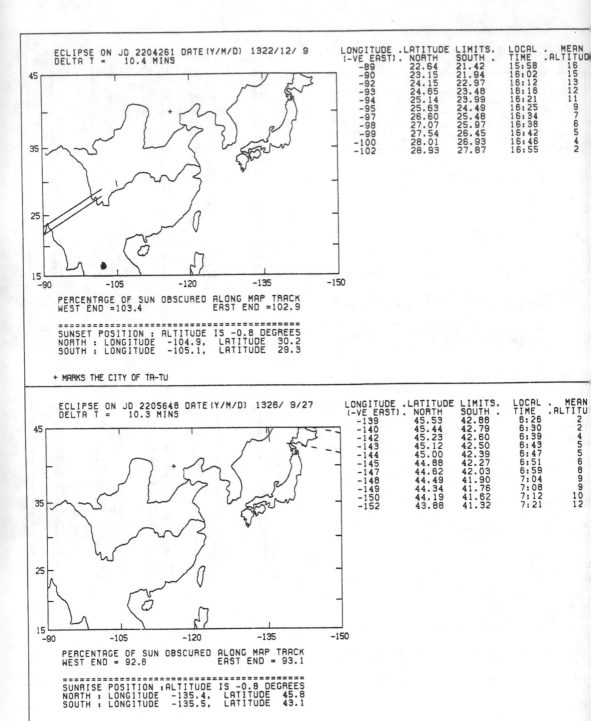

ECLIPSE ON JD 2204261 DATE(Y/M/D) 1322/12/ 9
DELTA T =   10.4 MINS

LONGITUDE (-VE EAST)	LATITUDE NORTH	LIMITS. SOUTH	LOCAL TIME	MEAN ALTITU
-89	22.64	21.42	15:58	16
-90	23.15	21.94	16:02	15
-92	24.15	22.97	16:12	13
-93	24.65	23.48	16:16	12
-94	25.14	23.99	16:21	11
-95	25.63	24.49	16:25	9
-97	26.60	25.48	16:34	7
-98	27.07	25.97	16:38	6
-99	27.54	26.45	16:42	5
-100	28.01	26.93	16:46	4
-102	28.93	27.87	16:55	2

PERCENTAGE OF SUN OBSCURED ALONG MAP TRACK
WEST END =103.4   EAST END =102.9

==
SUNSET POSITION : ALTITUDE IS -0.8 DEGREES
NORTH : LONGITUDE -104.9,   LATITUDE 30.2
SOUTH : LONGITUDE -105.1,   LATITUDE 29.3

+ MARKS THE CITY OF TA-TU

ECLIPSE ON JD 2205648 DATE(Y/M/D) 1326/ 9/27
DELTA T =   10.3 MINS

LONGITUDE (-VE EAST)	LATITUDE NORTH	LIMITS. SOUTH	LOCAL TIME	MEAN ALTITU
-139	45.53	42.88	6:26	2
-140	45.44	42.79	6:30	2
-142	45.23	42.60	6:39	4
-143	45.12	42.50	6:43	5
-144	45.00	42.39	6:47	5
-145	44.88	42.27	6:51	6
-147	44.62	42.03	6:59	8
-148	44.49	41.90	7:04	9
-149	44.34	41.76	7:08	9
-150	44.19	41.62	7:12	10
-152	43.88	41.32	7:21	12

PERCENTAGE OF SUN OBSCURED ALONG MAP TRACK
WEST END = 92.8   EAST END = 93.1

==
SUNRISE POSITION :ALTITUDE IS -0.8 DEGREES
NORTH : LONGITUDE -135.4,   LATITUDE 45.8
SOUTH : LONGITUDE -135.5,   LATITUDE 43.1

+ MARKS THE CITY OF TA-TU

ECLIPSE ON JD 2206683 DATE(Y/M/D) 1329/ 7/27
DELTA T =   10.2 MINS

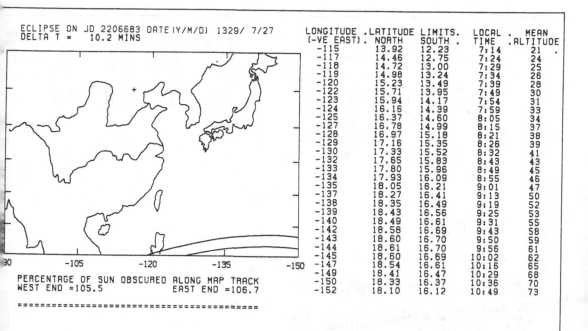

PERCENTAGE OF SUN OBSCURED ALONG MAP TRACK
WEST END =105.5   EAST END =106.7

===

+ MARKS THE CITY OF TA-TU

LONGITUDE (-VE EAST)	LATITUDE LIMITS. NORTH	SOUTH .	LOCAL TIME	MEAN ALTITUDE
-115	13.92	12.23	7:14	21
-117	14.46	12.75	7:24	24
-118	14.72	13.00	7:29	25
-119	14.98	13.24	7:34	26
-120	15.23	13.49	7:39	28
-122	15.71	13.95	7:49	30
-123	15.94	14.17	7:54	31
-124	16.16	14.39	7:59	33
-125	16.37	14.60	8:05	34
-127	16.78	14.99	8:15	37
-128	16.97	15.18	8:21	38
-129	17.16	15.35	8:26	39
-130	17.33	15.52	8:32	41
-132	17.65	15.83	8:43	43
-133	17.80	15.96	8:49	45
-134	17.93	16.09	8:55	46
-135	18.05	16.21	9:01	47
-137	18.27	16.41	9:13	50
-138	18.35	16.49	9:19	52
-139	18.43	16.56	9:25	53
-140	18.49	16.61	9:31	55
-142	18.58	16.69	9:43	58
-143	18.60	16.70	9:50	59
-144	18.61	16.70	9:56	61
-145	18.60	16.69	10:02	62
-147	18.54	16.61	10:16	65
-149	18.41	16.47	10:29	68
-150	18.33	16.37	10:36	70
-152	18.10	16.12	10:49	73

ECLIPSE ON JD 2207539 DATE(Y/M/D) 1331/11/30
DELTA T =   10.1 MINS

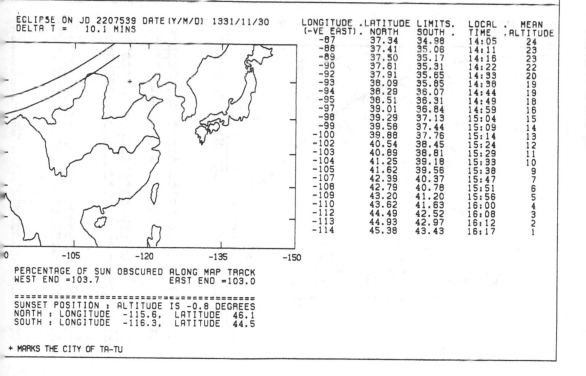

PERCENTAGE OF SUN OBSCURED ALONG MAP TRACK
WEST END =103.7   EAST END =103.0

===
SUNSET POSITION : ALTITUDE IS -0.8 DEGREES
NORTH : LONGITUDE -115.6, LATITUDE 46.1
SOUTH : LONGITUDE -116.3, LATITUDE 44.5

+ MARKS THE CITY OF TA-TU

LONGITUDE (-VE EAST)	LATITUDE LIMITS. NORTH	SOUTH .	LOCAL TIME	MEAN ALTITUDE
-87	37.34	34.98	14:05	24
-88	37.41	35.06	14:11	23
-89	37.50	35.17	14:16	23
-90	37.61	35.31	14:22	22
-92	37.91	35.65	14:33	20
-93	38.09	35.85	14:38	19
-94	38.29	36.07	14:44	19
-95	38.51	36.31	14:49	18
-97	39.01	36.84	14:59	16
-98	39.29	37.13	15:04	15
-99	39.58	37.44	15:09	14
-100	39.88	37.76	15:14	13
-102	40.54	38.45	15:24	12
-103	40.89	38.81	15:29	11
-104	41.25	39.18	15:33	10
-105	41.62	39.56	15:38	9
-107	42.39	40.37	15:47	7
-108	42.79	40.78	15:51	6
-109	43.20	41.20	15:56	5
-110	43.62	41.63	16:00	4
-112	44.49	42.52	16:08	3
-113	44.93	42.97	16:12	2
-114	45.38	43.43	16:17	1

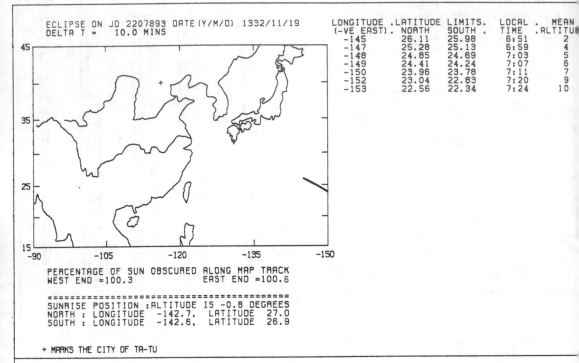

ECLIPSE ON JD 2207893 DATE(Y/M/D) 1332/11/19
DELTA T = 10.0 MINS

LONGITUDE (-VE EAST)	LATITUDE NORTH	LIMITS SOUTH	LOCAL TIME	MEAN ALTITU
-145	26.11	25.98	6:51	2
-147	25.28	25.13	6:59	4
-148	24.85	24.69	7:03	5
-149	24.41	24.24	7:07	6
-150	23.96	23.78	7:11	7
-152	23.04	22.83	7:20	9
-153	22.56	22.34	7:24	10

PERCENTAGE OF SUN OBSCURED ALONG MAP TRACK
WEST END =100.3   EAST END =100.6

===
SUNRISE POSITION :ALTITUDE IS -0.8 DEGREES
NORTH : LONGITUDE -142.7, LATITUDE 27.0
SOUTH : LONGITUDE -142.6, LATITUDE 26.9

+ MARKS THE CITY OF TA-TU

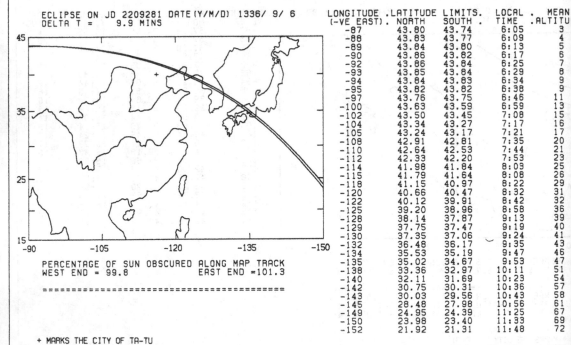

ECLIPSE ON JD 2209281 DATE(Y/M/D) 1336/ 9/ 6
DELTA T = 9.9 MINS

LONGITUDE (-VE EAST)	LATITUDE NORTH	LIMITS SOUTH	LOCAL TIME	MEAN ALTITU
-87	43.80	43.74	6:05	3
-88	43.83	43.77	6:09	4
-89	43.84	43.80	6:13	5
-90	43.86	43.82	6:17	6
-92	43.86	43.84	6:25	7
-93	43.85	43.84	6:29	8
-94	43.84	43.83	6:34	9
-95	43.82	43.82	6:38	9
-97	43.76	43.75	6:46	11
-100	43.63	43.59	6:59	13
-102	43.50	43.45	7:08	15
-104	43.34	43.27	7:17	16
-105	43.24	43.17	7:21	17
-108	42.91	42.81	7:35	20
-110	42.64	42.53	7:44	21
-112	42.33	42.20	7:53	23
-114	41.98	41.84	8:09	25
-115	41.79	41.64	8:08	26
-118	41.15	40.97	8:22	29
-120	40.66	40.47	8:32	31
-122	40.12	39.91	8:42	32
-125	39.20	38.96	8:58	36
-128	38.14	37.87	9:13	39
-129	37.75	37.47	9:19	40
-130	37.35	37.06	9:24	41
-132	36.48	36.17	9:35	43
-134	35.53	35.19	9:47	46
-135	35.02	34.67	9:53	47
-138	33.36	32.97	10:11	51
-140	32.11	31.69	10:23	54
-142	30.75	30.31	10:36	57
-143	30.03	29.56	10:43	58
-145	28.48	27.98	10:56	61
-149	24.95	24.39	11:25	67
-150	23.98	23.40	11:33	69
-152	21.92	21.31	11:48	72

PERCENTAGE OF SUN OBSCURED ALONG MAP TRACK
WEST END = 99.8   EAST END =101.3

===

+ MARKS THE CITY OF TA-TU

ECLIPSE ON JD 2211703 DATE(Y/M/D) 1343/ 4/25
DELTA T =   9.6 MINS

LONGITUDE (-VE EAST)	LATITUDE NORTH	LIMITS SOUTH	LOCAL TIME	MEAN ALTITUDE
-115	13.99	12.12	7:29	25
-118	15.39	13.46	7:45	29
-120	16.37	14.39	7:56	31
-123	17.88	15.85	8:12	35
-125	18.92	16.85	8:23	38
-128	20.51	18.41	8:40	42
-130	21.61	19.46	8:53	45
-133	23.24	21.08	9:10	50
-135	24.34	22.16	9:24	53
-137	25.43	23.24	9:36	55
-139	26.50	24.31	9:49	58
-140	27.03	24.84	9:56	59
-142	28.07	25.89	10:08	62
-144	29.08	26.90	10:22	64
-145	29.57	27.40	10:29	65
-147	30.51	28.36	10:42	67
-150	31.84	29.72	11:02	70

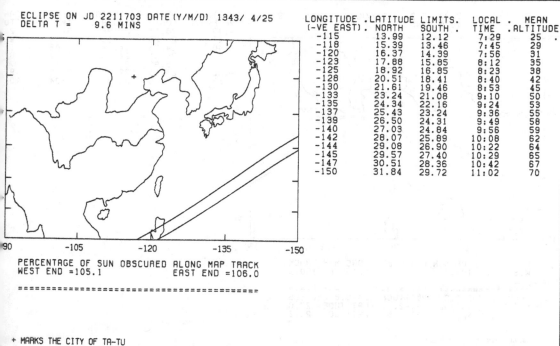

PERCENTAGE OF SUN OBSCURED ALONG MAP TRACK
WEST END =105.1   EAST END =106.0

==

+ MARKS THE CITY OF TA-TU

ECLIPSE ON JD 2212234 DATE(Y/M/D) 1344/10/ 7
DELTA T =   9.6 MINS

LONGITUDE (-VE EAST)	LATITUDE NORTH	LIMITS SOUTH	LOCAL TIME	MEAN ALTITUDE
-87	17.66	15.11	12:10	65
-89	16.44	13.89	12:27	65
-90	15.85	13.30	12:35	65
-92	14.70	12.18	12:52	64
-94	13.64	11.13	13:08	63

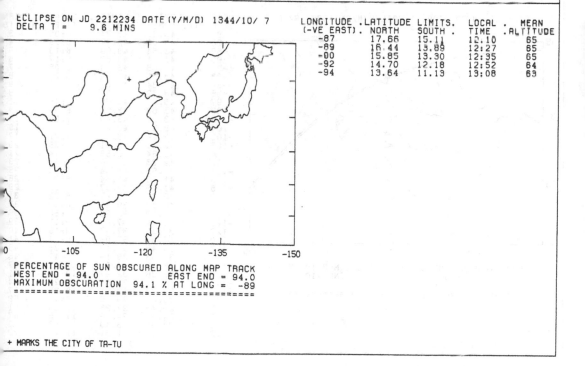

PERCENTAGE OF SUN OBSCURED ALONG MAP TRACK
WEST END = 94.0   EAST END = 94.0
MAXIMUM OBSCURATION 94.1 % AT LONG = -89
==

+ MARKS THE CITY OF TA-TU

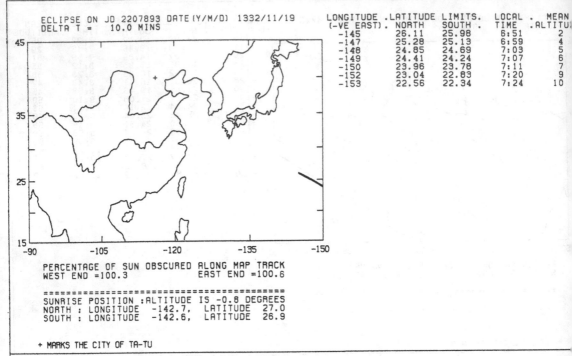

ECLIPSE ON JD 2207893 DATE(Y/M/D) 1332/11/19
DELTA T =   10.0 MINS

LONGITUDE (-VE EAST)	LATITUDE LIMITS. NORTH	SOUTH .	LOCAL TIME	MEAN .ALTITU
-145	26.11	25.98	6:51	2
-147	25.28	25.13	6:59	4
-148	24.85	24.69	7:03	5
-149	24.41	24.24	7:07	6
-150	23.96	23.78	7:11	7
-152	23.04	22.83	7:20	9
-153	22.56	22.34	7:24	10

PERCENTAGE OF SUN OBSCURED ALONG MAP TRACK
WEST END =100.3   EAST END =100.6

===
SUNRISE POSITION :ALTITUDE IS -0.8 DEGREES
NORTH : LONGITUDE -142.7, LATITUDE 27.0
SOUTH : LONGITUDE -142.6, LATITUDE 26.9

+ MARKS THE CITY OF TA-TU

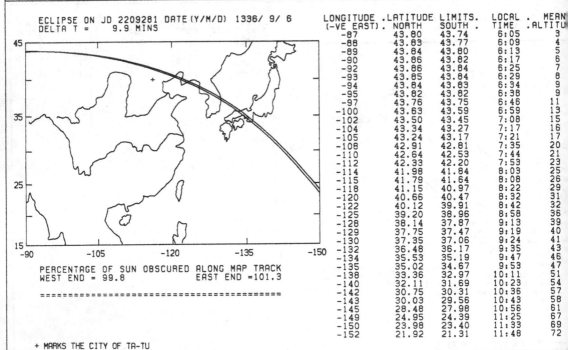

ECLIPSE ON JD 2209281 DATE(Y/M/D) 1336/ 9/ 6
DELTA T =   9.9 MINS

LONGITUDE (-VE EAST)	LATITUDE LIMITS. NORTH	SOUTH .	LOCAL TIME	MEAN .ALTITU
-87	43.80	43.74	6:05	3
-88	43.83	43.77	6:09	4
-89	43.84	43.80	6:13	5
-90	43.86	43.82	6:17	6
-92	43.86	43.84	6:25	7
-93	43.85	43.84	6:29	8
-94	43.84	43.83	6:34	9
-95	43.82	43.82	6:38	9
-97	43.76	43.75	6:46	11
-100	43.63	43.59	6:59	13
-102	43.50	43.45	7:08	15
-104	43.34	43.27	7:17	16
-105	43.24	43.17	7:21	17
-108	42.91	42.81	7:35	20
-110	42.64	42.53	7:44	21
-112	42.33	42.20	7:53	23
-114	41.98	41.84	8:03	25
-115	41.79	41.64	8:08	26
-118	41.15	40.97	8:22	29
-120	40.66	40.47	8:32	31
-122	40.12	39.91	8:42	32
-125	39.20	38.96	8:58	36
-128	38.14	37.87	9:13	39
-129	37.75	37.47	9:19	40
-130	37.35	37.06	9:24	41
-132	36.48	36.17	9:35	43
-134	35.53	35.19	9:47	46
-135	35.02	34.67	9:53	47
-138	33.36	32.97	10:11	51
-140	32.11	31.69	10:23	54
-142	30.75	30.31	10:36	57
-143	30.03	29.56	10:43	58
-145	28.48	27.98	10:56	61
-149	24.95	24.39	11:25	67
-150	23.98	23.40	11:33	69
-152	21.92	21.31	11:48	72

PERCENTAGE OF SUN OBSCURED ALONG MAP TRACK
WEST END = 99.8   EAST END =101.3

===

+ MARKS THE CITY OF TA-TU

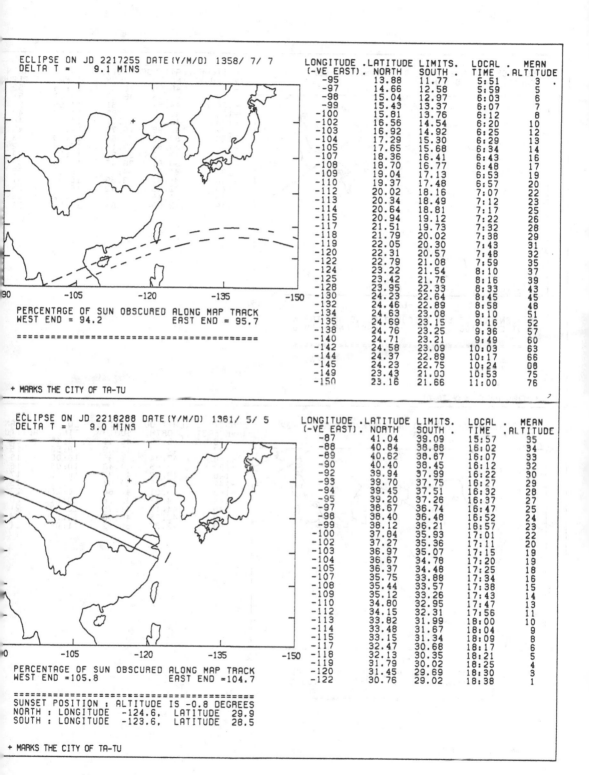

ECLIPSE ON JD 2217255 DATE (Y/M/D) 1358/ 7/ 7
DELTA T =   9.1 MINS

LONGITUDE (-VE EAST)	LATITUDE LIMITS. NORTH	SOUTH	LOCAL TIME	MEAN ALTITUDE
-95	13.88	11.77	5:51	3
-97	14.66	12.58	5:59	5
-98	15.04	12.97	6:03	6
-99	15.43	13.37	6:07	7
-100	15.81	13.76	6:12	8
-102	16.56	14.54	6:20	10
-103	16.92	14.92	6:25	12
-104	17.29	15.30	6:29	13
-105	17.65	15.68	6:34	14
-107	18.36	16.41	6:43	16
-108	18.70	16.77	6:48	17
-109	19.04	17.13	6:53	19
-110	19.37	17.48	6:57	20
-112	20.02	18.16	7:07	22
-113	20.34	18.49	7:12	23
-114	20.64	18.81	7:17	25
-115	20.94	19.12	7:22	26
-117	21.51	19.73	7:32	28
-118	21.79	20.02	7:38	29
-119	22.05	20.30	7:43	31
-120	22.31	20.57	7:48	32
-122	22.79	21.08	7:59	35
-124	23.22	21.54	8:10	37
-125	23.42	21.76	8:16	39
-128	23.95	22.33	8:33	43
-130	24.23	22.64	8:45	45
-132	24.46	22.89	8:58	48
-134	24.63	23.08	9:10	51
-135	24.69	23.15	9:16	52
-138	24.76	23.25	9:36	57
-140	24.71	23.21	9:49	60
-142	24.58	23.09	10:03	63
-144	24.37	22.89	10:17	66
-145	24.23	22.75	10:24	08
-149	23.43	21.03	10:53	75
-150	23.16	21.66	11:00	76

PERCENTAGE OF SUN OBSCURED ALONG MAP TRACK
WEST END = 94.2   EAST END = 95.7

===

+ MARKS THE CITY OF TA-TU

ECLIPSE ON JD 2218288 DATE (Y/M/D) 1361/ 5/ 5
DELTA T =   9.0 MINS

LONGITUDE (-VE EAST)	LATITUDE LIMITS. NORTH	SOUTH	LOCAL TIME	MEAN ALTITUDE
-87	41.04	39.09	15:57	35
-88	40.84	38.88	16:02	34
-89	40.62	38.67	16:07	33
-90	40.40	38.45	16:12	32
-92	39.94	37.99	16:22	30
-93	39.70	37.75	16:27	29
-94	39.45	37.51	16:32	28
-95	39.20	37.26	16:37	27
-97	38.67	36.74	16:47	25
-98	38.40	36.48	16:52	24
-99	38.12	36.21	16:57	23
-100	37.84	35.93	17:01	22
-102	37.27	35.36	17:11	20
-103	36.97	35.07	17:15	19
-104	36.67	34.78	17:20	19
-105	36.37	34.48	17:25	18
-107	35.75	33.88	17:34	16
-108	35.44	33.57	17:38	15
-109	35.12	33.26	17:43	14
-110	34.80	32.95	17:47	13
-112	34.15	32.31	17:56	11
-113	33.82	31.99	18:00	10
-114	33.48	31.67	18:04	9
-115	33.15	31.34	18:09	8
-117	32.47	30.68	18:17	6
-118	32.13	30.35	18:21	5
-119	31.79	30.02	18:25	4
-120	31.45	29.69	18:30	3
-122	30.76	29.02	18:38	1

PERCENTAGE OF SUN OBSCURED ALONG MAP TRACK
WEST END =105.8   EAST END =104.7

===
SUNSET POSITION : ALTITUDE IS -0.8 DEGREES
NORTH : LONGITUDE -124.6, LATITUDE 29.9
SOUTH : LONGITUDE -123.6, LATITUDE 28.5

+ MARKS THE CITY OF TA-TU

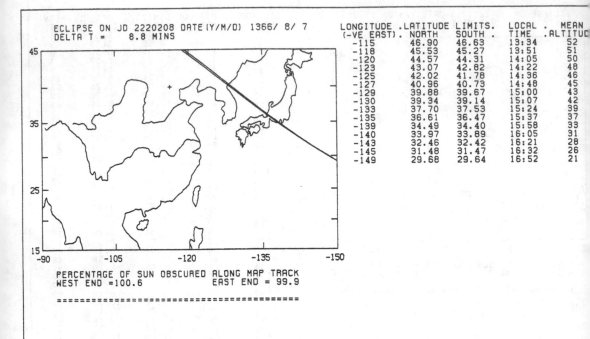

ECLIPSE ON JD 2220208 DATE(Y/M/D) 1366/ 8/ 7
DELTA T =   8.8 MINS

LONGITUDE (-VE EAST)	LATITUDE NORTH	LIMITS. SOUTH	LOCAL TIME	MEAN ALTITUDE
-115	46.90	46.63	13:34	52
-118	45.53	45.27	13:51	51
-120	44.57	44.31	14:05	50
-123	43.07	42.82	14:22	48
-125	42.02	41.78	14:36	46
-127	40.96	40.73	14:48	45
-129	39.88	39.67	15:00	43
-130	39.34	39.14	15:07	42
-133	37.70	37.53	15:24	39
-135	36.61	36.47	15:37	37
-139	34.49	34.40	15:58	33
-140	33.97	33.89	16:05	31
-143	32.46	32.42	16:21	28
-145	31.48	31.47	16:32	26
-149	29.68	29.64	16:52	21

PERCENTAGE OF SUN OBSCURED ALONG MAP TRACK
WEST END =100.6   EAST END = 99.9

+ MARKS THE CITY OF TA-TU

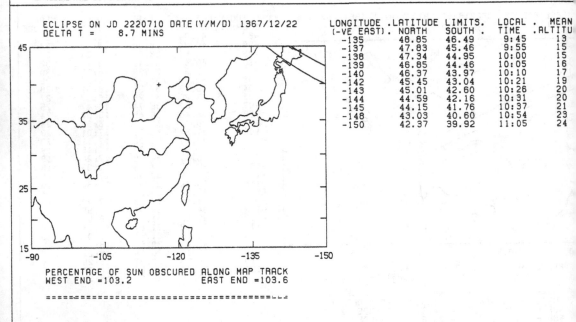

ECLIPSE ON JD 2220710 DATE(Y/M/D) 1367/12/22
DELTA T =   8.7 MINS

LONGITUDE (-VE EAST)	LATITUDE NORTH	LIMITS. SOUTH	LOCAL TIME	MEAN ALTITUDE
-135	48.85	46.49	9:45	13
-137	47.83	45.46	9:55	15
-138	47.34	44.95	10:00	15
-139	46.85	44.46	10:05	16
-140	46.37	43.97	10:10	17
-142	45.45	43.04	10:21	19
-143	45.01	42.60	10:26	20
-144	44.59	42.16	10:31	20
-145	44.15	41.76	10:37	21
-148	43.03	40.60	10:54	23
-150	42.37	39.92	11:05	24

PERCENTAGE OF SUN OBSCURED ALONG MAP TRACK
WEST END =103.2   EAST END =103.6

+ MARKS THE CITY OF TA-TU

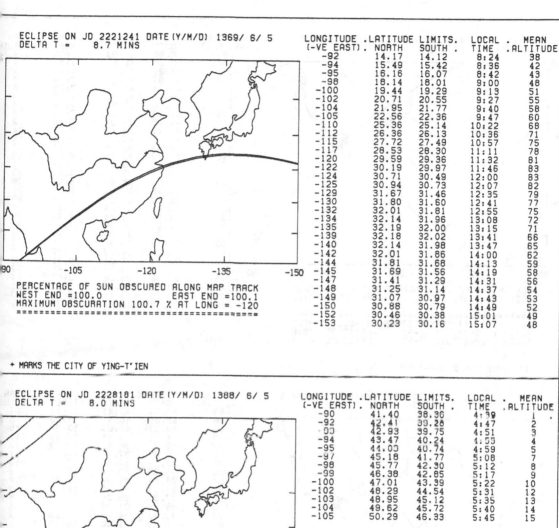

ECLIPSE ON JD 2221241 DATE (Y/M/D) 1369/ 6/ 5
DELTA T =   8.7 MINS

LONGITUDE (-VE EAST)	LATITUDE NORTH	LIMITS. SOUTH	LOCAL TIME	MEAN ALTITUDE
-92	14.17	14.12	8:24	38
-94	15.49	15.42	8:36	42
-95	16.16	16.07	8:42	43
-98	18.14	18.01	9:00	48
-100	19.44	19.29	9:13	51
-102	20.71	20.55	9:27	55
-104	21.95	21.77	9:40	58
-105	22.56	22.36	9:47	60
-110	25.36	25.14	10:22	68
-112	26.36	26.13	10:36	71
-115	27.72	27.49	10:57	75
-117	28.53	28.30	11:11	78
-120	29.59	29.36	11:32	81
-122	30.19	29.97	11:46	83
-124	30.71	30.49	12:00	83
-125	30.94	30.73	12:07	82
-129	31.67	31.46	12:35	79
-130	31.80	31.60	12:41	77
-132	32.01	31.81	12:55	75
-134	32.14	31.96	13:08	72
-135	32.19	32.00	13:15	71
-139	32.18	32.02	13:41	66
-140	32.14	31.98	13:47	65
-142	32.01	31.86	14:00	62
-144	31.81	31.68	14:13	59
-145	31.69	31.56	14:19	58
-147	31.41	31.29	14:31	56
-148	31.25	31.14	14:37	54
-149	31.07	30.97	14:43	53
-150	30.88	30.79	14:49	52
-152	30.46	30.38	15:01	49
-153	30.23	30.16	15:07	48

PERCENTAGE OF SUN OBSCURED ALONG MAP TRACK
WEST END =100.0   EAST END =100.1
MAXIMUM OBSCURATION 100.7 % AT LONG = -120
==

+ MARKS THE CITY OF YING-T'IEN

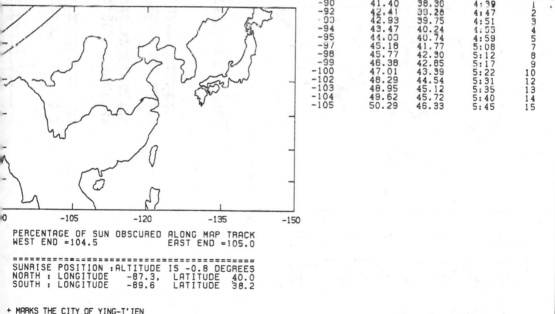

ECLIPSE ON JD 2228181 DATE (Y/M/D) 1388/ 6/ 5
DELTA T =   8.0 MINS

LONGITUDE (-VE EAST)	LATITUDE NORTH	LIMITS. SOUTH	LOCAL TIME	MEAN ALTITUDE
-90	41.40	38.36	4:39	1
-92	42.41	39.28	4:47	2
-93	42.93	39.75	4:51	3
-94	43.47	40.24	4:55	4
-95	44.03	40.74	4:59	5
-97	45.18	41.77	5:08	7
-98	45.77	42.30	5:12	8
-99	46.38	42.85	5:17	9
-100	47.01	43.39	5:22	10
-102	48.29	44.54	5:31	12
-103	48.95	45.12	5:35	13
-104	49.62	45.72	5:40	14
-105	50.29	46.33	5:45	15

PERCENTAGE OF SUN OBSCURED ALONG MAP TRACK
WEST END =104.5   EAST END =105.0

==
SUNRISE POSITION : ALTITUDE IS -0.8 DEGREES
NORTH : LONGITUDE -87.3, LATITUDE 40.0
SOUTH : LONGITUDE -89.6   LATITUDE 38.2

+ MARKS THE CITY OF YING-T'IEN

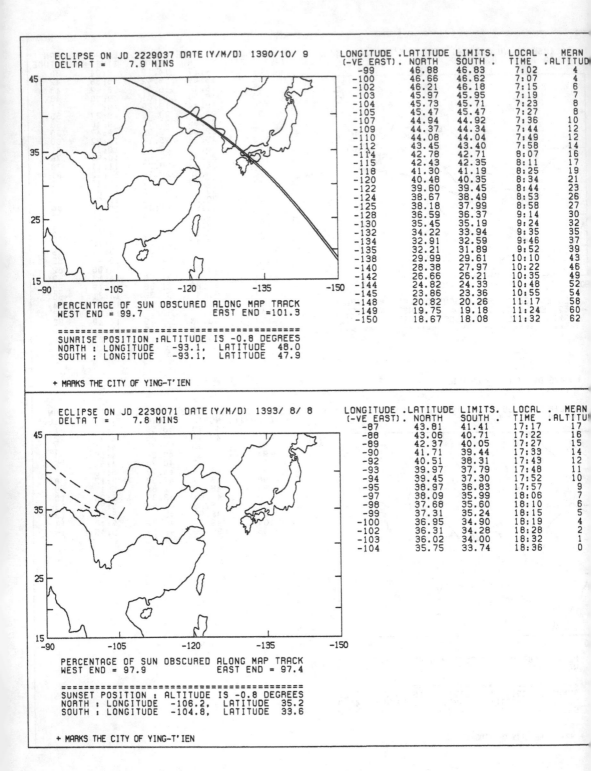

ECLIPSE ON JD 2229037 DATE(Y/M/D) 1390/10/ 9
DELTA T = 7.9 MINS

LONGITUDE (-VE EAST)	.LATITUDE NORTH	LIMITS. SOUTH	LOCAL TIME	MEAN .ALTITU
-99	46.88	46.83	7:02	4
-100	46.66	46.62	7:07	4
-102	46.21	46.18	7:15	6
-103	45.97	45.95	7:19	7
-104	45.73	45.71	7:23	8
-105	45.47	45.47	7:27	8
-107	44.94	44.92	7:36	10
-109	44.37	44.34	7:44	12
-110	44.08	44.04	7:49	12
-112	43.45	43.40	7:58	14
-114	42.78	42.71	8:07	16
-115	42.43	42.35	8:11	17
-118	41.30	41.19	8:25	19
-120	40.48	40.35	8:34	21
-122	39.60	39.45	8:44	23
-124	38.67	38.49	8:53	26
-125	38.18	37.99	8:58	27
-128	36.59	36.37	9:14	30
-130	35.45	35.19	9:24	32
-132	34.22	33.94	9:35	35
-134	32.91	32.59	9:46	37
-135	32.21	31.89	9:52	39
-138	29.99	29.61	10:10	43
-140	28.38	27.97	10:22	46
-142	26.66	26.21	10:35	49
-144	24.82	24.33	10:48	52
-145	23.86	23.36	10:55	54
-148	20.82	20.26	11:17	58
-149	19.75	19.18	11:24	60
-150	18.67	18.08	11:32	62

PERCENTAGE OF SUN OBSCURED ALONG MAP TRACK
WEST END = 99.7   EAST END =101.3

==
SUNRISE POSITION :ALTITUDE IS -0.8 DEGREES
NORTH : LONGITUDE -93.1, LATITUDE 48.0
SOUTH : LONGITUDE -93.1, LATITUDE 47.9

+ MARKS THE CITY OF YING-T'IEN

ECLIPSE ON JD 2230071 DATE(Y/M/D) 1393/ 8/ 8
DELTA T = 7.8 MINS

LONGITUDE (-VE EAST)	.LATITUDE NORTH	LIMITS. SOUTH	LOCAL TIME	MEAN .ALTITU
-87	43.81	41.41	17:17	17
-88	43.06	40.71	17:22	16
-89	42.37	40.05	17:27	15
-90	41.71	39.44	17:33	14
-92	40.51	38.31	17:43	12
-93	39.97	37.79	17:48	11
-94	39.45	37.30	17:52	10
-95	38.97	36.83	17:57	9
-97	38.09	35.99	18:06	7
-98	37.68	35.60	18:10	6
-99	37.31	35.24	18:15	5
-100	36.95	34.90	18:19	4
-102	36.31	34.28	18:28	2
-103	36.02	34.00	18:32	1
-104	35.75	33.74	18:36	0

PERCENTAGE OF SUN OBSCURED ALONG MAP TRACK
WEST END = 97.9   EAST END = 97.4

==
SUNSET POSITION : ALTITUDE IS -0.8 DEGREES
NORTH : LONGITUDE -106.2, LATITUDE 35.2
SOUTH : LONGITUDE -104.8, LATITUDE 33.6

+ MARKS THE CITY OF YING-T'IEN

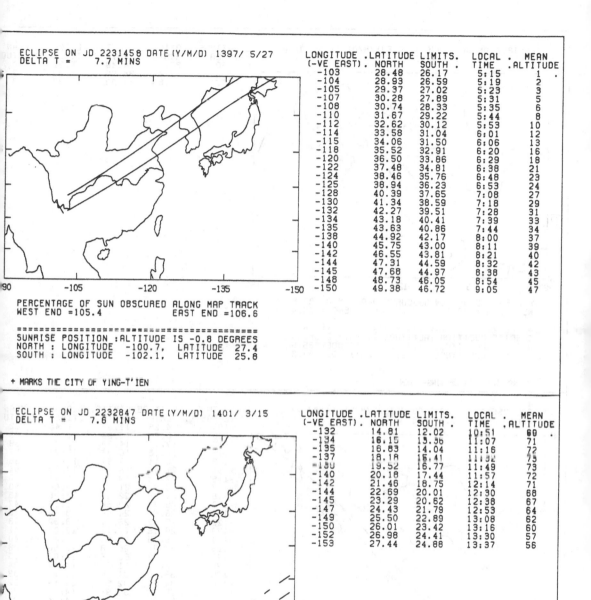

ECLIPSE ON JD 2231458 DATE (Y/M/D) 1397/ 5/27
DELTA T =   7.7 MINS

LONGITUDE (-VE EAST)	LATITUDE NORTH	LIMITS. SOUTH	LOCAL TIME	MEAN ALTITUDE
-103	28.48	26.17	5:15	1
-104	28.93	26.59	5:19	2
-105	29.37	27.02	5:23	3
-107	30.28	27.89	5:31	5
-108	30.74	28.33	5:35	6
-110	31.67	29.22	5:44	8
-112	32.62	30.12	5:53	10
-114	33.58	31.04	6:01	12
-115	34.06	31.50	6:06	13
-118	35.52	32.91	6:20	16
-120	36.50	33.86	6:29	18
-122	37.48	34.81	6:38	21
-124	38.46	35.76	6:48	23
-125	38.94	36.23	6:53	24
-128	40.39	37.65	7:08	27
-130	41.34	38.59	7:18	29
-132	42.27	39.51	7:28	31
-134	43.18	40.41	7:39	33
-135	43.63	40.86	7:44	34
-138	44.92	42.17	8:00	37
-140	45.75	43.00	8:11	39
-142	46.55	43.81	8:21	40
-144	47.31	44.59	8:32	42
-145	47.68	44.97	8:38	43
-148	48.73	46.05	8:54	45
-150	49.38	46.72	9:05	47

PERCENTAGE OF SUN OBSCURED ALONG MAP TRACK
WEST END =105.4   EAST END =106.6

===
SUNRISE POSITION : ALTITUDE IS -0.8 DEGREES
NORTH : LONGITUDE -100.7, LATITUDE   27.4
SOUTH : LONGITUDE -102.1, LATITUDE   25.8

+ MARKS THE CITY OF YING-T'IEN

ECLIPSE ON JD 2232847 DATE (Y/M/D) 1401/ 3/15
DELTA T =   7.6 MINS

LONGITUDE (-VE EAST)	LATITUDE NORTH	LIMITS. SOUTH	LOCAL TIME	MEAN ALTITUDE
-132	14.81	12.02	10:51	60
-134	16.15	13.56	11:07	71
-135	16.83	14.04	11:16	72
-137	18.18	15.41	11:32	75
-138	19.52	16.77	11:49	73
-140	20.18	17.44	11:57	72
-142	21.46	18.75	12:14	71
-144	22.69	20.01	12:30	68
-145	23.29	20.62	12:38	67
-147	24.43	21.79	12:53	64
-149	25.50	22.89	13:08	62
-150	26.01	23.42	13:16	60
-152	26.98	24.41	13:30	57
-153	27.44	24.88	13:37	56

PERCENTAGE OF SUN OBSCURED ALONG MAP TRACK
WEST END = 93.5   EAST END = 93.2
MAXIMUM OBSCURATION 93.5 % AT LONG = -137
===

+ MARKS THE CITY OF YING-T'IEN

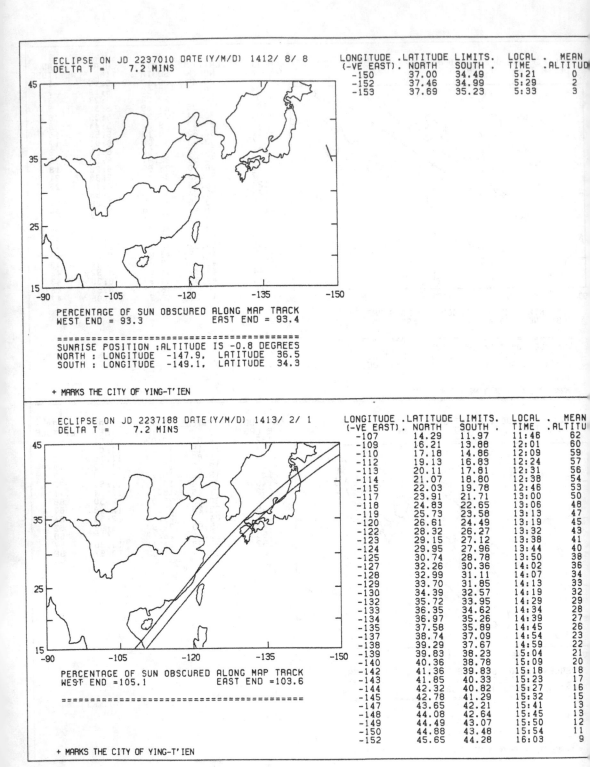

ECLIPSE ON JD 2237010 DATE(Y/M/D) 1412/ 8/ 8
DELTA T =   7.2 MINS

LONGITUDE	.LATITUDE	LIMITS.	LOCAL	. MEAN
(-VE EAST)	. NORTH	SOUTH .	TIME	.ALTITUD
-150	37.00	34.49	5:21	0
-152	37.46	34.99	5:29	2
-153	37.69	35.23	5:33	3

PERCENTAGE OF SUN OBSCURED ALONG MAP TRACK
WEST END = 93.3   EAST END = 93.4

===
SUNRISE POSITION :ALTITUDE IS -0.8 DEGREES
NORTH : LONGITUDE -147.9, LATITUDE   36.5
SOUTH : LONGITUDE -149.1, LATITUDE   34.3

+ MARKS THE CITY OF YING-T'IEN

ECLIPSE ON JD 2237188 DATE(Y/M/D) 1413/ 2/ 1
DELTA T =   7.2 MINS

LONGITUDE	.LATITUDE	LIMITS.	LOCAL	. MEAN
(-VE EAST)	. NORTH	SOUTH .	TIME	.ALTITU
-107	14.29	11.97	11:46	62
-109	16.21	13.88	12:01	60
-110	17.18	14.86	12:09	59
-112	19.13	16.83	12:24	57
-113	20.11	17.81	12:31	56
-114	21.07	18.80	12:38	54
-115	22.03	19.78	12:46	53
-117	23.91	21.71	13:00	50
-118	24.83	22.65	13:06	48
-119	25.73	23.58	13:13	47
-120	26.61	24.49	13:19	45
-122	28.32	26.27	13:32	43
-123	29.15	27.12	13:38	41
-124	29.95	27.96	13:44	40
-125	30.74	28.78	13:50	38
-127	32.26	30.36	14:02	36
-128	32.99	31.11	14:07	34
-129	33.70	31.85	14:13	33
-130	34.39	32.57	14:19	32
-132	35.72	33.95	14:29	29
-133	36.35	34.62	14:34	28
-134	36.97	35.26	14:39	27
-135	37.58	35.89	14:45	26
-137	38.74	37.09	14:54	23
-138	39.29	37.67	14:59	22
-139	39.83	38.23	15:04	21
-140	40.36	38.78	15:09	20
-142	41.36	39.83	15:18	18
-143	41.85	40.33	15:23	17
-144	42.32	40.82	15:27	16
-145	42.78	41.29	15:32	15
-147	43.65	42.21	15:41	13
-148	44.08	42.64	15:45	13
-149	44.49	43.07	15:50	12
-150	44.88	43.48	15:54	11
-152	45.65	44.28	16:03	9

PERCENTAGE OF SUN OBSCURED ALONG MAP TRACK
WEST END =105.1   EAST END =103.6

===

+ MARKS THE CITY OF YING-T'IEN

```
ECLIPSE ON JD 2238044 DATE(Y/M/D) 1415/ 6/ 7
DELTA T =    7.1 MINS
```

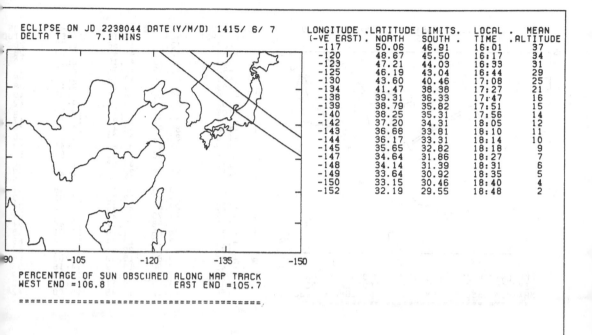

LONGITUDE (-VE EAST)	.LATITUDE NORTH	LIMITS. SOUTH	LOCAL TIME	. MEAN .ALTITUDE
-117	50.06	46.91	16:01	37
-120	48.67	45.50	16:17	34
-123	47.21	44.03	16:33	31
-125	46.19	43.04	16:44	29
-130	43.60	40.46	17:08	25
-134	41.47	38.38	17:27	21
-138	39.31	36.33	17:47	16
-139	38.79	35.82	17:51	15
-140	38.25	35.31	17:56	14
-142	37.20	34.31	18:05	12
-143	36.68	33.81	18:10	11
-144	36.17	33.31	18:14	10
-145	35.65	32.82	18:18	9
-147	34.64	31.86	18:27	7
-148	34.14	31.39	18:31	6
-149	33.64	30.92	18:35	5
-150	33.15	30.46	18:40	4
-152	32.19	29.55	18:48	2

```
PERCENTAGE OF SUN OBSCURED ALONG MAP TRACK
WEST END =106.8               EAST END =105.7

============================================

+ MARKS THE CITY OF YING-T'IEN
```

```
ECLIPSE ON JD 2239432 DATE(Y/M/D) 1419/ 3/26
DELTA T =    7.0 MINS
```

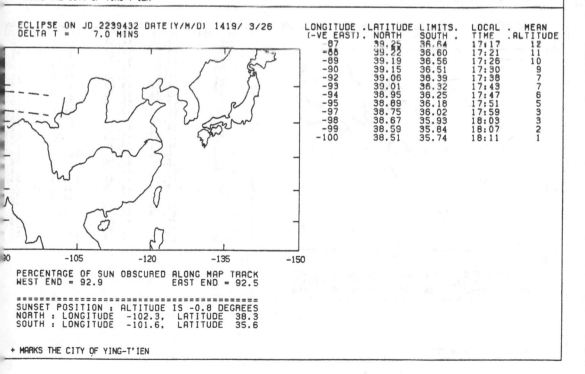

LONGITUDE (-VE EAST)	.LATITUDE NORTH	LIMITS. SOUTH	LOCAL TIME	. MEAN .ALTITUDE
-87	39.25	36.64	17:17	12
-88	39.22	36.60	17:21	11
-89	39.19	36.56	17:26	10
-90	39.15	36.51	17:30	9
-92	39.06	36.39	17:38	7
-93	39.01	36.32	17:43	7
-94	38.95	36.25	17:47	6
-95	38.89	36.18	17:51	5
-97	38.75	36.02	17:59	3
-98	38.67	35.93	18:03	3
-99	38.59	35.84	18:07	2
-100	38.51	35.74	18:11	1

```
PERCENTAGE OF SUN OBSCURED ALONG MAP TRACK
WEST END = 92.9               EAST END = 92.5

============================================
SUNSET POSITION : ALTITUDE IS -0.8 DEGREES
NORTH : LONGITUDE  -102.3,  LATITUDE   38.3
SOUTH : LONGITUDE  -101.6,  LATITUDE   35.6

+ MARKS THE CITY OF YING-T'IEN
```

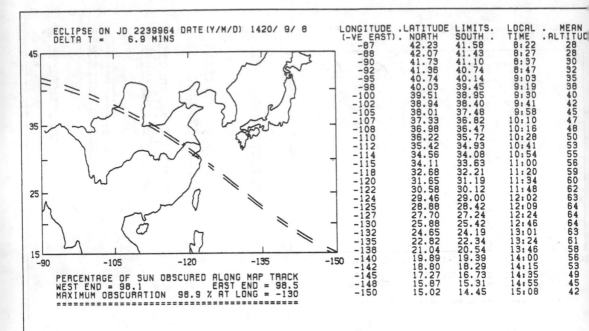

ECLIPSE ON JD 2239964 DATE(Y/M/D) 1420/ 9/ 8
DELTA T =   6.9 MINS

LONGITUDE (-VE EAST)	.LATITUDE . NORTH	LIMITS. SOUTH .	LOCAL TIME	. MEAN .ALTITUD
-87	42.23	41.58	8:22	28
-88	42.07	41.43	8:27	28
-90	41.73	41.10	8:37	30
-92	41.36	40.74	8:47	32
-95	40.74	40.14	9:03	35
-98	40.03	39.45	9:19	38
-100	39.51	38.95	9:30	40
-102	38.94	38.40	9:41	42
-105	38.01	37.48	9:58	45
-107	37.33	36.82	10:10	47
-108	36.98	36.47	10:16	48
-110	36.22	35.72	10:28	50
-112	35.42	34.93	10:41	53
-114	34.56	34.08	10:54	55
-115	34.11	33.63	11:00	56
-118	32.68	32.21	11:20	59
-120	31.65	31.19	11:34	60
-122	30.58	30.12	11:48	62
-124	29.46	29.00	12:02	63
-125	28.88	28.42	12:09	64
-127	27.70	27.24	12:24	64
-130	25.88	25.42	12:46	64
-132	24.65	24.19	13:01	63
-135	22.82	22.34	13:24	61
-138	21.04	20.54	13:46	58
-140	19.89	19.39	14:00	56
-142	18.80	18.29	14:15	53
-145	17.27	16.73	14:35	49
-148	15.87	15.31	14:55	45
-150	15.02	14.45	15:08	42

PERCENTAGE OF SUN OBSCURED ALONG MAP TRACK
WEST END = 98.1   EAST END = 98.5
MAXIMUM OBSCURATION 98.9 % AT LONG = -130
==

+ MARKS THE CITY OF YING-T'IEN

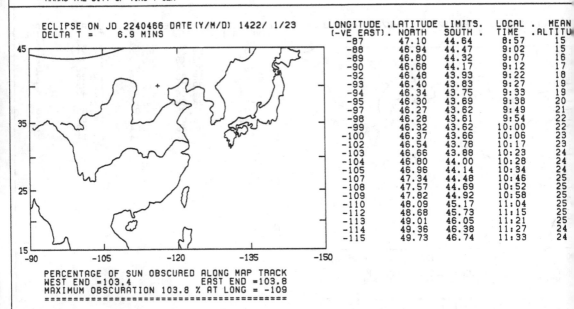

ECLIPSE ON JD 2240466 DATE(Y/M/D) 1422/ 1/23
DELTA T =   6.9 MINS

LONGITUDE (-VE EAST)	.LATITUDE . NORTH	LIMITS. SOUTH .	LOCAL TIME	. MEAN .ALTITU
-87	47.10	44.64	8:57	15
-88	46.94	44.47	9:02	15
-89	46.80	44.32	9:07	16
-90	46.68	44.17	9:12	17
-92	46.48	43.93	9:22	18
-93	46.40	43.83	9:27	19
-94	46.34	43.75	9:33	19
-95	46.30	43.69	9:38	20
-97	46.27	43.62	9:49	21
-98	46.28	43.61	9:54	22
-99	46.32	43.62	10:00	22
-100	46.37	43.66	10:06	23
-102	46.54	43.78	10:17	23
-103	46.66	43.88	10:23	24
-104	46.80	44.00	10:28	24
-105	46.96	44.14	10:34	24
-107	47.34	44.48	10:46	25
-108	47.57	44.69	10:52	25
-109	47.82	44.92	10:58	25
-110	48.09	45.17	11:04	25
-112	48.68	45.73	11:15	25
-113	49.01	46.05	11:21	25
-114	49.36	46.38	11:27	24
-115	49.73	46.74	11:33	24

PERCENTAGE OF SUN OBSCURED ALONG MAP TRACK
WEST END =103.4   EAST END =103.8
MAXIMUM OBSCURATION 103.8 % AT LONG = -109
==

+ MARKS THE CITY OF PEI-CHING

ECLIPSE ON JD 2243596 DATE(Y/M/D) 1430/ 8/19
DELTA T =   6.6 MINS

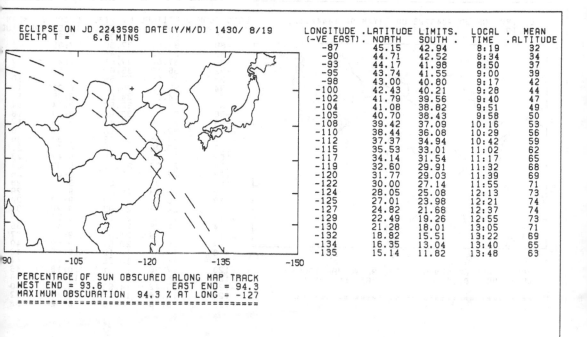

LONGITUDE (-VE EAST)	.LATITUDE NORTH	LIMITS. SOUTH	LOCAL TIME	. MEAN .ALTITUDE
-87	45.15	42.94	8:19	32
-90	44.71	42.52	8:34	34
-93	44.17	41.98	8:50	37
-95	43.74	41.55	9:00	39
-98	43.00	40.80	9:17	42
-100	42.43	40.21	9:28	44
-102	41.79	39.56	9:40	47
-104	41.08	38.82	9:51	49
-105	40.70	38.43	9:58	50
-108	39.42	37.09	10:16	53
-110	38.44	36.08	10:29	56
-112	37.37	34.94	10:42	59
-115	35.53	33.01	11:02	62
-117	34.14	31.54	11:17	65
-119	32.60	29.91	11:32	68
-120	31.77	29.03	11:39	69
-122	30.00	27.14	11:55	71
-124	28.05	25.08	12:13	73
-125	27.01	23.98	12:21	74
-127	24.82	21.68	12:37	74
-129	22.49	19.26	12:55	73
-130	21.28	18.01	13:05	71
-132	18.82	15.51	13:22	69
-134	16.35	13.04	13:40	65
-135	15.14	11.82	13:48	63

PERCENTAGE OF SUN OBSCURED ALONG MAP TRACK
WEST END = 93.6   EAST END = 94.3
MAXIMUM OBSCURATION 94.3 % AT LONG = -127
===

+ MARKS THE CITY OF PEI-CHING

ECLIPSE ON JD 2247936 DATE(Y/M/D) 1442/ 7/ 8
DELTA T =   6.2 MINS

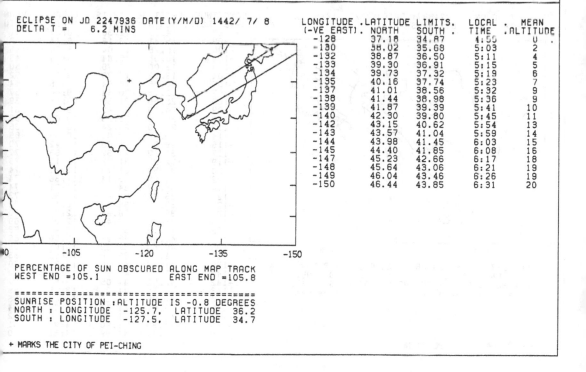

LONGITUDE (-VE EAST)	.LATITUDE NORTH	LIMITS. SOUTH	LOCAL TIME	. MEAN .ALTITUDE
-128	37.18	34.87	4:55	0
-130	58.02	35.68	5:03	2
-132	38.87	36.50	5:11	4
-133	39.30	36.91	5:15	5
-134	39.73	37.32	5:19	6
-135	40.16	37.74	5:23	7
-137	41.01	38.56	5:32	9
-138	41.44	38.98	5:36	9
-139	41.87	39.39	5:41	10
-140	42.30	39.80	5:45	11
-142	43.15	40.62	5:54	13
-143	43.57	41.04	5:59	14
-144	43.98	41.45	6:03	15
-145	44.40	41.85	6:08	16
-147	45.23	42.66	6:17	18
-148	45.64	43.06	6:21	19
-149	46.04	43.46	6:26	19
-150	46.44	43.85	6:31	20

PERCENTAGE OF SUN OBSCURED ALONG MAP TRACK
WEST END =105.1   EAST END =105.8

===
SUNRISE POSITION :ALTITUDE IS -0.8 DEGREES
NORTH : LONGITUDE -125.7, LATITUDE 36.2
SOUTH : LONGITUDE -127.5, LATITUDE 34.7

+ MARKS THE CITY OF PEI-CHING

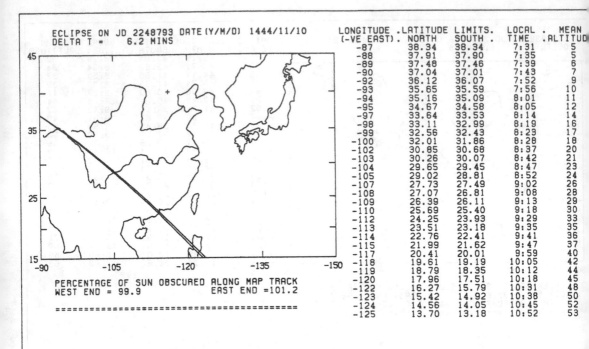

ECLIPSE ON JD 2248793 DATE (Y/M/D) 1444/11/10
DELTA T = 6.2 MINS

LONGITUDE	.LATITUDE	LIMITS.	LOCAL	. MEAN
(-VE EAST).	NORTH	SOUTH .	TIME	.ALTITUDE
-87	38.34	38.34	7:31	5
-88	37.91	37.90	7:35	5
-89	37.48	37.46	7:39	6
-90	37.04	37.01	7:43	7
-92	36.12	36.07	7:52	9
-93	35.65	35.59	7:56	10
-94	35.16	35.09	8:01	11
-95	34.67	34.58	8:05	12
-97	33.64	33.53	8:14	14
-98	33.11	32.99	8:19	16
-99	32.56	32.43	8:23	17
-100	32.01	31.86	8:28	18
-102	30.85	30.68	8:37	20
-103	30.26	30.07	8:42	21
-104	29.65	29.45	8:47	23
-105	29.02	28.81	8:52	24
-107	27.73	27.49	9:02	26
-108	27.07	26.81	9:08	28
-109	26.39	26.11	9:13	29
-110	25.69	25.40	9:18	30
-112	24.25	23.93	9:29	33
-113	23.51	23.18	9:35	35
-114	22.76	22.41	9:41	36
-115	21.99	21.62	9:47	37
-117	20.41	20.01	9:59	40
-118	19.61	19.19	10:05	42
-119	18.79	18.35	10:12	44
-120	17.96	17.51	10:18	45
-122	16.27	15.79	10:31	48
-123	15.42	14.92	10:38	50
-124	14.56	14.05	10:45	52
-125	13.70	13.18	10:52	53

PERCENTAGE OF SUN OBSCURED ALONG MAP TRACK
WEST END = 99.9   EAST END =101.2

===

+ MARKS THE CITY OF PEI-CHING

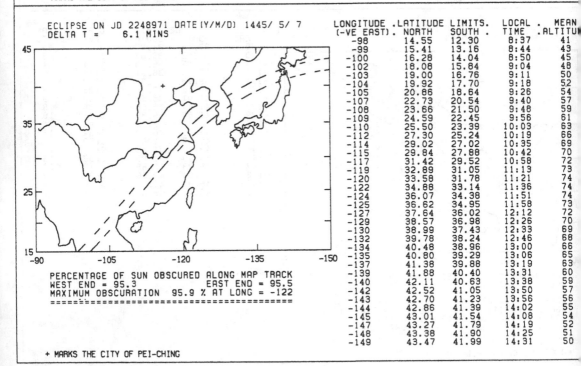

ECLIPSE ON JD 2248971 DATE (Y/M/D) 1445/ 5/ 7
DELTA T = 6.1 MINS

LONGITUDE	.LATITUDE	LIMITS.	LOCAL	. MEAN
(-VE EAST).	NORTH	SOUTH .	TIME	.ALTITUDE
-98	14.55	12.30	8:37	41
-99	15.41	13.16	8:44	43
-100	16.28	14.04	8:50	45
-102	18.08	15.84	9:04	48
-103	19.00	16.76	9:11	50
-104	19.92	17.70	9:18	52
-105	20.86	18.64	9:26	54
-107	22.73	20.54	9:40	57
-108	23.66	21.50	9:48	59
-109	24.59	22.45	9:56	61
-110	25.50	23.39	10:03	63
-112	27.30	25.24	10:19	66
-114	29.02	27.02	10:35	69
-115	29.84	27.88	10:42	70
-117	31.42	29.52	10:58	72
-119	32.89	31.05	11:13	73
-120	33.58	31.78	11:21	74
-122	34.88	33.14	11:36	74
-124	36.07	34.38	11:51	74
-125	36.62	34.95	11:58	73
-127	37.64	36.02	12:12	72
-129	38.57	36.98	12:26	70
-130	38.99	37.43	12:33	69
-132	39.78	38.24	12:46	68
-134	40.48	38.96	13:00	66
-135	40.80	39.29	13:06	65
-137	41.38	39.88	13:19	63
-139	41.88	40.40	13:31	60
-140	42.11	40.63	13:38	59
-142	42.52	41.05	13:50	57
-143	42.70	41.23	13:56	56
-144	42.86	41.39	14:02	55
-145	43.01	41.54	14:08	54
-147	43.27	41.79	14:19	52
-148	43.38	41.90	14:25	51
-149	43.47	41.99	14:31	50

PERCENTAGE OF SUN OBSCURED ALONG MAP TRACK
WEST END = 95.3   EAST END = 95.5
MAXIMUM OBSCURATION 95.9 % AT LONG = -122

===

+ MARKS THE CITY OF PEI-CHING

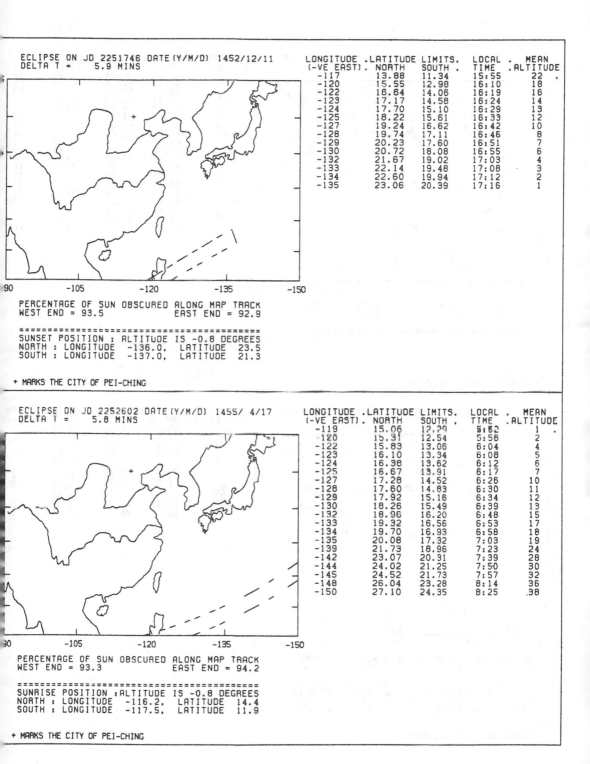

ECLIPSE ON JD 2251746 DATE (Y/M/D) 1452/12/11
DELTA T = 5.9 MINS

LONGITUDE	.LATITUDE LIMITS.		LOCAL	. MEAN
(-VE EAST).	NORTH	SOUTH .	TIME	.ALTITUDE
-117	13.88	11.34	15:55	22 .
-120	15.55	12.98	16:10	18
-122	16.64	14.06	16:19	16
-123	17.17	14.58	16:24	14
-124	17.70	15.10	16:29	13
-125	18.22	15.61	16:33	12
-127	19.24	16.62	16:42	10
-128	19.74	17.11	16:46	8
-129	20.23	17.60	16:51	7
-130	20.72	18.08	16:55	6
-132	21.67	19.02	17:03	4
-133	22.14	19.48	17:08	3
-134	22.60	19.94	17:12	2
-135	23.06	20.39	17:16	1

PERCENTAGE OF SUN OBSCURED ALONG MAP TRACK
WEST END = 93.5   EAST END = 92.9

==
SUNSET POSITION : ALTITUDE IS -0.8 DEGREES
NORTH : LONGITUDE -136.0, LATITUDE 23.5
SOUTH : LONGITUDE -137.0, LATITUDE 21.3

+ MARKS THE CITY OF PEI-CHING

ECLIPSE ON JD 2252602 DATE (Y/M/D) 1455/ 4/17
DELTA T = 5.8 MINS

LONGITUDE	.LATITUDE LIMITS.		LOCAL	. MEAN
(-VE EAST).	NORTH	SOUTH .	TIME	.ALTITUDE
-119	15.06	12.29	5:52	1 .
-120	15.31	12.54	5:58	2
-122	15.83	13.06	6:04	4
-123	16.10	13.34	6:08	5
-124	16.38	13.62	6:12	6
-125	16.67	13.91	6:17	7
-127	17.28	14.52	6:26	10
-128	17.60	14.83	6:30	11
-129	17.92	15.16	6:34	12
-130	18.26	15.49	6:39	13
-132	18.96	16.20	6:48	15
-133	19.32	16.56	6:53	17
-134	19.70	16.93	6:58	18
-135	20.08	17.32	7:03	19
-139	21.73	18.96	7:23	24
-142	23.07	20.31	7:39	28
-144	24.02	21.25	7:50	30
-145	24.52	21.73	7:57	32
-148	26.04	23.28	8:14	36
-150	27.10	24.35	8:25	.38

PERCENTAGE OF SUN OBSCURED ALONG MAP TRACK
WEST END = 93.3   EAST END = 94.2

==
SUNRISE POSITION : ALTITUDE IS -0.8 DEGREES
NORTH : LONGITUDE -116.2, LATITUDE 14.4
SOUTH : LONGITUDE -117.5, LATITUDE 11.9

+ MARKS THE CITY OF PEI-CHING

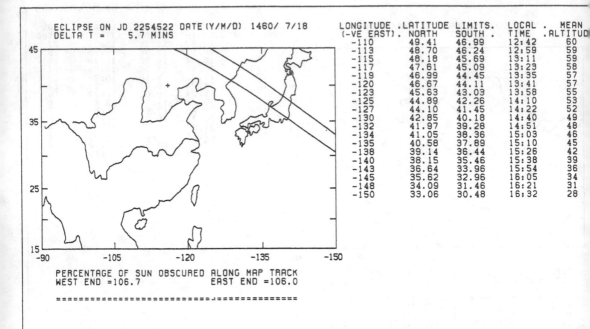

ECLIPSE ON JD 2254522 DATE (Y/M/D) 1460/ 7/18
DELTA T =   5.7 MINS

LONGITUDE (-VE EAST)	LATITUDE NORTH	LIMITS. SOUTH	LOCAL TIME	MEAN ALTITUD
-110	49.41	46.99	12:42	60
-113	48.70	46.24	12:59	59
-115	48.18	45.69	13:11	59
-117	47.61	45.09	13:23	58
-119	46.99	44.45	13:35	57
-120	46.67	44.11	13:41	57
-123	45.63	43.03	13:58	55
-125	44.89	42.26	14:10	53
-127	44.10	41.45	14:22	52
-130	42.85	40.18	14:40	49
-132	41.97	39.28	14:51	48
-134	41.05	38.36	15:03	46
-135	40.58	37.89	15:10	45
-138	39.14	36.44	15:26	42
-140	38.15	35.46	15:38	39
-143	36.64	33.96	15:54	36
-145	35.62	32.96	16:05	34
-148	34.09	31.46	16:21	31
-150	33.06	30.48	16:32	28

PERCENTAGE OF SUN OBSCURED ALONG MAP TRACK
WEST END =106.7   EAST END =106.0

==

+ MARKS THE CITY OF PEI-CHING

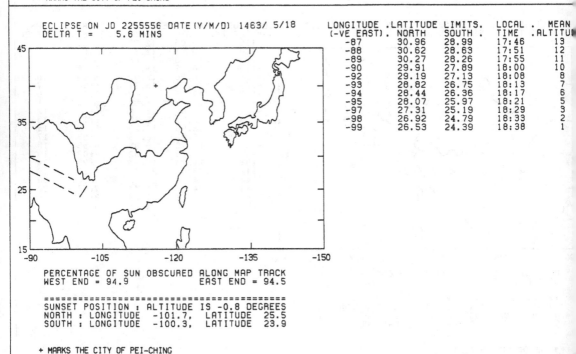

ECLIPSE ON JD 2255556 DATE (Y/M/D) 1463/ 5/18
DELTA T =   5.6 MINS

LONGITUDE (-VE EAST)	LATITUDE NORTH	LIMITS. SOUTH	LOCAL TIME	MEAN ALTITU
-87	30.96	28.99	17:46	13
-88	30.62	28.63	17:51	12
-89	30.27	28.26	17:55	11
-90	29.91	27.89	18:00	10
-92	29.19	27.13	18:08	8
-93	28.82	26.75	18:13	7
-94	28.44	26.36	18:17	6
-95	28.07	25.97	18:21	5
-97	27.31	25.19	18:29	3
-98	26.92	24.79	18:33	2
-99	26.53	24.39	18:38	1

PERCENTAGE OF SUN OBSCURED ALONG MAP TRACK
WEST END = 94.9   EAST END = 94.5

==
SUNSET POSITION : ALTITUDE IS -0.8 DEGREES
NORTH : LONGITUDE -101.7, LATITUDE 25.5
SOUTH : LONGITUDE -100.3, LATITUDE 23.9

+ MARKS THE CITY OF PEI-CHING

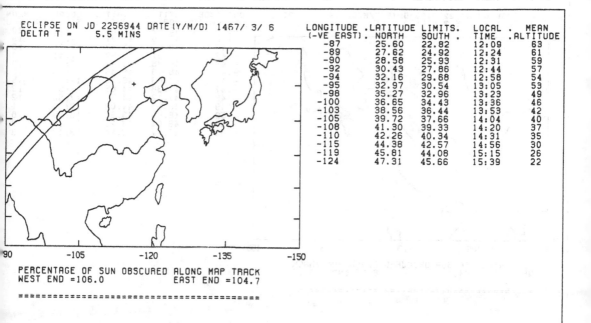

ECLIPSE ON JD 2256944 DATE (Y/M/D) 1467/ 3/ 6
DELTA T =   5.5 MINS

LONGITUDE (-VE EAST)	LATITUDE NORTH	LIMITS SOUTH	LOCAL TIME	MEAN ALTITUDE
-87	25.60	22.82	12:09	63
-89	27.62	24.92	12:24	61
-90	28.58	25.93	12:31	59
-92	30.43	27.86	12:44	57
-94	32.16	29.68	12:58	54
-95	32.97	30.54	13:05	53
-98	35.27	32.96	13:23	49
-100	36.65	34.43	13:36	46
-103	38.56	36.44	13:53	42
-105	39.72	37.66	14:04	40
-108	41.30	39.33	14:20	37
-110	42.26	40.34	14:31	35
-115	44.38	42.57	14:56	30
-119	45.81	44.08	15:15	26
-124	47.31	45.66	15:39	22

PERCENTAGE OF SUN OBSCURED ALONG MAP TRACK
WEST END =106.0   EAST END =104.7

==

+ MARKS THE CITY OF PEI-CHING

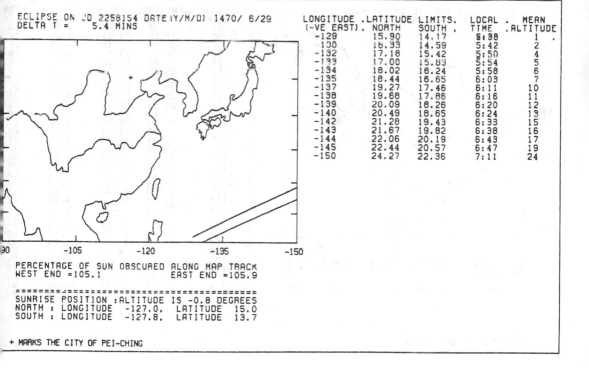

ECLIPSE ON JD 2258154 DATE (Y/M/D) 1470/ 6/29
DELTA T =   5.4 MINS

LONGITUDE (-VE EAST)	LATITUDE NORTH	LIMITS SOUTH	LOCAL TIME	MEAN ALTITUDE
-129	15.90	14.17	5:38	1
-130	16.33	14.59	5:42	2
-132	17.18	15.42	5:50	4
-133	17.00	15.83	5:54	5
-134	18.02	16.24	5:58	6
-135	18.44	16.65	6:03	7
-137	19.27	17.46	6:11	10
-138	19.68	17.86	6:16	11
-139	20.09	18.26	6:20	12
-140	20.49	18.65	6:24	13
-142	21.28	19.43	6:33	15
-143	21.67	19.82	6:38	16
-144	22.06	20.19	6:43	17
-145	22.44	20.57	6:47	19
-150	24.27	22.36	7:11	24

PERCENTAGE OF SUN OBSCURED ALONG MAP TRACK
WEST END =105.1   EAST END =105.9

==
SUNRISE POSITION : ALTITUDE IS -0.8 DEGREES
NORTH : LONGITUDE -127.0, LATITUDE 15.0
SOUTH : LONGITUDE -127.8, LATITUDE 13.7

+ MARKS THE CITY OF PEI-CHING

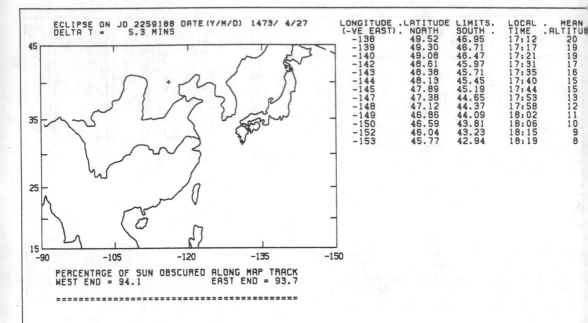

ECLIPSE ON JD 2259188 DATE(Y/M/D) 1473/ 4/27
DELTA T =   5.3 MINS

LONGITUDE (-VE EAST)	LATITUDE NORTH	LIMITS SOUTH	LOCAL TIME	MEAN ALTITU
-138	49.52	46.95	17:12	20
-139	49.30	46.71	17:17	19
-140	49.08	46.47	17:21	19
-142	48.61	45.97	17:31	17
-143	48.38	45.71	17:35	16
-144	48.13	45.45	17:40	15
-145	47.89	45.19	17:44	15
-147	47.38	44.65	17:53	13
-148	47.12	44.37	17:58	12
-149	46.86	44.09	18:02	11
-150	46.59	43.81	18:06	10
-152	46.04	43.23	18:15	9
-153	45.77	42.94	18:19	8

PERCENTAGE OF SUN OBSCURED ALONG MAP TRACK
WEST END = 94.1   EAST END = 93.7

==

+ MARKS THE CITY OF PEI-CHING

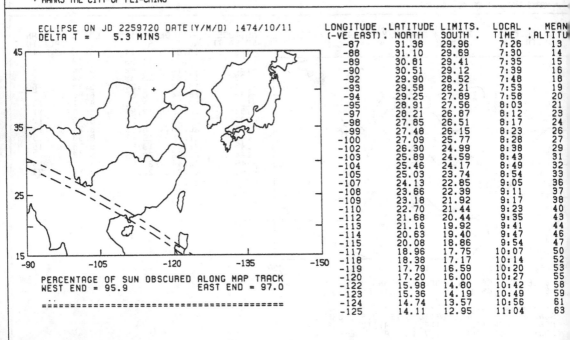

ECLIPSE ON JD 2259720 DATE(Y/M/D) 1474/10/11
DELTA T =   5.3 MINS

LONGITUDE (-VE EAST)	LATITUDE NORTH	LIMITS SOUTH	LOCAL TIME	MEAN ALTITU
-87	31.38	29.96	7:26	13
-88	31.10	29.69	7:30	14
-89	30.81	29.41	7:35	15
-90	30.51	29.12	7:39	16
-92	29.90	28.52	7:48	18
-93	29.58	28.21	7:53	19
-94	29.25	27.89	7:58	20
-95	28.91	27.56	8:03	21
-97	28.21	26.87	8:12	23
-98	27.85	26.51	8:17	24
-99	27.48	26.15	8:23	26
-100	27.09	25.77	8:28	27
-102	26.30	24.99	8:38	29
-103	25.89	24.59	8:43	31
-104	25.46	24.17	8:49	32
-105	25.03	23.74	8:54	33
-107	24.13	22.85	9:05	36
-108	23.66	22.39	9:11	37
-109	23.18	21.92	9:17	38
-110	22.70	21.44	9:23	40
-112	21.68	20.44	9:35	43
-113	21.16	19.92	9:41	44
-114	20.63	19.40	9:47	46
-115	20.08	18.86	9:54	47
-117	18.96	17.75	10:07	50
-118	18.38	17.17	10:14	52
-119	17.79	16.59	10:20	53
-120	17.20	16.00	10:27	55
-122	15.98	14.80	10:42	58
-123	15.36	14.19	10:49	59
-124	14.74	13.57	10:56	61
-125	14.11	12.95	11:04	63

PERCENTAGE OF SUN OBSCURED ALONG MAP TRACK
WEST END = 95.9   EAST END = 97.0

==

+ MARKS THE CITY OF PEI-CHING

ECLIPSE ON JD 2263352 DATE(Y/M/D) 1484/ 9/20
DELTA T = 5.0 MINS

LONGITUDE (-VE EAST)	LATITUDE NORTH	LIMITS. SOUTH	LOCAL TIME	MEAN ALTITUDE
-140	49.69	46.59	9:12	28
-143	48.42	45.28	9:27	30
-145	47.49	44.33	9:38	32
-148	45.97	42.73	9:54	35
-150	44.84	41.55	10:05	38

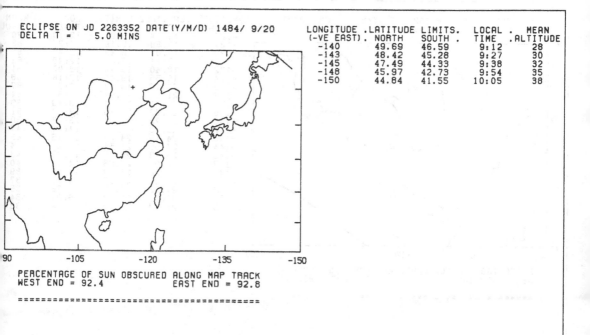

PERCENTAGE OF SUN OBSCURED ALONG MAP TRACK
WEST END = 92.4   EAST END = 92.8

==

+ MARKS THE CITY OF PEI-CHING

ECLIPSE ON JD 2264740 DATE(Y/M/D) 1488/ 7/ 9
DELTA T = 4.9 MINS

LONGITUDE (-VE EAST)	LATITUDE NORTH	LIMITS. SOUTH	LOCAL TIME	MEAN ALTITUDE
-87	31.19	28.98	11:51	81
-89	30.34	28.09	12:04	82
-90	29.89	27.62	12:11	82
-92	28.91	26.59	12:25	81
-94	27.84	25.48	12:38	79
-95	27.28	24.89	12:46	78
-97	26.08	23.64	13:00	76
-99	24.80	22.32	13:15	72
-100	24.13	21.63	13:22	71
-102	22.73	20.21	13:35	68
-104	21.28	18.73	13:50	64
-105	20.52	17.98	13:57	62
-107	19.00	16.44	14:10	59
-109	17.43	14.89	14:24	55
-110	16.64	14.11	14:31	54
-112	15.07	12.55	14:43	50
-114	13.49	11.01	14:56	47

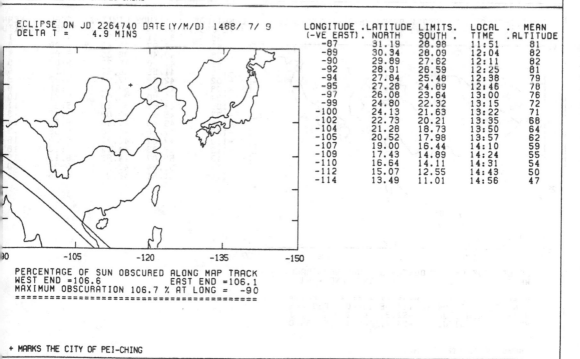

PERCENTAGE OF SUN OBSCURED ALONG MAP TRACK
WEST END =106.6   EAST END =106.1
MAXIMUM OBSCURATION 106.7 % AT LONG = -90

==

+ MARKS THE CITY OF PEI-CHING

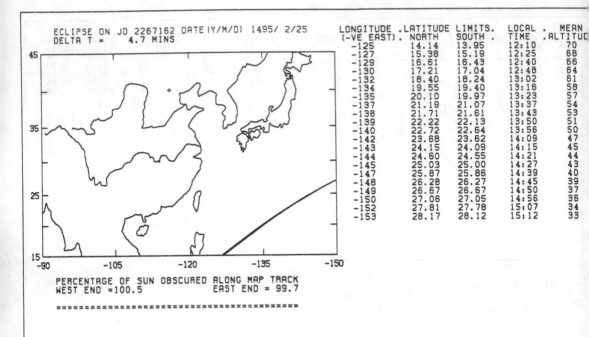

ECLIPSE ON JD 2267162 DATE(Y/M/D) 1495/ 2/25
DELTA T =   4.7 MINS

LONGITUDE (-VE EAST)	LATITUDE NORTH	LIMITS. SOUTH	LOCAL TIME	MEAN ALTITUDE
-125	14.14	13.95	12:10	70
-127	15.38	15.19	12:25	68
-129	16.61	16.43	12:40	66
-130	17.21	17.04	12:48	64
-132	18.40	18.24	13:02	61
-134	19.55	19.40	13:16	58
-135	20.10	19.97	13:23	57
-137	21.19	21.07	13:37	54
-138	21.71	21.61	13:43	53
-139	22.22	22.13	13:50	51
-140	22.72	22.64	13:56	50
-142	23.68	23.62	14:09	47
-143	24.15	24.09	14:15	45
-144	24.60	24.55	14:21	44
-145	25.03	25.00	14:27	43
-147	25.87	25.86	14:39	40
-148	26.28	26.27	14:45	39
-149	26.67	26.67	14:50	37
-150	27.06	27.05	14:56	36
-152	27.81	27.78	15:07	34
-153	28.17	28.12	15:12	33

PERCENTAGE OF SUN OBSCURED ALONG MAP TRACK
WEST END =100.5   EAST END = 99.7

==

+ MARKS THE CITY OF PEI-CHING

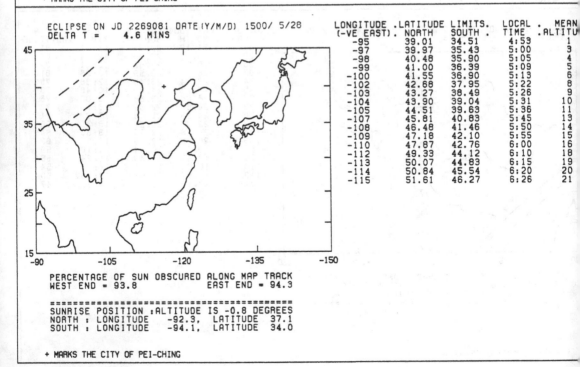

ECLIPSE ON JD 2269081 DATE(Y/M/D) 1500/ 5/28
DELTA T =   4.6 MINS

LONGITUDE (-VE EAST)	LATITUDE NORTH	LIMITS. SOUTH	LOCAL TIME	MEAN ALTITUDE
-95	39.01	34.51	4:53	1
-97	39.97	35.43	5:00	3
-98	40.48	35.90	5:05	4
-99	41.00	36.39	5:09	5
-100	41.55	36.90	5:13	6
-102	42.68	37.95	5:22	8
-103	43.27	38.49	5:26	9
-104	43.90	39.04	5:31	10
-105	44.51	39.63	5:36	11
-107	45.81	40.83	5:45	13
-108	46.48	41.46	5:50	14
-109	47.18	42.10	5:55	15
-110	47.87	42.76	6:00	16
-112	49.33	44.12	6:10	18
-113	50.07	44.83	6:15	19
-114	50.84	45.54	6:20	20
-115	51.61	46.27	6:26	21

PERCENTAGE OF SUN OBSCURED ALONG MAP TRACK
WEST END = 93.8   EAST END = 94.3

==
SUNRISE POSITION :ALTITUDE IS -0.8 DEGREES
NORTH : LONGITUDE   -92.3, LATITUDE   37.1
SOUTH : LONGITUDE   -94.1, LATITUDE   34.0

+ MARKS THE CITY OF PEI-CHING

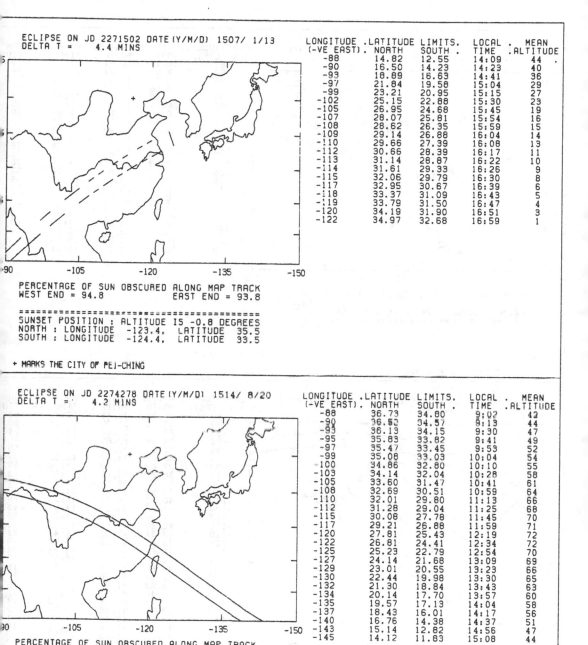

ECLIPSE ON JD 2271502 DATE(Y/M/D) 1507/ 1/13
DELTA T = 4.4 MINS

LONGITUDE (-VE EAST)	LATITUDE NORTH	LIMITS SOUTH	LOCAL TIME	MEAN ALTITUDE
-88	14.82	12.55	14:09	44
-90	16.50	14.23	14:23	40
-93	18.89	16.63	14:41	36
-97	21.84	19.58	15:04	29
-99	23.21	20.95	15:15	27
-102	25.15	22.88	15:30	23
-105	26.95	24.68	15:45	19
-107	28.07	25.81	15:54	16
-108	28.62	26.35	15:59	15
-109	29.14	26.88	16:04	14
-110	29.66	27.39	16:08	13
-112	30.66	28.39	16:17	11
-113	31.14	28.87	16:22	10
-114	31.61	29.33	16:26	9
-115	32.06	29.79	16:30	8
-117	32.95	30.67	16:39	6
-118	33.37	31.09	16:43	5
-119	33.79	31.50	16:47	4
-120	34.19	31.90	16:51	3
-122	34.97	32.68	16:59	1

PERCENTAGE OF SUN OBSCURED ALONG MAP TRACK
WEST END = 94.8   EAST END = 93.8

===
SUNSET POSITION : ALTITUDE IS -0.8 DEGREES
NORTH : LONGITUDE -123.4, LATITUDE 35.5
SOUTH : LONGITUDE -124.4, LATITUDE 33.5

+ MARKS THE CITY OF PEI-CHING

ECLIPSE ON JD 2274278 DATE(Y/M/D) 1514/ 8/20
DELTA T = 4.2 MINS

LONGITUDE (-VE EAST)	LATITUDE NORTH	LIMITS SOUTH	LOCAL TIME	MEAN ALTITUDE
-88	36.73	34.80	9:02	42
-90	36.52	34.57	9:13	44
-93	36.13	34.15	9:30	47
-95	35.83	33.82	9:41	49
-97	35.47	33.45	9:53	52
-99	35.08	33.03	10:04	54
-100	34.86	32.80	10:10	55
-103	34.14	32.04	10:28	58
-105	33.60	31.47	10:41	61
-108	32.69	30.51	10:59	64
-110	32.01	29.80	11:13	66
-112	31.28	29.04	11:25	68
-115	30.08	27.78	11:45	70
-117	29.21	26.88	11:59	71
-120	27.81	25.43	12:19	72
-122	26.81	24.41	12:34	72
-125	25.23	22.79	12:54	70
-127	24.14	21.68	13:09	69
-129	23.01	20.55	13:23	66
-130	22.44	19.98	13:30	65
-132	21.30	18.84	13:43	63
-134	20.14	17.70	13:57	60
-135	19.57	17.13	14:04	58
-137	18.43	16.01	14:17	56
-140	16.76	14.38	14:37	51
-143	15.14	12.82	14:56	47
-145	14.12	11.83	15:08	44

PERCENTAGE OF SUN OBSCURED ALONG MAP TRACK
WEST END =106.1   EAST END =106.2
MAXIMUM OBSCURATION 106.7 % AT LONG = -120
===

+ MARKS THE CITY OF PEI-CHING

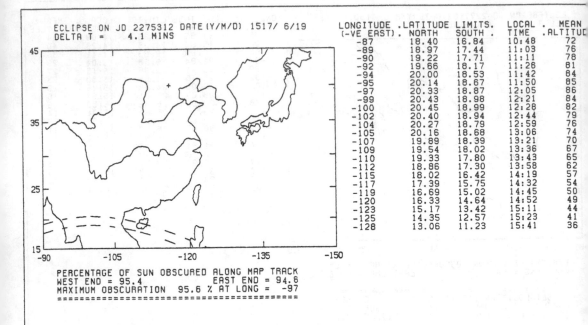

ECLIPSE ON JD 2275312 DATE(Y/M/D) 1517/ 6/19
DELTA T =   4.1 MINS

LONGITUDE (-VE EAST)	LATITUDE LIMITS. NORTH	SOUTH	LOCAL TIME	MEAN ALTITU
-87	18.40	16.84	10:48	72
-89	18.97	17.44	11:03	76
-90	19.22	17.71	11:11	78
-92	19.66	18.17	11:26	81
-94	20.00	18.53	11:42	84
-95	20.14	18.67	11:50	85
-97	20.33	18.87	12:05	86
-99	20.43	18.98	12:21	84
-100	20.45	18.99	12:28	82
-102	20.40	18.94	12:44	79
-104	20.27	18.79	12:59	76
-105	20.16	18.68	13:06	74
-107	19.89	18.39	13:21	70
-109	19.54	18.02	13:36	67
-110	19.33	17.80	13:43	65
-112	18.86	17.30	13:58	62
-115	18.02	16.42	14:19	57
-117	17.39	15.75	14:32	54
-119	16.69	15.02	14:45	50
-120	16.33	14.64	14:52	49
-123	15.17	13.42	15:11	44
-125	14.35	12.57	15:23	41
-128	13.06	11.23	15:41	36

PERCENTAGE OF SUN OBSCURED ALONG MAP TRACK
WEST END = 95.4   EAST END = 94.6
MAXIMUM OBSCURATION 95.6 % AT LONG = -97
===

+ MARKS THE CITY OF PEI-CHING

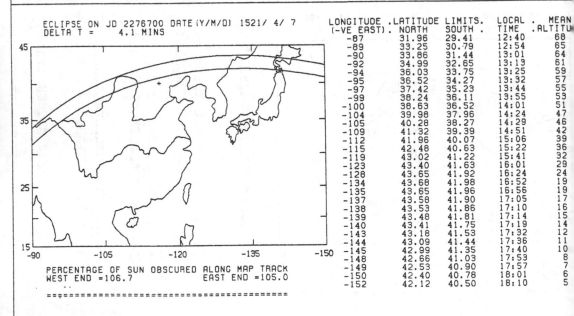

ECLIPSE ON JD 2276700 DATE(Y/M/D) 1521/ 4/ 7
DELTA T =   4.1 MINS

LONGITUDE (-VE EAST)	LATITUDE LIMITS. NORTH	SOUTH	LOCAL TIME	MEAN ALTITU
-87	31.96	29.41	12:40	68
-89	33.25	30.79	12:54	65
-90	33.86	31.44	13:01	64
-92	34.99	32.65	13:13	61
-94	36.03	33.75	13:25	59
-95	36.52	34.27	13:32	57
-97	37.42	35.23	13:44	55
-99	38.24	36.11	13:55	53
-100	38.63	36.52	14:01	51
-104	39.98	37.96	14:24	47
-105	40.28	38.27	14:29	46
-109	41.32	39.39	14:51	42
-112	41.96	40.07	15:06	39
-115	42.48	40.63	15:22	36
-119	43.02	41.22	15:41	32
-123	43.40	41.63	16:01	29
-128	43.65	41.92	16:24	24
-134	43.68	41.98	16:52	19
-135	43.65	41.96	16:56	19
-137	43.58	41.90	17:05	17
-138	43.53	41.86	17:10	16
-139	43.48	41.81	17:14	15
-140	43.41	41.75	17:19	14
-143	43.18	41.53	17:32	12
-144	43.09	41.44	17:36	11
-145	42.99	41.35	17:40	10
-148	42.66	41.03	17:53	8
-149	42.53	40.90	17:57	7
-150	42.40	40.78	18:01	6
-152	42.12	40.50	18:10	5

PERCENTAGE OF SUN OBSCURED ALONG MAP TRACK
WEST END =106.7   EAST END =105.0

===

+ MARKS THE CITY OF PEI-CHING

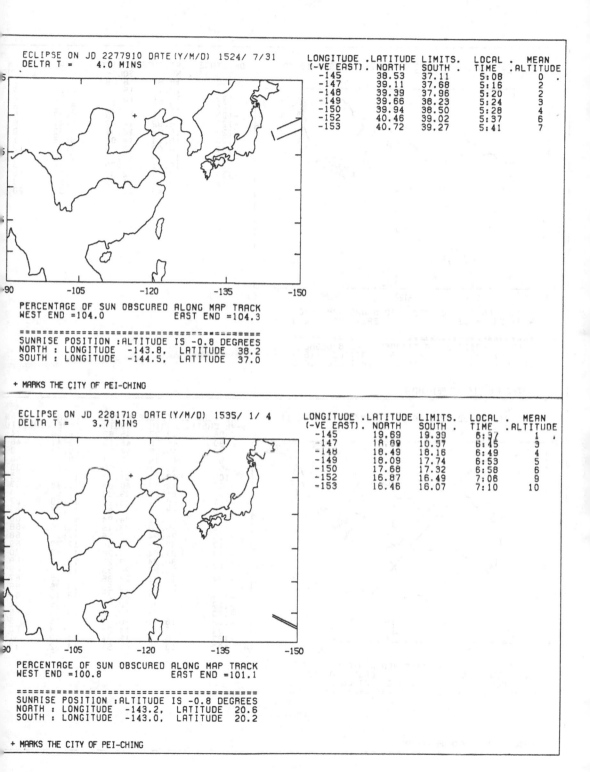

ECLIPSE ON JD 2277910 DATE(Y/M/D) 1524/ 7/31
DELTA T =   4.0 MINS

LONGITUDE (-VE EAST)	LATITUDE NORTH	LIMITS. SOUTH	LOCAL TIME	MEAN ALTITUDE
-145	38.53	37.11	5:08	0
-147	39.11	37.68	5:16	2
-148	39.39	37.96	5:20	2
-149	39.66	38.23	5:24	3
-150	39.94	38.50	5:28	4
-152	40.46	39.02	5:37	6
-153	40.72	39.27	5:41	7

PERCENTAGE OF SUN OBSCURED ALONG MAP TRACK
WEST END =104.0   EAST END =104.3

==
SUNRISE POSITION :ALTITUDE IS -0.8 DEGREES
NORTH : LONGITUDE -143.8, LATITUDE   38.2
SOUTH : LONGITUDE -144.5, LATITUDE   37.0

+ MARKS THE CITY OF PEI-CHING

ECLIPSE ON JD 2281719 DATE(Y/M/D) 1535/ 1/ 4
DELTA T =   3.7 MINS

LONGITUDE (-VE EAST)	LATITUDE NORTH	LIMITS. SOUTH	LOCAL TIME	MEAN ALTITUDE
-145	19.69	19.39	6:37	1
-147	18.89	10.57	6:45	3
-148	18.49	18.16	6:49	4
-149	18.09	17.74	6:53	5
-150	17.68	17.32	6:58	6
-152	16.87	16.49	7:06	9
-153	16.46	16.07	7:10	10

PERCENTAGE OF SUN OBSCURED ALONG MAP TRACK
WEST END =100.8   EAST END =101.1

==
SUNRISE POSITION :ALTITUDE IS -0.8 DEGREES
NORTH : LONGITUDE -143.2, LATITUDE   20.6
SOUTH : LONGITUDE -143.0, LATITUDE   20.2

+ MARKS THE CITY OF PEI-CHING

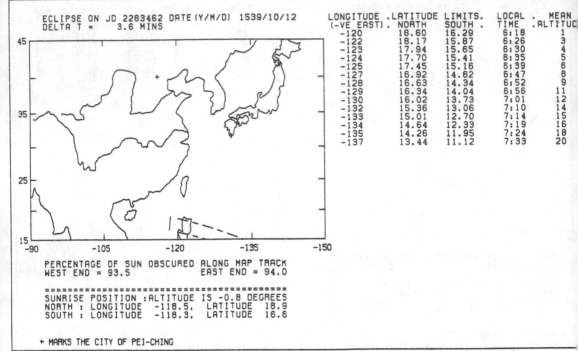

ECLIPSE ON JD 2283462 DATE(Y/M/D) 1539/10/12
DELTA T =   3.6 MINS

LONGITUDE (-VE EAST)	LATITUDE LIMITS. NORTH	SOUTH	LOCAL TIME	MEAN ALTITUDE
-120	18.60	16.29	6:18	1
-122	18.17	15.87	6:26	3
-123	17.94	15.65	6:30	4
-124	17.70	15.41	6:35	5
-125	17.45	15.16	6:39	6
-127	16.92	14.62	6:47	8
-128	16.63	14.34	6:52	9
-129	16.34	14.04	6:56	11
-130	16.02	13.73	7:01	12
-132	15.36	13.06	7:10	14
-133	15.01	12.70	7:14	15
-134	14.64	12.33	7:19	16
-135	14.26	11.95	7:24	18
-137	13.44	11.12	7:33	20

PERCENTAGE OF SUN OBSCURED ALONG MAP TRACK
WEST END = 93.5   EAST END = 94.0

==
SUNRISE POSITION :ALTITUDE IS -0.8 DEGREES
NORTH : LONGITUDE -118.5, LATITUDE 18.9
SOUTH : LONGITUDE -118.3, LATITUDE 16.6

+ MARKS THE CITY OF PEI-CHING

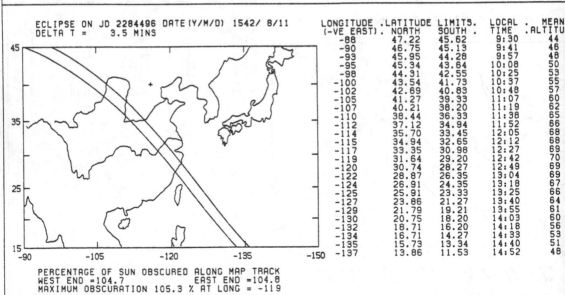

ECLIPSE ON JD 2284496 DATE(Y/M/D) 1542/ 8/11
DELTA T =   3.5 MINS

LONGITUDE (-VE EAST)	LATITUDE LIMITS. NORTH	SOUTH	LOCAL TIME	MEAN ALTITUDE
-88	47.22	45.62	9:30	44
-90	46.75	45.13	9:41	46
-93	45.95	44.28	9:57	48
-95	45.34	43.64	10:08	50
-98	44.31	42.55	10:25	53
-100	43.54	41.73	10:37	55
-102	42.69	40.83	10:48	57
-105	41.27	39.33	11:07	60
-107	40.21	38.20	11:19	62
-110	38.44	36.33	11:38	65
-112	37.12	34.94	11:52	66
-114	35.70	33.45	12:05	68
-115	34.94	32.65	12:12	68
-117	33.35	30.98	12:27	69
-119	31.64	29.20	12:42	70
-120	30.74	28.27	12:49	69
-122	28.87	26.35	13:04	69
-124	26.91	24.35	13:18	67
-125	25.91	23.33	13:25	66
-127	23.86	21.27	13:40	64
-129	21.79	19.21	13:55	61
-130	20.75	18.20	14:03	60
-132	18.71	16.20	14:18	56
-134	16.71	14.27	14:33	53
-135	15.73	13.34	14:40	51
-137	13.86	11.53	14:52	48

PERCENTAGE OF SUN OBSCURED ALONG MAP TRACK
WEST END =104.7   EAST END =104.8
MAXIMUM OBSCURATION 105.3 % AT LONG = -119
==

+ MARKS THE CITY OF PEI-CHING

ECLIPSE ON JD 2286918 DATE (Y/M/D) 1549/ 3/29
DELTA T = 3.4 MINS

LONGITUDE (-VE EAST)	LATITUDE LIMITS. NORTH	SOUTH	LOCAL TIME	MEAN ALTITUDE
-87	14.24	14.07	7:48	28
-90	15.33	15.20	8:04	32
-92	16.12	16.02	8:16	35
-94	16.97	16.89	8:27	37
-95	17.41	17.34	8:33	39
-98	18.80	18.78	8:52	43
-100	19.80	19.80	9:04	46
-102	20.87	20.84	9:17	48
-104	21.98	21.93	9:31	51
-105	22.54	22.48	9:37	52
-108	24.28	24.19	9:58	56
-110	25.47	25.36	10:12	58
-112	26.66	26.54	10:26	60
-114	27.85	27.71	10:41	62
-115	28.43	28.29	10:48	63
-119	30.72	30.57	11:17	64
-120	31.27	31.12	11:24	64
-124	33.35	33.21	11:52	64
-125	33.84	33.70	11:59	63
-129	35.67	35.54	12:26	61
-130	36.09	35.97	12:32	60
-133	37.27	37.16	12:52	58
-134	37.64	37.53	12:58	57
-135	37.99	37.89	13:04	56
-137	38.66	38.57	13:17	54
-138	38.97	38.89	13:23	53
-139	39.27	39.20	13:29	52
-140	39.56	39.49	13:35	51
-142	40.10	40.04	13:47	49
-143	40.35	40.30	13:53	48
-144	40.60	40.55	13:58	47
-145	40.83	40.79	14:04	46
-147	41.26	41.23	14:15	44
-148	41.46	41.44	14:21	43
-149	41.64	41.63	14:27	42
-150	41.82	41.82	14:32	42

PERCENTAGE OF SUN OBSCURED ALONG MAP TRACK
WEST END = 99.5   EAST END = 99.8
MAXIMUM OBSCURATION 100.4 % AT LONG = -122
===

+ MARKS THE CITY OF PEI-CHING

ECLIPSE ON JD 2288305 DATE (Y/M/D) 1553/ 1/14
DELTA T = 3.3 MINS

LONGITUDE (-VE EAST)	LATITUDE LIMITS. NORTH	SOUTH	LOCAL TIME	MEAN ALTITUDE
-87	14.34	13.48	13:58	46
-88	14.85	14.00	14:02	45
-89	15.37	14.53	14:09	43
-90	15.88	15.05	14:15	42
-92	16.92	16.11	14:27	39
-93	17.43	16.64	14:33	38
-94	17.95	17.17	14:39	36
-95	18.46	17.70	14:44	35
-97	19.48	18.75	14:56	32
-98	19.99	19.27	15:01	31
-99	20.49	19.78	15:06	30
-100	20.99	20.30	15:12	28
-102	21.97	21.31	15:22	26
-103	22.46	21.81	15:27	24
-104	22.94	22.30	15:32	23
-105	23.41	22.79	15:37	22
-107	24.34	23.75	15:47	20
-108	24.80	24.22	15:52	18
-109	25.26	24.69	15:57	17
-110	25.70	25.15	16:01	16
-112	26.58	26.05	16:11	14
-113	27.01	26.49	16:15	13
-114	27.44	26.93	16:20	12
-115	27.85	27.36	16:24	11
-117	28.67	28.20	16:33	9
-118	29.08	28.62	16:37	8
-119	29.47	29.02	16:41	7
-120	29.86	29.43	16:46	6
-122	30.62	30.21	16:54	4
-123	31.00	30.60	16:58	3
-124	31.37	30.97	17:02	2
-125	31.73	31.35	17:06	1

PERCENTAGE OF SUN OBSCURED ALONG MAP TRACK
WEST END =102.4   EAST END =101.0

===
SUNSET POSITION : ALTITUDE IS -0.8 DEGREES
NORTH : LONGITUDE -126.7, LATITUDE 32.5
SOUTH : LONGITUDE -126.9, LATITUDE 32.0

+ MARKS THE CITY OF PEI-CHING

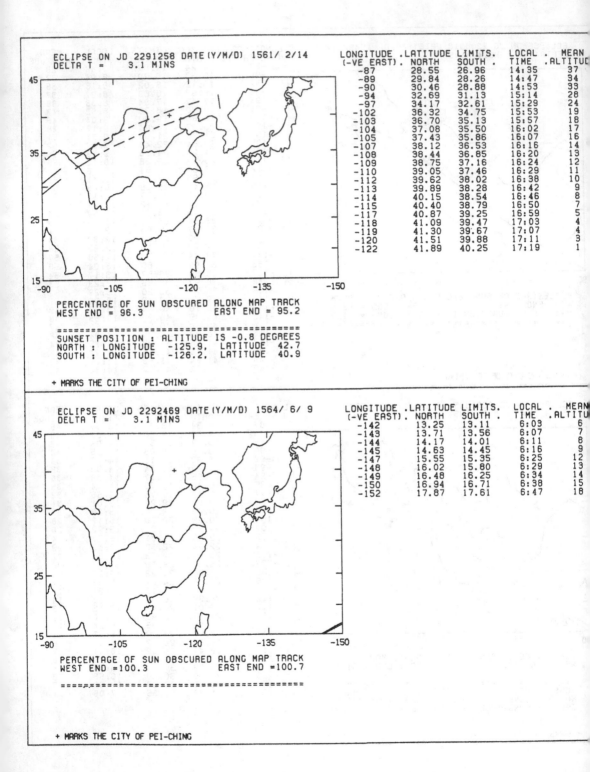

ECLIPSE ON JD 2291258 DATE (Y/M/D) 1561/ 2/14
DELTA T =   3.1 MINS

LONGITUDE (-VE EAST)	LATITUDE NORTH	LIMITS. SOUTH	LOCAL TIME	MEAN ALTITUDE
-87	28.55	26.96	14:35	37
-89	29.84	28.26	14:47	34
-90	30.46	28.88	14:53	33
-94	32.69	31.13	15:14	28
-97	34.17	32.61	15:29	24
-102	36.32	34.75	15:53	19
-103	36.70	35.13	15:57	18
-104	37.08	35.50	16:02	17
-105	37.43	35.86	16:07	16
-107	38.12	36.53	16:16	14
-108	38.44	36.85	16:20	13
-109	38.75	37.16	16:24	12
-110	39.05	37.46	16:29	11
-112	39.62	38.02	16:38	10
-113	39.89	38.28	16:42	9
-114	40.15	38.54	16:46	8
-115	40.40	38.79	16:50	7
-117	40.87	39.25	16:59	5
-118	41.09	39.47	17:03	4
-119	41.30	39.67	17:07	4
-120	41.51	39.88	17:11	3
-122	41.89	40.25	17:19	1

PERCENTAGE OF SUN OBSCURED ALONG MAP TRACK
WEST END = 96.3   EAST END = 95.2

==
SUNSET POSITION : ALTITUDE IS -0.8 DEGREES
NORTH : LONGITUDE -125.9, LATITUDE   42.7
SOUTH : LONGITUDE -126.2, LATITUDE   40.9

+ MARKS THE CITY OF PEI-CHING

ECLIPSE ON JD 2292469 DATE (Y/M/D) 1564/ 6/ 9
DELTA T =   3.1 MINS

LONGITUDE (-VE EAST)	LATITUDE NORTH	LIMITS. SOUTH	LOCAL TIME	MEAN ALTITUDE
-142	13.25	13.11	6:03	6
-143	13.71	13.56	6:07	7
-144	14.17	14.01	6:11	8
-145	14.63	14.45	6:16	9
-147	15.55	15.35	6:25	12
-148	16.02	15.80	6:29	13
-149	16.48	16.25	6:34	14
-150	16.94	16.71	6:38	15
-152	17.87	17.61	6:47	18

PERCENTAGE OF SUN OBSCURED ALONG MAP TRACK
WEST END =100.3   EAST END =100.7

==

+ MARKS THE CITY OF PEI-CHING

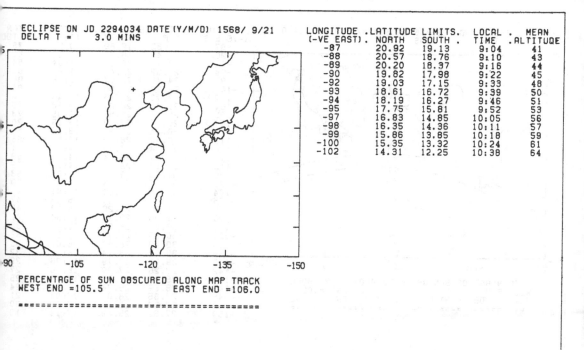

ECLIPSE ON JD 2294034 DATE (Y/M/D) 1568/ 9/21
DELTA T =   3.0 MINS

LONGITUDE (-VE EAST)	.LATITUDE NORTH	LIMITS. SOUTH .	LOCAL TIME	. MEAN .ALTITUDE
-87	20.92	19.13	9:04	41
-88	20.57	18.76	9:10	43
-89	20.20	18.37	9:16	44
-90	19.82	17.98	9:22	45
-92	19.03	17.15	9:33	48
-93	18.61	16.72	9:39	50
-94	18.19	16.27	9:46	51
-95	17.75	15.81	9:52	53
-97	16.83	14.85	10:05	56
-98	16.35	14.36	10:11	57
-99	15.86	13.85	10:18	59
-100	15.35	13.32	10:24	61
-102	14.31	12.25	10:38	64

PERCENTAGE OF SUN OBSCURED ALONG MAP TRACK
WEST END =105.5   EAST END =106.0

==

+ MARKS THE CITY OF PEI-CHING

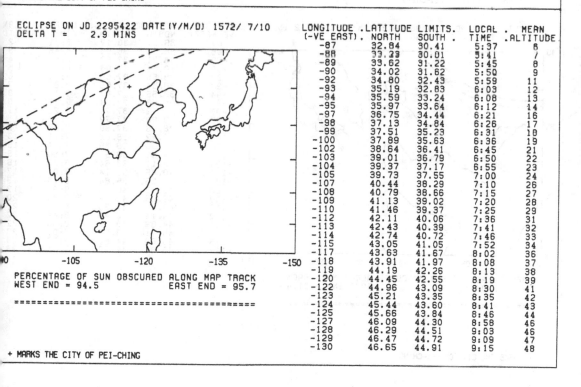

ECLIPSE ON JD 2295422 DATE (Y/M/D) 1572/ 7/10
DELTA T =   2.9 MINS

LONGITUDE (-VE EAST)	.LATITUDE NORTH	LIMITS. SOUTH .	LOCAL TIME	. MEAN .ALTITUDE .
-87	32.84	30.41	5:37	6
-88	33.23	30.01	5:41	7
-89	33.62	31.22	5:45	8
-90	34.02	31.62	5:50	9
-92	34.80	32.43	5:59	11
-93	35.19	32.83	6:03	12
-94	35.59	33.24	6:08	13
-95	35.97	33.64	6:12	14
-97	36.75	34.44	6:21	16
-98	37.13	34.84	6:26	17
-99	37.51	35.23	6:31	18
-100	37.89	35.63	6:36	19
-102	38.64	36.41	6:45	21
-103	39.01	36.79	6:50	22
-104	39.37	37.17	6:55	23
-105	39.73	37.55	7:00	24
-107	40.44	38.29	7:10	26
-108	40.79	38.66	7:15	27
-109	41.13	39.02	7:20	28
-110	41.46	39.37	7:25	29
-112	42.11	40.06	7:36	31
-113	42.43	40.39	7:41	32
-114	42.74	40.72	7:46	33
-115	43.05	41.05	7:52	34
-117	43.63	41.67	8:02	36
-118	43.91	41.97	8:08	37
-119	44.19	42.26	8:13	38
-120	44.45	42.55	8:19	39
-122	44.96	43.09	8:30	41
-123	45.21	43.35	8:35	42
-124	45.44	43.60	8:41	43
-125	45.66	43.84	8:46	44
-127	46.09	44.30	8:58	46
-128	46.29	44.51	9:03	46
-129	46.47	44.72	9:09	47
-130	46.65	44.91	9:15	48

PERCENTAGE OF SUN OBSCURED ALONG MAP TRACK
WEST END = 94.5   EAST END = 95.7

==

+ MARKS THE CITY OF PEI-CHING

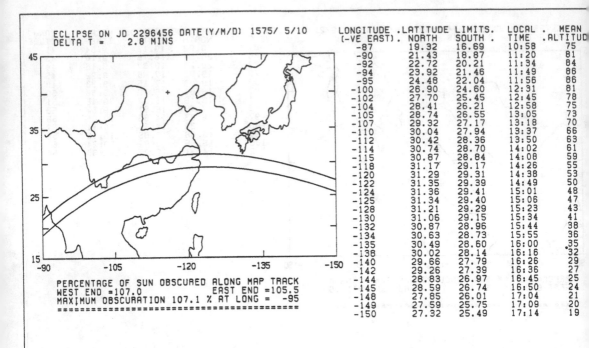

```
ECLIPSE ON JD 2296456 DATE(Y/M/D) 1575/ 5/10
DELTA T =    2.8 MINS
```

LONGITUDE (-VE EAST)	LATITUDE NORTH	LIMITS SOUTH	LOCAL TIME	MEAN ALTITUDE
-87	19.32	16.69	10:58	75
-90	21.43	18.87	11:20	81
-92	22.72	20.21	11:34	84
-94	23.92	21.46	11:49	86
-95	24.48	22.04	11:56	86
-100	26.90	24.60	12:31	81
-102	27.70	25.45	12:45	78
-104	28.41	26.21	12:58	75
-105	28.74	26.55	13:05	73
-107	29.32	27.17	13:18	70
-110	30.04	27.94	13:37	66
-112	30.42	28.36	13:50	63
-114	30.74	28.70	14:02	61
-115	30.87	28.84	14:08	59
-118	31.17	29.17	14:26	55
-120	31.29	29.31	14:38	53
-122	31.35	29.39	14:49	50
-124	31.36	29.41	15:01	48
-125	31.34	29.40	15:06	47
-128	31.21	29.29	15:23	43
-130	31.06	29.15	15:34	41
-132	30.87	28.96	15:44	38
-134	30.63	28.73	15:55	36
-135	30.49	28.60	16:00	35
-138	30.02	28.14	16:16	32
-140	29.66	27.79	16:26	29
-142	29.26	27.39	16:36	27
-144	28.83	26.97	16:45	25
-145	28.59	26.74	16:50	24
-148	27.85	26.01	17:04	21
-149	27.59	25.75	17:09	20
-150	27.32	25.49	17:14	19

```
PERCENTAGE OF SUN OBSCURED ALONG MAP TRACK
WEST END =107.0           EAST END =105.5
MAXIMUM OBSCURATION 107.1 % AT LONG =  -95
===================================================
```

+ MARKS THE CITY OF PEI-CHING

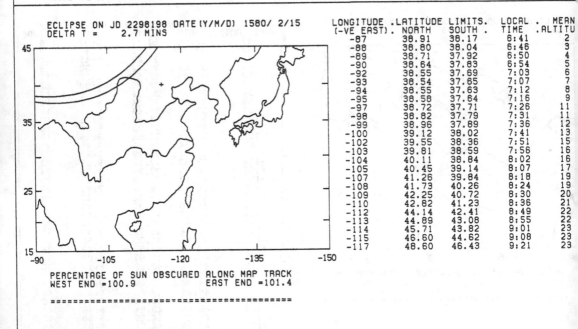

```
ECLIPSE ON JD 2298198 DATE(Y/M/D) 1580/ 2/15
DELTA T =    2.7 MINS
```

LONGITUDE (-VE EAST)	LATITUDE NORTH	LIMITS SOUTH	LOCAL TIME	MEAN ALTITUDE
-87	38.91	38.17	6:41	2
-88	38.80	38.04	6:46	3
-89	38.71	37.92	6:50	4
-90	38.64	37.83	6:54	5
-92	38.55	37.69	7:03	6
-93	38.54	37.65	7:07	7
-94	38.55	37.63	7:12	8
-95	38.58	37.64	7:16	9
-97	38.72	37.71	7:26	11
-98	38.82	37.79	7:31	11
-99	38.96	37.89	7:36	12
-100	39.12	38.02	7:41	13
-102	39.55	38.36	7:51	15
-103	39.81	38.59	7:56	16
-104	40.11	38.84	8:02	16
-105	40.45	39.14	8:07	17
-107	41.26	39.84	8:18	19
-108	41.73	40.26	8:24	19
-109	42.25	40.72	8:30	20
-110	42.82	41.23	8:36	21
-112	44.14	42.41	8:49	22
-113	44.89	43.08	8:55	22
-114	45.71	43.82	9:01	23
-115	46.60	44.62	9:08	23
-117	48.60	46.43	9:21	23

```
PERCENTAGE OF SUN OBSCURED ALONG MAP TRACK
WEST END =100.9           EAST END =101.4

===================================================
```

+ MARKS THE CITY OF PEI-CHING

ECLIPSE ON JD 2299054 DATE(Y/M/D) 1582/ 6/20
DELTA T =   2.7 MINS

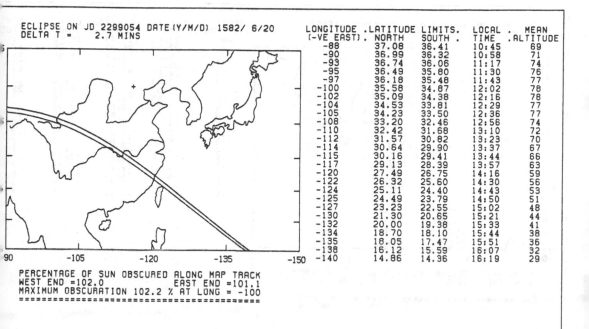

LONGITUDE	.LATITUDE	LIMITS.	LOCAL .	MEAN
(-VE EAST) .	NORTH	SOUTH .	TIME	.ALTITUDE
-88	37.08	36.41	10:45	69
-90	36.99	36.32	10:58	71
-93	36.74	36.06	11:17	74
-95	36.49	35.80	11:30	76
-97	36.18	35.48	11:43	77
-100	35.58	34.87	12:02	78
-102	35.09	34.38	12:16	78
-104	34.53	33.81	12:29	77
-105	34.23	33.50	12:36	77
-108	33.20	32.46	12:56	74
-110	32.42	31.68	13:10	72
-112	31.57	30.82	13:23	70
-114	30.64	29.90	13:37	67
-115	30.16	29.41	13:44	66
-117	29.13	28.39	13:57	63
-120	27.49	26.75	14:16	59
-122	26.32	25.60	14:30	56
-124	25.11	24.40	14:43	53
-125	24.49	23.79	14:50	51
-127	23.23	22.55	15:02	48
-130	21.30	20.65	15:21	44
-132	20.00	19.38	15:33	41
-134	18.70	18.10	15:44	38
-135	18.05	17.47	15:51	36
-138	16.12	15.59	16:07	32
-140	14.86	14.36	16:19	29

PERCENTAGE OF SUN OBSCURED ALONG MAP TRACK
WEST END =102.0   EAST END =101.1
MAXIMUM OBSCURATION 102.2 % AT LONG = -100
===

+ MARKS THE CITY OF PEI-CHING

ECLIPSE ON JD 2299586 DATE(Y/M/D) 1583/12/14
DELTA T =   2.7 MINS

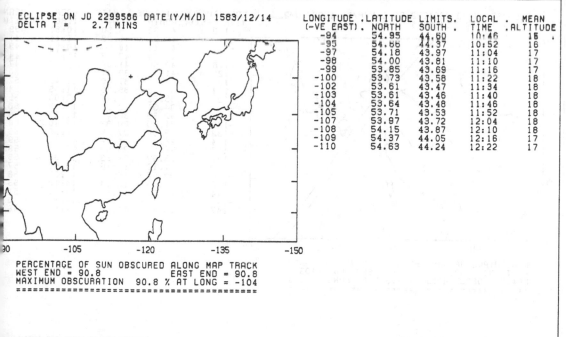

LONGITUDE	.LATITUDE	LIMITS.	LOCAL .	MEAN
(-VE EAST) .	NORTH	SOUTH .	TIME	.ALTITUDE
-94	54.95	44.60	10:46	16
-95	54.66	44.37	10:52	16
-97	54.18	43.97	11:04	17
-98	54.00	43.81	11:10	17
-99	53.85	43.69	11:16	17
-100	53.73	43.58	11:22	18
-102	53.61	43.47	11:34	18
-103	53.61	43.46	11:40	18
-104	53.64	43.48	11:46	18
-105	53.71	43.53	11:52	18
-107	53.97	43.72	12:04	18
-108	54.15	43.87	12:10	18
-109	54.37	44.05	12:16	17
-110	54.63	44.24	12:22	17

PERCENTAGE OF SUN OBSCURED ALONG MAP TRACK
WEST END = 90.8   EAST END = 90.8
MAXIMUM OBSCURATION 90.8 % AT LONG = -104
===

+ MARKS THE CITY OF PEI-CHING

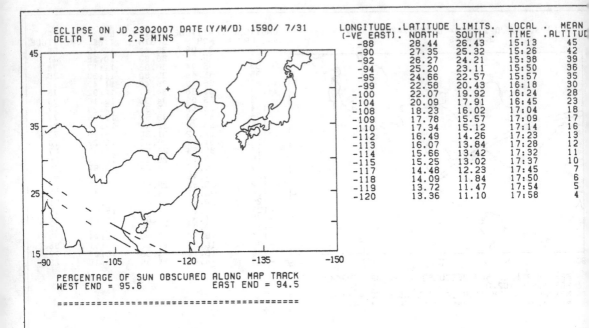

ECLIPSE ON JD 2302007 DATE(Y/M/D) 1590/ 7/31
DELTA T =   2.5 MINS

LONGITUDE (-VE EAST)	LATITUDE NORTH	LIMITS. SOUTH	LOCAL TIME	MEAN ALTITU
-88	28.44	26.43	15:13	45
-90	27.35	25.32	15:26	42
-92	26.27	24.21	15:38	39
-94	25.20	23.11	15:50	36
-95	24.66	22.57	15:57	35
-99	22.58	20.43	16:18	30
-100	22.07	19.92	16:24	28
-104	20.09	17.91	16:45	23
-108	18.23	16.02	17:04	18
-109	17.78	15.57	17:09	17
-110	17.34	15.12	17:14	16
-112	16.49	14.26	17:23	13
-113	16.07	13.84	17:28	12
-114	15.66	13.42	17:32	11
-115	15.25	13.02	17:37	10
-117	14.48	12.23	17:45	7
-118	14.09	11.84	17:50	6
-119	13.72	11.47	17:54	5
-120	13.36	11.10	17:58	4

PERCENTAGE OF SUN OBSCURED ALONG MAP TRACK
WEST END = 95.6   EAST END = 94.5

==

+ MARKS THE CITY OF PEI-CHING

ECLIPSE ON JD 2304252 DATE(Y/M/D) 1596/ 9/22
DELTA T =   2.4 MINS

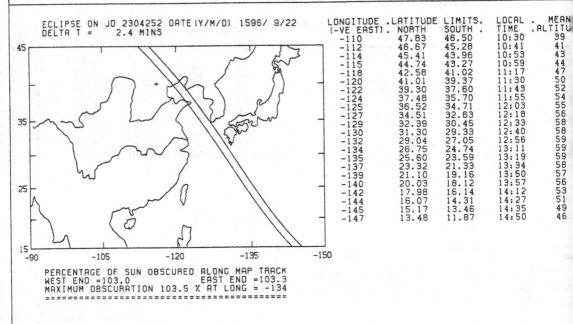

LONGITUDE (-VE EAST)	LATITUDE NORTH	LIMITS SOUTH	LOCAL TIME	MEAN ALTITU
-110	47.83	46.50	10:30	39
-112	46.67	45.28	10:41	41
-114	45.41	43.96	10:53	43
-115	44.74	43.27	10:59	44
-118	42.58	41.02	11:17	47
-120	41.01	39.37	11:30	50
-122	39.30	37.60	11:43	52
-124	37.48	35.70	11:55	54
-125	36.52	34.71	12:03	55
-127	34.51	32.63	12:18	56
-129	32.39	30.45	12:33	58
-130	31.30	29.33	12:40	58
-132	29.04	27.05	12:56	59
-134	26.75	24.74	13:11	59
-135	25.60	23.59	13:19	59
-137	23.32	21.33	13:34	58
-139	21.10	19.16	13:50	57
-140	20.03	18.12	13:57	56
-142	17.98	16.14	14:12	53
-144	16.07	14.31	14:27	51
-145	15.17	13.46	14:35	49
-147	13.48	11.87	14:50	46

PERCENTAGE OF SUN OBSCURED ALONG MAP TRACK
WEST END =103.0   EAST END =103.3
MAXIMUM OBSCURATION 103.5 % AT LONG = -134

==

+ MARKS THE CITY OF PEI-CHING

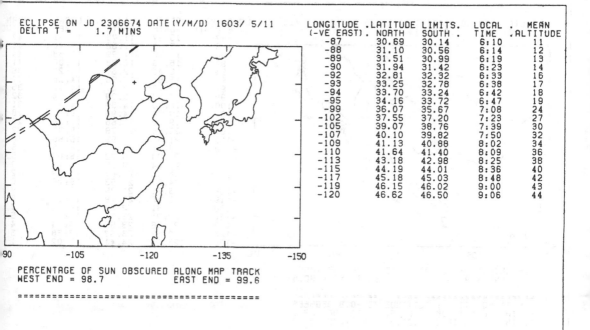

ECLIPSE ON JD 2306674 DATE(Y/M/D) 1603/ 5/11
DELTA T =   1.7 MINS

LONGITUDE (-VE EAST)	LATITUDE LIMITS. NORTH	SOUTH	LOCAL TIME	MEAN ALTITUDE
-87	30.69	30.14	6:10	11
-88	31.10	30.56	6:14	12
-89	31.51	30.99	6:19	13
-90	31.94	31.42	6:23	14
-92	32.81	32.32	6:33	16
-93	33.25	32.78	6:38	17
-94	33.70	33.24	6:42	18
-95	34.16	33.72	6:47	19
-99	36.07	35.67	7:08	24
-102	37.55	37.20	7:23	27
-105	39.07	38.76	7:39	30
-107	40.10	39.82	7:50	32
-109	41.13	40.88	8:02	34
-110	41.64	41.40	8:09	36
-113	43.18	42.98	8:25	38
-115	44.19	44.01	8:36	40
-117	45.18	45.03	8:48	42
-119	46.15	46.02	9:00	43
-120	46.62	46.50	9:06	44

PERCENTAGE OF SUN OBSCURED ALONG MAP TRACK
WEST END = 98.7   EAST END = 99.6

==

+ MARKS THE CITY OF PEI-CHING

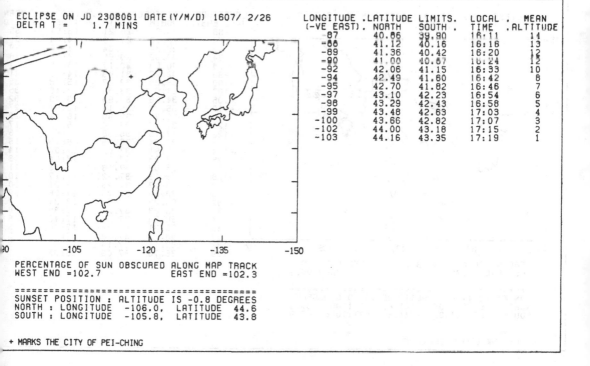

ECLIPSE ON JD 2308061 DATE(Y/M/D) 1607/ 2/26
DELTA T =   1.7 MINS

LONGITUDE (-VE EAST)	LATITUDE LIMITS. NORTH	SOUTH	LOCAL TIME	MEAN ALTITUDE
-87	40.65	39.90	16:11	14
-88	41.12	40.16	16:16	13
-89	41.36	40.42	16:20	12
-90	41.00	40.67	16:24	12
-92	42.06	41.15	16:33	10
-94	42.49	41.60	16:42	8
-95	42.70	41.82	16:46	7
-97	43.10	42.23	16:54	6
-98	43.29	42.43	16:58	5
-99	43.48	42.63	17:03	4
-100	43.66	42.82	17:07	3
-102	44.00	43.18	17:15	2
-103	44.16	43.35	17:19	1

PERCENTAGE OF SUN OBSCURED ALONG MAP TRACK
WEST END =102.7   EAST END =102.3

==

SUNSET POSITION : ALTITUDE IS -0.8 DEGREES
NORTH : LONGITUDE -106.0, LATITUDE 44.6
SOUTH : LONGITUDE -105.8, LATITUDE 43.8

+ MARKS THE CITY OF PEI-CHING

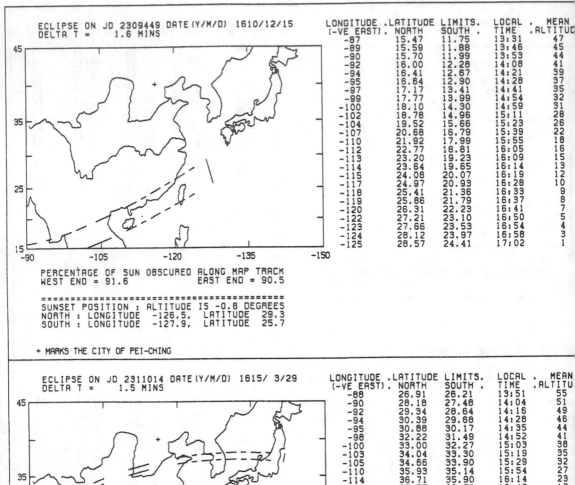

ECLIPSE ON JD 2309449 DATE (Y/M/D) 1610/12/15
DELTA T = 1.6 MINS

LONGITUDE (-VE EAST)	LATITUDE NORTH	LIMITS. SOUTH	LOCAL TIME	MEAN ALTITUDE
-87	15.47	11.75	13:31	47
-89	15.59	11.88	13:46	45
-90	15.70	11.99	13:53	44
-92	16.00	12.28	14:08	41
-94	16.41	12.67	14:21	39
-95	16.64	12.90	14:28	37
-97	17.17	13.41	14:41	35
-99	17.77	13.99	14:54	32
-100	18.10	14.30	14:59	31
-102	18.78	14.96	15:11	28
-104	19.52	15.66	15:23	26
-107	20.68	16.79	15:39	22
-110	21.92	17.99	15:55	18
-112	22.77	18.81	16:05	16
-113	23.20	19.23	16:09	15
-114	23.64	19.65	16:14	13
-115	24.08	20.07	16:19	12
-117	24.97	20.93	16:28	10
-118	25.41	21.36	16:33	9
-119	25.86	21.79	16:37	8
-120	26.31	22.23	16:41	7
-122	27.21	23.10	16:50	5
-123	27.66	23.53	16:54	4
-124	28.12	23.97	16:58	3
-125	28.57	24.41	17:02	1

PERCENTAGE OF SUN OBSCURED ALONG MAP TRACK
WEST END = 91.6   EAST END = 90.5

===
SUNSET POSITION : ALTITUDE IS -0.8 DEGREES
NORTH : LONGITUDE -126.5, LATITUDE 29.3
SOUTH : LONGITUDE -127.9, LATITUDE 25.7

+ MARKS THE CITY OF PEI-CHING

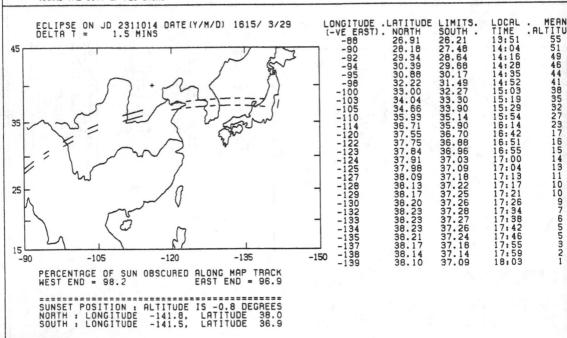

ECLIPSE ON JD 2311014 DATE (Y/M/D) 1615/ 3/29
DELTA T = 1.5 MINS

LONGITUDE (-VE EAST)	LATITUDE NORTH	LIMITS. SOUTH	LOCAL TIME	MEAN ALTITUDE
-88	26.91	26.21	13:51	55
-90	28.18	27.48	14:04	51
-92	29.34	28.64	14:16	49
-94	30.39	29.68	14:28	46
-95	30.88	30.17	14:35	44
-98	32.22	31.49	14:52	41
-100	33.00	32.27	15:03	38
-103	34.04	33.30	15:19	35
-105	34.66	33.90	15:29	32
-110	35.93	35.14	15:54	27
-114	36.71	35.90	16:14	23
-120	37.55	36.70	16:42	17
-122	37.75	36.88	16:51	16
-123	37.84	36.96	16:55	15
-124	37.91	37.03	17:00	14
-125	37.98	37.09	17:04	13
-127	38.09	37.18	17:13	11
-128	38.13	37.22	17:17	10
-129	38.17	37.25	17:21	10
-130	38.20	37.26	17:26	9
-132	38.23	37.28	17:34	7
-133	38.23	37.27	17:38	6
-134	38.23	37.26	17:42	5
-135	38.21	37.24	17:46	5
-137	38.17	37.18	17:55	3
-138	38.14	37.14	17:59	2
-139	38.10	37.09	18:03	1

PERCENTAGE OF SUN OBSCURED ALONG MAP TRACK
WEST END = 98.2   EAST END = 96.9

===
SUNSET POSITION : ALTITUDE IS -0.8 DEGREES
NORTH : LONGITUDE -141.8, LATITUDE 38.0
SOUTH : LONGITUDE -141.5, LATITUDE 36.9

+ MARKS THE CITY OF PEI-CHING

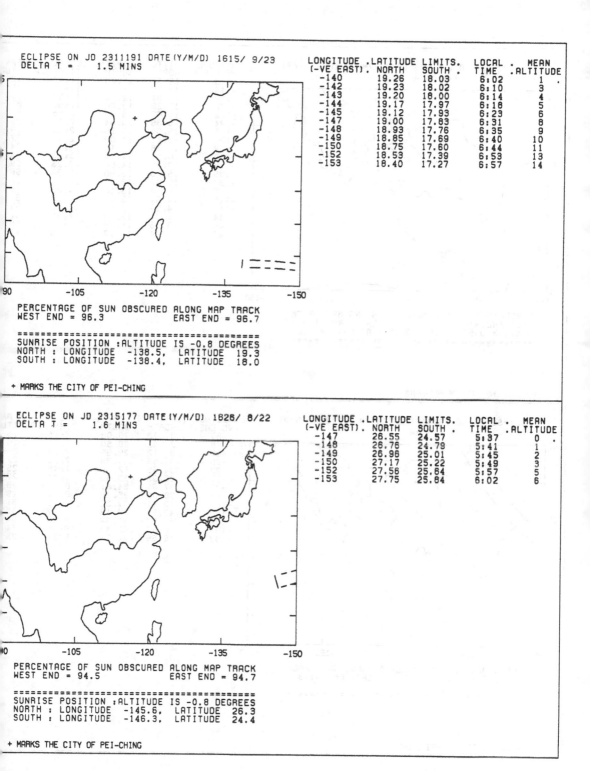

ECLIPSE ON JD 2311191 DATE(Y/M/D) 1615/ 9/23
DELTA T =   1.5 MINS

LONGITUDE (-VE EAST)	LATITUDE LIMITS. NORTH	SOUTH	LOCAL TIME	MEAN ALTITUDE
-140	19.26	18.03	6:02	1
-142	19.23	18.02	6:10	3
-143	19.20	18.00	6:14	4
-144	19.17	17.97	6:18	5
-145	19.12	17.93	6:23	6
-147	19.00	17.83	6:31	8
-148	18.93	17.76	6:35	9
-149	18.85	17.69	6:40	10
-150	18.75	17.60	6:44	11
-152	18.53	17.39	6:53	13
-153	18.40	17.27	6:57	14

PERCENTAGE OF SUN OBSCURED ALONG MAP TRACK
WEST END = 96.3   EAST END = 96.7

===
SUNRISE POSITION :ALTITUDE IS -0.8 DEGREES
NORTH : LONGITUDE -138.5, LATITUDE 19.3
SOUTH : LONGITUDE -138.4, LATITUDE 18.0

+ MARKS THE CITY OF PEI-CHING

ECLIPSE ON JD 2315177 DATE(Y/M/D) 1626/ 8/22
DELTA T =   1.6 MINS

LONGITUDE (-VE EAST)	LATITUDE LIMITS. NORTH	SOUTH	LOCAL TIME	MEAN ALTITUDE
-147	26.55	24.57	5:37	0
-148	26.76	24.79	5:41	1
-149	26.96	25.01	5:45	2
-150	27.17	25.22	5:49	3
-152	27.56	25.64	5:57	5
-153	27.75	25.84	6:02	6

PERCENTAGE OF SUN OBSCURED ALONG MAP TRACK
WEST END = 94.5   EAST END = 94.7

===
SUNRISE POSITION :ALTITUDE IS -0.8 DEGREES
NORTH : LONGITUDE -145.6, LATITUDE 26.3
SOUTH : LONGITUDE -146.3, LATITUDE 24.4

+ MARKS THE CITY OF PEI-CHING

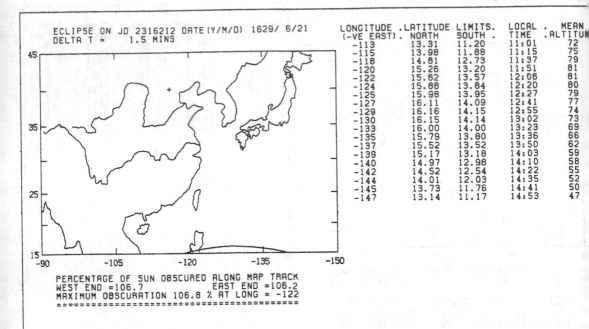

ECLIPSE ON JD 2316212 DATE(Y/M/D) 1629/ 6/21
DELTA T =   1.5 MINS

LONGITUDE (-VE EAST).	LATITUDE NORTH	LIMITS. SOUTH .	LOCAL TIME	MEAN ALTITU
-113	13.31	11.20	11:01	72
-115	13.98	11.88	11:15	75
-118	14.81	12.73	11:37	79
-120	15.26	13.20	11:51	81
-122	15.62	13.57	12:06	81
-124	15.88	13.84	12:20	80
-125	15.98	13.95	12:27	79
-127	16.11	14.09	12:41	77
-129	16.16	14.15	12:55	74
-130	16.15	14.14	13:02	73
-133	16.00	14.00	13:23	69
-135	15.79	13.80	13:36	66
-137	15.52	13.52	13:50	62
-139	15.17	13.18	14:03	59
-140	14.97	12.98	14:10	58
-142	14.52	12.54	14:22	55
-144	14.01	12.03	14:35	52
-145	13.73	11.76	14:41	50
-147	13.14	11.17	14:53	47

PERCENTAGE OF SUN OBSCURED ALONG MAP TRACK
WEST END =106.7   EAST END =106.2
MAXIMUM OBSCURATION 106.8 % AT LONG = -122
===

+ MARKS THE CITY OF PEI-CHING

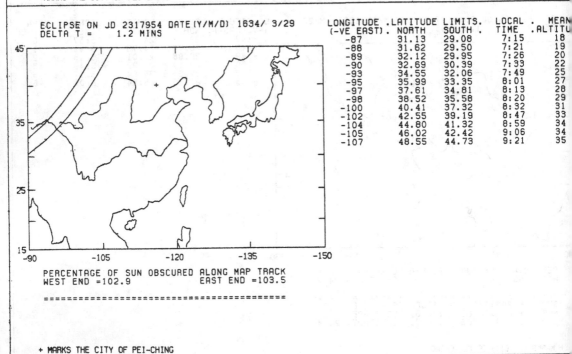

ECLIPSE ON JD 2317954 DATE(Y/M/D) 1634/ 3/29
DELTA T =   1.2 MINS

LONGITUDE (-VE EAST).	LATITUDE NORTH	LIMITS. SOUTH .	LOCAL TIME	MEAN ALTITU
-87	31.13	29.08	7:15	18
-88	31.62	29.50	7:21	19
-89	32.12	29.95	7:26	20
-90	32.69	30.39	7:33	22
-93	34.55	32.06	7:49	25
-95	35.99	33.35	8:01	27
-97	37.61	34.81	8:13	28
-98	38.52	35.58	8:20	29
-100	40.41	37.32	8:32	31
-102	42.55	39.19	8:47	33
-104	44.80	41.32	8:59	34
-105	46.02	42.42	9:06	34
-107	48.55	44.73	9:21	35

PERCENTAGE OF SUN OBSCURED ALONG MAP TRACK
WEST END =102.9   EAST END =103.5

===

+ MARKS THE CITY OF PEI-CHING

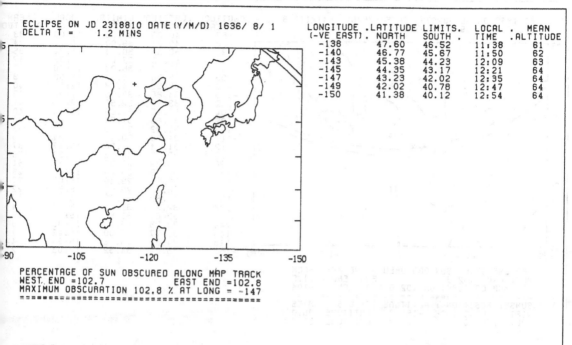

```
ECLIPSE ON JD 2318810 DATE(Y/M/D) 1636/ 8/ 1
DELTA T =    1.2 MINS
```

LONGITUDE (-VE EAST)	LATITUDE NORTH	LIMITS. SOUTH	LOCAL TIME	MEAN ALTITUDE
-138	47.60	46.52	11:38	61
-140	46.77	45.67	11:50	62
-143	45.38	44.23	12:09	63
-145	44.35	43.17	12:21	64
-147	43.23	42.02	12:35	64
-149	42.02	40.78	12:47	64
-150	41.38	40.12	12:54	64

```
PERCENTAGE OF SUN OBSCURED ALONG MAP TRACK
WEST END =102.7                    EAST END =102.8
MAXIMUM OBSCURATION 102.8 % AT LONG = -147
================================================

+ MARKS THE CITY OF PEI-CHING
```

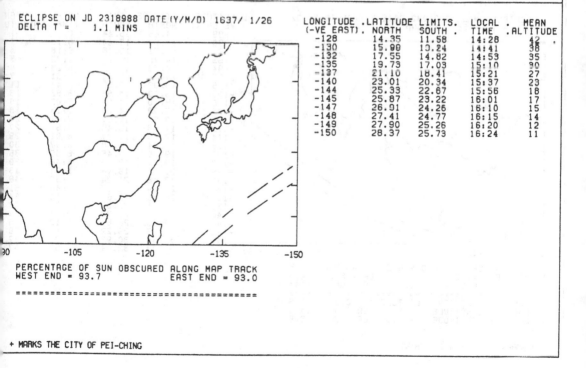

```
ECLIPSE ON JD 2318988 DATE(Y/M/D) 1637/ 1/26
DELTA T =    1.1 MINS
```

LONGITUDE (-VE EAST)	LATITUDE NORTH	LIMITS. SOUTH	LOCAL TIME	MEAN ALTITUDE
-128	14.35	11.58	14:28	42
-130	15.90	13.24	14:41	38
-132	17.55	14.82	14:53	35
-135	19.73	17.03	15:10	30
-137	21.10	18.41	15:21	27
-140	23.01	20.34	15:37	23
-144	25.33	22.67	15:56	18
-145	25.87	23.22	16:01	17
-147	26.91	24.26	16:10	15
-148	27.41	24.77	16:15	14
-149	27.90	25.26	16:20	12
-150	28.37	25.73	16:24	11

```
PERCENTAGE OF SUN OBSCURED ALONG MAP TRACK
WEST END = 93.7                    EAST END = 93.0

================================================

+ MARKS THE CITY OF PEI-CHING
```

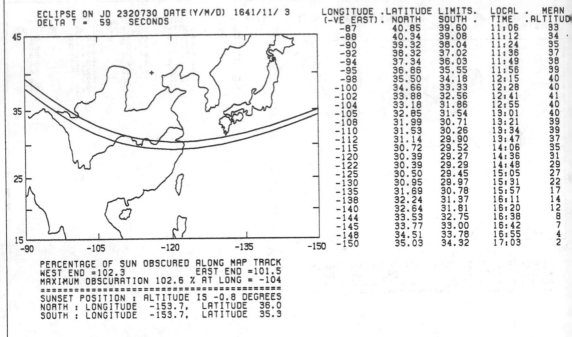

ECLIPSE ON JD 2320730 DATE(Y/M/D) 1641/11/ 3
DELTA T = 59 SECONDS

LONGITUDE (-VE EAST)	LATITUDE LIMITS. NORTH	SOUTH	LOCAL TIME	MEAN ALTITUD
-87	40.85	39.60	11:06	33
-88	40.34	39.08	11:12	34
-90	39.32	38.04	11:24	35
-92	38.32	37.02	11:36	37
-94	37.34	36.03	11:49	38
-95	36.86	35.55	11:56	39
-98	35.50	34.18	12:15	40
-100	34.66	33.33	12:28	40
-102	33.88	32.56	12:41	41
-104	33.18	31.86	12:55	40
-105	32.85	31.54	13:01	40
-108	31.99	30.71	13:21	39
-110	31.53	30.26	13:34	39
-112	31.14	29.90	13:47	37
-115	30.72	29.52	14:06	35
-120	30.39	29.27	14:36	31
-122	30.39	29.29	14:48	29
-125	30.50	29.45	15:05	27
-130	30.95	29.97	15:31	22
-135	31.69	30.78	15:57	17
-138	32.24	31.37	16:11	14
-140	32.64	31.81	16:20	12
-144	33.53	32.75	16:38	8
-145	33.77	33.00	16:42	7
-148	34.51	33.78	16:55	4
-150	35.03	34.32	17:03	2

```
PERCENTAGE OF SUN OBSCURED ALONG MAP TRACK
WEST END =102.3              EAST END =101.5
MAXIMUM OBSCURATION 102.6 % AT LONG = -104
===========================================
SUNSET POSITION : ALTITUDE IS -0.8 DEGREES
NORTH : LONGITUDE  -153.7,  LATITUDE  36.0
SOUTH : LONGITUDE  -153.7,  LATITUDE  35.3
```

+ MARKS THE CITY OF PEI-CHING

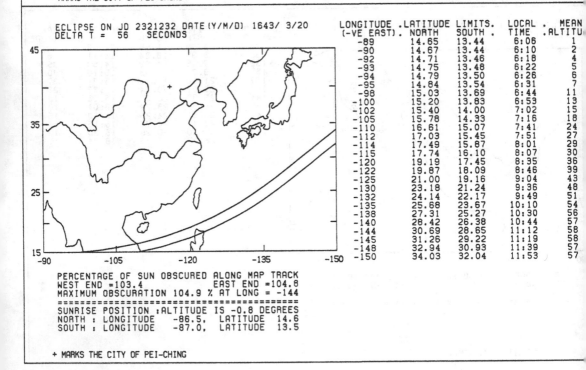

ECLIPSE ON JD 2321232 DATE(Y/M/D) 1643/ 3/20
DELTA T = 56 SECONDS

LONGITUDE (-VE EAST)	LATITUDE NORTH	LIMITS. SOUTH	LOCAL TIME	MEAN ALTITU
-89	14.65	13.44	6:06	1
-90	14.67	13.44	6:10	2
-92	14.71	13.46	6:18	4
-93	14.75	13.48	6:22	5
-94	14.79	13.50	6:26	6
-95	14.84	13.54	6:31	7
-98	15.03	13.69	6:44	11
-100	15.20	13.83	6:53	13
-102	15.40	14.00	7:02	15
-105	15.78	14.33	7:16	18
-110	16.61	15.07	7:41	24
-112	17.03	15.45	7:51	27
-114	17.49	15.87	8:01	29
-115	17.74	16.10	8:07	30
-120	19.19	17.45	8:35	36
-122	19.87	18.09	8:46	39
-125	21.00	19.16	9:04	43
-130	23.18	21.24	9:36	48
-132	24.14	22.17	9:49	51
-135	25.68	23.67	10:10	54
-138	27.31	25.27	10:30	56
-140	28.42	26.38	10:44	57
-144	30.69	28.65	11:12	58
-145	31.26	29.22	11:19	58
-148	32.94	30.93	11:39	57
-150	34.03	32.04	11:53	57

```
PERCENTAGE OF SUN OBSCURED ALONG MAP TRACK
WEST END =103.4              EAST END =104.8
MAXIMUM OBSCURATION 104.9 % AT LONG = -144
===========================================
SUNRISE POSITION :ALTITUDE IS -0.8 DEGREES
NORTH : LONGITUDE  -86.5,  LATITUDE  14.6
SOUTH : LONGITUDE  -87.0,  LATITUDE  13.5
```

+ MARKS THE CITY OF PEI-CHING

ECLIPSE ON JD 2321763 DATE(Y/M/D) 1644/ 9/ 1
DELTA T = 55   SECONDS

LONGITUDE (-VE EAST)	.LATITUDE LIMITS. NORTH SOUTH	LOCAL TIME	MEAN .ALTITUDE
-88	25.12	23.63	9:22	49
-90	24.70	23.22	9:35	52
-92	24.23	22.76	9:48	55
-94	23.70	22.22	10:01	58
-95	23.40	21.93	10:08	59
-97	22.77	21.29	10:22	63
-99	22.06	20.57	10:36	66
-100	21.68	20.19	10:43	68
-102	20.85	19.35	10:58	71
-104	19.95	18.44	11:13	74
-105	19.47	17.95	11:21	76
-107	18.46	16.91	11:36	79
-109	17.36	15.80	11:53	81
-110	16.79	15.21	12:01	82
-112	15.60	14.00	12:17	82
-114	14.36	12.73	12:33	80
-115	13.73	12.09	12:42	78

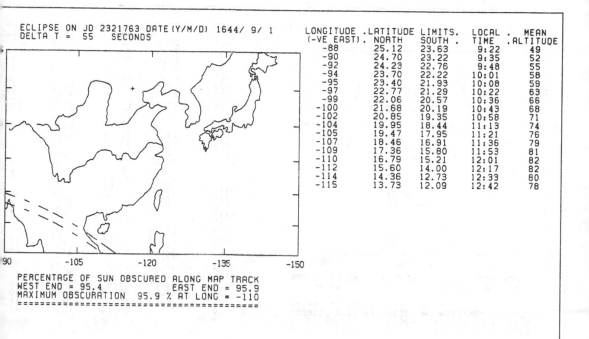

PERCENTAGE OF SUN OBSCURED ALONG MAP TRACK
WEST END = 95.4   EAST END = 95.9
MAXIMUM OBSCURATION 95.9 % AT LONG = -110
==

+ MARKS THE CITY OF PEI-CHING

ECLIPSE ON JD 2322619 DATE(Y/M/D) 1647/ 1/ 6
DELTA T = 51   SECONDS

LONGITUDE (-VE EAST)	.LATITUDE LIMITS. NORTH SOUTH	LOCAL TIME	MEAN .ALTITUDE
-145	36.91	32.69	7:11	1
-147	36.03	31.80	7:20	3
-148	35.58	31.36	7:24	4
-149	35.12	30.91	7:28	6
-150	34.66	30.46	7:32	6
-152	33.73	29.56	7:41	8
-153	33.26	29.11	7:45	9

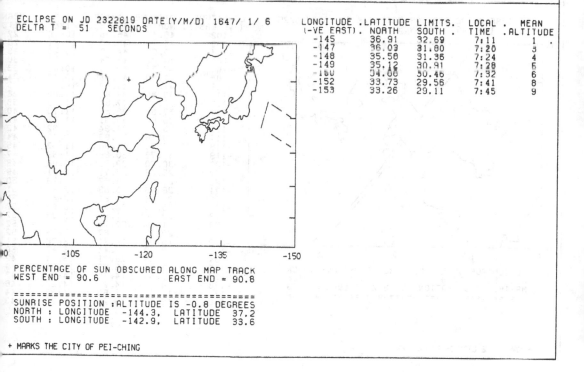

PERCENTAGE OF SUN OBSCURED ALONG MAP TRACK
WEST END = 90.6   EAST END = 90.8

==
SUNRISE POSITION :ALTITUDE IS -0.8 DEGREES
NORTH : LONGITUDE -144.3, LATITUDE 37.2
SOUTH : LONGITUDE -142.9, LATITUDE 33.6

+ MARKS THE CITY OF PEI-CHING

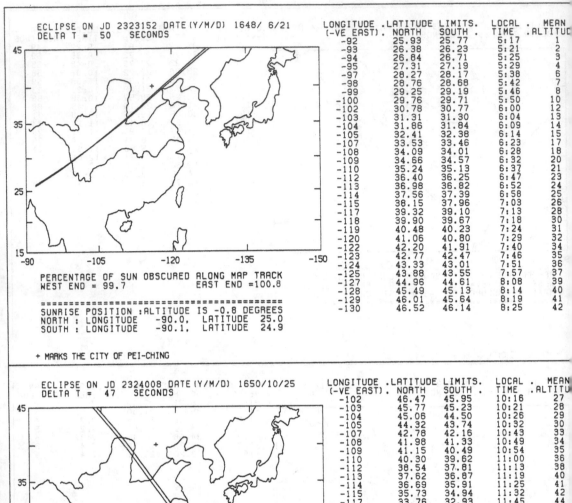

ECLIPSE ON JD 2323152 DATE(Y/M/D) 1648/ 6/21
DELTA T = 50 SECONDS

PERCENTAGE OF SUN OBSCURED ALONG MAP TRACK
WEST END = 99.7   EAST END =100.8

==
SUNRISE POSITION :ALTITUDE IS -0.8 DEGREES
NORTH : LONGITUDE   -90.0, LATITUDE   25.0
SOUTH : LONGITUDE   -90.1, LATITUDE   24.9

+ MARKS THE CITY OF PEI-CHING

LONGITUDE (-VE EAST)	LATITUDE NORTH	LIMITS. SOUTH	LOCAL TIME	MEAN ALTITUD
-92	25.93	25.77	5:17	1
-93	26.38	26.23	5:21	2
-94	26.84	26.71	5:25	3
-95	27.31	27.19	5:29	4
-97	28.27	28.17	5:38	6
-98	28.76	28.68	5:42	7
-99	29.25	29.19	5:46	8
-100	29.76	29.71	5:50	10
-102	30.78	30.77	6:00	12
-103	31.31	31.30	6:04	13
-104	31.86	31.84	6:09	14
-105	32.41	32.38	6:14	15
-107	33.53	33.46	6:23	17
-108	34.09	34.01	6:28	18
-109	34.66	34.57	6:32	20
-110	35.24	35.13	6:37	21
-112	36.40	36.25	6:47	23
-113	36.98	36.82	6:52	24
-114	37.56	37.39	6:58	25
-115	38.15	37.96	7:03	26
-117	39.32	39.10	7:13	28
-118	39.90	39.67	7:18	30
-119	40.48	40.23	7:24	31
-120	41.06	40.80	7:29	32
-122	42.20	41.91	7:40	34
-123	42.77	42.47	7:46	35
-124	43.33	43.01	7:51	36
-125	43.88	43.55	7:57	37
-127	44.96	44.61	8:08	39
-128	45.49	45.13	8:14	40
-129	46.01	45.64	8:19	41
-130	46.52	46.14	8:25	42

ECLIPSE ON JD 2324008 DATE(Y/M/D) 1650/10/25
DELTA T = 47 SECONDS

PERCENTAGE OF SUN OBSCURED ALONG MAP TRACK
WEST END =100.8   EAST END =101.4
MAXIMUM OBSCURATION 101.6 % AT LONG = -129
==

+ MARKS THE CITY OF PEI-CHING

LONGITUDE (-VE EAST)	LATITUDE NORTH	LIMITS. SOUTH	LOCAL TIME	MEAN ALTITU
-102	46.47	45.95	10:16	27
-103	45.77	45.23	10:21	28
-104	45.06	44.50	10:26	29
-105	44.32	43.74	10:32	30
-107	42.78	42.16	10:43	33
-108	41.98	41.33	10:49	34
-109	41.15	40.49	10:54	35
-110	40.30	39.62	11:00	36
-112	38.54	37.81	11:13	38
-113	37.62	36.87	11:19	40
-114	36.69	35.91	11:25	41
-115	35.73	34.94	11:32	42
-117	33.76	32.93	11:45	44
-118	32.75	31.90	11:52	45
-119	31.73	30.86	11:59	47
-120	30.70	29.81	12:06	48
-122	28.63	27.71	12:21	49
-123	27.59	26.67	12:29	50
-124	26.56	25.63	12:36	51
-125	25.54	24.61	12:44	51
-127	23.56	22.63	13:00	52
-128	22.60	21.68	13:08	52
-129	21.67	20.76	13:15	52
-130	20.77	19.87	13:23	52
-132	19.08	18.21	13:38	51
-133	18.29	17.44	13:46	50
-134	17.55	16.71	13:54	49
-135	16.84	16.03	14:01	49
-137	15.55	14.78	14:16	47
-138	14.96	14.21	14:23	46
-139	14.42	13.69	14:30	44
-140	13.91	13.20	14:37	43

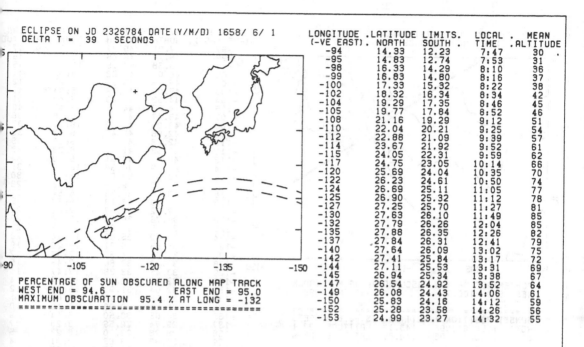

ECLIPSE ON JD 2326784 DATE(Y/M/D) 1658/ 6/ 1
DELTA T = 39   SECONDS

LONGITUDE (-VE EAST)	.LATITUDE LIMITS. NORTH	SOUTH	LOCAL TIME	MEAN .ALTITUDE
-94	14.33	12.23	7:47	30
-95	14.83	12.74	7:53	31
-98	16.33	14.29	8:10	36
-99	16.83	14.80	8:16	37
-100	17.33	15.32	8:22	38
-102	18.32	16.34	8:34	42
-104	19.29	17.35	8:46	45
-105	19.77	17.84	8:52	46
-108	21.16	19.29	9:12	51
-110	22.04	20.21	9:25	54
-112	22.88	21.09	9:39	57
-114	23.67	21.92	9:52	61
-115	24.05	22.31	9:59	62
-117	24.75	23.05	10:14	66
-120	25.69	24.04	10:35	70
-122	26.23	24.61	10:50	74
-124	26.69	25.11	11:05	77
-125	26.90	25.32	11:12	78
-127	27.25	25.70	11:27	81
-130	27.63	26.10	11:49	85
-132	27.79	26.26	12:04	85
-135	27.88	26.35	12:26	82
-137	27.84	26.31	12:41	79
-140	27.64	26.09	13:02	75
-142	27.41	25.84	13:17	72
-144	27.11	25.53	13:31	69
-145	26.94	25.34	13:38	67
-147	26.54	24.92	13:52	64
-149	26.08	24.43	14:06	61
-150	25.83	24.16	14:12	59
-152	25.28	23.58	14:26	56
-153	24.99	23.27	14:32	55

PERCENTAGE OF SUN OBSCURED ALONG MAP TRACK
WEST END = 94.6   EAST END = 95.0
MAXIMUM OBSCURATION 95.4 % AT LONG = -132
==

+ MHHKS THE CITY OF PEI-CHING

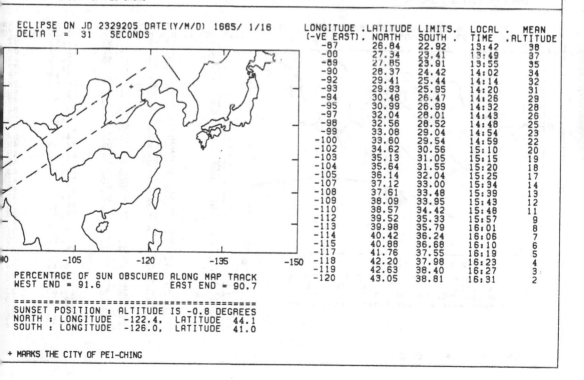

ECLIPSE ON JD 2329205 DATE(Y/M/D) 1665/ 1/16
DELTA T = 31   SECONDS

LONGITUDE (-VE EAST)	.LATITUDE LIMITS. NORTH	SOUTH	LOCAL TIME	MEAN .ALTITUDE
-87	26.84	22.92	13:42	38
-88	27.34	23.41	13:49	37
-89	27.85	23.91	13:55	35
-90	28.37	24.42	14:02	34
-92	29.41	25.44	14:14	32
-93	29.93	25.95	14:20	31
-94	30.46	26.47	14:26	29
-95	30.99	26.99	14:32	28
-97	32.04	28.01	14:43	26
-98	32.56	28.52	14:48	25
-99	33.08	29.04	14:54	23
-100	33.60	29.54	14:59	22
-102	34.62	30.56	15:10	20
-103	35.13	31.05	15:15	19
-104	35.64	31.55	15:20	18
-105	36.14	32.04	15:25	17
-107	37.12	33.00	15:34	14
-108	37.61	33.48	15:39	13
-109	38.09	33.95	15:43	12
-110	38.57	34.42	15:48	11
-112	39.52	35.33	15:57	9
-113	39.98	35.79	16:01	8
-114	40.42	36.24	16:06	7
-115	40.88	36.68	16:10	6
-117	41.76	37.55	16:19	5
-118	42.20	37.98	16:23	4
-119	42.63	38.40	16:27	3
-120	43.05	38.81	16:31	2

PERCENTAGE OF SUN OBSCURED ALONG MAP TRACK
WEST END = 91.6   EAST END = 90.7

==
SUNSET POSITION : ALTITUDE IS -0.8 DEGREES
NORTH : LONGITUDE -122.4, LATITUDE 44.1
SOUTH : LONGITUDE -126.0, LATITUDE 41.0

+ MARKS THE CITY OF PEI-CHING

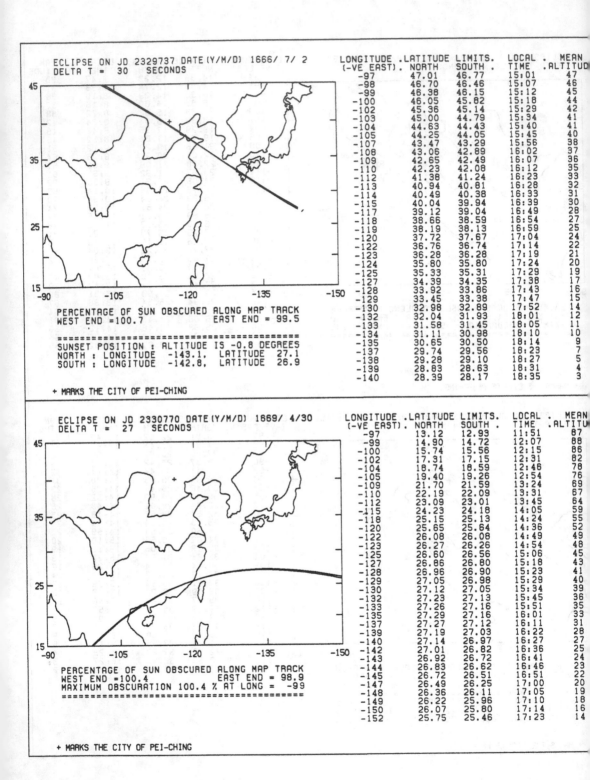

ECLIPSE ON JD 2329737 DATE(Y/M/D) 1666/ 7/ 2
DELTA T = 30 SECONDS

LONGITUDE (-VE EAST)	LATITUDE LIMITS. NORTH	SOUTH	LOCAL TIME	MEAN ALTITU
-97	47.01	46.77	15:01	47
-98	46.70	46.46	15:07	46
-99	46.38	46.15	15:12	45
-100	46.05	45.82	15:18	44
-102	45.36	45.14	15:29	42
-103	45.00	44.79	15:34	41
-104	44.63	44.43	15:40	41
-105	44.25	44.05	15:45	40
-107	43.47	43.29	15:56	38
-108	43.06	42.89	16:02	37
-109	42.65	42.49	16:07	36
-110	42.23	42.08	16:12	35
-112	41.38	41.24	16:23	33
-113	40.94	40.81	16:28	32
-114	40.49	40.38	16:33	31
-115	40.04	39.94	16:39	30
-117	39.12	39.04	16:49	28
-118	38.66	38.59	16:54	27
-119	38.19	38.13	16:59	25
-120	37.72	37.67	17:04	24
-122	36.76	36.74	17:14	22
-123	36.28	36.28	17:19	21
-124	35.80	35.80	17:24	20
-125	35.33	35.31	17:29	19
-127	34.39	34.35	17:38	17
-128	33.92	33.86	17:43	16
-129	33.45	33.38	17:47	15
-130	32.98	32.89	17:52	14
-132	32.04	31.93	18:01	12
-133	31.58	31.45	18:05	11
-134	31.11	30.98	18:10	10
-135	30.65	30.50	18:14	9
-137	29.74	29.56	18:23	7
-138	29.28	29.10	18:27	5
-139	28.83	28.63	18:31	4
-140	28.39	28.17	18:35	3

PERCENTAGE OF SUN OBSCURED ALONG MAP TRACK
WEST END =100.7   EAST END = 99.5

==
SUNSET POSITION : ALTITUDE IS -0.8 DEGREES
NORTH : LONGITUDE -143.1, LATITUDE 27.1
SOUTH : LONGITUDE -142.8, LATITUDE 26.9

+ MARKS THE CITY OF PEI-CHING

ECLIPSE ON JD 2330770 DATE(Y/M/D) 1669/ 4/30
DELTA T = 27 SECONDS

LONGITUDE (-VE EAST)	LATITUDE LIMITS. NORTH	SOUTH	LOCAL TIME	MEAN ALTITU
-97	13.12	12.93	11:51	87
-99	14.90	14.72	12:07	88
-100	15.74	15.56	12:15	86
-102	17.31	17.15	12:31	82
-104	18.74	18.59	12:46	78
-105	19.40	19.26	12:54	76
-109	21.70	21.59	13:24	69
-110	22.19	22.09	13:31	67
-112	23.09	23.01	13:45	64
-115	24.23	24.18	14:05	59
-118	25.15	25.13	14:24	55
-120	25.65	25.64	14:36	52
-122	26.08	26.08	14:49	49
-123	26.27	26.26	14:54	48
-125	26.60	26.56	15:06	45
-127	26.86	26.80	15:18	43
-128	26.96	26.90	15:23	41
-129	27.05	26.98	15:29	40
-130	27.12	27.05	15:34	39
-132	27.23	27.13	15:45	36
-133	27.26	27.16	15:51	35
-135	27.29	27.16	16:01	33
-137	27.27	27.12	16:11	31
-139	27.19	27.03	16:22	28
-140	27.14	26.97	16:27	27
-142	27.01	26.82	16:36	25
-143	26.92	26.72	16:41	24
-144	26.83	26.62	16:46	23
-145	26.72	26.51	16:51	22
-147	26.49	26.25	17:00	20
-148	26.36	26.11	17:05	19
-149	26.22	25.96	17:10	18
-150	26.07	25.80	17:14	16
-152	25.75	25.46	17:23	14

PERCENTAGE OF SUN OBSCURED ALONG MAP TRACK
WEST END =100.4   EAST END = 98.9
MAXIMUM OBSCURATION 100.4 % AT LONG = -99
==

+ MARKS THE CITY OF PEI-CHING

ECLIPSE ON JD 2337710 DATE(Y/M/D) 1688/ 4/30
DELTA T = 10.7 SECONDS

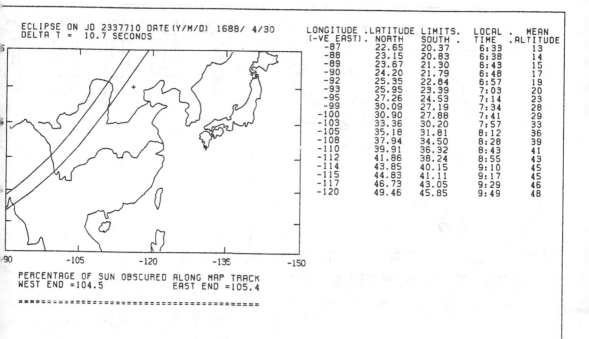

LONGITUDE (-VE EAST)	.LATITUDE NORTH	LIMITS. SOUTH	LOCAL TIME	. MEAN .ALTITUDE
-87	22.65	20.37	6:33	13
-88	23.15	20.83	6:38	14
-89	23.67	21.30	6:43	15
-90	24.20	21.79	6:48	17
-92	25.35	22.84	6:57	19
-93	25.95	23.39	7:03	20
-95	27.26	24.53	7:14	23
-99	30.09	27.19	7:34	28
-100	30.90	27.88	7:41	29
-103	33.36	30.20	7:57	33
-105	35.18	31.81	8:12	36
-108	37.94	34.50	8:28	39
-110	39.91	36.32	8:43	41
-112	41.86	38.24	8:55	43
-114	43.85	40.15	9:10	45
-115	44.83	41.11	9:17	45
-117	46.73	43.05	9:29	46
-120	49.46	45.85	9:49	48

PERCENTAGE OF SUN OBSCURED ALONG MAP TRACK
WEST END =104.5   EAST END =105.4

==

+ MARKS THE CITY OF PEI-CHING

ECLIPSE ON JD 2338744 DATE(Y/M/D) 1691/ 2/28
DELTA T = 9.7 SECONDS

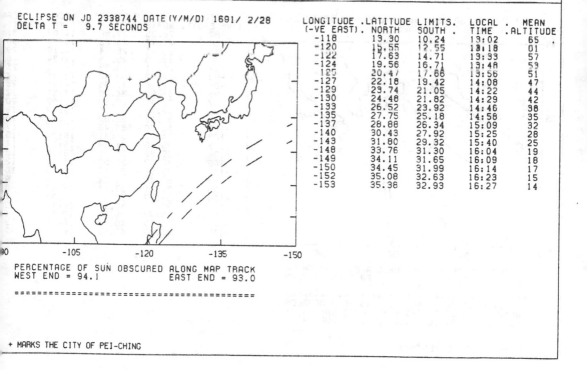

LONGITUDE (-VE EAST)	.LATITUDE NORTH	LIMITS. SOUTH	LOCAL TIME	. MEAN .ALTITUDE
-118	13.30	10.24	13:02	65
-120	15.55	12.55	13:18	61
-122	17.63	14.71	13:33	57
-124	19.56	16.71	13:48	53
-125	20.47	17.66	13:56	51
-127	22.18	19.42	14:08	47
-129	23.74	21.05	14:22	44
-130	24.48	21.82	14:29	42
-133	26.52	23.92	14:46	38
-135	27.75	25.18	14:58	35
-137	28.88	26.34	15:09	32
-140	30.43	27.92	15:25	28
-143	31.80	29.32	15:40	25
-148	33.76	31.30	16:04	19
-149	34.11	31.65	16:09	18
-150	34.45	31.99	16:14	17
-152	35.08	32.63	16:23	15
-153	35.38	32.93	16:27	14

PERCENTAGE OF SUN OBSCURED ALONG MAP TRACK
WEST END = 94.1   EAST END = 93.0

==

+ MARKS THE CITY OF PEI-CHING

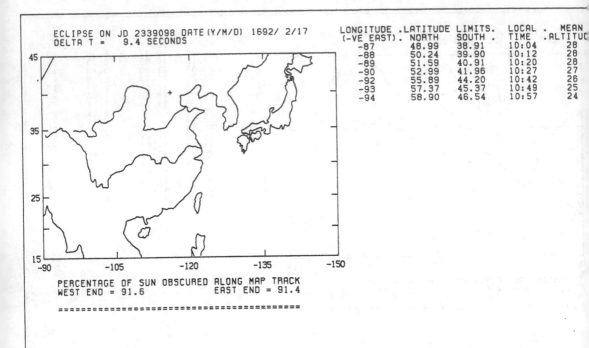

ECLIPSE ON JD 2339098 DATE (Y/M/D) 1692/ 2/17
DELTA T =   9.4 SECONDS

LONGITUDE	.LATITUDE	LIMITS.	LOCAL	. MEAN
(-VE EAST)	. NORTH	SOUTH .	TIME	.ALTITU
-87	48.99	38.91	10:04	28
-88	50.24	39.90	10:12	28
-89	51.59	40.91	10:20	28
-90	52.99	41.96	10:27	27
-92	55.89	44.20	10:42	26
-93	57.37	45.37	10:49	25
-94	58.90	46.54	10:57	24

PERCENTAGE OF SUN OBSCURED ALONG MAP TRACK
WEST END = 91.6   EAST END = 91.4

+ MARKS THE CITY OF PEI-CHING

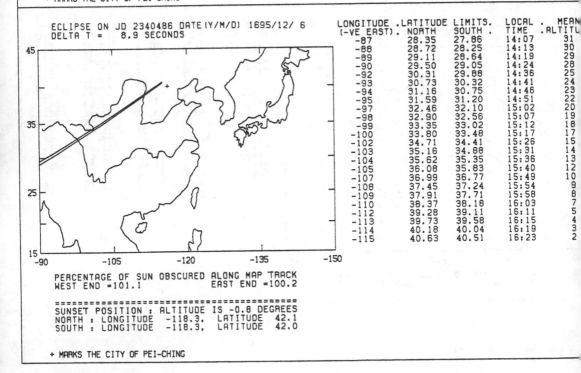

ECLIPSE ON JD 2340486 DATE (Y/M/D) 1695/12/ 6
DELTA T =   8.9 SECONDS

LONGITUDE	.LATITUDE	LIMITS.	LOCAL	. MEAN
(-VE EAST)	. NORTH	SOUTH .	TIME	.ALTITU
-87	28.35	27.86	14:07	31
-88	28.72	28.25	14:13	30
-89	29.11	28.64	14:19	29
-90	29.50	29.05	14:24	28
-92	30.31	29.88	14:36	25
-93	30.73	30.32	14:41	24
-94	31.16	30.75	14:46	23
-95	31.59	31.20	14:51	22
-97	32.46	32.10	15:02	20
-98	32.90	32.56	15:07	19
-99	33.35	33.02	15:12	18
-100	33.80	33.48	15:17	17
-102	34.71	34.41	15:26	15
-103	35.16	34.88	15:31	14
-104	35.62	35.35	15:36	13
-105	36.08	35.83	15:40	12
-107	36.99	36.77	15:49	10
-108	37.45	37.24	15:54	9
-109	37.91	37.71	15:58	8
-110	38.37	38.18	16:03	7
-112	39.28	39.11	16:11	5
-113	39.73	39.58	16:15	4
-114	40.18	40.04	16:19	3
-115	40.63	40.51	16:23	2

PERCENTAGE OF SUN OBSCURED ALONG MAP TRACK
WEST END =101.1   EAST END =100.2

SUNSET POSITION : ALTITUDE IS -0.8 DEGREES
NORTH : LONGITUDE -118.3, LATITUDE 42.1
SOUTH : LONGITUDE -118.3, LATITUDE 42.0

+ MARKS THE CITY OF PEI-CHING

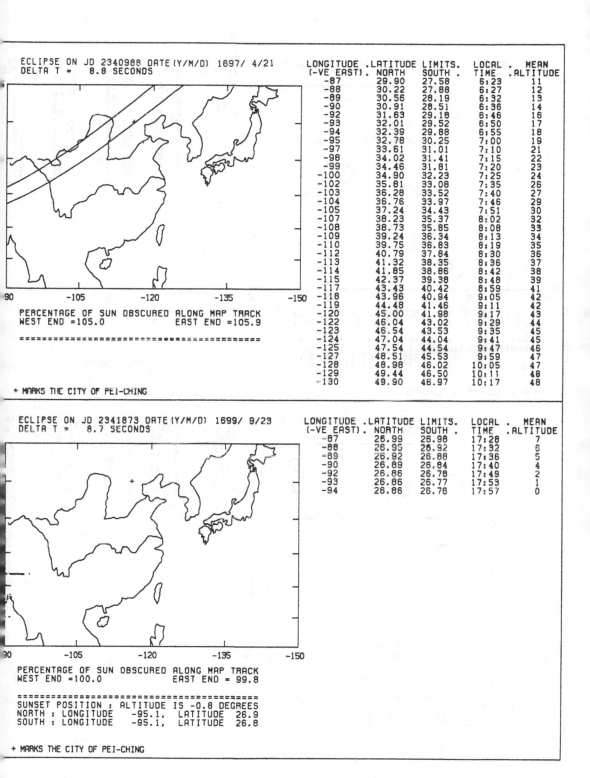

ECLIPSE ON JD 2340988 DATE(Y/M/D) 1697/ 4/21
DELTA T = 8.8 SECONDS

LONGITUDE (-VE EAST)	.LATITUDE LIMITS. NORTH	SOUTH .	LOCAL . TIME	MEAN .ALTITUDE
-87	29.90	27.58	6:23	11
-88	30.22	27.88	6:27	12
-89	30.56	28.19	6:32	13
-90	30.91	28.51	6:36	14
-92	31.63	29.18	6:46	16
-93	32.01	29.52	6:50	17
-94	32.39	29.88	6:55	18
-95	32.78	30.25	7:00	19
-97	33.61	31.01	7:10	21
-98	34.02	31.41	7:15	22
-99	34.46	31.81	7:20	23
-100	34.90	32.23	7:25	24
-102	35.81	33.08	7:35	26
-103	36.28	33.52	7:40	27
-104	36.76	33.97	7:46	29
-105	37.24	34.43	7:51	30
-107	38.23	35.37	8:02	32
-108	38.73	35.85	8:08	33
-109	39.24	36.34	8:13	34
-110	39.75	36.83	8:19	35
-112	40.79	37.84	8:30	36
-113	41.32	38.35	8:36	37
-114	41.85	38.86	8:42	38
-115	42.37	39.38	8:48	39
-117	43.43	40.42	8:59	41
-118	43.96	40.94	9:05	42
-119	44.48	41.46	9:11	42
-120	45.00	41.98	9:17	43
-122	46.04	43.02	9:29	44
-123	46.54	43.53	9:35	45
-124	47.04	44.04	9:41	45
-125	47.54	44.54	9:47	46
-127	48.51	45.53	9:59	47
-128	48.98	46.02	10:05	47
-129	49.44	46.50	10:11	48
-130	49.90	46.97	10:17	48

PERCENTAGE OF SUN OBSCURED ALONG MAP TRACK
WEST END =105.0   EAST END =105.9

==

+ MARKS THE CITY OF PEI-CHING

ECLIPSE ON JD 2341873 DATE(Y/M/D) 1699/ 9/23
DELTA T = 8.7 SECONDS

LONGITUDE (-VE EAST)	.LATITUDE LIMITS. NORTH	SOUTH .	LOCAL TIME	MEAN .ALTITUDE
-87	26.99	26.96	17:28	7
-88	26.95	26.92	17:32	6
-89	26.92	26.88	17:36	5
-90	26.89	26.84	17:40	4
-92	26.86	26.78	17:49	2
-93	26.86	26.77	17:53	1
-94	26.86	26.76	17:57	0

PERCENTAGE OF SUN OBSCURED ALONG MAP TRACK
WEST END =100.0   EAST END = 99.8

==
SUNSET POSITION : ALTITUDE IS -0.8 DEGREES
NORTH : LONGITUDE   -95.1, LATITUDE   26.9
SOUTH : LONGITUDE   -95.1, LATITUDE   26.8

+ MARKS THE CITY OF PEI-CHING

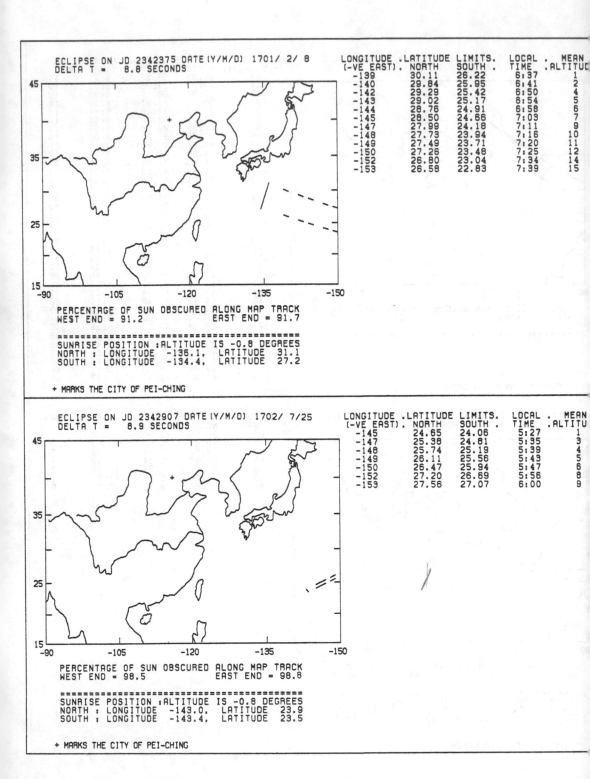

ECLIPSE ON JD 2342375 DATE(Y/M/D) 1701/ 2/ 8
DELTA T =   8.8 SECONDS

LONGITUDE (-VE EAST).	LATITUDE NORTH	LIMITS. SOUTH .	LOCAL TIME .	MEAN ALTITUDE
-139	30.11	26.22	6:37	1
-140	29.84	25.95	6:41	2
-142	29.29	25.42	6:50	4
-143	29.02	25.17	6:54	5
-144	28.76	24.91	6:58	6
-145	28.50	24.66	7:03	7
-147	27.99	24.18	7:11	9
-148	27.73	23.94	7:16	10
-149	27.49	23.71	7:20	11
-150	27.26	23.48	7:25	12
-152	26.80	23.04	7:34	14
-153	26.58	22.83	7:39	15

PERCENTAGE OF SUN OBSCURED ALONG MAP TRACK
WEST END = 91.2   EAST END = 91.7

==
SUNRISE POSITION :ALTITUDE IS -0.8 DEGREES
NORTH : LONGITUDE -136.1, LATITUDE   31.1
SOUTH : LONGITUDE -134.4, LATITUDE   27.2

+ MARKS THE CITY OF PEI-CHING

ECLIPSE ON JD 2342907 DATE(Y/M/D) 1702/ 7/25
DELTA T =   8.9 SECONDS

LONGITUDE (-VE EAST).	LATITUDE NORTH	LIMITS. SOUTH .	LOCAL TIME .	MEAN ALTITUDE
-145	24.65	24.06	5:27	1
-147	25.38	24.81	5:35	3
-148	25.74	25.19	5:39	4
-149	26.11	25.56	5:43	5
-150	26.47	25.94	5:47	6
-152	27.20	26.69	5:56	8
-153	27.56	27.07	6:00	9

PERCENTAGE OF SUN OBSCURED ALONG MAP TRACK
WEST END = 98.5   EAST END = 98.8

==
SUNRISE POSITION :ALTITUDE IS -0.8 DEGREES
NORTH : LONGITUDE -143.0, LATITUDE   23.9
SOUTH : LONGITUDE -143.4, LATITUDE   23.5

+ MARKS THE CITY OF PEI-CHING

ECLIPSE ON JD 2343764 DATE(Y/M/D) 1704/11/27
DELTA T =   9.0 SECONDS

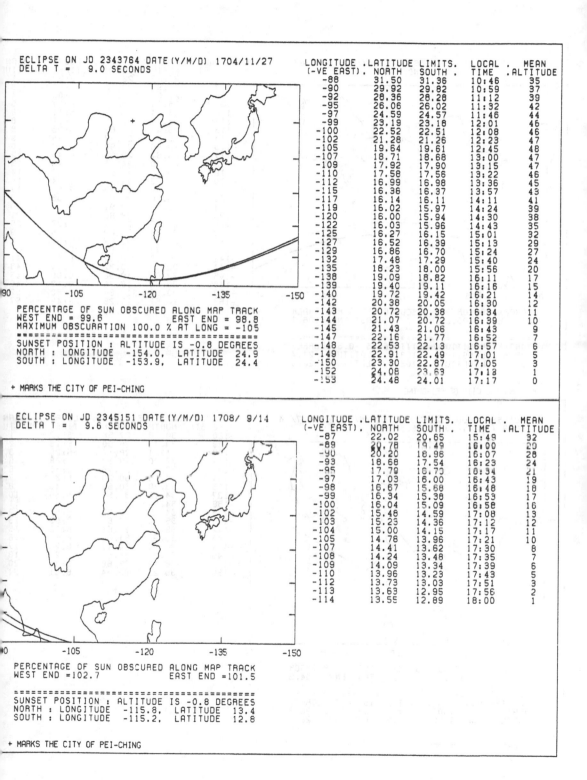

LONGITUDE (-VE EAST)	LATITUDE LIMITS. NORTH	SOUTH	LOCAL TIME	MEAN ALTITUDE
-88	31.50	31.36	10:46	35
-90	29.92	29.82	10:59	37
-92	28.36	28.28	11:12	39
-95	26.06	26.02	11:32	42
-97	24.59	24.57	11:46	44
-99	23.19	23.18	12:01	46
-100	22.52	22.51	12:08	46
-102	21.28	21.26	12:23	47
-105	19.64	19.61	12:45	48
-107	18.71	18.68	13:00	47
-109	17.92	17.90	13:15	47
-110	17.58	17.56	13:22	46
-112	16.99	16.98	13:36	45
-115	16.36	16.37	13:57	43
-117	16.14	16.11	14:11	41
-119	16.02	15.97	14:24	39
-120	16.00	15.94	14:30	38
-122	16.03	15.96	14:43	35
-125	16.27	16.15	15:01	32
-127	16.52	16.39	15:13	29
-129	16.86	16.70	15:24	27
-132	17.48	17.29	15:40	24
-135	18.23	18.00	15:56	20
-138	19.09	18.82	16:11	17
-139	19.40	19.11	16:16	15
-140	19.72	19.42	16:21	14
-142	20.38	20.05	16:30	12
-143	20.72	20.38	16:34	11
-144	21.07	20.72	16:39	10
-145	21.43	21.06	16:43	9
-147	22.16	21.77	16:52	7
-148	22.53	22.13	16:57	6
-149	22.91	22.49	17:01	5
-150	23.30	22.87	17:05	3
-152	24.08	23.63	17:13	1
-153	24.48	24.01	17:17	0

PERCENTAGE OF SUN OBSCURED ALONG MAP TRACK
WEST END = 99.6   EAST END = 98.8
MAXIMUM OBSCURATION 100.0 % AT LONG = -105
======================================
SUNSET POSITION : ALTITUDE IS -0.8 DEGREES
NORTH : LONGITUDE -154.0, LATITUDE 24.9
SOUTH : LONGITUDE -153.9, LATITUDE 24.4

+ MARKS THE CITY OF PEI-CHING

ECLIPSE ON JD 2345151 DATE(Y/M/D) 1708/ 9/14
DELTA T =   9.6 SECONDS

LONGITUDE (-VE EAST)	LATITUDE LIMITS. NORTH	SOUTH	LOCAL TIME	MEAN ALTITUDE
-87	22.02	20.65	15:49	32
-89	20.78	19.49	16:00	29
-90	20.20	18.96	16:07	28
-93	18.68	17.54	16:23	24
-95	17.79	16.73	16:34	21
-97	17.03	16.00	16:43	19
-98	16.67	15.68	16:48	18
-99	16.34	15.38	16:53	17
-100	16.04	15.09	16:58	16
-102	15.48	14.59	17:08	13
-103	15.23	14.36	17:12	12
-104	15.00	14.15	17:17	11
-105	14.78	13.96	17:21	10
-107	14.41	13.62	17:30	8
-108	14.24	13.48	17:35	7
-109	14.09	13.34	17:39	6
-110	13.96	13.23	17:43	5
-112	13.73	13.03	17:51	3
-113	13.63	12.95	17:56	2
-114	13.55	12.89	18:00	1

PERCENTAGE OF SUN OBSCURED ALONG MAP TRACK
WEST END =102.7   EAST END =101.5

======================================
SUNSET POSITION : ALTITUDE IS -0.8 DEGREES
NORTH : LONGITUDE -115.8, LATITUDE 13.4
SOUTH : LONGITUDE -115.2, LATITUDE 12.8

+ MARKS THE CITY OF PEI-CHING

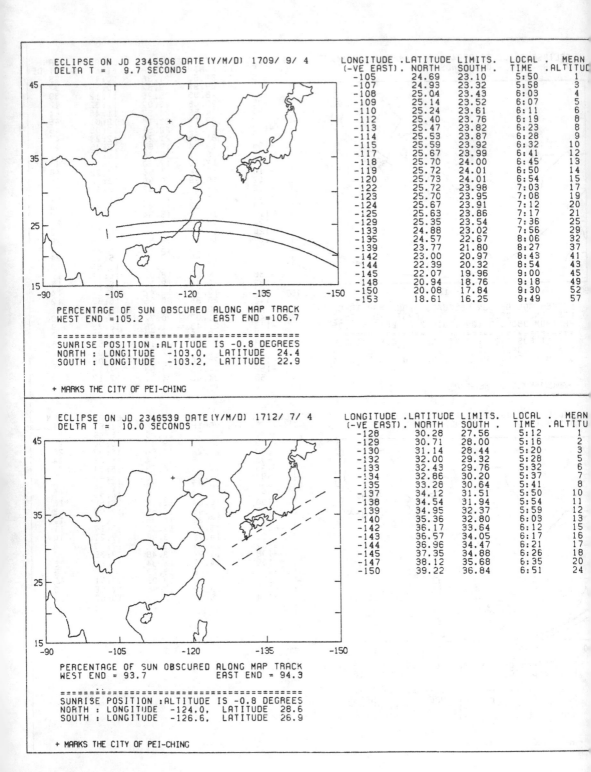

ECLIPSE ON JD 2345506 DATE(Y/M/D) 1709/ 9/ 4
DELTA T =   9.7 SECONDS

LONGITUDE (-VE EAST)	LATITUDE LIMITS. NORTH	SOUTH	LOCAL TIME	MEAN ALTITUD
-105	24.69	23.10	5:50	1
-107	24.93	23.32	5:58	3
-108	25.04	23.43	6:03	4
-109	25.14	23.52	6:07	5
-110	25.24	23.61	6:11	6
-112	25.40	23.76	6:19	8
-113	25.47	23.82	6:23	8
-114	25.53	23.87	6:28	9
-115	25.59	23.92	6:32	10
-117	25.67	23.99	6:41	12
-118	25.70	24.00	6:45	13
-119	25.72	24.01	6:50	14
-120	25.73	24.01	6:54	15
-122	25.72	23.98	7:03	17
-123	25.70	23.95	7:08	19
-124	25.67	23.91	7:12	20
-125	25.63	23.86	7:17	21
-129	25.35	23.54	7:36	25
-133	24.88	23.02	7:56	29
-135	24.57	22.67	8:06	32
-139	23.77	21.80	8:27	37
-142	23.00	20.97	8:43	41
-144	22.39	20.32	8:54	43
-145	22.07	19.96	9:00	45
-148	20.94	18.76	9:18	49
-150	20.08	17.84	9:30	52
-153	18.61	16.25	9:49	57

PERCENTAGE OF SUN OBSCURED ALONG MAP TRACK
WEST END =105.2   EAST END =106.7

===
SUNRISE POSITION :ALTITUDE IS -0.8 DEGREES
NORTH : LONGITUDE -103.0, LATITUDE 24.4
SOUTH : LONGITUDE -103.2, LATITUDE 22.9

+ MARKS THE CITY OF PEI-CHING

ECLIPSE ON JD 2346539 DATE(Y/M/D) 1712/ 7/ 4
DELTA T = 10.0 SECONDS

LONGITUDE (-VE EAST)	LATITUDE LIMITS. NORTH	SOUTH	LOCAL TIME	MEAN ALTITU
-128	30.28	27.56	5:12	1
-129	30.71	28.00	5:16	2
-130	31.14	28.44	5:20	3
-132	32.00	29.32	5:28	5
-133	32.43	29.76	5:32	6
-134	32.86	30.20	5:37	7
-135	33.28	30.64	5:41	8
-137	34.12	31.51	5:50	10
-138	34.54	31.94	5:54	11
-139	34.95	32.37	5:59	12
-140	35.36	32.80	6:03	13
-142	36.17	33.64	6:12	15
-143	36.57	34.05	6:17	16
-144	36.96	34.47	6:21	17
-145	37.35	34.88	6:26	18
-147	38.12	35.68	6:35	20
-150	39.22	36.84	6:51	24

PERCENTAGE OF SUN OBSCURED ALONG MAP TRACK
WEST END = 93.7   EAST END = 94.3

===
SUNRISE POSITION :ALTITUDE IS -0.8 DEGREES
NORTH : LONGITUDE -124.0, LATITUDE 28.6
SOUTH : LONGITUDE -126.6, LATITUDE 26.9

+ MARKS THE CITY OF PEI-CHING

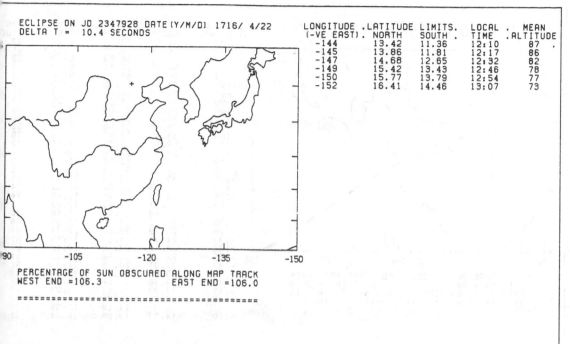

ECLIPSE ON JD 2347928 DATE(Y/M/D) 1716/ 4/22
DELTA T = 10.4 SECONDS

LONGITUDE	.LATITUDE	LIMITS.	LOCAL	. MEAN
(-VE EAST).	NORTH	SOUTH .	TIME	.ALTITUDE
-144	13.42	11.36	12:10	87
-145	13.86	11.81	12:17	86
-147	14.68	12.65	12:32	82
-149	15.42	13.43	12:46	78
-150	15.77	13.79	12:54	77
-152	16.41	14.46	13:07	73

PERCENTAGE OF SUN OBSCURED ALONG MAP TRACK
WEST END =106.3   EAST END =106.0

===

+ MARKS THE CITY OF PEI-CHING

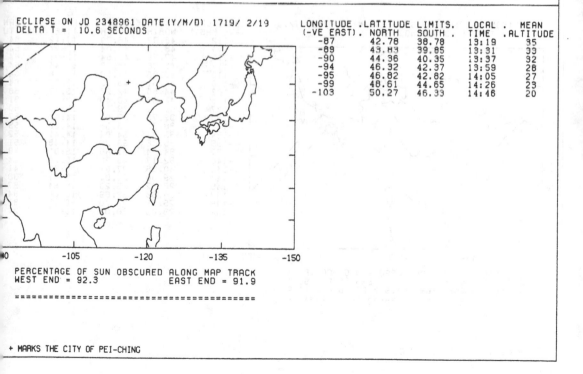

ECLIPSE ON JD 2348961 DATE(Y/M/D) 1719/ 2/19
DELTA T = 10.6 SECONDS

LONGITUDE	.LATITUDE	LIMITS.	LOCAL	. MEAN
(-VE EAST).	NORTH	SOUTH .	TIME	.ALTITUDE
-87	42.78	38.78	13:19	35
-89	43.83	39.85	13:31	33
-90	44.36	40.35	13:37	32
-94	46.32	42.37	13:59	28
-95	46.82	42.82	14:05	27
-99	48.61	44.65	14:26	23
-103	50.27	46.33	14:46	20

PERCENTAGE OF SUN OBSCURED ALONG MAP TRACK
WEST END = 92.3   EAST END = 91.9

===

+ MARKS THE CITY OF PEI-CHING

ECLIPSE ON JD 2349493 DATE (Y/M/D) 1720/ 8/ 4
DELTA T = 10.7 SECONDS

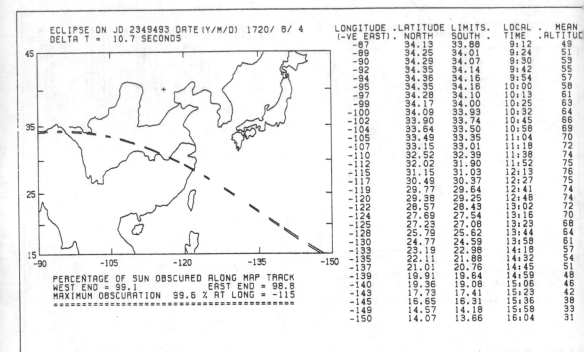

LONGITUDE (-VE EAST)	LATITUDE NORTH	LIMITS. SOUTH	LOCAL TIME	MEAN ALTITUDE
-87	34.13	33.88	9:12	49
-89	34.25	34.01	9:24	51
-90	34.29	34.07	9:30	53
-92	34.35	34.14	9:42	55
-94	34.36	34.16	9:54	57
-95	34.35	34.16	10:00	58
-97	34.28	34.10	10:13	61
-99	34.17	34.00	10:25	63
-100	34.09	33.93	10:32	64
-102	33.90	33.74	10:45	66
-104	33.64	33.50	10:58	69
-105	33.49	33.35	11:04	70
-107	33.15	33.01	11:18	72
-110	32.52	32.39	11:38	74
-112	32.02	31.90	11:52	75
-115	31.15	31.03	12:13	76
-117	30.49	30.37	12:27	75
-119	29.77	29.64	12:41	74
-120	29.38	29.25	12:48	74
-122	28.57	28.43	13:02	72
-124	27.69	27.54	13:16	70
-125	27.23	27.08	13:23	68
-128	25.79	25.62	13:44	64
-130	24.77	24.59	13:58	61
-133	23.19	22.98	14:18	57
-135	22.11	21.88	14:32	54
-137	21.01	20.76	14:45	51
-139	19.91	19.64	14:59	48
-140	19.36	19.08	15:06	46
-143	17.73	17.41	15:23	42
-145	16.65	16.31	15:36	38
-149	14.57	14.18	15:58	33
-150	14.07	13.66	16:04	31

PERCENTAGE OF SUN OBSCURED ALONG MAP TRACK
WEST END = 99.1   EAST END = 98.8
MAXIMUM OBSCURATION 99.6 % AT LONG = -115
==

+ MARKS THE CITY OF PEI-CHING

ECLIPSE ON JD 2353125 DATE (Y/M/D) 1730/ 7/15
DELTA T = 11.3 SECONDS

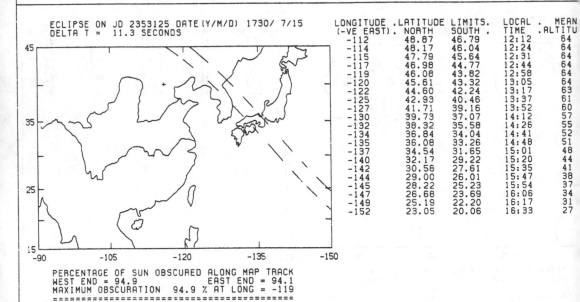

LONGITUDE (-VE EAST)	LATITUDE NORTH	LIMITS. SOUTH	LOCAL TIME	MEAN ALTITUDE
-112	48.87	46.79	12:12	64
-114	48.17	46.04	12:24	64
-115	47.79	45.64	12:31	64
-117	46.98	44.77	12:44	64
-119	46.08	43.82	12:58	64
-120	45.61	43.32	13:05	64
-122	44.60	42.24	13:17	63
-125	42.93	40.46	13:37	61
-127	41.71	39.16	13:52	60
-130	39.73	37.07	14:12	57
-132	38.32	35.58	14:26	55
-134	36.84	34.04	14:41	52
-135	36.08	33.26	14:48	51
-137	34.54	31.65	15:01	48
-140	32.17	29.22	15:20	44
-142	30.58	27.61	15:35	41
-144	29.00	26.01	15:47	38
-145	28.22	25.23	15:54	37
-147	26.68	23.69	16:06	34
-149	25.19	22.20	16:17	31
-152	23.05	20.06	16:33	27

PERCENTAGE OF SUN OBSCURED ALONG MAP TRACK
WEST END = 94.9   EAST END = 94.1
MAXIMUM OBSCURATION 94.9 % AT LONG = -119
==

+ MARKS THE CITY OF PEI-CHING

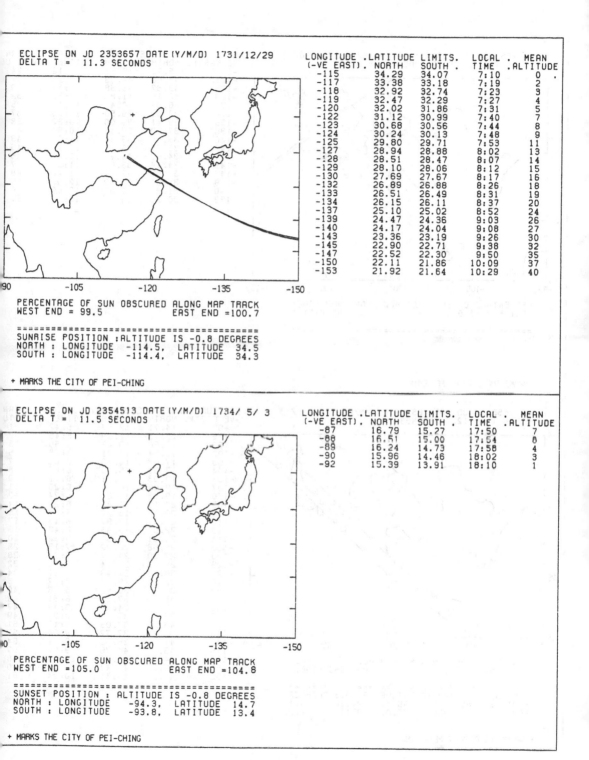

ECLIPSE ON JD 2353657 DATE (Y/M/D) 1731/12/29
DELTA T = 11.3 SECONDS

LONGITUDE (-VE EAST)	LATITUDE NORTH	LIMITS. SOUTH	LOCAL TIME	MEAN ALTITUDE
-115	34.29	34.07	7:10	0
-117	33.38	33.18	7:19	2
-118	32.92	32.74	7:23	3
-119	32.47	32.29	7:27	4
-120	32.02	31.86	7:31	5
-122	31.12	30.99	7:40	7
-123	30.68	30.56	7:44	8
-124	30.24	30.13	7:48	9
-125	29.80	29.71	7:53	11
-127	28.94	28.88	8:02	13
-128	28.51	28.47	8:07	14
-129	28.10	28.06	8:12	15
-130	27.69	27.67	8:17	16
-132	26.89	26.88	8:26	18
-133	26.51	26.49	8:31	19
-134	26.15	26.11	8:37	20
-137	25.10	25.02	8:52	24
-139	24.47	24.36	9:03	26
-140	24.17	24.04	9:08	27
-143	23.36	23.19	9:26	30
-145	22.90	22.71	9:38	32
-147	22.52	22.30	9:50	35
-150	22.11	21.86	10:09	37
-153	21.92	21.64	10:29	40

PERCENTAGE OF SUN OBSCURED ALONG MAP TRACK
WEST END = 99.5   EAST END =100.7

==
SUNRISE POSITION :ALTITUDE IS -0.8 DEGREES
NORTH : LONGITUDE -114.5, LATITUDE 34.5
SOUTH : LONGITUDE -114.4, LATITUDE 34.3

+ MARKS THE CITY OF PEI-CHING

ECLIPSE ON JD 2354513 DATE (Y/M/D) 1734/ 5/ 3
DELTA T = 11.5 SECONDS

LONGITUDE (-VE EAST)	LATITUDE NORTH	LIMITS. SOUTH	LOCAL TIME	MEAN ALTITUDE
-87	16.79	15.27	17:50	7
-88	16.51	15.00	17:54	6
-89	16.24	14.73	17:58	4
-90	15.96	14.46	18:02	3
-92	15.39	13.91	18:10	1

PERCENTAGE OF SUN OBSCURED ALONG MAP TRACK
WEST END =105.0   EAST END =104.8

==
SUNSET POSITION : ALTITUDE IS -0.8 DEGREES
NORTH : LONGITUDE   -94.3, LATITUDE 14.7
SOUTH : LONGITUDE   -93.8, LATITUDE 13.4

+ MARKS THE CITY OF PEI-CHING

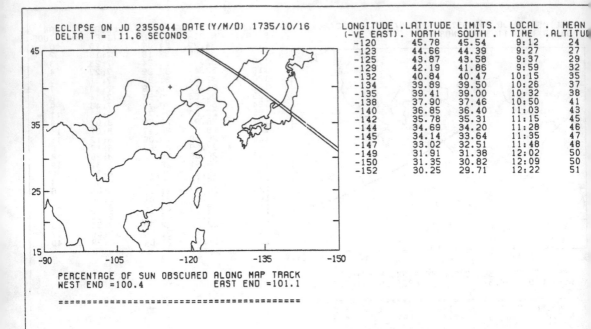

ECLIPSE ON JD 2355044 DATE(Y/M/D) 1735/10/16
DELTA T = 11.6 SECONDS

LONGITUDE (-VE EAST)	LATITUDE NORTH	LIMITS SOUTH	LOCAL TIME	MEAN ALTITUDE
-120	45.78	45.54	9:12	24
-123	44.66	44.39	9:27	27
-125	43.87	43.58	9:37	29
-129	42.19	41.86	9:59	32
-132	40.84	40.47	10:15	35
-134	39.89	39.50	10:26	37
-135	39.41	39.00	10:32	38
-138	37.90	37.46	10:50	41
-140	36.85	36.40	11:03	43
-142	35.78	35.31	11:15	45
-144	34.69	34.20	11:28	46
-145	34.14	33.64	11:35	47
-147	33.02	32.51	11:48	48
-149	31.91	31.38	12:02	50
-150	31.35	30.82	12:09	50
-152	30.25	29.71	12:22	51

PERCENTAGE OF SUN OBSCURED ALONG MAP TRACK
WEST END =100.4   EAST END =101.1

===

+ MARKS THE CITY OF PEI-CHING

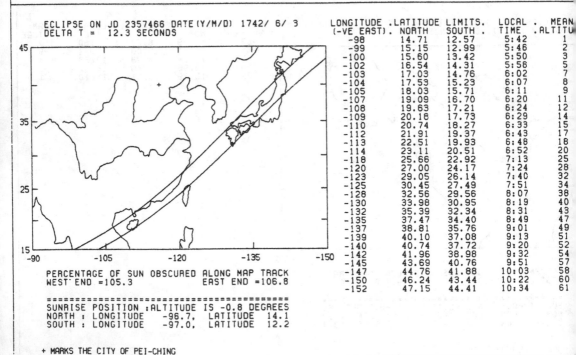

ECLIPSE ON JD 2357466 DATE(Y/M/D) 1742/ 6/ 3
DELTA T = 12.3 SECONDS

LONGITUDE (-VE EAST)	LATITUDE NORTH	LIMITS SOUTH	LOCAL TIME	MEAN ALTITUDE
-98	14.71	12.57	5:42	1
-99	15.15	12.99	5:46	2
-100	15.60	13.42	5:50	3
-102	16.54	14.31	5:58	5
-103	17.03	14.76	6:02	7
-104	17.53	15.23	6:07	8
-105	18.03	15.71	6:11	9
-107	19.09	16.70	6:20	11
-108	19.63	17.21	6:24	12
-109	20.18	17.73	6:29	14
-110	20.74	18.27	6:33	15
-112	21.91	19.37	6:43	17
-113	22.51	19.93	6:48	18
-114	23.11	20.51	6:52	20
-118	25.66	22.92	7:13	25
-120	27.00	24.17	7:24	28
-123	29.05	26.14	7:40	32
-125	30.45	27.49	7:51	34
-128	32.56	29.56	8:07	38
-130	33.98	30.95	8:19	40
-132	35.39	32.34	8:31	43
-135	37.47	34.40	8:49	47
-137	38.81	35.76	9:01	49
-139	40.10	37.08	9:13	51
-140	40.74	37.72	9:20	52
-142	41.96	38.98	9:32	54
-145	43.69	40.76	9:51	57
-147	44.76	41.88	10:03	58
-150	46.24	43.44	10:22	60
-152	47.15	44.41	10:34	61

PERCENTAGE OF SUN OBSCURED ALONG MAP TRACK
WEST END =105.3   EAST END =106.8

===
SUNRISE POSITION :ALTITUDE IS -0.8 DEGREES
NORTH : LONGITUDE -96.7, LATITUDE 14.1
SOUTH : LONGITUDE -97.0, LATITUDE 12.2

+ MARKS THE CITY OF PEI-CHING

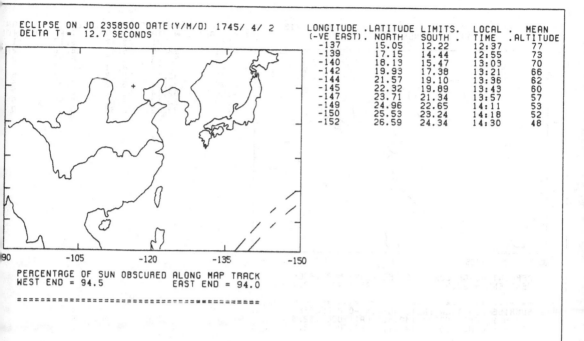

ECLIPSE ON JD 2358500 DATE(Y/M/D) 1745/ 4/ 2
DELTA T = 12.7 SECONDS

LONGITUDE (-VE EAST)	LATITUDE NORTH	LIMITS. SOUTH	LOCAL TIME	MEAN ALTITUDE
-137	15.05	12.22	12:37	77
-139	17.15	14.44	12:55	73
-140	18.13	15.47	13:03	70
-142	19.93	17.38	13:21	66
-144	21.57	19.10	13:36	62
-145	22.32	19.89	13:43	60
-147	23.71	21.34	13:57	57
-149	24.96	22.65	14:11	53
-150	25.53	23.24	14:18	52
-152	26.59	24.34	14:30	48

PERCENTAGE OF SUN OBSCURED ALONG MAP TRACK
WEST END = 94.5   EAST END = 94.0

===

+ MARKS THE CITY OF PEI-CHING

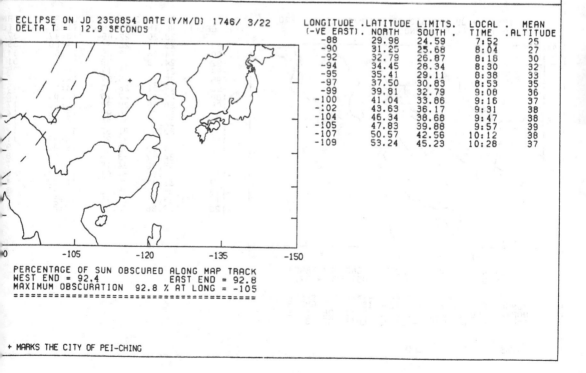

ECLIPSE ON JD 2358854 DATE(Y/M/D) 1746/ 3/22
DELTA T = 12.9 SECONDS

LONGITUDE (-VE EAST)	LATITUDE NORTH	LIMITS. SOUTH	LOCAL TIME	MEAN ALTITUDE
-88	29.98	24.59	7:52	25
-90	31.25	25.68	8:04	27
-92	32.79	26.87	8:18	30
-94	34.45	28.34	8:30	32
-95	35.41	29.11	8:38	33
-97	37.50	30.83	8:53	35
-99	39.81	32.79	9:08	36
-100	41.04	33.86	9:16	37
-:02	43.63	36.17	9:31	38
-104	46.34	38.68	9:47	38
-105	47.83	39.88	9:57	39
-107	50.57	42.56	10:12	38
-109	53.24	45.23	10:28	37

PERCENTAGE OF SUN OBSCURED ALONG MAP TRACK
WEST END = 92.4   EAST END = 92.8
MAXIMUM OBSCURATION 92.8 % AT LONG = -105

===

+ MARKS THE CITY OF PEI-CHING

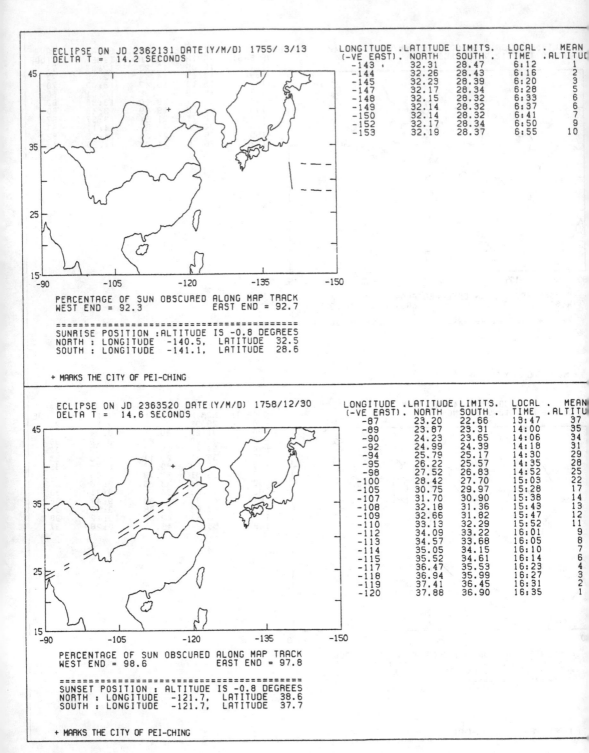

ECLIPSE ON JD 2362131 DATE(Y/M/D) 1755/ 3/13
DELTA T = 14.2 SECONDS

LONGITUDE (-VE EAST)	.LATITUDE .NORTH	LIMITS. SOUTH .	LOCAL . TIME	MEAN .ALTITU
-143 .	32.31	28.47	6:12	1
-144	32.26	28.43	6:16	2
-145	32.23	28.39	6:20	3
-147	32.17	28.34	6:28	5
-148	32.15	28.32	6:33	6
-149	32.14	28.32	6:37	6
-150	32.14	28.32	6:41	7
-152	32.17	28.34	6:50	9
-153	32.19	28.37	6:55	10

PERCENTAGE OF SUN OBSCURED ALONG MAP TRACK
WEST END = 92.3   EAST END = 92.7

==
SUNRISE POSITION :ALTITUDE IS -0.8 DEGREES
NORTH : LONGITUDE -140.5, LATITUDE 32.5
SOUTH : LONGITUDE -141.1, LATITUDE 28.6

+ MARKS THE CITY OF PEI-CHING

ECLIPSE ON JD 2363520 DATE(Y/M/D) 1758/12/30
DELTA T = 14.6 SECONDS

LONGITUDE (-VE EAST)	.LATITUDE .NORTH	LIMITS. SOUTH .	LOCAL . TIME	MEAN .ALTITU
-87	23.20	22.66	13:47	37
-89	23.87	23.31	14:00	35
-90	24.23	23.65	14:06	34
-92	24.99	24.39	14:18	31
-94	25.79	25.17	14:30	29
-95	26.22	25.57	14:35	28
-98	27.52	26.83	14:52	25
-100	28.42	27.70	15:03	22
-105	30.75	29.97	15:28	17
-107	31.70	30.90	15:38	14
-108	32.18	31.36	15:43	13
-109	32.66	31.82	15:47	12
-110	33.13	32.29	15:52	11
-112	34.09	33.22	16:01	9
-113	34.57	33.68	16:05	8
-114	35.05	34.15	16:10	7
-115	35.52	34.61	16:14	6
-117	36.47	35.53	16:23	4
-118	36.94	35.99	16:27	3
-119	37.41	36.45	16:31	2
-120	37.88	36.90	16:35	1

PERCENTAGE OF SUN OBSCURED ALONG MAP TRACK
WEST END = 98.6   EAST END = 97.8

==
SUNSET POSITION : ALTITUDE IS -0.8 DEGREES
NORTH : LONGITUDE -121.7, LATITUDE 38.6
SOUTH : LONGITUDE -121.7, LATITUDE 37.7

+ MARKS THE CITY OF PEI-CHING

ECLIPSE ON JD 2364051 DATE(Y/M/D) 1760/ 6/13
DELTA T = 14.9 SECONDS

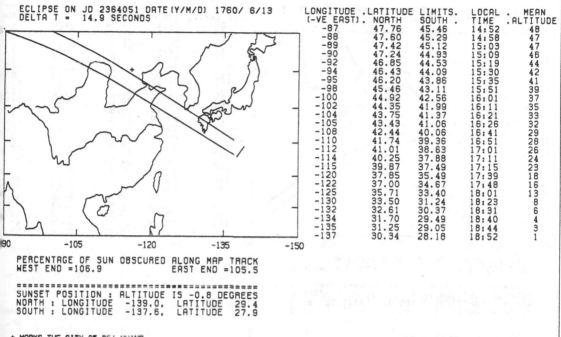

LONGITUDE (-VE EAST)	.LATITUDE LIMITS. NORTH	SOUTH .	LOCAL TIME	. MEAN .ALTITUDE
-87	47.76	45.46	14:52	48
-88	47.60	45.29	14:58	47
-89	47.42	45.12	15:03	47
-90	47.24	44.93	15:09	46
-92	46.85	44.53	15:19	44
-94	46.43	44.09	15:30	42
-95	46.20	43.86	15:35	41
-98	45.46	43.11	15:51	39
-100	44.92	42.56	16:01	37
-102	44.35	41.99	16:11	35
-104	43.75	41.37	16:21	33
-105	43.43	41.06	16:26	32
-108	42.44	40.06	16:41	29
-110	41.74	39.36	16:51	28
-112	41.01	38.63	17:01	26
-114	40.25	37.88	17:11	24
-115	39.87	37.49	17:15	23
-120	37.85	35.49	17:39	18
-122	37.00	34.67	17:48	16
-125	35.71	33.40	18:01	13
-130	33.50	31.24	18:23	8
-132	32.61	30.37	18:31	6
-134	31.70	29.49	18:40	4
-135	31.25	29.05	18:44	3
-137	30.34	28.18	18:52	1

PERCENTAGE OF SUN OBSCURED ALONG MAP TRACK
WEST END =106.9   EAST END =105.5

===
SUNSET POSITION : ALTITUDE IS -0.8 DEGREES
NORTH : LONGITUDE -139.0, LATITUDE 29.4
SOUTH : LONGITUDE -137.6, LATITUDE 27.9

+ MARKS THE CITY OF PEI-CHING

ECLIPSE ON JD 2364907 DATE(Y/M/D) 1762/10/17
DELTA T = 15.2 SECONDS

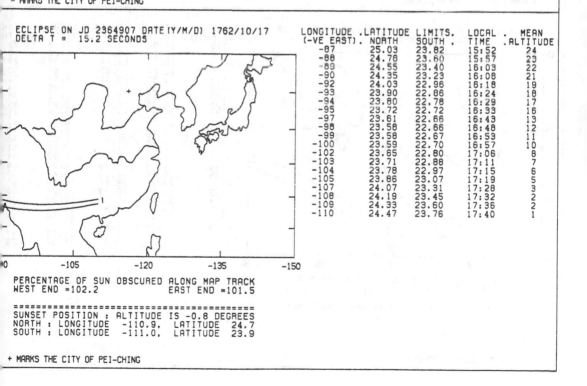

LONGITUDE (-VE EAST)	.LATITUDE LIMITS. NORTH	SOUTH .	LOCAL TIME	. MEAN .ALTITUDE
-87	25.03	23.82	15:52	24
-88	24.78	23.60	15:57	23
-89	24.55	23.40	16:03	22
-90	24.35	23.23	16:08	21
-92	24.03	22.96	16:18	19
-93	23.90	22.86	16:24	18
-94	23.80	22.78	16:29	17
-95	23.72	22.72	16:33	16
-97	23.61	22.66	16:43	13
-98	23.58	22.66	16:48	12
-99	23.58	22.67	16:53	11
-100	23.59	22.70	16:57	10
-102	23.65	22.80	17:06	8
-103	23.71	22.88	17:11	7
-104	23.78	22.97	17:15	6
-105	23.86	23.07	17:19	5
-107	24.07	23.31	17:28	3
-108	24.19	23.45	17:32	2
-109	24.33	23.60	17:36	2
-110	24.47	23.76	17:40	1

PERCENTAGE OF SUN OBSCURED ALONG MAP TRACK
WEST END =102.2   EAST END =101.5

===
SUNSET POSITION : ALTITUDE IS -0.8 DEGREES
NORTH : LONGITUDE -110.9, LATITUDE 24.7
SOUTH : LONGITUDE -111.0, LATITUDE 23.9

+ MARKS THE CITY OF PEI-CHING

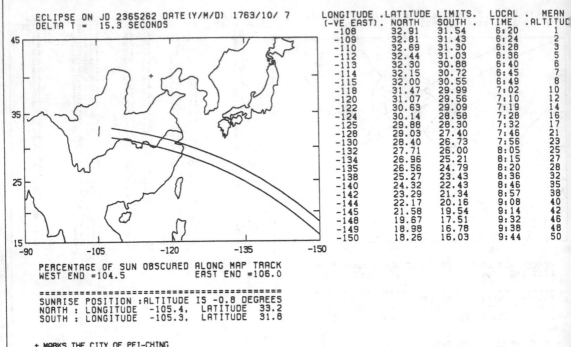

ECLIPSE ON JD 2365262 DATE (Y/M/D) 1763/10/ 7
DELTA T = 15.3 SECONDS

LONGITUDE (-VE EAST)	LATITUDE NORTH	LIMITS. SOUTH	LOCAL TIME	MEAN ALTITUDE
-108	32.91	31.54	6:20	1
-109	32.81	31.43	6:24	2
-110	32.69	31.30	6:28	3
-112	32.44	31.03	6:36	5
-113	32.30	30.88	6:40	6
-114	32.15	30.72	6:45	7
-115	32.00	30.55	6:49	8
-118	31.47	29.99	7:02	10
-120	31.07	29.56	7:10	12
-122	30.63	29.09	7:19	14
-124	30.14	28.58	7:28	16
-125	29.88	28.30	7:32	17
-128	29.03	27.40	7:46	21
-130	28.40	26.73	7:56	23
-132	27.71	26.00	8:05	25
-134	26.96	25.21	8:15	27
-135	26.56	24.79	8:20	28
-138	25.27	23.43	8:36	32
-140	24.32	22.43	8:46	35
-142	23.29	21.34	8:57	38
-144	22.17	20.16	9:08	40
-145	21.58	19.54	9:14	42
-148	19.67	17.51	9:32	46
-149	18.98	16.78	9:38	48
-150	18.26	16.03	9:44	50

PERCENTAGE OF SUN OBSCURED ALONG MAP TRACK
WEST END =104.5   EAST END =106.0

===
SUNRISE POSITION :ALTITUDE IS -0.8 DEGREES
NORTH : LONGITUDE -105.4, LATITUDE 33.2
SOUTH : LONGITUDE -105.3, LATITUDE 31.8

+ MARKS THE CITY OF PEI-CHING

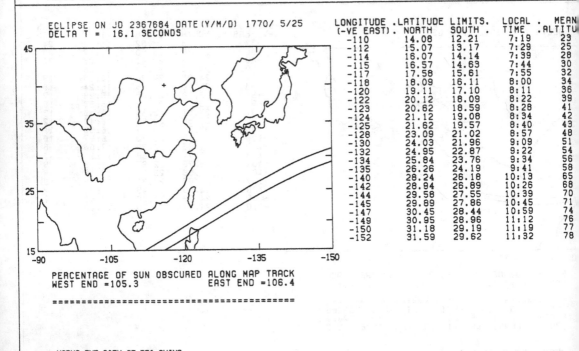

ECLIPSE ON JD 2367684 DATE (Y/M/D) 1770/ 5/25
DELTA T = 16.1 SECONDS

LONGITUDE (-VE EAST)	LATITUDE NORTH	LIMITS. SOUTH	LOCAL TIME	MEAN ALTITUDE
-110	14.08	12.21	7:19	23
-112	15.07	13.17	7:29	25
-114	16.07	14.14	7:39	28
-115	16.57	14.63	7:44	30
-117	17.58	15.61	7:55	32
-118	18.09	16.11	8:00	34
-120	19.11	17.10	8:11	36
-122	20.12	18.09	8:22	39
-123	20.62	18.59	8:28	41
-124	21.12	19.08	8:34	42
-125	21.62	19.57	8:40	43
-128	23.09	21.02	8:57	48
-130	24.03	21.96	9:09	51
-132	24.95	22.87	9:22	54
-134	25.84	23.76	9:34	56
-135	26.26	24.19	9:41	58
-140	28.24	26.18	10:13	65
-142	28.94	26.89	10:26	68
-144	29.58	27.55	10:39	70
-145	29.89	27.86	10:45	71
-147	30.45	28.44	10:59	74
-149	30.95	28.96	11:12	76
-150	31.18	29.19	11:19	77
-152	31.59	29.62	11:32	78

PERCENTAGE OF SUN OBSCURED ALONG MAP TRACK
WEST END =105.3   EAST END =106.4

===

+ MARKS THE CITY OF PEI-CHING

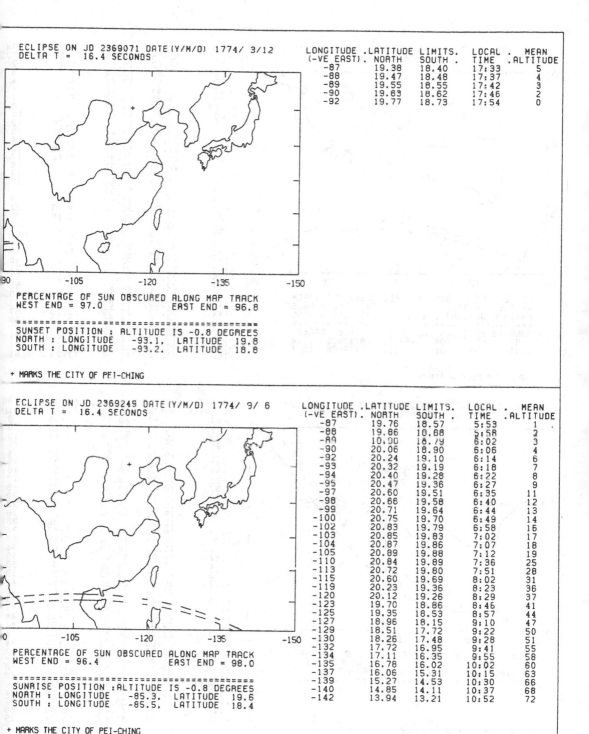

ECLIPSE ON JD 2369071 DATE(Y/M/D) 1774/ 3/12
DELTA T = 16.4 SECONDS

LONGITUDE (-VE EAST)	LATITUDE NORTH	LIMITS SOUTH	LOCAL TIME	MEAN ALTITUDE
-87	19.38	18.40	17:33	5
-88	19.47	18.48	17:37	4
-89	19.55	18.55	17:42	3
-90	19.63	18.62	17:46	2
-92	19.77	18.73	17:54	0

PERCENTAGE OF SUN OBSCURED ALONG MAP TRACK
WEST END = 97.0   EAST END = 96.8
===
SUNSET POSITION : ALTITUDE IS -0.8 DEGREES
NORTH : LONGITUDE   -93.1,   LATITUDE   19.8
SOUTH : LONGITUDE   -93.2,   LATITUDE   18.8

+ MARKS THE CITY OF PEI-CHING

ECLIPSE ON JD 2369249 DATE(Y/M/D) 1774/ 9/ 6
DELTA T = 16.4 SECONDS

LONGITUDE (-VE EAST)	LATITUDE NORTH	LIMITS SOUTH	LOCAL TIME	MEAN ALTITUDE
-87	19.76	18.57	5:53	1
-88	19.86	18.68	5:58	2
-89	19.90	18.79	6:02	3
-90	20.06	18.90	6:06	4
-92	20.24	19.10	6:14	6
-93	20.32	19.19	6:18	7
-94	20.40	19.28	6:22	8
-95	20.47	19.36	6:27	9
-97	20.60	19.51	6:35	11
-98	20.66	19.58	6:40	12
-99	20.71	19.64	6:44	13
-100	20.75	19.70	6:49	14
-102	20.83	19.79	6:58	16
-103	20.85	19.83	7:02	17
-104	20.87	19.86	7:07	18
-105	20.89	19.88	7:12	19
-110	20.84	19.89	7:36	25
-113	20.72	19.80	7:51	28
-115	20.60	19.69	8:02	31
-119	20.23	19.36	8:23	36
-120	20.12	19.26	8:29	37
-123	19.70	18.86	8:46	41
-125	19.35	18.53	8:57	44
-127	18.96	18.15	9:10	47
-129	18.51	17.72	9:22	50
-130	18.26	17.48	9:28	51
-132	17.72	16.95	9:41	55
-134	17.11	16.35	9:55	58
-135	16.78	16.02	10:02	60
-137	16.06	15.31	10:15	63
-139	15.27	14.53	10:30	66
-140	14.85	14.11	10:37	68
-142	13.94	13.21	10:52	72

PERCENTAGE OF SUN OBSCURED ALONG MAP TRACK
WEST END = 96.4   EAST END = 98.0
===
SUNRISE POSITION : ALTITUDE IS -0.8 DEGREES
NORTH : LONGITUDE   -85.3,   LATITUDE   19.6
SOUTH : LONGITUDE   -85.5,   LATITUDE   18.4

+ MARKS THE CITY OF PEI-CHING

413

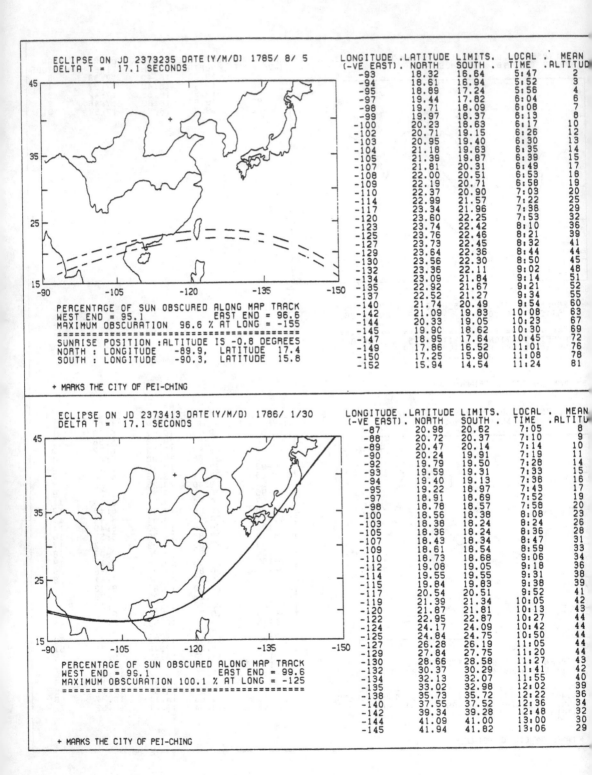

ECLIPSE ON JD 2373235 DATE (Y/M/D) 1785/ 8/ 5
DELTA T = 17.1 SECONDS

PERCENTAGE OF SUN OBSCURED ALONG MAP TRACK
WEST END = 95.1   EAST END = 96.6
MAXIMUM OBSCURATION 96.6 % AT LONG = -155
==
SUNRISE POSITION :ALTITUDE IS -0.8 DEGREES
NORTH : LONGITUDE -89.9, LATITUDE 17.4
SOUTH : LONGITUDE -90.3, LATITUDE 15.8

+ MARKS THE CITY OF PEI-CHING

LONGITUDE (-VE EAST)	LATITUDE NORTH	LIMITS. SOUTH	LOCAL TIME	MEAN ALTITU
-93	18.32	16.64	5:47	2
-94	18.61	16.94	5:52	3
-95	18.89	17.24	5:56	4
-97	19.44	17.82	6:04	6
-98	19.71	18.09	6:08	7
-99	19.97	18.37	6:13	8
-100	20.23	18.63	6:17	10
-102	20.71	19.15	6:26	12
-103	20.95	19.40	6:30	13
-104	21.18	19.63	6:35	14
-105	21.39	19.87	6:39	15
-107	21.81	20.31	6:49	17
-108	22.00	20.51	6:53	18
-109	22.19	20.71	6:58	19
-110	22.37	20.90	7:03	20
-114	22.99	21.57	7:22	25
-117	23.34	21.96	7:38	29
-120	23.60	22.25	7:53	32
-123	23.74	22.42	8:10	36
-125	23.76	22.46	8:21	39
-127	23.73	22.45	8:32	41
-129	23.64	22.36	8:44	44
-130	23.56	22.30	8:50	45
-132	23.36	22.11	9:02	48
-134	23.09	21.84	9:14	51
-135	22.92	21.67	9:21	52
-137	22.52	21.27	9:34	55
-140	21.74	20.49	9:54	60
-142	21.09	19.83	10:08	63
-144	20.33	19.05	10:23	67
-145	19.9C	18.62	10:30	69
-147	18.95	17.64	10:45	72
-149	17.86	16.52	11:01	76
-150	17.25	15.90	11:08	78
-152	15.94	14.54	11:24	81

ECLIPSE ON JD 2373413 DATE (Y/M/D) 1786/ 1/30
DELTA T = 17.1 SECONDS

PERCENTAGE OF SUN OBSCURED ALONG MAP TRACK
WEST END = 9S.1   EAST END = 99.6
MAXIMUM OBSCURATION 100.1 % AT LONG = -125
==

+ MARKS THE CITY OF PEI-CHING

LONGITUDE (-VE EAST)	LATITUDE NORTH	LIMITS. SOUTH	LOCAL TIME	MEAN ALTITU
-87	20.98	20.62	7:05	8
-88	20.72	20.37	7:10	9
-89	20.47	20.14	7:14	10
-90	20.24	19.91	7:19	11
-92	19.79	19.50	7:28	14
-93	19.59	19.31	7:33	15
-94	19.40	19.13	7:38	16
-95	19.22	18.97	7:43	17
-97	18.91	18.69	7:52	19
-98	18.78	18.57	7:58	20
-100	18.56	18.38	8:08	23
-103	18.38	18.24	8:24	26
-105	18.36	18.24	8:36	28
-107	18.43	18.34	8:47	31
-109	18.61	18.54	8:59	33
-110	18.73	18.68	9:06	34
-112	19.08	19.05	9:18	36
-114	19.55	19.55	9:31	38
-115	19.84	19.83	9:38	39
-117	20.54	20.51	9:52	41
-119	21.39	21.34	10:05	42
-120	21.87	21.81	10:13	43
-122	22.95	22.87	10:27	44
-124	24.17	24.09	10:42	44
-125	24.84	24.75	10:50	44
-127	26.28	26.19	11:05	44
-129	27.84	27.75	11:20	44
-130	28.66	28.58	11:27	43
-132	30.37	30.29	11:41	42
-134	32.13	32.07	11:55	40
-135	33.02	32.98	12:02	39
-138	35.73	35.72	12:22	36
-140	37.55	37.52	12:36	34
-142	39.34	39.28	12:48	32
-144	41.09	41.00	13:00	30
-145	41.94	41.82	13:06	29

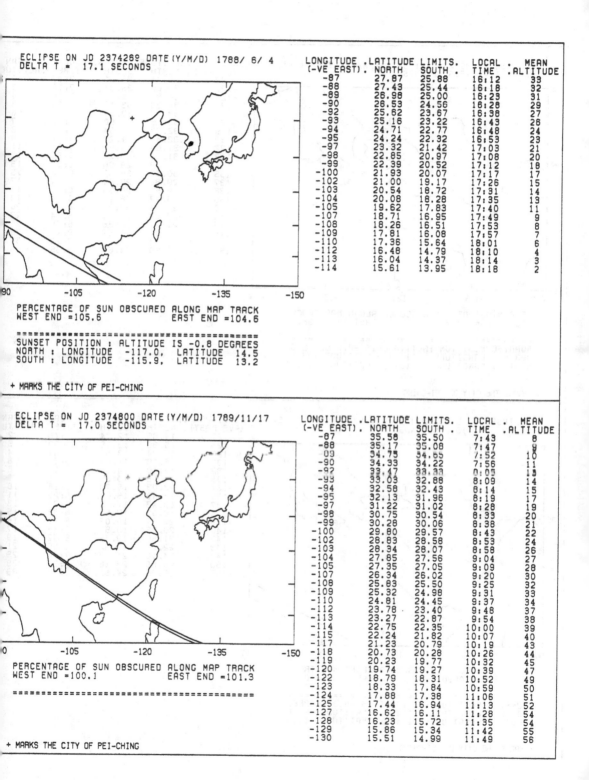

ECLIPSE ON JD 2374269 DATE(Y/M/D) 1788/ 6/ 4
DELTA T = 17.1 SECONDS

LONGITUDE (-VE EAST)	.LATITUDE LIMITS. NORTH	SOUTH	LOCAL TIME	.MEAN ALTITUDE
-87	27.87	25.88	16:12	33
-88	27.43	25.44	16:18	32
-89	26.98	25.00	16:23	31
-90	26.53	24.56	16:28	29
-92	25.62	23.67	16:38	27
-93	25.16	23.22	16:43	26
-94	24.71	22.77	16:48	24
-95	24.24	22.32	16:53	23
-97	23.32	21.42	17:03	21
-98	22.85	20.97	17:08	20
-99	22.39	20.52	17:12	18
-100	21.93	20.07	17:17	17
-102	21.00	19.17	17:26	15
-103	20.54	18.72	17:31	14
-104	20.08	18.28	17:35	13
-105	19.62	17.83	17:40	11
-107	18.71	16.95	17:49	9
-108	18.26	16.51	17:53	8
-109	17.81	16.08	17:57	7
-110	17.36	15.64	18:01	6
-112	16.48	14.79	18:10	4
-113	16.04	14.37	18:14	3
-114	15.61	13.95	18:18	2

PERCENTAGE OF SUN OBSCURED ALONG MAP TRACK
WEST END =105.6   EAST END =104.6

==

SUNSET POSITION : ALTITUDE IS -0.8 DEGREES
NORTH : LONGITUDE -117.0, LATITUDE 14.5
SOUTH : LONGITUDE -115.9, LATITUDE 13.2

+ MARKS THE CITY OF PEI-CHING

ECLIPSE ON JD 2374800 DATE(Y/M/D) 1789/11/17
DELTA T = 17.0 SECONDS

LONGITUDE (-VE EAST)	.LATITUDE LIMITS. NORTH	SOUTH	LOCAL TIME	.MEAN ALTITUDE
-87	35.58	35.50	7:43	8
-88	35.17	35.08	7:47	9
-09	34.75	34.65	7:52	10
-90	34.33	34.22	7:56	11
-92	33.47	33.33	8:05	13
-93	33.03	32.88	8:09	14
-94	32.58	32.43	8:14	15
-95	32.13	31.96	8:19	17
-97	31.22	31.02	8:28	19
-98	30.75	30.54	8:33	20
-99	30.28	30.06	8:38	21
-100	29.80	29.57	8:43	22
-102	28.83	28.58	8:53	24
-103	28.34	28.07	8:58	26
-104	27.85	27.56	9:04	27
-105	27.35	27.05	9:09	28
-107	26.34	26.02	9:20	30
-108	25.83	25.50	9:25	32
-109	25.32	24.98	9:31	33
-110	24.81	24.45	9:37	34
-112	23.78	23.40	9:48	37
-113	23.27	22.87	9:54	38
-114	22.75	22.35	10:00	39
-115	22.24	21.82	10:07	40
-117	21.23	20.79	10:19	43
-118	20.73	20.28	10:26	44
-119	20.23	19.77	10:32	45
-120	19.74	19.27	10:39	47
-122	18.79	18.31	10:52	49
-123	18.33	17.84	10:59	50
-124	17.88	17.38	11:06	51
-125	17.44	16.94	11:13	52
-127	16.62	16.11	11:28	54
-128	16.23	15.72	11:35	54
-129	15.86	15.34	11:42	55
-130	15.51	14.99	11:49	56

PERCENTAGE OF SUN OBSCURED ALONG MAP TRACK
WEST END =100.1   EAST END =101.3

==

+ MARKS THE CITY OF PEI-CHING

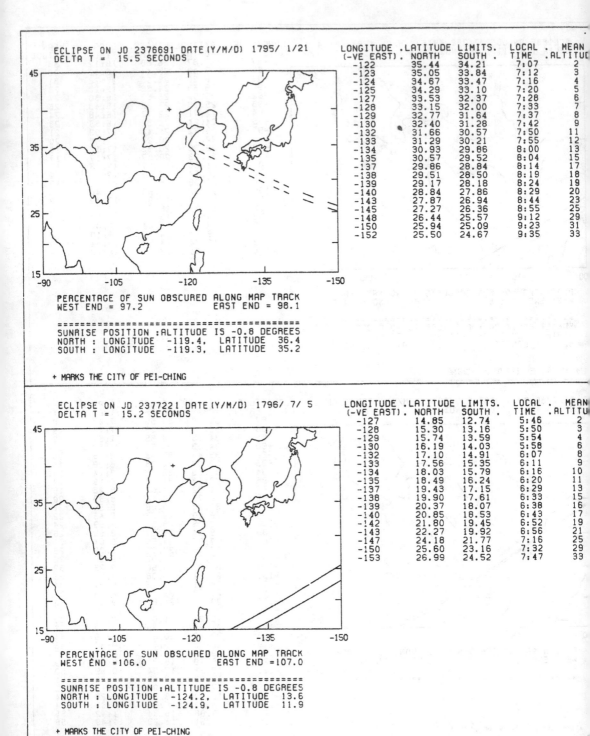

ECLIPSE ON JD 2376691 DATE(Y/M/D) 1795/ 1/21
DELTA T = 15.5 SECONDS

LONGITUDE (-VE EAST)	LATITUDE NORTH	LIMITS. SOUTH	LOCAL TIME	MEAN ALTITU
-122	35.44	34.21	7:07	2
-123	35.05	33.84	7:12	3
-124	34.67	33.47	7:16	4
-125	34.29	33.10	7:20	5
-127	33.53	32.37	7:28	6
-128	33.15	32.00	7:33	7
-129	32.77	31.64	7:37	8
-130	32.40	31.28	7:42	9
-132	31.66	30.57	7:50	11
-133	31.29	30.21	7:55	12
-134	30.93	29.86	8:00	13
-135	30.57	29.52	8:04	15
-137	29.86	28.84	8:14	17
-138	29.51	28.50	8:19	18
-139	29.17	28.18	8:24	19
-140	28.84	27.86	8:29	20
-143	27.87	26.94	8:44	23
-145	27.27	26.36	8:55	25
-148	26.44	25.57	9:12	29
-150	25.94	25.09	9:23	31
-152	25.50	24.67	9:35	33

PERCENTAGE OF SUN OBSCURED ALONG MAP TRACK
WEST END = 97.2   EAST END = 98.1

==
SUNRISE POSITION :ALTITUDE IS -0.8 DEGREES
NORTH : LONGITUDE -119.4,   LATITUDE   36.4
SOUTH : LONGITUDE -119.3,   LATITUDE   35.2

+ MARKS THE CITY OF PEI-CHING

ECLIPSE ON JD 2377221 DATE(Y/M/D) 1796/ 7/ 5
DELTA T = 15.2 SECONDS

LONGITUDE (-VE EAST)	LATITUDE NORTH	LIMITS. SOUTH	LOCAL TIME	MEAN ALTITU
-127	14.85	12.74	5:46	2
-128	15.30	13.16	5:50	3
-129	15.74	13.59	5:54	4
-130	16.19	14.03	5:58	6
-132	17.10	14.91	6:07	8
-133	17.56	15.35	6:11	9
-134	18.03	15.79	6:16	10
-135	18.49	16.24	6:20	11
-137	19.43	17.15	6:29	13
-138	19.90	17.61	6:33	15
-139	20.37	18.07	6:38	16
-140	20.85	18.53	6:43	17
-142	21.80	19.45	6:52	19
-143	22.27	19.92	6:56	21
-147	24.18	21.77	7:16	25
-150	25.60	23.16	7:32	29
-153	26.99	24.52	7:47	33

PERCENTAGE OF SUN OBSCURED ALONG MAP TRACK
WEST END =106.0   EAST END =107.0

==
SUNRISE POSITION :ALTITUDE IS -0.8 DEGREES
NORTH : LONGITUDE -124.2,   LATITUDE   13.6
SOUTH : LONGITUDE -124.9,   LATITUDE   11.9

+ MARKS THE CITY OF PEI-CHING

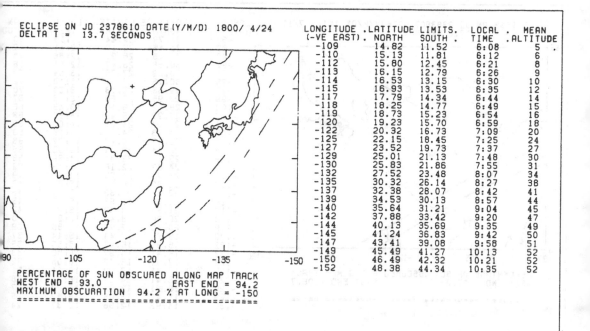

ECLIPSE ON JD 2378610 DATE(Y/M/D) 1800/ 4/24
DELTA T = 13.7 SECONDS

LONGITUDE (-VE EAST)	.LATITUDE NORTH	LIMITS. SOUTH	LOCAL TIME	. MEAN .ALTITUDE
-109	14.82	11.52	6:08	5
-110	15.13	11.81	6:12	6
-112	15.80	12.45	6:21	8
-113	16.15	12.79	6:26	9
-114	16.53	13.15	6:30	10
-115	16.93	13.53	6:35	12
-117	17.79	14.34	6:44	14
-118	18.25	14.77	6:49	15
-119	18.73	15.23	6:54	16
-120	19.23	15.70	6:59	18
-122	20.32	16.73	7:09	20
-125	22.15	18.45	7:25	24
-127	23.52	19.73	7:37	27
-129	25.01	21.13	7:48	30
-130	25.83	21.86	7:55	31
-132	27.52	23.48	8:07	34
-135	30.32	26.14	8:27	38
-137	32.38	28.07	8:42	41
-139	34.53	30.13	8:57	44
-140	35.64	31.21	9:04	45
-142	37.88	33.42	9:20	47
-144	40.13	35.69	9:35	49
-145	41.24	36.83	9:42	50
-147	43.41	39.08	9:58	51
-149	45.49	41.27	10:13	52
-150	46.49	42.32	10:21	52
-152	48.38	44.34	10:35	52

PERCENTAGE OF SUN OBSCURED ALONG MAP TRACK
WEST END = 93.0   EAST END = 94.2
MAXIMUM OBSCURATION 94.2 % AT LONG = -150
===

+ MARKS THE CITY OF PEI-CHING

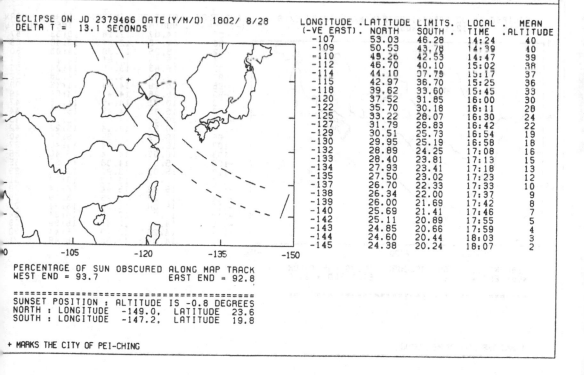

ECLIPSE ON JD 2379466 DATE(Y/M/D) 1802/ 8/28
DELTA T = 13.1 SECONDS

LONGITUDE (-VE EAST)	.LATITUDE NORTH	LIMITS. SOUTH	LOCAL TIME	. MEAN .ALTITUDE
-107	53.03	46.28	14:24	40
-109	50.53	43.78	14:39	40
-110	49.26	42.53	14:47	39
-112	46.70	40.10	15:02	38
-114	44.10	37.79	15:17	37
-115	42.97	36.70	15:25	36
-118	39.62	33.60	15:45	33
-120	37.52	31.85	16:00	30
-122	35.70	30.18	16:11	28
-125	33.22	28.07	16:30	24
-127	31.79	26.83	16:42	22
-129	30.51	25.73	16:54	19
-130	29.95	25.19	16:58	18
-132	28.89	24.25	17:08	16
-133	28.40	23.81	17:13	15
-134	27.93	23.41	17:18	13
-135	27.50	23.02	17:23	12
-137	26.70	22.33	17:33	10
-138	26.34	22.00	17:37	9
-139	26.00	21.69	17:42	8
-140	25.69	21.41	17:46	7
-142	25.11	20.89	17:55	5
-143	24.85	20.66	17:59	4
-144	24.60	20.44	18:03	3
-145	24.38	20.24	18:07	2

PERCENTAGE OF SUN OBSCURED ALONG MAP TRACK
WEST END = 93.7   EAST END = 92.8

===
SUNSET POSITION : ALTITUDE IS -0.8 DEGREES
NORTH : LONGITUDE -149.0, LATITUDE 23.6
SOUTH : LONGITUDE -147.2, LATITUDE 19.8

+ MARKS THE CITY OF PEI-CHING

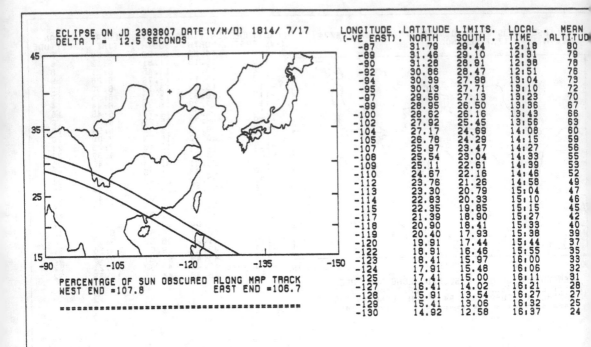

ECLIPSE ON JD 2383807 DATE(Y/M/D) 1814/ 7/17
DELTA T = 12.5 SECONDS

LONGITUDE (-VE EAST)	LATITUDE NORTH	LIMITS. SOUTH	LOCAL TIME	MEAN ALTITU
-87	31.79	29.44	12:18	80
-89	31.46	29.10	12:31	79
-90	31.28	28.91	12:38	78
-92	30.86	28.47	12:51	76
-94	30.39	27.98	13:04	73
-95	30.13	27.71	13:10	72
-97	29.56	27.13	13:23	70
-99	28.95	26.50	13:36	67
-100	28.62	26.16	13:43	66
-102	27.92	25.45	13:56	63
-104	27.17	24.69	14:08	60
-105	26.78	24.29	14:15	59
-107	25.97	23.47	14:27	56
-108	25.54	23.04	14:33	55
-109	25.11	22.61	14:39	53
-110	24.67	22.16	14:46	52
-112	23.76	21.26	14:58	49
-113	23.30	20.79	15:04	47
-114	22.83	20.33	15:10	46
-115	22.35	19.85	15:15	45
-117	21.39	18.90	15:27	42
-118	20.90	18.41	15:33	40
-119	20.40	17.93	15:38	39
-120	19.91	17.44	15:44	37
-122	18.91	16.46	15:55	35
-123	18.41	15.97	16:00	33
-124	17.91	15.48	16:06	32
-125	17.41	15.00	16:11	31
-127	16.41	14.02	16:21	28
-128	15.91	13.54	16:27	27
-129	15.41	13.06	16:32	25
-130	14.92	12.58	16:37	24

PERCENTAGE OF SUN OBSCURED ALONG MAP TRACK
WEST END =107.8   EAST END =106.7

+ MARKS THE CITY OF PEI-CHING

ECLIPSE ON JD 2384841 DATE(Y/M/D) 1817/ 5/16
DELTA T = 12.4 SECONDS

LONGITUDE (-VE EAST)	LATITUDE NORTH	LIMITS. SOUTH	LOCAL TIME	MEAN ALTITU
-88	14.07	12.16	13:33	67
-90	14.72	12.84	13:48	63
-92	15.25	13.39	14:03	60
-94	15.66	13.81	14:17	57
-95	15.83	13.98	14:24	55
-97	16.09	14.24	14:38	52
-98	16.19	14.34	14:45	50
-99	16.26	14.41	14:51	49
-100	16.31	14.45	14:58	47
-102	16.36	14.49	15:10	44
-103	16.35	14.47	15:16	43
-104	16.33	14.44	15:22	42
-105	16.29	14.39	15:28	40
-107	16.16	14.24	15:40	37
-108	16.07	14.14	15:46	36
-109	15.97	14.03	15:52	35
-110	15.85	13.90	15:57	33
-112	15.58	13.60	16:08	31
-113	15.42	13.44	16:14	29
-114	15.26	13.26	16:19	28
-115	15.08	13.07	16:24	27
-117	14.69	12.65	16:34	24
-118	14.48	12.43	16:39	23
-119	14.26	12.19	16:44	22
-120	14.04	11.95	16:49	21

PERCENTAGE OF SUN OBSCURED ALONG MAP TRACK
WEST END = 94.8   EAST END = 94.0

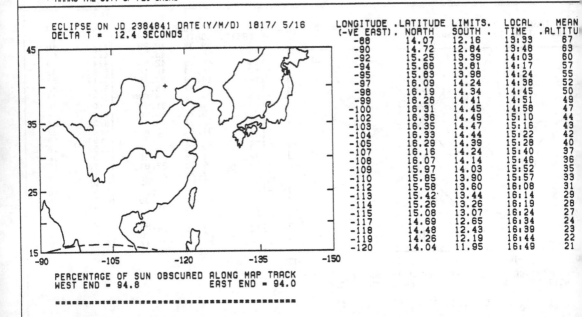

+ MARKS THE CITY OF PEI-CHING

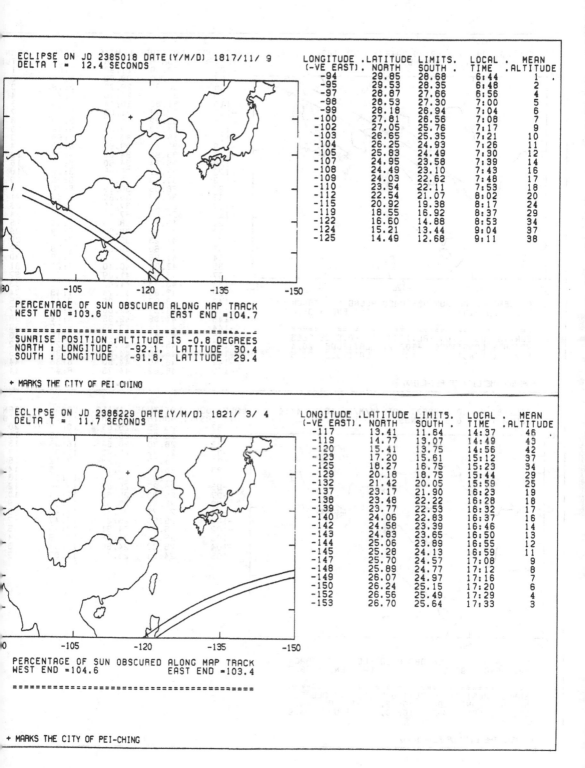

ECLIPSE ON JD 2385018 DATE(Y/M/D) 1817/11/ 9
DELTA T = 12.4 SECONDS

LONGITUDE (-VE EAST).	.LATITUDE NORTH	LIMITS. SOUTH .	LOCAL TIME	. MEAN .ALTITUDE
-94	29.85	28.68	6:44	1
-95	29.53	28.35	6:48	2
-97	28.87	27.66	6:56	4
-98	28.53	27.30	7:00	5
-99	28.18	26.94	7:04	6
-100	27.81	26.56	7:08	7
-102	27.05	25.76	7:17	9
-103	26.65	25.35	7:21	10
-104	26.25	24.93	7:26	11
-105	25.83	24.49	7:30	12
-107	24.95	23.58	7:39	14
-108	24.49	23.10	7:43	16
-109	24.03	22.62	7:48	17
-110	23.54	22.11	7:53	18
-112	22.54	21.07	8:02	20
-115	20.92	19.38	8:17	24
-119	18.55	16.92	8:37	29
-122	16.60	14.88	8:53	34
-124	15.21	13.44	9:04	37
-125	14.49	12.68	9:11	38

PERCENTAGE OF SUN OBSCURED ALONG MAP TRACK
WEST END =103.6   EAST END =104.7

===
SUNRISE POSITION :ALTITUDE IS -0.8 DEGREES
NORTH : LONGITUDE   -92.1, LATITUDE   30.4
SOUTH : LONGITUDE   -91.8, LATITUDE   29.4

+ MARKS THE CITY OF PEI CHING

ECLIPSE ON JD 2386229 DATE(Y/M/D) 1821/ 3/ 4
DELTA T = 11.7 SECONDS

LONGITUDE (-VE EAST).	.LATITUDE NORTH	LIMITS. SOUTH .	LOCAL TIME	. MEAN .ALTITUDE
-117	13.41	11.64	14:37	46
-119	14.77	13.07	14:49	43
-120	15.41	13.75	14:56	42
-123	17.20	15.61	15:12	37
-125	18.27	16.75	15:23	34
-129	20.18	18.75	15:44	29
-132	21.42	20.05	15:59	25
-137	23.17	21.90	16:23	19
-138	23.48	22.22	16:28	18
-139	23.77	22.53	16:32	17
-140	24.06	22.83	16:37	16
-142	24.58	23.39	16:46	14
-143	24.83	23.65	16:50	13
-144	25.06	23.89	16:55	12
-145	25.28	24.13	16:59	11
-147	25.70	24.57	17:08	9
-148	25.89	24.77	17:12	8
-149	26.07	24.97	17:16	7
-150	26.24	25.15	17:20	6
-152	26.56	25.49	17:29	4
-153	26.70	25.64	17:33	3

PERCENTAGE OF SUN OBSCURED ALONG MAP TRACK
WEST END =104.6   EAST END =103.4

===

+ MARKS THE CITY OF PEI-CHING

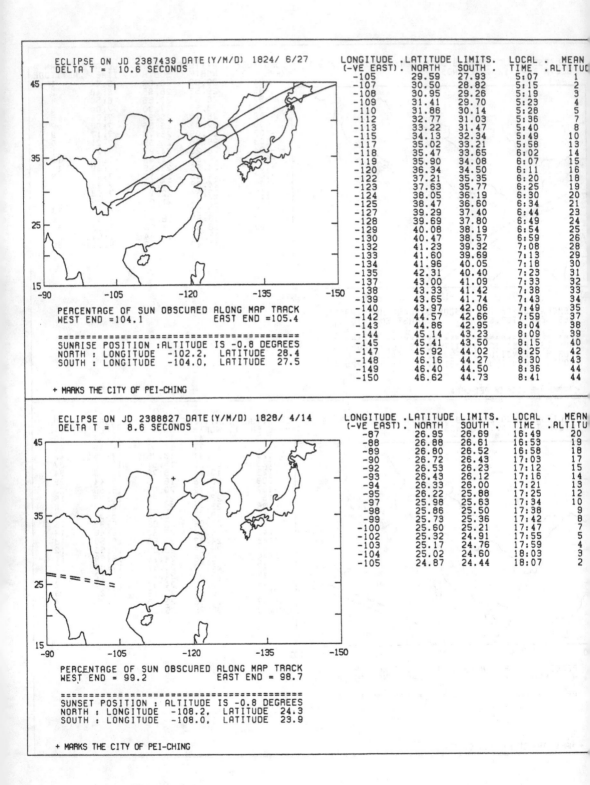

ECLIPSE ON JD 2387439 DATE (Y/M/D) 1824/ 6/27
DELTA T = 10.6 SECONDS

LONGITUDE (-VE EAST)	LATITUDE LIMITS. NORTH	SOUTH	LOCAL TIME	MEAN ALTITU
-105	29.59	27.93	5:07	1
-107	30.50	28.82	5:15	2
-108	30.95	29.26	5:19	3
-109	31.41	29.70	5:23	4
-110	31.86	30.14	5:28	5
-112	32.77	31.03	5:36	7
-113	33.22	31.47	5:40	8
-115	34.13	32.34	5:49	10
-117	35.02	33.21	5:58	13
-118	35.47	33.65	6:02	14
-119	35.90	34.08	6:07	15
-120	36.34	34.50	6:11	16
-122	37.21	35.35	6:20	18
-123	37.63	35.77	6:25	19
-124	38.05	36.19	6:30	20
-125	38.47	36.60	6:34	21
-127	39.29	37.40	6:44	23
-128	39.69	37.80	6:49	24
-129	40.08	38.19	6:54	25
-130	40.47	38.57	6:59	26
-132	41.23	39.32	7:08	28
-133	41.60	39.69	7:13	29
-134	41.96	40.05	7:18	30
-135	42.31	40.40	7:23	31
-137	43.00	41.09	7:33	32
-138	43.33	41.42	7:38	33
-139	43.65	41.74	7:43	34
-140	43.97	42.06	7:49	35
-142	44.57	42.66	7:59	37
-143	44.86	42.95	8:04	38
-144	45.14	43.23	8:09	39
-145	45.41	43.50	8:15	40
-147	45.92	44.02	8:25	42
-148	46.16	44.27	8:30	43
-149	46.40	44.50	8:36	44
-150	46.62	44.73	8:41	44

PERCENTAGE OF SUN OBSCURED ALONG MAP TRACK
WEST END =104.1   EAST END =105.4

SUNRISE POSITION :ALTITUDE IS -0.8 DEGREES
NORTH : LONGITUDE -102.2, LATITUDE 28.4
SOUTH : LONGITUDE -104.0, LATITUDE 27.5

+ MARKS THE CITY OF PEI-CHING

ECLIPSE ON JD 2388827 DATE (Y/M/D) 1828/ 4/14
DELTA T = 8.6 SECONDS

LONGITUDE (-VE EAST)	LATITUDE LIMITS. NORTH	SOUTH	LOCAL TIME	MEAN ALTITU
-87	26.95	26.69	16:49	20
-88	26.88	26.61	16:53	19
-89	26.80	26.52	16:58	18
-90	26.72	26.43	17:03	17
-92	26.53	26.23	17:12	15
-93	26.43	26.12	17:16	14
-94	26.33	26.00	17:21	13
-95	26.22	25.88	17:25	12
-97	25.98	25.63	17:34	10
-98	25.86	25.50	17:38	9
-99	25.73	25.36	17:42	8
-100	25.60	25.21	17:47	7
-102	25.32	24.91	17:55	5
-103	25.17	24.76	17:59	4
-104	25.02	24.60	18:03	3
-105	24.87	24.44	18:07	2

PERCENTAGE OF SUN OBSCURED ALONG MAP TRACK
WEST END = 99.2   EAST END = 98.7

SUNSET POSITION : ALTITUDE IS -0.8 DEGREES
NORTH : LONGITUDE -108.2, LATITUDE 24.3
SOUTH : LONGITUDE -108.0, LATITUDE 23.9

+ MARKS THE CITY OF PEI-CHING

ECLIPSE ON JD 2389359 DATE(Y/M/D) 1829/ 9/28
DELTA T =   8.0 SECONDS

LONGITUDE (-VE EAST).	LATITUDE NORTH	LIMITS. SOUTH .	LOCAL TIME	MEAN ALTITUDE
-137	50.03	46.82	9:58	33
-140	48.98	45.73	10:14	36
-142	48.21	44.95	10:26	37
-144	47.40	44.12	10:37	39
-145	46.97	43.70	10:43	40
-148	45.66	42.32	11:01	42
-150	44.70	41.36	11:14	44
-152	43.72	40.33	11:27	46

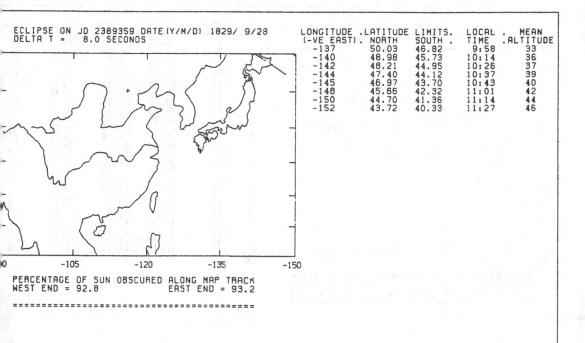

PERCENTAGE OF SUN OBSCURED ALONG MAP TRACK
WEST END = 92.8   EAST END = 93.2

===

+ MARKS THE CITY OF PEI-CHING

ECLIPSE ON JD 2392990 DATE(Y/M/D) 1839/ 9/ 8
DELTA T =   5.6 SECONDS

LONGITUDE (-VE EAST).	LATITUDE NORTH	LIMITS. SOUTH .	LOCAL TIME	MEAN ALTITUDE
-138	36.82	35.20	5:43	0
-140	37.01	35.41	5:51	2
-142	37.17	35.59	5:59	4
-143	37.24	35.67	6:03	4
-144	37.31	35.75	6:07	5
-145	37.36	35.82	6:12	6
-147	37.46	35.93	6:20	8
-148	37.49	35.97	6:24	9
-149	37.52	36.01	6:29	9
-150	37.54	36.04	6:33	10
-152	37.56	36.07	6:42	12
-153	37.55	36.08	6:46	13

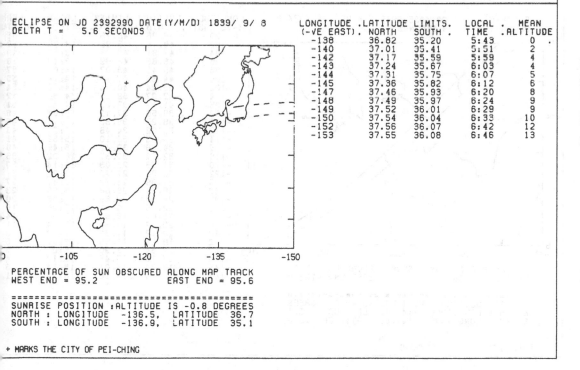

PERCENTAGE OF SUN OBSCURED ALONG MAP TRACK
WEST END = 95.2   EAST END = 95.6

===
SUNRISE POSITION :ALTITUDE IS -0.8 DEGREES
NORTH : LONGITUDE -136.5, LATITUDE 36.7
SOUTH : LONGITUDE -136.9, LATITUDE 35.1

+ MARKS THE CITY OF PEI-CHING

ECLIPSE ON JD 2393169 DATE(Y/M/D) 1840/ 3/ 4
DELTA T = 5.7 SECONDS

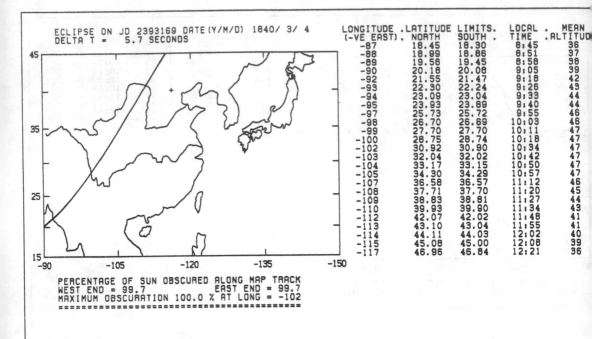

LONGITUDE (-VE EAST)	LATITUDE LIMITS NORTH	SOUTH	LOCAL TIME	MEAN ALTITUDE
-87	18.45	18.30	8:45	36
-88	18.99	18.86	8:51	37
-89	19.56	19.45	8:58	98
-90	20.18	20.08	9:05	39
-92	21.55	21.47	9:18	42
-93	22.30	22.24	9:26	43
-94	23.09	23.04	9:33	44
-95	23.93	23.89	9:40	44
-97	25.73	25.72	9:55	46
-98	26.70	26.69	10:03	46
-99	27.70	27.70	10:11	47
-100	28.75	28.74	10:18	47
-102	30.92	30.90	10:34	47
-103	32.04	32.02	10:42	47
-104	33.17	33.15	10:50	47
-105	34.30	34.29	10:57	47
-107	36.58	36.57	11:12	46
-108	37.71	37.70	11:20	45
-109	38.83	38.81	11:27	44
-110	39.93	39.90	11:34	43
-112	42.07	42.02	11:48	41
-113	43.10	43.04	11:55	41
-114	44.11	44.03	12:02	40
-115	45.08	45.00	12:08	39
-117	46.96	46.84	12:21	36

PERCENTAGE OF SUN OBSCURED ALONG MAP TRACK
WEST END = 99.7   EAST END = 99.7
MAXIMUM OBSCURATION 100.0 % AT LONG = -102
===

+ MARKS THE CITY OF PEI-CHING

ECLIPSE ON JD 2394025 DATE(Y/M/D) 1842/ 7/ 8
DELTA T = 5.9 SECONDS

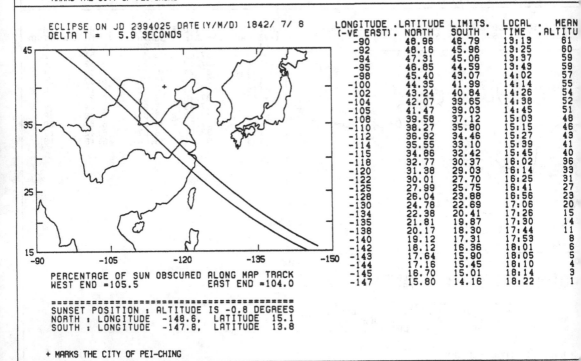

LONGITUDE (-VE EAST)	LATITUDE NORTH	SOUTH	LOCAL TIME	MEAN ALTITUDE
-90	48.96	46.79	13:19	61
-92	48.16	45.96	13:25	60
-94	47.31	45.06	13:37	59
-95	46.85	44.59	13:49	59
-98	45.40	43.07	14:02	57
-100	44.35	41.99	14:14	55
-102	43.24	40.84	14:26	54
-104	42.07	39.65	14:38	52
-105	41.47	39.03	14:45	51
-108	39.58	37.12	15:03	48
-110	38.27	35.80	15:15	46
-112	36.92	34.46	15:27	43
-114	35.55	33.10	15:39	41
-115	34.86	32.42	15:45	40
-118	32.77	30.37	16:02	36
-120	31.38	29.03	16:14	33
-122	30.01	27.70	16:25	31
-125	27.99	25.75	16:41	27
-128	26.04	23.88	16:56	23
-130	24.78	22.69	17:06	20
-134	22.38	20.41	17:26	15
-135	21.81	19.87	17:30	14
-138	20.17	18.30	17:44	11
-140	19.12	17.31	17:53	8
-142	18.12	16.36	18:01	6
-143	17.64	15.90	18:05	5
-144	17.16	15.45	18:10	4
-145	16.70	15.01	18:14	3
-147	15.80	14.16	18:22	1

PERCENTAGE OF SUN OBSCURED ALONG MAP TRACK
WEST END =105.5   EAST END =104.0

===
SUNSET POSITION : ALTITUDE IS -0.8 DEGREES
NORTH : LONGITUDE -148.6, LATITUDE 15.1
SOUTH : LONGITUDE -147.8, LATITUDE 13.8

+ MARKS THE CITY OF PEI-CHING

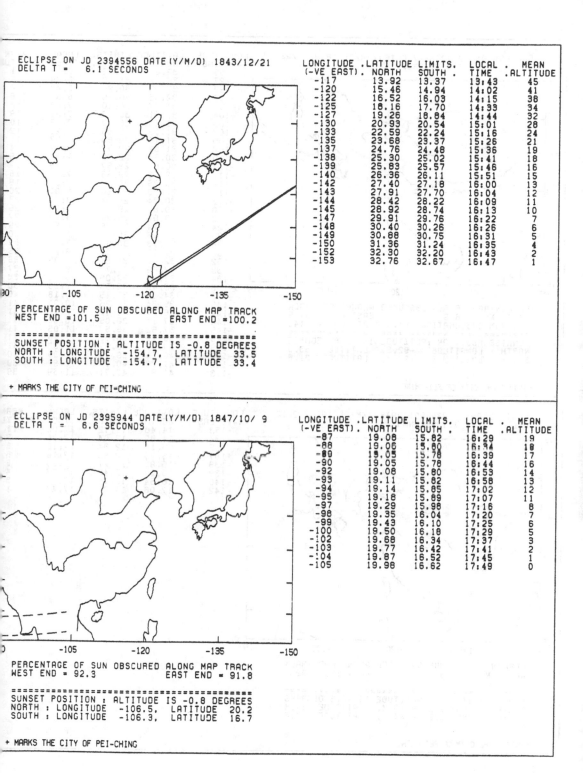

ECLIPSE ON JD 2394556 DATE(Y/M/D) 1843/12/21
DELTA T =   6.1 SECONDS

LONGITUDE (-VE EAST).	LATITUDE NORTH	LIMITS. SOUTH .	LOCAL TIME	MEAN .ALTITUDE
-117	13.92	13.37	13:43	45
-120	15.46	14.94	14:02	41
-122	16.52	16.03	14:15	38
-125	18.16	17.70	14:33	34
-127	19.26	18.84	14:44	32
-130	20.93	20.54	15:01	28
-133	22.59	22.24	15:16	24
-135	23.68	23.37	15:26	21
-137	24.76	24.48	15:36	19
-138	25.30	25.02	15:41	18
-199	25.83	25.57	15:46	16
-140	26.36	26.11	15:51	15
-142	27.40	27.18	16:00	13
-143	27.91	27.70	16:04	12
-144	28.42	28.22	16:09	11
-145	28.92	28.74	16:13	10
-147	29.91	29.76	16:22	7
-148	30.40	30.26	16:26	6
-149	30.88	30.75	16:31	5
-150	31.36	31.24	16:35	4
-152	32.30	32.20	16:43	2
-153	32.76	32.67	16:47	1

PERCENTAGE OF SUN OBSCURED ALONG MAP TRACK
WEST END =101.5   EAST END =100.2

==
SUNSET POSITION : ALTITUDE IS -0.8 DEGREES
NORTH : LONGITUDE -154.7, LATITUDE 33.5
SOUTH : LONGITUDE -154.7, LATITUDE 33.4

+ MARKS THE CITY OF PEI-CHING

ECLIPSE ON JD 2395944 DATE(Y/M/D) 1847/10/ 9
DELTA T =   6.6 SECONDS

LONGITUDE (-VE EAST).	LATITUDE NORTH	LIMITS. SOUTH .	LOCAL TIME	MEAN .ALTITUDE
-87	19.08	15.82	16:29	19
-88	19.06	15.80	16:34	18
-89	19.05	15.78	16:39	17
-90	19.05	15.78	16:44	16
-92	19.08	15.80	16:53	14
-93	19.11	15.82	16:58	13
-94	19.14	15.85	17:02	12
-95	19.18	15.89	17:07	11
-97	19.29	15.98	17:16	8
-98	19.35	16.04	17:20	7
-99	19.43	16.10	17:25	6
-100	19.50	16.18	17:29	5
-102	19.68	16.34	17:37	3
-103	19.77	16.42	17:41	2
-104	19.87	16.52	17:45	1
-105	19.98	16.62	17:49	0

PERCENTAGE OF SUN OBSCURED ALONG MAP TRACK
WEST END = 92.3   EAST END = 91.8

==
SUNSET POSITION : ALTITUDE IS -0.8 DEGREES
NORTH : LONGITUDE -106.5, LATITUDE 20.2
SOUTH : LONGITUDE -106.3, LATITUDE 16.7

+ MARKS THE CITY OF PEI-CHING

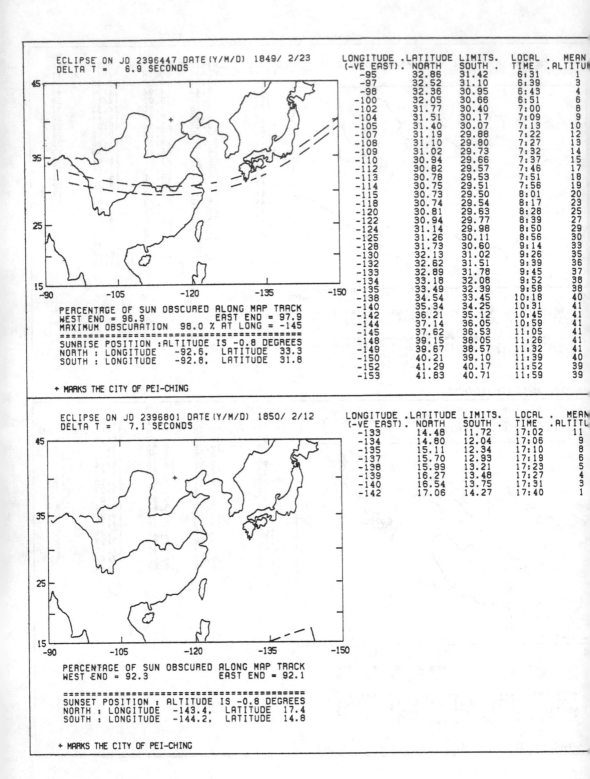

ECLIPSE ON JD 2396447 DATE(Y/M/D) 1849/ 2/23
DELTA T = 6.9 SECONDS

LONGITUDE (-VE EAST)	LATITUDE NORTH	LIMITS. SOUTH	LOCAL TIME	MEAN ALTITU
-95	32.86	31.42	6:31	1
-97	32.52	31.10	6:39	3
-98	32.36	30.95	6:43	4
-100	32.05	30.66	6:51	6
-102	31.77	30.40	7:00	8
-104	31.51	30.17	7:09	9
-105	31.40	30.07	7:13	10
-107	31.19	29.88	7:22	12
-108	31.10	29.80	7:27	13
-109	31.02	29.73	7:32	14
-110	30.94	29.66	7:37	15
-112	30.82	29.57	7:46	17
-113	30.78	29.53	7:51	18
-114	30.75	29.51	7:56	19
-115	30.73	29.50	8:01	20
-118	30.74	29.54	8:17	23
-120	30.81	29.63	8:28	25
-122	30.94	29.77	8:39	27
-124	31.14	29.98	8:50	29
-125	31.26	30.11	8:56	30
-128	31.73	30.60	9:14	33
-130	32.13	31.02	9:26	35
-132	32.62	31.51	9:39	36
-133	32.89	31.78	9:45	37
-134	33.18	32.08	9:52	38
-135	33.49	32.39	9:58	38
-138	34.54	33.45	10:18	40
-140	35.34	34.25	10:31	41
-142	36.21	35.12	10:45	41
-144	37.14	36.05	10:59	41
-145	37.62	36.53	11:05	41
-148	39.15	38.05	11:26	41
-149	39.67	38.57	11:32	41
-150	40.21	39.10	11:39	40
-152	41.29	40.17	11:52	39
-153	41.83	40.71	11:59	39

PERCENTAGE OF SUN OBSCURED ALONG MAP TRACK
WEST END = 96.9   EAST END = 97.9
MAXIMUM OBSCURATION 98.0 % AT LONG = -145
==
SUNRISE POSITION :ALTITUDE IS -0.8 DEGREES
NORTH : LONGITUDE -92.6, LATITUDE 33.3
SOUTH : LONGITUDE -92.8, LATITUDE 31.8

+ MARKS THE CITY OF PEI-CHING

ECLIPSE ON JD 2396801 DATE(Y/M/D) 1850/ 2/12
DELTA T = 7.1 SECONDS

LONGITUDE (-VE EAST)	LATITUDE NORTH	LIMITS. SOUTH	LOCAL TIME	MEAN ALTITU
-133	14.48	11.72	17:02	11
-134	14.80	12.04	17:06	9
-135	15.11	12.34	17:10	8
-137	15.70	12.93	17:19	6
-138	15.99	13.21	17:23	5
-139	16.27	13.48	17:27	4
-140	16.54	13.75	17:31	3
-142	17.06	14.27	17:40	1

PERCENTAGE OF SUN OBSCURED ALONG MAP TRACK
WEST END = 92.3   EAST END = 92.1

==
SUNSET POSITION : ALTITUDE IS -0.8 DEGREES
NORTH : LONGITUDE -143.4, LATITUDE 17.4
SOUTH : LONGITUDE -144.2, LATITUDE 14.8

+ MARKS THE CITY OF PEI-CHING

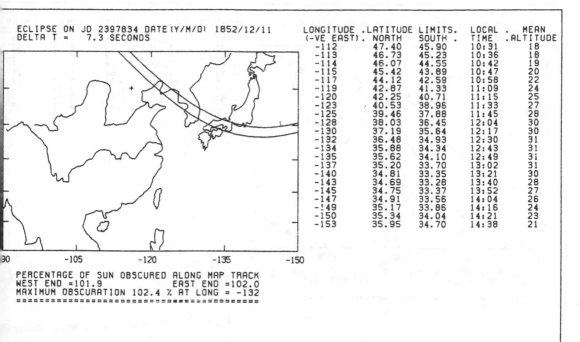

ECLIPSE ON JD 2397834 DATE(Y/M/D) 1852/12/11
DELTA T =   7.3 SECONDS

LONGITUDE (-VE EAST)	.LATITUDE NORTH	LIMITS. SOUTH .	LOCAL TIME	. MEAN .ALTITUDE
-112	47.40	45.90	10:31	18
-113	46.73	45.23	10:36	18
-114	46.07	44.55	10:42	19
-115	45.42	43.89	10:47	20
-117	44.12	42.59	10:58	22
-119	42.87	41.33	11:09	24
-120	42.25	40.71	11:15	25
-123	40.53	38.96	11:33	27
-125	39.46	37.88	11:45	28
-128	38.03	36.45	12:04	30
-130	37.19	35.64	12:17	30
-132	36.48	34.93	12:30	31
-134	35.88	34.34	12:43	31
-135	35.62	34.10	12:49	31
-137	35.20	33.70	13:02	31
-140	34.81	33.35	13:21	30
-143	34.69	33.28	13:40	28
-145	34.75	33.37	13:52	27
-147	34.91	33.56	14:04	26
-149	35.17	33.86	14:16	24
-150	35.34	34.04	14:21	23
-153	35.95	34.70	14:38	21

PERCENTAGE OF SUN OBSCURED ALONG MAP TRACK
WEST END =101.9   EAST END =102.0
MAXIMUM OBSCURATION 102.4 % AT LONG = -132
===

+ MARKS THE CITY OF PEI-CHING

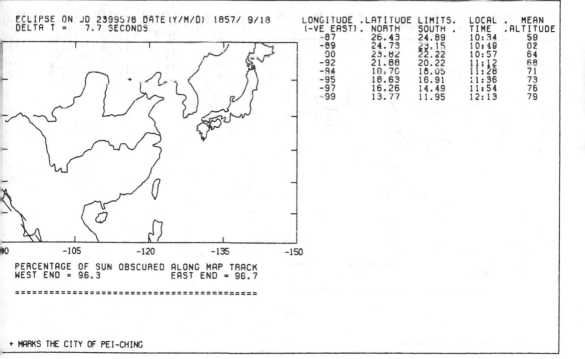

ECLIPSE ON JD 2399578 DATE(Y/M/D) 1857/ 9/18
DELTA T =   7.7 SECONDS

LONGITUDE (-VE EAST)	.LATITUDE NORTH	LIMITS. SOUTH .	LOCAL TIME	. MEAN .ALTITUDE
-87	26.43	24.89	10:34	59
-89	24.73	23.15	10:48	62
-90	23.82	22.22	10:57	64
-92	21.88	20.22	11:12	68
-94	19.70	18.05	11:28	71
-95	18.63	16.91	11:36	73
-97	16.26	14.49	11:54	76
-99	13.77	11.95	12:13	79

PERCENTAGE OF SUN OBSCURED ALONG MAP TRACK
WEST END = 96.3   EAST END = 96.7

===

+ MARKS THE CITY OF PEI-CHING

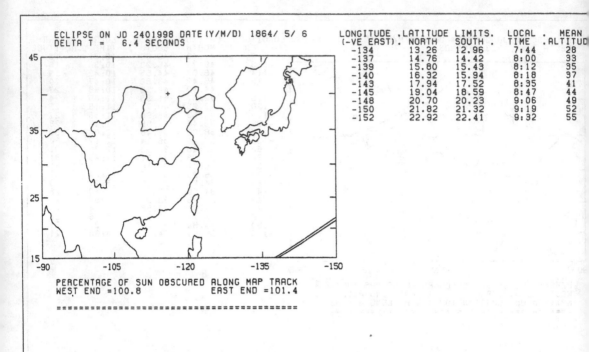

ECLIPSE ON JD 2401998 DATE (Y/M/D) 1864/ 5/ 6
DELTA T =   6.4 SECONDS

LONGITUDE (-VE EAST)	LATITUDE NORTH	LIMITS SOUTH	LOCAL TIME	MEAN ALTITUD
-134	13.26	12.96	7:44	28
-137	14.76	14.42	8:00	33
-139	15.80	15.43	8:12	35
-140	16.32	15.94	8:18	37
-143	17.94	17.52	8:35	41
-145	19.04	18.59	8:47	44
-148	20.70	20.23	9:06	49
-150	21.82	21.32	9:19	52
-152	22.92	22.41	9:32	55

PERCENTAGE OF SUN OBSCURED ALONG MAP TRACK
WEST END =100.8   EAST END =101.4

==

+ MARKS THE CITY OF PEI-CHING

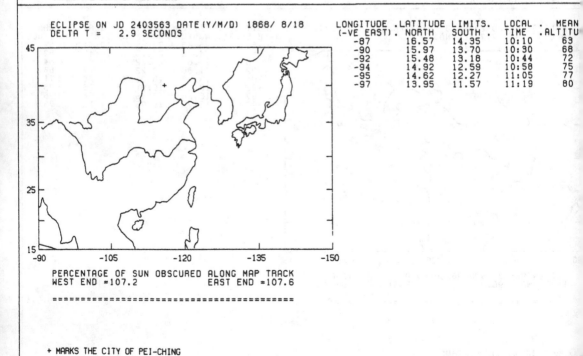

ECLIPSE ON JD 2403563 DATE (Y/M/D) 1868/ 8/18
DELTA T =   2.9 SECONDS

LONGITUDE (-VE EAST)	LATITUDE NORTH	LIMITS SOUTH	LOCAL TIME	MEAN ALTITU
-87	16.57	14.35	10:10	63
-90	15.97	13.70	10:30	68
-92	15.48	13.18	10:44	72
-94	14.92	12.59	10:58	75
-95	14.62	12.27	11:05	77
-97	13.95	11.57	11:19	80

PERCENTAGE OF SUN OBSCURED ALONG MAP TRACK
WEST END =107.2   EAST END =107.6

==

+ MARKS THE CITY OF PEI-CHING

426

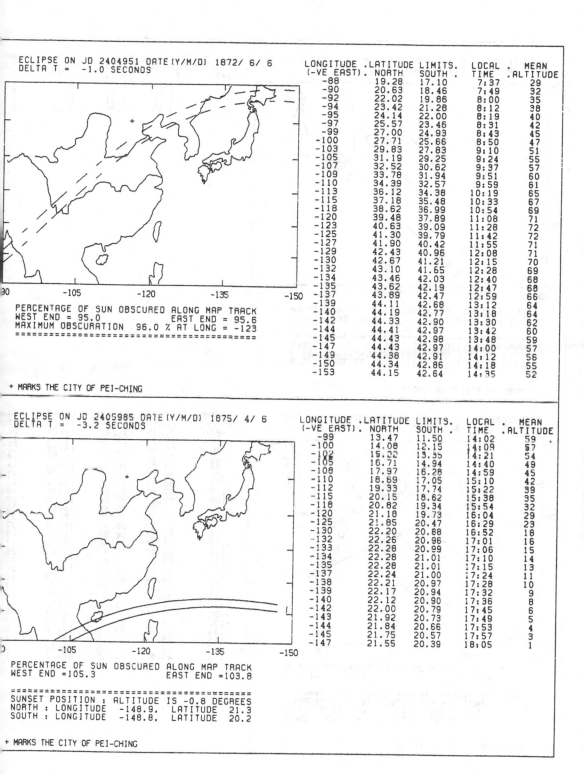

ECLIPSE ON JD 2404951 DATE(Y/M/D) 1872/ 6/ 6
DELTA T = -1.0 SECONDS

LONGITUDE (-VE EAST)	.LATITUDE LIMITS. NORTH	SOUTH .	LOCAL TIME	. MEAN .ALTITUDE
-88	19.28	17.10	7:37	29
-90	20.63	18.46	7:49	32
-92	22.02	19.86	8:00	35
-94	23.42	21.28	8:12	38
-95	24.14	22.00	8:19	40
-97	25.57	23.46	8:31	42
-99	27.00	24.93	8:43	45
-100	27.71	25.66	8:50	47
-103	29.83	27.83	9:10	51
-105	31.19	29.25	9:24	55
-107	32.52	30.62	9:37	57
-109	33.78	31.94	9:51	60
-110	34.39	32.57	9:59	61
-113	36.12	34.38	10:19	65
-115	37.18	35.48	10:33	67
-118	38.62	36.99	10:54	69
-120	39.48	37.89	11:08	71
-123	40.63	39.09	11:28	72
-125	41.30	39.79	11:42	72
-127	41.90	40.42	11:55	71
-129	42.43	40.96	12:08	71
-130	42.67	41.21	12:15	70
-132	43.10	41.65	12:28	69
-134	43.46	42.03	12:40	68
-135	43.62	42.19	12:47	68
-137	43.89	42.47	12:59	66
-139	44.11	42.68	13:12	64
-140	44.19	42.77	13:18	64
-142	44.33	42.90	13:30	62
-144	44.41	42.97	13:42	60
-145	44.43	42.98	13:48	59
-147	44.43	42.97	14:00	57
-149	44.38	42.91	14:12	56
-150	44.34	42.86	14:18	55
-153	44.15	42.64	14:35	52

PERCENTAGE OF SUN OBSCURED ALONG MAP TRACK
WEST END = 95.0   EAST END = 95.6
MAXIMUM OBSCURATION 96.0 % AT LONG = -123
===

+ MARKS THE CITY OF PEI-CHING

ECLIPSE ON JD 2405985 DATE(Y/M/D) 1875/ 4/ 6
DELTA T = -3.2 SECONDS

LONGITUDE (-VE EAST)	.LATITUDE LIMITS. NORTH	SOUTH .	LOCAL TIME	. MEAN .ALTITUDE
-99	13.47	11.50	14:02	59
-100	14.08	12.15	14:09	57
-102	15.22	13.55	14:21	54
-105	16.71	14.94	14:40	49
-108	17.97	16.28	14:59	45
-110	18.69	17.05	15:10	42
-112	19.33	17.74	15:22	39
-115	20.15	18.62	15:38	35
-118	20.82	19.34	15:54	32
-120	21.18	19.73	16:04	29
-125	21.85	20.47	16:29	23
-130	22.20	20.88	16:52	18
-132	22.26	20.96	17:01	16
-133	22.28	20.99	17:06	15
-134	22.28	21.01	17:10	14
-135	22.28	21.01	17:15	13
-137	22.24	21.00	17:24	11
-138	22.21	20.97	17:28	10
-139	22.17	20.94	17:32	9
-140	22.12	20.90	17:36	8
-142	22.00	20.79	17:45	6
-143	21.92	20.73	17:49	5
-144	21.84	20.66	17:53	4
-145	21.75	20.57	17:57	3
-147	21.55	20.39	18:05	1

PERCENTAGE OF SUN OBSCURED ALONG MAP TRACK
WEST END =105.3   EAST END =103.8

===
SUNSET POSITION : ALTITUDE IS -0.8 DEGREES
NORTH : LONGITUDE -148.9, LATITUDE 21.3
SOUTH : LONGITUDE -148.8, LATITUDE 20.2

+ MARKS THE CITY OF PEI-CHING

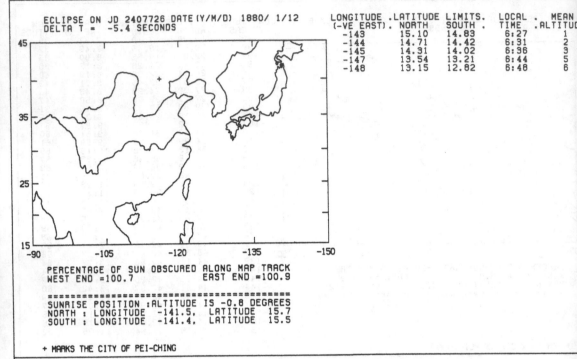

ECLIPSE ON JD 2407726 DATE (Y/M/D) 1880/ 1/12
DELTA T = -5.4 SECONDS

LONGITUDE	.LATITUDE LIMITS.		LOCAL	. MEAN
(-VE EAST).	NORTH	SOUTH .	TIME	.ALTITU
-143	15.10	14.83	6:27	1
-144	14.71	14.42	6:31	2
-145	14.31	14.02	6:36	3
-147	13.54	13.21	6:44	5
-148	13.15	12.82	6:48	6

PERCENTAGE OF SUN OBSCURED ALONG MAP TRACK
WEST END =100.7   EAST END =100.9

===

SUNRISE POSITION :ALTITUDE IS -0.8 DEGREES
NORTH : LONGITUDE -141.5, LATITUDE 15.7
SOUTH : LONGITUDE -141.4, LATITUDE 15.5

+ MARKS THE CITY OF PEI-CHING

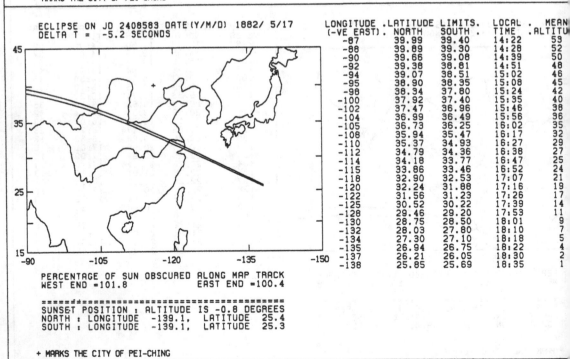

ECLIPSE ON JD 2408583 DATE (Y/M/D) 1882/ 5/17
DELTA T = -5.2 SECONDS

LONGITUDE	.LATITUDE LIMITS.		LOCAL	. MEAN
(-VE EAST).	NORTH	SOUTH .	TIME	.ALTITU
-87	39.99	39.40	14:22	59
-88	39.89	39.30	14:28	52
-90	39.66	39.08	14:39	50
-92	39.38	38.81	14:51	48
-94	39.07	38.51	15:02	46
-95	38.90	38.35	15:08	45
-98	38.34	37.80	15:24	42
-100	37.92	37.40	15:35	40
-102	37.47	36.96	15:46	38
-104	36.99	36.49	15:56	36
-105	36.73	36.25	16:02	35
-108	35.94	35.47	16:17	32
-110	35.37	34.93	16:27	29
-112	34.79	34.36	16:38	27
-114	34.18	33.77	16:47	25
-115	33.86	33.46	16:52	24
-118	32.90	32.53	17:07	21
-120	32.24	31.88	17:16	19
-122	31.56	31.23	17:26	17
-125	30.52	30.22	17:39	14
-128	29.46	29.20	17:53	11
-130	28.75	28.50	18:01	9
-132	28.03	27.80	18:10	7
-134	27.30	27.10	18:18	5
-135	26.94	26.75	18:22	4
-137	26.21	26.05	18:30	2
-138	25.85	25.69	18:35	1

PERCENTAGE OF SUN OBSCURED ALONG MAP TRACK
WEST END =101.8   EAST END =100.4

===

SUNSET POSITION : ALTITUDE IS -0.8 DEGREES
NORTH : LONGITUDE -139.1, LATITUDE 25.4
SOUTH : LONGITUDE -139.1, LATITUDE 25.3

+ MARKS THE CITY OF PEI-CHING

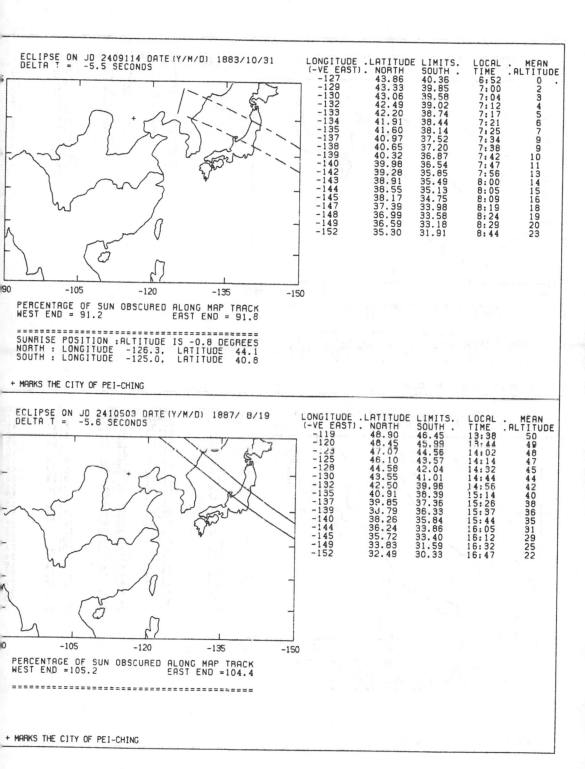

ECLIPSE ON JD 2409114 DATE(Y/M/D) 1883/10/31
DELTA T = -5.5 SECONDS

LONGITUDE (-VE EAST)	LATITUDE NORTH	LIMITS SOUTH	LOCAL TIME	MEAN ALTITUDE
-127	43.86	40.36	6:52	0
-129	43.33	39.85	7:00	2
-130	43.06	39.58	7:04	3
-132	42.49	39.02	7:12	4
-133	42.20	38.74	7:17	5
-134	41.91	38.44	7:21	6
-135	41.60	38.14	7:25	7
-137	40.97	37.52	7:34	9
-138	40.65	37.20	7:38	9
-139	40.32	36.87	7:42	10
-140	39.98	36.54	7:47	11
-142	39.28	35.85	7:56	13
-143	38.91	35.49	8:00	14
-144	38.55	35.13	8:05	15
-145	38.17	34.75	8:09	16
-147	37.39	33.98	8:19	18
-148	36.99	33.58	8:24	19
-149	36.59	33.18	8:29	20
-152	35.30	31.91	8:44	23

PERCENTAGE OF SUN OBSCURED ALONG MAP TRACK
WEST END = 91.2   EAST END = 91.8

==

SUNRISE POSITION :ALTITUDE IS -0.8 DEGREES
NORTH : LONGITUDE -126.3, LATITUDE   44.1
SOUTH : LONGITUDE -125.0, LATITUDE   40.8

+ MARKS THE CITY OF PEI-CHING

ECLIPSE ON JD 2410503 DATE(Y/M/D) 1887/ 8/19
DELTA T = -5.6 SECONDS

LONGITUDE (-VE EAST)	LATITUDE NORTH	LIMITS SOUTH	LOCAL TIME	MEAN ALTITUDE
-119	48.90	46.45	13:38	50
-120	48.45	45.99	13:44	49
-123	47.07	44.56	14:02	48
-125	46.10	43.57	14:14	47
-128	44.58	42.04	14:32	45
-130	43.55	41.01	14:44	44
-132	42.50	39.98	14:56	42
-135	40.91	38.39	15:14	40
-137	39.85	37.36	15:26	38
-139	38.79	36.33	15:37	36
-140	38.26	35.84	15:44	35
-144	36.24	33.86	16:05	31
-145	35.72	33.40	16:12	29
-149	33.83	31.59	16:32	25
-152	32.49	30.33	16:47	22

PERCENTAGE OF SUN OBSCURED ALONG MAP TRACK
WEST END =105.2   EAST END =104.4

==

+ MARKS THE CITY OF PEI-CHING

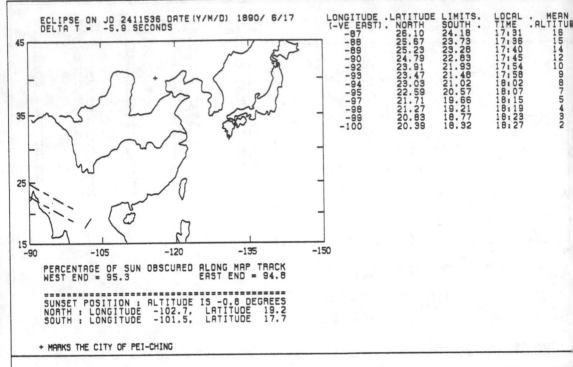

ECLIPSE ON JD 2411536 DATE (Y/M/D) 1890/ 6/17
DELTA T = -5.9 SECONDS

LONGITUDE (-VE EAST)	LATITUDE LIMITS NORTH	SOUTH	LOCAL TIME	MEAN ALTITU
-87	26.10	24.18	17:31	16
-88	25.67	23.73	17:36	15
-89	25.23	23.28	17:40	14
-90	24.79	22.83	17:45	12
-92	23.91	21.93	17:54	10
-93	23.47	21.48	17:58	9
-94	23.03	21.02	18:02	8
-95	22.59	20.57	18:07	7
-97	21.71	19.66	18:15	5
-98	21.27	19.21	18:19	4
-99	20.83	18.77	18:23	3
-100	20.39	18.32	18:27	2

PERCENTAGE OF SUN OBSCURED ALONG MAP TRACK
WEST END = 95.3   EAST END = 94.8

==
SUNSET POSITION : ALTITUDE IS -0.8 DEGREES
NORTH : LONGITUDE -102.7, LATITUDE 19.2
SOUTH : LONGITUDE -101.5, LATITUDE 17.7

+ MARKS THE CITY OF PEI-CHING

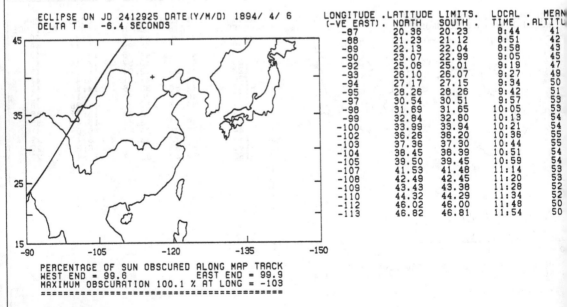

ECLIPSE ON JD 2412925 DATE (Y/M/D) 1894/ 4/ 6
DELTA T = -6.4 SECONDS

LONGITUDE (-VE EAST)	LATITUDE LIMITS NORTH	SOUTH	LOCAL TIME	MEAN ALTITU
-87	20.36	20.23	8:44	41
-88	21.23	21.12	8:51	42
-89	22.13	22.04	8:58	43
-90	23.07	22.99	9:05	45
-92	25.06	25.01	9:19	47
-93	26.10	26.07	9:27	49
-94	27.17	27.15	9:34	50
-95	28.26	28.26	9:42	51
-97	30.54	30.51	9:57	53
-98	31.69	31.65	10:05	53
-99	32.84	32.80	10:13	54
-100	33.99	33.94	10:21	54
-102	36.26	36.20	10:36	55
-103	37.36	37.30	10:44	55
-104	38.45	38.39	10:51	54
-105	39.50	39.45	10:59	54
-107	41.53	41.48	11:14	53
-108	42.49	42.45	11:20	53
-109	43.43	43.38	11:28	52
-110	44.32	44.29	11:34	52
-112	46.02	46.00	11:48	50
-113	46.82	46.81	11:54	50

PERCENTAGE OF SUN OBSCURED ALONG MAP TRACK
WEST END = 99.6   EAST END = 99.9
MAXIMUM OBSCURATION 100.1 % AT LONG = -103
==

+ MARKS THE CITY OF PEI-CHING

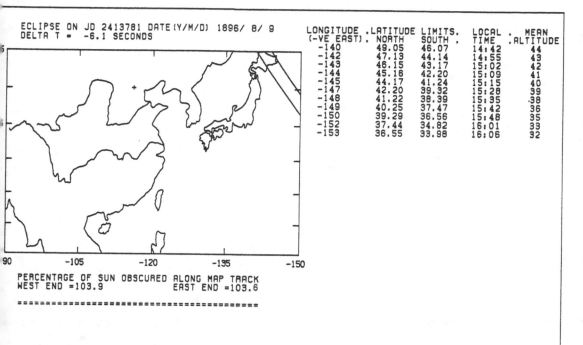

ECLIPSE ON JD 2413781 DATE(Y/M/D) 1896/ 8/ 9
DELTA T = -6.1 SECONDS

LONGITUDE (-VE EAST)	LATITUDE LIMITS. NORTH	SOUTH.	LOCAL TIME	MEAN ALTITUDE
-140	49.05	46.07	14:42	44
-142	47.13	44.14	14:55	43
-143	46.15	43.17	15:02	42
-144	45.16	42.20	15:09	41
-145	44.17	41.24	15:15	40
-147	42.20	39.32	15:28	39
-148	41.22	38.39	15:35	38
-149	40.25	37.47	15:42	36
-150	39.29	36.56	15:48	35
-152	37.44	34.82	16:01	33
-153	36.55	33.98	16:06	32

PERCENTAGE OF SUN OBSCURED ALONG MAP TRACK
WEST END =103.9   EAST END =103.6

===

+ MARKS THE CITY OF PEI-CHING

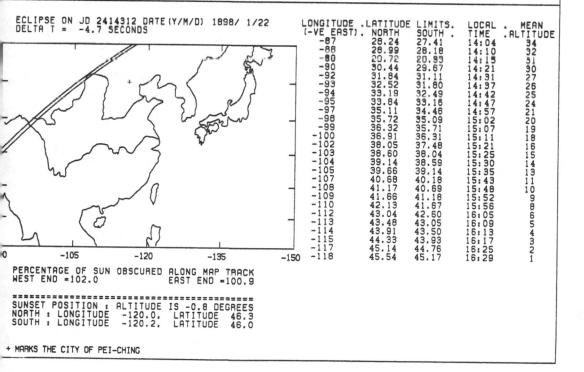

ECLIPSE ON JD 2414312 DATE(Y/M/D) 1898/ 1/22
DELTA T = -4.7 SECONDS

LONGITUDE (-VE EAST)	LATITUDE LIMITS. NORTH	SOUTH.	LOCAL TIME	MEAN ALTITUDE
-87	28.24	27.41	14:04	34
-88	28.99	28.18	14:10	32
-80	20.72	20.95	14:15	51
-90	30.44	29.67	14:21	30
-92	31.84	31.11	14:31	27
-93	32.52	31.80	14:37	26
-94	33.19	32.49	14:42	25
-95	33.84	33.16	14:47	24
-97	35.11	34.46	14:57	21
-98	35.72	35.09	15:02	20
-99	36.32	35.71	15:07	19
-100	36.91	36.31	15:11	18
-102	38.05	37.48	15:21	16
-103	38.60	38.04	15:25	15
-104	39.14	38.59	15:30	14
-105	39.66	39.14	15:35	13
-107	40.68	40.18	15:43	11
-108	41.17	40.69	15:48	10
-109	41.66	41.18	15:52	9
-110	42.13	41.67	15:56	8
-112	43.04	42.60	16:05	6
-113	43.48	43.05	16:09	5
-114	43.91	43.50	16:13	4
-115	44.33	43.93	16:17	3
-117	45.14	44.76	16:25	2
-118	45.54	45.17	16:29	1

PERCENTAGE OF SUN OBSCURED ALONG MAP TRACK
WEST END =102.0   EAST END =100.9

===
SUNSET POSITION : ALTITUDE IS -0.8 DEGREES
NORTH : LONGITUDE -120.0, LATITUDE 46.3
SOUTH : LONGITUDE -120.2, LATITUDE 46.0

+ MARKS THE CITY OF PEI-CHING